6th South East Asia Design Research International Conference 2018

Inspiring Students to Learn: Fostering Innovative Teaching and Learning of Science, Mathematics and Technology

Journal of Physics: Conference Series
Volume 1088

Banda Aceh, Indonesia
27 – 28 June 2018

ISBN: 978-1-5108-7372-8
ISSN: 1742-6588

Printed from e-media with permission by:

Curran Associates, Inc.
57 Morehouse Lane
Red Hook, NY 12571

Some format issues inherent in the e-media version may also appear in this print version.

Printed by Curran Associates, Inc. (2018)

For permission requests, please contact the Institute of Physics
at the address below.

Institute of Physics
Dirac House, Temple Back
Bristol BS1 6BE UK

Phone: 44 1 17 929 7481
Fax: 44 1 17 920 0979

techtracking@iop.org

Additional copies of this publication are available from:

Curran Associates, Inc.
57 Morehouse Lane
Red Hook, NY 12571 USA
Phone: 845-758-0400
Fax: 845-758-2633
Email: curran@proceedings.com
Web: www.proceedings.com

TABLE OF CONTENTS

THE EFFECTIVENESS OF STEM MENTORING PROGRAM IN PROMOTING INTEREST TOWARDS STEM 1
L Halim, T M T Soh, N M Arsad

HOW DO WE LET STUDENTS WORK AS 'YOUNG MATHEMATICIANS' IN THE CLASSROOM? 9
M Dolk

TEACHERS' COLLECTIVE KNOWLEDGE: THE CASE OF EQUIVALENT FRACTIONS 15
Z H Putra, C Winsløw

DEVELOPING SPATIAL REASONING ACTIVITIES WITHIN GEOMETRY LEARNING 21
D W Winarti

ANDROID BASED E-LEARNING TUTORIAL FOR MATHEMATICS TEACHERS 28
Y Roza, I Daqiqil, S N Siregar, S Salam, A Adnan

THE APPLICATION OF R AND ARCGIS SOFTWARE AS A LEARNING MEDIA TO ESTIMATE THE RETURN PERIOD OF THE DESTRUCTIVE EARTHQUAKE IN ACEH AREAS USING MAXIMUM LIKELIHOOD METHOD 34
H Sofyan, Jumila, L Rahayu

ENHANCING MANIPULATION OF ALGEBRAIC EQUATION THROUGH BALANCE METHOD 42
D N B Pg Badaruddin, K A Tengah, R C I Prahmana

DESIGNING AN INTEGRATED LEARNING STRATEGY TO DEVELOP STUDENTS' AWARENESS OF RIVER ENVIRONMENT AND SCIENCE PROCESS SKILLS 49
A Winarti, Sarbain, M Yamin

FORMATIVE ASSESSMENT IN SCIENCE EDUCATION: IS IT BEING PRACTICED? 56
M J B Espiritu, M Shahrill, J S H Q Perera, R C I Prahmana

UNCERTAINTY AND DATA CONTENT IN BOWLING: TASK DESIGN 62
R Permatasari, R I I Putri, Zulkardi

TEACHERS' REFLECTIONS ON STUDENTS' MATHEMATICAL PROBLEM SOLVING IN JUNIOR HIGH SCHOOL 68
Y Harisman, Y S Kusumah, K Kusnandi

DEVELOPMENT OF LEARNING TOOL WITH CONTEXTUAL TEACHING AND LEARNING (CTL) APPROACH TO IMPROVE STUDENT MATHEMATICAL CONNECTION ABILITY 76
Mauliana, M Ikhsan, M Subianto

LEARNING OF REPRODUCTION SYSTEM WITH AN INTEGRATIVE CURRICULUM APPROACH IN JUNIOR HIGH SCHOOL 81
Ibrahim, N Yusoff, M I Awang, Marwan

ASSESSING THE VALIDITY AND RELIABILITY OF QUESTIONNAIRES ON THE IMPLEMENTATION OF INDONESIAN CURRICULUM K-13 IN STEM EDUCATION 88
R Oktavia, Irwandi, Rajibussalim, M Mentari, I S Mulia

CHARACTERISING FORMATIVE ASSESSMENT PRACTICES IN THE MATHEMATICS CLASSES 95
J Low, M Shahrill, J S H Q Perera, R C I Prahmana

DESIGN LEARNING IN MATHEMATICS EDUCATION: ENGAGING EARLY CHILDHOOD STUDENTS IN GEOMETRICAL ACTIVITIES TO ENHANCE GEOMETRY AND SPATIAL REASONING 101
R Novita, M Putra, E Rosayanti, F Fitriati

ELEMENTARY STUDENTS' REPRESENTATIONS IN SOLVING WORD PROBLEMS 107
J T Manoy

THE DEVELOPMENT OF LEARNING INSTRUMENTS USING THE CREATIVE PROBLEM-SOLVING LEARNING MODEL TO IMPROVE STUDENTS' CREATIVE THINKING SKILLS IN MATHEMATICS 114
D M Sari, M Ikhsan, Z Abidin

PISA-LIKE MATHEMATICS PROBLEM: THE CONTEXT OF BASKETBALL IN ASIAN GAMES 119
R D Jannah, R I I Putri, Zulkardi

THE STUDY OF THE MATHEMATICAL PROBLEM SOLVING AND METACOGNITION STRATEGY ON A PAIRED HANDEP COOPERATIVE LEARNING MODEL 126
Demitra, D Sulisworo

INVESTIGATING STUDENTS' LEARNING TRAJECTORY: A CASE ON TRIANGLE 133
Anwar, I Rofiki
A SELF-EVALUATION TECHNIQUE IN IMPROVING TEACHER'S PROFESSIONAL DEVELOPMENT: THE USE OF "REALIA" MEDIA AND "WAIT TIME" STRATEGIES 139
W Artika, M Saputri
LEARNING FRACTION THROUGH THE CONTEXT OF ASIAN GAMES 2018 145
R I I Putri, Zulkardi
THE TEACHING AND LEARNING OF ADDITION AND SUBTRACTION OF INTEGERS THROUGH MANIPULATIVE IN BRUNEI DARUSSALAM 153
N Sahat, K A Tengah, R C I Prahmana
THE EFFECTIVENESS OF AN AUTOGRAPH-ASSISTED PROBLEM BASED LEARNING MODEL FOR IMPROVING HIGH SCHOOL STUDENTS' LEARNING OUTCOME ON GRAPHICS FUNCTIONS 160
M Ridha, R Johar, Marwan, Mailizar
DILEMMATIC SITUATIONS FOR LEARNING MATHEMATICS-RELATED BELIEFS 166
T Y E Siswono, A W Kohar, S Hartono, A H Rosyidi, P Wijayanti, R Ekawati
STUDENTS' COLLABORATIVE ABILITY IN LEARNING GEOMETRY TRANSFORMATION USING A SCIENTIFIC APPROACH BASED ON LEARNING COMMUNITY 174
Hobri, J Safitri, S Romlah, E Nazareth, Susanto
REALISTIC MATHEMATICS LEARNING BASED ON THE ETHNOMATHEMATICS IN BENGKULU TO IMPROVE STUDENTS' COGNITIVE LEVEL 182
W Widada, D Herawaty, A N M T Lubis
DESIGNING STUDENT WORKSHEET FOR RICH MATHEMATICAL TASKS 190
F Fitriati, R Novita
THE DEVELOPMENT OF LEARNING INSTRUMENTS THROUGH THE PROBLEM-BASED LEARNING MODEL TO ENHANCE STUDENTS' CREATIVITY 197
M Iqbal, Yusrizal, Z Abidin
ANALYSIS OF STUDENTS' ERRORS IN RESPONDING TO TIMSS DOMAIN ALGEBRA PROBLEM 203
B A Saputro, D Suryadi, R Rosjanuardi, B G Kartasasmita
TEACHING AND LEARNING STATISTICS IN COLLEGE: HOW LEARNING MATERIALS SHOULD BE DESIGNED 210
K Idris
THE DEVELOPMENT OF MATHEMATICS STUDENT WORKSHEET FOR SCHOOL LITERACY MOVEMENT 215
K N S Effendi, Zulkardi, R I I Putri, P Yaniawati
APPLICATION OF MICROSOFT EXCEL AS AN INTERACTIVE LEARNING MEDIA OF ACID-BASE TITRATION 222
I Khaldun, M Hasan, Nilawati
TEACHERS' USE OF LEARNING RESOURCES IN SPATIAL LEARNING 229
C Khairunnisak, E Elizar, R Johar, T P Utami
THE VALIDITY OF VECTOR ANALYSIS MODULE USING WXMAXIMA SOFTWARE 237
Khairina, M Ikhsan, Suhartati
DIFFICULTIES ANALYSIS OF MATHEMATICS EDUCATION STUDENTS ON THE REAL ANALYSIS SUBJECT 243
I Widiati, A Sthephani
THE DEVELOPMENT OF LEARNING INSTRUMENTS BASED ON AN OPEN-ENDED APPROACH TO IMPROVE STUDENTS' PROBLEM-SOLVING SKILL 248
N R Fandanu, M Ikhsan, Bahrun
THE DEVELOPMENT OF LEARNING MATERIALS USING CONTEXTUAL TEACHING LEARNING (CTL) APPROACH ORIENTED ON THE CHARACTER EDUCATION 254
R Johar, Agussalim, M Ikhsan, B Zaura
DEVELOPING LEARNING TRAJECTORY FOR TEACHING STATISTICS AT JUNIOR HIGH SCHOOL USING RME APPROACH 260
A Fauzan, E Musdi, J Afriadi
THE INTEGRATION OF SCIENCE AND MATH 267
Y Fitria, Y Helsa, H Nirwana, A P Zulkarnaini
THE PRACTICALITY OF THE QUADRATIC FUNCTION MODULE BY UTILIZING AUTOGRAPH SOFTWARE AND ANGRY BIRDS GAME 273
C M Zubainur, Suhartati, Iwannitona
DEVELOPMENT OF ALGEBRA TEST QUESTIONS BASED ON BLOOM'S TAXONOMY 280
M Husna, R Johar, Hajidin, Mailizar

DEVELOPING LTBI FOR ADDITION AND MULTIPLICATION RULES IN PROBABILITY THEORY WITH REALISTIC MATHEMATICS EDUCATION 287
H Julie
DEVELOPING A PHYSICS MODULE BASED ON THE LOCAL WISDOM OF HULU SUNGAI TENGAH REGENCY TO TRAIN THE MURAKATA CHARACTER 294
S Hartini, M F Isnanda, M Wati, M Misbah, S An'nur, S Mahtari
THE DEVELOPMENT OF LEARNING RESOURCES THROUGH BENTHIC SPECIES STUDY IN MANGROVE ECOSYSTEM REULEUNG LEUPUNG FOR INVERTEBRATE ZOOLOGY LEARNING 300
M Ali S, Supriyatno, M D Asiah, M Saputri, A Mursawal, Zulfikar
PISA-LIKE MATHEMATICS PROBLEMS USING THE CONTEXT OF ATHLETICS IN ASIAN GAMES 2018 308
I Pratiwi, R I I Putri, Zulkardi
LEARNING DIRECT PROPORTION BY USING THE CONTEXT OF TIMPAN RECIPES 315
S F Zuhra, C M Zubainur, T F Abidin
DEVELOPING MATHEMATICS TEACHING TOOL USING ELPSA 321
E Gradini, F Bahri
THE EFFECTIVENESS OF LEARNING INSTRUMENTS ON THE TOPIC OF THE SET USING PROBLEM-BASED LEARNING MODEL AT ISLAMIC JUNIOR HIGH SCHOOL IN PEKANBARU 328
Sakur, A Murni, R D Anggraini
THE TEACHER'S MATHEMATICAL LITERACY FOR THE CHANGE AND RELATIONSHIP PROBLEMS ON THE PISA ADAPTATION TEST 334
A Y Anggoro, H Julie, F Sanjaya, M A Rudhito
JUNIOR HIGH SCHOOL MATHEMATICS TEACHERS' PEDAGOGICAL CONTENT KNOWLEDGE IN TEACHING OF POLYHEDRA 341
Ma'rufi, M Ilyas, Salwah
THE EFFECT OF CREATIVE PROBLEM-SOLVING LEARNING MODEL USING GEOMETRY TRANSFORMATION BOOK BASED ON AL-QUR'AN ON STUDENTS' VAN HIELE THINKING LEVEL AND LEARNING OUTCOME 347
F Kristanti, C Ainy, S Shoffa
GENDER DIFFERENCES OF MATHEMATICAL CRITICAL THINKING SKILLS OF SECONDARY SCHOOL STUDENTS 354
Mawaddah, A Ahmad, M Duskri
ETHNOMATHEMATICS ANALYSIS ON JAMBI PLAIT ART AS THE MATHEMATICS LEARNING RESOURCES 360
Kamid, A Wandari, Rohati
THE FEASIBILITY OF AN ANDROID-BASED POCKETBOOK AS MATHEMATICS LEARNING MEDIA IN SENIOR HIGH SCHOOL 365
M Saputra, T F Abidin, B I Ansari, M Hidayat
INTEGRATED MODEL IN SCIENCE FOR ELEMENTARY SCHOOL 371
R Amini, Usmeldi, Y Helsa
STUDENTS' PERCEPTIONS ON THE IMPLEMENTATION OF E-LEARNING: HELPFUL OR UNHELPFUL? 376
L Vitoria, M Mislinawati, N Nurmasyitah
THE DEVELOPMENT OF AN INSTRUMENT TO EXPLORE NON-ROUTINE PROBLEM SOLVING STRATEGIES AMONG MATHEMATICS EDUCATION STUDENTS 382
R M Bambang S, R Salasi, M Hasbi, MZ Mardhiah
LIGHT EMITTING DIODE (LED) AS AN ESSENTIAL PROP COMPONENT FOR STEM EDUCATION IN THE 21ST CENTURY: A FOCUS FOR SECONDARY SCHOOL LEVEL 389
Irwandi, R Oktavia, Rajibussalim, A Halim, Melvina
THE ETHNOMATHEMATICS: EXPLORATION OF GAYO TRIBE LOCAL WISDOM RELATED TO MATHEMATICS EDUCATION 395
B Yustinaningrum, Nurliana, E Rahmadhani
THE DEVELOPMENT OF TEACHING MATERIAL: RIGOROUS MATHEMATICAL THINKING IN A GEOMETRY CLASSROOM 401
M Meilantifa, M T Budiarto
PISA-LIKE MATHEMATICS PROBLEM WITH KARATE CONTEXT IN ASIAN GAMES 406
H Nizar, R I I Putri, Zulkardi
STUDENTS' METACOGNITIVE ABILITY IN MATHEMATICAL PROBLEM-SOLVING LEARNING BASED ON LESSON STUDY FOR LEARNING COMMUNITY (LSLC) 413
Hobri, S Romlah, A C Prihandoko, J Safitri, E Nazareth

THE IMPACT OF USING LINE@ ON THE COOPERATIVE LEARNING TO IMPROVE THE CRITICAL THINKING SKILLS OF HIGH SCHOOL STUDENTS 421
D Sulisworo, S Daimah, M Toifur, A Suryadi
THE TEACHERS' MATHEMATICS LITERACY ABILITY FOR SOLVING UNCERTAINTY PROBLEMS ON A PISA ADAPTATION TEST 428
F Sanjaya, A Y Anggoro, H Julie, M A Rudhito
STUDENTS' CREATIVE THINKING ABILITY IN LEARNING MATHEMATICS THROUGH LEARNING MODEL OF LOGAN AVENUE PROBLEM SOLVING (LAPS) – HEURISTIC 434
U Husna, C M Zubainur, B I Ansari
THE IMPROVEMENT OF MATHEMATICS TEACHERS' PEDAGOGICAL CONTENT KNOWLEDGE (PCK) THROUGH MENTORING 440
T Zubaidah, R Johar, Suparno
CARTOGRAPHY IN DESIGNING DIGITAL MAP USING ADOBE FLASH CS6 447
Y Miaz, Y Helsa, Desyandri, R Febrianto
DIFFERENCES IN STUDENTS' LEARNING OUTCOMES BETWEEN DISCOVERY LEARNING AND CONVENTIONAL LEARNING MODELS 454
R D Anggraini, A Murni, Sakur
STUDENTS' ABSTRACTION IN SOLVING SYSTEM OF LINEAR EQUATIONS WITH TWO VARIABLES 459
F Pangaribuan
STUDENTS' ABILITY TO CONVERT A SITUATION INTO A MATHEMATICAL MODEL OR DIAGRAM USING PROBLEM SOLVING APPROACH 465
H Arianti, C M Zubainur, Hizir
INTEGRATING TECHNOLOGY AND MEDIA INTO MATHEMATICS LEARNING 471
U Rahmi, Y Helsa, Azrul
THE DEVELOPMENT OF STUDENT WORKSHEET BY USING DISCOVERY-BASED APPROACH: A CASE STUDY IN LEARNING MEDIA COURSE 475
E B Rahaju, P Wijayanti
GENERATIVE LEARNING MODEL TO IMPROVE MATHEMATICS PROBLEM SOLVING SKILLS ON POLYHEDRON 480
Andriani, M Ikhsan, S Munzir, C Khairunnisak
THE ANALYSIS OF JUNIOR HIGH SCHOOL STUDENTS' MATHEMATICAL ABSTRACTION ABILITY BASED ON LOCAL CULTURAL WISDOM 486
I Dewi, N Siregar, A Andriani
THE QUALITY OF LEARNING MATERIALS THROUGH MATHEMATICS REASLITIC TO IMPROVE STUDENTS' MATHEMATICAL COMMUNICATION ABILITY IN THE ELEMENTARY SCHOOL 492
Nurfadhillah, R Johar, A Ahmad
DESIGNING TEACHING INSTRUCTIONS CATERING STUDENTS' NEEDS: TEACHING SOLID GEOMETRY THROUGH PROBLEM-BASED LEARNING (PBL) 498
K Umam, S Maulina
CALISTUNG LITERACY THROUGH THE APPLICATION OF LECTORA 504
Masniladevi, Ritawati, Yullys Helsa
THE VALIDITY TEST OF THE LESSON PLAN TO REDUCE STUDENTS' MISCONCEPTIONS USING THE COGNITIVE CONFLICT STRATEGY 508
D Fitri, R Johar, A Ahmad
THE USE OF CIPP MODEL FOR EVALUATION OF COMPUTATIONAL ALGORITHM LEARNING PROGRAM 514
S J Hartati, N Sayidah, Muhajir
ANALYSIS OF SENIOR HIGH SCHOOL STUDENTS' EMOTIONAL INTELLIGENCE IN COOPERATIVE BASED MATHEMATICS LEARNING 520
M Ilyas, M Ma'rufi, F Fitriani, S Salwah
FAUZI'S COGNITIVE CONFLICT IN THE DEVELOPMENT OF GEOMETRY TEACHING MATERIAL: A CASE STUDY IN SHIFTING TRAPEZOIDAL DEFINITION 525
M T Budiarto
COGNITIVE CONFLICT STRATEGY TO MINIMIZE STUDENTS' MISCONCEPTION ON THE TOPIC OF ADDITION OF ALGEBRAIC EXPRESSION 531
Irawati, C M Zubainur, R M Ali
DEEPENING STUDENTS UNDERSTANDING OF TRIANGLE TOPIC THROUGH 'APPLICATION' COMPONENT OF ELPSA (EXPERIENCE, LANGUAGE, PICTORIAL, SYMBOL AND APPLICATION) FRAMEWORK 537
B R A Febrilia, D W Winarti

MATHEMATICAL PROBLEMS OF PISA-LIKE WITH THE 200M SWIMMING CONTEXTS IN ASIAN GAMES 543
D Yansen, R I I Putri, Zulkardi
DESIGNING LEARNING ACTIVITIES ON CONDITIONAL PROBABILITY 549
Benidiktus Tanujaya, Rully Charitas Indra Prahmana, Jeinne Mumu
THE USE OF THE WXMAXIMA LINEAR ALGEBRA MODULE ON GAUSS ELIMINATION LESSON FOR MATHEMATICS EDUCATION STUDENTS 555
D Afriza, C M Zubainur, A Veloo
STUDENTS' METACOGNITION ON MATHEMATICAL PROBLEM SOLVING THROUGH ETHNOMATHEMATICS IN REJANG LEBONG, INDONESIA 562
D Herawaty, W Widada, T Novita, L Waroka, A N M T Lubis
DEVELOPMENT OF LEARNING TOOLS USING TREFFINGER LEARNING MODEL TO IMPROVE CREATIVE THINKING 568
R Handayani, Hajidin, M Duskri, E Maidiyah
DEVELOPING WORKSHEETS THROUGH ISLE-BASED STEM APPROACH AND IMPLEMENTING THEM ON SENIOR HIGH SCHOOL STUDENTS 575
E Rahmayani, I Irwandi, R Rajibussalim
ASSESSING CONCEPTUAL AND ALGORITHMIC UNDERSTANDING OF STUDENTS IN SENIOR HIGH SCHOOL 583
Nuzulia, M Hasan, A Ismayani
THE DEVELOPMENT OF MATHEMATICS HIGHER ORDER THINKING SKILLS INSTRUMENT FOR GRADE VIII JUNIOR HIGH SCHOOL 588
Y Yunita, W Wahidin, A Tsurayya
THE REFLECTIVE ABSTRACTION PROFILE OF JUNIOR HIGH SCHOOL STUDENTS IN SOLVING MATHEMATICAL PROBLEMS BASED ON COGNITIVE STYLE OF FIELD INDEPENDENT AND FIELD DEPENDENT 594
B Panjaitan
MATHEMATICS LEARNING THROUGH PENDIDIKAN MATEMATIKA REALISTIK INDONESIA (PMRI) APPROACH AND ADOBE FLASH CS6 599
M Zainil, Y Helsa, Y Zainil, W T Yanti
DESIGNING LEARNING TRAJECTORY FOR TEACHING SEQUENCE AND SERIES USING RME APPROACH TO IMPROVE STUDENTS' PROBLEM SOLVING ABILITIES 604
E Gee, A Fauzan, A Atmazaki
THE QUALITY OF MATHEMATICS LEARNING MATERIAL USING A MODIFICATION OF THINK PAIR SHARE (TPS) MODEL 610
Maisyura, C M Zubainur, T F Abidin
STUDENT'S UNDERSTANDING OF NUMBERS THROUGH THE NUMBER SENSE STRATEGY 616
N F Helmy, R Johar, Z Abidin
PATTERNS PLAYING FOR EARLY CHILDHOOD EDUCATION: MATHEMATICS LEARNING FOR EARLY CHILDHOOD EDUCATION 622
Y Yuhasriati, A Yuriansa
TEACHERS' ACTIVITIES DURING DESIGNING HIGHER-ORDER THINKING SKILLS (HOTS) MATHEMATICAL QUESTIONS THROUGH TEACHER ASSISTING PROGRAM 627
Wahyuni, R Johar, M Duskri
THE DEVELOPMENT OF ALGEBRA TEACHING MATERIALS TO FOSTER STUDENTS' CREATIVE THINKING SKILLS IN HIGHER EDUCATION 634
M Mursalin, N L S Nuraini, H Purnomo, N W Damayanti, D Kristanti, A Rohim, R Widyastuti, Y O Wulandari, H Saleh, S N Mayangsari, M Fonna, R Rohantizani, I Muhammad, H Nufus, R Sulastri, R Amalia, N Nuraina, M Muliana
EFFECTIVENESS OF THE DEVELOPMENT OF THE INQUIRY-BASED LEARNING MODEL TO IMPROVE STUDENTS' PSYCHOMOTOR ACHIEVEMENT 640
R A Siregar
DEVELOPING MATHEMATICS TRAINING MODULES TO ENHANCE TEACHERS' CONTENT KNOWLEDGE 645
R Sulastri, R Johar, M Duskri, M Ikhsan, H Meutia
STUDENT'S REPRESENTATION OF FRACTION THROUGH ELPSA FRAMEWORK 652
E Juliangkary, R Johar
IMPROVING STUDENTS' HIGHER ORDER THINKING SKILLS IN THERMOCHEMISTRY CONCEPT USING WORKSHEETS BASED ON 2013 CURRICULUM 660
R Verdina, A Gani, Sulastri

THE EFFECTIVENESS OF GUIDED INQUIRY-BASED STUDENT WORKSHEETS ON STUDENTS' GENERIC SCIENCE SKILLS 666
M Faradilla, M Hasan, Sulastri

THE ASSOCIATION BETWEEN CONCEPTUAL UNDERSTANDING AND REASONING ABILITY IN MATHEMATICS: AN ANALYSIS OF DNR-BASED INSTRUCTION MODELS 672
MT Bakar, D Suryadi, Darhim, W S Tonra, M S Noto

STUDENTS' UNDERSTANDING OF CHARTS: THE STUDY OF PISA'S PROBLEM-SOLVING IN THE CONTENT OF DATA 677
I K Sari, A Nasriadi, M Salmina

QUESTIONING SKILL OF SCIENCE TEACHER FROM THE STUDENTS PERSCPECTIVE IN SENIOR HIGH SCHOOL 683
A Halim, Yusrizal, H Mazlina, Melvina, Zainaton

DESIGNING COMPUTER-BASED FRACTION WORKSHEETS FOR JUNIOR HIGH SCHOOL 689
R Amalia, S Saiman, S Sofiyan, M Mursalin

THE POTENTIAL OF JANGKA BEACHES AS A NATURAL LABORATORY FOR LEARNING THE CONCEPT OF BIODIVERSITY 696
A Abdullah, L Fitriana, M Ali Sarong, C Nurmaliah

STEM LEARNING IN REGULAR AND VOCATIONAL HIGH SCHOOLS ON THE TOPIC OF SCIENTIFIC MENU CARD FABRICATION 702
R P Sari, M Adlim, A Gani

VALIDATION OF PROTOTYPE INSTRUMENTS FOR IMPLEMENTING HIGHER ORDER THINKING LEARNING USING THE IMPROVE METHOD 708
BI Ansari, Saminan, R Sulastri

DEVELOPING GUIDED-INQUIRY-STUDENT WORKSHEETS TO IMPROVE THE SCIENCE PROCESS SKILLS OF HIGH SCHOOL STUDENTS ON THE HEAT CONCEPT 714
M Mahyuna, M Adlim, I Saminan

CHARACTERISTICS OF PRE-SERVICE TEACHERS' PERFORMANCE IN PROBLEM POSING 719
M Masriyah, I Kurniasari, E L W Palupi

STUDENTS' THINKING PROCESS IN SOLVING MATHEMATICAL PROBLEMS BASED ON THE LEVELS OF MATHEMATICAL ABILITY 724
A Sanjaya, R Johar, M Ikhsan, L Khairi

DESIGNING PISA-LIKE MATHEMATICS PROBLEMS USING THE CONTEXT OF KARAWANG 730
I N Aini, Zulkardi, R I I Putri, Turmudi

OPTIMALIZATION OF STUDENT'S LEARNING OUTCOMES AND LEARNING ACTIVITIES IN PHYSICS USING OPEN ENDED MODEL 735
A Hamid, Nofiza

THE DEVELOPMENT OF A MODULE WITH MICROSOFT EXCEL-BASED INTERACTIVE MEDIA ON THE TOPIC OF BUFFER SOLUTION 740
Anizar, A Gani, I Khaldun, M Bahi

THE DEVELOPMENT OF STUDENT WORKSHEETS BASED ON METACOGNITIVE APPROACH TO IMPROVE STUDENTS' MATHEMATICAL REPRESENTATION ABILITY 746
A Murni, R D Anggraini, Sakur

Author Index

The 6th South East Asia Design Research International Conference (SEA-DR IC) 2018

R Johar[1], C Morina[1], Anwar[1], Mailizar[1], Elizar[1], C Khairunnisak[1], R C I Prahmana[2], W Artika[1], L Vitoria[1], L Khairi[1], S Maulina[1] and M Ulfa[1]

[1]Syiah Kuala University, Banda Aceh, 23111, Indonesia
[2]Universitas Ahmad Dahlan, Jl. Pramuka Kav. 5, Yogyakarta 55161, Indonesia

E-mail: rahmahjohar@fkip.unsyiah.ac.id

Preface

The South East Asia Design Research (SEA-DR) as a forum of design research in collaboration with Master Program of Mathematics Education, Syiah Kuala University, organized the 6th SEA-DR conference with the theme “Inspiring students to learn: Fostering innovative teaching and learning of science, mathematics and technology”.

This conference was an excellent opportunity for academics, researchers, teachers and students to share knowledge, experiences and research findings as well as to inspire the best practice of development research in the field of teaching mathematics, science, and technology.

We had four keynote speakers that were Prof. Berinderjeet Kaur, Prof. Maarten Dolk, Prof. Lilia Halim, and Dr. John Willison. We also had ten invited speakers and four keynote speakers in workshop sessions. Furthermore, there were 181 papers, including 148 oral presentation and 33 posters presentations. The 6th SEA-DR conference successfully attracted delegates from many countries. There were seven countries participating in this conference, including: Singapore, the Netherlands, Denmark, Australia, Malaysia, Brunei Darussalam, and Indonesia.

Finally, we would like to extend our gratitude for everyone involved for their contribution in the conference.

IOP Conf. Series: Journal of Physics: Conf. Series **1088** (2018) 011001 doi:10.1088/1742-6596/1088/1/011001

The Committee of 6th South East Asia Design Research International Conference (SEA-DR IC) 2018

Steering Committee
Prof. Dr. Ir. Samsul Rizal, M. Eng (Rector of Syiah Kuala University)
Prof. Dr. Djufri, M. Si (Dean of FKIP Syiah Kuala University)
Prof. Dr. Maarten Dolk
Prof. Dr. R.K Sembiring
Dr. Y. Marpaung
Prof. Dr. Zulkardi, M. I. Kom, M.Sc.
Prof. Dr. Sutarto Hadi, M.Si., M. Sc.
Prof. Dr. Dian Armanto, M.Pd, M.Sc.
Prof. Dr. Ahmad Fauzan, M.Pd, M. Sc.
Prof. Dr. Turmudi, M.Ed, M.Sc.
Dr. M. Ikhsan, M.Pd. (Head of Mathematics Education Department of Syiah Kuala Univesity)
Prof. Berinderjeet Kaur
Prof. Lilia Halim

Organizing Committee
Conference Chair : Dr. Rahmah Johar, M.Pd.
Conference Co-Chair : Dr. Anwar, M.Pd.
Secretary : Dr. Mailizar, M.Ed.
Vice Secretary : Elizar, Ph.D.
Treasurer : Dr. Cut Morina Zubainur, M.Pd.
Vice Treasurer : Suhartati, M.Pd.

Scientific Committee
Scientific Committee Chair : Prof. Dr. Marwan, M. Si
Scientific Committee Member :
Prof. Maarten Dolk
Prof. Rohaida Mohd. Saat
Prof. Dr. Musri Musman, M.Sc.
Prof. Dr. Ratu Ilma Indra Putri, M.Si.
Prof. Dr. Muchlisin Z.A, M.Sc.
Prof. Dr. Adlim, M.Sc.
Prof. Dr. Ahmad Fauzan, M.Sc.
Dr. Cathy Wissehr
Dr. Wanty Widjaja
Dr. Abdul Halim Abdullah
Dr. Ariyadi Wijaya, M.Sc.
Dr. Al Jupri, M.Sc.
Dr. Tatag Yuli Eko Siswono, M.Pd.
Dr. Yenita Roza, M.Sc.
Dr. Shintia Revina, M.Sc.
Dr. Neni Mariana, M.Sc.
Dr. Rooselina Ekawati, M.Sc.
Dr. Nasrullah Idris, M.Eng.
Dr. Hongki Julie, M.Pd.
Dr. Taufik Fuadi Abidin, M.Tech.
Dr. Rully Charitas Indra Prahmana

Keynote Speakers

Prof. Berinderjeet Kaur	National Institute of Education, Singapore
Prof. Maarten Dolk	Utrecht University, the Netherlands
Prof. Lilia Halim	University Kebangsaan Malaysia, Malaysia
Dr. John Willison	the University of Adelaide, Australia

Figure 1. Prof Berinderjeet Kaur from National Institute of Education, Singapore, delivering her keynote talk

IOP Conf. Series: Journal of Physics: Conf. Series **1088** (2018) 011001 doi:10.1088/1742-6596/1088/1/011001

Figure 2. Dr. John Willison from the University of Adelaide, Australia, delivering his keynote talk

Figure 3. Opening ceremony

Figure 4. Participants attending one of the keynote sessions

Figure 5. One of the participants delivering her talk in a parallel session

Figure 6. The chair and one of the steering committees announcing the next SEA-DR conference

Peer review statement

All papers published in this volume of *Journal of Physics: Conference Series* have been peer reviewed through processes administered by the proceedings Editors. Reviews were conducted by expert referees to the professional and scientific standards expected of a proceedings journal published by IOP Publishing.

IOP Conf. Series: Journal of Physics: Conf. Series **1088** (2018) 012001 doi:10.1088/1742-6596/1088/1/012001

The effectiveness of STEM mentoring program in promoting interest towards STEM

L Halim, T M T Soh and N M Arsad
Unversiti Kebangsaan Malaysia, Bangi Selangor, 43600 UKM, Malaysia

E-mail: lilia@ukm.edu.my

Abstract. STEM mentoring program were introduced in order to overcome the declining interest in STEM subjects in schools. This study aims to measure the effectiveness of STEM mentoring programs on the students' interest towards STEM. Comparisons between girls and boys on their interest towards STEM were also conducted. The STEM Mentoring programs in this study consisted of an array of the program from various faculties namely (i) BITARA STEM and STEM Career Interest from Faculty of Education (FPEND), (ii) CRYSTAL from Faculty of Engineering and Built Environment (FKAB), and (iii) Science Camp and SEE Nature from the Faculty of Science and Technology (FST). This study used post-test control group design involved 198 students in one of the states in Malaysia. Data garnered from the study was described by means of descriptive analysis, which later followed by inferential analysis using independent sample t-test. Overall, the findings revealed that there is significant difference between girls in treatment group in interest towards mathematics. This study showed that research should be done more thoroughly on mentoring program, need to find out the suitable framework in implementing STEM mentoring programs and possible implementation of Blended Mentor Mentee programs for further research to get a corroborative evidence.

1. Introduction

Interest in science and mathematics learning among school students is declining worldwide including Malaysia [1-3]. Reasons for the decline include learning in silo and not integrated, unable to see the relevance of science and mathematics in everyday situation, subject matter too difficult and career is uninteresting [4]. One of the ways to address this phenomenon is by promoting STEM education in that STEM integrated education brings real world problems and contexts into schools. At the same time, [5] argued that another underlying issue to the decline, which has not been addressed, is concerned with the students' own identity to the discipline- i.e. unwillingness to opt for science relate studies and careers. Both the pedagogical and psychosocial factors could be addressed through a mentoring system between STEM higher education (HE) students (commonly known as facilitators) and school students. Often courses in higher education involve integration of more than one discipline and deal with complex and real world problems. STEM facilitators serve as role models are able to project positive STEM identity - such as projecting images whereby STEM researchers and their work is actually helping the community e.g. Device methods to overcome famine or engineers develop method to produce alternative energy sources with less pollution consequences. While mentoring is seen important, the effectiveness of mentoring program varies in results and effect size. Traditional face to face program is better than online mentoring and vice versa [6]. The fact that mentoring studies lack of treatment group which is a limitation of any mentoring program [6]. Thus, this study addresses the problem of declining interest in science and mathematics and promoting positive STEM identity through mentoring system. In addition, this study determines the effectiveness of the program by using a post-test control design approach.

1.1. Mentoring system

Improving the state of science education in Malaysia has become a national priority. In the USA, one response to this problem has been the implementation of STEM outreach programs designed to reignite the interest and awareness of students and parents with the aim to increase the number of students that enter graduate programs in STEM at school and tertiary levels [7]. Studies such as [8] showed that outreach programs in Malaysia, which are gaining in numbers and participations from industries, universities and NGOs, are able to excite and engage students in STEM and students are also able to develop the STEM knowledge and skills. The common underlying thread in outreach or enrichment programs (e.g. robotic clubs, science camps, STEM camps) by STEM graduates and academia is the mentoring process between a mentor and mentee during and after the program. Research shows that students are inspired to involve in STEM when they see the relevance in STEM namely impact of STEM research in the improvement the world that they associate with [5].

Research suggests providing access to faculty mentoring through enrichment programs may also increase the degree to which students identify themselves as scientists [9]. In a review of the impact of afterschool programs, [1] found that university students' current interest in STEM is associated with their participation in afterschool programs. They also argued that students with average achievement in science and mathematics but with an early interest in pursuing STEM fields at a higher level is a better predictor compared to those who are high achievers but do not show interest in the same careers. Thus, mentoring could create a positive STEM identity that leads to better predictor for the student to pursue STEM fields in the future. Mentoring is a relationship between a less experienced individual, called a mentee or protégé and a more experienced individual known as the mentor [10]. Various models of mentorships exist- the more traditional ones such as dyadic, face-to-face, long-term connection, including those at the opposite end of the spectrum-short term and exclusively electronic. Mentors support their mentees in setting appropriate and challenging goals. They motivate mentees for STEM activities and increase their interest in STEM topics through conversations and learning activities [6]. STEM mentoring leads to greater STEM achievement than their counterparts who lacked such opportunities. At the same time, face to face mentoring program such as internship where students learn through immersion in the authentic environment is also meaningful to students socially and intellectually [11], especially those from rural schools who might not have access to internet. Aspects of effective mentoring is also essential. [10] suggest based on literature that with mentors should provide (1) psychological or emotional support, (2) a role model, (3) assistance in goal setting and career paths, and (4) subject-specific expertise. Thus, mentors provide psychosocial and academic support. Psychosocial support is most closely linked to identity formation.

The STEM mentoring programs in UKM has been provided outreach programs since 2013. Thus, one can assume that the programs are equipped with experienced facilitators. Some programs such as BITARA STEM conduct training of trainers' program for their own facilitators so that a more effective mentoring process is provided to the schools. The STEM Mentoring Program was design based on the multidisciplinary STEM activities from three faculties in UKM. The scope of the activities in the program varies from learning embedded system engineering, the nature, insect studies and chemical processes. The hands-on activities gave participants the opportunities to touch and knowledge and products. Such activities aim to create an understanding and awareness that multidisciplinary STEM have an impact on students' everyday life and industrialist in the future.

1.2. Objectives and research questions of the study

The objectives of this study were to assess the impact of students' interest towards STEM through multidiscipline STEM projects in mentoring STEM program. The research questions were:

- What is the level of students' interest towards STEM after they were involved in STEM mentoring program?
- Was there any significant different of interest towards STEM between groups (participated and non-participated in STEM mentoring program)?
- Was there any significant interaction of interest towards STEM between groups and gender?

2. Research design
Posttest-Only Design with Non-equivalent Groups was used in this study. Based on this type of design, both groups (control and treatment) were only given post-test on a range knowledge of STEM and interests towards STEM.

3. Method

3.1. STEM mentoring program
UKM STEM mentoring program comprised of an array of STEM based programs namely CRYSTAL, See Nature, STEM Career, BITARA STEM and Science Camp (See Table 1). These programs are common in terms of: (a) types of activities- which are hands- on, (b) outside of the classrooms, (c) application of STEM knowledge, and (d) related to formal school curriculum. Most programs were conducted in one day at respective schools.

Table 1. UKM STEM mentoring program and activities.

Programs	Modules	Example of activities
CRYSTAL	SCRATCH	Create animation and character using SCRATCH software.
See Nature	Forest Insect Bird Fireflies Nature documentation	Students seek insects in their surroundings. Study the structure and morphology of the insects. Shared their findings
BITARA STEM	Energy Transportation Wireless Communication Urban infrastructure	Developing mini SMART communities
STEM Career	Career personality	Explore students profile of STEM career
Science Camp	Energy Transportation Wireless communication Urban infrastructure	Develop mini SMART communities

3.2. Participants
A total of 400 students (mentee) from six secondary schools in Selangor district, Malaysia participated in STEM mentoring program. 25% from the total of mentees were randomly selected as a treatment group (N = 99). Non-participants in the STEM mentoring program from the same schools were randomly selected as a control group (N = 99). In addition, lecturers as a mentor (N = 20) and graduate students as a facilitator (N = 70) from three faculties (Faculty of Education, Faculty of Science and Technology and Faculty of Engineering & Built Environment) in UKM also participated in this program.

3.3. Instrument and procedure

A set of questionnaire of interest towards STEM was used in this research and employed 5-point agreement of Likert type response (1= strongly not agree to 5= strongly agree). The instrument was adapted from [13] which takes into consideration three main interests of STEM namely: (i) interest towards science, (ii) interest towards mathematics, and (iii) interest towards science & technology. This instrument has been verified by experts in term of construct and content validation. The reliability of the instrument from the pilot study indicates high Cronbach's Alpha value for all constructs (Cronbach alpha > 0.07) (See Table 2). This instrument (post-test) then was distributed to the participants (control and treatment group) after the mentoring program ends.

Table 2. Items and Cronbach alpha value for interest towards STEM.

Element	Cronbach alpha	Example of items	Number of items
Interest towards science	0.733	- I enjoyed learning science - I love doing challenging science activities	4
Interest towards mathematics	0.910	- I enjoyed learning mathematics - I love solving difficult mathematical problem	4
Interest towards engineering & technology	0.815	- I love designing product activities - I am interested to find out how the machine works	7
Interest towards STEM	0.854		17

3.4. Data analysis

The data that have been collected from the post-test was analysed using SPSS 2.10 at the significant level $p < 0.05$. The analysis involved statistic descriptive to determine the level of interest towards STEM using the mean score and standard deviation. Meanwhile, inferential analysis involves MANOVA and independent sample t-test were used to identify the significant difference and interaction between group (control and treatment) and gender.

4. Result and discussion

4.1. Level of interest towards STEM

Table 3 shows the mean score and standard deviation value following by the level of interest towards STEM. The table indicates, the level of all three aspects of interest towards STEM are moderately high and high with the mean scores between 3.5 to 4.3. The interpretation of the level of the mean score was based on [14]. The level of the interest towards science is 'high' for both group and gender. This high score mean of interest towards science probably because in the early ages of learning, in this context were form one student (age 13 years old), the students have more desire or interest to know more about science. But it is feared that their interest in science decreases as their education level increases [15,16]. The current issues that showed interest towards science among school students is declining worldwide [1-3]. Hence, research have suggested [7,9], access of mentoring as an enrichment program or afterschool program can help to increase and maintain their interest.

Meanwhile, the level of interest towards mathematics indicated the different of level in both group and gender. Boys in treatment group shows moderate level of interest towards mathematics compared to the boys in control group with high level of interest towards mathematics. This finding could not be clarified further. It may be explain clearly if the pre-test was conducted as well as a qualitative study to learn more about the cause. On the contrary, girls in the treatment group indicate high level of interest towards mathematics compared to the girls in control group. Some research shows girls interest towards mathematics declined starting in the middle school [17] and worldwide including Malaysia [1-3]. However, one possibility which is contributing to increase the interest among girls is the usability of mentoring group approach [18]. It is shows that mentoring help students to hinder from losing interest towards mathematics and provide them with the scaffold they need to develop greater interest [6].

Based on the Table 3, interest toward engineering and technology indicate boys in the treatment group shows higher interest compared to the girls and control group. One possible reason for the low level interest among girls compared to boys is that that engineering and technology are stereotyped linked to a masculine field and machines are related to the quality valuable for boys compared to girls [19]. However, both girls and boys in the treatment group shows higher score mean compared to the control group. The mentoring programs which involved hands-on activities and multidisciplinary STEM project in creating product, together with the support and motivation form the facilitator, may help both girls and boys to cope with the challenge in accomplished their goals of projects.

Overall, boys and girls in the treatment group shows higher level of interest towards STEM compared to control group. The finding of this study does support previous finding on the positive effect of the STEM Mentoring programs on students' interest towards STEM subject [8]. Findings showed that students' current interest in STEM is associated with their participation in afterschool programs [7]. STEM mentoring leads to a better STEM achievement than their peers who lacked such opportunities. Besides, according to [11], involving in face to face mentoring program is also meaningful to students socially and intellectually.

Table 3. Level of interest towards STEM.

Element	Group	Gender	Mean score	Standard deviation	Level
Interest towards Science	Control	Boys	4.129	0.646	High
		Girls	4.214	0.595	High
	Treatment	Boys	4.250	0.490	High
		Girls	4.388	0.433	High
Interest towards mathematics	Control	Boys	4.000	0.756	High
		Girls	3.620	1.055	Moderate high
	Treatment	Boys	3.885	0.956	Moderate high
		Girls	4.108	0.826	High
Interest towards engineering & technology	Control	Boys	3.961	0.637	Moderate high
		Girls	3.538	0.631	Moderate high
	Treatment	Boys	4.004	0.689	High
		Girls	3.743	0.620	Moderate high
Interest towards STEM	Control	Boys	4.030	0.496	High
		Girls	3.791	0.614	Moderate high
	Treatment	Boys	4.046	0.492	High
		Girls	4.080	0.451	High

Mean score interpretation [14]:
1.00 - 1.99 (low)
2.00 - 2.99 (moderate low)
3.00 - 3.99 (moderate high)
4.00 - 5.00 (high)

4.2. Different and interaction between group and gender

Table 4 shows there are no significant group effect on the three elements of interest in STEM; interest towards science, interest towards mathematics and interest towards engineering and technology. Whereas, the main effect of gender is significant only for element interest towards engineering & technology, $F(3,194) = 8.504$, $p < 0.05$; Pillai's Trace = 0.118; partial $\eta 2 = 0.118$. However, only boys showed the 'significant' effect from 'moderately high' to 'high' as the girls only showed slightly significant different of mean score (refer Table 3). While the effect of interaction between group and gender was significant only for element interest towards mathematic, F (3,194) = 1.724, p = 0.163; Pillai's Trace = 0.026; partial η2 = 0.026.

Table 4. Multivariate test.

Source	Element	Sum of square	df	Mean square	F	Sig.	Partial eta squared
Group	Science	0.950	1	0.950	3.278	0.072	0.017
	Math	1.527	1	1.527	1.768	0.185	0.009
	E & T	0.673	1	0.673	1.639	0.202	0.008
Gender	Science	0.540	1	0.540	1.863	0.174	0.010
	Math	0.269	1	0.269	0.031	0.577	0.002
	E & T	5.127	1	5.127	12.491	0.001	0.061
Group*gender	Science	0.031	1	0.031	0.106	0.745	0.001
	Math	3.999	1	3.999	4.629	0.033	0.023
	E & T	0.288	1	0.288	0.701	0.403	0.004

Based on the results of the independent sample t-test as shown in Table 5 and Table 6 it was found that there was a significant difference between girls for interest in mathematics in control and treatment group (t = -2.897, df = 127, p = 0.004) . Mean score of interest towards mathematics for girls in treatment groups is higher compared to girls in control group (treatment: mean=4.108; control: mean=3.620. (Refer Table 3).

Table 5. Independent sample t-test by group.

Element	Group	Gender	Mean	Standard deviation	t	df	Sig.
Interest towards mathematics	Control	Boys	4.000	0.756	1.759	97	0.082
		Girls	3.620	1.055			
	Treatment	Boys	3.885	0.956	-1.237	97	0.219
		Girls	4.108	0.826			

Table 6. Independent sample t-test by gender.

Element	Group	Gender	Mean	Standard deviation	t	df	Sig.
Interest towards mathematics	Boys	Control	4.000	0.756	0.537	66	0.593
		Treatment	3.885	0.956			
	Girls	Control	3.620	1.055	-2.897	127	0.004
		Treatment	4.108	0.826			

Based on the findings above, UKM STEM mentoring program is effective in improving interest towards mathematics especially for girls (see Table 6). This finding shows, negative stereotype with maths [19] or interest declined among girls can be overcome through STEM mentoring program with multidisciplinary STEM project. This mentoring program offers psychology and academic support by trained and experienced facilitators from different faculties. Because of this factor, therefore, the process of mentoring becomes effective [12].

5. Conclusion

UKM STEM mentoring program was conducted to explore students' interest towards science, mathematics and engineering and technology. Participants consists of mentees (N=198), mentors and facilitators. The descriptive analysis showed that the level of the interest towards science is 'high' for both group and gender. Meanwhile, the inferential analysis showed that there was a significant difference between girls for interest in mathematics in control and treatment group. However, there were no significant group effect on the three elements of interest in STEM; interest towards science, interest towards mathematics and interest towards engineering and technology. The study indicated that the program was effective at modifying students; interest level as the result revealed positives changes for both group and gender but the data still not enough to suggest that the positive changes due to the interaction in this STEM mentoring program. Nevertheless, students who participate in a mentoring program tend to be more satisfied with their academic experience than their non-mentored peers [20,7]. The limitation of the study that there was no pre-test data thus future research should include a pre-post-test controlled design. In addition, followed up interviews would illuminate the effects of mentoring.

Acknowledgments

Authors wishing to acknowledge that this research were funded by Universiti Kebangsaan Malaysia (UKM) (Code Project: GUP-2017-016) and Ministry of Education (MOE) and under National STEM Movement. This project also in collaboration with Selangor State Education Department (JPNS) and Schools (SMK Sg. Rawang, SMK Sepang Putra, SMK Engku Husain, SMK Darul Ehsan, SMK Gombak Setia, and SMK Sg. Ramal).

References

[1] Jayarajah K, Saat R M, Rauf A and Amnah R 2014 A Review of Science, Technology, Engineering & Mathematics (STEM) Education Research from 1999-2013: A Malaysian Perspective *Eurasia Journal of Mathematics, Science & Technology Education* **10** 155

[2] Osman K, Soh T M T and Arsad N M 2010 Development and validation of the Malaysian 21st century skills instrument (M-21CSI) for science students *Procedia Social and Behavioral Sciences* **9** 599

[3] Halim L 2013 *Science education and the development of a scientific literate society* (Malaysia: National University of Malaysia Press)

[4] English L D 2017 Advancing elementary and middle school STEM education *International Journal of Science and Mathematics Education* **15** 5

[5] Schreiner C and SjØberg S 2006 Science Education and Young People's Identity Construction - Two Mutually Incompatible Projects? *Science Education and Youth's Identity Construction* 1–17

[6] Stoeger H, Duan X, Schirner S, Greindl T and Ziegler A 2013 The effectiveness of a one-year online mentoring program for girls in STEM *Computers and Education* **69** 408

[7] Krishnamurthi A, Ballard M and Noam G G 2014 Examining the Impact of Afterschool STEM Programs *Afterschool Alliance* 1-36

[8] Shahali E H M, Halim L, Rasul M S, Osman K and Zulkifeli M A 2017 STEM Learning through Engineering Design: Impact on Middle Secondary Students' Interest towards STEM *Eurasia Journal of Mathematics, Science and Technology Education* **13** 1189

[9] Zaniewski A M and Reinholz D 2016 Increasing STEM success: a near-peer mentoring program in the physical sciences *International Journal of STEM Education* **3** 14

[10] Stumbo N J, Martin J K, Nordstrom D, Rolfe T, Burgstahler S, Whitney J, Turnbaugh S L, ... Misquez E 2010 Evidence-based Practices in Mentoring Students with Disabilities: Four Case Studies *Journal of Science Education for Students with Disabilities* **14** 33

IOP Conf. Series: Journal of Physics: Conf. Series **1088** (2018) 012001 doi:10.1088/1742-6596/1088/1/012001

[11] National Academies of Sciences 2017 *Undergraduate Research Experiences for STEM Students: Successes, Challenges, and Opportunities* (Washington D C: The National Academies Press)
[12] Hudson P 2013 Strategies for mentoring pedagogical knowledge *Teachers and Teaching: Theory and Practice* **19** 363
[13] Unfried A, Faber M, Stanhope D S and Wiebe E 2015 The Development and Validation of a Measure of Student Attitudes Toward Science, Technology, Engineering, and Math (S-STEM) *Journal of Psychoeducational Assessment* **33** 622
[14] Nunnally J C and Bernstein I H 1994 *Psychometric Theory (McGraw-Hill Series in Psychology)* (USA: McGraw-Hill)
[15] Barmby P, Kind P M and Jones K 2008 Examining Changing Attitudes in Secondary School Science *International Journal of Science Education* **30** 1075
[16] Osman K 2007 Sikap terhadap sains dan sikap saintifik di kalangan pelajar sains *Jurnal Pendidikan Malaysia (Malaysian Journal of Education)* **32** 39
[17] Hayden K, Ouyang Y, Scinski L, Olszewski B and Bielefeldt T 2011 Increasing student interest and attitudes in STEM: Professional development and activities to engage and inspire learners *Contemporary Issues in Technology and Teacher Education* **11** 47
[18] Stoeger H, Hopp M and Ziegler A 2017 Online Mentoring as an Extracurricular Measure to Encourage Talented Girls in STEM (Science, Technology, Engineering, and Mathematics): An Empirical Study of One-on-One Versus Group Mentoring *Gifted Child Quarterly* **61** 239
[19] Cheryan S, Master A and Meltzoff A N 2015 Cultural Stereotypes as gatekeepers: Increasing girls' interest in computer science and engineering by diversifying stereotypes *Frontiers in Psychology* **6** 49
[20] Bearman S, Blake B S, Hunt L and Crosby F J 2007 *New directions in mentoring* In T D Allen and L T Eby (Eds.) *The Blackwell handbook of mentoring: A multiple perspectives approach* (New York: Blackwell) p 375

IOP Conf. Series: Journal of Physics: Conf. Series **1088** (2018) 012002 doi:10.1088/1742-6596/1088/1/012002

How do we let students work as 'young mathematicians' in the classroom?

M Dolk
Inspiration 4 Learning, Utrecht, the Netherlands / Freudenthal Institute, Utrecht University, the Netherlands

Email: m.dolk@uu.nl

Abstract. We must encourage students to solve mathematical problems and-in my opinion, more importantly- to develop and justify mathematical theories. If that succeeds, mathematics becomes more than finding the right answer or repeating a teacher's argument. But how do you do that in the class room? The problems we pose to students and the questions we ask largely determine the image of mathematics that students develop. If we constantly ask for the 'right' answer, students will think that mathematics is primarily a mathematical activity and that mathematical thinking stops when they have found the 'teacher's answer'. However, when we ask students to defend their approach and encourage students to generalize the strategy used, a different, richer picture of mathematics arises.

1. What is the mathematics in the investigation?

There are basically two ways to look at a mathematical investigation that teachers ask children to work on. On the one hand, we can ask ourselves what we need to teach our children so that they can get the right answer to that problem. On the other hand, we can analyze what mathematics our children can develop while working on this problem, and we can try to support our children to develop and understand this mathematics?

Teachers who use a textbook that contains series of small calculation tasks without any internal connection, often follow the first approach. They want to help the children by teaching them how to get the answer. As the problems in those textbook series are unrelated (see fig 1), children will not develop any new insights in mathematics; they might, however, practice fluency in one or more mathematical operations.

Teachers who focus more on the development and understanding of mathematical theory will be more inclined to see investigations as an opportunity for the children to develop mathematical theory. They will look for entry points to discuss the mathematics behind the investigation and therefore go beyond the answer to the problem. They will support children by asking questions that are at the edge of their understanding and

Children will also behave differently. When they focus on getting a correct answer, children will play safe. They will not feel challenged to experiment with strategies they do not fully understand, however they will use lower order strategies that will give them *the* answer. When children get space to experiment, they will not always find the correct answer. They will be extremely interested in discussing their approach and-if needed-adjust it.

Consider children who are still developing subtraction strategies, trying solve 36 - 19. If pressed for the correct answer they will probably count three times: they tally 36, they strike through 19 tallies and count the remaining ones. However, if they are giving space to explore this and other problems, some might try subtracting 20 in two jumps of 10: 36, 26, 16. They realize they have to compensate 1, but they might not be sure if it is 15 or 17.

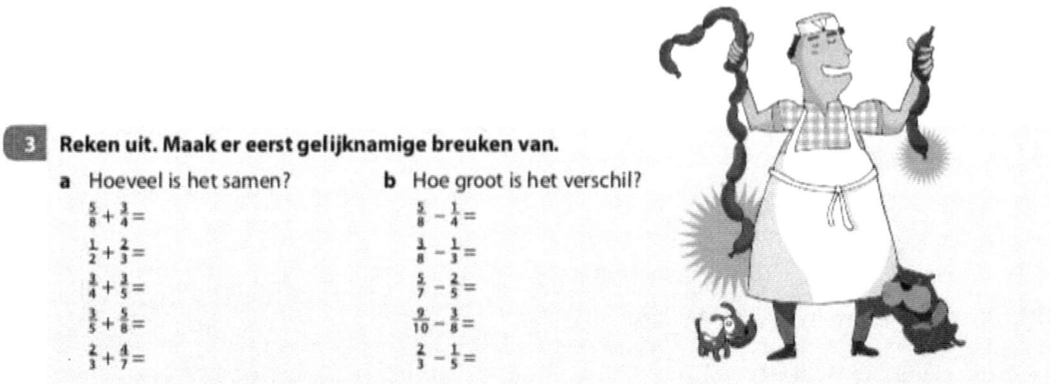

Figure 1 A series of unrelated problems, taken from a Dutch mathematics textbook. The problem states: “Calculate. First, convert the fractions so that they have common denominators. (a) How much is it together? (b) How much is the difference?”[1]

Peter’s method, as describe by Max Stephens [2], is a great example that shows how children’s inventions can be a starting point for vertical mathematizing - for the further development of mathematical. Peter says, “I do these by first adding 5 and then subtracting 10, like 32 - 5 = 32 + 5 - 10. Working it out this way is easier.” Peter is using a strategy that not many children will use initially. So, the first question might be whether Peter’s method even works: “will it always give the correct answer?” Tim (9 years, 1 month) shows how powerful it can be to discuss Peter’s method with children: “Here is an explanation for all numbers. Whatever number he (Peter) is taking away (assumed to be less than 10), you plus the number that would make a ten, and you take away ten. The bigger the number you are subtracting, the smaller the number you are plussing. They all make a ten together” [2]. Tim generalizes over all minuends and all subtrahends. In algebraic terms, he says: ‘a - b is the same as a + (10 - b) - 10.’

In the next paragraph, we will look at an example of an investigation that will allow children to develop mathematical theory. We will not only look at the investigation, but we will also focus on the mathematics the students can invent.

2. One investigation: many problems at many levels

SEAMEO Regional Centre for QITEP in Mathematics (SEAQiM) in Yogyakarta (Ind), organizes yearly a Southeast Asia Realistic Mathematics Education (SEA-RME) course which is based on the regional culture, nature, and characteristics of Southeast Asia. This course gives teachers from SEAMEO Member Countries the opportunity to work collaboratively, celebrate diversity, and bring these experiences to their mathematics class rooms [3]. During the RME course in 2017, a group of 30 teachers worked on a number of mathematical investigations in number explorations, algebra, geometry, and data handling; they observed and discussed video-examples of RME investigations in primary schools; they read and discussed articles about RME in Indonesia; and they designed and implemented an investigation for primary school students.

In the design of an investigation, the course followed a lessons study approach. Groups of participants, supported by SEAQiM staff members, designed an investigation. Each investigation was first given to all other participants. Given the experiences with this try-out, the investigation was further improved. Next, the group who designed the investigation implemented it in a class in Yogyakarta. Other participants observed and made notes. These experiences were discussed with all participants. Subsequently, the designers redesigned the investigation. This final design was given to all participants to implement in their school and in their country.

One of the investigations was inspired by work of Wahid Yunianto, head of the 'Research and Development, Capacity Building and Training' division [4,5]. Let me summarize the investigation in my words: a farmer wants to fence off a small coral. He has 25 fence pieces; each fence piece was one-meter long. He wants to know what shape would give him the largest enclosed area. As support, the students were given 25 yellow squares (the fence) and a larger set of green squares (the field).

During the try-out in a class in Yogyakarta, the children started to create rectangles with the green squares and a fence with the yellow squares. All groups started with a rectangle, most groups started with a square, a few used a non-square rectangle. Once all yellow squares were used, the children counted the number of enclosed greens squares and gave that total as their answer. If we only focus on the answer, this would have been the endpoint of the investigation. The students would say "25" and a happy teacher would say "great, next problem." However, so much more could be at stake. I will discuss five issues that are possible discussion points in this investigation: the structure of the array model, the shape of the largest enclosed area, the relationship between perimeter and area, the transition from counting to skip-counting, and the transition from skip-counting to multiplication.

2.1. Understanding the array model

To go beyond counting, children need to understand the structure of an array as Figure 2 below. They need to see an array as a series of connected rows or columns to start using skip counting or multiplication. Seeing an array as a combination of rows or columns is not something children will automatically do; they need to construct it [6,7]. Therefore, it is not supporting when a teacher tells children that they need to count the number of tiles in one row and count the number of rows. We can assess if children have constructed this idea by asking them — the next day when the green and yellow tiles are not around— to draw the field the farmer has built.

Figure 2. Children are investigating what the size is of the largest field enclosed by 25 (yellow) fence pieces.

2.2. What border shape will create the largest enclosed area?

Starting with a square does not mean that the children are convinced that a square will give the largest area, nor that they will that be able to justify why the square will lead to the largest number of enclosed tiles. It is not too hard to challenge them to explore if another rectangular shape, say 4 by 6, would create a larger area. Figure 2 shows a group that also explores a concave, non-rectangular shape. They might even explore octagons or other more circular shapes; however, we need to keep in mind that the green squares will hinder the children to go in this direction. Once the children are convinced that the square will give a larger area compared to non-square rectangles, they will need to prove that statement.

2.3. Rectangles with the same perimeter do not need to have the same area

While they are exploring what shape will enclose the largest area, they will find that different rectangles with a perimeter of 25 will not always have the same area. So, not only will they start to discuss why the square has the largest area, in the end, they might also be able to discuss a big idea: "when the perimeter of two rectangles are the same, the area does not have to be the same", and also: "when the area of two rectangles are the same, the perimeter does not have to be the same".

2.4. From counting-all to skip counting.

When children see the array as a series of connected rows or columns, they will understand that counting the number of tiles in one row and the number of rows will help to find the total number of tiles. However, something odd happens. While counting the children know that you have to count each tile once. Yet, when children count the number of tiles in a row and count the number of rows, it appears as if they count at one tile twice. They point at that particular tile counting the number of tiles in the row, and they point at that same tile again when they count the number of rows. Being able to justify that they do not count that tile double, requires a deep understanding of the structure of the array.

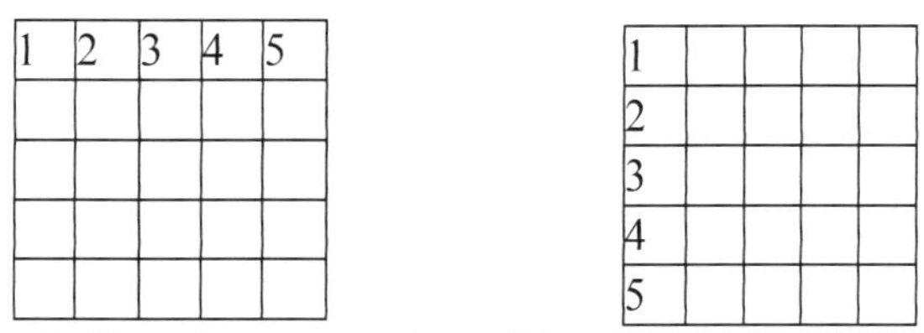

Figure 3. Counting the number of tiles in one row (left) and counting the number of rows (right). Did we count the top-left tile twice?

The size of the array matters. For children, there is not much difference in skip counting 5 fives or counting 25 tiles. Larger numbers will make it more rewarding to skip count. Imagine that the field is 10 by 16. Counting 160 tiles is a lot of work, skip counting 10, 20, ..., 160 will be rewarding and intriguing. In designing an investigation, we have to decide between using numbers that make the investigation doable or choosing numbers that will entice children to look for more advanced approaches.

2.5. From skip counting to multiplication

A deeper understanding will be developed when students start to talk about multiplication. When students skip count, they count the number of tiles. However, what does it mean when they say the area is 5×5. Clearly, they do not mean that the area is 5 tiles by 5 tiles or 25 square tiles. In the multiplication, the 5 does not refer the number of tiles, but to the length of one side of the rectangle.

2.6. An accessible investigation

The children realize what they can do in this context and with the given materials. They start laying the tiles down, building the fences, and find the enclosed area. They can work on this investigation at a very concrete level, therefore this investigation is accessible for all. It clearly allows for natural differentiation, as it also offers many challenging problems at many levels. Children can be challenged to continue working on an issue that is at the edge of their understanding. Here, the support of the teacher is very important. The teacher needs to be able to challenge each group of children to work as young mathematicians.

3. How do we change classrooms into communities of mathematicians?

Changing classroom practice is not easy. This is not so much a problem for the children, it is foremost a problem for the teachers. They have to change their habits, teaching styles, believes, and split-second behavior. They have to provide children with interesting and challenging investigations. A good start is to:

- Present an investigation in a clear context. It can be a real-world context, it can be a fantasy context, it might even be the world of mathematics. Presenting a context means more than handing out a worksheet. Tell a story that connects to the children's experience. It should allow all children to work at their own level: they can work at a very concrete level, or they work at a more abstract level.
- Let the children work on the investigation in small groups. I would suggest creating an optimal mismatch when you group the children. When you create a group with a child who works concrete and a student who works abstract, they might not understand what the other is talking about. When all children are working in an identical way, there is nothing to discuss. It is better to create groups of children who are close but not too closely aligned in their approach to the investigation.
- Support the children during their investigation. Do not tell them how they can solve their problems but support them to work on the mathematics themselves. Cathy Fosnot [8] talks about three stages of a conferral: clarification, celebration, and challenge. When you start a conversation or a conferral with children, you might start by listening to the children talking and clarifying what you have heard and seen the children do. This gives them a chance to explain their understanding and their wonderments. Next, you celebrate their findings. And you end by posing some further investigations at the edge of their understanding
- When most groups have solved the problem, ask each group to create a poster that shows and justifies their thinking. Creating a poster that only shows a solution is not informative as all groups will have solved the investigation. All children will know *the answer*; however, they might not understand the approach of other groups. A poster should contain a justification of their solution, a viable mathematical argument [9], open questions, wonderments... This will force the children to reflect on their work and thinking, to develop mathematical argumentation, and to be ready to defend their thinking.
- Ask a few of the groups to explain their poster to the other children in the class. Let other children ask questions and clarifications. It is not a share, where all groups have to tell their story, it is a mathematical congress where the community of learners defends, questions, develops, and explores mathematical understanding. You could ask questions that might lead to vertical mathematizing, to the further development of mathematical theory. This might happen when you contrast two approaches, or when you build on an invention by one of the groups, or when you explore one of their wonderments.

References

[1] Malmberg 2010 De Wereld in Getallen: voorproefje [Online] Available: http://www2.malmberg.nl/BAO/WIG/wig-8-lb-lessen/files/513830-01_lessen_180dpi1.pdf. [Accessed 9 5 2018]

[2] Stephens M 2007 Using number sentences to introduce the idea of variable [Online] Available: https://nzmaths.co.nz/2007-national-numeracy-facilitators-conference [Accessed 9 5 2018]

[3] SEAQiM Southeast Asia Realistic Mathematics Education (SEA-RME) [Online] Available: http://www.qitepinmath.org/en/programmes/regular-courses/southeast-asia-realistic-mathematics-education-sea-rme/. [Accessed 18 5 2018]

[4] Yunianto W Supporting 7th grade students' understanding of the area measurement of quadrilaterals and triangles through reallotment activities [THESIS]. Palembang, Sriwijaya University, 2014

[5] W Yunianto "Developing understanding of area through reallotment activities," in Examples of Mathematics Investigations for Primary Schools in Indonesia (working title) ed F van Galen, D van Eeerde (Utrecht, Freudenthal Institute) (Under development) Available: http://www.fisme.science.uu.nl/en/impome/investigations

[6] Battista M T, Clements D H, Arnoff J and Van Auken Borrow C 1998 Students' Spatial Structuring of 2D Arrays of Squares *Journal for Research in Mathematics Education* **29** 503

[7] Outhred L and Mitchelmore M 2004 Students' Structuring of Rectangular Arrays Proceedings of the 28th Conference of the International Group for the Psychology of Mathematics Education (Bergen, Norway)

[8] Fosnot c 2016 *Conferring with Young Mathematicians at Work* (New London: New Perspectives on Learning LLC)

[9] Stylianou D and Blanton M 2018 *Teaching with Mathematical Argument* (Portsmouth, NH: Heinemann)

[10] Gravemeijer K, Stephan M, Julie C, Lin F L and Ohtani M 2017 What Mathematics Education May Prepare Students for the Society of the Future? *Int J of Sci and Math Educ* **15** 115

IOP Conf. Series: Journal of Physics: Conf. Series **1088** (2018) 012003 doi:10.1088/1742-6596/1088/1/012003

Teachers' collective knowledge: the case of equivalent fractions

Z H Putra[1,2] and C Winsløw[1]
[1]Department of Science Education, University of Copenhagen, Øster Voldgade 3, 1350 Copenhagen, Denmark
[2]Faculty of Teacher Training and Education, University of Riau, Kampus Bina Widya Km 12,5 Simpang Baru, Pekanbaru, Riau 28293, Indonesia

E-mail: zetra.putra@ind.ku.dk

Abstract. Research on teachers' mathematical knowledge has grown significantly over the last few decades. Many studies concern teachers' knowledge by using written tests, as for students. But in reality, teachers do not work in isolation but in institutions where professional knowledge is shared. How can this shared or collective knowledge be studied systematically and precisely? With this in mind, we designed so-called hypothetical teacher tasks (HTTs) which teachers solve in pairs. Each HTT can be used to investigate teachers' knowledge of some specific mathematical piece of knowledge (like how to add fractions) and knowledge about how to teach it. We present a case from Danish and Indonesian pre-service teachers' collaborative work on one HTT about equivalent fractions. We analyse how the shared mathematical and didactical knowledge of equivalent fractions differ between the two groups. Indeed, there are visible differences. The Danish pair proposed several didactical ideas to teach equivalent fractions, while the Indonesian pair focused on the direct instruction of standard algorithm.

1. Introduction

Research on teachers' knowledge has grown significantly over the last few decades especially after Shulman's seminal work introduced a new perspective on content knowledge in teaching [1]. This knowledge consists of subject matter content knowledge, pedagogical content knowledge, and curricular knowledge. These categories are closely related to learning and teaching practices in classrooms, and they become the primary focus of many studies including in mathematics education such as mathematical knowledge for teaching (MKT) [2].

Most studies on teachers' knowledge focus on teachers' knowledge by using written tests [e.g., 3], as for students, in which it is not in line with Shulman's view on how to study teachers' content knowledge and pedagogical content knowledge. For instance, he argued that content knowledge is not only about teachers' capability to define a correct answer in a specific subject, but "teachers must also be able to explain why a particular proposition is deemed warranted, why it is worth knowing, and how it relates to other propositions, both within discipline and without, both in theory and in practice" [1]. Therefore, to study teachers' knowledge one should let teachers share their knowledge that is related to what and how the teachers work in institutions where professional knowledge should not be isolated but shared. We try to address this issue in the present study.

Teachers' collective knowledge has been studied by Winsløw and colleagues [4,5] by designing so-called *hypothetical teacher tasks* (HTTs) which teachers solve in pairs. Each HTT aims to investigate teachers' mathematical knowledge, and how the teachers share this knowledge during their collaborative work. The mathematical task involved in the HTTs is both standard and elementary, that means in principle be addressed in many institutions [4]. But, it should give several possibilities, both practices, and theory, for teachers to solve the mathematical task and propose adequate didactical ideas. We adopt the idea of HTTs to study pre-service teachers' mathematical knowledge of rational numbers, specifically the case of equivalent fractions. We investigate how HTTs can be a useful tool to study and compare teachers' collective knowledge from two different institutions, Denmark and Indonesia.

In designing and studying teachers' collective knowledge, this study is developed based on the anthropological theory of the didactic (ATD) introduced by Chevallard [6], specifically the notion of *praxeology*. We apply the praxeology because it provides an epistemological model to study human practice and theory. The praxeology comes from a Greek word, *praxis,* and *logos*. The praxis or practical block is unified by *a type of tasks* (T) and *techniques* (τ). In the case of rational numbers, a mathematical type of task is to find a fraction that is equal to $\frac{a}{b}$. To solve it, one can employ several possible mathematical techniques such as multiplying both numerator and denominator by a positive integer or doubling each numerator and denominator. While, in the case of teachers' knowledge, the type of tasks is not only about mathematics but also didactics. An example of didactical tasks is to organise a teaching activity related to the equivalent fractions to support pupils learning, and the didactical technique could be to instruct pupils the standard mathematical technique of equivalent fraction $\left(\frac{a}{b} = \frac{na}{nb}, n \text{ is an integer}\right)$. While the logos or theoretical block is also unified by two components called a *technology* (θ) and a *theory* (Θ). The technology functions to justify or explain several techniques. An example of technological discourse related to the technique for the equivalent fraction task is that two fractions are equal if the value or size of the two fractions is the same. To show them, one can change both fractions into decimals. There is also a didactical technology associated with the didactical technique. For instance, a teacher who directly instructs pupils the standard mathematical technique may believe that the pupils need to receive the direct instruction on how to solve a mathematical task, and then use it to the similar task type. The second part of logos is the theory used to be a basis and support of the technology. Equivalent value of different representations of rational numbers can be seen as a theory to justify that mathematical technology, and the direct instruction theory is the example of didactic theory to justify the didactical technology. We use the mathematical and didactical praxeology as an analytical framework to the present study of teachers' mathematical and didactical knowledge of rational numbers, especially in the case of equivalent fractions, in the setting of collaborative work, because it provides us with a more detailed model to investigate teachers' knowledge, and the collaborative work provides opportunities for teachers to share and critically re-examine how they make claims from facts, and how they build on one another's ideas to construct more sophisticated ways of reasoning [7].

2. Method

The present study focuses on one among five HTTs designed for studying teachers' mathematical and didactical knowledge of rational numbers [8]. The design of HTT about equivalent fractions follows a similar way to the other HTTs. It is started by reviewing some previous studies on equivalent fractions that pupils have difficulties and misconceptions to find and justify the equivalent values of two fractions. One of the main misconceptions of finding an equivalent fraction is that pupils tend to add both numerator and denominator of a fraction by a positive integer $\left(\frac{a}{b} = \frac{a+n}{b+n}, \boldsymbol{n} \textbf{ is a positive integer}\right)$. By proposing this situation to PsTs, they may consider different mathematical and didactical praxeologies. The concrete HTT of equivalent fractions given to PsTs is presented in the following figure:

> You ask fourth grade pupils to find fractions which are equal to $\frac{3}{4}$.
> *A pupil claims that $\frac{3}{4} = \frac{8}{9}$ because if you add 5 to both the top and the bottom of a fraction, the new fraction must be equal to the original.*
> a. What do you think about this answer? Please explain! (*to be solved individually within 3 minutes*).
> b. What would you do as a teacher to help the pupils from this case to understand the concept of equal fractions better? (*to be discussed and solved in pair within 5 minutes*).

Figure 1. HTT about equivalent fractions

The HTT about equivalent fractions along with other four HTTs of rational numbers has been tested to 31 Danish and 32 Indonesian PsTs. Almost all of them worked in pairs except for one Danish group consisting of three PsTs, so there were 15 Danish and 16 Indonesian pairs. However, in this particular study, we focus on presenting and analysing one pair from each group because the primary aim of this study is showing how HTTs can be used to study teachers' collective knowledge, and how this knowledge can be developed during their discussion.

From figure 1, we can describe two possible mathematical types of tasks and one didactical type of tasks. The first mathematical type of tasks (T_{m1}) is to find a fraction that is equal to $\frac{a}{b}$. The second mathematical type of tasks (T_{m2}) is to evaluate the two fractions $\frac{a}{b}$ and $\frac{a+n}{b+n}$ are equal, and it is based on the pupil's claim. Then, from question b the didactical task type (T_d) is to propose didactical praxeologies to help pupils to solve the task type T_{m1}, but some PsTs may relate it to T_{m2} when they discuss during their collaborative work. For each type of task, we can describe some possible techniques, and those can be found in Putra and Winsløw [8].

The data consists of PsTs' written answers from question a. and video recordings from question b. Especially for the video recordings, their utterances during the discussion were transcribed from time to time to produce detailed mathematical and didactical ideas. The transcripts were analysed in term of mathematical and didactical praxeologies (techniques, technologies, and a theory), both on mathematics and didactics. The first author did the principal analysis by reading the transcript several times and interpreting based on the praxeologies. Then some questionable points found in the data were discussed with the second author.

3. Findings

Now, we present our praxeological analysis of one Danish and one Indonesian pair working on the HTT about equivalent fractions. We select the pairs that can represent a commonality for their groups.

The Danish pair (D_1 and D_2). The written answer given by D_1 focused on mathematical technologies to clarify the pupil's incorrect mathematical technique of adding the positive integer 5 to the numerator and the denominator (τ_{m1}). He wrote "The size of a fraction depends solely on the relationship between the numerator and the denominator. If you add 5 to the numerator and the denominator, the size of the fraction increases because the numerator will be a larger percentage". D_1 had the mathematical technology to justify τ_{m1} based on the ratio, or a relationship between two quantitates, the numerator and the denominator. This is indicated that adding a number to both quantities does not give the same size of the two fractions. D_1 also wrote, "If the pupil had extended the fraction by multiplying the same factor in the numerator and the denominator, the relationship would still be the same, and the fraction size would be the same." This means that the appropriate mathematical technique to find an equivalent fraction is to use multiplication instead of addition (τ_{m1}) because it will give the same ratio between the two quantities. While D_2 also proposed a mathematical technology to justify τ_{m1} that was a little bit similar to what has been suggested by D_1. He said that pupils mix between addition and multiplication, but when one affects the numerator, the same goes to the denominator. Besides, D_2 also proposed a didactical technique that he would like to show to pupils that $\frac{8}{9}$ give a bigger number than $\frac{3}{4}$ by finding the quotients of the two fractions.

At the collaborative work, the Danish pair concerned on how to teach the pupils. They proposed three different didactical techniques to support pupils to understand the equivalent fractions. The first didactical technique was to present pupils with a common and straightforward fraction. It was stated by D_1 when they started the discussion. He said "Perhaps something about simplifying fractions. Perhaps [we] present them with the fractions of $\frac{10}{100}$, and then somehow to teach them, that is the same as a tenth". The didactical technology behind this technique was to have something easy for the pupils to grasp the meaning of equal fractions.

The second didactical technique was to support pupils to check the quotients of the two fractions $\frac{3}{4}$ and $\frac{8}{9}$. D2 explained it based on what he wrote in his written answer.

D_2: It's. They are on the track of the correct thing since they have the understanding that. They are just mixing up addition and multiplication, and they know that when you influence the numerator, that also works in the denominator, since [they] can prolong a fraction. In that sense, they are quite close, and I think that the only thing they are doing now is to test some hypotheses that they think, well earlier experiences in some way. Because, if you have two numbers then, and you add 5 to both then the difference should be the same. It just isn't when calculation [with] fractions. That's a shame, but what they have tested, which they could earlier.... and therefore, I think (..) almost right, to give the pupils an understanding... I think it's ok to make the pupils calculate these two numbers. Find out what it is as a decimal [number]. Perhaps do it together on the blackboard, have some pupils' approach [on the board], and use the calculator, so they will see the difference, and they see hmm, what it is then, that we perhaps did wrong.

D_1: Yes, perhaps start there, instead of giving them the answer to make them wonder.

D_2: Yes, because it is a little hypothesis they have [gotten] wrong. But it's something that is fine enough because if you have the number 5 and the number 10, and you have to prolong, you know make those numbers bigger, but it doesn't matter how there has to be the same value in between them. The proportion should be the same, well then it is okay to add the same in both. But it is just a little mistake when considering the rules of [operating] fractions.

D_1 : Yes. There it is the proportion between the numerator and the denominator which is important.

From the excerpt, D_2 supported the pupils to change both fractions into decimals because these could help them to see that the two fractions did not give the same value, the same decimal representation. The use of a calculator to check the value of the two fractions can be seen as a didactical tool to support pupils' understanding, but it also provides a challenge for teaching and learning activities, such as $\frac{8}{9}$ gives a result of repeated decimals. Besides, they also provided a didactical technology based on the mathematical idea of proportion. This means that the two fractions are equivalent when they have the same proportion.

The last didactical technique was to teach pupils using pizza representations, and the main purpose was to show that the two fractions $\frac{3}{4}$ and $\frac{8}{9}$ are not equal. During the discussion, D_1 drew the two diagrams to represent both fractions (figure 2). He suggested this didactical technique to let pupils find by themselves what went wrong. He believed that the visual representation could help the pupils learn a lot about the equivalent fractions especially for the pupils who were still in the fourth grade. So, it seems that the nature of teaching elementary school pupils based on contextual situations becomes a didactical technology to justify the didactical technique based on the pizza representations.

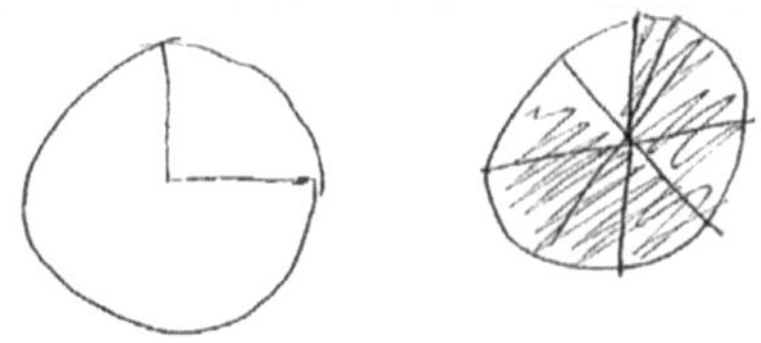

Figure 2. Pizza representations to support pupils' conception of equivalent fractions

The Indonesian pair (S_1 and S_2). The written answers given by both PsTs indicated that they recognised that the pupil's mathematical technique, adding both numerator and denominator by a positive integer (τ_{mi}), was incorrect. So, they proposed a correct mathematical technique, multiplying/dividing both numerator and denominator by a positive integer (τ_{m1}). S_1 also provided a technological discourse to justify the mathematical technique τ_{m1}. He argued that the two fractions are equivalent when the quotient of both fractions was the same (θ_{m1}). Besides, S_1 also stated that the pupil's explanation could be directed to find the least common multiple (LCM) because the pupil has understood that it has to be done an arithmetic operation to both numerator and denominator to get an equivalent fraction. This means that the pupil just chose an inappropriate operation, the addition instead of the multiplication. This explanation could be interpreted as a didactical technology to justify the pupil's mathematical thinking, and the mathematical idea of finding LCM seems less appropriate because the right term is just "multiple" or to multiply each numerator and each denominator.

At the collaborative work to address the task type T_d, S_1 dominated the discussion. He started it by asking S_2's opinion and continued by re-reading question b (T_d). Instead of giving an opportunity to S_2 to explain her thoughts, S_1 just proceeded to explain the technological discourse behind the pupil's answer based on what he has written on question a. He argued that the pupil has already understood how to find an equivalent fraction that it has to be done with the numerator and the denominator (adding the same number to both numerator and denominator), but it was incorrect. He continued the explanation by proposing a correct mathematical technique τ_{m1} that the pupil should apply to solve the task type T_{m1}, and the technique was correct if the quotient of the two fractions gave the same result (re-explain the technology θ_{m1}). While S_2 just seemed to agree with that explanation that it was indicated by her short answers "hmm" during the discussion. So, we can say that both PsTs agreed to instruct pupils with the correct and standard mathematical technique τ_{m1} to solve the task type T_{m1} as the main didactical technique for finding equivalent fractions.

The discussion was continued by proposing further possible justification behind the incorrect mathematical technique (τ_{mi}) given by the pupil. S_1 argued "[the arithmetic] operations needed [for the task] are multiplication and division," but when S_2 asked him a reason behind these two operations, S_1 claimed that it was based on "the rule." Although he did not elaborate on it, he tried to define the definition of equivalent fractions from the meaning of a fraction itself. He stated, "A fraction [$\frac{a}{b}$] is the division [*a* by *b*], and equivalent value to the division is the multiplication and the division itself." From this statement, we can interpret that S_1 tried to justify the mathematical technique τ_{m1} is correct from the concept of fraction as a division of two positive integers. They continued the discussion, but they still discussed the idea behind the reason to choose the multiplication or the division. For instance, we present a further discussion of the two Indonesian pair as follows:

S_2 : I think it is better to explain why it has to be multiplied not to be added. Do you understand it?

S_1 : [I] understand [it].Yes, it is like that, [but] why it should be multiplied. It is not only to be multiplied, but it is also possible to be divided.

S_2: Yes, it is divided when the fraction involves [two] bigger [digits]. Yes, it has to be divided by the same [positive] integer. If the case like this. The pupil asks to find a fraction that is equal to $\frac{3}{4}$. It involves [two] smaller [digits 3 and 4].

S_1: Maybe like this. A fraction is equal to the division. $\frac{3}{4}$ means that 3 is divided by 4. Which is equivalent to the division, ... division, multiplication, subtraction, and addition...the same level to the division is multiplication, so only the multiplication can be used to find the equivalent fraction. Hmm, wait a minute, … How to explain it? Yes, it can be like that.

The excerpt shows that S_2 tried to propose an idea to teach pupils by explaining why the mathematical technique τ_{m1} should be used instead of $\tau_{m1'}$. This kind of didactical technique may support pupils to develop their technological discourse to accept the correct mathematical technique. Although S_1 tried to convince S_2 using the concept of fraction as a division of two positive integers, this technological discourse may be beyond the mathematical concept that could be accepted by S_2 and also by the pupils. It seems that S_2 was not satisfied with S_1, but she also could not provide some justification to the mathematical technique τ_{m1}. Then, at the end of their discussion S_1 argued "if we use addition to find [an equivalent fraction of $\frac{a}{b}$], the numerator and the denominator are added by the same positive integer and [got a new fraction $\frac{c}{d}$], and when we divide [*a* by *b* and *c* by *d*], we can show the quotients are different". So, this is one of didactical technology linked to mathematical technology to justify why such the technique proposed by the pupils on the HTT is incorrect.

4. Discussion and concluding remarks

The findings of the study show manifest differences between the Danish and Indonesian mathematical and didactical praxeologies for addressing the task of equivalent fractions. The main difference is that the Danish pair suggested three different possible didactical techniques, while the Indonesian pair only focused on one didactical technique based on the direct instruction of the mathematical technique τ_{m1}. That discrepancy is caused by their personal mathematical praxeologies. For instance, both Indonesian PsTs provided almost the same written answers to question a, and they just discussed the idea based on what they have written. The Danish pair, on the other hand, seems to have different personal mathematical praxeologies, so when they bring these praxeologies into the discussion, they come to different possible didactical praxeologies to teach the pupils. The differences between the two pairs can be caused by their mathematical technologies or theories. For instance, the Danish pair tends to justify their techniques based on the ratio or proportion. They tend to argue that the two fractions are equivalent when the ratio between the numerators and the denominators of the two fractions remains the same, and understanding a fraction as ratio between two numbers is considered a prerequisite for equivalent fractions [9]. While, the Indonesian pair focuses on the linguistic meaning of "fraction" as the division between the numerator and the denominator stated by S_1, and uses this technology to justify τ_{m1}. Their difficulty in explaining this makes them discuss the same thing for several times and could not achieve more advanced mathematical and didactical praxeologies.

Acknowledgment

We thank the Ministry of Research, Technology, and Higher Education (RISTEKDIKTI) of the Republic of Indonesia for funding the first author PhD research under the grant no.102.7/E4.4/2015.

References

[1] Shulman L 1986 *Educational Researcher* **15** 4
[2] Ball D L, Thames M H and Phelps G 2018 *Journal of Teacher Education* **59** 389
[3] Depaepe et al. 2015 *Teaching and Teacher Education* **47** 82
[4] Winsløw C and Durrand-Guerrier V 2007 *Nordic Studies in Mathematics Education* **12** 5
[5] Durrand-Guerrier V, Winsløw C and Yoshida H 2010 *Annales de Didactique et de Sciences Cognitives* **15** 147
[6] Chevallard Y 2006 *Proceedings CERME 4 (Sant Feliu de Guíxols, Spain)* 21
[7] Francisco J M 2013 *Educational Studies in Mathematics* **82** 417
[8] Putra Z H and Winsløw C 2016 *Presented in CITAD 5 (Castro-Urdiales, Spain)*
[9] Charalambous C Y and Pitta-Pantazi D 2007 *Educational Studies in Mathematics* **64** 293

Developing spatial reasoning activities within geometry learning

D W Winarti
University of Canberra, 11 Kirinari St, Bruce ACT 2617, Canberra, Australia

E-mail: Destina.Winarti@canberra.edu.au

Abstract. Spatial ability is a capability that is used in everyday life as well as one of the keys to be successful in the fields such as science, technology, engineering and mathematics (STEM). Numbers of research show the positive correlation between spatial ability and mathematics performance. This study has an aim to develop spatial reasoning activities within mathematics curriculum to support student's learning in mathematics, particularly in geometry. Two lessons (80 minutes each) for learning surface area and volume of a prism were designed. The implementation was conducted in two female classes with the total of 57 grade 8 students in West Nusa Tenggara, Indonesia. The data were obtained through video recording, observation form, students' worksheets and interview with some students. The data shows that the spatial reasoning activities within the geometry lesson such as drawing on isometric paper, forming a prism from connected cubes, give an opportunity for students to thinking spatially by visualising, predicting and checking the object and any manipulation involved. It helps students to contextualise surface area and volume of a prism and understand the concept of it before they go further to the formulation of the area and volume of a prism.

1. Introduction

Spatial ability can be described as the ability to mentally understand, manipulate, reorganise and interpret visual relationship [1]. Some research has revealed the positive correlation between spatial ability and mathematics performance [2,3]. Those who perform better in spatial tasks tend to have a better performance on a test of mathematical ability [4,5]. To date, some countries have recognised the importance of developing spatial ability during the school years such as Turkey, the United States and Australia [6,7,8,9]. Students from those countries did not perform in the top 10 countries with the highest mathematics performance in international tests such as Programme for International Student Assessment (PISA) and Trends in International Mathematics and Science Study (TIMSS). Nevertheless, these students performed far better than students from many other countries including Indonesia.

Numbers of studies support the idea that spatial ability can be developed through training because of the malleability of spatial thinking. The spatial thinking was explored to demonstrate spatial thinking's malleability and its importance in STEM education [10]. Even though the opinions on the result of spatial ability improvement are diverse, research indicates that spatial training can help to improve spatial ability when the appropriate material is given [11]. To date, even though only in a small number of studies, the enhanced spatial ability has been shown to improve students' performance in STEM subjects. Six studies have been identified for its finding to improve STEM achievement through spatial training [12]. Also, a recent study from [13] found that embedded spatial reasoning into an existing elementary mathematics curriculum showed promising results that both spatial ability and mathematics performance improve after spatial training.

The study from [13] gives the possibilities that students in higher level will have the support for their mathematics learning through spatial reasoning intervention as well.

This present study, therefore, investigated the effect of spatial reasoning activities to support high school student's mathematics learning, particularly geometry. Considering that spatial ability is an essential factor in geometry achievement and geometry problem solving [14], this study analyses grade 8 students' learning process on the subchapter of a flat surface 3D shape (i.e., prism). For achieving the aim of this study, spatial activities embedded within mathematics classrooms will be designed which involves several spatial activities involving spatial visualisation and spatial orientation.

2. Methods

2.1. Participants

This developmental research involved two classes of grade 8 in one of the female boarding school in West Lombok, West Nusa Tenggara, Indonesia. The school covered mainly middle-low socio-economic demography. The two classes that participated were chosen among six classes based on the availability of the teachers to join the spatial training before conducting the classroom implementation. In total, there are 57 students with the mean age in years = 13.56 and one mathematics teacher who participate in this study.

2.2 Data collection and analysis

Data was collected from the implementation of lesson design during classroom intervention through classroom observation, video recording, interviews, and students' worksheets then analysed in a retrospective analysis. Classroom intervention consisted of spatial activities embedded within mathematics classroom particularly in a subchapter of surface area and volume of a prism. The classroom activities were videotaped, and the author transcribed it herself. The interview was aimed to get more understanding of students thinking about specific activities which cannot be explained only by watching the video recording. Students' works were also utilised to analyse the way students employ their understanding on the discussion of surface area and perimeter of a prism.

Table 1. Lesson design.

Lesson	Activity	Spatial Construct	Mathematics strand	General Capabilities
A	Form a prism from connected cube blocks: 1. Visualize on adding up some cubes on a prism 2. Draw some different possible prisms that can be formed from 4 cubes (isometric drawing) 3. Game: form a prism	Spatial Visualisation & Spatial Orientation	- Define the surface area and volume of a prism and how to find it - The ability to form a prism from cube blocks - The ability to represent/visualise 3D object into the 2D isometric drawing	- Use the spatial visualisation and the spatial orientation skills in decoding and encoding the information - Use the spatial visualisation skill in constructing a prism from the unit cubes.
B	Form some prism with the same surface area but different volume and vice versa: 1. Form different possible prisms with the same volume but with the different surface area (isometric drawing) 2. Form different possible prisms with the same surface area but with the different volume (isometric drawing) 3. Find the surface area and volume of a prism	Spatial Visualisation	- Define the surface area and volume of a prism - Different prism can have the same volume but the different surface area and vice versa	Use the spatial visualisation skill in constructing the prism

IOP Conf. Series: Journal of Physics: Conf. Series **1088** (2018) 012004 doi:10.1088/1742-6596/1088/1/012004

2.3 Lesson design
This lesson design is included in the designing stage of developmental research. The lesson was designed collaboratively with the participating teacher in a series of lesson plan where spatial activities are embedded in mathematics lessons. It is expected that, while learning mathematics, students will also develop their spatial reasoning through the activities designed. The activities that were designed will also help students to understand the mathematics lesson better. The brief description of the lessons is presented in table 1.

3. Results and discussion

3.1. Results
The results of this study are presented in the retrospective analysis of students learning process during the lesson. The finding shows that the spatial activities within mathematics lesson help students to improve their spatial understanding as well as their understanding of surface area and volume of a prism.

In the first activity of lesson A, the teacher shows a model made from 4 x 3 x 3 unit cubes (see figure 1) then asked the student to name the model as well as to find the surface area and volume of it. Prior the lesson A, the students have learned about the characteristics of the prism as well as the surface area of it and volume of 3D shape has been learned in brief during their elementary years.

Figure 1. A model of a prism.

Some students mention the model as cuboid and some as a prism, and many of them use their gesture to find the surface area and the volume of the prism since the teacher only hold the model and did not hand it to students. In this activity, students started using their spatial orientation when they were counting the surface area; it was shown from the gesture of student's palm as it open downward, upward, left-right, front and back or when they count the volume, the finger keeps pointing to the prism while counting.

On the second activity, the teacher asked the students to close their eyes and imagine if they were asked to place three more cubes on the top left of the prism, where will it be? In this activity, since the students are asked to close their eyes, then their option was visualising the model in their mind and where to place the additional three cubes. Students then orientate themselves as the teacher position as one of the students' explanation that she placed the three cubes on her right side because her right side was the left side of the teacher. In this activity, the teacher gave enough time for students to orientate and visualise then predict where to locate the cubes. In the end, the class checked together whether the placing was right.

Teachers then asked the students whether the new shape was still named as a prism (see figure 2). This question provoked students to think back on the definition of the prism. The teacher gave an uncommon form of a prism so that students would not glue on the visualisation of the typical form of a prism with the regular base. The students knew that they need to find at least one pair of parallel and congruence side and it is not necessarily on the top or bottom part of the shape.

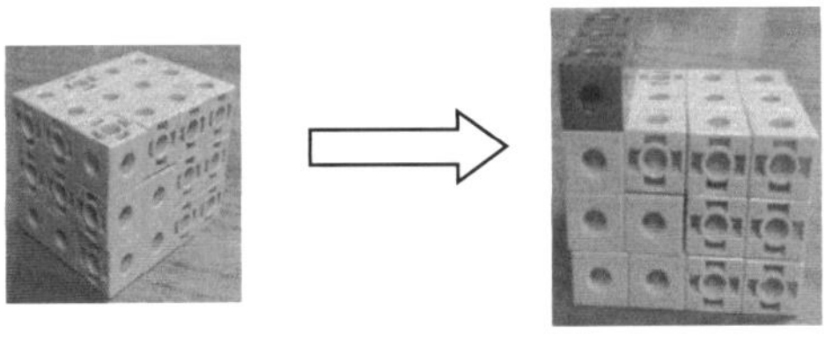

Figure 2. Two models of the prism.

Along with the activity in the first part of the lesson, after forming the new prism, the teacher also asked the students whether the surface area and volume of the prism remained the same. The class agreed that the surface area and volume were changing. Students might use the sense that adding up three cubes also change the surface area and volume.

In the next activity, students were asked to form a prism from 4 cube blocks, as many different prisms as possible. The students did not practically have four cube blocks in their hands, but they had it in mind then drawing all the possible arrangement on isometric paper. The example of students' drawing is shown in figure 3.

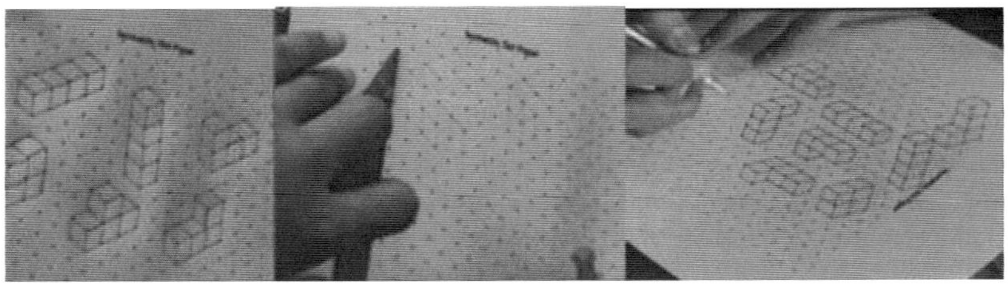

Figure 3. Examples of students work on the isometric papers.

The idea that students did not use real unit cubes was that they were employing their spatial thinking while drawing a different kind of prism on isometric paper. First, they visualise the arrangement of those four cubes then predicted it on isometric paper so that it will be seen as the stacks they have in mind. In the end, they checked all their design by using real cube blocks. Unfortunately, there is no further data on how actually student thought while drawing, but from the interview, we find that student found difficulties in drawing on the paper and how to connect between dots. "*To visualise or imagine the arrangement of 4 cubes were easy but to draw it on a paper was hard*", said one of the students.

The next activity is a game of cubes arrangement (see figure 4). In this game students alternately instructed each other to form a prism from 6 unit cubes by first wrote down the instruction. This activity might not directly correlate to the concept of area and perimeter, but it built the students awareness of spatial orientation while receiving the instruction, despite using their spatial visualisation skill while writing the instruction. Also, the student used the concept of the prism to make sure that the shape they formed was a prism, not just merely arranging different forms of 4 unit cubes. For the one who gave instruction, she/he must think of the steps in building the blocks so that her/his friend who received the instruction could understand it easily. On the other hands, the one who received the instruction should be able to orientate he/herself as the person who gives the instruction.

Figure 4. A game of cubes arrangement.

The following activity was arranging 6 unit cubes to form different prism with the same volume but with the different surface area and vice versa. The same as previous activity, students, did the arrangement in mind and drew it on isometric paper. The idea of this activity was getting students to understand the concept of area and volume better, as well as deepening their spatial visualisation through this activity.

The teacher gave the initial prism made from 6 unit cubes (see figure 5) then asked the students to form different prism from that model. The new prism should have the same volume as the initial prism but with the different surface area and vice versa. Some example of students' drawing is shown in figure 6.

Figure 5. The initial form of prism given by the teacher as a model.

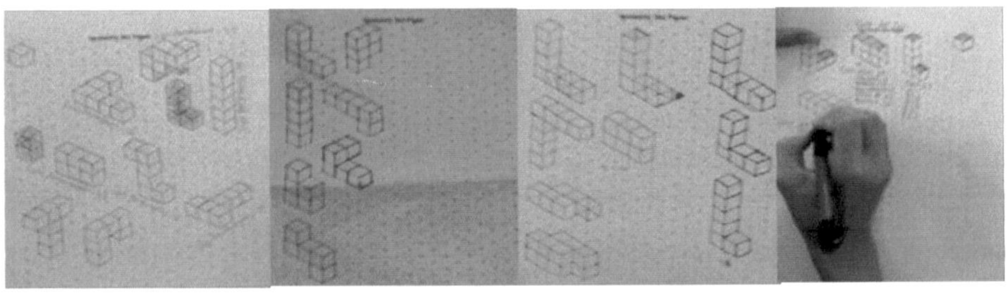

Figure 6. Examples of students work on the isometric papers

In these activities, students thought in two ways, first was when students start thinking about the different arrangement so that the surface area would be different but the volume remained the same. The second was when the students tried to add or reduce the number of the cube so that the prism would have different volume but the surface area remained the same. The students were mainly used spatial visualisation in this activity.

In the last activity, students were given a picture of 3D shape without dimension, as shown in figure 7.

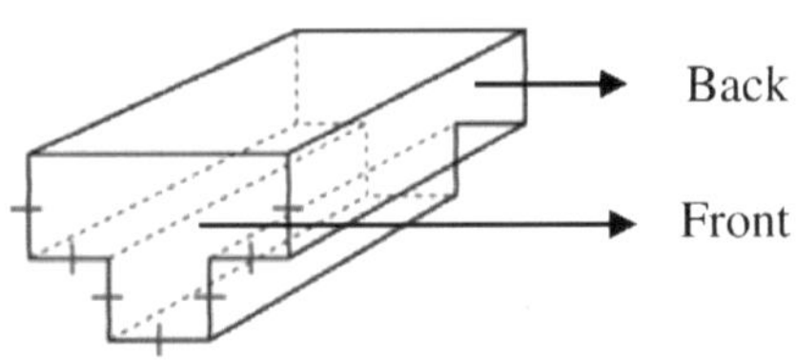

Figure 7. A composite 3D shape.

The teacher asked whether the shape of the picture is a prism. Students, in fact, could explain that the shape is a prism by giving the justification that the shape has a pair of parallel congruent sides in its front part and back part. The student could also explain how to find the surface area of the shape, one of the students explain during the interview that to find the total surface area she needed to add all side planes of the prism which all were in rectangle form except the front and the back side of the solids. When she was asked how to find the area of the front and back sides, she said she would divide the shape into rectangle and square as shown in figure 8.

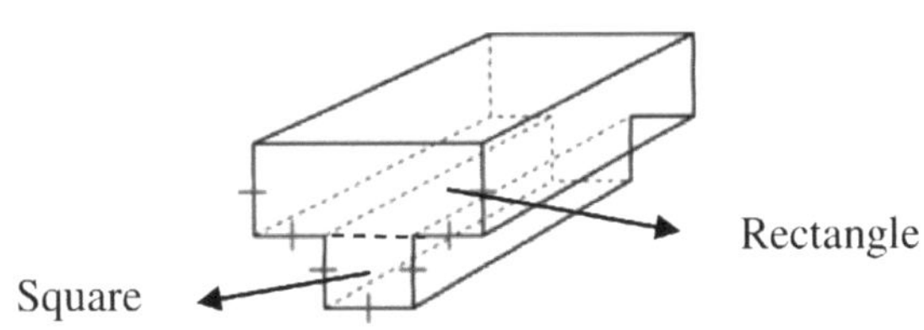

Figure 8. Illustration in splitting the front side of the 3D shape

The student also explained that he needs to divide the shape into two prisms as shown in figure 9. Students explain that she needed to add up the volume of prism 1 and prism 2 by using the formula of length × width × height for each prism.

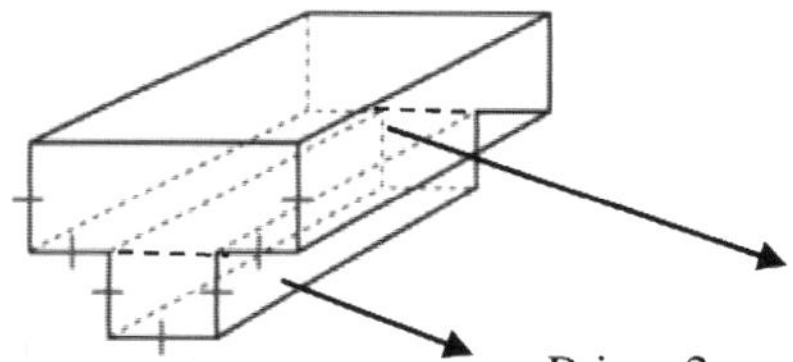

Figure 9. Illustration in splitting the front side of the 3D shape.

In this activity, the student explained clearly every step on how to calculate the surface area and volume of the prism. Since there is no dimension given, then students were urged to use their spatial skills in describing the steps. Such as mentioning all parts of the sides planes and how to do the partition for the front and back sides of the solid as well as the partition of the solids to find the volume of it.

3.2. Discussion

Mathematics problem solving, often, requires students to reason about spatial information [12]. In this instance, for learning surface area and volume of a 3D shape, students require to be able to identify which parts of the 3D shape that was called as the surface and which part is the volume before they know the formula of that measurement. The series of activities in this study was designed to promote the spatial thinking of students in various ways to support their understanding of surface and volume of the 3D shape. The activities involved can be grouped into three main activities namely manipulating the unit cubes, drawing on isometric paper, and at the end, there was a review regarding the surface area, volume, and a prism shape. All the activities designed gave a chance for students to visualise, use their spatial reasoning as well as construct a geometric modelling to solve a problem. It is in line with one of the principles and standards for school mathematics from National Council of Teachers of Mathematics [15] where students are expected to learn mathematics with understanding, actively building new knowledge from experience and previous knowledge.

The first group of activities (Lesson A), manipulating unit cubes into a different type of prisms, let the students use a lot of spatial visualisation and somewhat spatial orientation. Despite sharpening students' spatial reasoning, this activity also deepens students' mathematical understanding. Through the activities of forming a prism from 4 unit cubes and a game of forming a prism, the students were expected to keep on repeating the idea of a prism so that they will also know the shapes which are not a prism. Meanwhile, from the activity of forming a prism from 6 unit cubes let the students understand the concept of surface area and volume. Students also get the idea that two prisms with the same volume do not necessarily have the same surface area and vice versa.

The second group of activities (Lesson B), drawing on isometric paper, was a novel activity for all students and they found it somewhat challenging at the beginning. Besides the reasons for novelty, the challenge most probably came from the urgency of using spatial visualisation and or spatial orientation while drawing on a paper. In this activity, students need to use spatial visualisation frequently as well as spatial orientation. Students need to think about how to connect the dots so that the shape they draw was the same as the shape they have in mind. The ability to visualise the shape, know the whole part of the shape and how it was constructed help students to get the concept of surface area and volume of the 3D shape.

In the closing activities, the students were given a picture of a composite figure without dimensions. In this activity, students were provoked to not glued into the formula of surface area and volume but more to the exploration on how they give their reasoning on finding the surface area and volume of the shape when the shape is not a regular prism. In this activity, students recalled back their knowledge about prism to make sure that the shape they saw was a prism. Regarding the counting of surface area, the students have the idea that side planes need to be added together, but for the front side and back side, they need to think on how to count it. Students use their mental cutting of the front part so that it comes into square and rectangle. This activity is known as partitioning and [16] described partitioning as one of the five basic concepts in learning to measure area other than unit iteration, conservation, structuring the array and linear measurement.

4. Conclusion

The series of activities in this study was designed to promote the spatial thinking of students in various ways to support their understanding of surface and volume of the 3D shape. All the activities designed gave a chance for students to visualise, use their spatial reasoning as well as construct a geometric modelling to solve a problem. Current study contributes to the possibility to embed spatial activities within the mathematics classroom through various activities related to the topic given (i.e., geometry). Because this study only gave examples on geometry topic, further study is needed to see the possibilities of other mathematics topics to be embedded with spatial activities.

5. References

[1] Tartre L A 1990 Spatial orientation skill and mathematical problem solving. *Journal for Research in Mathematics Education* **21** 216

[2] Cheng Y and Mix K S 2014. Spatial training improves children's mathematics ability *Journal of Cognition and Development* **15** 2

[3] Pittalis M and Christou C 2010 Types of reasoning in 3D geometry thinking and their relation to spatial ability *Educ. Stud Math,* **75** 191

[4] Holmes J, Adams J W and Hamilton C J 2008 The relationship between visiospatial sketchpad capacity and children's mathematical skills *European Journal of Cognitive Psychology* **20** 272

[5] Rasmussen C and Bisanz J 2005 Representation and working memory in early arithmetic *Journal of experimental child psychology* **91** 137

[6] Cakmak S, Isiksal M and Koc Y 2014. Investigating the effect of origami-based instruction on elementary students' spatial skills and perceptions *The Journal of Educational Research* **107** 59

[7] National Academy of Science 2012 *Learning to Think Spatially: GIS as a Support System in the K-12 Curriculum* (Washington DC: National Academy of Science)

[8] Foundation N S 2010 *Preparing the next generation of stem innovators: Identifying and developing our nation's human capital* (National Science Foundation)

[9] Australian Curriculum, Assessment, and Reporting Authority 2010 *Australian F-10 Curriculum* Retrieved from https://www.australiancurriculum.edu.au/f-10-curriculum/general-capabilities/numeracy/

[10] Miller D I, Newcombe N S and Uttal D H 2013 Exploring and Enhancing Spatial Thinking: Links to Achievement in Science, Technology, Engineering, and Mathematics *Current Directions in Psychological Science* **22** 367

[11] Olkun S 2003 Making connections: Improving engineering drawing activities *International Journal of Mathematics Teaching and Learning*, **3** 1

[12] Stieff M and Uttal D 2015 How much can spatial training improve STEM achievement? *Educational Psychology Review* **27** 607

[13] Lowrie T, Logan T and Ramful A 2017 Visiospatial training improves elementary students' mathematics performance *British Journal of Educational Psychology*

[14] Battista M T 2007 The development of geometric and spatial thinking *Second handbook of research on mathematics teaching and learning* **2** 843

[15] National Council of Teachers of Mathematics 2000 *Principles and standards for school mathematics* (Reston: NCTM)

[16] Clements D H and Stephan M 2004. Measurement in Pre-K to grade 2 mathematics *Engaging young children in mathematics: Standards for early childhood mathematics education* 299-317

Android based e-learning tutorial for mathematics teachers

Y Roza, I Daqiqil, S N Siregar, S Salam and A Adnan
Universitas Riau, Jl Kampus Bina Widya Km 12,5 Simpang Baru Pekanbaru 28293 Indonesia

E-mail: yenita.roza@lecturer.unri.ac.id

Abstract. A large number of teachers and the scattered location due to geography condition is one of the causes the slow process of teacher training in Indonesia. The current national curriculum (Kurikulum-13) had been launched in 2013 but the implementation of the curriculum has the problem in socialization as well as teacher training in using this curriculum. This development research produces android- based E-tutorial material of the K-13 Curriculum Implementation for Mathematics Subjects that can be used by Mathematics teachers. The product contains of the curriculum K-13, the Competency analysis, design Lesson Plan and design Evaluation for the knowledge and skills aspect as well as video using a scientific approach in Problem Based Learning and of Discovery Learning Model. This e-tutorial is developed by using Waterfall method and used android based. The product has been used by mathematics teachers for testing product practicality. The product is assessed based on the aspect of cosmetics and program. The product then revised based on input from the teacher. The average score on program aspect is 3.8 and 3.5 on aspect of cosmetics. Therefore, the results of this study will be one of the solutions for preparing teachers to implement the curriculum.

1. Introduction

The 2013 curriculum focuses on the acquisition of contextual knowledge related to the subjects and its learning environment. The curriculum seeks to develop students' skills in three areas: attitude (honesty, politeness, and discipline), technical skills (through practical work/school projects), and scientific knowledge. At the elementary level, the curriculum emphasizes on the development of attitudes and functional skills over scientific knowledge, which receives more attention at higher educational levels. At the junior and senior high school levels, the academic rigor is increased since the students' personalities were emphasized at the primary level [1].

The curriculum has been implemented nationally at the beginning of the 2013-2014 academic year. In the first year of the implementation, the curriculum was only implemented nationally in grade 1, 4, 7, and 10 The government has conducted various training for teachers, however, not all teachers have understood this curriculum well [2]. Many teachers who are appointed as curriculum implementers feel not ready to implement this curriculum. As a result, they still use the previous curriculum in their teaching. Teachers encounter a various problem in implementing the curriculum including; preparing learning media, integrating lessons in thematic learning, and mastery of information technology [3]. Teacher training is very important to note because teachers play an important role in the process of curriculum implementation [4]. Nurmalasari [5] said the role of teachers in the implementation of current Curriculum, namely: a) teachers discuss in the process of preparing lesson plan; b) teachers prepare lesson plan; c) teachers deliver the lesson d) teachers as facilitators; e) teachers provide character education; f) teachers guide students according to a scientific approach; g) teachers choose and use varied methods, media, and learning resources, h) teachers perform authentic assessments; i) teachers choose and use varying assessment techniques; and j) teachers provide remedial teaching.

The current curriculum is an improvement from the previous curriculum with an addition to some aspects of implementation. For the implementation of the curriculum of 2013, students' books and teachers' books are provided by the central government as compulsory learning resources in schools. Teachers' books and government-provided student books are designed for national purposes, teachers need to be aware of them in order to adapt to the characteristics of each school.

To maximize the implementation of the 2013 curriculum every year the government revises both the content and the socialization process to prepare teachers using the 2013 curriculum. The government is targeting that by 2019 all teachers are ready to learn using this curriculum. The new problem that arises is the way of socialization for all teachers who spread throughout the territory of Indonesia with very varied geography conditions.

In the previous year, researchers have developed a desktop-based e-tutorial, teachers can use this material if they get the file directly from the researchers or get files from colleagues. However, when the government revises the contents of the curriculum and other tools the teacher cannot access the revised results directly from the revised e-tutorial. To overcome these problems researchers developed a desktop-based e-tutorial into an android-based e-tutorial. Android is an open-source, Linux-based operating system for mobile devices such as smartphones and tablet computers [6]. By using android based e-tutorials every researcher to upgrade the contents of the program so the user e-tutorial teachers can upgrade their materials. Seppala [7] found that e-tutorial as flexible teaching solutions, which will enable access to information using different devices, and support learning in a variety of situations.

The aim of this research is to produce the android-based e-tutorial on the implementation of curriculum 2013 for mathematics teachers. Setiabudi [8] said android based learning is contributing to every day learning more people are able to access the program to complete their education. This research is useful for junior and senior high school mathematics teachers in preparing the implementation of the 2013 curriculum. Teachers can learn this product themselves without the help of instructors. This product is also useful for resource persons and Instructors of the 2013 curriculum to be used for material in the 2013 curriculum training for targeted teachers. Overall, this e-tutorial is useful for the government, especially the education offices at various levels to facilitate the implementation of the 2013 curriculum so that the curriculum can be fully implemented in 2019.

2. Methodology

The research activities at the design stage and develop activities was focus on the design and development of the Android-based Curriculum E-Tutorial system 2013. System development method used was Waterfall method. The Waterfall Model was the first Process Model to be introduced. It is also referred to as a linear-sequential life cycle model. It is very simple to understand and use [9]. Wells [10] said development is discrete, and do not allow for jumping back and forth, or any overlap between the phases. The steps in Waterfall can be seen in Figure 1.

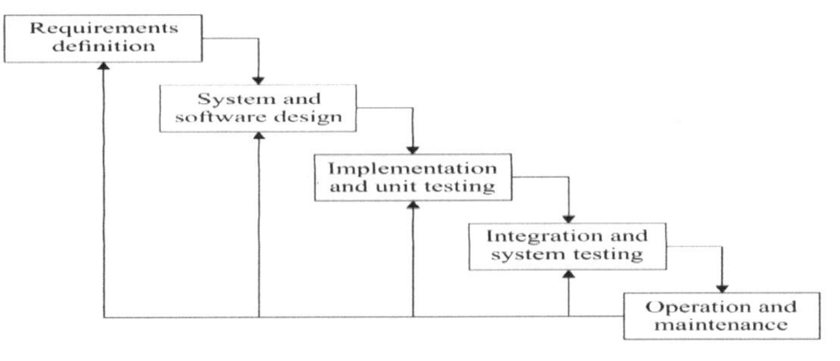

Figure1. Waterfall method

The waterfall method is often called the classic life cycle, which illustrates a systematic and sequential approach to software development, starting with the user needs specification and continuing through the planning, modelling, (construction), as well as delivery of the system to the customers/users (deployment), which ends with support in the complete software generated. Stages of activity can be seen in the following description:

Requirements definition, at this stage the needs analysis process is intensified and focused on software. The first step is to identify the nature of the program to be created, and then the software engineer must understand about the information domain of the software, such as required functions, user interface, and user character. The three activities (search system and software needs) were documented and shown to customers. Needs are collected by interviews, field surveys, questionnaires, and surveys.

System and Software Design process is used to transform the above needs into a representation into the "blueprint" of the software before the coding. The design should be able to implement the requirements already mentioned in the previous stage. On the stage of Implementation and Unit Testing, to be understood by the machine, in this case is the computer, the design had to be transformed into a real form and can be understood by the machine, that is into the programming language through the process of coding. This stage is an implementation of the design stage that technically will be done by the programmer. Integration and system testing are done to ensure that a product must be tested. Likewise with software, all software functions should be tested, so that the software is free from errors, and the results should be perfectly in accordance with the needs that have been defined previously. Operation and maintenance of software is conducted including in its development, because the software need to be revised periodically.

Data collected in this research is a training material published by the government for teacher training in implementation of curriculum 2013 such as guideline for designing lesson plan, evaluation and various learning model. Some of these materials are used as they are, but most of them are edited into difference appearance. Other data used for this e-tutorial product are the teachers' and students' books uses for mathematics learning of year 7 up to class 12. An instructional instrument such as lesson plan, student worksheet, evaluation instrument are developed by researcher and revise based on the result of validation by 20 mathematics teachers. Those teachers are members of the mathematics teacher group from Regency of Siak. They were given a questionnaire for assessing the product based on the aspects of program and cosmetics.

3. Results and Discussion

In the previous year, researchers have developed a desktop-based e-tutorial, teachers can use this material if they get the file directly from the researchers or get files from colleagues. However, when the government revises the contents of the curriculum and other tools the teacher cannot access the revised results directly from the revised e-tutorial. To overcome these problems researchers developed a desktop-based e-tutorial into an android-based e-tutorial. By using android based e-tutorials every researcher to upgrade the contents of the program so the user e-tutorial teachers can upgrade their materials. E-learning expect will promote more efficient learning technique [11].

The first result of the research proses is the navigation structure of the program. The e-tutorial program consists of 7 part of information, they are the introduction, the dynamics of curriculum development, teaching material, instructional instrument, learning assessment, implementation of curriculum 2013 and related government regulation. The user can surf into the program through those seven choices of the main menu as shown in Figure 2.

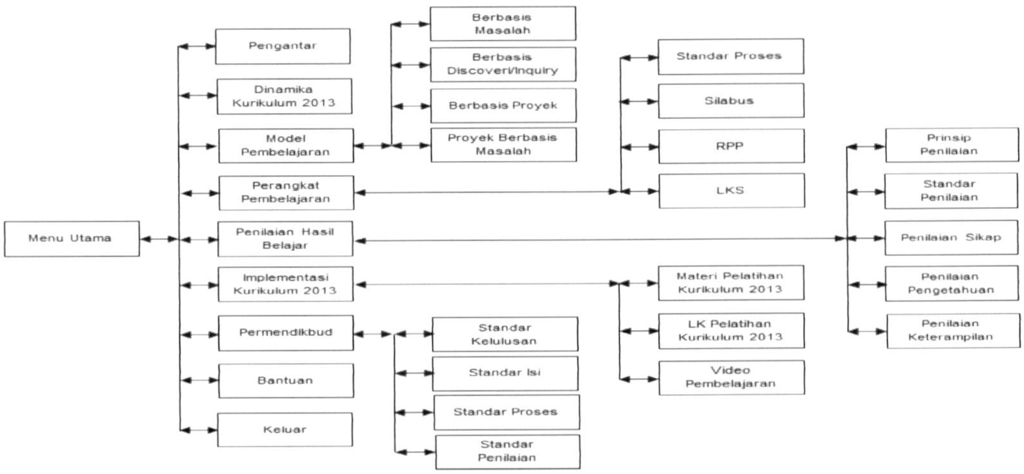

Figure 2. Structure navigation of program

This e-tutorial product uses android hybrid application. This app utilizes Webview as a the main component. The framework used is Phonegap. The product design can be seen in Figure 3 and Figure 4:

Figure 3. Cover design

Figure 4. Design of list content

The prototype of android-based e-tutorials was tested to the group of junior high and high school teachers who are members of Mathematics Teachers Group Discussion *(MGMP)* Mathematics. Programs are given to teachers to be learned and they are given a questionnaire for assessment and feedback on the program. Assessment is given for two aspects they are program and cosmetic aspects. Assessments of the program aspects are presented in Figure 4:

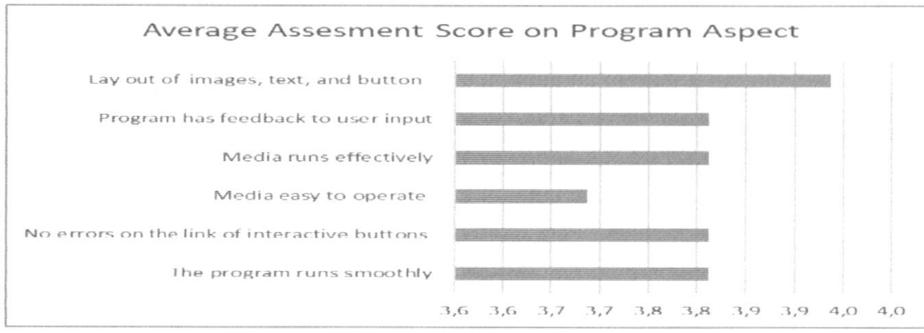

Figure 5. The average assessment score on program aspect

The picture above shows the average score of the six points questioned on the program aspect. All aspects get very good scores on a 1-4 scale, all above 3.7. The best scores are given on the layout of the image, text and interactive buttons used in the program. The lowest scores are given on the ease of using the program, this is considered reasonable as they are the first to use the given program.
Scoring assessment for cosmetic aspects is presented in Figure 5.

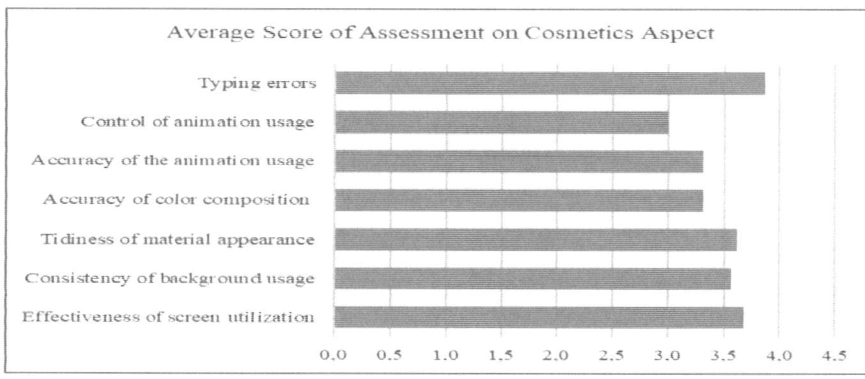

Figure 6. The average score of assessment on cosmetics aspect

From the figure above can be seen there are seven things that are questioned on the cosmetic aspect, this is to ensure that the program appears attractive and fun to read. The best judgment is given on typing points, almost no typos found in the created program, but there are some differences in the writing style so it does not get the full score. The points that get the lowest score are about the use of animation. Teachers expect to find more animation so it is more fun to use the program. For the use of color, some of the input given is for the program to use a brighter color, so more uplifting to learn.

4. Conclusion

Implement e-tutorial can be used as an alternative method in the learning process or training. The results show that the use of android based e-tutorials increases the mastery of knowledge and teachers consider learning more fun, faster and easier than traditional tutorials. The advantages of android based e-tutorials because it can be accessed anywhere and anytime, thus increasing participants' control of learning. This new e-tutorial contains two examples of learning models, so there is still an opportunity to develop examples of other learning models for different levels of school.

Acknowledgment

This research is fully supported by University of Riau research grant scheme *Bidang Ilmu*. Appreciation is given to mathematics teachers of Mathematics Teachers Group Discusion (MGMP) Siak Regency who give the input for the lesson plan and student worksheet.

References

[1] Kaimuddin 2014 Implementasi Pendidikan Karakter Dalam Kurikulum 2013 *Din Ilmu* **14** 47
[2] Alawiyah F 2015 Kesiapan Guru dalam Implementasi Kurikulum 2013 Kaji_Singk
[3] Krissandi A D S and Rusmawan R 2015 Kendala guru sekolah dasar dalam implementasi Kurikulum 2013 *J Cakrawala Pendidik* **34** 457
[4] Ingvarson L, Beavis A, Bishop A, Peck R, & Elsworth, G 2004 Investigation of Effective Mathematics Teaching and learning in Australian secondary schools *Melbourne: Australian Council for Educational Research*
[5] Nurmalasari R, Wati R D P, Puspitasari P, Diana W and Dewi N K 2013 PERAN GURU DALAM IMPLEMENTASI KURIKULUM 2013 *J Berk Progr Pascasarj UM Malang* p 722
[6] Tutorial Point 2014 Android Tutorial *Android Tutor Point* pp 1–2
[7] Seppälä P and Alamäki H 2003 Mobile learning in teacher training *Journal of Computer Assisted Learning* **19** 330
[8] Setiabudi D H and Tjahyana L J 2013 Mobile learning application based on hybrid mobile application technology running on Android smartphone and Blackberry *Int. Conf ICT Smart Soc* pp 1–5
[9] N d 2013 SDLC Waterfall Model *Tutorialspoint*
[10] Wells C J 2014 Systems Analysis and Design - The Waterfall Model*TechnologyUK* Available: http://www.technologyuk.net/computing/sad/waterfall_model.shtml.
[11] Rana H, Rajiv, and Lal M 2014 E-Learning: Issues and Challenges *Int. J. Comput. Appl. Technol* **97** 20.

IOP Conf. Series: Journal of Physics: Conf. Series **1088** (2018) 012006 doi:10.1088/1742-6596/1088/1/012006

The application of R and ArcGIS software as a learning media to estimate the return period of the destructive earthquake in Aceh areas using maximum likelihood method

H Sofyan, Jumila and L Rahayu
Syiah Kuala University, Jl. Teuku Nyak Arief Darussalam, Banda Aceh 23111, Indonesia

E-mail: hizir@unsyiah.ac.id

Abstract. Aceh is one of the most vulnerable areas for earthquakes. The earthquakes occurred has caused much loss including lives. Therefore, earthquake occurrence time estimation is needed to mitigate preparedness to the earthquake disaster. The purpose of this research was to explain the measurement result of destructive earthquake repetitive period in Aceh and other surrounding areas using Likelihood Maximum method toward the utilization of R and ArcGIS software instructional media. The data was obtained on trial utilization of R and ArcGIS software on Spatial Statistics course. The participants are 18 students taking the spatial statistics class. The research instrument was the observation sheet used to evaluate the implementation of student worksheets in the class. The data sources the students used for the measurement were obtained from USGS and BMKG. The results found that the students were able to predict the earthquake correctly using student worksheet assisted by the R and ArcGIS software. That is the earthquake would occur in the next 5.2 years at the marine territory of the southern region of Aceh Singkil. Based on this result, it can be concluded that the student worksheet could be used in the spatial statistics class assisted by the R and ArcGIS software.

1. Introduction

The province of Aceh is a region that coincides with the meeting of the Eurasian plate with the Indo-Australian plate along the Indian Ocean. In this zone, there have been a number of major earthquakes that have caused severe damage in some areas around the Indian Ocean; one of which occurred in December 26, 2004 that had triggered the tsunami with a magnitude of 9.1 and the earthquake that occurred in Pidie Jaya on December 7, 2016 with a magnitude of 6.5.

Research on seismotectonic potential as a precursor of earthquake level in Aceh has analyzed the level of seismicity and the period of earthquake re-occurring in Aceh region using the relationship between the frequencies with earthquake magnitude using the Maximum Likelihood method with *ZMAP* software. The study concludes that between two to five years there would be an earthquake for the 6 Mw scale in Simeulue Island, Northwest Aceh and North Aceh (Andaman) [1]. The results of this research can be used as the context of the problem in Spatial Statistics course. Students are assigned to look for relevant data sourced from USGS and BMKG, then, analyze the period of the destructive earthquake in Aceh and surrounding areas using the Maximum Likelihood method with a learning media, the R and ArcGIS software.

Maximum Likelihood aims to estimate statistics of random variables by maximizing probabilities with existing samples. This method is often used in different applications and studies in various situations [2]. In this study this method is used to analyze the relationship of frequency (N) and magnitude (M) earthquake which is known by Gutenberg Richter [3].

$$log\ N(M) = a - bM \quad (1)$$

Where:
N = the number of earthquakes
a = a determination that depends on the period, total area and area of observation activity
b = seismotectonic parameter of an area in which in an earthquake occur and depending on the local rock.
M = magnitude

The value of the tectonic parameters in the research area can be seen by looking at b-value. The b-value can show the stress level of the rock. If b-value is low, the stress condition in the area is high [4]. b-value is said to be low if b-value has a smaller value, the average value of b-value in the entire study area [5]. The b-value approximation is correct if the standard deviation of b-value is less than one [6]. The equation for b-value is presented as follows [7].

$$b = \frac{\log e}{\bar{M} - M_0} \quad (2)$$

Where:
$\bar{M}$ = Average magnitude
M_0 = Minimum magnitude
$log\ e = log\ 2{,}71828 = 0{,}434$

The error rate or standard deviation of the b-value calculation with the magnitude can be estimated using the equation in (3).

$$\sigma_b = 2{,}30b^2\sqrt{\frac{\sum_{i=1}^{N}(M_i - \mu)^2}{N(N-1)}} \quad (3)$$

In addition, there is an a-value that shows the level of seismicity in a region. The smaller the a-value, the smaller the seismic activity [8]. The equation for *a-value is* shown in [1].

$$a = \log N(M \geq M_0) + \log(b \ln 10) + M_0 b \quad (4)$$

Thus, to obtain the average value of earthquake return periods in a region can be calculated using the following equation.

$$P(\Theta) = \frac{1}{N_1(M \geq M_0)} \quad (5)$$

Where Θ is the period of repeating earthquake for magnitude $M \geq M_0$ [1].

The probability of earthquakes occurring is the possibility of destructive earthquakes in one area over a period of time. Assuming the time interval follows the exponential form e^{-NT}, the probability of earthquake occurrence is provided as follows.

$$P(M,T) = 1 - e^{-N_1(M \geq M_0)T} \quad (6)$$

Where $P(M,T)$ is the probability of an earthquake with magnitude M and period T.

The condition of Aceh that is often stricken by an earthquake disaster can be used as the context of the problems to be solved by students in spatial statistics class that is one of the courses in Statistics Department of Mathematics and Science Faculty at Syiah Kuala University. To analyze the calculation of earthquake return period required a learning media that can facilitate the students. Learning media is a tool in teaching process that is a physical means to deliver learning content/material [9]. Learning media is also a means of communication in the form of print and hearing-sight, including hardware technology [10].

Over time, there are many learning media in the form of software that students can use to solve various problems that occur in the universe, such as QGIS that has been used as Development of Web GIS for information of renewable energy in Aceh province after the rehabilitation and reconstruction process [11]. Besides, Xplore, GRASS, and ERDAS have also been used to classify satellite imagery data using the fuzzy clustering method [12].

Learning media used in this research were R and ArcGIS software. R is one of the languages in statistical programming based on the S programming language developed by Bell Laboraties. R was first introduced in 1996 by two statistics professors, Ross Ihaka and Robert Gentlemen, from the University of Auckland in New Zealand. They released the code as a free software package. ArcGIS is the most widely used software by users of Geographic Information System (GIS). One of the packages in ArcGIS is ArcMap, which is the main application for most GIS process and computer mapping. ArcMap has major capabilities for visualization, developing new spatial databases, selecting (querying), editing, creating map designs, analyzing and creating the final look in activity reports. In this study, ArcMap is used to visualize the results obtained from calculations with the R software into map form.

The spatial statistic learning with the utilization of R and ArcGIS software could maximize by using student worksheet that guides students in predicting the return period of an earthquake. Thus, this study aims to investigate the implementation of the worksheet assisted by R and ArcGIS software in spatial statistic class.

2. Method

This research generally aims to develop learning tools in spatial statistics class on Statistic Study Program of Mathematics and Natural Sciences Faculty that train students to solve problems. The learning tools were developed by following the research phases of controlling the Plomp Model (1999). This article focuses on instructional media used to help solve the given problem of "estimate return period earthquake." The learning media used is R and ArcGIS software. The explanation given is limited to the results obtained by the students in analyzing the repeated periods of earthquakes in Aceh and surrounding areas by the Maximum Likelihood method based.

The data used were earthquake occurrences from January 1907 to December 2017. The data obtained from the Agency of Meteorology, Climatology, and Geophysics (BMKG) were the data in 2009 while the data in 1907 to 2008 were obtained from the catalog of earthquakes in the United States Geological Survey (USGS). The selected data were earthquake data located on longitude 92° - 98.5° East Longitude and latitude 0° - 8° North latitude which coincide in Aceh province and at the maximum depth of 80 km. Research area is divided into 1° x 1° grid to obtain 48 grids with earthquake incidences as much as 9202.

The result of students' calculation when solving the worksheet based on the learning materials used was collected. Trials were conducted to measure the practicality of the learning materials. The implementation of the worksheet with R and ArcGIS software in this trial was examined based on students' calculation. Students are expected to predict the repeated period of earthquake correctly, which is 5.2 years. If the students get the expected result, it can be said that the worksheet can be implemented in the spatial statistic class.

3. Result and discussion

At the beginning of the lesson, students were instructed to learn information about earthquakes through BMKG and USGS website. The detailed instruction is written on the worksheet. The data obtained were earthquake data from The USGS and BMKG earthquake catalog, and the largest magnitude earthquake ever recorded in that period was in 2004 and 2012 in Simeulue. Earthquakes continue to occur from 1907 to 2017 with the cumulative incidence annually seen in Figure 1.

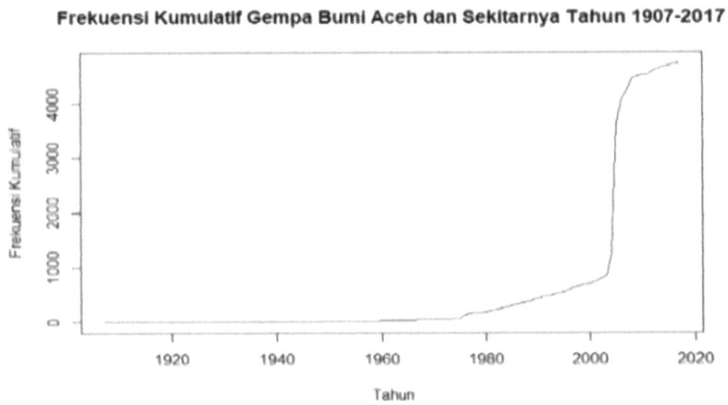

Figure 1. Cumulative frequency of earthquake events.

Figure 1 is a graph visualized by the students using R software. The graph shows that the earthquake increased significantly after 2000, where there was a major earthquake that caused the tsunami in 2004, and there had been many subsequent earthquakes.

Furthermore, the students created illustrations of earthquake distribution in Aceh and surrounding areas in 1907 to 2017, presented in Figure 2, visualized students with ArcGIS software using data obtained from BMKG and USGS.

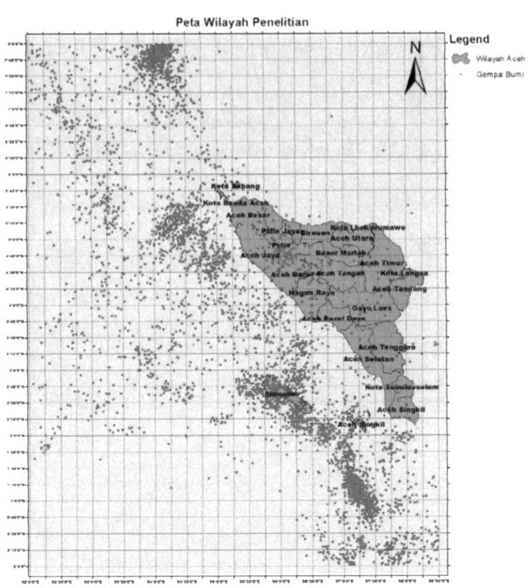

Figure 2. Map of spreading earthquakes.

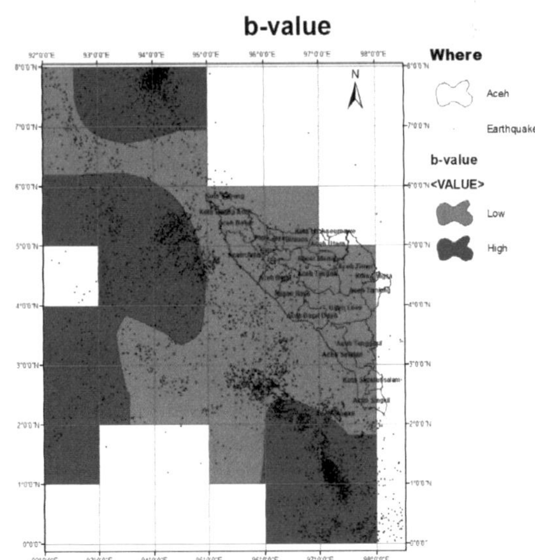

Figure 3. Map of *b-value* spatial variation.

The red dots in Figure 2 represent 4750 points of shallow earthquakes in Aceh and surrounding areas in 1907 to 2017 with a minimum magnitude of 5 amounting to 4750 events.

The next learning activity was b-value analysis to know the tectonic condition of the research area. The spatial b-value analysis conducted by the students in the research area with the number of earthquakes of 4750 events obtained the b-value value of 0.720 and the standard deviation obtained for 0.316 which is smaller than 1. This finding suggests that the calculation of b-value was close to the true value.

Furthermore, the students calculated spatial variation analysis of b-value. Spatial variation of b-value was determined by dividing the research area located at $92^0 - 98^0$ East Longitude and $0^0 - 8^0$ West Longitude into 832 parts with grid $1^0 \times 1^0$. The result of the analysis shows that b-value ranges from 0,208 - 1,717 with average 0.705 . 25 grids that had a low b-value is a region with b-value smaller than the average. In addition, there were 19 grids that had a high b-value, i.e., an area with b-value more than the average. Furthermore, 6 grids could not be analyzed because of the absence of earthquake data. However, for mapping, the area with an earthquake frequency of more than 10 was used, resulting in 34 areas with an average b-value of 0.7066. The variation of b-value can be seen in Figure 3.

Figure 3 is a contour map visualized by students with ArcGIS *software*, which shows that the blue region is a low b-value area. The region had high levels of stress and rare earthquakes, but in case of an earthquake, there will be earthquakes with a large magnitude. This is due to the buildup of energy in the region and when energy was released; it resulted in large-scale earthquakes. The red area is a region with high b-value which indicates that earthquakes were frequent in the area, with small magnitude due to the earthquake immediate energy released. The level of accuracy of the calculation can be seen based on the standard deviation value of b-value. Overall the standard deviation of b-value was less than 1, which means the calculation of b-value was close to the true value.

After the spatial analysis of b-value, the students conducted a temporal analysis of -value. The temporal b-value analysis was performed by distributing earthquake occurrence time. The data used were the data from the BMKG earthquake catalog. Data grouping was obtained by combining the earthquake data on a quarterly basis each year. The first quarter starts from January to March, the second quarter from April to June, the third quarter from July to September, and the fourth quarter from October to December. The purpose of the temporal variation analysis of b-value was to look at random signs before a major earthquake. The values obtained for temporal variation b-value is visualized in Figure 4.

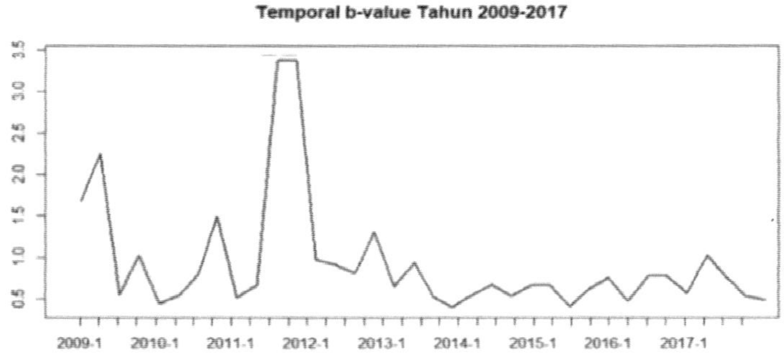

Figure 4. Temporal variation of b-value.

Figure 4 illustrates that temporarily in the event of a b-value earthquake decreased until a major earthquake occurred. This can be seen in 2012 and 2016 where there had been a major earthquake in

Simeulue and Pidie Jaya. In 2011, b-value increased after the earthquake in Simeulue in 2010. Also, b-value increased in 2017 after the earthquake in Pidie Jaya at the end of 2016.

After obtaining the value of b-value, the students calculated the return period of earthquakes using R software and mapped them using ArcGIS software. The return period of earthquakes in Aceh and surrounding areas with $M_w \geq 5$ is illustrated in the following contour maps.

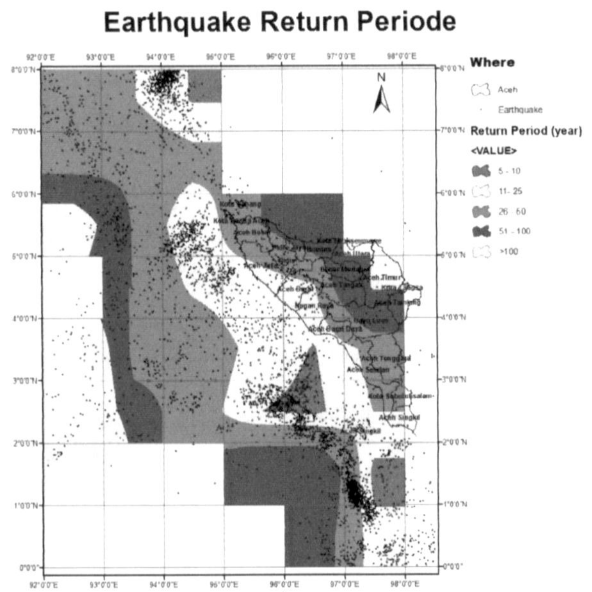

Figure 5. Map of contours of return periods of earthquakes.

Figure 5 shows that red areas are areas with return periods of earthquakes in the shortest period of five to ten years when the occurrence of repeated periods of earthquakes continued to increase every change in red, yellow, green to blue. While the white color indicates the area that there is an earthquake frequency of less than 10; thus, no mapping was obtained. The period of the earthquake can be observed in Table 1.

Table 1. The period of return period of destructive earthquakes.

Grid	Period (years)	Year estimates of return period	Occurrence probability
12	5,2	2023	0,632
17	6,3	2024	0,633
33	8,2	2026	0,633

Table 1 shows that the most rapidly estimated earthquake will occur in the 12th grid that coincides in the sea area of Southern area of Aceh Singkil with a repeat period of 5.2 years, in the 12th grid that coincides in Simeulue with a repeat period of 6.3 years, and in the 33rd that coincides in sea areas of Northern area of Aceh of with a return period of approximately 8.2 years. The probability of earthquake occurrences in the grid to 12, 15 and 33 is about 63%. The estimated year the earthquake repeated period was calculated after 2017 where the last earthquake event data were used, so earthquakes are expected to occur in the 12th grid in 2022, the 17th grid in 2023, and the 33rd grid by 2026.

As the validity of return period analysis, an analysis of the probability of earthquake occurrence in the near-estimated time, earthquakes will occur again in the same areas. The probability of earthquake occurrence can be seen in Table 2, which is the result of student calculation using the R software.

Table 2. The probability of the occurrence of destructive earthquakes.

Grid	Probability			
	5 years	10 years	30 years	50 years
12	0.61809	0.85414	0.9969	0.99993
17	0.55073	0.79815	0.99178	0.99967
33	0.45656	0.70467	0.97424	0.99775

Table 2 shows that the probability of an earthquake re-occurrence increased by increasing time. In the 12th grid, it is estimated that the probability of an earthquake is 61.8% in the next five years and 85.4% in the next ten years. The results are in accordance with the estimated return period of the earthquake that occurred within 5.2 years with the probability of occurrence of 63.2% (Table 1).

In addition, Table 2 shows that the highest probability of earthquakes with an estimated 12th grid within 30 years will occur with a probability of 99.69% and within 100 years with a probability of 100%.

Based on the students' answer, it can be concluded that the students successfully predicted the return period of earthquakes as expected. This finding implies that the worksheet with the utilization of R and ArcGIS software could help students achieve the learning objective. In another word, the worksheet utilized by R and ArcGIS software could be implemented in spatial statistics class.

4. Conclusion

Based on the estimation of return period of the destructive earthquake in Aceh area using the maximum likelihood method, the most rapid destructive earthquake is expected to occur in the sea area of Southern of Aceh Singkil in 5.2 years, which means that the earthquake is expected to occur in 2023. This estimation was obtained by students through the worksheet assisted by the R and ArcGIS software. It can be concluded that the R and ArcGIS software are ones of learning media for students to solve the case of spatial statistics. The use of the software needs to be assisted by the worksheet. The worksheet developed in this study meets the criteria of practicality in the learning on spatial statistics class.

References

[1] Tauladani R 2015 *Kajian Seismisitas dan Periode Ulang Gempa Bumi di Aceh* (Skripsi)
[2] Wang J P, Xu Y and Wu Y M 2017 Earthquake probability in Taipei based on non-local model with limited local observation: Maximum likelihood estimation *Soil Dynamics and Earthquake Engineering* **99** 150
[3] Gutenberg B and Richter C F 1936 Magnitude and energy of earthquakes *Science* **83** 183
[4] Nuannin P Kulhánek O and Persson L 2012 Variations of b-values preceding large earthquakes in the Andaman-Sumatra subduction zone *J Asian Earth Sci* **61** 237
[5] Kulhánek O Nuannin P and Persson L 2018 Variation of b-values Proceding Large Earthquake in The Shallow Subduction Zones of Cocos and Nazca Plates S*outh Am Earth Sci* **82** 207
[6] Happrobo S *Variasi b-value terhadap Ruang dan Waktu sebagai Prekursor Gempa Bumi Daerah Andaman Sumatera* [Skripsi] 2016
[7] Han Q, Wang L, Xu J, Carpinteri A and Lacidogna G 2015 A robust method to estimate the b-value of the magnitude-frequency distribution of earthquakes *Chaos, Solitons and Fractals* **81** 103
[8] Kanamori H 1981 The nature of seismic patterns before large earthquakes. In earthquake prediction: An international review *Maurice Ewing Ser* 1981 **4** 1

[9] John B B, Kevin F and Collis 1982 *Evaluating the Quality of Learning: the solo Taxonomy* (New York: Academic Press)
[10] National Education Association 1969 *Audiovisual Instruction Department New Media and College Teaching* (Washington D C: NEA)
[11] Nizamuddin, Hizir, Ardiansyah, Pertiwi D and Handayani P 2017 *Development of Web GIS for information of renewable energy in Aceh Province after rehabilitation and reconstruction process* In: Earth and Environmental Science
[12] Sofyan H, Said MAM, Affan M and Bawahidi K 2014 *The Application of Fuzzy Clustering to Satellite Images Data* In: Remote Sensing. Venice

IOP Conf. Series: Journal of Physics: Conf. Series **1088** (2018) 012007 doi:10.1088/1742-6596/1088/1/012007

Enhancing manipulation of algebraic equation through Balance Method

D N B Pg Badaruddin[1], K A Tengah[1] and R C I Prahmana[2]

[1]Sultan Hassanal Bolkiah Institute of Education, Universiti Brunei Darussalam, Jalan Tungku Link, Gadong, BE 1410, Brunei Darussalam
[2]Universitas Ahmad Dahlan, Jl. Pramuka Kav. 5, Yogyakarta 55161, Indonesia

E-mail: khairul.tengah@ubd.edu.bn

Abstract. This action research study measured the effectiveness of the Balance Method as an intervention in teaching and learning of manipulation of the algebraic equation. The lesson intervention involved twenty-five students from two Year 9 classes in one all-girls secondary school in Brunei. Data collection involved pre-test and post-test attempted by all students, followed by the interview of three selected students. Three lessons designed explicitly as the intervention in the learning of manipulation of algebraic equation, which transition from the use of manipulatives to written work, making use of Balance Method concept. Any improvement or beneficial outcome of the previous lesson was used to adapt the following experience as part of the Design Research process. A paired sample t-test revealed that there was a significant improvement in students' performance on the topic after the intervention lesson using Balance Method. More answers that are correct were obtained when students used the Balance Method, compared to their previous failed attempts or incorrect attempts when using another method. Whereas interview revealed that whether students prefer Balance Method or others, they correctly attempted manipulation of the algebraic equation through Balance Method, particularly with equations that involved the combination of operations. It provides evidence that Balance Method should be introduced and taught as the correct concept of manipulating algebraic equation at students' first encounter of the topic.

1. Introduction

Algebra is an important topic in mathematics curriculum and agreed by teachers as one that is difficult to teach and make students understand. This topic brings about an abstract idea, which is difficult for the learners to grasp at first [1-2]. Students find it difficult to learn algebra as they fail to relate why they should learn this particular topic and how it relates to real world [1].

In Brunei, algebra is introduced to students in Year 6 where they are taught basic concepts. These basic concepts involve simple algebraic terms compared to the secondary level syllabus, such as adding and subtracting like terms, substituting values and solving equation of one unknown. Students who encountered problems in algebra at primary level will face difficulties at higher level, which require deeper understanding and mastery of the concept and skills [3-4].

The National Education System for the 21st Century, SPN21, was first implemented in Brunei in 2008 and proposed to raise the students' achievements in core subjects including mathematics as one of the rationale for the changes in the education system [5]. The new curriculum listed the expected learning outcomes of students for each learning area in which they need to accomplish the specific objectives identified by subject content. Students are expected to develop learning skills, conceptual understanding and attitudes to handle mathematics in everyday life, intended to achieve learning outcomes that includes developing the ability to interpret and communicate mathematical ideas both orally and written, stimulate students' interest in learning mathematics and nurture good learning habits. It is important that students improve their understanding on the mathematics concept by displaying good mathematical thinking and reasoning in solving mathematical problem [5]. As part of the objectives of SPN21, equipping students with necessary skills, such as collaboration, creative thinking, problem solving and critical reasoning and ICT skills is crucial for future encounter in university and career [5]. Thus, this intervention lesson aimed to improvise students' way of presenting their procedures to solutions both verbally and written as this could reflect their conceptual understanding and thinking process. This may also provide assurance whether students has fully understand the concept of equal sign in an equations. The research question used to guide this study: *How is the balancing equation method affecting students' performance?*

Mathematics is considered as an essential and fascinating subject because it can be learnt without complications if only the students are manage to produce understanding and fluency of what they need to learn [6]. Mathematical concepts may be difficult to understand or grasp at the beginning of lesson due to unfamiliarity of the concept especially algebra. Most of mathematical concepts can only be understood fully if procedural skills of the concept are acquired by the learners' themselves [7]. Students are still struggling with mathematical concepts even at their upper secondary level of education and that misconceptions learnt by them appeared to be the cause of their poor achievements in test and examinations [8].

In algebra, some of the errors and misconceptions that can be found include errors in forming an equation based on a given word problem or statement [9]. Students often misinterpret the statement or directly translated the words into an equation, making errors in expansion and factorization of algebraic expressions or equation and errors in rearranging or manipulating algebraic equations [10, 11]. Other factors that may result in poor performance in mathematics algebra include low English proficiency in interpreting word problems. When a mathematics question becomes difficult to understand, students will most likely not attempt the question or use inappropriate approach in solving the problem [10].

Algebraic manipulation involves the rearrangement of variables in an equation to obtain the desired outcome. For this section of algebra, these basic skills in manipulating an equation are important and foundation to other higher level concepts algebra. Manipulation of algebraic equation is also applicable in other topics such as function, geometry, and trigonometry. Because of students' inability to manipulate algebraic equation properly, it has led them to have difficulties in solve algebraic problems such as formulating word problem [11].

Students were found to encounter difficulties when solving problems that involve manipulation of algebraic equation [12]. Therefore, enriching and enhancing students' understanding in manipulating equations might increase their capability to solve other problems not only in algebra but also in other topics in mathematics. With the students having better understanding of mathematical concept, it may realize Brunei's aims to increase the students' performance in mathematics.

Balance Method Model takes into consideration that in a mathematics equation, a state of equilibrium achieve in any variables or constants in the left hand side are equal to those in the right hand side. In Balance Method approach, you aim to leave the unknown variable to one side by applying any operation on both side of the equation, while still maintain equilibrium or balance. If we imagine this as seesaw at a balance state, adding or subtracting the same amount from both side will maintain balance, although there is changes value in both sides of the balance. Similarly, multiplying or dividing is the same amount will result in similar outcome. This concept of manipulating of quadratic equation through the Balance Method was successful in several studies [13-15]. Balance Method does have limitation where negative numbers cannot visually be represented [15-17].

2. Method

The initial participants involved in this quantitative study consisted of convenient sampling of thirty-four female students of Year 9 International General Certificate of Secondary Education (IGCSE) students, from one of the all-girl secondary schools in Brunei-Muara District. The abilities of these students ranged from average to low since these students were from two of the IGCSE classes in the school. Since there were some absentees during the pre-test and post-test, only results from twenty-five students were analysed for this study.

The test instruments consist of pre-test and post-test. Both the tests contained the same ten items with questions designed to bring about possible errors and misconceptions that students might commit. Items 1 to 4 involved individual addition, subtraction, multiplication and division operation, followed by two items containing square and square root operation and the remaining four items involving equations that contained combination of any of the four operations with square and fraction (Table 1). Correct attempts in question 1-4 each carried one mark, while the rest carried two marks each. The questions for the test were adapted from past year papers of IGCSE mathematics and Mathematics IGCSE textbook for Year 9. Students were expected to complete the test within 30 minutes and the use of calculators was permitted.

Table 1. List of items included in the test.

Item number	Item question
1	Given that $\boldsymbol{x = y - 1}$. Express $\boldsymbol{y}$ in terms of $\boldsymbol{x}$.
2	Given that $\boldsymbol{p = m + n}$. Express $\boldsymbol{m}$ in terms of $\boldsymbol{p}$ and $\boldsymbol{n}$.
3	Given that $\boldsymbol{C = 2\pi r}$. Express $\boldsymbol{r}$ in terms of $\boldsymbol{C}$.
4	Given that $\boldsymbol{Q = \frac{P}{3}}$. Express $\boldsymbol{P}$ in terms of $\boldsymbol{Q}$.
5	Given that $\boldsymbol{y = a + bc}$, make $\boldsymbol{c}$ as the subject.
6	Given that $\boldsymbol{v = \frac{w-7}{x}}$. Express $\boldsymbol{w}$ in terms of $\boldsymbol{x}$ and $\boldsymbol{y}$.
7	Given that $\boldsymbol{t = m^2}$, make $\boldsymbol{m}$ as the subject.
8	Given that $\boldsymbol{T = 2\sqrt{n}}$. Express $\boldsymbol{n}$ in terms of $\boldsymbol{T}$.
9	Given that $\boldsymbol{C = \frac{5}{F} - 3}$. Express $\boldsymbol{F}$ in terms of $\boldsymbol{C}$.
10	Given that $\boldsymbol{A = \frac{5\pi r^2}{2}}$, make $\boldsymbol{r}$ as the subject.

Prior to the actual study, the test instrument was administered to a group of year 9 IGCSE students of the same school not involved in the main study to check for the reliability of the test questions. A test-retest approach of one week apart under same conditions produced high positive correlation coefficient, indicating reliable instrument. Moreover, students were able to complete the test within 30 minutes during the first and second time taking the test. Three teachers validated the test itself with at least five years' experience teaching mathematics at this level.

Pre-test was carried out at the beginning of the research where students were not informed in advance about the test as to anticipate their current understanding of algebraic manipulation. Post-test was given to students after conducting all the intervention lessons as to measure the impact of the intervention lessons. A paired sample t-test comparing the means of pre-test and post-test was done using Statistical Package for Social Sciences or SPSS to determine if there is the significant difference; hence determine whether the intervention lessons have any impact towards students' performance in the topic.

Interviews were conducted after the post-test, which aimed to gain more feedback from the students regarding their opinions on the method introduced during the intervention. Three randomly selected students were interviewed with the designed questions related to the research question. The selection for the interview was also based on the availability of the students at the time of the interview. The interview was audio recorded and later transcribed for analysis.

For the intervention, three lessons of one hour each were conducted separately in both classes. The three lessons were planned to ensure that each part in rearranging algebraic equation or formula would be covered with sufficient work examples and exercises for each and combined operation. The lessons were designed in a sequence of increasing difficulties and hence students were expected to enhance their understanding in each operation so that they will be able to proceed to the next part of the intervention. (Table 2). In addition, any improvement or useful outcome of the prior lesson was added or adapted to the following lesson as part of the Design Research process.

Table 2. The intervention lessons conducted in this research study

Lesson	Focus	Materials used	Duration (hour)
1	Introduction of balance concept and use this concept, with the aid of manipulatives, to manipulate equations involving addition, subtraction, multiplication and division.	Manipulatives, activity sheets, exercise sheet	1
2	Applying balance concept to continue with manipulation of equation, expanding to square, square roots, fractions and combination of operations.	Manipulatives, Activity sheets, exercise sheets	1
3	Reinforcement questions on of manipulation of equation using combinations of operations	Work sheet	1

The first lesson introduced the concept of balance to resemble the equation and any addition, deletion, manipulation of number or variable in the equation needed to be done on both side. Students also used Manipulative to support the learning concepts (Figure 1).

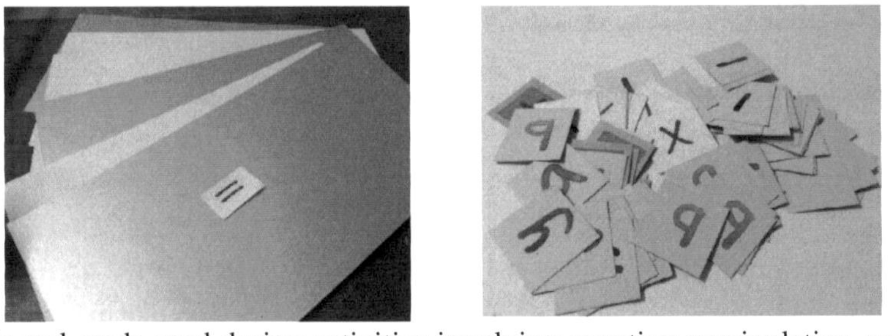

Figure 1. Boards and cards used during activities involving equation manipulation using balancing method.

Four operations were focused in the first lesson, which correspond to specific actions on manipulatives (Table 3).

Table 3. The intervention lessons conducted in this research study

Operation	Action with manipulative in Lesson 1 activity
Addition	Adding the corresponding number or unknown on both side
Subtraction	Removing the corresponding number or unknown on both side
Multiplication	Repeated addition of the corresponding number and unknown on both side
Division	Regrouping into divisor set on both side, follow by removal/leaving only one set.

The second lesson expanded the manipulation of equation to include operation of square, square root and fractions, in addition to reinforcement of the four operations previously covered (single and combined). Initially, the manipulative was only planned in the first lesson, but was kept in the following two lessons so that teacher could use as quick reminder to students the concept related to Balance Method to help and reinforce students visualise the concept of actual balanced equation. Students could also use manipulatives to self-check and clarify any confusion. The lesson moves on to more manipulation via written work. The final intervention lesson stressed is on the reinforcement of all aspects in previous lessons.

3. Result and discussion

The descriptive statistics (Table 4) of the pre-test results indicated that the mean score of the pre-test was 2.68 with standard deviation (S.D.) of 1.82. The lowest score obtained was 0 and the highest score was 8. Surprisingly, only one student out of the total participants passed the test (scored 8). In the post-test, the mean score was 10.40 (S.D. = 2.65). The lowest score gained by students in the post-test was 6 while the highest score obtained was 16, with twenty-one students passed the test (scored 8 and above).

Table 4. Descriptive statistics of the pre-test and post-test result

	Total students	Score				
		Total	Mean	Standard deviation	Lowest	Highest
Pre-test	25	16	2.68	1.82	0	8
Post-test	25	16	10.40	2.65	6	16

Prior to conducting the paired sample t-test, the conditions for suitability for t-test were determined if fulfilled. The dependent variables are continuous (marks of 0 to 20), the independent variable consists of related groups, no significant outliers marks, and the distribution of the differences in the pre-test and post-test is approximately normally distributed. Thus, paired-sample t-test was carried out with the pre-test and post-test results.

Table 5. Paired sample t-test for mean difference of pre-test and post-test.

	Mean difference	Standard Deviation	t	df	Sig (2 tailed)
Post-test – Pre-test	7.720	2.072	18.629	24	0.000

A paired sample t-test (Table 5) produced mean score increase of 7.720, with pre-test having mean of 2.68 (SD = 1.82) and the post-test reaching mean of 10.40 (SD = 2.65), with $t = 18.629$, $df = 24$ and $p = 0.000$ ($p < 0.05$ indicates significance changes). Since the p-value was less than 0.05, analysis indicated that there was significant improvement of the post-test compared to pre-test, thus giving evidence that intervention lessons based on Balance Method has improved students' performance in that topic.

Table 6. Number of students got correct responses for each question in the both tests

		Item number									
		1	2	3	4	5	6	7	8	9	10
Total students	Pre-test	15	16	3	17	1	2	1	1	0	0
	Post-test	24	25	24	22	14	6	25	17	6	11

From Table 6, in the pre-test, Question 1, 2 and 4 had high correct responses with 15, 16 and 17 students respectively. This coincided with questions involving addition, subtraction and division. Low correct attempts in pre-test for question 3 indicated students are having difficulty in successfully attempting multiplication question. Similarly, the rest of the questions indicated extremely low correct attempt (no students correctly responded question 9 and 10 at all), proving that participants are unable to attempt combination and complex questions of manipulating equations. Comparing to the post-test responses, the number of correct responses of every questions increased; with question 2 and 7 had correct response from all students. Question 7 had the most increase in the number of correct responses with only one correct response in the pre-test to 25 correct responses in the post-test. This question is on rearranging equation that involve square in which that Balance Method greatly facilitated correct attempts on such question.

Question 3 which involved multiplication, also increased by large amount of correct responses in the post-test from 3 to 24 correct responses. For the small increase in the correct responses, question 6 only increased by four correct responses in the post-test. This small increase might due to possibly students having the lack of practice in questions involving mixed of operations as well as fraction.

From the interview with the selected students, two of them revealed that they prefer to use balancing equation method in manipulating algebraic equation, which evidently was reflected in their big increase in test's performance. The other student who still preferred 'move to the other side' method also performed better in the post-test by using balancing equation method. It was observed that students were able to manipulate equation correctly using Balance Method with consistence use of the method and more practices, particularly with equations that involved combination of operations.

The significant results from paired sample t-test proved that there was a significant improvement in students' performance on the topic after the intervention of Balance Method. Students attempting questions using Balance Method in post-test produce more correct answers compared to 'move to the other side' method majority of students applied during the pre-test. Students committed fewer errors in the post-test using Balance Method, even for those complex and combination equations [18].

These findings showed that teaching the right mathematical concept did enhanced students' performance on the topic, and hopefully, understanding of the topic. With these, it is important for teachers to develop both conceptual and procedural understanding for the students in learning mathematics, which should be implied when designing a lesson [19-21].

4. Conclusion

Manipulating algebraic equation is introduced with Balance Method in most textbooks, yet teachers tend to use to keep using alternative or short-cut method such as the 'move to the other side' method, which consistently show misconceptions and fail attempts by students. Some students do not even know 'move to the other side' method itself is a consequence of Balance Method and on its own such concept does not exist. Thus, teachers should apply the right mathematical concept and procedure to deliver such topic to the students to ensure not only strengthens students' understanding of this basic concept but also as to avoid teaching accidental misconception to begin with.

References

[1] Revina S and Leung F K S 2018 Educational Borrowing and Mathematics Curriculum: Realistic Mathematics Education in the Dutch and Indonesian Primary Curriculum *International Journal on Emerging Mathematics Education* **2** 1
[2] Purnami A S, *et al* 2018 *J. Phys.: Conf. Ser.* **948** 012020
[3] Ahmad S, *et al* 2018 The instruments of higher order thinking skills *J. Phys.: Conf. Ser.* **943** 012053
[4] Risdiyanti I and Prahmana R C I 2018 *J. Phys.: Conf. Ser.* **943** 012032
[5] Ministry of Education 2013 *The National Education System for the 21st Century SPN21* (Brunei: Ministry of Education)
[6] French D 2002 *Teaching and learning algebra* (London: A&C Black)
[7] Angle D 2007 *What is conceptual understanding?* Retrieved from http://www.maa.org/external_archive/devlin/devlin_09_07.html
[8] Sarwadi H R and Shahrill M 2014 *Math. Ed. Trends and Research* **2014** 1
[9] Clement J 1982 *J. For Research In Math.* Ed. **13** 16
[10] Kani N H and Shahrill M 2015 *Asian Journal of Management Sciences and Education* **4** 20

[11] Matzin E S and Shahrill M 2015 In Pursuit of Quality Math. Ed: *Proc. of the 7th ICMI-East Asia Regional Conf. on Math. Ed.* 2015 (pp. 233-239)
[12] Nickson M 2004 *Teaching and Learning Mathematics: A teacher's guide to recent research and its application* (London: A&C Black)
[13] Gavin M K and Sheffield L J 2015 *Math Teaching in the Middle School* **20** 460
[14] Suh J and Moyer P S 2007 *J. of Computers in Math. and Sci. Teaching* **26** 155
[15] Vlassis J 2002 *Educational Studies in Mathematics* **49** 341
[16] Aczel J 1998 Eds A Oliver and K Newstead *Proc. of the 22nd conf. of the Int. Group for the Psych. of Math. Ed.* (Stellenbosch, South Africa: PME) **2** 2
[17] Warren E and Cooper T J 2005 *Math. Ed. Res. J.* **17** 58
[18] Booth J L, Lange K E, Koedinger K R and Newton K J 2013 Using example problems to improve student learning in algebra: Differentiating between correct and incorrect examples *Learning and Instruction* **25** 24
[19] Widodo S A, *et al* 2018 Teaching materials of algebraic equation *J. Phys.: Conf. Ser.* **943** 012017
[20] Permatasari D and Harta I 2018 The gap between the beginning and the end of algebraic thinking transition period *International Journal on Emerging Mathematics Education* **2** 79
[21] Tanujaya B, *et al* 2017 Mathematics instruction, problems, challenges, and opportunities: A case study in Manokwari regency, Indonesia *World Transactions on Engineering and Technology Education* **15** 287

Designing an integrated learning strategy to develop students' awareness of river environment and science process skills

A Winarti, Sarbain and M Yamin
Universitas Lambung Mangkurat, Jl. H. Hasan Basry, Banjarmasin, Indonesia

E-mail: atiekwin_kimia@ulm.ac.id

Abstract. The river plays a vital role in the daily life of the Banjarese people, Indonesia. Unfortunately, the awareness of the people to keep the rivers clean is still lacking. Hence, the number of rivers in Banjarmasin is decreasing, and the water quality is getting poorer. The study aimed to improve students' awareness of river environment and science process skills by designing learning strategies integrating four subjects, namely: Science, Indonesian language, English, and Civics education. This research employed Research and Development with the 4-D model. The pilot study of the learning design was administered for 100 junior high school students. The data were collected by questionnaires and a science process skill test and were analyzed using descriptive analysis. The results showed that the integrated learning strategy developed in this research was valid, practical, and effective to develop students' awareness of river environment and science process skills. Four weeks after the learning was completed, students' awareness of river environment remains high. Students' process skills of observing, collecting data, and concluding developed higher compared to other skills. Further research is needed to provide more evidence by implementing the strategy involving other subjects.

1. Introduction

Banjarmasin is known for its wetlands as it is surrounded by many rivers required everyone's attention and care. In fact, some Banjarese houses are located on the river. Garbage is the current problem of the river. According to the data, among 117 existing rivers in Banjarmasin, there are now only 60 rivers left [1]. The number will surely continue decreasing if the government does not take this problem into account. Nowadays, these rivers are not only facing the threat of sitting up and narrowing but also the threat of pollution hence the water quality is not worth consuming. According to the pick test results `conducted by the Regional Environment Agency (BLHD) of South Kalimantan, the general condition of river water in Banjarmasin has many heavy metals and E. Coli bacteria.

The results of the preliminary study carried out by Winarti, Sarbaini and Yamin [2] showed that in general junior high school students do not fully understand the importance of environmental awareness yet. Besides, the habit of littering is still prevalent. Teacher responses in some schools also reinforce that students still look down on the spirit of environmental care. Research by Manase [3] found that environmental education methods are not satisfactory enough to increase environmental conditions. Different environmental education methods need to be developed to create sustainable learning outcomes within environmental education.

Treatment implemented in an environmental education course can be effective to change student perception toward the environment and increased students' environmental attitudes [4,5]. Based on these findings, this study designed a learning strategy that can build students' environmental awareness. Different from the previous studies, the learning strategy developed in this study is presented in the thematic form in which the material is supported by science subjects, Indonesia language, English, and Civics. Environmental awareness will be improved through listening and science process activities. The main purpose of listening is to get facts, analyze, evaluate, get inspired, and to improve the ability to speak. In the context of this study, listening is not only hearing, but also giving attention and care. Therefore, through listening habituation, students' awareness of the environment will be developed.

Science process skills are the things done by scientists in learning and investigating [6]. The science process skills are the skills underlying a person in scientific thinking and making decisions. Concerning environmental awareness, practicing process skills will enable students to discover, develop facts and concepts, and cultivate and develop positive attitudes and values for the environment. By training both listening and science process skills, it is expected to develop a positive attitude towards the environment and raise awareness as well as students' critical attitude in decision-making related to the environment.

To be implemented, the developed learning strategy must meet the valid, practical and effective criteria [7]. What is meant by valid is fulfilling the validator judgment, the practical criterion is practical according to experimental results of learning device implementation, and the effectiveness criterion is effective in building environmental awareness and science process skills. Generally, this study aims to develop webbed integrated learning strategy which is valid, practical, and effective to build students' environmental awareness in wetland areas.

2. Method

This study employed a Research and Development with 4-D design. The research stages were Define, Design, Develop, and Disseminate. The pilot product applied the Tessmer formative test [7]. The study was conducted in schools located near the rivers in Banjarmasin. Through purposive sampling technique, three junior public schools in Banjarmasin, Indonesia were selected for the pilot test. This study involved 100 Year 7 students.

This research data was in the form of worksheets readability test results, the validity of learning tools, science process skills, and environmental awareness of students. The worksheet readability, the learning tools validity, and the students' environmental awareness data were obtained through a questionnaire technique. The learning implementation data was obtained through observation, while the science process skill data were obtained through the test. Analyze of the readability student worksheet data, the learning devices validity, learning effectiveness, and students' environmental awareness were conducted by using percentage technique, while the validity of instructional learning was analyzed using the category of validity. There are four categories of validity, namely highly valid, valid, invalid, and highly invalid [7]. This criterion states the learning device has a good degree of validity if the minimum validity reaches a valid level. Practical data on instructional devices are viewed based on the results of the observer's assessment.

The practicality test of learning strategy was conducted three times in each school. Revisions are made to each one-time lesson based on iterative principles. Practical data on instructional devices are viewed based on the results of the assessment observer.

The environmental awareness was measured four times. The first measurement was done before the learning, and the second measurement was done immediately after the learning, the third and fourth measurements were each conducted on the second and the fourth week after the learning was completed. The aim was to see whether the expected environmental awareness has been internalized on the students, even though the learning has been completed. Meanwhile, the science process skills were trained three times. Then, at the end of the learning, the science process skills test was given to the students.

3. Results and discussion

Validation results on learning devices that include the lesson plan, student worksheet, and textbook showed that the developed learning devices are valid, with a mean score of 4.2. The lesson plan reached the score of 4.4, the student worksheet score was 4.1, and 4.1 was in the textbook. Even though this result was declared valid, some small revisions on learning devices related to writing and goal formulation according to the advice of the validator were done.

Based on the results of the student worksheet legality test, it is concluded that the worksheet material is useful, the presentation of the drawing is impressive, and the question and the writing are clear and interesting. Revision of student worksheet only changes the type of letters, from times new romance to arial because it is considered preferable to students. The low score of the practicality of lesson plan 1 on aspects of student activity listening to the text in the cause of the text used is too long, so students do not focus and find it difficult. In the second study, the text was revised so that it did not happen again.

Based on worksheet readability test, several revisions were conducted namely (1) simplifying working procedures; (2) replacing some images with other relevant images; (3) adding some new questions and

IOP Conf. Series: Journal of Physics: Conf. Series **1088** (2018) 012008 doi:10.1088/1742-6596/1088/1/012008

eliminating irrelevant questions; (4) adding color variation. The sample of the student worksheet can be viewed in Figure 1.

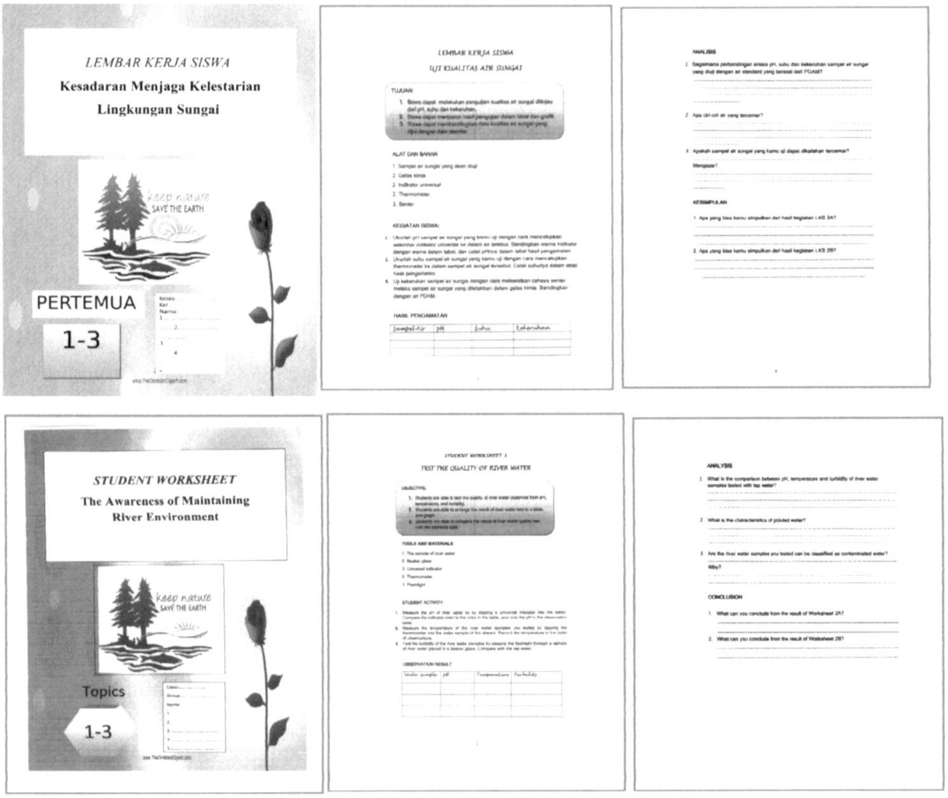

Figure 1. Sample pages of the student worksheet

Figure 1 shows the cover of student worksheet and part of student worksheet containing students' activities of checking the sample of river water. Different from other learning strategies in science which focus on laboratory activities only, the laboratory activities in this worksheet are combined with environmental observation and reading analysis. By doing these activities, students awareness of the environment will be built.

Table 1. The implementation of learning devices

No.	The observed activity	Learning			Average
		1	2	3	
1	The teacher motivates students by showing pictures, etc.	3.9	4.4	4.4	4. 2
2	The teacher explains the English reading passage	3.8	4.1	4.3	4.1
3	The teacher guides students to listen to the content of the passage	3.9	4.0	4.3	4.1
4	The teacher provides guidance on the environmental observation/ practicum activities	4.2	4.4	4.5	4.4
5	The teacher guides the discussion of the preparation of the observation report	4.0	4.2	4.4	4.2
6	The teacher guides the discussion of the presentation of the observation report	4.0	4.2	4.4	4.2
7	The teacher guides students to conclude the lesson.	4.0	4.2	4.4	4.2
	Average	4.0	4.2	4.4	4.2

The practical test results of the developed learning strategy are viewed from the implementation of the learning devices and student activities presented in Table 1 and 2.

Table 2. Students learning activities

No	The observed activity	Material learning 1 (%)	Material learning 2 (%)	Material learning 3 (%)	Average (%)
1	Listen to the teacher's explanation.	91.3	92 .5	96.7	93.5
2	Listen to the English reading passage.	74.3	7 9.2	81.7	78.4
3	Actively search for the data on the environmental/ practicum observation.	90.0	90 .2	92.0	90.7
4	Actively inquire about the environmental/ practicum observation.	76.7	82.5	90.7	83.3
5	Be active in discussion activity preparation of the report.	85.0	85.2	86.7	85.6
6	Be active in the presentation and discussion of the report.	80.0	80.0	83.3	81.1
7	Actively conclude the lesson.	85.0	85.0	85.0	85 .0
	Average	83.2	84.9	88.0	85.4

The data in Table 1 and Table 2 show the high mean score of the used learning devices by 4.2 with the good category, and students activities were done well with the average percentage of 85.4 %. The teacher's teaching activities also went well. It means that the teacher did not have difficulty in understanding the instructions in the lesson plan and student worksheet or in other words, the learning strategy is easy to implement. The most challenging activity for the students was listening to the passage, and the students' most preferred activity was finding data on the current activities of the observation and lab work.

The students' prior ability in listening was still low. They were not very interested in discussing the English discourse as it is a bit too long. The teacher solved this problem by increasing the interaction at the listening time so that students understand the purpose of the activity. This teacher's effort seemed successful as can be seen in the following excerpt of the conversation.

Teacher: ... use one short sentence to express the meaning of this discourse.
Student: The plastic waste contaminates the river.
Teacher: Exactly. What do you think about this condition?
Student: If the water is used for bathing, it would be itchy, ma'am.

It is seen in this excerpt that the teacher's effort to divert the students' boredom when listening to the passage was successful. One of the students responded in line with the habits of the population around the area that is using the river for bathing and washing purposes.

This situation continued with the observation of the river and sampling activities, in which students appeared to be enthusiastic. This outdoor learning activity attracted the students' attention; therefore, they were very active. This relates to the good teacher's ability when guiding the observation (Table 1 point 4). Good teacher guidance keeps students motivated to do the activity. Another reason for the students' interest in observation activity was because this activity is outdoor. Frequent outdoor learning experiences implemented by a trained teacher in a familiar setting can result in greater engagement for students [9,10,12]. In this research, outdoor learning also gives students the opportunity to contact with the environment directly. Therefore, it provides a balance between the application of knowledge and attitude formation, thus making learning feels effective for students. However, while feeling happy with the observational activity, the students rarely asked the teacher questions during observation. In the second lesson, this concern was improved by the teacher by asking more questions to provoke the students' curiosity.

The developed learning strategy effectiveness is seen from the development of the environmental awareness up to 4 weeks after the learning as well as the increase in the students' science process skills. The environmental awareness was built on how to practice listening and science process skills. The listening abilities were developed by introducing environment-related vocabulary, analyzing poems about nature's destruction, and discussing efforts to overcome the natural damage. The skills of the science process were carried out in the student worksheet activities, including observing activities, formulating hypotheses, collecting data, analyzing data, and summarizing integrated into river observation activities, comparing

IOP Conf. Series: Journal of Physics: Conf. Series **1088** (2018) 012008 doi:10.1088/1742-6596/1088/1/012008

polluted and uncontaminated river conditions, cleaning up plastic waste around the rivers, taking samples of the river water, and conducting physical and chemical tests on water samples. The results of the effectiveness test of the developed learning strategy are presented in Table 3.

Table 3. Percentage of the students' ecological awareness level

Indicator	Level of ecological awareness before learning (%)			Level of consciousness Ecological after learning (%)								
	High	Moderate	Low	High			Moderate			Low		
	(1)[a]	(1)[a]	(1)[a]	(2)[b]	(3)[c]	(4)[d]	(2)[b]	(3)[c]	(4)[d]	(2)[b]	(3)[c]	(4)[d]
Concern for the condition of the river.	20	40	40	100	90	90	0	10	10	0	0	0
Awareness of the importance of maintaining the environment.	20	35	45	100	100	100	0	0	0	0	0	0
Awareness of the importance of education to build the environmental care.	35	35	30	100	100	90	0	0	10	0	0	0
Average	*31.7*	*36.7*	*31.7*	*100*	*96.7*	*93.3*	*0*	*3.3*	*6.7*	0	0	0

[a]= the first test; [b] = the second test; [c] = the third test; [d]= the fourth test

It can be seen in Table 3 that before the learning, the percentage of students with a high level of awareness is still lacking (mean of 31.7%). After learning, the environmental awareness was on the third indicator, reached 100% or increased by 56.7%. After two to four weeks of learning was completed, the percentage of the students with high environmental awareness slightly decreased (3.3% to 6.7%)' however, this reduction is minimal. During this period of 1 to 4 weeks, the teacher also continued to monitor and reminded students not to throw garbage into the rivers.

The results showed that the awareness of the waking student environment appears to be internalized. This increased awareness is related to the high interest of students in outdoor learning activities, as well as the involvement of the four subjects that make learning more interesting. This corresponds to the findings stating that the use of nature-based outdoor activities is effective in increasing the students' environmental awareness and sensitivity to the environment. In contrast to the previous studies [4,5] which involving one subject only, the student perception toward the environment in this study involving four subjects, thus the student activities were more varied.

The students' rising environmental awareness after using this strategy was recognized by the students on the student worksheet readability test. The students were very enthusiastic about the second student worksheet material. As much as 68.18 % of students stated strongly agree and 22.73% agreed that the student worksheet material raises their environmental awareness. Thus, the students were convinced that the student worksheet material has a good impact on the development of their environmental awareness. In addition to the environmental awareness, the science process skills after learning also developed to the height category (the mean 3.4 to 3.7). This is possible because of the exercises of the science process skills steps performed on the three previous worksheets. The data on the students' science process skills are presented in Figure 2.

Figure 2. Science process skills of students after learning

As shown in Figure 2, the mean score of the science process skills across all aspects of the three schools ranged from moderate to high. On the observing, data collecting, and concluding aspects in all three schools are high, while the hypothesizing and data analyzing aspects are quite high. It shows that the exercises and laboratory activities applying the science process skills for learning can play a role in the development of these science process skills. The findings are relevant to previous research [12-14] which found that applying science process skill activities in learning will increase the science process skills of students. Even though the previous researches using different methods to train science process skills, however, all efforts provided the same results, science process skills improvement. I-diagram implemented by Karamustafaoglu [14], computer simulation applied by Siahaan, Suryani, Kaniawati, Suhendi, and Samsudin [12], and five stage learning cycle used by Safaah [13] train science process skill in different ways but have the same purposes. Same as the previous researches, applying laboratory work integrated with reading and environment observation in this research made students train their science process skill frequently, hence the skills develop well, especially observation, data collection, and conclusion. At first, the students were not accustomed to doing the science process skills. The observing skill was still low, and there was a confusion between observing and referencing as it is known through the following students' excerpt.

Student A: What are the characteristics of a polluted river?
Student B: ...the dirty river.
Student A: So, write "The river is polluted" in the observation table.
Student B: Ehmmm (pause) I don't think that is what the teacher said. Write the results of the observation according to what we see. It means we write it there that this river has a lot of garbage.
Student A: The Polluted river is the same as the *dirty river.*
Student B: (Hesitant)...I don't think it is the same. Let's ask the teacher then.

Such confusion did not happen again in further learning. From the results of this study, it can be concluded that the developed webbed model integrative thematic learning strategy proved valid, practical to be implemented, and effective in building students' environmental awareness and science process skills.

4. Conclusion

In general, the integrative learning strategy developed in this study has met the valid, practical and effective criteria for building environmental awareness and develop science process skills of the junior high school students in river areas. Specifically, it can be concluded that the developed integrative learning strategy is valid according to the validator's assessment result. The developed learning strategy is practically used based on the teacher's activity and students' activity. The developed integrative learning strategy is effective in building the students' environmental awareness and science process skills. Students' process skills in observing, collecting data, and concluding develop higher compared to the other skills. Two to four weeks after the learning was completed, students' awareness of river environment remains high. Further research is needed to find out more justified evidence by implementing the strategy by involving other subjects.

Acknowledgment
This study was conducted with the fund from *Penelitian Unggulan Perguruan Tinggi*. We would like to thank the Rector of Universitas Lambung Mangkurat, LPPM, and Ditlitabmas Dirjendikti who have facilitated this study.

References

[1] The Statictical Center Agency of Banjarmasin City 2015 *Daftar Nama Sungai di Kota Banjarmasin beserta Data Panjang dan Lebarnya.* Available at: https://banjarmasinkota.bps.go.id/statictable/2015/12/09/487/daftar-nama-sungai-di-kota-banjarmasin-beserta-data-panjang-dan-lebarnya-2014.html
[2] Winarti A, Sarbaini and Yamin M 2016 *Pengembangan Keterampilan Menyimak di Daerah Lahan Basah* (Unpublished: Banjarmasin)
[3] Manase J 2016 The Adequacy of Environmental Education Techniques and Strategies Employed in Uluguru Mountains Hotspot *International Journal of Education and Literacy Studies* **4** 65

[4] Cho Y and Lee D 2018 Love Honey, Hate Honey Bees: Reviving Biophilia of Elementary School Students through Environmental Education Program *Environmental Education Research* **4** 445

[5] Saribas D, Kucuk Z D and Ertepinar H 2017 Implementation of an Environmental Education Course to Improve Pre-Service Elementary Teachers' Environmental Literacy and Self-Efficacy Beliefs *International Research in Geographical and Environmental Education* **26** 311

[6] James H, Fiel R, Okey J R and Jaus H H F 1979 *Learning Science Process Skill* (Paperback)

[7] Nieveen N 1999 Prototyping to Reach Product Quality in *Design Approaches and Tools in Education and Training* (Doordrecht, Kluwer) 125

[8] Bloom B S, Madaus G F and Hastings J T 1981 *Evaluation to Improve Learning* (New York: Mc Graw Hill)

[9] Rios J M and Brewer J 2014 Outdoor education and Science Achievement *Applied Environmental Education and Communication* **13** 234

[10] Okur-Berberoglul E, Ozdilek H G and Yalcin-Ozxilek 2015 The Short Ter Effectiveness of an Outdoor Environmental Education on Environmental Awareness and Sensitivity of In-service Teachers *International Electronic Journal of Environmental Education* **5** 1

[11] Sukhontapatipak C and Srikosamatara 2012 The Role of Field Exercises in Ecological Learning and Values Education: Action Research on the Use of Campus Wetlands *Journal of Biological Education* **46** 36

[12] Siahaan P, Suryani A, Kaniawati I, Suhendi E and Samsudin A 2017 Improving Students' Science Process Skills through Simple Computer Simulations on Linear Motion Conceptions *Journal of Physics: Conference Series* **812** 012017

[13] Safaah E S, Muslim M and Liliawati W 2017 Teaching Science Process Skills by Using the 5-Stage Learning Cycle in Junior High School *Journal of Physics: Conference Series* **812** 012106

[14] Karamustafaoglu S 2017 Improving the Science Process Skills Ability of Science Student Teachers Using I Diagrams *Eurasian J. Phys. Chem. Educ* **3** 26

Formative assessment in science education: Is it being practiced?

M J B Espiritu[1], M Shahrill[1], J S H Q Perera[1] and R C I Prahmana[2]

[1]Sultan Hassanal Bolkiah Institute of Education, Universiti Brunei Darussalam, Bandar Seri Begawan, Brunei Darussalam
[2]Universitas Ahmad Dahlan, Jl. Pramuka Kav. 5, Yogyakarta 55161, Indonesia

E-mail: masitah.shahrill@ubd.edu.bn

Abstract. This study investigated the formative assessment practices in Science education in relation to the enhancement of student learning. The sample comprised of three teachers and 114 students at the Year 9 to 11 levels at one of the private secondary schools in Brunei Darussalam. The study was carried out in three phases: classroom observations, dissemination of the teacher/student self-reflection surveys and individual interviews with the teachers. The findings revealed that the weaknesses, such as the partial use of formative assessment and few student-led discussions that outweigh the strengths of the current formative assessment practices at this particular secondary school. Despite the teachers' attempts to improve the process of teaching and learning, there is a need to change their mindsets and upgrade their skills so that the formative assessment in Science education can be implemented successfully.

1. Introduction

A number of researchers have asserted that articulating formative assessment methods can support student development of inquiry competences in the classroom [1-7]. Teachers may use portfolios, journals, rubrics, checklists, graphic organisers, written reports, the Socratic questioning or other evidence-eliciting techniques to best stimulate and assess the students' cognitive processes [8-15]. Thus, formative assessment practices are slowly emerging to bridge the gap between classroom teaching and learning. When formative assessment is integrated into classroom instruction, it is a planned process to and it does not happen by coincidence [16]. Formative assessment occurs when ongoing and collaborative feedback is provided to the students to improve learning and teaching [17-19]. Moreover, formative assessment underpins student involvement both as evaluators of their own learning and as resources to other classmates [20]. Through providing descriptive feedback, the students gain an understanding of their strengths and weaknesses in connection with classroom learning. However, limited feedback to the students may not lead to improved student learning as suggested by a significant body of research.

This study aims to investigate and report the strengths and weaknesses of formative assessment practices in Science education in one of the private secondary schools in Brunei Darussalam. The student learning was not measured in this study since it only focused on the participants' perceptions of how formative assessment is being practiced in Science education. The formative assessment practices to be examined are limited to the ten dimensions or criteria as identified by the 'Formative

Assessment of Students and Teachers State Collaborative on Assessment and Student Standards' also known as the FAST SCASS [21, 22], namely: (1) learning goals, (2) criteria for success, (3) tasks and activities that elicit evidence of student learning, (4) questioning strategies that elicit evidence of student learning, (5) feedback loops during questioning, (6) individualised descriptive feedback, (7) peer assessment, (8) self-assessment, (9) collaboration, and (10) use of evidence to inform instruction.

2. Method

This study investigated the formative assessment practices in Science education in relation to the enhancement of student learning. The sample comprised of three teachers and 114 students at the Year 9 to 11 levels at one of the private secondary schools in Brunei Darussalam. The teachers were selected purposefully so that they can contribute specific knowledge about the topic of the study. On the other hand, the students were randomly chosen with the use of cluster sampling based on their subject teachers. All participants were made aware of the study and they were willing to participate.

The study was carried out in three phases, namely: classroom observations, dissemination of the teacher/student self-reflection surveys and individual interviews. Mixed quantitative and qualitative methods were employed to investigate the topic presented in this paper. In the first two phases, a quantitative method was applied to develop an assessment tool for classroom observation and subsequently in developing the survey. For the classroom observations of the first phase, the lessons were observed in order to determine if the teachers were indeed consistently practicing the formative assessment. In the second phase, two surveys were performed to validate the classroom observations with the participants' perceptions. In the final phase, applications of the qualitative method from the recorded structured interviews were done with the teachers to explicate the survey results. Descriptive statistics was used to describe the quantitative data gathered from the classroom observations and surveys. And finally, thematic analysis was employed to present the qualitative data that emerged from the individual interviews.

3. Results and discussions

3.1. Demographic profile of the teachers

Table 1 shows the demographic profile of the three female teachers who participated in the study. The teachers came from two different nationalities and the subjects taught are Physics and Combined Science. Note that the subject Combined Science generally covers a basic combination of the three pure science subjects of Biology, Chemistry and Physics. From Table 1, the mean scores of the teachers' age and the years of teaching are 31 and 9 respectively.

Table 1. Frequency distribution of teachers' demographic profile.

Nationality	Age	Years of Teaching Experience	Subjects
Indian	34	12	Combined Science
Indian	29	11	Physics
Chinese	30	4	Combined Science and Physics

3.2. Demographic profile of the students

Table 2 displays the student gender distribution for each year level in the first phase. The male students for each year level mainly dominate the sample population with 59%. Meanwhile, Year 10 had the most number of participants (47%) and Year 11 with 25% of the sample.

Table 2. Student gender and year level distribution for the classroom observation (N = 114).

Year Levels	Male	Female	Total
9	18 (56%)	14 (44%)	32 (28%)
10	30 (56%)	24 (44%)	54 (47%)
11	19 (68%)	9 (32%)	28 (25%)
Total	67 (59%)	47 (41%)	114

Table 3 illustrates the student gender distribution for each subject taught by the teachers in the second phase of the study. The females are more than 20% than males who answered the survey on student self-reflection. Almost three-quarters of the respondents were enrolled in Physics.

Table 3. Student gender and subject distribution for the survey (N = 15).

Subjects	Male	Female	Total
Combined Science	2 (40%)	3 (60%)	5 (33%)
Physics	4 (40%)	6 (60%)	10 (67%)
Total	6 (60%)	9 (40%)	15

3.3. First phase of data collection

The results of the classroom observation based on the ten criteria identified by the FAST SCASS [14, 15] are shown in Table 4. The highest mean score is for item 3. Meanwhile, the lowest mean score is for item 7.

Table 4. Frequency distribution of respondents based on the levels of implementation for the classroom observation (N = 3).

Dimensions	1	2	3	4	5	6	7	8	9	10
Beginning (1)	0	0	0	0	0	0	0	0	0	0
Developing (2)	1	1	0	1	1	0	3	0	2	1
Progressing (3)	2	2	1	1	1	2	0	2	1	2
Extending (4)	0	0	2	1	1	1	0	1	0	0
Mean Score	2.7	2.7	3.7	3	3	3.3	2	3.3	2.3	2.7

Note: 1 – Learning goals, 2 – Criteria for success, 3 – Tasks and activities that elicit evidence of student learning, 4 – Questioning strategies that elicit evidence of student learning, 5 – Feedback loops during questioning, 6 – Individualised descriptive feedback, 7 – Peer assessment, 8 – Self-assessment, 9 – Collaboration, and 10 – Use of evidence to inform instruction.

3.4 Second phase of data collection

Table 5 presents the survey results of the teachers' responses for every criterion of formative assessment. The highest mean score is for item 10. By contrast, the lowest mean score is for item 9.

Table 5. Survey results of teacher self-reflection (N = 3).

Items	1	2	3	4	5	6	7	8	9	10
Extensively (4)	1	2	1	0	0	1	1	0	0	2
Reasonably (3)	2	0	1	2	3	2	1	3	1	1
Partial (2)	0	1	1	1	0	0	1	0	2	0
Limited (1)	0	0	0	0	0	0	0	0	0	0
Mean Score	3.3	3.3	3	2.7	3	3.3	3	3	2.3	3.7

Note: 1 – Learning goals, 2 – Criteria for success, 3 – Tasks and activities that elicit evidence of student learning, 4 – Questioning strategies that elicit evidence of student learning, 5 – Feedback loops during questioning, 6 – Individualised descriptive feedback, 7 – Peer assessment, 8 – Self-assessment, 9 – Collaboration, and 10 – Use of evidence to inform instruction.

Table 6 exhibits the frequency distribution of the students' responses in the second phase of the study. The highest mean score is for item 3. On the contrary, the lowest mean score is 2.9 for both items 7 and 10.

Table 6. Survey results of student self-reflection (N = 15).

Items	1	2	3	4	5	6	7	8	9	10
Extensively (4)	8	5	10	7	8	9	4	7	9	8
Reasonably (3)	4	7	4	3	5	2	6	7	2	1
Partial (2)	2	1	1	3	2	3	4	1	3	2
Limited (1)	1	2	0	2	0	1	1	0	1	4
Mean Score	3.3	3	3.6	3	3.4	3.3	2.9	3.4	3.3	2.9

Note: 1 – Learning goals, 2 – Criteria for success, 3 – Tasks and activities that elicit evidence of student learning, 4 – Questioning strategies that elicit evidence of student learning, 5 – Feedback loops during questioning, 6 – Individualised descriptive feedback, 7 – Peer assessment, 8 – Self-assessment, 9 – Collaboration, and 10 – Use of evidence to inform instruction.

3.5 Third phase of data collection

Table 7 presents the thematic analysis results of the structured interviews with the teachers to elicit how they understood and applied formative assessment in Science education. The interviewees had varied viewpoints about formative assessment. The brief responses expressed in concordance are, providing feedback to advance learners, the purpose of formative assessment is checking understanding of the topic, and there are challenges in planning the assessment.

Table 7. Thematic analysis results from the teacher interviews (N = 3).

Themes	Summarised responses (% responded)
1. Purpose of formative assessment	Checking understanding of topic (67%); Testing knowledge (33%)
2. Tools used in formative assessment	Giving short quizzes (33%); Giving questionnaire (33%); Analysis of knowledge (33%); Providing written examinations (33%)
3. Benefits of formative assessment	Differentiate level of understanding (33%); As a self-reflection (33%); To assess quickly the strengths and weaknesses (33%)
4. Evidence of learning	Students asked questions (33%); Students' satisfaction (33%); From the students' language and facial expressions (33%); By asking the students directly (33%); By getting clarifications from the students (33%); Admitting to limitations (33%)
5. Strategies for advance learning	Giving extended topics (33%); Providing feedback to the students (100%)
6. Challenges in planning the assessment	Assessment Planning (67%); Lacking motivation (33%); Students not responding to some assessment questions (33%)

The findings obtained from this study provided an overview of the strengths and weaknesses of the formative assessment practices in Science education at a private school in Brunei. Based on the classroom observations, surveys and interviews, the weaknesses, such as the partial use of formative assessment and few student-led discussions that outweigh the strengths of the current formative assessment practices at this particular secondary school. In analysing the reported qualitative and quantitative data, there are six points that needs to be highlighted.

1. Formative assessment was only practiced partially by the teachers in the classroom setting. One probable reason is that the teachers' understanding of formative assessment was parallel to summative assessment.
2. The students were able to understand the learning goals at the start of the lesson with the added presentation using visual aids. For example, in one of the Physics lessons, a thermal expansion was brought to the class to stimulate the students' interest in the topic.
3. The students perceived that their tasks and activities were connected to the learning goals and these goals will be able to elicit evidence of learning. They had opportunities to ask questions but only a few students were involved in the discussions. Furthermore, it was observed that the students preferred to copy and write the notes rather than being involved in the student-led activities. This observation implied that the students relate copying and writing notes as one of the task activities thereby enabling them to learn the topic.
4. In reporting the participants' self-reflection of formative assessment, item 4 on the use of evidence to inform instruction had the highest mean score for the teachers (3.7 or 93%) but it is the lowest mean score for the students (2.9 or 73%). There were opposing views on collecting evidences of student learning connected to the learning goals. Thus, the teachers' efforts to modify their teaching styles in accommodating learning had not been executed extensively to impress upon the minds of the students.
5. Although the teachers assessed the students' mastery of the topic, peer assessment, as discerned by the students, had not been explored thoroughly. The absence of peer assessment restricted the students' exposure to a variety of work that may help them incite their cognitive processes. This insight validated the teachers' self-reflection on having less collaboration to deepen the learning experience in the classroom.
6. The teachers gauge learning experience based on the students' reactions to the words and body language. Using portfolios, journals, rubrics, checklists or graphic organisers had been overlooked in the lessons. Importantly, based on the findings from previous studies [8-15], the use of such tools may assist in making learning experience more meaningful by allowing and supplementing teachers to impart evidence-based descriptive feedback to their students before, during, and after the lesson.

Generally, despite the teachers' attempts to improve the process of teaching and learning, there is a need to change their mindsets and upgrade their skills to implement the formative assessment successfully in Science education. Exploring the differences in the understanding and implementation of formative assessment practices by secondary teachers in Science education has revealed important facets for further research undertakings. Based on the findings of the study, there are several suggestions and recommendations that may be taken into consideration, such as the need to conduct school-wide investigation of the current formative assessment practices in the secondary school section to identify gaps between teaching and learning; to review existing curriculum to determine how to embed formative assessment aligned with the learning goals, content, and outcomes; to establish mentoring and coaching programme to assist the teachers in making significant transitions in mindset and skills for effective implementation of the formative assessment; to develop an evaluation tool to be used for to ensure that criteria for success in the implementation of formative assessment practices are achieved; and to organise student-led conferences to allow the students to have reflection and take responsibility for their learning as they collect, compile, and present them using a portfolio or graphic organiser.

4. Conclusions
Formative assessment has its advantages and disadvantages when it comes to validating student success. Therefore, teachers must determine how to blend formative assessment with other approaches in order to have a meaningful evaluation of student learning. As a consequence, teachers need to keep abreast with emerging trends that shape the future of teaching and learning.

References

[1] Black P, *et al* 2004 *Phi Delta Kappan* **86** 8
[2] Hume A and Coll R K 2010 *Res. in Sci. & Tech. Ed.* **28** 43
[3] Yatab R S and Shahrill M 2014 *Int. J. of Sci. and Res.* **3** 685
[4] Nor H N H M and Shahrill M 2014 Incorporating the use of poster and oral presentations as an alternative assessment in the teaching of secondary mathematics *Proc. of the 2nd Int. Conf. on Soc. Sci. Res.* (Malaysia: ICSSR 2014, WorldConferences.net) pp 369–78
[5] Botty H M R H and Shahrill M 2015 *Int. J. of Ed. and Prac.* **3** 17
[6] Damit A H, *et al* 2015 *Med. J. of Soc. Sci.* **6** 214
[7] Andersson C and Palm T 2017 *Ed. Inq.* **8** 104
[8] Pinchok N and Brandt W C 2009 *Connecting Formative Assessment Research to Practice: An Introductory Guide for Educators* (Chicago, IL: Learning Point Associates)
[9] Khoo J S, *et al* 2016 *J. on Math. Ed.* **7** 83
[10] Idris S H, *et al* 2017 *Adv. Sci. Lett.* **23** 901
[11] Suhaimi Z, *et al* 2016 *J. of Math. Ed. at Teachers College* **7** 11
[12] Suhaimi Z, *et al* 2017 *Int. J. on Emerging Math. Ed.* **1** 39
[13] Mohammad H H, Jawawi R, Matzin R, Shahrill M, Jaidin J H, Mundia L and Shamsu L S 2017 *Turkish Online J. of Ed. Tech. Dec. Sp. Iss.* **12** 281
[14] Sai F L, *et al* 2018 *J. of Phys.: Conf. Ser.* **948** 012057
[15] Tan E H Y, *et al* 2018 *Adv. Sci. Lett.* **24** 5401
[16] Popham W J 2008 *Transformative Assessment* (Alexandria: ASCD)
[17] Rawlins P and Leach L 2014 Questions in Assessment for Learning and Teaching *Facing the Big Questions in Teaching: Purpose, Power and Learning* eds A St George, S Brown, and J O'Neill (Auckland: Cengage Learning)
[18] Smith K 2010 *Assessment Matters* **2** 6
[19] Wiliam D 2011 *Stud. in Ed. Eval.* **37** 3
[20] Clark I 2012 *Ed. Psyc. Rev.* **24** 205
[21] Heritage M 2010 *Formative Assessment and Next-Generation Assessment Systems: Are We Losing an Opportunity?* (Washington: Council of Chief State School Officers)
[22] Wylie C and Lyon C 2013 *Using the Formative Assessment Rubrics, Reflection and Observation Tools to Support Professional Reflection on Practice* (Washington: Council of Chief State School Officers)

Uncertainty and data content in bowling: Task design

R Permatasari, R I I Putri and Zulkardi

Universitas Sriwijaya, Jl. Srijaya Negara, Bukit Besar, Palembang, Indonesia

E-mail: ratu.ilma@yahoo.com

Abstract. The students still difficulties in solving the PISA-like mathematics problem on uncertainty and data content. It is because the students were never solving context-based problems such as PISA. This research aimed to produce the valid and practical of PISA-like problems using bowling context. The research involved tenth-grade students of the senior high school in Palembang. The research method used is design research with development studies type through two phases: the preliminary and formative evaluation. In the preliminary phase have been reviewed some of literature and design. Meanwhile, the formative evaluation phase only consists of self-evaluation, expert reviews, one-to-one, and small group. The data collection techniques used were walkthrough, document, observation, and interview. This research produced two valid and practical problems using bowling context. From the expert reviews and one-to-one, the problems using bowling context have been well based on content, constructs, and language. From the small group, the problems could be understood and easy to be used by students. In addition, the bowling context that used could involve students' mathematics literacy ability such as communication, representation, reasoning and argument abilities to find the concept, give perception, and can interpret the problems.

1. Introduction

Uncertainty and data are part of the content of the Program for International Student Assessment (PISA) that very important and useful to solve the problems which are closely related to daily life [1, 2]. However, many students in Indonesia are still difficulties in solving PISA problems. It's because students never solving context-based problems such as PISA which is seen in the evaluation system in Indonesia is still using routine-problems [3-6]. Accordance with PISA's result, Indonesia is ranked 63 out of 65 countries with a score of 384 on uncertainty and data content [7].

Therefore, the low result of PISA's Indonesia students is the basis for the development of the 2013 Curriculum. Also, the teachers required to forming the students to be creative in thinking, solving problems, making decisions, reasoning, and conveying ideas in activities through PISA problems [8]. In PISA, the problems were presented mostly in real-world situations, so students can feel the benefits of mathematics in solving the problems of daily life [1, 9].

Furthermore, the use of real-world situations or context in learning mathematics can present an abstract mathematical concept in the form of representation that is easy to understand by students, so the mathematical concept becomes more meaningful [10]. One of the contexts that can be used is sports because it's one of the fields that use probability or in the PISA framework called uncertainty and data content [2]. Besides that, on August 8, 2018, Indonesia is a host of the 18th Asian Games precisely in Jakarta and Palembang. Also, one of the sports in this competition is bowling.

In addition, several previous studies have shown good results in learning mathematics by using a context, namely the research by [11] which used Jambi context in developing PISA-like mathematics problem. A study conducted by [12] using swimming context in mathematics learning. Then, [13] shown that the use of sports rowing context can enlarge the students' thought and can make the students understand the mathematical concept. In this case, the use of bowling context is a new experience for students in solving PISA-like mathematics problems and one of the efforts to enliven the 18th Asian Games among the students. Therefore, researchers are interested in using bowling as a context for solving the PISA-like mathematics problems on uncertainty and data content with the aim to produce the valid and practical problems.

2. Method

This research was conducted tenth-grade students of the senior high school in Palembang by using design research method with development studies type through two phases: preliminary and formative evaluation [14]. In the preliminary that includes the preparation and design, the researchers reviewed some of the literature related to this research, then designed instruments such as lattices, question cards, and rubric assessment by the 2015 PISA framework.

The next phase is a formative evaluation that includes the phase of self-evaluation, expert reviews, one-to-one, small group, and field test. But in this research, the researchers only discussed until small group by the problem statements of this research.

In self-evaluation, the researchers have analyzed the designed problems by ourselves and produce prototype I. After that it used in the expert reviews phase. The prototype I was validated by experts, in this phase the researchers using three methods of validation process: mails review with PISA expert from Brunei Darussalam University; panel review with lecturers of mathematics education in Sriwijaya University and some of the peers that had experience in developing of PISA-like problems; face-to-face with mathematics teacher. Along with validations with experts, a one-to-one phase performed. It's done to know the validity of PISA-like mathematics problems. This phase involving three students of senior high school with various abilities (high, medium, low). The validation result was produced prototype II and used in the next phase.

Small group phase was conducted to find out the practicality of the problems developed by involving six students with high-ability, medium-ability, and low-ability. The results of this phase produce prototype III that valid and practical.

Data collection techniques used was the walkthrough, document, observation, and an interview. Then the data were analyzed by the qualitative descriptive method to describe the result of each step of development.

3. Result and discussion

This research produced two items of PISA-like mathematics problems on uncertainty and data content using bowling context in Asian Games which was valid and practical. However, in this paper, the researchers only covered one problem using bowling context at sixth level.

3.1. Preliminary

In preliminary, the researchers analyzed the students with a mathematics teacher by determining the level of mathematical ability of the students in choosing the research subject. Therefore, the researchers got three tenth-grade students (low, medium, high-ability) for one-to-one phase and six tenth-grade students with the different ability for the small group phase. Next, curriculum analysis by determining indicators of problem achievement, and the strategies students use in mathematical literacy ability. After that, the results of this phase were designed lattices, some of the PISA-like mathematics problems using bowling context on uncertainty and data content, and rubric assessment.

3.2. Formative evaluation

3.2.1. Self-evaluations

In self-evaluation, the PISA-like problems that designed, the evaluated and examined are owned by the researchers. The revision results in this phase called prototype I. Then, it used in the next step. One of the PISA-like mathematics problems using bowling context in Asian Games can be seen in Figure 1.

Question 1
Pin formation in the bowling game can be seen below:

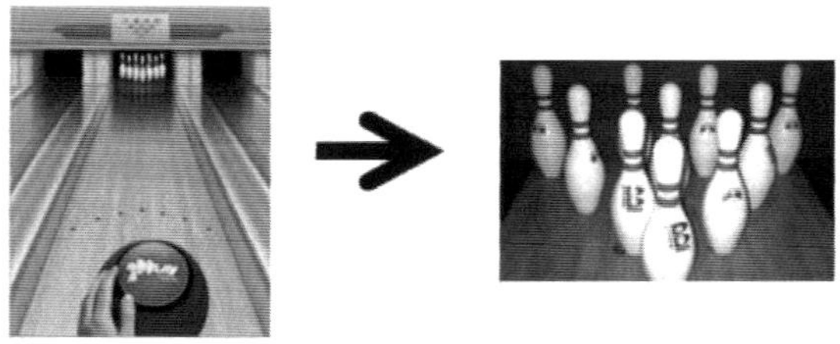

What position of the ball in order to fall all the pin?

Figure 1. The PISA-like problem before revision

3.2.2. Expert reviews and one-to-one

Expert reviews and one-to-one phase were conducted in parallel to see the validity of problems. The prototype I was validated by experts based on existing criteria both regarding content, construct, and language. While the process of one-to-one involved three tenth-grade students of the senior high school, namely BHIP (high-ability), RR (medium-ability), and CAP (low-ability).

The students are asked to read and examine the problem so that the researchers can find out the responses, constraints faced, and understanding of students in solving the problem using bowling context. The responses and obstacles that observed focus on legibility and clarity of problems. Table 1 shows the validation result from expert reviews and one-to-one phases.

Table 1. Comments of experts and students

Validation	Comments/Response	Revise
Expert reviews	• The question is not clear, so change it to "What position should the ball be shot by the player in order to strike all the pin?"	• Change the question according to the suggestion from the expert
	• Change figure of pin formation that seen from in front of to figure that seen from the top	• Change the figure
Students	• How many are the strategy of ball positions asked this question? • I don't understand what the meaning of this problem	• Make the problem clearly

Based on the results of expert validation and one-to-one, the researchers revised it. Furthermore, it can be concluded that the problems have designed are categorized as valid problems. It's reflected in the comment provided by expert validation, then responses, constraints, and understanding of the students when solving the problem [14].

The validity of the problems in terms of content (according to the domain of mathematics literacy in PISA, also according to probability learning in tenth-grade); in terms of construct, the problems have been in accord with the characteristics of PISA problems level and abilities of the tenth-grade students; and in terms of language, the problems use of enhancing spelling, could be understood, and didn't have variety of meanings [15]. Then, the revised result called prototype II and used in the next phase. The problem in prototype II can be seen in Figure 2.

Question 1
Pin formation in the bowling game can be seen below:

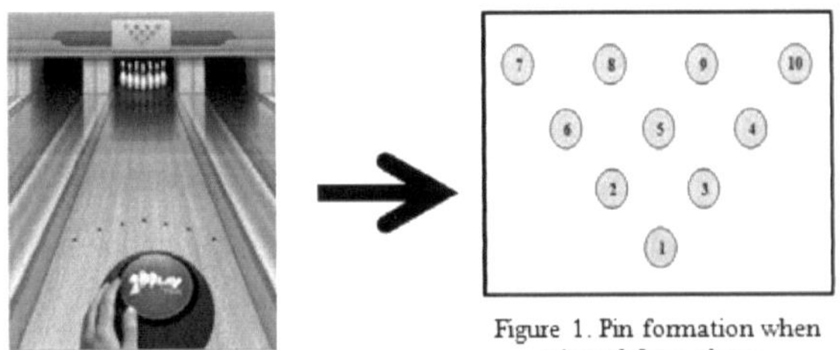

Source: www.google.com

Figure 1. Pin formation when viewed from above

What position should the ball be shot by the player in order to strike all the pin?

Figure 2. PISA-like problem after revision

3.2.3. Small group

Small group phase was conducted to see the practicality of problems for students in the implementation. This phase involved six tenth-grade students of the senior high school, namely DBA (high-ability), MA (high-ability), TCBS (medium-ability), AFP (medium-ability), DG (low-ability), PUR (low-ability).

The students were first asked to solve the problem individually. After that, students asked to discuss in their group. Almost all of the student were able to understand and solve the problem well in the learning process. From the analysis result, the students use different strategies to solve the problem (see Figure 3 and Figure 4).

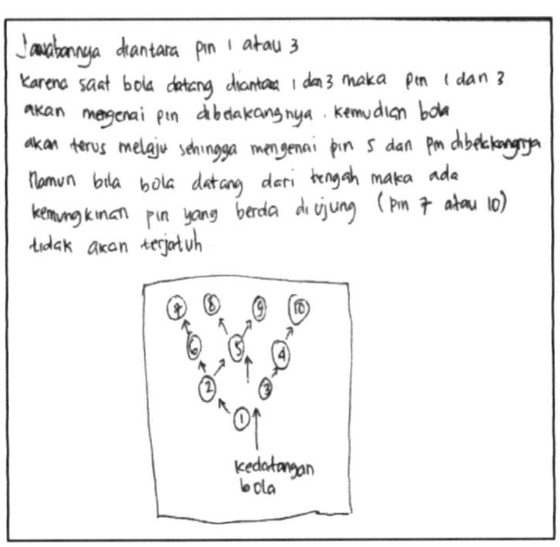

Figure 3. DG's answer to the problem using bowling context.

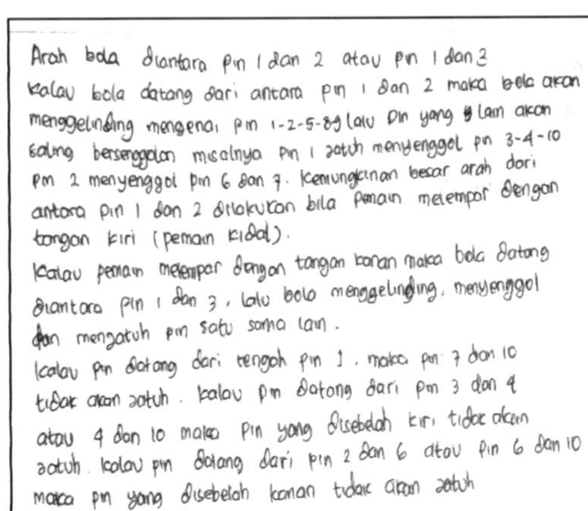

Figure 4. DBA's answer to the problem using bowling context.

Figure 3 shows the students explain the ball should be shot between the 1st and 3rd pin because it will hit the pin of the behind and the pin will fall each other. The student also said that if the ball shot in the middle, so pin on the end namely the 7th or 10th pin will not fall. Then, they can draw the process of pinfall by the picture. It means that the student can use their representation ability. While from the figure 4, it appears that students can explain in detail of the probabilities that will happen. Based on their analysis, there are two ways to get the strike: between the 1st and 2nd pin or between the 1st and 3rd pin. Students are also able to communicate the process of pinfall by their self-sentences.

Figure 4 also shows the analysis of student from various sides and the reasons that support their arguments such as they explain if the players use the left hand so that the ball will shot between the 1st and 2nd pins. It will hit the 1st, 2nd, 5th, 8th, and 9th pins. Also, the 1st pin will hit another pin on end. Then the 2nd pin will hit the 6th and 7th pins. Not only that but also the students explain the process of pinfall if the players use right hand and shot between the 1st and 3rd pins. Furthermore, they clarify if the players shoot between the 3rd and 4th pins or between the 4th and 10th pins, the pins on the left will not fall. Besides, if the ball shot between the 2nd and 6th pins or between the 6th and 10th pins, also the pins on the right will not fall.

The analysis result of students' worksheet shows that most students can solve the problems well that using bowling context. Students also involving the communication, representation, reasoning and argument abilities to compare the probabilities of each other in solving the problems. Accordance with the results of research by [16] which states that students with good reasoning ability can understand, formulate, and solve problems correctly and adequately.

Based on the result of expert reviews, one-to-one, and small group, it can be concluded that the problems using bowling context that has designed are categorized valid and practical. It's not only reflected from the result of expert reviews, but also from student's response and understanding in solving the problem [17, 18]. In this case, almost students can solve the problems well that using bowling context. Students also involving the communication, representation, reasoning and argument abilities to compare the probabilities of each other in deciding the position should the ball shot. The practicality illustrated by the result of small group phase which is the problems could be understood for learning probability, easy to be used because they can solve the problems well by imagining the situations and probabilities that can happen [15, 19, 20].

In addition, the interview result of students' answer sheet, it can be concluded that the PISA-like mathematics problems using bowling context on uncertainty and data can improve students' mathematics literacy ability such as communication, representation, reasoning and argument abilities. Also, students said that they felt happy to do such problems because it is interesting to them. The implemented context in problems can help students to find the concept, help students to give the perception and can interpret information of problems easily, also can stimulate informal knowledge to formal knowledge [21-24].

4. Conclusion

This research produced two valid and practical of PISA-like mathematics problems using bowling context on uncertainty and data content. Valid theoretically can be seen from the validation results where all validators state that the items have been well based on content, constructs, and language. In terms of content, the problems that have been designed according to the domain of mathematics literacy in PISA, also according to probability learning in tenth-grade; in terms of construct, the problems have been in accord with the characteristics of PISA problems level and abilities of the tenth-grade students; and in terms of language, the problems use of enhancing spelling, could be understood, and didn't have variety of meanings. Practically illustrated by the results of small group phase where the problems could be understood as learning probability, easy to use because students can solve the problems well by imagining the situations and probabilities that can happen. In addition, the bowling context could help students involving their communication, representation, reasoning and argument abilities to find the concept, give perception, and interpret the problems.

Acknowledgments

The researchers would like to express gratitude to the Directorate General of Higher Education Indonesia who has funded Pasca-grant research in 2018, and to Ms. Dian Fitriana and her students for participating in this research. As well as those who have helped the researchers in developing the research and writing this article.

IOP Conf. Series: Journal of Physics: Conf. Series **1088** (2018) 012010 doi:10.1088/1742-6596/1088/1/012010

References

[1] Johar R 2012 Domain soal PISA untuk literasi matematika *Jurnal Peluang* **1** 30
[2] Yanti W, Nusantara T, and Qohar A 2016 Analisis kesalahan dalam menyelesaikan soal pada materi permutasi dan kombinasi *Proc. Nas. Sem. on Mathematics* (Malang: Indonesia) p 97
[3] Stacey K 2010 Mathematical and scientific literacy around the world *Journal of Science and Mathematics Education in Southeast Asia* **33** 1
[4] Wu M 2011 Using PISA and TIMSS mathematics assessments to identify the relative strengths of students in western and asian countries *Journal of Research in Education Sciences* **56** 67
[5] Novita R, Zulkardi, and Hartono Y 2012 Exploring primary student's problem-solving ability *Journal on Mathematics Education (IndoMS-JME)* **3** 133
[6] Wijaya A 2016 Students' information literacy: A perspective from mathematical literacy *Journal on Mathematics Education (IndoMS-JME)* **7** 73
[7] OECD 2014 *PISA 2012 Results: What Students Know and Can Do – Student Performance In Mathematics, Reading and Science* **1** (Paris: OECD Publishing)
[8] Kemendikbud 2013 *Pengembangan kurikulum 2013* (Jakarta: Kementerian Pendidikan dan Kebudayaan)
[9] R I I Putri and Z Zulkardi 2018 *J. Phys.: Conf. Ser.* **948** 012056
[10] Putri R I I and Zulkardi 2017 Fraction in shot-put: A learning trajectory *AIP Conference Proceedings* **1868** 050005
[11] Charmila N, Zulkardi, and Darmawijoyo 2016 Pengembangan soal matematika model PISA menggunakan konteks jambi *Jurnal Penelitian dan Evaluasi Pendidikan* **20** 198
[12] Gunawan M S, Putri R I I, and Zulkardi 2017 Learning fractions through swimming context for elementary school students *Proc. Int. Conf. on 5th South East Asia Development Research (SEA-DR)* (Banjarmasin: Universitas Lambung Mangkurat) **100** p 61
[13] Nasution M F, Putri R I I, and Zulkardi 2017 Dayung Context in Fraction *Proc. Int. Conf. on 5th South East Asia Development Research (SEA-DR)* (Banjarmasin: Universitas Lambung Mangkurat) **100** p 1
[14] Zulkardi 2002 *Developing a learning environment on realistic mathematics education for Indonesian student teachers* (Enschede: Universiteit Twente)
[15] Zulkardi 2006 *Formative Evaluation: What, why, when, and how*
[16] Ahyan S, Zulkardi, and Darmawijoyo 2014 Developing mathematics problems based on PISA level *Journal on Mathematics Education (IndoMS-JME)* **5** 47
[17] Marwan, Ikhsan M, and Marwan 2016 Meningkatkan kemampuan berpikir kritis matematis siswa SMK melalui model pembelajaran berbasis masalah *Jurnal Didaktik Matematika* **3** 9
[18] S Arifin *et al* 2017 *J. Phys.: Conf. Ser.* **943** 012038
[19] Hendroanto A, *et al* 2018 Photography activities for developing students' spatial orientation and spatial visualization *J. Phys.: Conf. Ser.* **943** 012029
[20] Setyawan F, *et al* 2018 Visualizer's representation in functions *J. Phys.: Conf. Ser.* **943** 012004
[21] Rahayu C, Putri R I I, and Zulkardi 2017 Multiplication of fraction with natural number by using hurdles *Proc. Int. Conf. on 5th South East Asia Development Research (SEA-DR)* (Banjarmasin: Universitas Lambung Mangkurat) 100 pp 43
[22] Roni A, Zulkardi, and Putri R I I 2017 Sprint Context of Asian Games in the Division of Fractions *Proc. Int. Conf. on 5th South East Asia Development Research (SEA-DR)* (Banjarmasin: Universitas Lambung Mangkurat) 100 pp 22
[23] R I I Putri *et al* 2017 *J. Phys.: Conf. Ser.* **943** 012035
[24] S Rejeki and R I I Putri 2018 *J. Phys.: Conf. Ser.* **948** 012058

IOP Conf. Series: Journal of Physics: Conf. Series **1088** (2018) 012011 doi:10.1088/1742-6596/1088/1/012011

Teachers' reflections on students' mathematical problem solving in junior high school

Y Harisman[1,2], **Y S Kusumah** [1] **and K Kusnandi** [1]
[1]Universitas Pendidikan Indonesia, Jl. Dr. Setiabudi No. 229, Bandung 40154, Indonesia
[2]STKIP PGRI Sumatera Barat, Jl. Gunung Pangilun Padang Utara Kota Padang Sumatera Barat 25000, Indonesia

E-mail: yulyantiharisman_math@student.upi.edu

Abstract. Reflection will help teachers improve their learning that has been done, especially in the problem-solving process. Therefore, this research was designed to answer the question of how to picture teacher reflection in mathematical problem solving learning process. This study adopted qualitative research with grounded theory and systematic design. Three teachers from three junior high schools with different clusters were selected as research subjects. Teaching and learning process were documented during three meetings. The selected subject was geometry because it requires a pattern of thinking in applying concepts and skills in solving problems. The learning video was analysed to create open questions that could explore how teachers reflect on the learning process they have conducted. The teachers and the researcher watched the video to reflect the teaching and learning process. The teachers were given several questions displayed in the video. The study revealed three categories of teacher resulted in their reflections process. The teacher categories are excellent, very good, and good. These categories of reflection can be a reference for teachers in studies that have a reflection phase such as lesson study and classroom action research.

1. Introduction

Teacher reflection is how a teacher reviews the learning process that has been conducted. The reflection will help them to fix any mistake they have made in order to avoid the same mistake to occur again in the next learning process. Expert researchers on teacher reflection in mathematics such as Muir and Beswick's [1] have categorized three levels of teacher reflection on mathematics learning process. The category was obtained based on the case studies of several school teachers. The measurement of the reflection levels was accomplished by interviewing research subjects after observing their learning process videos. The reflection levels achieved by Muir and Beswick are presented in Table 1. Muir and Beswick's research was followed by Geiger [2]. The research conducted by Geiger [2] developed by adding dimension aspect, that is: aspect of self, practice, and students. The reviewed reflection level from Geiger's reflection response object technical is explained in Table 2.

Table 1. Teacher's reflection levels

Teacher Level Reflections	
Level 1: *Technical Reflection*	A teacher describes general accounts of classroom practice, often with a focus on technical aspects, with no consideration of the value of the experiences For example: The lesson went well. I did not ask enough questions. The students could all do the task
Level 2: *Deliberate Reflection*	A teacher identifies 'critical incidents' and offers a rationale or explanation for the action or behavior For example: Johnny was really off task today. I think the question was too hard for him; the way he was working out the area showed me he was confusing it with perimeter I wanted them to use the concrete materials
Level 3: *Critical Reflection*	The teachers move beyond identifying 'critical incidents' and providing explanations to considering others' perspectives and offering alternatives For example: I should not have put Jack on the spot by asking him to explain what a square number was. He was obviously uncomfortable. Perhaps I could incorporate a 'think-pair-share' strategy whereby the students could talk with each other before sharing more publicly. .

Table 2. Levels of reflection against object of reflective response

Reflection Levels	Object of reflective response		
	Self	Practice	Students
Technical	Personal role is described during a teaching event. The description is factual rather than personally insightful	Teaching activity is described in terms of technical aspects. Focus is on consequences or outcomes of their practice	Students' responses to teaching activity are described in terms of technical aspects. The focus is on consequences or outcomes of teaching practice
Deliberate	Personal role is described during a teaching event. A rationale or explanation for the personal behavior is provided	Critical incidents are related to teaching practice, and a rationale or explanation for the practice is articulated	Students' responses to teaching activity are noted, and a rationale or explanation for the response or behavior is constructed
Critical	Personal role is described during a teaching event. The behaviour is critically analysed and alternative behaviours discussed	The purpose of an activity is clearly articulated, and a judgment is made about the success or otherwise of a teaching practice. When unsuccessful, an alternative practice or activity is suggested	Students' responses to teaching activity are noted, and a rationale or explanation for the response or behaviour is constructed. Potential improvements to the activity are related to an anticipated student response

However, the measurement of those levels was held in general teaching and learning mathematics. There is still not much research that highlights how teachers reflect on the learning process of problem-solving. Therefore, this research paper will answer a question of how the relationship of the category of teacher reflection on the problem-solving learning process is.

2. Method

This research is a preliminary analysis of the development research. The product of this research is the theory about category of teacher reflection in problem-solving. The method used is a qualitative approach and the research design is grounded theory. According to Creswell [3], the grounded theory design is a systematic qualitative procedure that is used to bring up a general explanation of researchers based on the views of the participants, explaining the process, act, or the interaction among participants. The type of grounded theory research design used in this research is a systematic design, which is widely used in the field of educational research. The design is composed of basic-obedient research, detailed, and predictable compared to the initial concept of grounded theory designs. Three secondary schools (Junior High School) in Bandung in Indonesia with a different cluster (level) were selected as the site of the research.

Table 3. Reflection techniques

Reflection Techniques	Purposes
Pre-Review of Reflection (prior to reflection): Making reflection questions	In order to understand the depth of teachers-done reflection through playing learning videos done by teachers.
Viewing of video footage (Watching recorded video together with teachers)	In order to enable teachers to take a role as an observer, not a member.
Professional Conversation	In order to observe further aspects of teaching behaviour through discussion.
Comparison of two students' mathematical behaviours (Highlighting some students intentionally on learning videos)	In order to encourage teachers to do critical reflection towards mathematical behaviours of students individually.
Professional Reading	In order to understand if teachers are able to recognize students' strength and weakness on mathematical problem-solving

A cluster is a group of schools based on the school rankings. According to information obtained from the Department of Education in Bandung, there were three clusters for Junior High Schools in Bandung. The clusters were A, B, and C. The purpose of selecting a school with various levels is an effort to avoid the problems of reliability and validity that may exist on a small sample Patton [4]. A voluntary teacher (volunteer) at eighth grade of each school served as the subject of the observation in the learning process of mathematical problem-solving. What was needed in order to obtain the category of teachers in the learning process can respectively be seen in Table 3 of reflection technique, and reflection analysis can be seen in Table 4.

Table 4. Analysis of Teachers' Reflection Data

Data gathering	Used instruments	Analyzed unit	Data interpretation
Transcript of teachers' interview	Semi-structured questions corresponding to teachers' reflection technique	Object of reflection response: Mathematical problem-solving learning process (understanding problems, planning strategies, implementing strategies and verifying solutions). Level of reflection: 1. *Technical; 2. Deliberate; 3. Critical*	Qualitative interpretation

IOP Conf. Series: Journal of Physics: Conf. Series **1088** (2018) 012011 doi:10.1088/1742-6596/1088/1/012011

3. Result and discussion

Based on the results of interview transcripts analysis, as well as analysis of learning videos conducted by the three teachers, the theory of teacher reflection categories on the learning process of problem-solving was obtained and presented in Table 5.

Table 5. Categories of teacher reflection on the learning process of problem solving

Aspect	Aspect description	Categories		
		Good	*Very Good*	*Excellent*
Teacher's reflection on the learning process of mathematical problem solving	Teacher reflection on how to provide understanding to understand problems in the learning process of mathematical problem solving	Reflecting on the process of how to provide understanding to understand the problem in the learning process of solving the problem is limited to revealing the facts that occurred in class	Able to reflect on the process of how to provide understanding to understand problems in the learning process of problem-solving, but cannot provide an alternative how to overcome them	Able to reflect on the process of how to provide understanding to understand problems in the learning process of problem-solving, and provide an alternative how to overcome them
	Teacher's reflection on the selection of strategies in the learning process on mathematical problem solving	Reflecting the process of selecting strategies in learning about mathematical problem solving is limited to revealing the facts that occur in the classroom	Able to reflect on the strategy selection process in learning about mathematical problem solving, but cannot give an alternative how to overcome it	Able to reflect on the strategy selection process in learning about mathematical problem solving, and provide an alternative how to overcome it
	Teacher's reflection on the use of strategies in the learning process on mathematical problem solving	Reflecting on the use of strategies in learning about mathematical problem solving is limited to revealing the facts that occur in the classroom	Able to reflect the process of using strategies in learning about mathematical problem solving, but cannot provide an alternative how to overcome them	Able to reflect on the use of strategies in learning about mathematical problem solving, and provide an alternative how to overcome them
	Teacher's reflection on the verification of solutions in the learning process on mathematical problem solving	Reflecting the process of verifying solutions in learning about mathematical problem solving is limited to revealing facts that occurred in the classroom	Able to reflect the process of verifying solutions in learning about mathematical problem solving, but cannot provide an alternative how to overcome them	Able to reflect the process of verifying solutions in learning about mathematical problem solving, and providing an alternative how to overcome them
	Reflection of teachers on students in the learning process on mathematical problem solving	Reflecting student behavior during the learning process on mathematical problem solving is limited to revealing the facts that occur in the classroom	Able to reflect student behavior during the learning process on mathematical problem solving, but cannot provide an alternative how to overcome it	Able to reflect on student behavior during the learning process on mathematical problem solving, and provide an alternative how to overcome it

The subject selected in this research is the process of geometry. Based on Rusyida' research [5], it is clear that students' mathematical problem solving ability in geometry is still low. The students have difficulties to solve a problem measuring problem-solving ability in all of the indicators in geometry. It may be because the teacher does not reflect on the learning process. This research provides information on how the categories of teachers' reflection on their learning process. Many studies have looked at how the teacher reflects on their learning process such as the study done by Wice [6]. Wice's research emphasizes the importance of evaluation after learning. The same argument is also expressed by Gore [7] in his research entitled "Reflecting on Reflective Teaching". There has been no research focusing on reflection of teachers on the learning process of mathematical problem-solving. One form of illustration of how to obtain teacher reflection category on the process of understanding the problem is described as follows.

3.1 T-1 teacher's (teacher who comes from school one) beliefs about mathematical problem solving learning

The process undertaken to observe the teacher's reflection category on how to provide an understanding to understand the problem in the mathematical problem-solving process done to teacher T-1 (teacher from school one) was by showing the learning video about the mathematical problem solving that has been done. The subject observed in the video was the problem-solving process of how to find the cube nets. Video was served with an initial duration starting from 00.02- 01.15 (Pre-Review Reflection). Several reflection questions have been made regarding the process of providing an understanding of the problem-solving process (Pre-Review Reflection). In that duration, the teacher T-1 gave an understanding of what the cube nets are.

The researcher along with the teacher T-1 observed the video recording. This was done to provide teacher T-1 in taking the role of an observer, not as a participant. The video shows the T-1 teacher did not do the process of providing an understanding of finding the shape of the cube nets but directly planned the chosen strategy to find the cube nets.

The researcher then asked questions and conducted discussions related to the learning video to see teacher T-1's reflection category on how to provide an understanding that students are able to understand the problem in the learning process of professional problem solving (Professional Conversation). Based on the answers given, the teacher T-1 can be categorized as an excellent teacher in reflecting the process of providing understanding in solving mathematical problems. This can be seen from the following interview footage.

Interviewer	: At the beginning of the lesson teacher do not give the process of understanding to find the shape of the cube nets; is it not required by the students?
Teacher (T-1)	: Oh yes it seems I forgot to inform students
Interviewer	: If necessary, according to you what is the easiest way for students to explain how to find the cube nets?
Teacher (T-1)	: It can be conveyed in various ways, can be by drawing a series of adjoining rectangles, which, if assembled again, will form a cube build. The next way can be by giving the intact cube then opened, it will form a flat wake with the condition that no side is loose, side by side and not overlapping, well that's called the cube nets. It is a universal definition and agreement.
Interviewer	: Is it necessary to involve students in explaining how to understand the problem?
Teacher (T-1)	: Yes, teacher should possibly ask students to bring various cube-shaped spaces from their houses or pastry boxes and others or handed them over to students, so that they can construct their own what is meant by the cube nets

Based on the results of the interview, T-1 teacher has been able to reflect on the process of how to provide an understanding to understand the problem in the learning process of problem solving. The teacher has also provided an alternative way of providing an understanding of what is meant by the cube nets and how to understand it such as depicting a rectangle on the board and removing the intact cube in order to be opened and form a flat wake.

3.2 T-2 teacher's (teacher who comes from school two) reflections about mathematical problem solving learning

Furthermore, the same process is also conducted to see the reflection of teacher T-2 (teacher coming from school two). Based on the answers given, the Teacher T-2 can be categorized as an excellent teacher in reflecting the process of providing understanding in solving mathematical problems. This can be seen from the following interview footage:

Interviewer: Based on the video presentation the process of providing an understanding of finding the shape of the cube nets committed by Mother, it is to draw a series of squares on the board and explains if the squares are folded and form a cube build, it is a web of cubes and vice versa.
Is that easy for students to understand?

Teachers (T-2) : Before going there, I want to reveal in advance that the lesson that started just before the media is not ready, it should be ready and the tools and materials have been distributed to each student, then students should be given instructions through LKS or LJK in order to not spend time.
It is easy because so far children have understood it from a series of rectangles to form cube nets. But maybe the way that you did was less interesting for students; maybe most children also do not feel interesting to pay attention to it. It is seen from the video, there is still a child who tells the story with his friend.

Interviewer: If so, do you have any alternative to explain how to give an understanding of what a cube's web is?

Teachers (T-2) : It would be easier to understand if students bring their own concrete objects in the shape of a cube later like a cake box that can be removed from its sides. Later students take off their sides. In order for students to understand what the cube nets are with any real or concrete objects it might be easier. Next, ask the students to release the cake box to form a flat wake, and then say to this child is the cube nets, 'why would it form a cube when we assemble?'. Only the constraints later in the cake box have a few more sides to stick it, do not let the child affected by it, later thought the child is part of the cube nets. We emphasize to the students that it does not include or ask the students to remove the more sides first

Interviewer: Is it necessary to involve students in explaining how to understand the problem

Teachers (T-2) : Yes that is, you should not explain it in the same direction. Students are asked to bring themselves wake-up cube-shaped space, so they can automatically understand what the cube nets with activities they do.

Based on the results of interviews, teacher T-2 have been able to reflect on the process of how to provide understanding to understand the problem in the learning process about problem solving and provide an alternative how to overcome them. Teacher T-2 revealed that the process of giving comprehension was not interesting for students. The video shows that there are still students who talked with friends during the class.

Furthermore, teacher T-2 has also provided an alternative way of providing understanding to be more interesting for students by asking the students to bring themselves concrete objects in the form of a cube like a box of cakes where the sides can be removed. Then, students were asked to remove the sides, so that students easily understood what the cube nets are.

3.3 T-3 teacher's (teacher who comes from school three) reflections about mathematical problem solving learning

In the next fact, based on the answers given, the teacher T-3 (teacher who comes from school three) can be categorized as a good teacher in reflecting the process of providing understanding in solving mathematical problems. This can be seen from the following interview footage:

Interviewer	:	The process of providing an understanding of what the cube's nets are performed by a teacher is by using the power point media. while the media is shown, it is seen that all of cube sides are open. Then Mother gives an explanation when all the sides open and form a flat cube, then that's called the cube nets. Is that easy for students to understand?
Teachers (T-3)	:	According to my vision, it has been easy and effective, but sometimes there are pupils who do not like the power point, so there is still my handwriting on the board. On the board, I explain the other again, not about what the webs are. For example in the power point, there are nets and I asked this child what is it? the answer of the son is square, how many? There are 6 (six) square children, children's answer. I made it a bridge to find the surface area of the cube. It is also explained that the webs will form a cube, but that is just a short, which they need to understand is the surface area of the cube.
Interviewer	:	Do you have any alternative to explain how to find the cube nets?
Teachers (T-3)	:	No, this is an effective way. This is related to the concept. This concept should be quickly understood and the application will be used later to find the surface area formula
Interviewer	:	Is it necessary to involve students in explaining how to understand the problem?
Teachers (T-3)	:	Yes, but they make it at home only. I ask them to make cube nets at home. In this section, it does not have to be long time because the ultimate goal is the surface area of the cube, so this is just a bridge only. Teacher likes to ask the child how many square, there are 6 (six), it will be associated with surface area of a cube, later.

Based on the results of the interview, the T-3 teacher was only able to reflect on the process of how to provide understanding to understand the problem in the process of learning about solving the problem that is only limited to reveal the facts that occurred in the class, and did not provide an alternative strategy in understanding the problem. Teacher T-3 revealed that the process of providing comprehension that had been done was effective for students and the time is convenient.

Furthermore, teacher T-3 did not provide insights on alternative ways of giving comprehension to be more interesting for students, for example by asking students to bring their own concrete objects in the form of a cube like a box of cakes that can be removed from the sides.

The same process was also performed on every process of learning made by the mentioned teacher in every learning process on students who attended the learning process. The results of the analysis of each interview process, which were coded, resulted category in Table 5.

Based on the categories of reflection, there are different gradations of each teacher. In order to reflect on the learning process in problem solving teachers can use these categories as standardization. The process of reflection is related to the professionalism of teachers in teaching according to Donald [8] and Danielson[9]'s opinion that a professional teacher must reflect on his or her actions. Lin [10] also conducted a study in professional development where teachers reflect on themselves with peers. Zeichner [11] has used his years of experience to reflect on the world's teacher learning process. According to Calderhead [12], teachers should be aware that the role of reflection is very important in learning.

4. Conclusion

Teacher reflection on the learning of mathematics in general is also the same as the reflection of teachers on the learning process of problem solving. The study revealed three categories of teacher resulted in their reflections process after they finished teaching and learning process. The resulted category consists of excellent, very good, and good teacher. An excellent teacher in doing reflections on the learning process of problem solving is a teacher who can anticipate and give solutions for better learning of Polya's stages. A very good teacher in doing reflections on the learning process of problem solving is a teacher who notices her/his mistake but cannot give solutions for better learning of Polya's stages. A good teacher in doing reflections on the learning process of problem solving is a teacher who only focuses in revealing the facts that occur in the classroom of Polya's stages. This category of reflection can be a reference for teachers in studies that have a reflection phase, for example, lesson study or classroom action research. Furthermore, when the teacher misses their mistake in learning, the teacher can improve for the next learning process.

References

[1] Muir T and Beswick K 2007 *Teacher Education and Development Special Issue* **8** 74
[2] Geiger V, Muir T and Lamb J 2015 *J Math Teacher Educ* **47** 4
[3] Creswell J W 2015 *Penelitian Kualitatif dan Desain Riset* (Yogyakarta: Pustaka Pelajar)
[4] Patton M 1990 *Qualitative Evaluation and Research Methods (2nd ed.)* (Newbury Park, CA: Sage Publication)
[5] Rusyda N A, Kusnandi K and Suhendra S 2017 *Journal of Physics: Conference Series* **895** 012081
[6] Wice, Viky L, Spiegel, Amy N and Bruning, Roger H 1999 *Journal of Teacher Education* **50** 1
[7] Gore J M 1987 *Journal of Teacher Education* **38** 2
[8] Schon D A 1987 *Educating the Reflective Practitioner. Toward a New Design for Teaching and Learning in the Professions. The Jossey-Bass Higher Education Series* (Francisco: Jossey-Bass Publishers)
[9] Danielson and Lana M 2009 *Fostering Reflection* (Alexandria: Educational leadership)
[10] Lin H S, Hong Z R, Yang K K and Lee S T 2013 *International Journal of Science Education* **35** 18
[11] Zeichner K 2008 *Educ. Soc journal* **29** 103
[12] Calderhead J 1989 *Teaching and Teacher Education Journal* **5** 43

Development of Learning Tool with Contextual Teaching and Learning (CTL) Approach to Improve Student Mathematical Connection Ability

Mauliana, M Ikhsan and M Subianto
Syiah Kuala University, Jl. Teungku Chik Pante Kulu No. 5, Banda Aceh 23111, Indonesia

E-mail: subianto@unsyiah.ac.id

Abstract. The ability of mathematical connections must be possessed by students when studying mathematics. However, the ability of students' mathematical connections is still very poor and restricted among mathematical topics only. Therefore, it is necessary to develop learning tools that can improve students' ability of mathematical connections by using a meaningful learning approach. This study aimed to produce learning tools by using Contextual Teaching and Learning (CTL) approach which are valid, practical, and effective. This development research only focussed on valid aspect and used 4-D model. This research involved 35 Year 7 students from one of the high schools in Banda Aceh. The developed learning tools included lesson plans, student worksheets, and a test on quadrilateral topics. The development of learning tools was validated by four experts until the learning tools were considered eligible for use in the learning activities. The results showed that the learning tool with CTL approach by using 4-D development model could be used on quadrilateral topics with minor revision. Therefore, the developed learning tools with CTL approach in this research can be used by teachers in the quadrilateral topic to foster students' mathematical connection. Also, the result of this study may be used for further analysis on practicality and effectiveness phases.

1. Introduction

Mathematics plays a vital role in elementary and junior high school students since it contributes positively to intellectual development to respond to an increasingly progressive change. There are five basic mathematical skills that become the standard of learning, including problem-solving, reasoning and proof, communication, connection, and representation [1]. Accordingly, this study used one of the basic capabilities, which is making the connection. The ability of mathematical connections is one of the abilities that must be possessed by students when studying mathematics. This ability helps students to understand the relationships among mathematical topics, between mathematics with other sciences, and mathematics with real life. Nevertheless, the reality showed that geometry was not fully mastered by some students. There were still many students who had difficulty in learning geometry. Similarly, student achievement in geometry was still not satisfactory. Students have a lack of understanding on geometry topics, such as in recognizing and understanding the geometry and its elements [2]. Many high school students have difficulty in completing the task of writing evidence, in solving standard geometry test and final geometry test of the program [3].

In one of the tasks of the Program for International Student Assessment (PISA) in 2012, only around 20% of Indonesian students answered the task correctly. The task required students'ability to understand the questions and prior knowledge to solve the problem, which was the knowledge to measure the circumference of a square, rectangle, and parallelogram. Besides, students needed to be able to apply mathematical knowledge in a new situation. Students who failed to respond to the task probably could not connect mathematics topics with real life [4]. Similarly, [5] stated that some students could not understand or solve problems that include more than one mathematical concept. Also, some students found difficulties when dealing with mathematical problems in daily life. Another study also showed that students had a lack of mathematical connection ability [6]. To improve students' mathematical connection abilities, suitable learning tools and learning methods should be used to meet the students' needs. Therefore, teachers need to arrange learning tools that can construct students' knowledge, in which students will be trained to discover their concepts and knowledge [7].

Successful mathematics learning can be achieved by applying the learning approach. The potential approach needs to support students' ability to think contextually, in which mathematical context is associated with students' real life. In the learning process, students need to find out by themselves the concept of the topic learned so that the understanding will be embedded well. In addition, the cooperation between students can also support them to have good strategic competence. All the learning practices above can be provided by Contextual Teaching and Learning (CTL) method [7]. In relation to the aspects of mathematical connection capability as well as the characteristics of the CTL approach, it is believed that the use of CTL approach can improve students' ability of mathematical connection. Therefore, this study aimed to a valid develop learning tools that use CTL approach to improve students' ability of mathematical connection.

2. Research method

This study was designed as a developmental research. The product of this research was the development of learning tool to improve students' mathematical connection ability through CTL approach. The learning tool development model was by modifying the 4-D model from Thiagarajan. This research focused only on the development stage, which was until the experimental activities of learning tools. Learning tools that were developed included lesson plans, student worksheets, and tasks or questions related to students' mathematical connection ability. The developed learning tools emphasized the quality of the product.

The experiment involved 35 Year 7 students in one of MTsN in Banda Aceh on the topic of area and circumference of quadrilateral and was conducted twice. Three aspects were used to assess the developed learning quality, including valid, practical and effective [8]. However, this research only focused on validity aspect. The validity was assessed by four validators, including 1 mathematics education lecturers, two experts on learning tools development, and a mathematics teacher. Data to assess quadrilateral learning tools were obtained from research instruments. A validation sheet was used as the research instrument, consisting of lesson plan validation sheet, student worksheet, and mathematical connection capability test. The criteria used to assess the module developed are based on Nieveen's [9].

Table 1. Criteria of validity score

$4 \leq$	VS	≤ 5	Highly valid
$3 \leq$	VS	≤ 4	Valid
$2 \leq$	VS	≤ 3	Less valid
$1 \leq$	VS	≤ 2	Invalid

3. Result and discussion

The development stage is the stage of product realization. Activity undertaken at this stage was to develop learning tools with CTL approach based on the design framework on the design stage. The products have met the valid criteria, based on the results of the assessment by experts, teachers, and peers. Stages of composing learning tools include a lesson plan, students worksheet, and test on mathematical connection capability. Furthermore, this learning tool product was validated by experts and practitioners. The instruments to validate the learning tools consisted of validation sheets for each type of the learning tools. Overall validation of all instruments can be categorized well. The average of lesson plan validation is 2.9 with a very good category. Students worksheet is 2.7 with a good category, and mathematical connection capability test is 2.8 with a very good category. In other words, the developed learning tools could be used after minor revision. The results of the validation conducted on the three quadrilateral learning tools in Table 2.

Table 2. Revision results on lesson plan and worksheet

Mistakes	Validators' Suggestions
Lesson plan The lesson plan designed was not in accordance with the CTL approach and did not involve contextual issues; thus, students were less motivated in learning.	The validators suggested to connect with contextual issues so that the lesson plan becomes eligible to use.
Worksheet The student worksheet did not engage students actively in a contextual learning process	The validators provided suggestions and improvements to make student worksheet more contextual in order to connect students with real life and make the learning process becomes more enjoyable and meaningful. Student worksheet aims to help the teacher in teaching and learning process and help students to understand the mathematics concept. Therefore, the worksheet needs to be designed carefully, so the task given could meet the criteria of learning objectives.

The initial draft of the student worksheet consisted of one question only, which was: "Write down the quadrilateral traits: Rectangle, Square, Parallelogram, Rhombus, Trapezoid, and Kite!". After the revision as suggested by the validators, the question was displayed in a table, as shown in Table 2. The complete worksheet was changed as: "Look at each quadrilateral in Figure 1, then note the things associated with the shape such as sides, corners, and diagonals. After that, please complete the following table, which indicates the properties of each quadrilateral. Hint: write (√) if any statements below fulfill the properties of each quadrilateral. "

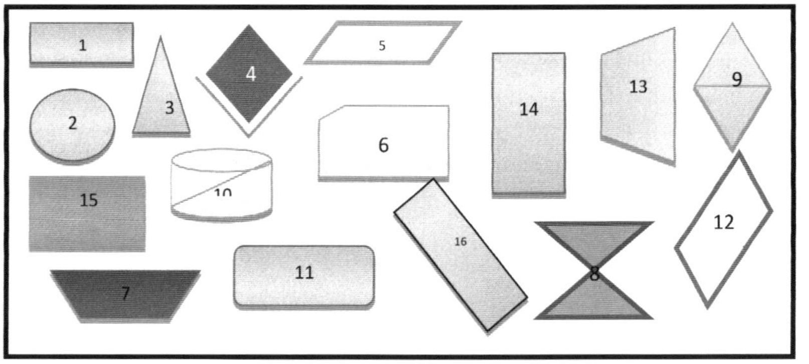

Figure 1. Quadrilateral figures

Table 3. The revision of student worksheet

No	Quadrilateral Trait	PP	P	JG	BK	TR	LL
1.	Opposite sides are parallel	√				X	
2.	Opposite sides have equal length.						
3.	All sides have equal length.						
4.	Opposite angles are equal.						
5.	All angles are equal.						
6.	Each diagonal divides the quadrilateral into two						
7.	The diagonals bisect each other.						
8.	The diagonals formed a right angle.						
9.	Only a pair of sides is parallel.						

Explanation:

√ = exist X = do not exist
JG = Parallelogram LL = Kite
PP = Rectangle P = Square
BK = Rhombus TR = Trapezoid

Furthermore, the connection test is designed according to mathematical connection indicators, which are (1) connected between mathematical topics, (2) connected with the real world, (3) connected with other disciplines. The validators provide an assessment and suggestion or input based on the mathematical connection test. Below is the example of the revision made by the researcher based on validator suggestions.

Table 4. Revision results of mathematical connection test

Initial Plan (Problem no.1)	**Revision** (Problem no.1)
A piece of paper is rectangular, with its length and width is 110 cm and 90 cm. How many sheets of paper are needed to make 70 kites with diagonal size 30 cm and 45 cm?	Mr.Mamat wants to make 70 kites for sale. Each kite has a diagonal size of 30 cm and 45 cm. If the available paper is rectangular, each sheet of paper is 110 cm in length and 90 cm wide. How many pieces of paper does Mr.Mamat need to make those 70 kites?

In the test of mathematical connection ability, the revision of the task was done only in question number 1. The number 1 hint given in the initial draft was not appropriate to assess students' ability of mathematical connection related to the rectangular material. There was still an ambiguous language, so the researcher considered the advice given by the validators and discussed with the mentor to change the form of the problem. Another revision was the lack of complete guidance for scoring, so the researcher must complete it with the alternative scoring guideline that provides the solution of the problems. The study concludes that scoring guidelines are required, in order to see to what extent students' grade is assessed. This is in accordance with the explanation which states that students' grades should be based on the student's effort in the form of test or other activities if there is no grade, then the students feel the activities that have been done not something important or useful [10]. Therefore, the scoring system should be clear in the workforce because it reflects students' abilities. Description of activities in the learning of quadrilateral materials through CTL approach [11].

During the treatment, the students' learning motivation was developed in which the average motivation was 2.8. This score means that students had excellent motivation in learning quadrilateral topics. Based on the consideration and comments from the experts and practitioners on the development of learning tools through CTL approach, the tools were valid. After the treatment, it was found that the learning tool based on CTL approach in this study was feasible to be used after some input from experts, practitioners, and students.

4. Conclusion

This research produced a learning tool on quadrilateral topics by using CTL approach that was valid. This validity was indicated by the average presentation of the validity which was very good. In addition, there was an increase in the mathematical connection ability of the students, indicating that the learning tool was appropriate to be used. There was also an increase in student's motivation when they learned using the learning tool. Therefore, it could be concluded that the development of learning tools through CTL approach has met the valid criteria.

References

[1] NCTM 2000 *Principle and standards fos school mathematic* (Virginia)
[2] Soedjadi R. 1999-2000 *Kiat pendidikan matematika di Indonesia* (Direktorat Pendidikan Tinggi Depdikbud Jakarta)
[3] Senk S L 1989 Van hiele levels an achievement in writing geometry proofs *Journal for Research in Mathematics Education* **20** 309
[4] PPPPTK 2012 *Instrumen penilaian hasil belajar matematika SMP Belajar dari PISA dan TIMSS* (Yogyakarta: Kementerian Pendidikan Nasional)
[5] Muhidin R 2016 *Peningkatan kemampuan koneksi matematis siswa SMP dengan model pembelajaran Connecting Organizing Reflecting Extending (CORE)*
[6] Rahmawati U N, Sugiatno & Hamdani 2017 Kesulitan koneksi matematis siswa dalam menyelesaikan soal cerita materi peluang di sekolah menengah atas *Jurnal Untan dan Pembelajaran Untan*
[7] Sanjaya W 2006 *Strategi pembelajaran berorientasi standar proses pendidikan* (Jakarta: Kencana)
[8] Nieveen N 1999 Prototyping to Reach Product Quality Dalam Plomp T Nieveen N Gustafson K Branch RM dan van den Akker J (eds) *Design Approaches and Toolsin Education and Training* (London: Kluwer Academic Publisher)
[9] Widoyoko S E P 2015 *Teknik penyusunan instrumen penelitian* (Yogyakarta: Pustaka Belajar)
[10] Van De Walle 2006 *Matematika sekolah dasar dan menengah pengembangan pengajaran* Edisi Keenam (Jakarta: Erlangga)
[11] Mauliana 2018 *Peningkatan Kemampuan Koneksi Matematis dan Motivasi Siswa melalui Pendekatan CTL berbantuan Peta Konsep* Tesis tidak diterbitkan (Universitas Syiah Kuala Banda Aceh)

IOP Conf. Series: Journal of Physics: Conf. Series **1088** (2018) 012013 doi:10.1088/1742-6596/1088/1/012013

Learning of reproduction system with an integrative curriculum approach in junior high school

Ibrahim[1], N Yusoff[2], M I Awang[2] and Marwan[3]
[1]Serambi Mekkah University, Banda Aceh, Indonesia
[2]College of Art and Science CAS, Universiti Utara Malaysia, Kedah Darul Aman, Malaysia
[3]Al Muslem University, Bireun, Aceh, Indonesia

E-mail: himsufi@gmail.com

Abstract. The education system in Aceh has not implemented the Qanun of Aceh's Education integrating Islamic values in the learning process. Integrative Curriculum (IC) is an effective approach to implement Islamic values in biology learning. Learning materials on reproduction topic with the IC approach have been developed using Plomp development model. This paper describes the learning activities on reproduction topic implemented in a trial learning. The trial conducted in this study aims to investigate the implementation of the learning materials with the IC approach. The implementation is then used as an indicator of the practicality of the learning materials. This trial involved 28 students of grade 8 in a junior high school in Indrapuri, Aceh Besar regency. This study utilized an observation sheet to observe the learning implementation with the IC approach. The results show that the learning on reproduction topic with the IC approach could be implemented in a biology classroom. The learning process integrated the Islamic values based on the verses of the Qur'an that describe the process of human creation. Qur'an Surah Al-Hajj verse 5 was assigned for Group 1, Surah Al-Mu'minuun verses 12-14 for Group 2, Surah Al-Mu'miin verse 67 for Group 3, and Surah Al-Qiyaamah verses 37-38 for Group 4. This finding suggests that the learning on reproduction topic integrating Islamic values needs further investigation on evaluating the practicality and effectivity of the learning with the IC approach.

1. Introduction

National education curriculum in Indonesia emphasizes that teachers should integrate the values of life in the learning process by incorporating knowledge, skills, and attitudes. The attitude aspect needs to accommodate religious and social attitudes [1-4]. Social and religious attitudes need to be in accordance with the local and religious wisdom of the students [5]. Education can integrate life values, such as curiosity and affection while teaching the competencies demanded by the curriculum [6]. The integration of life values in education is known as the Integrative Curriculum (IC) [7]. IC is done by synchronizing conceptual contents, skills, and attitudes. The attitude aspect in IC is based on the belief, behavior, and character education [8].

IC combines basic knowledge with the learning experience that fits students' needs [9]. IC can help students understand topics taught through the problems that relate to students' needs [10] and offers a range of possibilities of strategies to integrate and relate daily activities to experience taught at school [11]. Some countries have been implemented the integration of life values in learning and the implementation shows positive results. For example, curriculum in the United States engages students through social and humanitarian activities [12]. In addition to that, it integrates history values in science learning [13], such as biology and physics [14]. While in Nigeria, the integration of life values in the learning proccess is applied to physics and chemistry classroom [15,16]. In contrast, Japan emphasizes the integration of hygiene, health and healthy food in learning. Curriculum in Japan also emphasized education that integrates preparations for reducing disasters [17,18]. Meanwhile, New Zealand emphasizes education that incorporates culture and respects the life of Maori society [19]. Some *tahfiz* institutions in Malaysia integrate the local community culture in the learning to improve the quality of students [20].

Also, the IC approach has been implemented in Aceh by integrating the value of water and sanitation, local wisdom, and cultural diversity. All three values are embedded in learning without additional lesson time [21]. The IC approach with the integration of Islamic values also needs to be included in curriculum in Aceh to support the Aceh Education Qanun which mandates that education in Aceh implements Islamic education based on *Al-Quran* and *Al-Hadith*, *Pancasila* (Indonesian philosophy), the Indonesian Constitution of 1945, and Aceh culture [22-24].

The integration of Islamic values in learning reflects that Aceh is excellent in implementing an Islamic Sharia-based education to support the ideals of generations that are qualified, superior, competitive, faithful, and cautious [25,26]. Islamic values also help teachers grow character as characteristic of Indonesia's national education curriculum [5]. Besides, learning can combine religious knowledge and general knowledge through the materials being taught [27].

The learning process with the IC approach has not been implemented thoroughly. There has been no significant progress in Islamic sharia-based education [28]. This is due to the lack of learning implementation guidance that facilitates teachers to implement learning that integrates Islamic values [29]. This is a challenge in the implementation of Islamic education in Aceh, and this challenge makes teachers reluctant to implement IC with Islamic values. Also, the unavailability of ideas, teaching materials, and assessment instruments for learning makes the implementation more challenging. Teachers are still looking for and building ideas of lessons that they will teach. Therefore, this study intends to develop biology learning materials that integrate Islamic values through the IC approach. This paper will discuss biology learning activities that integrate Islamic values through the IC approach on reproduction topic taught in grade 8 in a junior high school.

2. Method

This paper describes biology learning activities integrating Islamic values through the IC approach. These activities are based on a pilot study on learning materials developed through the development model [30]. The materials with the integration of the Islamic values used in the biology classroom have met valid criteria based on recommendations from experts and practitioners. Trials conducted in this study aims to examine the implementation of the learning materials. This implementation becomes the practicality criteria of the learning materials [31]. The learning activities used in the learning materials focus on reproduction topic.

3. Result and discussion

The biology learning process integrating Islamic values through the IC approach on reproduction topic was developed to achieve indicators of competencies in describing the reproductive system in men and women, explaining the function of the reproductive system in men, describing the reproductive hormone system in women, and describing venereal diseases in the reproductive organs. This paper will describe the results of a learning trial in the first meeting of the reproductive system in men.

The learning process in the first meeting began with reviewing the previous material and student experience related to the material. That is about the importance of the reproduction of living things. Next, students were motivated to know the importance of doing activities in everyday life including reproduction that follows the provisions of God. The students were given an understanding of the term of faithful that refers to doing all the commandments of God and avoiding all His prohibitions. Furthermore, the teacher delivers the learning objectives in the first meeting. Then, the students are expected to understand the reproduction system in men and their functions through reading references, torso media observations, pictures, and discussions.

The main activity in the first meeting began with the activity of students in groups that observe the verses of the Qur'an associated with the reproductive system in human beings. Students read and examined the meaning of the verses of the Qur'an through the reading material provided. Each group was responsible for discussing and understanding the meaning of assigned verses of the Qur'an. Qur'an Surah Al-Hajj verse 5 was assigned for Group 1, Qur'an Surat Al-Mu'minuun verses 12, 13 and 14 for Group 2, Qur'an Surah Al-Mu'miin verse 67 for Group 3, and Qur 'an Surah Al-Qiyaamah verses 37-38 for Group 4. The sentence, interpretation, and meaning of the verses were provided in the reading material.

Chapter 1 of the reading material provides verse 5 from Surat Al-Hajj. The interpretation of the verse is shown as follows.

"Oh Human, if you are in doubt about the resurrection, then We have made you from clay, then from a drop of semen, then from a clot of blood, then from a clot of flesh perfect and incorruptible, that We may make clear to you, and establish in the womb, what We will, until the appointed time; and then We bring you forth as a baby, and then come to maturity, and between You are the one who has been raised up and has been raised among you for a length of time, so that he does not know anything that he had known before. And when you see the earth dry, then when We have sent down water on it, the earth lays its life and grows and produces all sorts of beautiful plants."

Students discussed the meaning of verse 5 of Surah Al-Hajj. This activity aims to foster students' understanding that the verse describes the creation process of humans and plants as clear evidence of truth and the coming of the Day of Resurrection. This means that as God's *khalifah* on the earth, humans must always be vigilant and not fall into sinful deeds.

The reading material given for Group 2 is Surah Al-Mu'minun verses 12, 13, and 14. The interpretation of verse 12 of Surah Al-Mu'minuun is *"And verily We created man from a quint (origin) from the ground."* Then, it is followed by Surah Al-Mu'minuun verse 13 with the interpretation of the verse: *"Then We made a semen-drop (stored in a firm place)."* Next, the meaning of Surah Al-Mu'minuun verse 14 is *"Then the semen we made a clot of blood, and clothed it We made a clot of flesh, and clothed it We made the bones, then bones We covered with flesh. Then We made him another (form). So Blessed is Allah, the Best Creator."*

The students in Group 2 read and discussed the reading material provided. This activity aims to build students' understanding of Surah Al-Mu'minuun verses 12, 13, and 14 which explains that the long process of human creation that originated from food as a source of life. The source of human food comes from the soil which then becomes the essence of semen formation in men and ovaries in women. The semen (spermatozoa) that meets the ovaries will become the fetus through a long process. The students need to understand that after God commanded his servant to guard and cultivate his faith.

In the teaching material it is also explained that through the verse, God reminds us of the origin of human creation, from where it came, how he lives and where he will return. This will eliminate the arrogant attitude in the students as God's creation. In verses 12, 13, and 14 of the Surah Al-Mu'minuun, it is said that all the food that God bestowed upon humans contain all sorts of substances needed for the survival of human beings. There are iron, egg whites, vitamins, calories, hormones and so on. The benefits of these foods help drain blood in the human body. In a healthy body, blood flow centering on eagles and from the heart flows through the whole body. In the blood, there is a substance that will become semen. A drop of semen has thousands and even the "look" that will be human, stored in the male *shulbi* and female *taroib*. Subsequently, God's will pertains to the appearance of a man who appears to be a tiny worm, in combination with semen in a woman who is a very small egg. The combination of both, it is called *nutfah*. The *nutfah* is growing within 40 days.

During the 40-day the semen that has been mixed, it gradually turns into a clot of blood. After 40 days in the form of a lump of semen, she turned into a lump of blood. It then freezes up into a clump of meat and continues to change its nature into bones. The bones are surrounded by water which then becomes the meat to cover the bones. Bones then grow into a head, legs, hands and all the bones in the body was covered by meat. At that moment, it was conferred the spirit and the formation of a respiratory system.

Students are motivated to see the reality of God's greatness through the process of human creation. It aims to instill a sense of faith in students. Group 3 discussed Qur'an Surah Al-Mu'miin verse 67. This verse motivates the students about the importance of studying the reproduction system as a part of worship to God. The interpretation of the verse is presented as follows.

"He who created you from the earth then from a drop of semen, after that from a clot of blood, then born of you as a child, then (you are left alive) that you may reach the age (then) until you are old, between you was what had been taken before. (That we may do so) that you may come to an appointed term and that you may understand."

Group 4 discusses verse 37 of Qur'an Surah Al-Qiyaamah. The meaning of the verse is "*Did not he first drop a seminal spilled (into the womb).*" Then, it is continued with verse 38 of the Qur'an Surat Al-Qiyaamah with the interpretation: "*Then the semen becomes a clot of blood, then God created it, and perfected it.*"

The students' understanding of the meaning of the Qur'an verses during the discussion was re-written in the student worksheet. In the worksheet, the students also wrote in their own language the relationship between the verse and the reproductive system in men.

The next activity was that students observed the problem about the development of reproductive organs in men and the obligation of circumcision in men regarding Islamic Sharia, Acehnese culture and local wisdom, and cleanliness. This activity referred to observing activities that are the first activities of a scientific approach.

The students were given the opportunity to ask questions on the problems they had not understood. This activity is a questioning activity on a scientific approach. Next, the students discussed the reading material or notes on the key points of the lecture. These activities are literacy reading activities that are emphasized in the Curriculum 2013 and also data collection activities on a scientific approach.

The students identified reproduction type and function in men through Torso media, image media and clipping. The information the students obtained from their investigation was used to answer questions on the worksheet. The activity of identifying reproduction type and function in men through image media is shown in Figure 1.

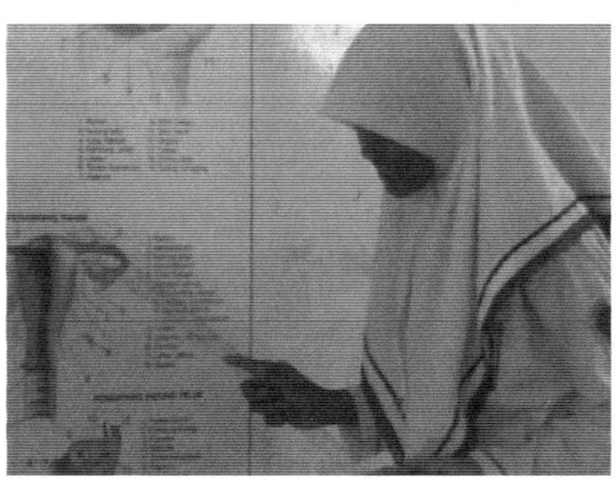

Figure 1. A student identified reproduction type and function in men through image media.

The learning activities were continued by discussing and writing down the results obtained. Furthermore, each group made a display of male reproduction and function in their respective groups. This activity is shown in Figure 2.

Figure 2. Students prepared their group displays.

Furthermore, groups of students present their work. Presentation activity is shown in Figure 3 below.

Figure 3. Students presented their work result.

In the group visits, the students classically raised the problems encountered during the visit and discussed the solution. The teacher guided the discussion to find solutions to student problems.

In closing activities, the students were asked to make the conclusions about the reproductive system in humans and their functions. The teacher provided reinforcement to the conclusions formulated by the students related to the Aceh culture and Islamic Sharia. Next, the students completed the individual tasks. Scores of each student contributed to their group scores. The group that got the highest average score earned an award from the teacher. Then, the teacher provided follow-up information as preparation for further material.

Based on the implementation of reproduction topic learning with IC approach, it is found that learning materials that have been developed can be implemented in the biology classroom. Students were enthusiastic in discussing the meaning of Qur'an verses. The students were passionate in interpreting and relating the meaning of the verses to the to reproduction system topic. The integration of Islamic values through the verses of the Qur'an could connect the topic to the students' life. This happened because the students in Aceh, especially participants in this study, implement the Islamic values based on the Qur'an. This condition can be utilized by teachers to achieve learning objectives [31]. Student enthusiasm on the Islamic value integrated into the learning process because the value is close to students' lives [32].

The conformity between classroom learning and learning materials developed shows that Islamic values can be integrated into the learning [11,33,34]. Furthermore, the learning materials developed and tested in the classroom were carried out with an IC approach. This was done with the consideration that the integration of Islamic values through an IC approach will give an emphasis not only on the learning materials but also on the students' ability to understand and solve problems related to daily life and religious matter [7-9].

4. Conclusion

Based on the experiment conducted, it can be concluded that the learning materials of reproduction topic with Integrated Curriculum (IC) approach can be implemented in the classroom. The evidence is showed by the implementation of learning in the classroom relevant to the lesson plan. Every aspect of IC, i.e., knowledge, skill and attitude, can be integrated without additional lesson time. In the learning, the students were trained to solve problems related to reproduction system based on Islamic values. Also, the religious attitude learned in the learning is understanding the reproductive system based on Islamic values.

References

[1] Nasution A 2003 *Model Pembelajaran Terpadu dari Kegiatan Profesional Guru* (Jakarta: Gramedia Press)
[2] Hamalik O 2010 *Kurikulum dan Pembelajaran* (Jakarta: Bumi Aksara)
[3] Hadi S 2005 *Pendidikan Matematika Realistik dan Omplementasinya* (Banjarmasin: Tulip)
[4] Subhan 2007 *Perjalanan Kurikulum Nasional (pada Pendidikan Dasar dan Menengah)* (Jakarta: BSNP)
[5] Hariyanto M S 2011 *Konsep dan Model Pendidikan Karakter* (Bandung: Remaja Rosda Karya)
[6] Levine D U, Levine R F and Ornstein A C 1985 Guidelines for Change and innovation in the Secondary School Curriculum *NASSP Bulletin* **69** 9
[7] Drake S M and Burns R C 2004 *Meeting standards through integrated curriculum* (Alexandria VA: Association for Supervision and Curriculum Development)
[8] Drake S M and Reid J 2010 Integrated curriculum: Increasing relevance while maintaining accountability *What Works* **28** 1
[9] Fogarty R 1991 *The mindful school: How to integrate the curriculum* (Illinois: IRI/Skylight Pub. Inc.)
[10] Johnson E B 2002 *Contextual teaching and learning: What it is and why its here to stay* (Corwin Press Inc.)
[11] Sabda S 2006 *Model Kurikulum Terpadu IPTEK dan IMTAQ: Desain, Pengembangan dan Implimentasi* (Jakarta: PT. Ciputat Press Group)
[12] Brooks J G and Brooks M G 1999 In search of understanding: The case for constructivist classroom. Alexandria: Association for Supervision and Curriculum Development *The Case for Constructivist Classrooms, Virginia: ASCD Alexandria*
[13] Kuhn S T 1996 *The structure education of scientific revolutions* (California: The University of Chicago Press.)
[14] Panasuk R M and Horton L B 2012 Integrating history of mathematics into curriculum: What are the chances and constraints *International Electronic Journal of Mathematics Education* **7** 125
[15] Avison 1980 *Traditional education versus realistic education: Hoping for changes* (Chicago: University of Chicago)
[16] Haggis H 1980 Thinking about thinking, learning about learning: Constructivism in physics education *Journal Physics Education* **21** 326
[17] Jalaluddin and Azwir 2012 *Management Berbasis Sekolah pada SMA dalam Provinsi Aceh (konsep dan aplikasi)* (Banda Aceh: Bandar Publishing)
[18] Susetyo 2006 *Pendidikan Kebencanaan dalam Mengurangi Resiko Kerusakan Masyarakat* (Yogyakarta: Kanisius)
[19] Agarwal R, Sambamurthy V and Stair R M 2000 The evolving relationship between general and the specific computer self-efficacy—An empirical assessment *Information and Systems Research* **11** 418
[20] Rahman, Azizul R A, Mohamad, Mohamad S 2010 *Strategi Pengajaran Cemerlang* (Kuala Lumpur: Majlis Guru Cemerlang Malaysia)
[21] SEAMEO SEAMOLEC 2007 *Bahan Pelatihan Pembelajaran untuk Guru dalam Kurikulum Integratif Pasca Tsunami di Provinsi Aceh* (Banda Aceh: DIKTI Unsyiah)
[22] Undang-undang Nomor 44 Tahun 1999 *Tentang Penyelenggaraan Provinsi Daerah Istimewa Aceh*

[23] Qanun Pendidikan Aceh 2002 *Qanun No 23 Tentang Pelaksanaan sistem pendidikan berasaskan syariah Islam di Provinsi Aceh*

[24] Qanun Pendidikan Aceh 2008 *Qanun No 5 tentang pendidikan berasaskan syariat Islam Peraturan Daerah Provinsi Aceh*

[25] Walidin W 2005 *Discussion Panel Recovery Pendidikan NAD* (Malang: Forum Mahasiswa Aceh)

[26] Puteh Y 2012 *Peran Pendidikan Agama sebagai Upaya Mencegah Pendangkalan Aqidah* (Banda Aceh: Bandar Publising)

[27] Abuddin N 2009 *Filsafat pendidikan Islam dan orientasi kepada pemikiran modern* (Jakarta: Gaya Media Pratama)

[28] Banta T A 2005 *Discussion Panel Recovery Pendidikan NAD* (Malang: Forum Mahasiswa Aceh)

[29] Rahmatsyah 2014 *Pendidikan Islami Aceh* (Aceh: Kompasiana)

[30] Plomp T 2010 *Educational Design Research: An Introduction* (Netherlands Institute for Curriculum Development)

[31] van den Akker J, Branch R M, Gustafson K, Nieveen N and Plomp T 2012 Design Approaches and Tools in Education and Training *Springer Science and Business Media*

[32] Thiessen D 2006 Student knowledge, engagement, and voice in educational reform *Curriculum Inquiry* **36** 345

[33] Bonotto C 2007 *How informal out-of school mathematics can help students make sense of formal in-school mathematics: The case of multiplying by decimal numbers* (Lawrence Erlbaum Associates, Inc.)

[34] Achsin 2007 *Introduction to research in educations. (6thed)* (United States: Wadswork Thomson Learning)

[35] Zainuddin S 1996 *Jihad intelektual, merumuskan parameter-parameter sains Islam* (Surabaya: Risalah Gusti)

IOP Conf. Series: Journal of Physics: Conf. Series **1088** (2018) 012014 doi:10.1088/1742-6596/1088/1/012014

Assessing the validity and reliability of questionnaires on the implementation of Indonesian curriculum K-13 in STEM education

R Oktavia, Irwandi, Rajibussalim, M Mentari, and I S Mulia
Syiah Kuala University, Jl. Teuku Nyak Arief Darussalam, Banda Aceh, 23111, Indonesia

E-mail: rini_oktavia@unsyiah.ac.id

Abstract. This study aims to investigate how educators of STEM-related subjects implemented K-13 in their classrooms. A forum group discussion (FGD) on the implementation of Indonesian curriculum known as *Kurikulum 2013* (K-13) in STEM (Science, Technology, Engineering, and Mathematics) Education was held in March 2018 by STEM Research Center Unsyiah (STEM.id). Participants of the FGD consist of 60 educators of STEM-related subjects in Banda Aceh and Aceh Besar Districts. Since K-13 was implemented in July 2013, there is no survey study to evaluate its implementation in Aceh has been conducted. This study initiates to develop survey instruments to measure participants' understanding and opinions regarding the implementation of K-13 in STEM Education and their satisfaction of the FGD. Eight items in participants' satisfaction questionnaire are strongly valid with validity values range from 0.632 to 0.824. For the questionnaires on the implementation of K-13, nine items are valid with validity values range from 0.584 to 0.821. For internal consistency reliability, the Cronbach's alpha of the FGD satisfaction responses is 0.838; meanwhile the Cronbach's alpha of implementation of K-13 responses is 0.882. These results showed that the questionnaires have good internal consistency reliability and validity.

1. Introduction

STEM Research Center Unsyiah (STEM.id) is a center in Syiah Kuala University that initiate the activities to advocate the development of integrated instructions of Science, Technology, Engineering, and Mathematics (STEM) so that the learning processes took place in STEM-related subject classrooms become more interesting and enjoyable for students. The activities include conducting surveys and Focus Group Discussion (FGD) to gather information on how to integrate *Investigative Science Learning Environment* (ISLE)-Based STEM lessons in K-13, developing ISLE-Based STEM learning modules, conducting workshops for teachers and students on ISLE-Based STEM lessons, developing websites that provides resources to implement ISLE-Based STEM modules, summer camps for students etc. The slogan "*learning is fun*" becomes a goal in developing integrated instructional approaches for STEM-related subjects. As one of the programs in its current project titled "*Integrated ISLE in Integrated Science Instructions to Improve Science Teachers' Abilities in Science Education*" funded by the United States of America (USAID-PEER) Awards, STEM.id held an FGD on the implementation of Kurikulum 2013 (K-13) in STEM Education on March 2018 that was participated by 60 STEM-related educators and educational policy makers from the city of Banda Aceh and Aceh Besar District.

The FGD aims to investigate how well educators of STEM-related subject understand and are able to implement the standards mandated by Indonesian national school curriculum known as Kurikulum 2013 or K-13. More importantly, it is intended to investigate the challenges that the secondary school teachers have in implementing K-13 in STEM related subjects such as Physics, Chemistry, Biology, and Mathematics. Since ISLE was developed for Physics Lessons by Prof. Dr. Eugenia Etkina [1], the USG-Partner of the USAID-PEER project, some of the information gathered will specifically relate to Physics courses.

Since it was implemented in July 2013, there is no survey study to evaluate the implementation of K-13 in Aceh has been conducted. Therefore, as one of the data collection techniques applied in this study, surveys on the implementation of K-13 in STEM education and on the satisfaction of participants of the FGD were administered at the end of the FGD. The survey instruments were developed to measure participants' understanding and opinions regarding the implementation of K-13 in STEM Education and their satisfaction of the FGD. In this article, the results of classical test theory (CTT) analyses on the validity and reliability of the instruments will be reported.

2. Literature Review

The psychometric analyses reported in this article only involve the classical test theory (CTT) analyses. CTT is based on the underlying principle that the scores that respondents got on responding to an instrument (observed scores) is the sum of the true scores and some errors:

$$X = T + E \tag{1}$$

where X is the observed score; T is the true score; and E is the error [2]. Two common measures of the quality of instrument in CTT are validity and reliability. A discussion regarding these two measures is described in the following paragraphs.

The importance of validity and reliability as criteria of the quality of an instrument has been addressed in many studies [3,4,5]. The term validity denotes the degree to which an instrument measures what it purports to measure. The validity of an instrument specifies how well the instrument met the standards that were set by certain criteria to judge the quality of an instrument. An instrument that has a poor validity indicates that there are items in the instrument that do not measure what the instrument intends to measure. In other words, the validity of an instrument represents how accurate the instrument measure what it is intended to measure. There are several types of validity of a measurement, including face validity, content validity, and construct validity. Descriptions of these types of validity are presented below.

2.1. Face validity

Face validity represents the estimation regarding the clarity of the purpose of an instrument. It means a questionnaire has a high face validity if its purpose or construct is clear by the respondents taking it. Researchers will let some experts to review whether the instrument has met the criteria of measuring the traits purpotedly measured.

2.2. Content validity

Content validity represents the estimation regarding the propriety of content and how it is presented. In this study, content validity is ensured by establishing the connections between items and the implementations of K-13 in science lessons and the effectiveness of the FGD. Using Dillman's Tailored Design Method [6], the questionnaire blueprint was created. The content validity can be estimated by asking experts in the content area to review the blueprint and the items in the instrument. In this study, we asked the experts of K-13 and STEM education to review the contents regarding the implementation of K-13 in STEM education. These experts were given the list of content areas specified in the test blueprint along with the test items. The experts were then asked to indicate whether they agree that each item is appropriately matched to the contents indicated in the blueprint and whether the ways the items were represented purportly measure the intended contents.

2.3. Construct validity

Construct validity is the measure of the propriety of the constructs in the instrument. According to Westen & Rosenthal [7], despite its strongly important measure of the quality of an instrument, there is no simple formula to represents the extent that an instrument can be considered as construct valid. The most common calculation used by researchers to obtain the construct validity metrics is by presenting correlations between a measure of a construct and a number of other measures that are theoretically associated with the construct known as the convergent validity or vary independently of the construct known as the discriminant validity [7]. In this article, the construct validity of the instruments used in our survey is described by calculating the Pearson's correlation coefficient r of the scores of respondents' responses to an item with their total scores. Pearson's correlation coefficient is also known as Pearson's product-moment correlation coefficient (PPMCC) and calculated as below:

$$r = \frac{n\sum xy - \sum x \sum y}{\sqrt{[n\sum x^2 - (\sum x)^2][n\sum y^2 - (\sum y)^2]}} \quad (2)$$

where r is the Pearson's correlation coefficient, n is the number of valid responses, x represents the score of an item, and y represents the total score of each respondent with valid responses; with an assumption that both variables (x and y) are normally distributed. The criteria of interpreting a validity coefficient of an item are presented in Table 1.

Table 1. Interpretations of validity coefficients [8]

Validity coefficient values	Interpretation
Above 0.35	Very beneficial (Strongly Valid)
0.21-0.35	Likely to be useful
0.11-0.20	Depends on circumstances
Below 0.11	Unlikely to be useful

If validity of a measurement represents the accuracy of the measurement, the reliability of an instrument indicates the consistency of the scores acquired from the instrument. The term reliability was first introduced by Spearman in 1904 who defined it as the ratio between the variances of the true scores with the variance of the observed scores [9]. By estimating the reliability of an instrument, several criteria of consistency of the scores can be explained. There are different ways to estimate the reliability of a questionnaire including: (1) Test-Retest reliability that is estimated by calculating the correlations between scores of two or more administrations of the questionnaire with the same participants; (2) Parallel-Forms reliability that is estimated by creating two or more forms of questionnaire that have the same content and then administering the forms to the same subjects at the same time; (3) Inter-Rater reliability that is estimated by comparing the scores of two or more independent judges on the responses of the participants to the instrument; the comparison is used to determine whether the raters' estimates are consistent; and (4) Internal Consistency Reliability that is estimated to judge the consistency of results across items on the same test. There are different perspectives regarding the internal consistency reliability as explained by Tang, Cui and Babenko [9]. Without ignoring the other interpretations of the internal consistency reliability, in this CTT analyses the definition of internal consistency reliability by Haertel [10] will be used, that is the reliability estimate of an instrument based on a single administration.

According to Tang, Cui and Babenko [9] there are several measures of internal consistency reliability under CTT including Cronbach's alpha, Revelle's beta, McDonald's ω_h, and Sijtsma's ECV. Among those measures, the Cronbach's alpha is the most common measure used. In our analysis, the Cronbach's alpha coefficient will be used as the measure of the internal consistency reliability of the instrument. The mathematical expression of the coefficient alpha introduced by Cronbach's [11] is presented below:

$$\alpha = \left[\frac{n}{n-1}\right]\left[1 - \frac{\sum_{i=1}^{n}\sigma_i^2}{\sigma_X^2}\right] \tag{3}$$

where α is a lower-bound estimate of the true reliability, n is the number of items in test X, σ_X^2 is the observed score variance of test X, and σ_i^2 is the variance of item i. The criteria of interpreting an internal consistency reliability coefficient of an instrument are presented in Table 2.

Table 2. Interpretations of internal consistency reliability coefficients [8]

Internal consistency reliability coefficient value	Interpretation
Greater than or equal to 0.90	Excellent
0.80-0.90	Good
0.70-0.79	Adequate
Below 0.7	Less Applicable

3. Method

The questionnaire for the survey administered during the FGD was developed based on the blueprint created to determine the constructs and contents to be measured. There are two constructs measure in this survey: (1) the implementation of K-13 (K-13 construct), and (2) the effectiveness of the FGD to raise concerns regarding STEM education and ISLE (FGD construct). There are 10 items developed for each construct that measure contents including: (1) How well the participants understand and implement K-13 in their lessons; (2) Whether the support system is available to implement K-13 for their lessons; (3) How easy the implementation of K-13 to be executed; (4) How they understand STEM education; (4) How they integrate STEM in K-13 based lessons; (5) How they understand ISLE; (6) How they think about using ISLE in physics lessons; and (7) How effective the FGD enlighten their understanding about STEM and ISLE. To ensure the content validity and the face validity of the instrument, we requested two experts in STEM education and ISLE to evaluate and provide critiques and suggestions to improve the instrument. After being reviewed, the revised version of the instrument was evaluated by the experts. The final version is the one that has been declared to be construct and content valid.

The participants of this study were 32 teachers and STEM educators attending the FGD held by STEM.id in March 2018. The survey was administered during the FGD. The administration took place at the end of the discussions session of the FGD. Each item in the questionnaire is given five pre-coded responses (Likert's scale): 1 for strongly disagree, 2 for disagree, 3 for neutral, 4 for agree, and 5 for strongly agree. Items that measured the same construct were arranged in the same group of items. Therefore there are two groups of items that later will be analyzed separately. Some examples of the items (translated into English from Bahasa Indonesia) are presented in Table 3 below.

Table 3. Examples of items in the instrument

Item	Statement
FGD_3	Before attending this FGD, I have known about STEM
FGD_6	I have known about ISLE learning approach before attending this FGD
K13_	I think the available laboratory equipment has supported me in conducting practicum activities
K13_9	I have mastered the use of the available laboratory equipment

Before analyzing the internal consistency validity and reliability, a descriptive analysis was conducted, including the normality assessment of the variables. Since the sample size is smaller than 50, the Shapiro-Wilk Test is more appropriate [12]. The results of CTT analysis, especially the internal consistency reliability and validity of our instrument will be discussed below.

4. Results and Analysis

From the distribution of the scores of each item and the total scores, we found that the average total score is 36.41 with a standard deviation of 6.257 for FGD construct and the average total score is 35.56 with a standard deviation of 8.758 for K-13 construct. It was also found that a little bit more than half of the students (56.2 %) scored 38-45 (maximum of 50) for FGD construct; meanwhile 54.1% of the participants scored 39-47 (maximum of 50) for K-13 construct. In evaluating the tendency of respondents in the survey the average responses of each item ranges from 2.7813 to 4.1875. These results indicate that most participants tend to agree with the items. Only two items fall in the neutral category (item FGD_3 and FGD_6). To calculate the internal consistency validity of the items in the instrument, the Shapiro-Wilk test statistics is presented in Table 4 and Table 5 below.

Table 4. Shapiro-Wilk test statistics for FGD items

Item	Stat	df	Sig
FGD_1	.854	26	.002
FGD_2	.839	26	.001
FGD_3	.885	26	.007
FGD_4	.828	26	.001
FGD_5	.710	26	.000
FGD_6	.828	26	.001
FGD_7	.815	26	.000
FGD_8	.671	26	.000
FGD_9	.737	26	.000
FGD_10	.624	26	.000

Table 5. Shapiro-Wilk test statistics for K-13 items and total scores

Item	Stat	df	Sig
FGD_1	.854	26	.002
FGD_2	.839	26	.001
FGD_3	.885	26	.007
FGD_4	.828	26	.001
FGD_5	.710	26	.000
FGD_6	.828	26	.001
FGD_7	.815	26	.000
FGD_8	.671	26	.000
FGD_9	.737	26	.000
FGD_10	.624	26	.000

The Saphiro-Wilk statistics of the items are all above 0.5 with p-values less than 0.05. The results indicate that all variables are normally distributed, therefore the use of Pearson's correlation coefficients are appropriate to measure the internal consistency construct validity. The values of the Pearson's correlation coefficients of each item with their total scores are presented in Table 6 and Table 8.

From Table 6 it can be seen that there are two items that fail to be valid items. Analyzing the items, we found that these two items have different nature than the other FGD items. When we exclude these two items from the analysis, the results still supported that the other items have strong validity values. Reviewing the statements of the items, we found that these two items measure respondents' knowledge about STEM and ISLE, which are new terms for the respondents; meanwhile other items purely asked their opinion about the information that they obtain from the FGD. The Pearson's correlation coefficients and the p-values showed that respondents, who tend to agree with the other statements on the questionnaire, disagree with the statements on these two items. To fix this problem, we have revised the items as showed in Table 7.

Table 6. Internal consistency validity of FGD items

Item	*r*	Significance	Interpretation
FGD_1	.726	.000	Strongly Valid
FGD_2	.759	.000	Strongly Valid
FGD_3	.081	.665	Unlikely to be useful
FGD_4	.824	.000	Strongly Valid
FGD_5	.781	.000	Strongly Valid
FGD_6	.264	.144	Likely to be useful
FGD_7	.632	.000	Strongly Valid
FGD_8	.859	.000	Strongly Valid
FGD_9	.750	.000	Strongly Valid
FGD_10	.788	.000	Strongly Valid

Table 7. Revised items in for FGD construct

Revised Item	Statement
FGD_3	After attending this event, I have a better understanding about STEM learning approach
FGD_6	After attending this event, I have understood ISLE learning model

Table 8. Internal consistency validity of K-13 items

Item	*r*	Significance	Interpretation
K13_1	.720	.000	Strongly Valid
K13_2	.652	.000	Strongly Valid
K13_3	.584	.000	Strongly Valid
K13_4	.794	.000	Strongly Valid
K13_5	.780	.000	Strongly Valid
K13_6	.727	.000	Strongly Valid
K13_7	.821	.000	Strongly Valid
K13_8	.790	.000	Strongly Valid
K13_9	.441	.017	Strongly Valid
K13_10	.719	.000	Strongly Valid

Since all items in K-13 construct are strongly valid following the rule of thumb in Table 1, then there are no revisions applied to the items in K-13 construct. For internal consistency reliability, it was found that the Cronbach‘s alpha of the FGD instrument is 0.838; meanwhile for the K-13 instrument, the response data have Cronbach’s alpha of 0.882. If we combine all responses of the 20 items, we found that the Cronbach’s alpha coefficient of the instrument is 0.848. Based on the interpretation of internal consistency reliability value in Table 8, from these results, we can confirm that the internal consistency reliability of the instrument is good.

5. Conclusions

Based on the CTT analysis conducted in this study, there are two items that are not strongly valid and need to be revised. The internal consistency coefficients of validity and reliability of the questionnaire showed that the instrument has good quality as a measurement tool to measure participants' knowledge and opinion regarding the implementation of K-13 in STEM education and the effectiveness of the FGD in enlightening participants about STEM and ISLE. The revised survey instrument will be administered in the future survey that will be conducted a few weeks after this report was written. The comparison among the initial and revised instruments will be reported in our next article.

Acknowledgment
This study is funded entirely by the NAS and USAID under the USAID-PEER Grant with USAID Prime Award Number AID-OAA-A-11-00012, and that any opinions, findings, conclusions, and recommendations stated in the article are from the authors only, and do not always reflect the views of USAID or NAS.

References:
[1] Etkina E 2015 Millikan award lecture: Students of physics-Listeners, observers, or collaborative participants in physics scientific practices? *American Journal of Physics* **83** 669
[2] Kline T J 2005 *Psychological testing: A Practical Approach to Design and Evaluation* (Canada: Sage)
[3] Cohen L, Manion L and Morrison K 2011 *Research Methods in Education (7th ed)* (London: Routledge)
[4] Cook D A and Beckman T J 2006 Current concepts in validity and reliability for psychometric instruments: Theory and application *The American Journal of Medicine* **119** 166
[5] Moskal B M and Leydens J A 2000 Scoring rubric development: Validity and reliability *Practical Assessment, Research and Evaluation* **7** 71
[6] Dillman D A, Smyth J D and Christian L M 2014 *Internet, Phone, Mail, and Mixed-mode Surveys: The Tailored Design Method (4th ed)* (Hoboken, NJ: John Wiley and Sons)
[7] Westen D and Rosenthal R 2003 Quantifying construct validity: Two simple measures *Journal of Personality and Social Psychology* **84** 608
[8] Saad S, Carter G W, Rothenberg M and Israelson E 1999 *Testing and Assessment: An Employer's Guide to Best Practices* (Washington DC: Employment and Training Administration)
[9] Tang W, Cui Y and Babenko O 2014 Internal consistency: Do we really know what it is and how to assess it *Journal of Psychology and Behavioral Science* **2** 205
[10] Haertel E H 2013 *Reliability and Validity of Inferences about Teachers based on Student Test Scores* (Princeton, NJ: Educational Testing Service)
[11] Cronbach L J 1951 Coefficient Alpha and The Internal Structure of Tests *Psychometrika* **16** 297
[12] Hanusz Z, Tarasinzka J and Zielinski W 2012 Adaptation of Shapiro-Wilk test to the case of known mean *Colloquium Biometricum* **42** 43

IOP Conf. Series: Journal of Physics: Conf. Series **1088** (2018) 012015 doi:10.1088/1742-6596/1088/1/012015

Characterising formative assessment practices in the mathematics classes

J Low[1], M Shahrill[1], J S H Q Perera[1] and R C I Prahmana[2]

[1]Sultan Hassanal Bolkiah Institute of Education, Universiti Brunei Darussalam, Bandar Seri Begawan, Brunei Darussalam
[2]Universitas Ahmad Dahlan, Jl. Pramuka Kav. 5, Yogyakarta 55161, Indonesia

E-mail: masitah.shahrill@ubd.edu.bn

Abstract. This study investigated how formative assessment is conducted in one of the secondary schools Brunei Darussalam. A framework that consisted of the five key strategies was utilised to observe two mathematics teachers teaching at the Year 7 and Year 10 levels. The qualitative approach was used in analysing the data in which two classroom observations and interviews with the teachers and randomly selected students were conducted. From the lesson observational findings, only a few of the characteristics from the modified key strategies were observed. The interview findings revealed contrasting findings with one teacher being aware of using formative assessment in her lesson and using activities to access her students' understandings of the lesson. However, the other teacher was not aware of using formative assessment in her lesson and prefers her students to think independently. In order to improve the teachers' performance to teach formatively, more time should be given for the teachers to plan, reflect and design formative assessment activities.

1. Introduction

The use of formative assessment practices is sometimes overlooked in the classrooms, as teachers typically tend to measure students' performance through summative assessments [1, 2]. However, formative assessment could also improve students' achievements [3, 4]. Formative assessment in Mathematics is an important tool to improve the level of learning in the Mathematics classroom. It is a deliberate instructional process utilised by teachers and students to provide specific, actionable and immediate feedback to students through daily, ongoing instructional strategies that are more focused towards student-and-classroom-centered [5]. Accordingly, Black et al. [6] suggested a few criteria for teachers in order for formative assessment to be practiced and implemented in the classrooms, such as teachers need to discuss and share the learning objectives with their students, teachers are encouraged to ask higher order thinking questions that encourage students to explore and make connections from their previous lesson, and to give constructive feedback to students by letting them know how they progress during the lesson [7-9], for example, what they have done well and how to improve their learning.

This study investigated how formative assessment is carried out in a secondary school in Brunei Darussalam. A framework from Wiliam and Thompson [10] that consisted of the five key strategies was utilised to observe two teachers teaching Year 7 and Year 10 respectively. Table 1 displays how each of the key strategies are related to define the features of formative assessment. Several characteristics of formative assessment were observed with reference to this framework and a checklist was used to identify if the two teachers had achieved the stated criteria. The chosen framework was used because Anderson and Palm [3] discovered the effectiveness of the key strategies based on the professional development given to the teachers in implementing the formative assessment activities in their lessons. From their findings, it was reported that the students' academic performance had improved when compared to the control group. In addition, the framework encompassed the approaches on the use of students' learning as evidences that adjusted the mode of instructions in order to suit the students' learning ability.

Table 1. The framework comprised of the five key strategies (taken from [10, p 107]).

Key Strategies and Characteristics
KS1 – Clarifying, sharing and understanding learning intentions and the criteria for success
KS2 – Engineering effective classroom discussions, questions and tasks that elicit evidence of learning
KS3 – Providing feedback that moves learners forward
KS4 – Activating students as instructional resources for one another
KS5 – Activating students as the owners of their own learning

Ruiz-Primo [11] suggested another form of informal formative assessment where teachers used instructional dialogues such as assessment conversation that are dialogic and interactive in nature. The main role of the teachers is to encourage students to ask and then answer the questions, subsequently, the students learn how to participate in reasoned argumentation with the requirement to be investigative, be independent by thinking critically and learn how to solve problems with their peers [12]. However, regardless of the teachers' teaching experiences, not all the teachers practiced and implemented the skills that were shared during the professional development [13]. The teachers preferred the direct instructional method.

Only recently, the Mathematics teachers from Brunei government schools were given professional development in dialogic teaching so that the students will be asked higher order thinking skills questions to assess their level understanding during the lessons. Therefore, this study aims to compare the classroom practices of two mathematics teachers. Teacher A who was the Program for International Student Assessment (PISA) learning partner, trained by the PISA International Coach to implement formative assessment strategies in her lessons, whereas Teacher B is an experienced teacher who had been teaching for more than 20 years who mainly prefers the direct method of instruction. The investigations centred to whether both teachers had shown the stated key strategies that were derived from the framework by Wiliam and Thompson [10], and to compare, if any, differences with the mathematics performance of their students in the respective classrooms.

2. Method

The two participating Mathematics teachers in the study were conveniently selected. Teacher A is in her late twenties, possessed a Master qualification with two years of teaching experience. Teacher A taught the Year 7 Mathematics topic on 'Congruency and Similarity'. Meanwhile, Teacher B is in her late fifties, with a Certificate qualification but had more than 25 years of teaching experience, and she taught the Year 7 Mathematics topic on 'Directed Numbers'. Each of their Year 7 mathematics lessons were observed and the teachers were subsequently interviewed. In total, six students (three from each of the teachers' classes) were also interviewed to seek their views on formative assessments that were used in their respective lessons. These students were randomly selected from both classes. Due to ethical considerations, the identity of these students will not be revealed.

This study used the qualitative approach in which two classroom observations and interviews were conducted. The two lessons were video recorded in order to analyse the respective teachers' assessment strategies. Semi-structured interview questions were used and the interviews lasted approximately 10 minutes. The conversations with each teacher from the interview recordings were then analysed for themes that focused on the teachers' views on the use of feedbacks and their opinions on formative assessment activities. A checklist, modified from Andersson and Palm [3, p 113] that consisted of the five key strategies with detailed formative assessment description activity used in the classrooms. The checklist was also used to report on the lesson observational findings that focused on the strengths and weaknesses for both teachers.

3. Results and discussions

3.1. Lesson observational findings

Few of the characteristics mentioned from the key strategies [10] were observed during the lesson observations. The formative assessment activity for each teacher are reported in Table 2, which gives the overall report based on the five key strategies adapted from the framework [3, p 113].

Table 2. Details of the overall report for the two observed teachers.

Formative Assessment Activity	TA	TB
(AN1) Clarifies learning goals by specifying sub-goals or success criteria	✓	✓
(AN2) Clarifies relevant mathematical procedures as learning goals by specifying sub-goals or success criteria	✓	×
(AN3) In oral and written communications, the teacher emphasises learning as a goal in contrast to completing a task	✓	×
(AN4) Clarifies relevant mathematical communications as learning goals by specifying sub-goals or success criteria, in most cases using rubrics	×	×
(AN5) Teacher uses mini-whiteboards to produce evidence of students learning	✓	×
(AN6) Teacher uses exit passes to produce evidence of students learning	×	×
(AN7) Teacher uses tests to produce evidence of students learning	✓	✓
(AN8) Encourage students to ask questions and allow students to inform their misconceptions	✓	×
(AN9) Teacher uses feedback	✓	✓
(AN10) Provides more thoughtful feedback	×	×
(AN11) Gives students comments instead of grades	✓	×
(AN12) Encourages students to help each other	✓	×
(AN13) Provides students with the descriptions of how to act as a resource for one another	×	×
(AN14) Provides students with the descriptions of how to regulate their learning	×	×
(AN15) Provides students with descriptions of how to handle situations when they do not know to solve tasks	×	×
(AN16) Organises activities for students to monitor and evaluate their use of time.	×	×

Note 1: KS1 = AN1, AN2, AN3 & AN4; KS2 = AN5, AN6, AN7 & AN8; KS3 = AN9, AN10 & AN11; KS4 = AN12 & AN13; KS5 = AN14, AN15 & AN16. Note 2: KS denotes Key Strategy; AN denotes Activity Number; TA denotes Teacher A and TB denotes Teacher B

From the Table 2, Teacher A was observed to implement 11 of the 16 formative assessment activities in her classroom, whereas Teacher B only managed three of the stated activities. It seems that Teacher B did not strengthen the first dimension of conducting formative assessment, such as clarifying mathematical procedures as the learning goals and the oral and written communication. Teacher A was observed to strengthen the learning goals clearly by communicating with her students at the beginning of the lesson and what to achieve at the end of the lesson. However, she did not use rubrics as a tool to interpret and grade the students' work against criteria and standards for the particular tasks. For the individual work, she also assigned two tasks targeting the particular learning goals, allowed student discussions so that answers were shared, and mistakes and questions accepted.

To produce evidence of students' learning (KS2), Teacher A had used mini-whiteboards to access individual students' understanding of the given task, such as in identifying the differences and similarities of the two given shapes. The students wrote their answers on the mini-whiteboards and held them up for the teacher's view. Teacher B, on the other hand did not strengthen this key strategies (KS2) but she tend to give a lot of direct questions as a review from the previous topic. The students were given a time limit of 15 minutes to answer the questions on the whiteboard and then exchanged their books to mark their peer's work. Answers were discussed directly with the students but they were not given any opportunity to talk and to discuss the corrections among themselves. However, both teachers did not use exit passes to provide evidence of students' understanding of the lessons. For the KS3 dimensions of learning, Teacher A was observed to give oral feedback to individual students as she monitors the students' work during the individual tasks. She replaced scores with comments. Meanwhile, Teacher B did not give any comments for her students to improve their understanding of directed numbers. However, she did ask the students verbally if they need more questions as a review for the current topic. Additionally, the female students were seen to respond more compared to the male students who were silent throughout the lesson.

For the second dimension involving all agents in the classroom, Teacher A encouraged her students to assist one other (KS4), however she did not provide any descriptions, or develop the application techniques of formative assessment to the students in supporting one another. Teacher B was not observed to carry any of these (KS4) strategies. In KS5, Teacher A organised an activity through peer assessment where students discussed and corrected their peers' work within the given time limit. She monitored the time limit for each activity, however the students were not allowed to monitor and reflect their time usage for the activity. Although Teacher A discussed the answers with her students at the end of the lesson, no encouragements were given for the students to do self-assessment of their learning with respect to the given goals. For the KS5, Teacher B was not seen instructing her students to regulate their own learning. The students were drilled to answer all the 17 questions from the whiteboard and they were not given any opportunities to reflect and self-assess their own learning as they were not informed of the goals needed to be achieved at the end of the lesson.

3.2. Interview findings

From the interview sessions, the teachers were asked on their views on formative assessment. Teacher A was aware of it but not for Teacher B. Due to her education background, Teacher A had no difficulties in describing the purpose of using formative assessment during her lesson, but not for Teacher B. Teacher A's opinion on formative assessment is an on-going assessment and it can also be used after a summative assessment, such as giving feedbacks to the students. She personally thought that formative assessment is more important after the summative assessment where it reflects the performance of the students. Teacher A expressed her opinions on the sharing of objectives of the lesson, in which she stated that it depending on the flow of the lesson. She sometimes shared it at the beginning of the lesson and if the students had achieved the goal of the lesson, she will let the students know at the end of the lesson. Teacher B was not aware of sharing the objectives of the lesson with the students and she disregarded it. She explained that she would only know how much her students understood after answering her questions through classwork. Nonetheless if many of her students made mistakes, then she would use mental arithmetic to tackle her students' mathematical skills. She also asked her students if they needed more questions from the previous topics and she would drill them with more questions before beginning a new topic. Teacher B preferred to assess her students through summative assessment.

Teacher A explained that using the mini-whiteboard is an efficient way to access her students' understanding of the lesson where it could help her to correct the students' mistakes immediately. She would correct the students' misconceptions immediately before giving the classwork or else it would be too late to give the students' feedback after marking the classwork. She would also use pre-designed activities that targeted the student's misconceptions and sometimes she gave a short test from the past year examination questions for 10 minutes, which she immediately discussed the answers with them. Teacher B did not believe in giving group work or activities to access her students formatively. She preferred her students to sit down, solve the problem and to think individually. Nonetheless, she allowed her students to ask her questions if they did not understand. When asked on giving feedbacks, she told the students to see her during her free time and she retaught the whole lesson once again. She would instruct the students to do the corrections immediately and if the students did not attempt the corrections, the students would face punishment. From the interview findings, it seems that Teacher A was aware of using formative assessment in her lesson and used activities to access her students' understandings of the lesson. However, Teacher B was not aware of using formative assessment in her lesson and prefers her students to think independently. She also did not want to practice any activities that is used to access students formatively. She preferred to access her students by giving more extra questions and instructs students to do the corrections in order to avoid repeating the same mistakes.

The six students were interviewed on their views of being formatively assessed by their teachers in class and the use of peer feedbacks. The familiar themes that emerged were: All preferred to discuss their answers with their peers; Through the activities done by Teacher A and the use of mini-whiteboards, the students were aware of being assessed formatively and it helped them to understand better; Teacher B's students were not aware of being assessed formatively mainly because the teacher instructed and discussed the answers with them and; They are used to being assessed through tests and preferred to be given numerical grades in order to see if they had done well in the examinations.

3.3. Discussions

The Wiliam and Thompson [10] framework is well structured, as it comprised of the three dimensions of learning within the five key strategies. The first dimension consisted of the three processes in teaching, learning and formative assessment and it establishes where the learners are at a specific stage in their learning, where the learner is currently and what required to be done in order to get there [10]. The first dimension clearly states what the teacher needs to be aware of in her teaching, which focuses more on where and how to move the learners from one stage to another. As observed from the framework, Teacher A had covered the first and second dimensions of the framework where she gets her students going by clarifying her learning objectives and the goals of her lesson. This framework is structured to acknowledge the importance of teachers and students to reach a common understanding of the specific goals, the processes focused on eliciting, interpreting and to use the information to make decisions on what needs to be done in the next steps in teaching. The second dimension of the framework involves the three agents in the classroom, which is the teacher, peer and the learner who participated in the formative assessment processes. This framework touches on the importance of students that need to act as both learners and peers, thus indicating where students need to help each other as well as themselves. Teacher A had involved her students to work cooperatively and collaboratively [14-17], her students were actively engaged during the discussions, and she also motivates her students to keep trying the questions.

The third dimension of this framework is used to differentiate the different ways of managing formative assessment, which is the time frame of the adjustment cycle. The framework is virtuous as it includes the length of the adjustment cycle where the teacher will adjust her teaching strategies according to the students' feedback and students will adjust their learning behaviour after receiving the feedback from the teacher. During instruction time, teachers should be aware of the students' feedback while assessing them formatively and at certain time, teachers need to make changes in planning their lesson, such as differentiating different tasks for different students' abilities. Perhaps, to review the topic before moving to the next lesson if the objectives of the lesson were not achieved. The three dimensions of the formative assessment from this framework are useful in order to structure a checklist for analysis of any teachers' formative assessment classroom practice.

4. Conclusions

According to the study conducted by Andersson and Palm [3] who had reviewed Wiliam and Thompson [10] framework, students' achievements improved when teachers changed their method of instruction through the professional development that consisted of the five key strategies. So far, professional development that was conducted by the PISA International Coaches did not contain certain elements of these key strategies. In order to improve the teachers' performance to teach formatively, more time should be given for the teacher to plan, reflect and design formative assessment activities in order to make it work.

It is recommended that teachers should be given time to focus more on their professional development. In addition, the teachers should be monitored frequently and given guidance to improve their instructional strategies. As observed from this study, Teacher A, who is the PISA learning partner had implemented most of the key strategies from the framework. However, she did not use the rubrics, which is a comprehensible set of criteria for students' work that included descriptions of levels of performance quality on the criteria [18, 19]. She should also allocate more time for the students to do peer assessment like checking their peers' work, encourage students to make decisions, to converse more and ask questions in class. In contrast, Teacher B who is an experienced teacher did not want to apply the knowledge from the professional development. Through the use of summative assessment, Teacher B viewed that it is the best method for students to improve their academic performance, thereby overlooking the formative assessment strategies. She should put in more effort in making use of the formative assessment activities so as to assist her students to better learn the skills rather than studying through rote learning without understanding what to achieve at the end of the lesson. However, Teacher B did not want to change her method of instructions because she believed that she had produced a number of students who had excelled in their examinations.

References

[1] Yatab R S and Shahrill M 2014 *Int. J. of Sci. and Res.* **3** 685
[2] Botty H M R H and Shahrill M 2015 *Int. J. of Ed. and Prac.* **3** 17
[3] Andersson C and Palm T 2017 *Ed. Inq.* **8** 104
[4] Nor H N H M and Shahrill M 2014 Incorporating the use of poster and oral presentations as an alternative assessment in the teaching of secondary mathematics *Proc. of the 2nd Int. Conf. on Soc. Sci. Res.* (Malaysia: ICSSR 2014, WorldConferences.net) p 369
[5] Wisconsin Department of Public Instruction 2017 Formative Assessment Retrieved from https://dpi.wi.gov/strategic-assessment/cycles-assessment/formative
[6] Black P, *et al* 2004 *Phi Delta Kappan* **86** 8
[7] Rawlins P and Leach L 2014 Questions in Assessment for Learning and Teaching *Facing the Big Questions in Teaching: Purpose, Power and Learning* eds A St George, S Brown, and J O'Neill (Auckland: Cengage Learning)
[8] Smith K 2010 *Assessment Matters* **2** 6
[9] Wiliam D 2011 *Stud. in Ed. Eval.* **37** 3
[10] Wiliam D and Thompson M 2008 Integrating Assessment with Learning: What Will it Take to Make it Work? *The Future of Assessment: Shaping Teaching and Learning* ed C A Dwyer (Mahwah: Lawrence Erlbaum Associates) p 53
[11] Ruiz-Primo M A 2011 *Studies in Ed. Eval.* **37** 15
[12] Gillies R M 2016 *Int. J. of Ed. Res.* **76** 178
[13] Wood K, *et al* 2017 *Int. J. for Lesson and Learning Studies* **6** 151
[14] Damit A H, Shahrill M and Roslan R M 2015 *Med. J. of Soc. Sci.* **6** 214
[15] Sulaiman N D and Shahrill M 2015 *Med. J. of Soc. Sci.* **6** 544
[16] Lim M T L, *et al* 2016 *Adv. Sci. Lett.* **22** 1725
[17] Lee C, *et al* 2018 *International Journal on Emerging Mathematics Education* **2** 49
[18] Brookhart S M 2013 *How to Create and Use Rubrics for Formative Assessment and Grading* (Alexandria: ASCD)
[19] Mumu J, *et al* 2018 *J. Phys.: Conf. Ser.* **943** 012011

Design learning in mathematics education: *Engaging early childhood students in geometrical activities to enhance geometry and spatial reasoning*

R Novita[1], M Putra[2], E Rosayanti[3] and F Fitriati[1]
[1]Mathematics Department, STKIP Bina Bangsa Getsempena Jl. Tanggul Krueng Aceh No.34, Rukoh, Darussalam, Banda Aceh 23112, Indonesia.
[2]Hiroshima University, Higashihiroshima, 〒739-0046 Hiroshima Prefecture, Higashihiroshima, Kagamiyama, 1 丁目 3-2, Japan
[3]TK Al-Washliyah Alue Naga, Banda Aceh, Indonesia

E-mail: rita@stkipgetsempena.ac.id

Abstract. Understanding geometry and spatial reasoning is an essential area of mathematics learning. Besides, geometry and spatial reasoning serve as the foundation for most learning of mathematics and other subjects, especially for early childhood students. Unfortunately, these aspects are often ignored or minimized in early education and the professional development of early childhood teachers. This case may be due to a lack of activities which involve students to reason about geometric phenomena in their surrounding world. Therefore, this study aimed to design an instructional activity to guide the early childhood students in understanding geometry and spatial reasoning by doing some activities related to shapes. Therefore, this study used van den Akker et al. design research model including three phases, namely: Preparing for the experiment, Experiment in the classroom, and Retrospective analysis. The study involved 13 children ages 4-5 in one of the kindergartens in Banda Aceh, Indonesia. Teaching experiment results showed that the learning trajectory developed is potential to make a significant difference in the geometry learning of early childhood helps children to connect their experience about various types of form in their life with mathematical knowledge (i.e., geometry and spatial reasoning) that allows them to reinvent and reinforce the geometry concept.

1. Introduction

Learning mathematics is not only oriented to the mastery of numbers, but some other aspects are equally important to be understood by young children. The aspects are geometric concepts and relationships. Understanding geometry and spatial reasoning is an essential area of mathematics learning. Besides, for early childhood students, geometry and spatial reasoning serve as the foundation for the learning of mathematics and other subjects. Furthermore, knowledge of geometry concepts is essential in many real-life contexts [1]; not only in and of themselves but also for the supporting roles they play in learning algebra and the concepts and skills of number and arithmetic [1,2,3]. Geometric models also contribute to learn and solve problems in measurement, graphing, ratio, and proportion, probability, as well as algebra. Other studies have also mentioned that high achieving students' numerical ability connect to their spatial and measurement abilities. Furthermore, low achieving students in mathematics generally show little growth in geometry. Therefore, educators should give full attention to the development and understanding of geometry and spatial reasoning [2,4,5,6].

Geometry includes a wide range of skills - from recognizing simple shapes to create complicated proofs, from understanding above and beside navigating complex spaces with maps and graphing algebraic functions. By learning geometric principles, students develop reasoning and justification skills and learn how to interpret and describe physical environments [3]. Therefore, it is essential to introduce geometry at every age, grade, and year of education [1], especially for early childhood. In the early childhood education, understanding of geometry is defined as the ability of children to recognize, point, mention and collect objects around them based on geometric shapes [7].

Thus, the construction of the geometry concepts in children starting from identifying forms, investigating shapes and separating the usual pictures such as, quadrilateral, circle, and triangle; learning the concepts of location, such as below, above, left, right are the basics of understanding geometry [8]. Furthermore, NCTM's Curriculum Focal Points [9] described areas of geometry that children should learn by age and education level from pre-kindergarten to 12-year-olds.

In the Indonesia standard curriculum of early childhood education, it also described the indicators of geometry learning achievement for children aged 4-6 years [10]. It expected that children could: (1) Know and show geometric shapes based on the size of the objects; (2) Classify more objects into similar groups, or groups in pairs; (3) Classifying objects by colour, shape, and size; (4) Sorting objects from the smallest size to the largest or vice versa. Young children have already recognized shapes before they come to school because they see all kind of shapes in their surroundings. They might experience with shapes, shadow or other geometric phenomena in informal ways. For example, they know that the form of a door is rectangular and the wheel is round. However, having that knowledge does not guarantee that they understand geometry and can explain why such things happen. Therefore, the learning activities focusing on understanding geometry and spatial reasoning need to be given and taught to the children.

However, some studies [2, 4, 5, 8, 11, 12] show that students are less involved in geometry learning that is rich in concepts and depictions and students are less well-versed with geometric phenomena in their surrounding world. Also, some international studies like TIMSS and PISA also indicate a weakness in student's geometry skills [1]. These deficits have been identified in the earliest years of life which caused by the teachers who do not always provide with adequate preparation in geometry as well as the teaching and learning of geometry [1,2,4]. Besides, it could be due to the lack of teachers' abilities in developing activities that appeal to students to learn geometry.

Therefore, this study aimed to design an instructional activity to guide the early childhood students in understanding Geometry and spatial reasoning related to shape. The idea of this research was to design a learning trajectory in geometry domain that giving children the meaningful experience in geometry and providing them the opportunity to develop geometric and spatial reasoning in particular shapes. The learning-teaching trajectory description for geometry in the lower grades emphasizes three aspects of geometry including orienting, constructing and operating with shapes and figures [8]. This research focused on learning trajectory in orienting or recognizing the variety of shapes. Utilizing the constructing activities, children will develop their skills to know and describe types of shapes in term of their color, form, and size. Consequently, the primary goal of this research was to develop children's learning process in geometry properties of shapes by constructing activities and providing meaningful learning for children by discussion. We hope this research can provide input for teachers in developing activities on geometry and spatial learning.

2. Research method

This research used *design research methods* as a tool to achieve the research objective of developing the *Local Instructional Theory (LIT)*. Design research is a research method aiming to develop local learning theory or Local Instruction Theory (LIT) that can be done by researchers together with teachers to improve the quality of learning [13]. Furthermore, this study used *authentic classroom assessment* in the process of developing LIT that has been prepared. Research subjects were 13 children aged 4-5 years in one of the kindergartens in Banda Aceh, Indonesia.

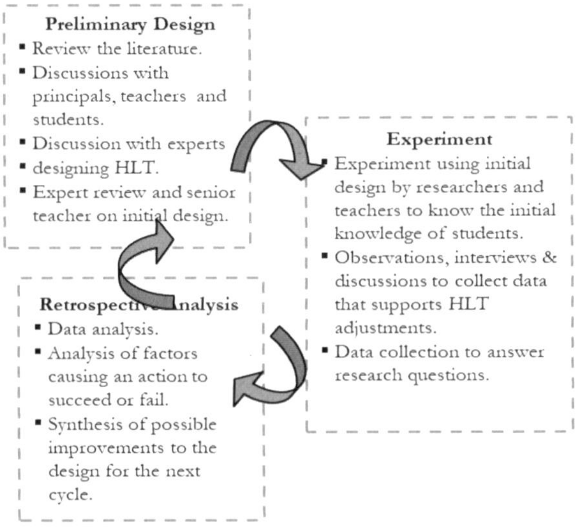

Figure 1. The design research procedure [14].

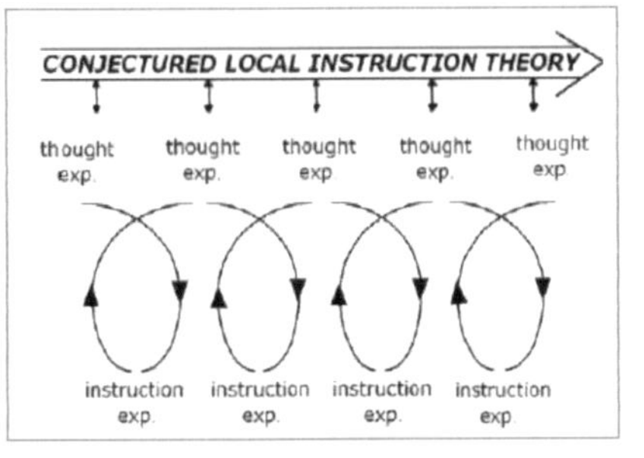

Figure 2. Developmental research, a cumulative cyclic process [15].

According to van Den Akker [15], research design is divided into three main phases, namely *preparing for the experiment*, *teaching experiment*, and *retrospective analysis*. Those can repeatedly be conducted to discover a new theory as the result of a revision of the learning theory tested. It also related to Gravemeijer's idea in [14] arguing that design research is a cyclical process (Figure 1) of thought and instruction experiments. The cycle process *(recurring)* is intended from the experiment of thought (*thought experiment*) and then experimental learning (*instruction experiments*). So the procedure of this study is described as Figure.2

Data collection in this study was conducted using documentation, interviews, and observation. The data were analyzed descriptively by considering the validity and reliability of the data. The validity of the research took into account the Hypothetical Learning Trajectory and the trackability. Furthermore, a reliability test was conducted in two ways, namely data triangulation technique and cross interpretation.

The research begins with a study of literature on the concept of LIT geometry that will be developed before designing the learning activities. Based on the review of the literature, the first Hypothetical Learning Trajectory (HLT) which is the hypothesis of the LIT will then be designed. According to Gravemeijer & Cobb [12], HLT consists of three components, including (a) the purpose of mathematics learning for students; (b) the learning activities and tools or media used in the learning process; (c) the conjecture learning process, how to examine the understanding and strategies of students arisen and developed when the learning activities are conducted in the classroom. At the preliminary design stage, HLT serves as a guideline for instructional materials to be developed. While, at the teaching test stage, HLT serves as a guideline for teachers and researchers in teaching, interviewing, and observing [9]. While LIT aims at "... the description of, the rationale for, the envisioned learning route as it relates to a set of instructional activities for a specific domain." [15]. Therefore, we can define LIT as a theory of the process in which students learn a mathematical topic and theory about media or learning material used in helping the learning process of the topic. In addition, Gravemeijer & Van Eerde [2] said that LIT also contains conjecture about the possibilities occurred in the learning process and the chances of ways that can be done to support the learning process.

3. Result and discussion

The first step of this research was to design learning trajectory and Hypothetical Learning Trajectory (HLT) of geometry learning based on literature study of LIT for geometry and spatial reasoning in early childhood (4-5 years). The developed HLT contains three components, namely: (a) the learning objectives of the geometry for the children; (b) The learning activities that are accompanied by devices or media used in the learning process of geometry and spatial reasoning; and (c) The conjecture learning process occurs in the classroom. The preparation of this HLT was based on preliminary study results on the child's ability and limited discussion with the classroom teachers about the experiences of teaching the concepts of geometric and spatial reasoning. The results of this discussion showed that the teacher only introduced geometric forms classically in front of the class during the geometric learning. This activity was conducted by indicating if this object was a circle or a rectangle or just a triangle without continuing with directional activity. On the other hand, students were directed to construct blocks and play with blocks; however, it was unclear what skills and competencies they should master. Teachers made the lesson plans before teaching geometry yet often the planning they made was useless because they did not implement them in their teaching where the teachers only write a lesson plan to address a requirement by the school's principal. Besides, based on interviews conducted with teachers it was found that the lack of teachers' knowledge about the concept of geometry was also a constraint in designing the learning activities for geometric and spatial reasoning. Consequently, the research concerning activities related to geometric and spatial reasoning in school is insufficient.

Table 1. Geometry learning activities for early childhood students along with the competencies.

No	**Learning objectives/competencies**	**Activities**
1	Children can interpret the physical world with geometric ideas and describe it with corresponding vocabularies	***Nature exploration***
2	Children can find shapes in their environments and describe them in their own words.	***Recognize shape***
3	Children can build figures and designs by combining two- and three-dimensional shapes	***Create the picture*** ***Shape picture***
4	Children can discuss the relative positions of objects with vocabulary such as "above," "below," and "next to."	***Put the ball***
5	Children can identify names and describe a variety of shapes with different sizes and orientations.	***Identify the shape***
6	Children can classify more objects into similar groups, or groups in pairs	***Find the similar shape*** ***Coloring the similar shape***

The HLT developed refers to four geometric and spatial reasoning competencies that must be acquired by children aged 4-5 years based on NCTM [14] and Indonesian curriculum [10]. Table 1 shows a series of geometry learning activities along with the competencies to be achieved after the revision was done (based on the validators' feedbacks and the findings of the pilot experiment), while the learning trajectory stage can be seen in Figure 3.

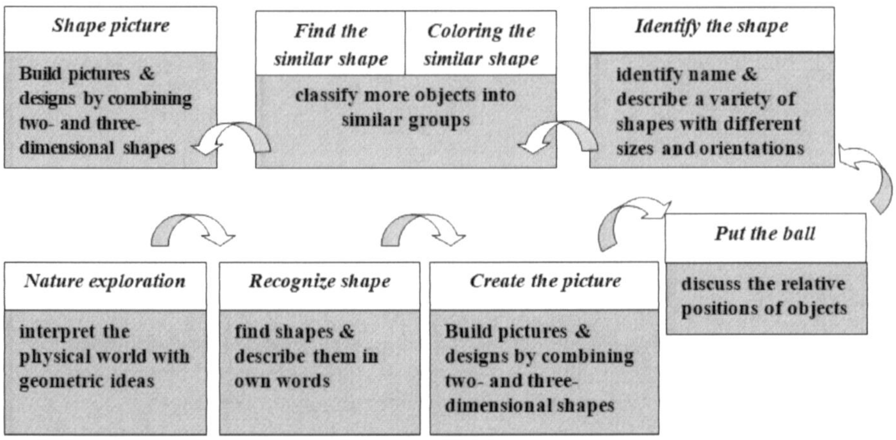

Figure 3. Learning trajectory in geometry & spatial reasoning for early childhood.

In ***Nature exploration*** activity, the children were taken out of the classroom to explore and observe the school environment. During the activity, children were asked to name objects similar to geometric shapes they can find in the school environments, such as windows and doors that are rectangular, tires of cars and motorcycles that are circular, the roof of a house that is the triangular shape. In this activity, the researcher found that the children are familiar with geometric objects, but they cannot explain correctly like saying a circle on the ball (meaning a circle shape on the ball), side three for a triangle.

In the ***Recognize shape*** & ***create the picture*** activity, the children were asked to draw the objects they found during their browsing activities. The children were free to draw any object, and then they must describe what they have drawn (Figure 4). This activity is excellent to build children's creativity and imagination. They can explore their ability in drawing all the objects that they see.

Figure 4. Children's attempt to draw the table in their class as a rectangular object.

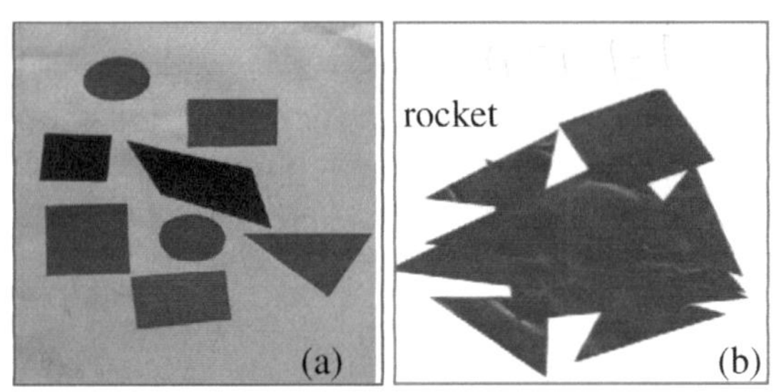

Figure 5. Children design their imaginary picture.

In ***Put the ball*** *activity*, the children were given a ball. Teachers asked them to put the ball based on the instructions given such as place the ball on the table, next to the cupboard, under the seat and so on. The other directions were given to the children, such as, "I want Syifa to put her ball inside a .... ." "Nila should put his ball under a table." And this was continued until all the children have had their turn. The reflection gained from this activity is that teachers should use a larger area than in a classroom because it has limited places to play.

In ***shape picture*** activities, the teacher gives children pieces of color paper cut into basic shapes: circles, squares, triangles, and rectangles. Then ask them to glue the various shapes onto a piece of paper to create their shape picture. Appearing conjecture in this activity was that children were able to create their imaginary objects creatively, although the color paper they used was coinciding with each other. However, the interesting finding of this activity is that some students tend to be less creative when given shape in a colorful paper (Figure 5a) because they assume that the shapes are separate images from each other.

4. Conclusion

This research created a Local Instructional Theory (LIT) on the learning of geometry and spatial reasoning of early childhood students that can be used by teachers and parents to teach the concept and develop geometry and spatial reasoning abilities of early childhood students, especially on the aspect of orienting or recognizing the variety of shapes for 4-5 year olds. Then, the activities that have been promoted in this LIT can be a good start to help children in developing their understanding about geometry and spatial reasoning for the future.

Acknowledgment

We would like to thank many people who have contributed to this research and supported us through the process. We are very grateful to our institution of STKIP BBG and Hiroshima University that supported and funded us to conduct this research.

References

[1] Sarama J, Clements D H, Parmar R S and Garrison R 2006 *Achieving Fluency in Special Education and Mathematics*

[2] Clement D H 2008 *Geometric and Spatial Thinking in Early Childhood Education* (State University of New York: University at Buffalo)

[3] National Mathematics Advisory Panel 2008 *Foundations for Success: the Final Report of the National Mathematics Advisory Panel* (Washington D.C.: U.S. Department of Education, Office of Planning, Evaluation, and Policy Development)

[4] Clement D H and Sarama J 2011 *J. Mathematics Teacher Education.* ***14*(2)** 133-48.

[5] Cheng Y L and Mix K S 2014 *J. Cognition and Development.* **15** 2–11.

[6] Cross C T, Taniesha A W and Heidi S 2009 *Mathematics Learning in Early Childhood Paths Toward Excellence and Equity* (United States of America: Committee on Early Childhood Mathematics)

[7] Lestari, KW 2011 *Konsep Matematika Untuk Anak Usia Dini* (Jakarta: Direktorat Jendral Pendidikan AUD Nonformal dan Informal)

[8] van den Heuvel-Panhuizen, M. 2005 *Young Children Learn Measurement and Geometry: Measurement and Geometry in Line (*The Netherlands: Freudenthal Institute, Utrecht University)

[9] National Council of Teachers of Mathematics (NCTM) 2006 *Curriculum focal points for prekindergarten through grade 8 mathematics: A quest for coherence* (Reston, Va.: NCTM)

[10] *Permendikbud 146 Tahun 2014* Kurikulum 2013 PAUD dan Lampirannya. (Jakarta: Peraturan Menteri Pendidikan Dan Kebudayaan)

[11] Acharya A and Ghose A 2015 *J. IJEPR.* ***4*(4)** 91-8

[12] Jardine R, and Martin N S 1983 Spatial ability and throwing accuracy *Behavior Genetics* **13** 331–40.

[13] Gravemeijer K and Van Eerde D 2009 *The Elementary School Journal* **109** 5

[14] Novita R and Putra M 2017 *J. Pendidikan Matematika* **11(1)** 43-56

[15] Gravemeijer K and Cobb P 2006 Design research from a learning design perspective eds Van Den Akker, K Gravemeijer, S, McKenney, & N. Nieveen *Educational Design Research* pp. 17 – 51 (New York: Routledge)

Elementary students' representations in solving word problems

J T Manoy
Universitas Negeri Surabaya, Jl. Ketintang, Ketintang, Gayungan, Kota SBY, Jawa Timur 60231

E-mail: janetmanoy@unesa.ac.id

Abstract. The mathematical representation is essential for students as a means of communicating their ideas so it will be easier for them to solve the problems that they may face. In solving word problems, students learn how to develop their models because, in principle, every individual has their way of understanding things. The purpose of this study was to describe the representation of third-grade elementary school students in solving mixed arithmetic operations word problems. Students' representation data were obtained from the results of the test related to word problems, and task-based interviews with two students. The results showed that the students used images, mathematical expressions, and word representations. This research implies that the teachers need to develop tasks for the students along with the answer alternatives by taking into consideration various representations that might be provided by the students

1. Introduction

The mathematical representation of this research is the mathematical expression of the idea in the form of images, symbols, mathematical expressions, words or written text that are used to show the students' thinking in solving problems in a certain way [1,2,3]. The students' ability in solving mathematical problems is closely related to the way how the students understand math knowledge. According to Bruner [4,5], to understand the knowledge, a person will go through three stages of representation, namely: enactive, iconic and symbolic stage. In the enactive *stage,* it will be easier for the students to use concrete objects that are related to the concept being studied, so students learn to use or to manipulate objects directly. This is called concrete model stage. In the iconic stage, students attempt to comprehend knowledge related to mental images, which are the description of concrete objects that have been manipulated. Furthermore, students do not directly manipulate concrete objects in this stage but only using visual images, drawings or diagrams outlining concrete activities or situations enacted in the enactive stage. In the symbolic stage, students change the shape of the images into symbolic systems, so the students directly manipulate symbols and no longer associated with the objects. Bruner's stages then refined by Lesh [6]. Lesh [6] splitted the iconic stage into manipulative and pictorial models, the symbolic stage into the spoken language and written symbols. On the other hand, Ishida [7] and Manoy [8] classified representation into: (a) the realistic model, which is the representation with realistic or real-world situation, (b) manipulative model, the representation with manipulative aids, (c) Pictorial model, the representation with static figures or pictures, (d) language model, the representation with written everyday language, and (e) symbolic model, that is the representation with mathematical symbols. In solving word problems related to mixed arithmetic operations, students learn to represent their ideas in different ways.

The question in this research is how the representation of third-grade elementary school students in solving word problem related to mixed arithmetic operations, while the purpose of the research is to describe the representation of third-grade elementary school students in solving word problems related to arithmetic operations.

2. Methods

This study includes qualitative research because it examines the phenomenon in its real-life context and tries to describe the phenomenon. Subjects of this study were third-grade primary school students, with the following criteria: (a) They have studied mixed arithmetic operation lesson consisting of addition, subtraction, multiplication, and division; and (b) They can communicate ideas/ opinions or thoughts orally and in written simply and clearly. Two chosen subjects who meet the criteria were interviewed by employing task-based interviews. Data were collected using the test and interview, while the instruments were word problems and interview guidelines. The word problems provided presented two mathematical problems. Results of interviews were analyzed using the following five steps:

- Selecting and identifying data related to ways of solving the problems by the subjects, and then generating code in many ways presented.
- Writing data set related to the ways of solving the problems by the subjects.
- Checking the validity of the data, this was done by comparing and examining written answers and clarified them with interview responses. If there is a match between the written answers and the response data, so it can be concluded that the data obtained are valid. In case of the data obtained are not valid, then interviews need to be re-conducted until valid data were obtained.
- Establishing the data that is writing the data set related to the ways the subjects solved the problems so conclusions can be drawn from the data. The data summarized were the subjects' responses regarding the ways they solve the problems.
- Drawing conclusions from the collected data.

3. Result

The research subjects were given two word problems to describe their representations related to mixed arithmetic operations which were the word problem 1 (W1) and the word problem 2(W2). Next, it was explored in more detail the mathematical representations of the two subjects –the subject 1 (S1) and the subject 2 (S2) Below are the word problems used in this research.

> W1: Mrs. Janet bought a box of eggs containing 20 eggs. The eggs were placed into several plastic bags. Each plastic bag contained three eggs. How many plastic bags should Mrs. Janet prepare for all eggs? Provide as many possible ways to answer this problem.
>
> W2: Today Dinda had a birthday party, and she invited 25 friends. Dinda set up some tables, and four people sit at each table. How many tables were needed to accommodate all her friends? Provide as many possible ways to answer this problem.

Figure 1. The problem given to students

The representations by S1 and S2 in solving the word problems involved mathematical expressions. This is in line with Hwang et al. [9] that state that representation means the process to make abstract models, symbols or concepts to describe inter-sentence relations. S1 and S2 presented their representations because every individual has different representations [10,11,12]. S1 and S2 also

IOP Conf. Series: Journal of Physics: Conf. Series **1088** (2018) 012017 doi:10.1088/1742-6596/1088/1/012017

used another representation called image representation. In solving the problem, drawing an image might appear as one of the representations [13,14,15].

3.1 Representations by S1

Data or information provided by S1 of W1 and W2 show the consistency of information (valid) so that the data displayed to answer the research question can use the W1 data. Here Some representations of solving W1 by S1are presented in Figure 2.

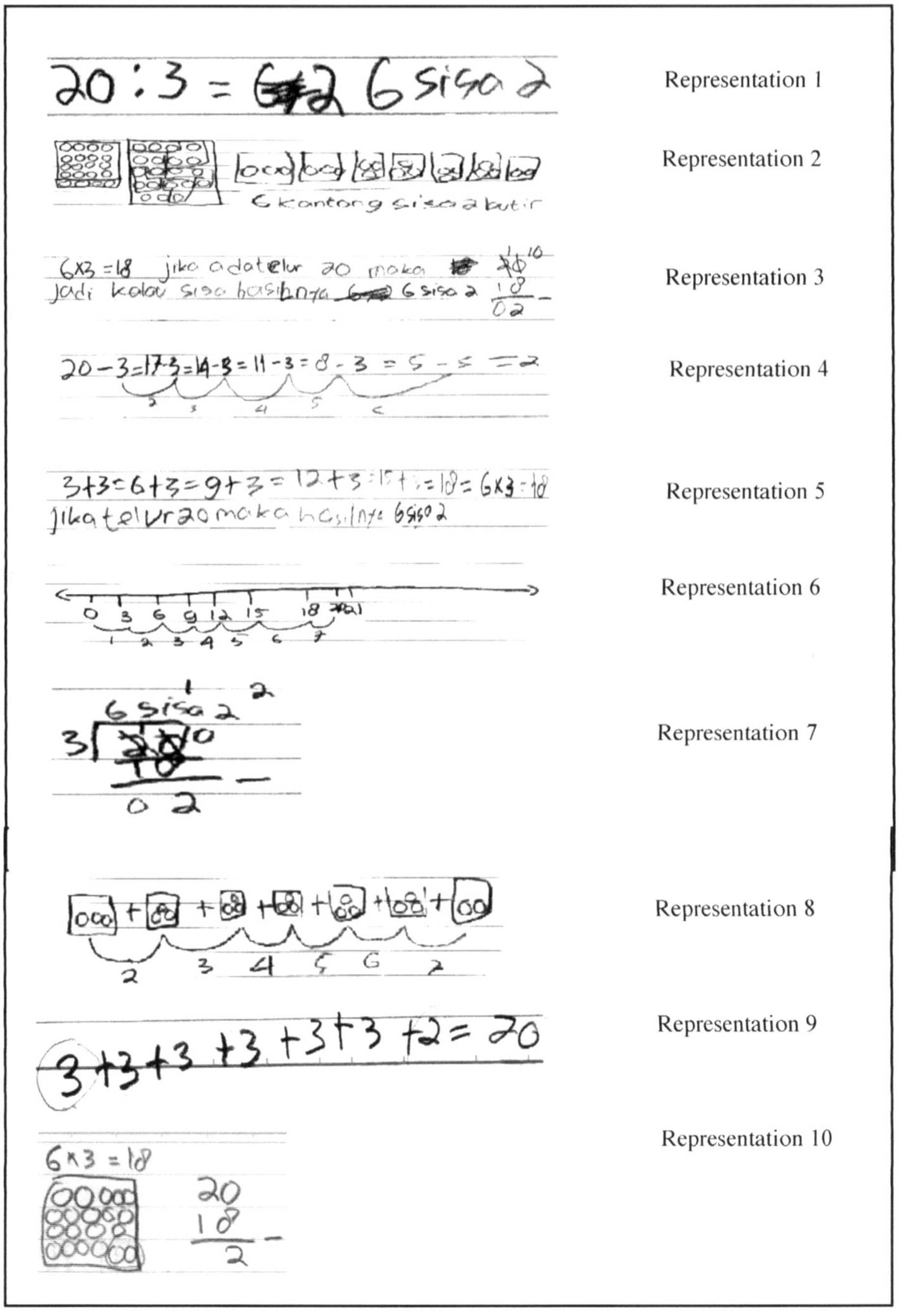

Figure 2. Representations by S1

Based on the interview data, the first that came into S1's thinking was to do the multiplication of 3 that the result was close to 20. S1 mentioned: 3 × 5 = 15, 3 × 6 = 18 but if 7 × 3 = 21, then what is used was 3 × 6 because 3 × 6 = 18 then the result will be 6.2, but the first way provided by S1 was by using the division. After completing the division, then S1 drew the image as the next step. First, S1 drew 20 circles in one box. According to SI, the image shows that there are 20 eggs in a box. In the next picture, every three circles were encircled into one except for the two last circles. According to S1, he gave mark for every 3 eggs that were put in the bags made in the next image, while the last two circles were still included in the bag (representation 1). In the next picture, S1 put every three circles in one box while the seventh box contained only two circles. According to S1, there was no relation between the first and the second way. The representation 2 provided by S1 was correct. In the next representation, S1 wrote 6 × 3 = 18. if there are 20 eggs, then 20-18 = 2. So, if there is the remainder, it was 6.2. According to S1, this idea was obtained by applying the multiplication that has been learned previously. The third way used is related to the first and second way. It can be concluded that the representation 3 was correct, S1 confirmed that S1 did the multiplication and remainder was 2. So, 2 at 6.2 means the remainder (representation 4). The researcher suggested the way to make it easier for the readers (peers) to understand. The researcher gave the option of writing 6.2 or 6 with a remainder of 2 and S1 chose the second so that all previous writing used 6.2 crossed without being given the cue by the researcher. S1 changed it to 6 with a remainder of 2. Next, clarification on representation 4 was done. In doing this, S1 20 subtracted from 3 for several times and the result still show that there was a remainder of 2. Looking at the results in the representation 4, there was a mistake in the way of writing ≠ 20-3 17-3, and so on, which should have been written as 20-3 = 17 and comma was needed to be placed or in the next line S1 should write 17-3 = 14, and so on. Based on the explanation of the completion, representation 4 was then declared incorrect. The following clarification is regarding representation 5 which was done by way of summation. S1 summed 3 for six times and the result was 18. S1 wrote 6 × 3 = 18. According to S1, all the summations made was the same as 6 × 3. Based on the written result by S1 (representation 5), there was a way of writing error because 3 + 3 ≠ 6 + 3 and so on, which should have been written as 3 + 3 = 6 and place a comma or 6 + 3 = 9 and so on should be written in the next line. Based on the explanation above, representation 5 was then considered incorrect. In the written reply before the clarification, S1 only answered until representation 5. The next answers raised when clarification was in progress. Consider the following clarification. When the researcher tried asking clarification on a number line, it turned out that S1 understood what is going to implement. S1 made the number line and wrote the skip correctly (skip counting by three) as representation 6. Although the line made was not straight and the dots placement, was not symmetrical in the distance, but it can be concluded that S1 had a good understanding of what he was doing. Completion of the representation 6 was correct.

The researcher was still asking the possibility of other ways that can be made by S1, and S1 immediately responded with *"porogapit"* method and directly did the long division as the representation 7, and it was correct. The researcher then asked for other possible answers by joining the existing answers. To respond to this question, S1 provided an answer as representation 8 and answered any questions from the researcher correctly. As seen on the representation 8, there was a mistake in drawing it, for in the image there should be no operation symbol. Based on the explanation, completion of the representation 8 was wrong. As the attempt to have another representation, S1 used the same way as in representation 5 but with the different writing, as representation 9, and the result was correct. Researcher still asked about the possibility of other ways that can be made. S1 answered perhaps by combining addition and subtraction. SI made 20 circles in one box and wrote 6 × 3 = 18. By way of long reduction, SI subtracted 20 from 18, and the remainder was 2 (representation 10). S1 the explained that the answer is still seven bags. Researcher still gave an opportunity for S1 to provide any other alternatives. However, S1 responded "no" which means that S1 had no more idea to answer that question. It turned out that S1 could do ten different representations.

Based on the analysis of written answers and clarifications (wrong answers excluded), the researcher concluded that in completing W1, S1 used seven correct ways which were ordinary division

(Representation 1) and the long division (Representation 7), drawing (Representation 2), mixed multiplication (Representation 3and 10), the addition (Representation 9), and the number line (Representation 6). There were some images as well as long divisions that appear. Ideas generated came from the previous ideas, so the way of the completion made by S1 was interrelated.

3.2 Representations by S2

The data or information provided by S2 of W1 and W2 show the consistency of information (valid) so that the data displayed to answer the research question could use the W1 data. Several representations made by S2 were presented in Figure 3.

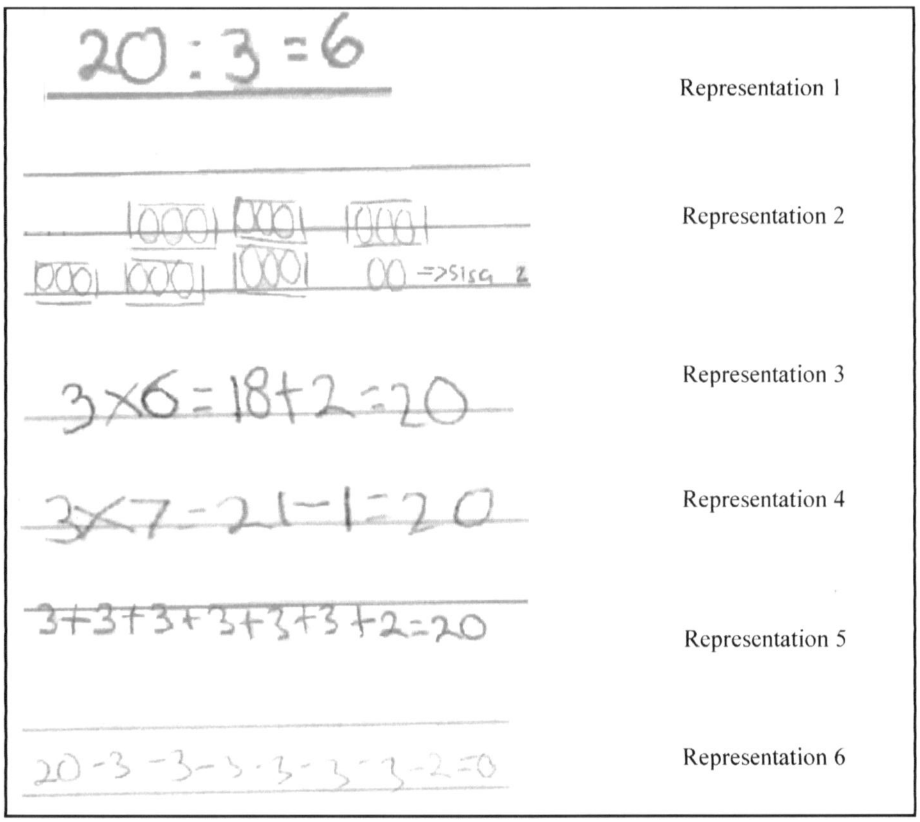

Figure 3. Representations by S2

In completing W1, S2 directly put three eggs and insert them into each plastic bag. When doing so, S2 just counted and did not use props prepared by the researcher. After collecting the eggs, the first thing to do that S2 divided twenty by three and the result was six with a remainder of 2. Witten result of Representation 1shows that there was a mistake in writing because S2 did not write the remainder of 2. Based on the explanation, completion on representation one was declared incorrect. S2 finished second way by drawing. According to S2, there are seven images which were six plastic bags containing three eggs each and the image of 2 eggs which are not in the bag (Representation 2). Both representations made were correct. The researcher asked S2 to write the results of his thinking. S2 was silent and took some time to think, and later researcher intervened with the question of how many as containing three eggs and how many are the rest. Finally, S2 wrote $3 \times 6 = 18 + 2 = 20$ (S41096). S2 used a mix of multiplication and addition. Based on the results written by S2 in the representation 3, there was a mistake in writing $3 \times 6 \neq 18 + 2$, which should have been written as $3 \times 6 = 18$ and placed a coma or on the next line should be written as 18 + 2 = 20. Based on this, the representation 3 was declared incorrect. The researcher asked S2 about how S2 did 3 × 7. The answer was because the result was 21. Next, the researcher asked S2 again why 3 × 7 = 21. S2 thought that if it was added, but later S2 revised the answer that when subtracted from one the result is 20. This means that S2 used a mix of multiplication and subtraction in the representation 4. base on the written result for Representation 4, there was mistake in the way of writing 3 × 7 ≠ 21-1, which should have been written as 3 × 7 = 21 and put a coma or on the next line, it should be written as 21-1 = 20. Based on the explanation, completion in representation 4 was declared false.

Furthermore, the researcher tried to ask S2 other ways to resolve the problem by asking "how else it can be done?" S2 spontaneously answered by addition- 3 + 3 + 3 + 3 +3 + 3 + 2 = 20 (Representation 5). Completion for representation 5 was correct. Next answer provided by S2 was by subtraction. S2 wrote 20 – 3 – 3 – 3 – 3 – 3 – 3 – 2 = 0 (Representation 6). Completion of the representation 6 was correct. The next question was focused on the ideas that have been raised by S2. According to S2, there is a relation between the first and second ideas concerning the way of thinking to get answers which were from the eggs that had been grouped.

Based on the analysis of the written answers and the clarifications (wrong answers excluded), the researcher concluded that in completing W1, S2 used three representations to solve the problem: drawing (Representation 2), addition (Representation 5) and subtraction (Representation 6). Ideas generated were from the previous ideas of eggs that had been grouped. S2 solve the problem in an interrelated way.

4. Conclusion

Students solved word problems used images, mathematical expressions, and word representations. Students used pictorial representations to illustrate the completion of word problems, mathematical expressions to write arithmetic operations, while the representation of words or written text used by students to write down the information that was known and being asked. There were several representations of the images that appear as well as long division. Ideas generated came from the previous ideas, so the ways to solve the problem provided by the subject were inter-related.

References

[1] Pape S J and Tchoshanov M A 2001 The Role of Representation(s) in Developing Mathematical Understanding *Theory into Practice* **40** 118

[2] NCTM 2000 *Principles and Standards for School Mathematics* (Reston, VA: NCTM)

[3] Rosengrant D 2005 *An Overview of Recent Research on Multiple Representations* (Rutgers: The State University of New Jersey)

[4] Hudojo H 1988 *Mengajar Belajar Matematika* (Jakarta: Depdikbud Dirjen Dikti PPLPTK)

[5] Tall D 1990 *A Versatile Theory of Visualisation and Symbolisation in Mathematics*

[6] Lesh R 1981 Applied Mathematical Problem Solving *Educational Studies in Mathematics* **12**

[7] Ishida T 1984 A Study on the Representational System in Mathematics Teaching *International Conference for Psychology of Mathematics education* (Australia: Sydney)

[8] Manoy J T *Variasi Model Pemecahan Masalah Matematika yang Terkait dengan Operasi Hitung Campuran oleh Siswa Sekolah Dasar* [Dissertation] Surabaya: Unesa 2011

[9] Hwang W Y, Chen N S, Dung J J and Yang Y L 2007 Multiple Representation Skills and Creativity Effects on Mathematical Problem Solving using a Multimedia Whiteboard System *Journal of Educational Technology & Society* **10** 191

[10] Rowe H L and Trickett E J 2017 *Student Diversity Representation and Reporting in Universal School-Based Social and Emotional Learning Programs: Implications for Generalizability* (USA: Springer)

[11] Krawec J L *Problem Representation and Mathematical Problem Solving of Students of Varying Math Ability* [Dissertation] Miami: University of Miami 2010

[12] Boonen A J H et al. 2014 The Role of Visual Representation Type, Spatial Ability, and Reading

Comprehension in Word Problem Solving: An Item-Level Analysis in Elementary School Children *International Journal of Educational Research* **68** 15

[13] Carreira S 2015 Mathematical Problem Solving Beyond School: Digital Tools and Students Mathematical Representations *Selected Regular Lectures from the 12th International Congress on Mathematical Education* 93

[14] Rubin A, Storeygard J and Koile K 2015 Supporting Special Needs Students in Drawing Mathematical Representations *The Impact of Pen and Touch Technology on Education* 57

[15] Earnest D 2015 From Number Lines to Graphs in the Coordinate Plane: Investigating Problem Solving Across Mathematical Representations *Cognition and Instruction* **33** 46

The development of learning instruments using the creative problem-solving learning model to improve students' creative thinking skills in mathematics

D M Sari[1], M Ikhsan[1] and Z Abidin[2]
[1]Syiah Kuala University, Jl. Teungku Chik Pante Kulu No. 5, Banda Aceh 23111, Indonesia
[2]Ar-Raniry State Islamic University, Jl. Syeikh Abdul Rauf Darussalam, Banda Aceh 23111, Indonesia

Email: ikhsan@unsyiah.ac.id

Abstract. Students should have creative thinking skill in solving mathematics problems by using various alternatives solutions. Unfortunately, many students still have poor quality in creative thinking skill because they have less attention to develop their ability when learning mathematics. One of the efforts to improve students' creative thinking skills is by applying creative problem solving (CPS) learning model. This research aimed to develop the valid and effective CPS learning instruments to improve the students' creative thinking skill by using the development method used a 4D model. There are four stages in the 4D model, including defining, designing, developing and disseminating. However, this research only involved the first three stages. The learning instruments developed were lesson plan, worksheet, and test that covered creative thinking skills. Four validators validated these instruments. The findings showed that all instruments were in a highly valid category, with an average score of 4.1, 4.4, and 4.2 respectively and can be used with minor revisions. Furthermore, the test of creative thinking skills was also classified into the highly valid category with an average score of 4.2 and was ready to be used minor revision.

1. Introduction

Mathematics is taught in each level of education. Mathematics is a universal science underlying the development of modern technology and has an important role in various disciplines and advances the human mind. Mathematics is a branch of systematically organized science [1]. Mathematics needs to be provided to all students starting from elementary school to equip students with logical, analytical, systematic, critical, innovative and creative thinking skills, as well as the ability to work together in a team [2]. However, this research focussed on the creative thinking skill only.

The creative thinking skill plays an important role in the learning of all children and is part of the higher order thinking skills that need to be developed. Creative thinking has had little concern in the mathematics learning. Many teachers only prioritize the logic and computational skill, while creativity is not considered important in mathematics teaching and learning process [3]. However, when students pay attention to the learning process, creativity will enhance understanding and foster the development of students' learning outcomes [4]. Students need to explore their creative potential to deal with various problems of contextual mathematics, which require reasoning, argumentation and creative thinking skills. Both critical and creative thinking skills are crucial to achieve the goals of mathematics learning. These skills must be nurtured and cultivated. A more meaningful mathematics learning allows students to ask questions and express their ideas [5]. Recognizing the importance of creative thinking skills, teachers should develop teaching instructions that could help students to improve their higher order thinking skills in mathematics learning.

One of the efforts to improve students' creative thinking skills is by applying creative problem-solving learning model, which refers to the learning model focusing on teaching and problem-solving skills, followed by strengthening the skills. This learning model aims to foster students' interest and motivation in learning mathematics so that students can obtain maximum benefits of learning outcomes [6]. The creative problem-solving model is a framework in which individuals or groups can use to formulate, challenge, generate and analyze new ideas and plan to be implemented to find the solutions [7]. The implementation of creative problem solving allows students to improve thinking skills, including creative and critical thinking [8]. The creative problem solving is one of the operational model problem solving that require students to apply creativity to complete the given task [9]. It can be inferred that creative problem solving is one of the learning models that can be used to improve students' problem-solving skills by using new ideas and different approaches to solve the problem and plan the effective procedures to find the solutions. This study aimed to develop the valid learning instruments of CPS learning model. Learning instruments developed were lesson plan, student worksheet, and tests of creative thinking skills.

2. Method

In this study, the development of learning instruments used a scientific approach that is relevant to the national curriculum of 2013. This research is classified as the type of development research and used the modified 4D model, including to define, to design, to develop, and to disseminate. However, this study was only conducted in the first three stages of the model. The developed instruments in this study included the lesson plans, student worksheets, and test of creative thinking skill. The subjects of this study were Year 8 students of one of the junior high schools in Banda Aceh.

It is crucial to analyze the quality of the instruments that were developed. The analysis could be assessed from three aspects, namely validity, practicality and effectiveness [10]. This research only focussed on the validity aspect, which involved four validators. The validators were the experts of mathematics teaching and learning, including the two lecturers of mathematics education department and two junior high school mathematics teachers. The used research instruments were validation sheets for the lesson plan, student worksheet, and test. The data the developed learning instruments were analyzed by determining the average of the validity level for each criterion. This analysis involved several phases, including the recapitulation of statements obtained from the validator, finding the average of validity level for each criterion, calculating the average score of each aspect, calculating the total averages, and matching the total averages with the category of validity of a predefined category as can be seen in Tabel 1 [11]. KV is the average score obtained from the experts when validating the learning instruments.

Table1. The validity criteria of developed learning instruments

Criteria	Average
Highly Valid	$4 \leq KV < 5$
Valid	$3 \leq KV < 4$
Less Valid	$2 \leq KV < 3$
Invalid	$1 \leq KV < 2$

The learning instruments could be used in the teaching practices once they show either "highly valid" or "valid" criteria. However, if the validation results show "less valid" or "invalid" levels, then the revisions on the developed learning instruments are needed.

3. Result and discussion

The recapitulation of the validity scores of the developed learning instruments obtained from the validators is presented in Table 2.

Table2. The validation results of learning instruments

Learning instrument	Average	Criteria
Lesson plan	4.1	Highly Valid
Student worksheet	4.4	Highly Valid
Creative thinking skills test	4.2	Highly Valid

Based on Table 2, it indicated that the lesson plan was categorized as highly valid with the average score of 4.1 and could be used with minor revision. Similarly, the student worksheet was also highly valid (average score of 4.4) and could be used with minor revision. Also, the creative thinking skills test was highly valid with the average score of 4.4 and could be used with minor revision. These results suggested that all the developed learning instruments were are highly valid; thus they are eligible to be implemented in future teaching instructions. Nevertheless, the validators also provided some feedback that was used to revise the learning instruments. The revisions made for the learning instruments are presented in Table 3.

Table 3. The revision results of the instruments based on validators' suggestions

Aspect	Before validation	After validation
Lesson plan	- The revision of language should be made to the lesson plan. - Students should be as the subject rather than as the object on the lesson plan. - The chosen students' name should be based on the common names used in Aceh - Teacher needed to encourage the learning awareness of the relevant knowledge so that students were ready to receive the new material.	- It has been revised, and students are no longer the object but the subject of the lesson plan - Students' names have been revised - The teacher showed the picture and related the picture to the learning objectives of the topics relation and function so that students could identify the concepts of linear equalities and inequalities of two variables
Student worksheet	- The colors of the pictures had been similar - The language used had not been communicative - The required students' activities had not been clear	- The colors of the picture were various - The used language was revised - Students' activities were clear
Test questions	- The pretest and posttest should be made slightly different not all the pretest, and post-test questions could be understood by using the operation process and graph. - The five items in the test should be checked for their difficulty level due to the limited time. - Item 5 in the post-test should be replaced with a word problem.	- The test questions have been revised, and the pretest and post-test were made slightly different. Not all the pretest and posttest can be understood using the operation process and graph. - The test was revised, and the level of difficulty was evaluated. - Item 5 in the post-test has been replaced with the word problem

Based on Table 3, the results of the revision before and after validation showed the visible improvement. Lesson plan, student worksheet, and test questions were assessed on the indicators suggested by the validators. As suggested in an earlier study, the preparation of learning instruments that can effectively enhance the ability of students' critical and creative thinking is not easy, which would need a research development to retrieve the research-based instruments in improving students' creative thinking skills [12]. The students who can think creatively may have an awareness of the need to develop the sense of curiosity and to stay practicing in solving mathematical problems. This awareness cannot be separated from the learning context that the students engage in which is crucial to develop students' creative thinking and cognitive domains [13]. Another study suggested that to improve students' learning performance and reasoning mathematical skill, the development of the learning instruments that are highly valid, practical, and effective is needed. The products of the development could be in forms of the lesson plans, student worksheets, or the assessment instrument that are worth using in teaching and learning activities [14].

4. Conclusion

The development of learning instruments in this study that employed creative problem solving (CPS) model to improve the creative thinking skills resulted in the highly valid learning instruments. This result is indicated by the average validation scores of the experts that showed that: the lesson plan was a highly valid category (the score was 4.1) and can be used with minor revisions. Also, both student worksheet and the test of creative thinking skills had highly valid criteria (with the scores of 4.4 and 4.2 respectively) and could be used with minor revisions. This research implies that creative problem-solving model could also be used in other research to develop learning instruments. Besides, this research is limited to the topics of relation and function, which means that the development of learning instruments to improve creative thinking skill by CPS model could also be applied to other topics.

References

[1] Soedjadi R 2000 *Kiat Pendidikan Matematika di Indonesia (Konstatasi Masa Kini Menuju Harapan Masa Depan)* (Jakarta Direktorat Jendral Pendidikan Tinggi Departemen Pendidikan Nasional)
[2] Permendikbud 2013 *Salinan Lampiran Permendikbud No. 58 Tahun 2014 Tentang Pembelajaran pada Pendidikan Menengah*
[3] Saefuddin A A 2012 Pengembangan Kemampuan Berpikir Kreatif dalam Pembelajaran Matematika dengan Pendekatan Pendidikan Matematika Realistik Indonesia (PMRI) *Universitas PGRI Yogyakarta* **4**
[4] Beetlestone F 2012 *Creative Learning* (Bandung: Nusa Media)
[5] Suherman dkk 2010 *Strategi Pembelajaran Matematika Kontemporer* (Bandung: Refika Aditama)
[6] Pepkin K L 2004 *Creative Problem Solving In Math* (Online)
[7] Kwon J &Ahn S 2014 A study on creative problem solving founded on computational thinking *International Journal of Applied Engineering Research* **9** 9185
[8] Tseng K H, Chang C C, Lou S J & Hsu P S 2013 Using creative problem solving to promote students' performance of concept mapping*International Journal of Technology and Design Education*, **23**1093
[9] Isaksen S &Parnes S J 1985 Curriculum planning for creative thinking and problem solving *The Journal of Creative Behavior* **19**1
[10] NieveenN1999Prototyping to Reach Product Quality Dalam van den AkkerJ BranchRM GustafsonKNieveen N Plomp T (Eds) *Design Approaches and Tools in Education and Training*(London: Kluwer Academic Publisher)
[11] Khabibah S 2006 Pengembangan Model Pembelajaran Matematika dengan Soal Terbuka untuk Meningkatkan Kreativitas Siswa SD (Surabaya: UNESA)

[12] Syahrir S 2018 Pengembangan Perangkat Pembelajaran Matematika SMP untuk Meningkatkan Kemampuan Berfikir Kreatif *Jurnal Ilmiah Mandala Education (JIME)* **2** 436
[13] Tan L S, Lee S S, Ponnusamy L D, Koh E R, and Tan K C K 2016 Fostering Creativity in the Classroom for High Ability Student: Context Does Matter *Education Sciences* **6**1
[14] Sulistyowati Yand Sugiman 2014 Pengembangan Perangkat Pembelajaran Bangun Ruang di SMP dengan Pendekatan Creative Problem Solving *Jurnal Pendidikan Matematika* **9**1

IOP Conf. Series: Journal of Physics: Conf. Series **1088** (2018) 012019 doi:10.1088/1742-6596/1088/1/012019

PISA-like mathematics problem: The context of basketball in Asian Games

R D Jannah, R I I Putri and Zulkardi

Universitas Sriwijaya, Jl. Srijaya Negara, Bukit Besar, Palembang, Indonesia

E-mail: ratu.ilma@yahoo.com

Abstract. The research background is based on the low achievement of Indonesian students on PISA mathematics literacy. The research aims to produce the valid and practical PISA-like mathematics problem that included uncertainty and data content by using basketball sports conducted in the Asian Games. The research subject was tenth-grade students of Senior High School. The researchers developed a problem based on two stages in the design research with development studies consisting of preliminary and formative evaluation stage. The data collection techniques used were a walkthrough, documentation, an observation, and an interview. The results showed that the developed problem met the valid and practical criteria. Valid criteria came from validator's assessment of two PISA expert teams from Australia, one expert from Indonesia regarding content, constructs, and language and the students' comments on the clarity (readability) of the phase one-to-one. Practical criteria obtained in the small group phase. It finds out that students had been able to comprehend the purpose of the problem. Besides, the basketball context used could make students actively in the discussion and help students to solve the problem and provide arguments from the given issues.

1. Introduction

Having good mathematical literacy skills is one of the most important things for everyone in this globalization era to face and solve real-life problems. Mathematical literacy is interpreted as the ability to use knowledge and mathematical understanding effectively in facing challenges in daily life [1]. Also in line with Stacey and Turner [2] opinion, that the main idea in mathematical literacy is the use of mathematics in daily life. In the PISA mathematical literacy is explained about four contents in which one of them is uncertainty and data content. Uncertainty and data is the heart of mathematical analysis of many problem situations, and the theory of probability and statistics as a technique of data representation and description [3]. However, in the results of the PISA survey in 2012, Indonesian students got a score of 384, which is still below the average score of other countries for uncertainty and data content [4].

Moreover, the result from the survey of the Programme for International Student Assessment (PISA) on mathematical literacy in the last two periods showed that in 2012, Indonesia is ranked 64 out of 65 countries with the score 375 from the OECD countries' average score is 494 [5,6]. Meanwhile, in 2015, Indonesia is ranked 62 out of 70 countries with the score 386 from the OECD countries' average score is 490 [7]. Based on these results, the importance of mathematical literacy is not in line with the results and achievements obtained by Indonesian students in the International PISA survey. The facts showed that Indonesia is still ranked below with the average score below the OECD's average score.

Indonesian students have difficulty in solving PISA-like mathematics problems using the context and in turning them into mathematical problems [8]. The effort which can be done is to design problem by the characteristics of PISA to familiarize students practicing to solve the problems of PISA-like and teachers are also required to be able to design problems by using context which closes to the students' lives [9]. The use of context in the learning of mathematics can make the concept of mathematics more meaningful because its context can present an abstract mathematical concept in the form of representation that is easy to understand by students [10-14]. The learning which is emphasizing on the context or situation is PMRI Approach PMRI approach is one of the learning approaches that will lead students to understand the concept of mathematics by constructing by them through the previous knowledge related to their daily lives, by finding their concept, it is hoped that students' learning will be more meaningful [15]. In addition, *Pendidikan Matematika Realistik Indonesia* (PMRI) is a learning innovation that adapts from a similar learning approach as RME (Realistic Mathematics Education) [15, 16].

One of the interesting contexts to be implemented is the sports in Asian Games which will be held in Indonesia in 2018, among which is basketball. It is such a common sport played and watched by students, so they are familiar with this context. The previous research by using the context of sports at the Asian Games can help students in understanding mathematics learning such as using sprint context [17], *dayung* context [18], swimming context [19], and hurdles context [20]. They showed that the context of the Asian Games sports could be used to attract students' learning interests.

The research about the development of mathematics problems of PISA-like has been done. However, no one has used basketball context in the Asian Games. The previous research developed the PISA-like problems level 6 for Junior High students [21] and the context of Indonesia's natural and cultural heritage to measure students' mathematical literacy [22]. Therefore, the researchers are interested to develop mathematics problem of PISA-like mathematics by using basketball context in Asian Games on uncertainty and data content.

2. Method

The method in this research was design research with the type of development studies with two stages, preliminary and formative evaluation that includes self-evaluation, expert reviews, one-to-one, small group, and field tests [23, 24]. The subject of this research was tenth-grade students of Senior High School.

In the preliminary stage, the researchers determined the place and subject of the research. Researchers not only conducted a curriculum analysis, analysis of mathematical problems based on PISA framework but also recognized PISA-like mathematics problems that had been developed previously as a reference material in developing the problem. Then, the researchers designed the problem's device that contained the problem grid, problem card, and scoring rubric so that it produced the initial prototype. In self-evaluation phase, the researchers evaluated and reviewed the initial prototype and to form prototype 1. The prototype I was then examined and judged by the validator at the expert reviews phase and tested to the students in the one-to-one phase. Expert reviews phase are the steps undertaken to evaluate the problems conducted by experienced experts. The experts assessed prototype I by evaluating it based on the content, constructs, and language. In expert reviews, the panel discussion was also employed. According to Turner [25], panel discussion or item panelling was one of some essential steps in the development of high-quality test items.

Along with the expert reviews phase, prototype I was tested to three students with different abilities (high, medium, and low ability) at one-to-one phase. The focus of this phase was to get students' comments on the clarity and readability of the problem, suggest improvements, and investigate the reasons for the difficulty found in solving the created problem. From the expert reviews and one-to-one phase would produce a valid prototype II.

The next phase was small group. In the small group phase, prototype II was tested to six students with high, medium, and low abilities that were involved to solve the problem and provide suggestions and comments to know the practicality of the developed problems. In the small group phase, the problem was used in the learning process. The comments and findings at this small group phase were taken into consideration for revising the prototype II. From the small group phase, would produce a valid and practical prototype III. The data collection techniques used were walk-through, document, observation, and an interview. Then the data were analyzed by the qualitative descriptive to describe the result of each step of development.

3. Result and discussion

This research generated PISA-like mathematics problem that included the context of basketball in Asian Games that included uncertainty and data which are valid and practical. The PISA-like mathematics problem had been through some development studies stages including preliminary stage and formative evaluation stage consisting of self-evaluation, expert reviews, one-to-one, and small group.

3.1. Preliminary

In the preliminary stage, the researchers conducted student analysis, curriculum analysis, PISA framework, and design an instrument that included the problem grid, problem card, and scoring rubric. Next, researchers designed the PISA-like mathematics problem. The researchers developed the PISA-like mathematics problem from an original PISA mathematics problem. In designing the problem using the context of the basketball, the researchers were inspired by the problem of PISA in 2006 which included uncertainty and data content that is "test scores". This problem asked students to use their reasoning to provide a reason or argument in proving and solving the problem. Indonesian students are weak in solving problems that require giving arguments or sense-making [26-30]. It is, therefore, necessary to develop students' reasoning in expressing their opinions and arguments by showing appropriate evidence and steps [28-30]. Thus, the researchers developed the problem using this basketball context by asking students' arguments that could be given to support the Lebanese basketball team as the winners in the game to able to assist educators in the learning process for the students in the school. Then, the next stage was the formative evaluation.

3.2. Formative Evaluation

In the formative evaluation stage, the first phase was self-evaluation. The researchers evaluated and re-examined the PISA-like mathematics set prepared previously. Researchers found out errors or shortcomings that had been designed with attention to the characteristics that became the focus of the prototype (i.e. content, constructs, and language). The results obtained at this stage were called prototype I. Next, the prototype I was assessed or validated by the validator at the expert content stage regarding terms of content, constructs, and language. The validators involved are experts; KS and RT from Australia, including HJ from Indonesia. The researchers also conducted panel discussion to get suggestions and comments on the developed problems. Panel discussion was done with nine colleagues who were students of Masters of mathematics Education of Sriwijaya University who also in the meantime conducted research on the development of PISA-like mathematics problems and the three lecturers of Sriwijaya University as well. In conjunction with expert reviews phase, the prototype I was tested to three tenth-grade students of Senior High School with different abilities (high, medium, and low). During the process, the researchers observed and found out how each student answered and understood the problem. The researchers asked the students if there are words, instructions, and information that was not clear on the matter. The following shows some suggestions/comments on the prototype I in expert reviews and one-to-one phase in Table 1.

Table 1. Comments/suggestions from experts and students

Validation	Comments/Suggestions	Revision
KS	Basically good problems but there is a miscalculation in the answer contained in the answer key.	1. Revise some words or sentences in the problem - "Winning score is 2" becomes "The team gets two points for every winning game." - "Winning score is 1" becomes "The team gets one point for every losing game." - "Score results" becomes " Score of the Match Results " 2. Revise the data of the competition scores by using the original data obtained from the real match
RT	1. This is a nice problem, reasoning testing, and strategy devising. 2. There are some sentences on the instructions that should be fixed, they are; - The winning score is 2 should be changed to the team gets two points for each winning match. - The value of defeat is 1 should be changed to the team gets one point for every lost match.	
HJ	Problem received and no comments	
Panel discussion	Data presented should be the actual data	
Students	Tables and instructions or rules contained in the problems are clear enough.	

The prototype I was then revised based on the validator suggestions from expert review phase and one-to-one phase. The result of this revision was called prototype II that had been valid. The validity of the problem from the validator's assessment of the content, construct, and language and from suggestions and comments one-to-one phase to the clarity/readability of the problem [23]. The following was prototype II in Figure 1.

BASKETBALL

Basketball is one of the games which will be conducted at the 2018 Asian Games. To determine the rank in a basketball game, it can be viewed from the following criteria:

1. The team gets two points for every winning game.
   The team gets one point for every losing game.
2. The team collects the highest score, first and so on.
3. If there are teams that have the same score, the right of victory is given to the team that is already compatible with both of them competition.
4. If there are more than two teams that have the same score, then the victory is determined by the average score.

Here are the results of basketball games from the six teams that had competed.

Score of the Match Results		
Iran	76 – 54	Lebanon
South Korea	107 – 56	Indonesia
Japan	98 – 78	Kazakhstan
Iran	83 – 86	South Korea
Kazakhstan	80 – 61	Indonesia
Japan	51 – 73	Lebanon
Japan	81 – 77	Iran
Lebanon	103 – 35	Indonesia
South Korea	105 – 66	Kazakhtan
Iran	90 – 59	Kazakhstan
Lebanon	80 – 67	South Korea
Indonesia	54 – 113	Japan
Indonesia	40 – 106	Iran
Japan	72 – 92	South Korea
Kazakhstan	58 – 59	Lebanon

Source : http://www.scoreboard.com/id/bola-basket/asia/kejuaraan-asia-u18/

Based on these results, the committee chairman decided that South Korea team became the top ranked team. The Lebanese team disagreed and tried to convince the committee chairman that the Lebanese team was the top-ranked team. Write down the reasons that could support the Lebanese team in convincing the committee chairman.

Figure 1. Prototype II

The valid prototype II was then tested to six students at the small group phase. The six students were two students with high abilities, two students with medium abilities, and two students with low abilities. Based on the results of the work, most students had been able to answer the problem with the right steps to completion. Moreover, in the learning process, it found that students were very enthusiastic in solving the problem and active during the discussion it also in line with the result of previous research [17-19] in which found that the Asian Games context could help students in understanding mathematics and improve their motivation throughout the learning process. The following is an example of student answers that researchers found in the small group phase in Figure 2.

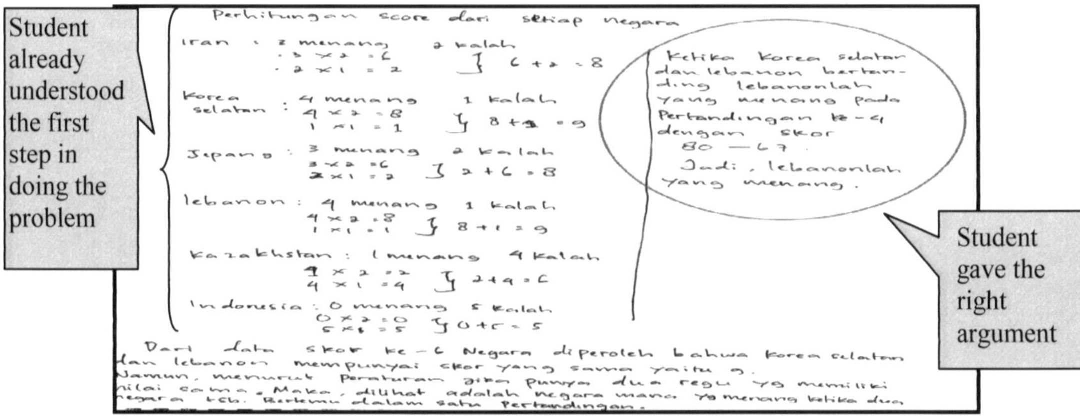

Figure 2. Student's answer

Figure 2 indicates that it can be seen that the student was able to understand the problem well. Student wrote complete and appropriate completion steps that were to calculate the score of each country based on the number of victories and defeats. Therefore, the student was able to draw conclusions and provide the correct argument to the given problem that was the Lebanese basketball team and the South Korean team has the same highest score compared with the other teams. However, based on the game rules stated on the problem, if both teams get the same score then the winner is determined by the result of the match when the two teams compete. When the Lebanese and South Korean teams competed, Lebanese team became the winner.

There was no revision that the researchers did from prototype II to prototype III. Based on the small group phase, it obtained valid and practical prototype III. Practicality was seen from the ease of students in understanding the problem [23, 24, 31, 32].

4. Conclusion

The developed PISA-like mathematics problem that included uncertainty and data content using the basketball context in Asian Games had fulfilled the valid and practical criteria and had been in accordance with the PISA framework. The validity of the problem was obtained through the validation of validators who validated in terms of content, constructs, and language as well as comments and suggestions from students about the clarity/readability of the matter at the one-to-one phase. Then, the practicality of the problem was evidenced in the small group phase; students could understand and use PISA-like mathematics problem well. In addition, the basketball context that was used could make students actively in the discussion and help students to solve the problem and provide arguments from the problems given.

Acknowledgments

The researchers would like to express their deepest gratitude to the Directorate General of Higher Education Indonesia who has funded Postgraduate Grant research in 2017-2018, and to Ms. Eriga and her students for participating in this research. As well as those who have helped the researchers in developing the research and writing this article.

References

[1] Steen L and Turner R 2007 *Developing Mathematical Literacy In Blum, W., Galbraith, P., Henn, H-W., & Niss, M (Eds), Modeling and Application in Mathematics Education- The 14th ICMI Study* (New York: Springer)

[2] Stacey K and Turner R 2015 *Assessing Mathematical Literacy: The PISA Experience* (Australia: Springer)

[3] OECD 2016 *PISA 2015 Assessment and Analytical Framework: Science, Reading, Mathematics and Financial Literacy* (Paris: OECD)

[4] OECD 2014 *PISA 2012 Results in Focus: What 15-year-old know and what they can do with what they know* (Paris: OECD)

[5] OECD 2013 *PISA 2012 Assessment and Analytical Framework: Mathematics, Reading, Science, Problem solving and Financial Literacy* (Paris: OECD)

[6] Putri R I I and Zulkardi Z 2018 Higher-order thinking skill problem on data representation in primary school: A case study *J. of Physics: Conf. Series* **948** 012056

[7] OECD 2016 *PISA 2015 Result (Volume 1): Excellence and Equity in Education* (Paris: OECD)

[8] Wijaya A, Heuvel-Panhuizen, M v d, Doorman M and Robitzsch A 2014 Difficulties in Solving Context-Based (PISA) *J. The Mathematics Enthusiast* **11** 555

[9] Zulkardi and Putri R I I 2006 *Mendesain Sendiri Soal Kontekstual Matematika Proc. KNM13* (Semarang: UNNES) p 1

[10] Arifin S, Zulkardi, Putri R I I, Hartono Y, and Susanti E 2017 Developing Ill-defined problem solving for the context of "South Sumatera" *J. of Physics: Conf. Series* **943** 012038

[11] Haris D and Putri R I I 2011 The role of context in third grader's learning of area measurement *IndoMS J. Mathematic Education* **2** 55

[12] Fauziah A, Putri R I I , Zulkardi, and Somakim 2017 Primary school student teachers' perception to Pendidikan Matematika Realistik Indonesia (PMRI) instruction *J. of Physics: Conf. Series* **943** 012044

[13] Putri, R I I and Zulkardi 2017 Fraction in shot-put: a learning trajectory AIP Conference Proceedings **1868** 050005

[14] Rejeki S and Putri R I I 2018 Models to support students' understanding of measuring area of circles *J. of Physics: Conf. Series* **948** 012058

[15] Putri R I I 2011 Pembelajaran materi bangun datar melalui cerita menggunakan pendekatan Pendidikan Matematika Realistik Indonesia (PMRI) di sekolah dasar *J. Pendidikan dan Pembelajaran* **18** 234

[16] Risma D A, Putri R I I, and Hartono Y 2013 On developing students' spatial visualisation ability *J. International Education Studies* **6** 12

[17] Roni A, Zulkardi and Putri R I I 2017 *Sprint Context of Asian Games in the Division of Fractions Proc. 5th South East Asia Development Research (SEA-DR) International Conference 2017*, (Banjarmasin: ULM) p 22

[18] Nasution M F, Putri R I I and Zulkardi 2017 *Dayung Context in Fraction Proc. 5th South East Asia Development Research (SEA-DR) International Conference 2017*, (Banjarmasin: ULM) p 1

[19] Putri R I I , Gunawan M S, and Zulkardi 2017 Addition of fraction in swimming context *J. of Physics: Conf. Series* **943** 012035

[20] Rahayu C, Putri R I I and Zulkardi 2017 *Multiplication of Fraction With Natural Number by Using Hurdles Proc. 5th South East Asia Development Research (SEA-DR) International Conference 2017* (Banjarmasin : ULM) p 43

[21] Kamaliyah, Zulkardi and Darmawijoyo 2013 Developing the sixth level of PISA-like Mathematics problems for secondary school students *J. on Mathematic Education* **4** 9

[22] Oktiningrum W, Zulkardi and Hartono Y 2016 Developing PISA-like Mathematics Task with Indonesia Natural and Cultural Heritage as Context to Assess Students' Mathematical Literacy *J. on Mathematics Education*, **7** 1

[23] Zulkardi 2002 *Developing A Learning Environment on Realistic Mathematics Education for Indonesian Student Teachers* Published Dissertation (Enschede: University Of Twente)

[24] Tessmer M 1998 *Planning and Conducting Formative Evaluations* (London: Kogan Page)

[25] Turner R 2000 *Item panelling, or cognitive walk-through* Available from https://works.bepress.com/ross_turner/21/ (Accessed 11th August 2017)

[26] Firdaus, Kailani I, Bakar M. N bin and Bakry 2015 Developing Critical Thinking Skills of Students in Mathematics Learning *J. of Education and Learning* **9** 226
[27] Simpol N S H, *et al* 2018 Implementing thinking aloud pair and Pólya problem solving strategies in fractions *J. Phys.: Conf. Ser.* **943** 012013
[28] Sundayana R, *et al* 2017 Using ASSURE learning design to develop students' mathematical communication ability *World Transactions on Engineering and Technology Education* **15** 245
[29] Simpol N S H, *et al* 2018 Implementing thinking aloud pair and Pólya problem solving strategies in fractions *J. Phys.: Conf. Ser.* **943** 012013
[30] Sukirwan, *et al* 2018 The students' mathematical argumentation in geometry *J. Phys.: Conf. Ser.* **943** 012026
[31] Riyanto B, Zulkardi, and Putri R I I 2017 Mathematical modeling in realistic mathematics education *J. of Physics: Conf. Series* **943** 012049
[32] Shahrill M, *et al* 2017 *J. Phys.: Conf. Ser.* **943** 012001

The study of the mathematical problem solving and metacognition strategy on a paired *Handep* cooperative learning model

Demitra[1] **and D Sulisworo**[1,2]
[1]University of Palangka Raya, Jl. Yos Sudarso, Palangka Raya, Kalimantan Tengah, Indonesia.
[2]Ahmad Dahlan University, Jl. Pramuka No. 42, Yogyakarta, Indonesia

E-mail: dwi.suliswor@uad.ac.id

Abstract. Mathematics problem-solving skill is essential in mathematics learning. However, the difficulty to solve mathematics problem still occur. The model of learning created based on cooperation culture of Dayak tribe developed to facilitate the learning. The research aimed to create the draft of *Handep* cooperative learning model as single path prototype level, to evaluate the feasibility of model components and to assess the problem-solving and metacognition questioning strategy fulfillment, to obtain the model of mathematical problem solving and metacognition strategy based on the local wisdom. Research and development used were the Recursive, Reflective, Design, and Development-Dissemination (R2D2). The development focused on *Handep* cooperative learning model construction. The learning technology expert reviewed the feasibility of the model, and the mathematics-learning expert validated the fulfillment of problem-solving and metacognition questioning strategy. Instruments were included questionnaire and rubrics. The result of the paired *Handep* cooperative learning model and experts' validation result concluded that learning technology experts agreed that the components of the model are feasible as a model of cooperative learning. The mathematics-learning experts found that the model has feasibility for mathematical problem-solving strategies. For advanced expert validation, it is recommended to assess the fulfillment of critical elements of the *Handep* cooperative learning model.

1. Introduction

Mathematics problem-solving skill in K-13 is a competency to develop students' mathematics ability [1]. The mathematics problem-solving skill is essential to develop students' ability [2]. The mastery of mathematics problem-solving skill is a cornerstone, which facilitated the improvement of reasoning and critical thinking.

The objective for mathematics problem-solving skill is a challenge for teachers and students in Indonesia. Difficulties in solving the mathematics problems in schools still occur. Results of research on mathematical problem solving explained that students still solve mathematics problems [3]. Students have the challenge to remember facts and concepts, understand issues, understand ideas and use them, as well as difficulty communicating meta-cognition in solving math problems [4]. Mathematical difficulties found in formulating equations, schemes, diagrams to create mathematical models of algebra materials in junior high school [5]. One of the contributing factors is the lack of effort to make innovations in learning and still using rigid learning models that sometimes are outside of the Indonesian learning context. Therefore, this research will study the feasibility of the local wisdom as an idea for developing a cooperative learning model.

The innovation of learning method has done in Indonesia but it is still limited to some areas. Indeed, Indonesia has diverse ethnics, various local wisdom; which teachers potentially used as a basis for innovation in education. Cultural factors are believed to affect the thinking process of understanding mathematical concepts [6,7]. Mathematics is the result of the cultural or human thought that is dominated by the western world [8,9]. When students from other cultural contexts learn math, students experience the conflict of thinking. Students construct mathematical knowledge when they overcome the friction; where learners have been able to accept, modify understanding in the context of thinking. The cultural background of students influences this process. Mathematics as a result of culture will experience enculturation in diverse contexts of various cultural backgrounds of students.

The pattern of cooperation in Dayak, Central Kalimantan is a form of local or indigenous knowledge. It is known as *Handep*. *Handep* cooperation has inspired the creation of cooperative learning model [10]. *Handep* is a word in Dayak, which means 'help each other.' The uniqueness of cooperation with *Handep* is the mutual collaboration between community members who agree to cooperate, by mutual give *Handep* (assistance) to other members and receives *andep* (who receive support) from other members in turn. The benefit of the *andep* is to enhance the ability of each community member [11,12].

The implementation of *Handep* learning model with group member 3-4 people for learning mathematics at SMPN-1 Palangka Raya conducted by Demitra and Wulandari [13] showed mastery of cube and block concept on students taught by *the Handep* model were higher than Student Team Achievement Division (STAD) model. However, in the aspect of motivation, there is no difference. The results of the classroom observation showed that the students had difficulties in carrying out group cooperation, in which the number of group members was 3-4 students. After reflecting on the encountered challenges and addressing the faced problems in front of friends in the group, the more talented students or smarter student did not want to help their friends who had difficulty solving math problems. The process of collaboration in the group is less smoothly. One of the causative factors observed was the number of 3-4 group members making the students reluctant to assist friends in groups. Cooperation with the number of group members of 3-4 people seems less suitable for students in junior high school. The researcher solved this problem by elaborating on the development of the *Handep* model, to reduce the group member to two or a paired group.

The purpose of this research is to explore the *Handep* model of a paired group and to evaluate the feasibility of component model and validation of fulfillment of mathematical problem solving and metacognition question strategy. The research problem was, firstly, how is the structure of the paired *Handep* model in single path prototype?. Second, how is the feasibility of the cooperative learning paired *Handep* model as a proven model of learning to found the model in the alpha version?

2. Method

The research is development research the development model of Recursive, Reflective, Design, and Development-Dissemination (R2D2) according to Willis [14]. The study used the descriptive method in ways of qualitative and quantitative approach. The activity on define focus involves creating and supporting a participatory team. It is a core team building activity consisting of two people (research team) and then involving experts as supporting groups during the development process. So the subject is supporting group, that they are experts in the field of learning technology there are two people and experts in the area of mathematics learning there are two people.

Progressive problem-solving activities in the design of cooperative paired cooperative learning model, beginning with formulating problems of learning mathematics and defining goals, and seek solutions to these problems. The researcher examines the components of the model that are suitable for constructing the learning model. Development activities prognosis or learning context is done by studying the context of learning mathematics in junior high school. The researcher explores the cultural context of the Dayak ethnic of Central Kalimantan. The reference syntax of the paired cooperative learning model related to this culture.

The focus of design and development involves creating *Handep* cooperative learning model. The researcher has done the process of designing cooperative paired learning in the early stages of the activity, by creating the model through the synthesis process toward the components of mathematical problem-solving procedures [15,16], metacognition thinking strategy [17,18,19] and the syntax of *Handep* model [10], and *Handep* cooperative procedures originating in of Dayak culture [11]. The researcher conducted the model development process by building and elaborating *Handep* model to get various types of the cooperative paired learning model. From the previous research, the *Handep* model procedure covers four steps, i.e.:

- Presenting the new material. Teachers recall material prerequisites and present new material to students. Then ask students to relearn by working on the worksheet.
- Troubleshooting individually. Students learn the material in the worksheet and work on solving problems in the questions contained in the worksheets facilitated by metacognition questions. Four stages of mathematical problem solving include understanding the problem, planning the solution, executing the problem-solving plan, and re-checking the solution.
- Solve problems using *Handep*. Students form groups in pairs and sit in opposite positions to make face-to-face interactions and convey the difficulty to solve problems with their partners in turn. Couples agree on a turn to solve the problem, and one student helps his/ her partner in solving the problem-solving. The partner who has received helping, assist other friends in solving the problem. All members did this stage until all the members can solve the problem.
- Present the results of solutions and celebrate success. Students post a problem-solving sheet on the wall or the board. Alternatively, present the results of split-ups on the board and explain if anyone asks from other groups. Each group presents their solutions. The teacher assesses the student's work. After completing the problem-solving results, students sing a song or yell for their success.

Stages of model elaboration, as follows (1) development of a detailed and descriptive learning model draft [20]; (2) expert test; (3) field test (small group); and (4) field test (in one class at school). The development of this learning model uses the R2D2 development model [14]. Expert validation is used to get the alpha version. The alpha version is the model development result after validated by experts. Expert validated the model by requesting a learning technology expert to assess the feasibility of the component of the paired cooperative learning model. Expert validation was also sought from mathematics problem-solving learning experts to assess the fulfillment of problem-solving strategies and meta-cognition questioning strategy [17,18,19]. The reference to revising the model based on the result of the assessment of mathematics learning and learning technology expert. The data are narratives opinion and percentage fulfillment of mathematical problem-solving component and metacognition question strategy. The data analyzed through meaning the expert opinion and constructing presented tabulation of the result of the expert assessment of the model.

3. Result and discussion

3.1. Single path prototype level

The product result of the prototype of the *Handep* model is the textbook. The first step in developing the prototype was the basic theory construction as the basis of the learning model. This model is based on teaching and learning theories in the view of constructivism, especially on mathematics. The next basis is the theory of cooperative elements, useful learning principles, and quantum learning approach. Teaching and learning mathematics contains problem-solving strategies and meta-cognition questioning strategies. The book title is *Model Pembelajaran Kooperatif Handep Berpasangan* [Paired *Handep* Cooperative Learning Model]. The prototype of the draft model is already at the single path prototype level.

IOP Conf. Series: Journal of Physics: Conf. Series **1088** (2018) 012020 doi:10.1088/1742-6596/1088/1/012020

3.2. Expert validation for the alpha version

The draft of the model at the level of single path prototype needs to be refined to get the excellent model. At this level, the draft model would be validated by two people who the learning technology expert (who will review on the constructivism theories) and two people whom mathematics was learning expert (review on the mathematics problem-solving strategies). Table 1 shows the result of the review of learning technology experts.

Table 1. The result from learning technology expert

Question items		Review result	
		Expert 1	Expert 2
A.	*Syntax*		
1.	Does the primary step of learning refer to the paired *Handep* mechanism?	Yes. The learning step referred to the paired *Handep* mechanism	Already
2.	Is the learning step relevant to the cooperative/ collaborative learning characteristics?	As a prescriptive, the steps already have cooperative/ collaborative learning	Almost all the steps are relevant to the primary procedure of collaborative activities.
3.	Based on the learner characteristics at what education level this model syntax will be fit or suitable?	Probably at all levels, but it requires some modification at some level.	This model will be suitable for student upper of junior high school.
4.	Is the learning step refer to the *Handep* cooperative components, mathematics problem solving, and metacognition questioning?	Already	Already
5.	At what part of this syntax should be improved or revised?	Nothing to be revised	The syntax parts are already detail
B.	*Instructional effect and nurturing effect*		
1.	That is the possibility to achieve the learning goals if this model is applied?	This model can be used to achieve the learning goals	The learning goals can be met because the procedure of this model is student oriented.
2.	If this model is applied, what social skill would the students acquire?	Patient, appreciate other, self-esteem, communication skill, reduce ego	Collaboration skill on problem-solving, not judgment someone
C.	*Overall components*		
1.	Does the prescription on the components give the complete explanation of the model?	The prescription is comprehensive and self-explanation	The components are detailed enough to describe the overall expected model

3.3. Assessment by mathematics learning experts

The results of the mathematics-learning expert on mathematical problem solving and meta-cognition questioning strategies are figure out in Table 2. Both experts assessed that paired *Handep* cooperative learning models had met the problem-solving stage to understand the problem and implement problem-solving (87.5%). In both aspects, it still needs to be added explanations on how to math problems and solve problems.

While aspects of planning problem solving and re-examining, paired *Handep* cooperative learning model have fulfilled (100%). *Similarly*, in the element of giving metacognition questioning strategy has achieved 100%. Fulfilling aspects of mathematical problem-solving strategy is 90.63%. This result means that cooperative paired cooperative learning model has fulfilled the stages of solving mathematical [15]. There are four stages: understanding the problem, making the plan of solving, executing the plan of solving, and re-checking the result. With the component of mathematical problem-solving strategy in paired *Handep* cooperative learning model able to facilitate the process of reasoning, mastery of problem-solving skills of mathematics to students. Results of the previous research indicate that the mathematical problem-solving strategy [15] can develop mathematical proficiency [16,21,22]. However, it still needs a prescriptive explanation on the stages of creating a plan for solving and examining the results.

Table 2. The result from mathematics learning experts

Aspects	Maximum Score	Item Score	Expert 1		Expert 2	
			Freq.	%	*Freq.*	%
A. Mathematics problem solving						
		3	2	50	2	50
1. Understand the problem	12	4	1	33.3	1	33.3
		Sum	10	87.5	14	87.5
2. Plan to solve the problem	8	4	2	100	2	100
		Sum	8	100	8	100
		3	2	37.5	2	37.5
3. Execute the problem solving	16	4	2	50	2	50
		Sum	14	87.5	14	87.5
		3	2	75	0	0
4. Re-check the solution	8	4	0	0	2	100
		Sum	6	75	8	100
Average				87.5		93.75
Average of both experts					**90.63**	
B. Metacognition Questioning	12	4	3	100	3	100
		Sum	12	100	12	100

The results of the assessment on the meta-cognition questioning strategies aspect were 100%. It means that this model has fulfilled all aspects of the strategy. The fulfillment of this meta-cognition questioning strategy aspect provides an excellent opportunity for this model to facilitate the students' mathematical reasoning process. Previous research [7,18,19] showed that the use of meta-cognition questioning strategies in mathematics learning could improve the learning outcomes of mathematics. The results of the expert's validation indicate that the *Handep* cooperative learning model has been eligible to be declared a model of learning.

The validation results of the feasibility of model components and the fulfillment of mathematics problem-solving strategy and meta-cognition questioning strategy in the *Handep* cooperative learning model become the reference for revising the model. To revise the model, the researcher needs to asses from aspects of the fulfillment of elements of the cooperative learning model [23], the leading principles of active learning and the principles of quantum teaching [24].

4. Conclusion

Based on the results of elaboration of paired *Handep* cooperative learning model and expert validation result, it is concluded that learning technology experts agreed that the components of the model are feasible as a model of cooperative learning. The mathematics-learning experts find that the model has components feasibility for mathematical problem-solving strategies. For advanced research, it is recommended to assess the fulfillment of critical elements of the *Handep* cooperative learning model in the real implementation in the classroom.

References

[1] Voyer D 2011 Performance in mathematical problem solving as a function of comprehension and arithmetic skills *International Journal of Science and Mathematics Education* **9** 1074

[2] As'ari A R, Tohir M, Valentino E, Imron Z and Taufiq I 2017 *Buku guru matematika SMP/MTs kelas VIII* (Jakarta: Ministry of Education and Culture)

[3] Surya E, Sabandar J, Kusumah Y S and Darhim 2013 Improving junior high school visual thinking representation ability in mathematical problem solving by CTL *Journal of Mathematics Education* **4** 113

[4] Novferma N 2016 Analisis kesulitan dan self-efficacy siswa SMP dalam pemecahan masalah matematika berbentuk soal cerita *Jurnal Riset Pendidikan Matematika* **3** 76

[5] Jupri A and Drijver P 2016 Student difficulties in Mathematizing word problems in Algebra *Eurasia Journal of Mathematics, Science & Technology Education* **12** 2481

[6] Bose A and Kantha V K 2014 Influence of socio-economic background and cultural practices on mathematics education in India: a contemporary overview in historical perspective *ZDM Mathematics Education* **46** 1073

[7] d'Entremont Y 2015 Linking mathematics, culture and community. In İşman, A. (Ed.) *International Conference on New Horizons in Education, INTE 2014* (Paris: France Sakarya, Turkey: Sakarya University) 2818

[8] Bishop A J 1997 *Mathematical enculturation, a culture perspective on mathematics education* (Netherland: Kluwer Academic Publishers)

[9] Prediger S 2004 Intercultural perspective on mathematics learning-developing a theoretical framework *International Journal of Science and Mathematics Education* **2** 377

[10] Demitra, Sarjoko and Uda S K 2012 Pengembangan model pembelajaran kooperatif Handep untuk pembelajaran Matematika *Jurnal Pendidikan dan Pengajaran* **19** 15

[11] Bunu H Y 2011 *Pendidikan dalam konstruksi masyarakat suku Dayak Ot Danum: Kajian fenomenologi di Kabupaten Gunung Mas, Kalimantan Tengah* (Surabaya: Jenggala Pustaka Utama)

[12] Yanarita, Naiem M, Budiadi and Sukarna 2014 Development of the Dayak Ngaju community forest in the forest and peatland area, Central Kalimantan, Indonesia *IOSR Journal of Environmental Science, Toxicology and Food Technology* **8** 40

[13] Demitra and Wulandari J 2016 Hasil belajar matematika dan motivasi siswa menggunakan model pembelajaran kooperatif *Handep* dan STAD *Indonesian Journal of Curriculum and*

Educational Technology Studies **3** 1

[14] Willis J 2009 *Constructivist instructional design: foundations, models, and examples* (Manhattan: Information Age Publishing)

[15] Polya G 1973 *How to solve it, a new aspect of mathematical method* (Princeton: Princeton University Press)

[16] Yuan S 2013 Incorporating Pólya's problem-solving method in remedial math *Journal of Humanistic Mathematics* **3** 96

[17] Mevarech Z R and Kramarski B 1997 IMPROVE: a multidimensional method for teaching mathematics in heterogeneous *American Educational Research Journal* **34** 365

[18] Sarjoko Implementasi model IMPROVE dalam pembelajaran pangkat rasional dengan setting kooperatif di SMAN-2 Palangkaraya [Tesis] Malang; PPS-Universitas Negeri Malang: 2003

[19] Kramarski B and Mevarech Z R 2003 Enhancing mathematical reasoning in the classroom: the effects of cooperative learning and metacognitive training *American Educational Research Journal* **40** 281

[20] Degeng I N S 2013 *Ilmu pembelajaran: klasifikasi variable untuk pengembangan teori dan penelitian* (Bandung: Kalam Hidup)

[21] Bayat S and Tarmizi R A 2010 Assessing cognitive and metacognitive strategies during Algebra problem solving among university students *Procedia Social and Behavioral Science* **8** 403

[22] Luthfiananda I M A, Mardiyana and Saputro D R S 2016 Analisis proses berpikir reflektif siswa dalam memecahkan masalah matematika non-rutin di kelas VIII SMP IIS PSM Magetan ditinjau dari kemampuan awal *Jurnal Elektronik Pembelajaran Matematika* **4** 812

[23] Jacob G M, Lee G S and Ball J 1997 *Learning cooperative learning via cooperative learning* (Singapore: Kagan Cooperative Learning)

[24] De Porter B, Reardon M and Singer-Nourie 2001 *Quantum teaching: mempraktikkan quantum learning di ruang-ruang kelas* [Quantum teaching: practicing quantum learning in the classrooms] translated by Ary Nilandari (Bandung: Kaifa)

Investigating students' learning trajectory: a case on triangle

Anwar[1] and I Rofiki[2]
[1]Syiah Kuala University, Banda Aceh, Aceh, Indonesia, 23373.
[2]Universitas Islam Negeri Maulana Malik Ibrahim, Jl. Gajayana No. 50, Malang, Indonesia, 65144.

E-mail: imam.rofiki@uin-malang.ac.id

Abstract. Learning trajectory becomes the main issues in mathematics education research. However, there have been limited studies of students' learning trajectory for triangle construction. Therefore, this design research was conducted to investigate students' learning trajectory for the topic of the triangle. The study involved 22 students from 7th grade in Malang, Indonesia. Data were collected through a videotaped, a student's worksheet, and a classroom observation. The results showed that students discovered the requirement of forming a triangle given three side lengths. In this condition, the starting point of students' learning trajectory was the drawing of a line segment from the given three side lengths. Students examined two side lengths whether these side lengths can be joined to the line segment as a triangle or not. Students used rulers for determining those three side lengths that could form a triangle. They made a statement that the sum of any two sides of a triangle must be greater than the third side. Furthermore, teachers should consider students' learning trajectory for achieving successfully the learning goal.

1. Introduction

Learning trajectory has been studied by researchers. Many studies on this issue have been conducted for several mathematics topics such as measurement [1,2], statistics [3], geometry [4,5], logic [6], functional relationships [7], fraction [8], and rational number [9]. The facts showed that the learning trajectory became an interesting topic to be further investigated in learning of mathematics.

Clements & Sarama [10] states that learning trajectory is a description of students' thinking and learning in particular mathematical domains and associated route guesses through a set of learning tasks designed to induce an action or mental process moving toward the progression of the thinking level. Based on the definition, there are two important aspects in learning trajectory, namely developmental progression and learning tasks. The developmental progression includes students' thinking and learning process in the mathematical domain. Students' thinking and learning process relate to a conjectured route on the benchmark of students' thinking levels through a hypothesised set of learning tasks to promote the development of a concept. Thus, learning trajectory can be defined as a student's thinking process in the process of learning created with the intention to achieve goals in certain mathematical domains.

Learning trajectory is based on a constructivist paradigm for the development of students' understanding. There are three components in a learning trajectory, namely the goal for students' learning, the learning activity, and students' thought and learning process [11]. The learning trajectory may not progress linearly. There are two types of learning trajectory, namely hypothetical learning trajectory and actual learning trajectory. The phrase of hypothetical learning trajectory was first

IOP Conf. Series: Journal of Physics: Conf. Series **1088** (2018) 012021 doi:10.1088/1742-6596/1088/1/012021

defined by Simon [11] as the teacher's prediction of the learning paths produced by students from less sophisticated to more sophisticated ideas and accumulated understandings. The characteristic of the hypothetical learning trajectory is flexible. Teachers may modify the learning trajectory that has been previously designed when students fail to understand the topic. The design of alternative learning trajectory can be used to help all students successfully achieving the learning goal. In the hypothetical learning trajectory, the actual learning trajectory is not knowable beforehand. Actual learning trajectory is obtained from implementing the hypothetical learning trajectory in the teaching experiment and analyzing students' learning path. Battista [12] focuses the idea of learning trajectory on the framework of a cognition-based assessment. This notion emphasizes the level of the model for a topic that not only describes a student's cognitive process, but also what things students can or cannot do, students' reasoning and conceptualization, a cognitive obstacle, and mental processes for progressing to higher levels.

Investigating students' learning trajectory in the mathematical learning process is crucial. The detection of the learning trajectory is the core of research projects, curricula, and professional development [10,11]. Learning trajectories have become an essential study in teaching and learning mathematics [10,11]. Knowing students' learning trajectory can assist educators in designing instructional model/strategy in response to students' mathematical thought process.

The construction task in geometry became one of the most studied issues in mathematics education [13], but there have been limited studies of students' learning trajectory for the triangle topic, especially triangle construction. Therefore, the purpose of this study is to investigate students' learning trajectory as the progressions of students' thinking about a triangle. The reason underlying the selection of the triangle topics because it is very widely used in everyday life. Yet, the triangle construction task becomes a problematic condition for some students. Students faced difficulty in constructing an equivalent triangle [14,15]. Graham & Chick [16] reported that some students were perplexed in determining all possible triangles that can be made from 20 matchsticks. Hence, teachers can use the result of this study to enhance the comprehension of students' learning progression about the triangle concept. This study makes a significant contribution related to students' learning trajectory. Teachers also can overcome students' misconceptions or mistakes.

2. Method

As a part of design research, this study focuses on students' learning trajectory. Design research is a design experiment that involves three phases: (1) preparing for the experiment (design), (2) implementing teaching experiment, and (3) carrying out retrospective analysis [17]. The design experiment aims to develop both learning processes theory and learning environment that supports the learning processes [17,18]. It is noteworthy that design research is not an experimental study. Design research is also labelled as developmental research [19] or design experiments [18].

Participants involved in this study were 22students of 7th grade. They were selected purposively from one of Islamic Junior High School in Malang, Indonesia. The study used a qualitative approach. Data were collected through videotape, learning artifacts (student's work), and a classroom observation. In the teaching experiment, the second author was the teacher and the first author was the observer. We also focused on students' gesture and group discussion. Cobb et al. [18] emphasizes the usage of various data sources (e.g., gesture, social interaction, artifact, and classroom discourse) and technological support (e.g., audio-recording tools) for obtaining a good data.

In this study, we designed learning tasks to support instructional experiment and foster students' learning about a triangle. The teaching experiment process was recorded using a digital camera and was transcribed for detailed analysis. In addition, for data triangulation step, various types of data were collected including students' thinking about the triangle, field note, and student's work. We further transcribed the recording of the learning process, especially the conversation between students and teacher. Finally, we analyzed the data of students' learning trajectory.

3. Results and discussion

In the teaching experiment, the teacher divided the students into 5 groups with details of 3 groups (Group 1, Group 2 and Group 3) each of 4 students, and 2 groups (Group 4 and Group 5) each of 5 students. In the learning process, 4 other groups completed all the triangle construction tasks correctly, but there was one remaining group (Group 3) who answered with a construction duties D. All students in Group 3 determined 3 side lengths in Task D could not form a triangle. They did not realize that the 3 side lengths actually could form a triangle. Distribution of students' answer is described completely in Table 1.

Table 1. Distribution of students' answer from each group

Construction task	Forming a triangle				
	Group 1	Group 2	Group 3	Group 4	Group 5
Task A (4 cm, 4.5 cm, and 6 cm)	Yes	Yes	Yes	Yes	Yes
Task B (3 cm, 2 cm, and6 cm)	No	No	No	No	No
Task C(4 cm, 3 cm, and 4 cm)	Yes	Yes	Yes	Yes	Yes
Task D (3 cm, 4 cm, and 2 cm)	Yes	Yes	No	Yes	Yes
Task E (2 cm, 3 cm, and 5 cm)	No	No	No	No	No

In this article, the discussion about students'learning trajectory is represented by Group 2. The reason for choosing this group is because the students were very confident in expressing their ideas.They also cooperated solidly. They explored the construction tasks well. They made the first step by drawing a line segment. Then, they used two rulers to represent the other two line segments. They joined the two line segments to the line segment as shown in Figure 1. If 3 side lengths are connected to each other, they decided that it could form a triangle. However, if it does not happen, they decided that 3 side lengths did not form a triangle. Students' learning trajectory can be seen in Figure 2.

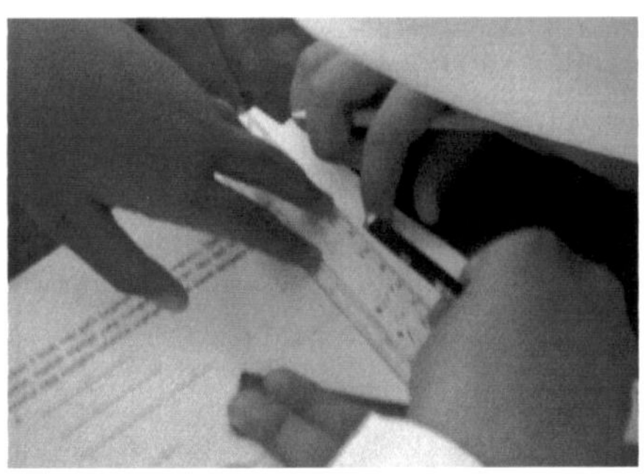

Figure 1. Students' activity in constructing a triangle

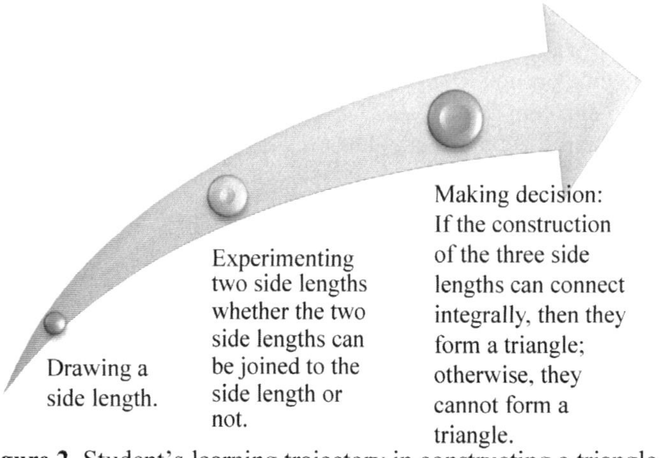

Figure 2. Student's learning trajectory in constructing a triangle

Students were enthusiastic in the discussion. In the construction task, each student gave a significant contribution. When students faced a problem/difficulty, they discussed again to find a solution. The teacher also gave some hints or questions to them. The following is the conversations protocol between teacher and students in group discussion.

Teacher : How do you solve the task of triangle construction?

Student 1: We used pencils, pens, and rulers. We drew a line segment first. We continued to use two rulers to check whether two line segments can be joined with one previous line segment. If it can connect, it means the triangle is formed. If it does not connect, yes…it cannot be made a triangle. Of the 5 tasks that have been done, Tasks A, C, and D could form a triangle, but Task B and E could not form a triangle.

Teacher : How do you guarantee that the three side lengths can or cannot form a triangle?

Student 1: Ya … ya all line segments in Task A, C, and D connect each other forming a triangle when we construct the tasks. For Task B and E, if it is drawn it cannot form a triangle. Maybe like that, Sir. How are my arguments, friends?

Student 2: Wait a minute ... Let's discuss this.

Student 3: Yes, that's right.

Student 4: But, how to solve the problem?

Teacher : Let's pay attention to the problems. Noticing the task that can or cannot form a triangle. Is there a specific pattern?

Student 2: (Student think for a long time) Oh ... ya we can investigate the group of triangle tasks that can form a triangle or not. There seems to be something unique on the sides of the triangle. Maybe we can find the pattern.

Student 1: Aha… I found the idea. That is a good point, friend. Let's check all the side lengths. For side lengths 2 cm, 3 cm, and 5 cm (Task E), all possible experiments cannot form a triangle. For example, the length of the first line segment drawn is 5 cm. This is the 2cm side and 3cm side connected to the 5cm side, it is not possible to be a triangle. When both sides (2 cm and 3 cm) are joined, it is parallel to the 5 cm side. Next, for side lengths of 3 cm, 2 cm, and 6 cm (Task B), if 3 cm and 2 cm sides are joined, the total length is 5 cm. The total side is parallel to the third side and will not be able to form a triangle. The 3cm and 6cm sides are connected. When the result is tried to be joined with the third side, it still does not connect. For side lengths of 4 cm, 4.5 cm, and 6 cm (Task A), all combinations of two side lengths are always more than the third side. Similarly, this is also valid for Task C and D.

Teacher : Wow… Great idea. So, what is the condition of the existence of a triangle?

Student 2: It is sought all possibilities, Sir. If all possibilities sum of two side lengths is more than the third side, then the triangle exists.

Student 1: It is definitely. The sum of any two side lengths has to greater than another side.

In another activity, a student in Group 2 used their index finger to represent two segments of a fused line. The student demonstrated the formation of a triangle whose sides have lengths 4 cm, 3 cm, and 4 cm. The student revealed that the triangle was anisosceles triangle. The studentconvincedthe teachers related to her argumentation and clarified her reasoning.The finding shows that the student used gestures to support and explain her reasoning. The fact is in accordance with the existing studies [20, 21] which reported that gestures have the contribution to students' reasoning.

In the process of learning, students in Group 2 initially had difficulty to determine the conclusion about the requirement of the forming a triangle given threeline segments. However, by the scaffolding of the teacher, they ccould ultimately make a conclusion that the sum of all possible two sides is more than the third side. The teacher also provided scaffolding to Group 3 regarding errors in Task D. The teacher instructed students to reconstruct TaskDcarefully. Scaffolding is very effective in overcoming thestudents' difficulties or errors. Some scholars stated that scaffolding can help students to gain success in problem-solving [22,23].

The success of teachers in providing scaffolding is inseparable from a good pedagogical content knowledge (PCK). PCK has a positive effect on students' understanding and learning process [24,25].In an effective instruction, teachers should not only master the material but also be able to apply appropriate learning strategies by using contingent dominants caffolding [26]. Teachers can previously design hypothetical learning trajectory, respond to events that occur in the classroom, overcome students' difficulties/misconceptions, and arouse students' thinking processes.

4. Conclusion

To conclude, this study found students' learning trajectory about forming a triangle given three side lengths as shown in Figure 2. Students were able to determine whether the three side lengths can be made into a triangle or not. The learning process was significant because the design of instruction supported the invention of triangle theorem. Students discovered the theorem by the scaffolding of their teachers. Students also controlled their thinking when they faced a difficulty or a mistake. Therefore, in the instruction teachers can help students to realise, regulate, and evaluate their thought process as the component of metacognition in achieving learning goals. For further studies, it is imperative to examine students' metacognition and characteristics of teachers' scaffolding in the mathematics teaching.

References

[1] Cobb P 2003 Investigating students' reasoning about linear measurement as a paradigm case of design research *Journal for Research in Mathematics Education[Monograph]* **12** 1

[2] Sarama J, Clements D H, Barrett J E, Van Dine D W and McDonel J S 2011 Evaluation of a learning trajectory for length in the early years *ZDM Mathematics Education* **43** 667

[3] McGatha M, Cobb P and McClain K 2002 An analysis of students' initial statistical understandings: developing a conjectured learning trajectory *Journal of Mathematical Behavior* **21** 339

[4] Clements D H, Wilson D C and Sarama J 2004 Young children's composition of geometric figures: a learning trajectory *Mathematical Thinking and Learning* **6** 163

[5] Kim D and Ju M-K 2012 A changing trajectory of proof learning in the geometry inquiry Classroom *ZDM Mathematics Education* **44** 149

[6] Yopp D A 2017 Eliminating counterexamples: a Grade 8 student's learning trajectory for contrapositive proving *Journal of Mathematical Behavior* **45** 150

[7] Blanton M, Brizuela B M, Gardiner A M, Sawrey K and Newman-Owens A 2015 A learning trajectory in 6-year-olds' thinking about generalizing functional relationships *Journal for Research in Mathematics Education* **46** 511

[8] Steffe L P 2004 On the construction of learning trajectories of children: the case of commensurate fractions *Mathematical Thinking and Learning* **6** 129

[9] Simon M A and Tzur R 2004 Explicating the role of mathematical tasks in conceptual learning: an elaboration of the hypothetical learning trajectory *Mathematical Thinking and Learning* **6** 91

[10] Clements D H and Sarama J 2004 Learning trajectories in mathematics education *Mathematical Thinking and Learning* **6** 81

[11] Simon M A 1995 Reconstructing mathematics pedagogy from a constructivist perspective *Journal for Research in Mathematics Education* **26** 114

[12] Battista M T 2011 Conceptualizations and issues related to learning progressions, learning trajectories, and levels of sophistication *The Mathematics Enthusiast* **8** 507

[13] MarinkovićV2015 On-line compendium of triangle construction problems with automatically generated solutions *The Teaching of Mathematics* **18** 29

[14] Kordaki M 2003 The effect of tools of a computer microworld on students' strategies regarding the concept of conservation of area *Educational Studies in Mathematics***52** 177

[15] Kordaki M and Balomenou A 2006 Challenging students to view the concept of area in triangles in a broad context: exploiting the features of Cabri-II *International Journal of Computers for Mathematical Learning* **11** 99

[16] Graham P and Chick H 2015 The 20 matchsticks triangle challenge: an activity to foster reasoning and problem solving *Australian Mathematics Teacher* **71** 9

[17] Gravemeijer K and CobbP 2006 *Design research from a learning design perspective Educational Design Researched J van den Akker et al* (New York: Routledge)

[18] Cobb P, Confrey J, DiSessa A, Lehrer R and Schauble L 2003 Design experiments in educational research *Educational Researcher* **32** 9

[19] Gravemeijer K 1994 Educational development and educational research in mathematics education. *Journal for Research in Mathematics Education* **25** 443

[20] Rofiki I, Nusantara T, Subanji and Chandra, T D 2017 Local plausible reasoning: the case of inequality tasks *Journal of Physics: Conference Series* **943** 012002

[21] Williams-Pierce C C, Pier E L, Walkington C, Boncoddo R, Clinton V, Alibali M W and Nathan MJ 2017 What we say and how we do: action, gesture, and language in proving *Journal for Research in Mathematics Education* **48** 248

[22] Bell C V and PapeS J 2012 Scaffolding students' opportunities to learn mathematics through social interactions *Mathematics Education Research Journal* **24** 423

[23] Kilic H 2018 Pre-service mathematics teachers' noticing skills and scaffolding practices *International Journal of Science and Mathematics Education* **16** 377

[24] Kellner E, Gullberg A, Attorps I, Thoren I and Tarneberg R 2011 Prospective teachers' initial conceptions about pupils' difficulties in science and mathematics: a potential resource in teacher education *International Journal of Science and Mathematics Education* **9** 843

[25] Campbell P F, Nishio M, Smith T M, Clark L M, Conant D L, Rust A H, … Choi Y 2014 The relationship between teachers' mathematical content and pedagogical knowledge, teachers' perceptions, and student achievement *Journal for Research in Mathematics Education* **45** 419

[26] Anwar, Ipung Y, Edy B I, and Abdur R A 2016 Investigation of Contingency Patterns of Teachers' Scaffolding in Teaching and Learning Mathematics *Journal on Mathematics Education,* **8** 65

IOP Conf. Series: Journal of Physics: Conf. Series **1088** (2018) 012022 doi:10.1088/1742-6596/1088/1/012022

A self-evaluation technique in improving teacher's professional development: The use of "realia" media and "wait time" strategies

W Artika[1,2] and M Saputri[2]
[1]Curriculum and Instruction Department, College of Education and Health Profession, University of Arkansas, Fayetteville, AR, Unites States
[2]Biology Education Department, Education and Teacher Training Faculty, Syiah Kuala University, Aceh, Indonesia

E-mail: wartika@unsyiah.ac.id

Abstract. Personal teaching evaluation is an influential way to improve teacher's professional development and student achievement. The purpose of this study was to assess the teaching performance of the use of "realia" media and wait time strategy found in daily life to teach science for elementary school students. The anatomy of a blue crab was selected as the topic in this study. The teaching performances were analyzed qualitatively using the modification of Florida Teacher Performance Measurement system. Students' perceptions (n=32) of teacher teaching practice were used as extra information for this personal evaluation task. The result showed a positive response to class activity that has been designed using the realia media. Also, some students suggested that the wait time should have been implemented to show a sign of the willingness to give elaboration answer. Based on the analysis, although the teaching is less effective, it could be concluded that being confident (in relation to self-efficacy) was essential to support teaching activities such as showing enthusiasm, respecting the students, and other positive attitudes. Preparing a preserved specimen of a blue crab can be an alternative way to prevent fear and unpleasant scent in observing a blue crab for early grade school students.

1. Introduction

The independent study (teaching internship) is one of core requirement that offers an opportunity to have experiences as an instructor in a CIED 3143 class (teaching science in elementary grade). Attending to a semester class schedule, getting involved in teaching process (such as class preparation and class discussion), preparing lesson plans and teaching in the class, and writing an internship report, were the overall activities that have been done for this teaching internship program. In this case, writing report of teaching activities was one of the self-assessment techniques that was used to evaluate personal knowledge and skills got during graduate study. Personal teaching evaluation is an important way to improve teacher's professional development and student achievement. Self-evaluation is a powerful technique for self-improvement as a theory of teacher practice for facilitating professional growth and teaching effectiveness that can be done through self-monitoring, audio, and video recording, and students' feedback on teaching [1] [2].

Moreover, Airasian and Gullickson stated that teacher self-assessment is the process of teacher's expertise (including scientific knowledge and social understanding) in classroom settings including (1) a teacher's reflection on classroom action, beliefs, and outcome; (2) a resulting self-critique of one's practice; and (3) if appropriate, a change in teacher understanding or future practice [3]. Besides, Tuckman considered effective teaching is the way that causes students' improvement which might be assessed using performance-based assessment approach to evaluate teacher behavior [4].

From a review of the literature, *realia* are objects or activities related to classroom teaching to the real life that can promote hands-on interactions and involvements [5][6]. While wait time was defined as the interval time (in seconds) between an instructor's question and a student response (wait time with response/wait time 1) or a decision by the instructor to move on to another topic (wait time without response/wait time 2)" (p.6) [7][8][9]. Therefore, there were little reports of the use of "creature" as real media and wait time strategy as the supporting tools to do self-evaluation in the teaching process. The purpose of these topics was to analyze teaching performance of the use of "realia" media enrichment in teaching science that is based on common things found in daily life or around us (such as home yard, grocery stores, or parks). The teaching performances have been analyzed qualitatively using the modification of Florida Teacher Performance Measurement system. In general, my teaching practice evaluation was analyzed to address the following research questions:

1. To what extent is this teaching practice effective?
2. How does the involvement in the classroom interact questioning and feedback to students' responses?
3. How does the "wait time" concept apply in this teaching practice?

2. Methods

Introduction to the external anatomy (morphology) of blue crab (*Callinectes sapidus*) was selected. The purpose of this topic was to introduce the external anatomy of one of crustacean subfamily as a basic way for the classification system. External anatomy (Dorsal and Ventral) characteristics were done through observation and discussion during the class.

There were six group of students (pre-service teachers) who took "science teaching for elementary student" course. Each group consisted of five to six students. One female and one male of blue crabs were planned to be provided for each group. Since there were not enough number of male and female crabs, each table was just provided with one female blue crab only. The only one male blue crab was only shown in the class through the table to table explanation. The class opening was done by showing the power point presentation, asking several questions and short explaining about what the students should have done for the class activity. No specific teaching instruction method applied to teach the topic. Additionally, the observation was emphasized to teach the topic. Instructional plan, including teaching material and student worksheet, was arranged in a lesson plan. Moreover, the teaching practice was recorded using two different video cameras to get more views of the classroom activities. Many researches found the effectiveness of using video analysis as teachers' teaching reflection across several studies focus including "type of reflection tasks, the guiding or facilitation of reflection, individual and collaborative reflection, video length, number of reflections and ways of measuring reflection" (p.3) [10]. The teaching activity project was analyzed qualitatively using the Florida Teacher Performance Measurement system (Table.1) [4]. Students' perceptions toward teacher teaching practice were gathered from students' notes (optional) collected at the end of classroom instruction.

3. Results and discussion

Observing personal teaching video and analyzing it without unbiased perceptions are not easy. Thus, students' evaluation towards the strengths and weaknesses of this practice teaching might be used as the second view for this personal teaching evaluation. In general, it was less than 50 minutes teaching time because it was started in the middle of class introduction and discussion, which leads to a misunderstanding of the amount of time that needs to be spent in the class. There was a technical problem in the recorder, so we missed a couple of minutes of lesson opening at the time to engage student using a book of Eric Carle "A house for a hermit crab."

Table 1. The Florida teacher performance measurement system

Domain	Effective indicators		Blue crab anatomy	Ineffective indicators
Organization and development instruction	1. Begins instruction promptly		On Time	Delays
	2. Handles materials in an orderly manner		handled	Does not handle materials systematically
	3. Orients students to classwork/ maintain academic focus		focus	Allows talk/activity unrelated subject
	4. Conducts beginning/ending review		yes	
	5. Questions: academic comprehension/lesson development	Asks single factual	Asks single factual	Process multiple questions asked as one
		Requires analysis/ reasons	recall and analysis type of questions	Process non-academic questions/ procedural questions
	6. Recognizes response/amplifies/gives corrective feedback		Giving appropriate feedback with the instructor's help	Ignores student or response/express sarcasm, disgust, harshness
	7. Gives specific academic praise		Giving praise	Uses general, nonspecific praise
	8. Provides for practice		Hands-on activities: macroscopic observation	Extends discourage, changes topic/ no practice
	9. Gives directions/assigns/checks comprehension of assignment/gives feedback		Students instruction sheet/ worksheet	Gives inadequate directions/ no homework/ no feedback
	10. Circulates and assists students		Assisting all tables with the help of the class instructor	Remains at desk/circulates inadequately
Presentation of subject matter	11. Treats attributes/examples/non-examples		Definitions were attributed with examples and pictures	Gives definition or examples only
	12. Wait time 1		3-10 seconds	
	13. Wait time 2		none	
Verbal and non-verbal communication	14. Emphasizes important points		emphasized the external anatomy of blue crab	
	15. Expresses enthusiasm/challenges students		Yes	
	16. Uses body behavior that shows interest/smiles/gestures		yes	Frowns/deadpan/lethargic
Management of student misconduct	17. Stops misconducts		Yes, with the help of a class instructor	Does not stop misconduct/desists punitively
	18. Maintains instructional momentum		yes	Loses momentum/fragments/over dwells

The analysis of teaching effectiveness based on the Florida teacher performance measurement system is described and discussed based on Table 1. The first measurement was related to organization and development of instruction. In this teaching practice case, the instruction promptly was begun by the instructor since there are two or three topic sessions to be taught in one meeting. Teaching blue crab topic was conducted in the second session. The session was opened by engaging student to recall the example of crab that the might recognize from their kid time story telling. Next, the objective of the lesson has been introduced to maintain the academic focus. The teaching materials did not handle in an orderly manner purposively. It has been planned for handling worksheet and general information of the blue crab first; then it would be ended by giving the morphology of the crab "cheating sheet" at the end of the observation. Instead of distributing materials in two steps, all teaching materials had been handled at one time because the teacher was feeling nervous. Moreover, several students criticized the quality and attractiveness of the powerpoint that the teacher made. In addition, there was no precise instruction given to students about the activities. This fact was supported by a statement from a student as follow:

> "I was sort of lost with what to do with the worksheet..., a little more instruction with that would be good" (Std-12).

Regarding questioning, academic praise, and feedback or responses, there was no application of the concept of wait-time-2 at all. Even though the wait time 1 (3-10 seconds) was applied, there was no "postponing time" that students might want to give more responses or more explanation for the question. Researchers found that wait time 2 have powerful in growing the probability of elaboration [11] and promote critical thinking skills [12]. In addition, it is found that 4-6 seconds of wait-time-1 has shown the highest frequency and depth of responses towards teacher' question [7]. Four of the thirty-two students have realized and gave suggestions regarding the "wait time" while asking questions. Some of the student's statements as follow:

> "maybe when you ask questions, allow more wait time for the students to respond!" (Std-5)
> ". . . and giving us a little bit wait time to think after the question" (Std-20)

The questions asked to students at the beginning of activity were expected to see the relationship between the observation and the classification of organisms. Those questions are categorized as "interpretation" type of question according to Blosser [13]. In addition, Shahrill reviewed that asking effective questions, applying "wait few seconds time," structuring the teacher questions, being flexible, being open and non-judgmental, and being honest in showing a keen interest of students' response towards questions were strategies to be an effective teacher in a classroom [14]. Another significant weakness of this teaching practice was an uncertain assessment that has been assigned to this class. The observation towards students' practical activity, a comprehensive task given, and feedback given was not enough to provide the evidence of learning assessment.

The second measurement was related to a presentation of subject matter. There were none of the criteria that suggested by the Florida Teacher Performance Measurement system has been done in the class for this domain. In addition, students suggested that teacher could give more explanation or story about the blue crab such as the habitat, mating behavior, or type of food the blue crab eats. Students also needed to have more "space" to discuss their observation result and discuss further some information that might be useful for them. The last measurement was related to verbal and non-verbal communication and management of student misconduct. For this point, the enthusiasm for doing the class activity and stopped misconduct when it was needed has been done. Yet, the low English proficiency teacher had, especially speaking with correct grammar, might influence the success of the communication during the teaching practice. English proficiency is also needed to choose the appropriate words to explain certain information to avoid misconception.

In general, students appreciated and welcomed to this teaching practice time. Most of them gave positive responses towards the topic and class activity that have been designed. Several interesting responses from students such as:

> "this activity was great! I've never looked at the anatomy of a sea creature, so it was a fun, new experience for me!" (Std-14)
> "good at explaining things and walking around the classroom to ask us further questions making us think more into the activity" (Std-2)

Therefore, based on the analysis result, it might say that this teaching practice was still far to be categorized as effective teaching. Cultural differences might also be a limit to understand the language (ea. student responses) and to have effective communication between the students and teacher such as the time to get students' attention to focus on the lesson. It also needs to improve questioning behavior such as suggested in McComas [15]: asking a variety of questions, using appropriate pauses and wait time, wording question clearly, encouraging students to expand on theirown responses and asking questions that permit multiple responses. Questioning also plays an important role in the instructional strategy to assess learning progress, to enhance conceptual understanding, to recognize prior conceptions and misconceptions, to help the teacher in probing and redirecting, and to empower students in active learning [16]. Nevertheless, this teaching practice and self-evaluation have given a lot of feedback to improve further teaching practice.

Furthermore, concerning the revision of this teaching, the class instructor suggested a well-prepared alternative strategy when a certain realia media could not be provided due to any unpredicted conditions. She also suggested (to those pre-service teachers) that a teacher should not show any negative expressions while they are using unpleasant teaching tools in front of their students. In addition, an expert in marine invertebrate and ecology has provided a suggestion for another way to use crab as the realia media as follow:

> "The introduction of a blue crab anatomy can be done through direct observation. However, the difference in student's school level should also be taken into account, especially related to the types of media used. As an alternative, for elementary school children at 4 -6 years of age, teachers can provide a specimen that has been preserved with Alcohol 70% to reduce students' fears, unpleasant scent, and the observation time concern. Students will be able to observe a blue crab easily since it has a very prominent carapace compared with its abdomen and has a bluish color and the bright white spots on the males, while the females are greenish-green with a rather dull white patch".

4. Conclusion

Based on this evaluation, being confident in relation to self-efficacy is important to support teaching activities such as by showing enthusiasm, respecting the students, and other positive attitudes. Yet, some students suggested that the wait time should have been implemented to show a sign of the willingness to give elaboration answer. Although the teaching is less effective, a teacher should also consider other essential components in teaching such as questioning, providing feedback, wait time, and instructional design planning that is important to support more profound understanding and long-life learning.

References

[1] Rose J A and Bruce C D 2007 Teacher self-assessment: A mechanism for facilitating professional growth *Teaching and Teacher Edu.* **23** 146
[2] Schwartz M nd Self-evaluation of teaching Assessed at https://www.ryerson.ca/content/dam/lt/resources/handouts/ SelfEvaluation.pdf

[3] Airasian P W & Gullickson A 1994 Examination of teacher self-assessment *Journal of Personnel Evaluation in Education* **8** 195

[4] Tuckman B W 1995 Assessing effective teaching Peabody *Journal of Education* **70** 127

[5] "realia" *Merriam-Webster.com 2018* https://www.merriam-webster.com (1 May 2018)

[6] Onasanya S A 2004 Selection and utilization of instructional media for effective practice teaching Institute *journal of studies in education* **2** 127

[7] Larson L R & Lovelace M D 2013 Evaluating the efficacy of questioning strategies in lecture-based classroom environments: Are we asking the right questions? *Journal on Excellence in College Teaching* **24** 1

[8] Tofade T, Elsner J & Haines S T 2013 Best practice strategies for effective use of questions as a teaching tool American *journal of pharmaceutical education* **77** 155

[9] Maroni B 2011 Pauses, gaps and wait time in classroom interaction in primary schools *Journal of Pragmatics* **43** 2081

[10] Tripp T and Rich P 2012 Using video to analyze one'sown teaching. *British Journal of Educational Technology British Journal of Educational Technology* **43** 678

[11] Rowe M B 1986 Wait time: Slowing down may be a way of speeding up! J *Teacher Edu* **37** 43

[12] Saeed T, Khan S, Ahmed A, Gul R, Cassum S & Parpio Y 2012 Development of students' critical thinking: the educators' ability to use questioning skills in the baccalaureate programmes in nursing in Pakistan *Journal of the Pakistan Medical Association* **62** 200

[13] Blosser P E 1991 How to Ask the Right Questions (Washington, DC: NSTA)

[14] Shahrill M 2013 *International Journal of Humanities and Social Science* **3** 224

[15] McComas W F nd Questioning skills (compilation of comments from Good and Brophy "Teacher-Student Relationships: Causes and Consequences," and Dillon "Asking Better Questions"

[16] Crowe M and Stanford P 2010 Questioning for quality *Delta kappa gamma bulletin* **76** 36

Learning fraction through the context of Asian Games 2018

R I I Putri and Zulkardi

Universitas Sriwijaya, Bukit Besar, Palembang, Indonesia

E-mail: ratu.ilma@yahoo.com

Abstract. In August 2018, Indonesia will host Asian Games in two cities namely Jakarta and Palembang. As the researchers stay in the city of Palembang, we use this event as the context for designing mathematics lesson materials. This paper shares the results of the design research method, which aims to developing Learning Trajectory for primary school students. Three phases of design research were used namely preparing for an experiment, teaching experiment, and finally retrospective analysis. As the result, this research produces Local Instructional Theory (LIT) on fraction through the contexts of Asian Games 2018 sports namely swimming, shot put, rowing, running, and hurdles. The swimming context could stimulate students' informal knowledge about the meaning of fractions. By using shot-put context can assist students in learning the subtraction of fractions. Furthermore, running context can help students understand the division of fractions. Meanwhile, rowing context can help students understand addition and subtraction of fractions. Lastly, the context of hurdles can help students to realize their knowledge about the concept of multiplication of fractions with natural numbers. Therefore, this study contributes positively to the various parties on how to teach fractions using the contexts in the learning process of Indonesian Realistic Mathematics Education.

1. Introduction

Mathematics is not only material transferred by the teacher to the students [1, 2]. Students should not be treated as passive recipients who only receive materials with using formula and procedure to solve a problem, but students should be given opportunities and guided to the situation to reinvent mathematics concept with their ways. One of the branches of mathematics that deal with the number is the fraction. Fraction is one of the complex concepts, but students need it in the primary school [3]. Many students in the world have difficulties in learning problems that relate to the fraction [4]. In the many cities, the average of students did not have conceptual knowledge about concept, and the lesson was lack of daily lives of students [5].

According to Streefland [6], there are five levels on learning fractions based on the student's level, managing the strategy of teaching fraction, ordering the way of operation of the equation infraction, students operate fraction on their ideas with accuracy. These are relevant with the curriculum that learning not only about textual but more on the development of thinking. This skill will help students use thinking mathematically that transferable in every condition. As a professional teacher, she has to use various contexts, starting from informal to the things more formal. Instructional activity will interest in and meaningful for students and started with the realistic context which experientially real to the students.

In Curriculum 2013 stated that mathematics subject is important to give to the students starting from Primary School in order to make students competence in thinking in logically, analytics, systematic, critics and creative and working collaborative [7]. These are relevant to the focus of the learning process. Students were expected to develop their understanding of the material that there are learning so they will more understand the given materials. Consistent with that, PMRI has character using students contribution [8]. PMRI was adapted from learning theory called Realistic Mathematics Education (RME) founded by Frudenthal [9] in the Netherlands. He suggested that mathematics should be connected to students daily lives [10]. Use of context and PMRI in designing lesson Indonesia has been conducted since 2001 [11]. It has been used in many researchers in efforts to improve attitude, motivation and knowledge of students either in the form of research by lecturers, student teachers and master students [12-16].

Learning mathematics in the school suggested connecting learning materials to the daily life. This suggestion good in the family environment, school, social with aims at finding the meaning of the materials for their daily life [17]. The role of context will make mathematics more accessible for students. Therefore, learning materials which support students improve their competencies and communication skills are needed. The learning process is not only an independent process or individual but also as social process [18, 19]. Here, interactivity is one of the five characteristics of RME or PMRI [20]. Interactivity stresses social interaction among students to support an individual process of each student. Therefore, the social norm is needed in the learning process.

The method of communication in mathematics is also supporting to develop meaning, publish ideas, and giving students opportunities to improve their understanding [17]. Therefore, activities for students are needed that support students to connect abstract thinking to more realistic so they can connect abstract thinking of students into the more real so mathematics understanding can be reached. Also, activities which used Indonesian local cultures are also has been shown in supporting students understand mathematics concept [21, 22]. However, contexts of sports in the event such as Asian Games are not used so far for research especially in mathematics for learning activities in primary school.

Asian Games 2018 is an international event which will be conducted in Jakarta and Palembang. It will be conducted from August 18th to September 2nd, 2018. This event supported by the president Republic of Indonesia with Instruction No.2 in 2016 with the aims success prestige and success in hosting the event. One of the multiplier effects are the Minister of Education and Culture to facilitate the students to join the game and to socialize the event. Therefore, activities for students using the context of Asian Games is significant to be designed. In Asian Games 2018, 37 sports will have competed. Since 2016, the citizen of South Sumatera have to make people Asian Games fever [23]. It shows that how important to socialize the event Asian Games 2018 through learning in the schools including mathematics in Primary Schools.

Based on the description above, therefore, the researcher interested into design learning realistic materials using Asian Games 2018 context in the primary school. The results of this research are expected as meaningful learning materials that support students develop mathematical thinkings from informal to formal. Therefore, the purposes of this paper are to know the role of Hypothetical Learning Trajectory realistic mathematics for primary school students using context sports in Asian Games 2018 and to produce Learning Trajectory on realistic mathematics for primary school students which are valid and practice using context sports in Asian Games 2018.

2. Method

This research used design research with the type of validation study as the research method. In collaboration with PMRI Primary Schools, data were collected using observation, interview, video camera, and test. Using design research aims at developing local instruction theory based on the theory-driven and empirically based through collaboration between researcher and teacher to improve relevance between research and regulation in educational practice [24].

This research produced five learning trajectory materials on fraction. Three phases of design research used namely preparing for the experiment, the experiment and the retrospective analysis [25, 26]. The research subjects are five PMRI schools with total of 170 students get involved.

IOP Conf. Series: Journal of Physics: Conf. Series **1088** (2018) 012023 doi:10.1088/1742-6596/1088/1/012023

3. Results and discussion

Using all steps of the design research, this research produced five Learning Trajectory (LT) on the materials subtraction with four operations that is the division, addition, subtraction, and multiplication of fractions. The following are the explanation of how to design the LT.

3.1 Preparing for the experiment

In this phase, the researcher discusses characteristics of students with mathematics teacher who performs as a teaching model. Based on the interview with the teacher model, the fraction is learned in the grade four. Also, the researcher explores literature on the fraction concepts and the 2013 curriculum edition revision. In addition, five contexts used namely swimming, shot-put, rowing, running and hurdles. Therea are several context implemented in learning process, such as the introduction of the fraction through rowing context (Figure 1), the addition of fraction using swimming context (Figure 2), the subtraction of fraction through shot-put context (Figure 3), the multiplication of fraction using integer and hurdles' context (Figure 4), and worth fraction through Atletic's context (Figure 5). Furthermore, this paper only explains the topic of adding fraction by using swimming context.

Figure 1. Rowing context [27].

Figure 2. Shot-put's context.

Figure 3. Swimming's context.

Figure 4. Hurdles' context [28]

Figure 5. Running context [29]

3.2 The design experiment

3.2.1 Pilot experiment

The pilot experiment aims at as a bridge between the first phase and teaching experiment. This first cycle of the experiment tried-out the first prototype of the HLT. The experiment was followed by six students in the non-research class, which have three levels of performance. The researcher discussed the characteristics of students. Besides discussion, students were interviewed about their opinions on the learning process that was used for the restropcctive analaysis. This paper focuses on the addition of fraction using the context of swimming.

3.2.2 Teaching experiment

This phase is the implementation of the revised HLT. The goal is to explore strategies and student thinking in the learning process. During the investigation, student's conjectures can be used for the next lesson. During this phase, the role of the researcher is as both a teacher and an observer. At the end of the teaching and learning process, the researcher and the teacher do a reflection.

Results of the first phase are five topics with some activities on each. In swimming's context, three activities are designed that is understanding the fraction, adding the fraction, and fractions with same denominators or not. They do not understand the meaning of 'part' of a whole. In addition, they do not understand that two fractions with different denumerator should be made the same before operated [16, 30]. The following shows a student has difficulties in solving a task that is related to adding of fraction. The task is as follow.

Indah and Silvy swim in the pool, which has length, is 30 meter. After 20 seconds, Indah has been swimming 12 meters while Silvy 15 meters. What is the total distance both of them?

From the task, students use model and understand the concept of addition of fraction with the same as well as the different denominator. Then, adding the fractions with different denominator using the context of swimming. In solving the problem, the activity used fraction bar as the media as well as the use of number line. In the activity with the swimming context, students have to add the distances reached by the athletes. First, students used fraction bar. Then, in the second activity, students used the number line. Students related to the addition operation of fractions with the different denominator.

Using swimming context, students have to add the distance of both students. In the first activity, students use the fraction bar. Then, in the second, students use the number line. Students related to the addition operation with different denominator. According to Van De Walle, *et al* [31], the long model with bar can support the student to invent components of the fractions. Figure 6 shows the right solution of the student in adding the different denominator of two fractions.

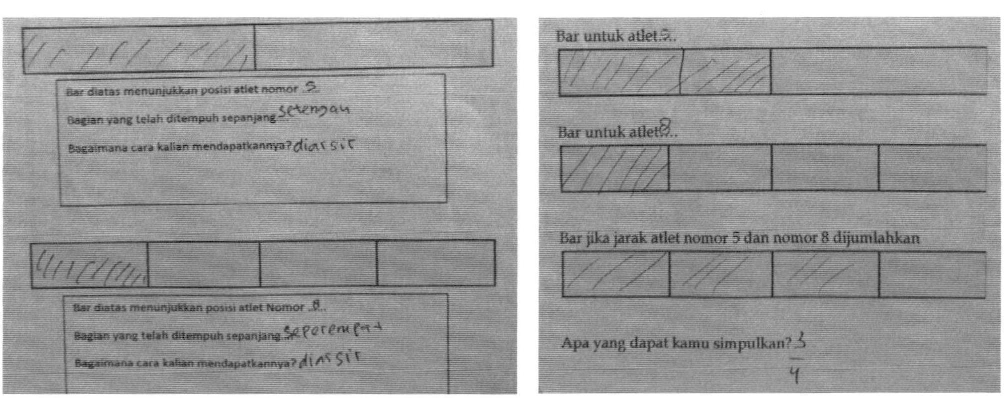

Figure 6. A students's solution.

The student adds both fractions using the given fraction bar. Then, he moves the results to the given worksheet. This activity is relevant with the HLT. The following discussion shows the communication among students and the teacher about the fraction with different denominator.

Teacher : Kenapa nomor 5 [why number 5?]
Student 1 : Karena dia mau duluan sampai [because he the leader]
Student 2 : 50 meter
Teacher : Berapa bagian kolam? [how much the part of pool]
Student 2 : 100 meter... 50 meter.. eh $\frac{1}{2}$
Teacher : Dapat darimana?[how to get it?]
Student 2 : Dari 100 meter itukan 50 meter. Setengah dari 100, 50 meter.
Teacher : Bagaimana mendapatkannya? [how to get it?]
Student 2 : 100 meter kan dibagi 2 sama dengan 50 sama dengan $\frac{1}{2}$
Teacher : Lalu bar kedua menunjukkan atlet nomor? [Then the second bar for who?]
Student 1 : 8
Student 2 : Atlet Jepang juga [Japan Athlete too]
Student 2 : Korea
Teacher : Korea, pinter! Berarti nomor berapa?[Korea,smart! Then what number?]
Student 2 : 8
Teacher : Bagiannya berapa?[So, what is the part]
Student 1 : 25 meter
Student 2 : 100 dibagi 4 (100 divided by 4]
Student 2 : 25 meter
Teacher : Dapatnya darimana? [hwo to get?]
Student 2 : 50 dibagi 2 lagi (50 divided by 2 again]
Teacher : Berarti bagiannya berapa?[So, what is the part?)
Student2 : $\frac{1}{4}$

In the next activity, students solve the different task, yet still the swimming context. Students added the two distances. The distances of both athletes were known using parts of the pool. In this task, th part of the pool using different fraction with different denominator Then, students use the number line. It stated in the solution sheet. Students only have to measure. Example of student solution can be seen in the Figure 7.

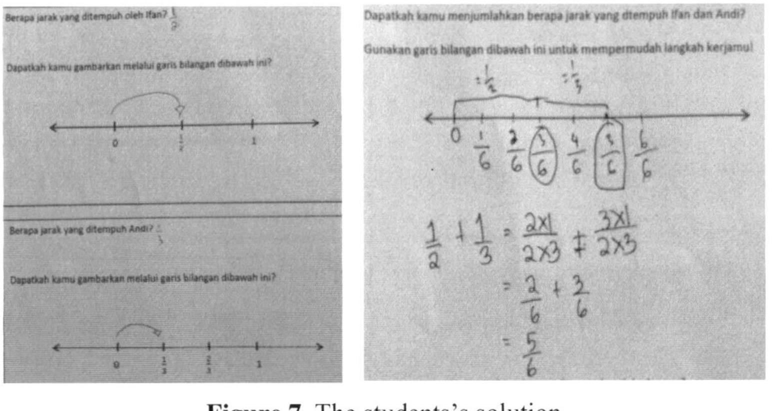

Figure 7. The students's solution.

Figure 7 shows students are able in doing addition of fractions using the number line. They are able to add fractions using the number line. In addition, a group has been understood that make the denominator are similar that can be founded by using the Least Common Multiple (LCM) of both fractions. Students knew that addition of fraction with different denominators could not directly add. One has to make the denominator are the similar. It is consistant with the HLT. For more clear, it can be seen in the following transcript.

Student$_1$: *Ifan* $\frac{1}{2}$
Student$_2$: $\frac{1}{2}$ *panjang kolam [length of the pool]*
Student$_3$: *Sama seperti yang tadi itu [same with the previous]*
Student$_2$: *Garis! [line!]*
(Student menggaris pada garis bilangan- Student make a line in the number line)
Teacher : *Lalu? [then]*
Student$_3$: *Andi itu* $\frac{1}{3}$ *[Andi that 1/3]*
Student$_1$: *Seperti itu? [Like that?]*
Teacher : *Oke... Jadi itu dapat* $\frac{5}{6}$ *dari mana? [Oke....then where 5/6 comes from]*
Student$_2$: *Jarak yang ditempuh Ifan* $\frac{1}{2}$, *jarak Andy* $\frac{1}{3}$, *dijumlahkan [the distance of Ifan]*
Teacher : $\frac{1}{2}$ *sama dengan* $\frac{3}{6}$ *itu maksudnya?*
Student$_1$: $\frac{2}{6}$ *ditambah* $\frac{3}{6}$ *[2/6 add by 3/6]*
Student$_1$: *Kalau penyebutnya beda harus disamakan dulu penyebutnya [if the denominator is different then they have to be the same first]*

In the second transcript, the students are able in pointing a point in a number line. When students have to add the two distances, students confused. This can be happened due to students have to find a value of the fraction in the decimal. Then, students order the fractios in the number line and add the two distances of the two atlhetes.

Then, students present their solutions in the front of the class. After classoom discussion, the teacher support students to conclude the lesson that adding fractions with different denominators cannot directly be added but one has to make the denominator are the same. All strategies can be using bar, the number line or LCM. Finally, the teacher closes the lesson.

The student's solutions are consistent with the HLT. Here one can see that sequences of instructional activities using PMRI theory shows how the characteristics of PMRI as basic of the learning process on each its activity.

There are three principles of RME or PMRI and five characteristics of RME or PMRI that are used in designing lesson materials [13, 15, 20, 32, 33]. Based on the principle of Guided reinvention and Didactical phenomenology, students were guided by the teacher using swimming's context in the Asian Games 2018. As a result, students reinvent the concept of fractions such as nominator and denominator.

3.3 The retrospective analysis

In the topic fraction using swimming context all data that were collected in the experiment were analyzed. Learning process in the first activity is consistent with the learning process in the first HLT. In other words, we can say many HLT are consistent with ALT. There are some differences such as to the students should be stressed about the meaning of fraction as a part of a whole. Students have difficulty in differentiating between part and distance. In HLT, the expected of the solution is the form of the fraction, but a number of students have the solution in the form of only the distance. This is not consistent with the assumption of the research in the beginning about adding fraction.

The first problem is about addition of problem with different denominators. Students asked to add the two distances reached by the two athletes using fraction bar. Then, students conclude their solutions. In the beginning, students are confused in using fraction bar with different denominators. However, after reread the guidelines in the worksheet, finally, students are able solve the problems. The results are consistent with the plan of HLT.

4. Conclusion

From activities can be shown that students can solve problem of adding fraction. Students work right on the track consistent with the hypothesized conjecture. Using PMRI, various solutions emerged in the student's solutions. Use of bar model in PMRI has a role in the learning of addition of fraction. Using bar model in the swimming context supports students in solving problem on fraction.

Acknowledgements

Researchers expressed acknowledgement to DIKTI-Indonesian Directorate General of Higher Education that has funded research this grants graduate.

References

[1] Gravemeijer K 1994 *Developing Realistic Mathematics Education* (Utrecht: CD--β Press)
[2] Gravemeijer K 2010 Realistic mathematics education theory as a guideline for problemcentered, interactive mathematics education *A Decade of PMRI in Indonesia* eds R K Sembiring, K Hoogland, and M Dolk (Utrecht: APS International) pp 41-50
[3] Mamede E 2010 Issues On Children's Ideas of Fraction when Quotient Interpretation is Used 1-10
[4] Siegler R S, Lisa K F, Drew H B, and Xinlin Z 2013 Fraction: The new frontier for theoris of numerical development *Trend in Cognitive Sciences* **17** 13
[5] Haris D, Putri R I I 2011 The role of context in third grader's learning of area measurement *Journal on Mathematic Education* **2** 55
[6] Streefland L 1991 *Fractions in realistic mathematics Education* (Dordrecht: Kluwer Academic Publishing)
[7] Kemendikbud 2013 *Implementasi kurikulum 2013* (Jakarta: Kementerian Pendidikan dan Kebudayaan)
[8] Soedjadi R 2007 Inti dasar-dasar pendidikan matematika realistik Indonesia *Jurnal Pendidikan Matematika* **1** 1
[9] Frudenthal H 1991 *Revisiting Mathematics Education-China Lectures* (Dordrecht: Kluwer Academic Publisher)
[10] Sembiring R, Hoogland K, and Dolk M 2010 *A Decade of PMRI In Indonesia* (Utrecht: APS)
[11] Zulkardi and Putri R I I 2006 Mendesain sendiri soal kontekstual matematika *Prosiding Konferensi Nasional Matematika ke-13 (KNM13)* (Semarang: Indonesia) p 1
[12] Zulkardi and Kohar A W 2018 Designing PISA-Like mathematics tasks in Indonesia: Experiences and challenges *J. Phys.: Conf. Ser.* **947** 012015
[13] Putri R I I and Zulkardi 2018 Higher-order thingking skill problem on data representation in primary school: a case study *J. Phys.: Conf. Ser.* **948** 012056
[14] Putri R I I and Zulkardi 2017 Fraction in shot-put: A learning trajectory *AIP Conference Proceedings* **1868** 050005
[15] Sumarto S N, Van Galen F, Zulkardi, and Darmawijoyo 2014 Proportional reasoning: How do the 4th graders use their intuitive understanding? Int. Educ. Studies **7** 69
[16] Gunawan M S, Putri R I I, and Zulkardi 2017 Learning fractions through swimming context for elementary school students *Proceeding in the Fifth South East Asia Design/Development Research (SEA-DR) International Conference 2017* (Banjarmasin: Universitas Lambung Mangkurat) p 61
[17] Putri R I I 2011 Improving Mathematics Comunication Ability of Students in Grade 2 Through PMRI Approach *Paper presented in International Seminar and The Fourth National Conference on Mathematics Education* (Yogyakarta: Universitas Negeri Yogyakarta)

[18] Cooke B D and Buchholz D 2005 Mathematical communication in the classroom: Teacher makes a difference *Early Childhood Education Journal* **32** 365

[19] Lopez L M and Allal L 2007 Sociomathematical norms and the regulation of problem solving in classroom multicultures *Int. J. of Educ. Research* **46** 252

[20] Zulkardi 2002 Developing a learning environment on realistic mathematics education for Indonesian student teachers *Ph.D Thesis University of Twente* (Enschede, the Netherlands)

[21] Putri R I I, Dolk M, and Zulkardi 2015 Professional development of PMRI teachers for introducing social norms *IndoMs-JME* **6** 1

[22] Mumu J, *et al* 2018 Construction and reconstruction concept in mathematics instruction *J. Phys.: Conf. Ser.* **943** 012011

[23] Tribun Sumsel 2016 Masyarakat sumsel wajib demam asian games 2018 Available online at http://http://sumsel.tribunnews.com/2015/02/26/tahun-2016-masyarakat-sumsel-wajib-demam - asian-games-2018

[24] Gravemeijer K and van Eerde D 2009 Design research as a means for building a knowledge base for teachers and teaching in mathematics education *The Elementary School Journal* **109** 510

[25] Gravemeijer K and Cobb P 2006 Design research from the learning design perspective *Educational Design R esearch* eds Van den Akker J, Gravemeijer K, McKenney S, and Nieveen N (London: Routledge) p 15

[26] Bakker A 2004 *Design Research in Statistics Education on Symbolizing and Computer Tools* (Amersfoort: Wilco Press)

[27] Nasution M F, Putri R I I, and Zulkardi 2017 Rowing sport in learningfractions of the fourthgrade students *Journal on Mathematics Education (IndoMS-JME)* **8** 69

[28] Rahayu C, Putri R I I, and Zulkardi 2017 Multiplication of fraction with natural number by using hurdles *Proceeding in the Fifth South East Asia Design/Development Research (SEA-DR) International Conference 2017* (Banjarmasin: Universitas Lambung Mangkurat) p 43

[29] Roni A, Zulkardi, and Putri R I I 2017 Sprint context of asian games in the division of fractions *Proceeding in the Fifth South East Asia Design/Development Research (SEA-DR) International Conference 2017* (Banjarmasin: Universitas Lambung Mangkurat) p 22

[30] Halim N L A, *et al* 2018 Teaching strategies in the learning of highest common factor and lowest common multiple *J. Phys.: Conf. Ser.* **943** 012041

[31] Van De Walle J A, Karp K S, and Williams J M B 2013 *Elementary and Middle School athematics Teaching Developmentally* (United States of America: Pearson Educations)

[32] Putri R I I and Zulkardi 2018 Higher-order thingking skill problem on data representation in primary school: a case study *J. Phys.: Conf. Ser.* **948** 012056

[33] Castle K and Needham J 2007 First Graders' Understanding of Measurement *Early Childhood Education Journal* **35** 315

IOP Conf. Series: Journal of Physics: Conf. Series **1088** (2018) 012024 doi:10.1088/1742-6596/1088/1/012024

The teaching and learning of addition and subtraction of integers through manipulative in Brunei Darussalam

N Sahat[1], K A Tengah[2] and R C I Prahmana[3]

[1]Rimba Secondary School, Ministry of Education, Perumahan Negara Kampong Rimba, Gadong, BE 3319, Brunei Darussalam
[2]Sultan Hassanal Bolkiah Institute of Education, Universiti Brunei Darussalam, Jalan Tungku Link, Gadong, BE 1410, Brunei Darussalam
[3]Universitas Ahmad Dahlan, Jl. Pramuka Kav. 5, Yogyakarta 55161, Indonesia

E-mail: khairul.tengah@ubd.edu.bn

Abstract. This action research studied on how the use of manipulative in the teaching and learning of integers could affect students' performances in adding and subtracting integers, involving Year 9 students from two mixed ability classes of the same government secondary school in Brunei. The intervention lessons used different colored counters to represent different signs of integers, followed by the adding and removal actions of these counters to mirror the addition and subtraction operations, with the added concept of zero-pair. Any improvement or beneficial outcome of the prior lesson was used to adapt the initially prepared following lesson as part of the Design Research process. The data collection consisted of pre-test and post-test. With the use of descriptive statistics and Wilcoxon Signed Ranked Test, this study concluded that there was the significant improvement on students' performances in the post-test and the effect size showed that the intervention gave big impact towards students' learning. Item analysis indicated that the intervention improved correct attempts in questions involving both addition and subtraction, with the more significant increase in the latter. The overall mean scores of correct attempts in the post-tests also showed that students scored higher on the addition of integers (89.7%) compared to subtraction of integers (81.0%).

1. Introduction

The poor performance of students in mathematics has become a source of worry to many. Many experts proposed several solutions to this problem, yet the problem persisted, believing that poor knowledge in basic mathematics among students is the key root to the problem [1-3].

One problem that students are facing when dealing with integers is that they are confused with signs and operations of the integers, which make them struggle in computing the integers [3]. Students taught using number line model can easily get confused with which direction to move when adding or subtracting integers since students always thought that adding means going forward and subtracting means going backward. It is not always true when the second addend involved negative integers [4]. However, it is unfortunate that this model has been widely used by teachers in Brunei [4]. Using algorithm as a teaching pedagogy is also popular yet ineffective, as some students could not remember the algorithm [4].

Researches revealed that students faced problems in understanding the concept in adding and subtracting integers [5-9]. A lack of foundation in mathematics, which includes the computations of integers, caused many Brunei upper secondary students unable to solve algebraic equations [10]. Moreover, a local study discovered that more than 50% of the Year 7 students in one government school failed in the computation of integers' pre-test [4].

The instructional approach that emphasised only on the mathematics procedural skills was one of the causes to the failure of computing integers among Bruneian students [11-13]. Brownell stated that repetition or computational drill does not lead to understanding [14]. Conceptual understanding plays a vital role in developing procedural skills. Through constructivist perspective, using manipulatives can actually enhance students' mathematics conceptual understanding. Through concrete representation that the manipulatives offer, it provides students the concrete experience in building the abstract concepts of mathematics [15].

Several researches, which made use of counters and zero-pair effect in the teaching and learning of integers, have proven to be effective in improving students' achievement [4, 16, 17]. Zero-pair effect involves neutralization or cancellation of positive with negative integers. Take red and blue counters as an example to represent positive and negative integers respectively. If a red counter and a blue counter come together, the positive and the negative signs cancel each other, creating zero. This means that if there are six red and six blue counters, it will result in zero, despite total twelve counters.

When doing subtraction, it involves the removal of positive or negative integers. If there are not enough positive or negative integers to be subtracted, zero-pair needs to be added in since zero pair will not cause any effect to the equation. For example, in $-3-(-9)$, there are initially three blue counters but since there are not enough blue counters to remove 9 of the blue counters, 6 zero-pairs need to be added in order to have enough blue counters to be removed. If students are already familiar with the manipulation of integers, they can advance themselves in computing the integers [18].

A previous research on the teaching and learning on the four computations of integers (addition, subtraction, multiplication and division) was done to a lower secondary school in Brunei [4]. The study used jar model as a teaching pedagogy in computing integers, which is similar concept of using counters, and made use of zero-pair effect. The result revealed that students significantly improved in adding and subtracting integers but not for multiplying and dividing integers. This study, stemmed from that research, is fine-tuned and expanded as follows:

- The study involved upper secondary students concentrating only on addition and subtraction of integers.
- It focused on the computation of single and two digit integers.
- It differentiated and emphasised on the terms 'positive' and 'negative' for signs, and 'plus' and 'minus' for operations.

The general purpose of this study is to investigate the effectiveness of using manipulatives in the teaching and learning of integers by measuring students' performances on the addition and subtraction of integers. Hence, the research question used to guide this study: *How does the teaching and learning of integers using manipulatives affect students' performances on addition and subtraction of integers*?

2. Method

This study adopted quantitative method, with convenient sample of fifty students from two Year 9 classes from a co-ed Brunei government school in Brunei-Muara district. All students attempted the pre-test and post-test, and attended all the intervention lessons to be part of the data in this study.

Both pre-test and post-test consisted the same open-ended twenty-five items, covering different combinations of operation and signs of the two numbers (Table 1). However, addition and subtraction of zero integer were not included in the test item but covered in the intervention exercise. Calculators were not allowed in both tests and essential workings needed to be shown. A score of one would be given to correct response and zero for incorrect answer or unattempt question.

Table 1. Different combination of questions for addition and subtraction of integers

Operation	Sign of numbers	Item No.
Addition	positive & positive	1, 13
	positive & negative	2, 3, 4, 5, 14, 16, 17, 18
	negative & negative	6, 19
Subtraction	positive & positive	7, 8, 20, 21
	positive & negative	9, 10, 22, 23
	negative & negative	11, 12, 24, 25

The test instrument was analysed for reliability via test-retest to a group of Year 9 students of a mixed ability class not involve in the main study from the same school. Pearson's correlation to test for reliability obtained value of 0.866. This indicated that the test items were reliable, for correlation greater than 0.8, the test items were considered reliable [19]. The pre-test and post-test was also validate by three experienced teachers teaching that level. Since the test instrument was reliable and valid, no modification the items were necessary and all items used in main study.

The intervention lessons for this study consisted of seven lessons, focusing on introduction of and discovering rule of addition and subtraction of different combinations of integers using manipulatives (Table 2). Notes and supporting exercises were also included in each lesson. In addition, any improvement or useful outcome of the prior lesson was added or adapted to the following initially prepared lesson as part of the Design Research process. The lesson intervention lasted 5 hours in total.

Table 2. The intervention lessons conducted in this research study

Lesson	Focus	Materials used	Duration (minutes)
1	Differentiation and reinforce difference between Signs and Operations	Activity Worksheets	30
2	Introduce manipulatives (single and base ten) and reinforce concept of Zero Pairs Effect	Manipulatives, Activity sheets,	30
3	Introduction and activities using manipulative to Addition of Integers	Manipulatives, Task and Questions sheets.	60
4	Guided discovery of rule in Adding Integers	Manipulatives, Question sheets	30
5	Introduction and activities using manipulative to subtraction of integers	Manipulatives, Task and Questions sheets.	60
6	Guided discovery of rule in subtracting integers	Manipulatives, Question sheets	30
7	Reinforcement activities and exercises related to addition and subtraction of integers	Work sheet	60

The intervention lesson involved the adding and removal of manipulatives to represent the concept in this topic. Two different colored Lego (called counters) to represent the positive and negative signs of numbers, and single counter to represent unit and a 10-stacked counter to represent base-ten (Figure 1). For the activities, adding action to starting counter is used to represent addition, while removal represented subtraction. Zero-pair effect is used and reinforced throughout the intervention.

(a) (b)

Figure 1. Single cube counters (a) and base-ten cube counters (b)

Initially, manipulatives was planned to be used throughout the intervention lesson. However, some students attempted some questions using drawings instead of manipulatives in the addition and subtraction operation lessons, thus drawings were included and encouraged as transition phase of manipulatives to written work.

In this study, the pre-test results act as a baseline performance of students' knowledge prior to the intervention; while post-test measure the impact of interventions on students' performances. Paired sample t-test was initially intended to compare the pre-test and post-test mean scores. However, since the students' results did not satisfy one condition of t-test measure (difference between the pre-test and post-test scores' were not normally distributed), Wilcoxon Signed Ranked Test was used instead to determine whether the two mean scores of the pre-test and post-test were significantly different in students' overall performance. The effect size of the intervention was also measured. In addition, item-by-item analysis was also done to compare overall percentage mean scores of six different combinations of test items to determine which combination of questions were greatly improved by the intervention lessons.

3. Results and discussions

Simple descriptive statistics (Table 3) confirmed there was difference in pre-test and post-test performance before and after the intervention. The mean score increased from 13.18 (*SD*=5.706) in the pre-test to 21.42 (*SD*=5.706) in the post-test.

Table 3. Overall pre-test and post-test scores of the sample based on total marks of 25

Test type	Total	Mean ($\overline{X}$)	Std. dev.(*SD*)	Lowest score	Highest score
Pre-test	50	13.18 (51.75%)	5.706	4	24
Post-test	50	21.42 (84.79%)	4.536	10	25

Out of fifty students in the sample, the highest pre-test mark was 24, scored by one student only. After the intervention, sixteen students obtained a full mark of 25. The lowest mark obtained in the pre-test was 4, whilst in the post-test, 10 was the lowest mark. Only four of the fifty participants scored lower in their post-test compared to pre-test, and one scored the same in both the pre-test and the post-test. Overall, the results suggest that students performed better in the post-test than the pre-test.

The Wilcoxon Signed Rank Test result (Table 4) revealed significance level, $p \leq 0.05$, which indicated that the difference in score between the pre-test and the post-test was significant, with large effect size (*ES* = 0.5788). For effect size, criteria of 0.1 indicates small effect, 0.3 indicates medium effect and 0.5 above indicates large effect [20]. Therefore, these tests provided evidences that students scored significantly better in their post-test as compared to pre-test, and the intervention had a large impact on students' performance and hence the intervention was suitable for all participants.

Table 4. Wilcoxon Signed Rank Test and effect size of the overall results

No. of students (N)	Median		Z	Asymp. Sig. (2-tailed) (p)	Effect size ($ES = \frac{\lvert Z \rvert}{\sqrt{2N}}$)
	Pre test	Post-test			
50	12	23	− 5.788	0.000	0.5788

Bar graph (a) in Figure 2 indicated that addition of negative integer with negative integer improved the most (overall increase by 36.0%) followed by addition of positive integer with negative integer (from 55.1% to 89.1%). Most students did not have problem in adding the positive integers to begin with and after intervention, as reflected in high score in both pre-test and post-test scores. Overall, there was improvement in students' performance on the addition of integers in which the overall percentage increased significantly from average of 60.2% in the pre-test to average of 89.7% in the post-test.

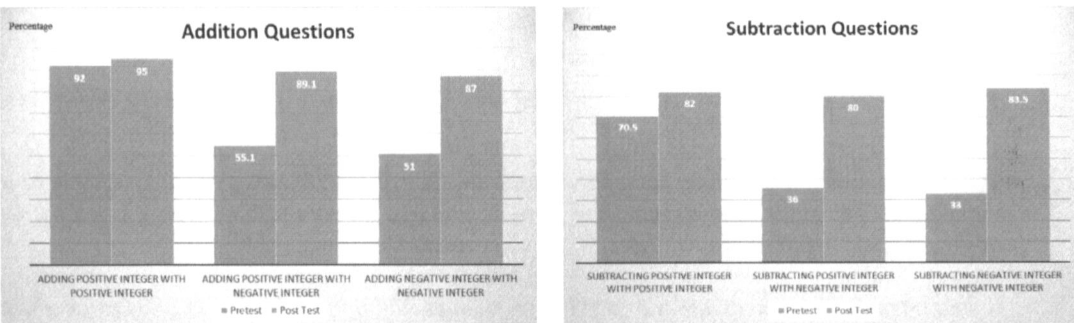

(a) (b)

Figure 2. Percentage of correct attempts in the pre-test and post-test for the different combinations of integers in test items involving addition (a) and subtraction (b)

With respect to the subtraction of integers, shown in Figure 2(b), students improved most significantly when subtracting negative integer with negative integer, in which only 33.0% got the answers correct in the pre-test to 83.5% of the total participants who responded correctly in the post-test. The second most improved was when subtracting positive integer with negative integer, which increased by 44.0%. Small improvement was also observed when students subtracted positive integer with positive integer, which started relatively high in pre-test to begin with. Generally, there was a considerable improvement in students' performance in the subtraction of integers, from overall 46.5% (pre-test) to 81.0% (post-test).

Overall result revealed that there was an increased in the mean score of 50 participants from 51.75% in the pre-test to 84.79% in the post-test. The Wilcoxon Ranked Test Analysis of the overall result revealed that the significance level, $p = 0.000 \leq 0.05$ signified that the difference in scores between the pre-test and post-test was significantly different. The large effect size of $r = 0.58$ implied that the intervention lessons gave a huge impact in their post-test score.

This study also discovered that overall, the intervention in general improved on students' performance in any combinations of signs and operations. The overall means score of correct attempts in the post-tests showed higher on addition of integers (89.7%) compared to subtraction of integers (81.0%). Nonetheless, when comparing their performance improvement from pre-test to post-test, students improved more on the subtraction of integers (with percentage increase of 34.5%) than the addition of integers (with percentage increase of 29.5%).

The findings in this study cannot be generalized to the whole population (since based on relatively small sample size of only 50 students) but it could provide a guideline in planning for remediation. A significant improvement in students' achievement in the computation of integers was observed, largely due to the carefully designed instruction with meaningful use of manipulatives. With this positive indication, teachers are encouraged to use manipulatives in their teaching to promote students' understanding in mathematics [21-23]. This concrete representation truly aids in clarifying mathematical concepts [24]. It also provided meaningful learning experiences for students.

4. Conclusion

In conclusion, in view of the importance of incorporating the use of manipulatives in mathematics teaching and learning as outline in the SPN21 mathematics curriculum framework, there is a need for continuous professional development to equip teachers with necessary content knowledge and pedagogical skills in using various manipulatives [25]. This will ensure that teachers become more competent in teaching using manipulatives, not only to integer topics but to other basic mathematics topics as well.

References

[1] Gningue S M, Menil V C and Fuchs E 2014 *The Electronic J. Math. Tech.* **8** 159
[2] Ismail S F Z, Shahrill M and Mundia L 2015 *Procedia – Soc. and Behav. Sci.* **186** 474
[3] Rubin R J, Marcelino J, Mortel R and Lapinid M R C 2014 Activity-based teaching of integer concepts and its operations *DLSU Research Congress.*
[4] Badarudin B R 2006 *Investigating teaching and learning of integers at form 1 level in Brunei Darussalam* (Brunei: Universiti Brunei Darussalam)
[5] Bofferding L 2010 Addition and subtraction with negatives: Acknowledging the multiple meanings of the minus sign *Proc. of the 32nd Annual Meeting of the North American Chapter of the Int. Grp. for the Psyc. of Math. Ed.* (Columbus, Ohio: Purdue University).
[6] Bolyard J J 2005 *A comparison of the impact of three virtual manipulatives on student achievement and conceptual understanding of integer addition and subtraction* (Fairfax, VA: George Mason University)
[7] Fagnant A, Vlassis J and Crahay M 2005 *Powerful environments for promoting deep conceptual and strategic learning*, eds L Verschaffel, E De Corte, G Kanselaar and M Valcke (Leuven, Belgium: Leuven University Press) p 81
[8] Terao A, Sawaki R, Hasegawa M and Murohashi H 2005 *Improving Skills of Addition and Subtraction Involving Negative Numbers Based on Cognitive Task Analysis and Assessment of Mental Representations of Negative Numbers: A Case Study of a Seventh-Grade Student.* (Troy, NY: Rensselaer Polytechnic Institute)
[9] Vlassis J 2004 *Learning and Instruction* **14** 469
[10] Hing L T 2000 *The teaching and learning of algebraic equations and factorisation in O-level mathematics: A case study* (Brunei: Universiti Brunei Darussalam)
[11] Clements M A 2002 Multiple perspectives and multiple realities of school mathematics *Seventh Annual International Conference of the Department of Science, Mathematics and Technical Education* (Brunei: Universiti Brunei Darussalam)
[12] Khoo S C 2001 *The teaching and learning of geometry* (Brunei: Universiti Brunei Darussalam)
[13] Lim T H 2000 *The teaching and learning of algebraic equations and factorization in O-level Mathematics: A case study* (Brunei: Universiti Brunei Darussalam)
[14] Brownell W A 1935 *The Teaching of Arithemetic, 10th Yearbbook of National Council of Teachers of Mathematics* 1
[15] Khalid M 2006 *Mathematical Thinking in Brunei Darussalam: Implementation Issues and Challenges. 2nd APEC-Tsukuba Int. Conf. Innovation of Classroom Teaching and Learning through Lesson Study II-Focusing on Math. Thinking, CRICED* (Japan: University of Tsukuba)
[16] Charalambous C Y, Hill H C and Mitchell R N 2012 *J. of Curriculum Studies* **44** 489
[17] Hunt A W, Nipper K L and Nash L E 2011 *Current Issues in Middle Level Education* **16** 1
[18] Tanner J and Hale K 2007 *Research and Teaching in Dev. Ed.* **23** 78

[19] Price P C, Jhangiani R S and Chiang I A 2015 Reliability and validity of measurement *Research Methods in Psyc.* (British Columbia, Canada: Kwantlen Polytechnic University)
[20] Cohen J 1992 A power primer *Psychological Bulletin* **112** 155
[21] Khalid M and Tengah M K A 2007 Communication in mathematics: The role of language and its consequences for English as second language students *3rd APEC-Tsukuba International Conf. Innovation of Classroom Teaching and Learning through Lesson Study III—Focusing on Math. Communication, CRICED* (Japan: University of Tsukuba)
[22] Zainil M, *et al* 2018 *J. Phys.: Conf. Ser.* **943** 012046
[23] Runisah, *et al* 2017 Using the 5E learning cycle with metacognitive technique to enhance students' mathematical critical thinking skills *International Journal on Emerging Mathematics Education* **1** 87
[24] Setyawan F, *et al* 2018 *J. Phys.: Conf. Ser.* **943** 012004
[25] Curriculum Development Department 2011 *Framework and guidelines for curriculum and assessment: Mathematics year 7 and year 8* (Brunei Darussalam: Ministry of Education)

The effectiveness of an autograph-assisted problem based learning model for improving high school students' learning outcome on graphics functions

M Ridha, R Johar, Marwan, and Mailizar
Universitas Syiah Kuala, Jl. Teuku Nyak Arief, Banda Aceh, 23111, Indonesia

E-mail: marwan.math@unsyiah.ac.id

Abstract. Students' learning outcomes can be enhanced through tasks that require students to relate images in their daily life to mathematical graphs. However, students find it difficult when they draw mathematical graphs by using pens or pencils. Therefore, they need a digital tool to assist them. This study aimed at developing a digital tool for teaching and learning of graphics functions. It adopted Plomp's model of design research that consists of three phases: (1) preliminary research (2) prototyping phase, and (3) assessment phase. It is important to highlight that this paper presents the assessment phase of the study. Data were collected to determine the effectiveness of Autograph-assisted problem-based learning model. Participants of this study were 17 year-ten students from a high school in Banda Aceh city, Indonesia. Data were obtained through student's worksheet and questionnaire, and it was analysed descriptively. The findings showed that 88% of the students achieve learning objectives and 90% of them had positive responses toward the leaning.

1. Introduction

Graph function is one of the mathematics topics at Year 10 and this subject is applied to many other fields such as physics. However, students find this topic difficult to master. Most students have difficulties in this subject as they have lack understanding of the subject particularly graph function [1]. Furthermore, based on the interview conducted by the researchers to one of the mathematics teachers in a high school in Banda Aceh, it was found that students did not fully understand the concept of graph function which resulted most of them were not able to solve mathematical problems associated with graph functions. In addition, students have difficulties in applying formulas and solving problems related to applications of graph function in everyday life.

Several studies have been conducted to develop learning tools using Autograph software for learning various topics [2,3] such as linear programming [4] and calculus [5]. However, there is no research that developed the learning tool for teaching graph functions through Project-Based Learning (PBL) assisted with Autograph. Thus, it is necessary to develop such teaching tools to assist students in the learning process.

In the learning, students are required to explore their ideas and strategies in solving mathematical problems by using Problem Based Learning (PBL) model. In the view of constructivism, learning aims to help students to understand and apply knowledge. They must work together, find concepts by themselves, and try with their ideas [6]. The PBL model is one of the constructivism learning models, which builds the concepts of students' understanding with real experience and through the process of collaboration in building their understanding and being able to convey the ideas that they find through the collaboration process [7]. PBLs are designed to help students develop their thinking skills, problem-solving skills, intellectual skills, and the model also stimulate students to become self-sustaining, as well as autonomous learners [8]. The PBL model can develop specific skills, including the ability to think critically, to analyze and solve complex and real-world problems [9]. Moreover, this model can help students to find, evaluate and use resources appropriately as well as to work together, to demonstrate effective communication skills, and to use intellectual knowledge and skills in order to keep the students highly motivated to follow the learning process [9]. One way to maximize student motivation is by using software as a learning media. One of the appropriate mathematical software is the Autograph software.

Through Autograph-assisted learning, students can do an exploration, investigation, and discovery [10]. Students can test more samples in a short time to prove the results that they obtained by pen and paperwork. As a result of their experiments, students are able to find, construct and draw a conclusion about the concepts of the Graph Function to determine the equation, and finally understand how to draw the equations of a graph correctly. Autograph Software provides quite complete features including geometry and algebra. The software also has 2D and 3D graphics capabilities for topics such as transformation, vector, and derivatives [11]. In addition, it can be utilized to draw statistical graphs, functions, and vectors as well as to alter the shape of an object or geometry shape.

Based on the above description, the problem formulation in this research is how effective the use of autograph assisted problem-based learning model for high school students' learning outcome on graphics functions.

2. Method

This study adopted Plomp's model of design research that consists of (1) preliminary research (2) prototyping phase, and (3) assessment phase. We collected data to examine the effectiveness of Autograph-assisted problem based learning model on learning graph functions. In total, 17 students from grade 10 of a high school in Banda Aceh participated in this study. Instruments used in this were tasks and questionnaire, and data were analysed descriptively. There were three lessons delivered during this study. In the first lesson, students were asked to draw graphs of a linear function while in the second lesson students were assigned to create graphs of quadratic functions. Moreover, in the last lesson, students were asked to draw graphs of trigonometric functions. In order to determine the effectiveness of the tool, we set the following criteria. Learning tools are considered effective when 85% of the students get a score of completeness criterion (75). On the other hand, the developed learning instruments fall into category positive when 80% of students respond happy, new, interested, obvious, attracted categories.

3. Result and discussion

In this section, we elaborate on the results and discuss them with related literature. We present student learning outcomes and their perceptions of learning activities respectively. Regarding their learning outcome, first, we present students learning outcome in groups, which is then followed by a discussion of individual students learning outcomes.

3.1. Students' learning outcome

As mentioned earlier, this study looked at students' learning outcome in groups and students' individual learning outcome. Both aspects are presented in the following subsections.

3.1.1. Students' learning outcome in groups

Assessment of student learning outcomes in groups is by investigating students during group work. The results of student learning in groups are presented in Table 1.

Table 1 shows that students in groups I, III and IV satisfied very good criteria whereas students in group II met good criteria. Overall, from the student's worksheet, it showed that the students were able to use and completed the developed worksheet.

Table 1. Scores on student's worksheet

Group	Scores of student's worksheet			Mean Score	Criteria
	1	2	3		
I	90	89	95	91.33	Very Good
II	85	90	93	89.33	Good
III	92	95	95	94.00	Very Good
IV	87	93	95	91.67	Very Good

The learning outcomes falls into the good category that indicates a positive result. This result is consistent with [12] that suggest that problem-based learning is more effective in improving learning achievement. This is due to PBL has the following advantages: (a) Challenges students 'ability as well as provides satisfaction to find new knowledge for students, (b) Increases student's motivation and learning activities, (c) Helps students in transferring their' knowledge to understand real-world problems, (d) Assists students to develop new knowledge and be responsible for their learning, (e) Develops students' critical thinking skills and develop their ability to adapt to new knowledge, (f) Provides opportunities for students to apply the knowledge they have in the world (g) Develops students' interest in continuous learning even when learning in formal education has completed, (h) Facilitates students in mastering the concepts to solve real-world problems [13]. Students' learning outcomes in groups are required when students complete the student's worksheet. In this study, at the learning stage, it helps independent and group mobilization.

3.1.2. Students' individual learning outcome

Students' individual learning outcomes were obtained through tasks at the end of the lesson. Regarding students' individual learning outcomes, 88% of the students received passing grades and 12% were unable to reach the standardized grade. It means that students were able to relate the given images to their daily life in order to find the graphics function. Therefore, it can be concluded that students' learning outcome on graphs function fall into the good category. Examples of students' works are presented in Figure 1 and Figure 2.

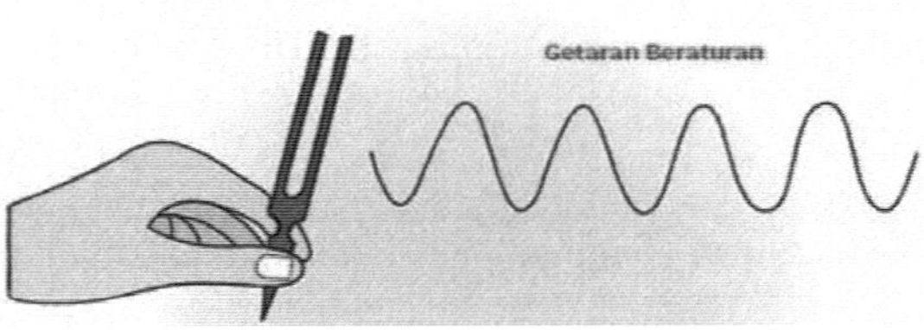

Figure 1. Graph of a trigonometric function

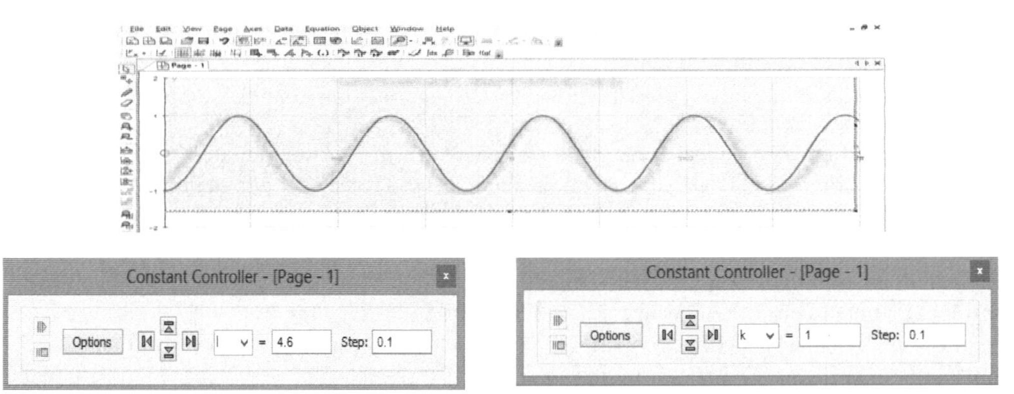

Figure 2. Student's answer

A function that matches the image is $y = -\cos 4{,}6x$. Furthermore, the minimum value on the graph is -1 while the maximum value on the graph is 1. In addition, period of the graph is 2π and amplitude is 1. Amplitude = 1

It can conclude that students were able to determine the trigonometric function of the image that they find in their daily life, determine the maximum and the minimum value of the trigonometric function, determine the period and determine the amplitude of the graph. Furthermore, for another problem, we can see in Figure 3.

Figure 3. Graph of quadratic functions

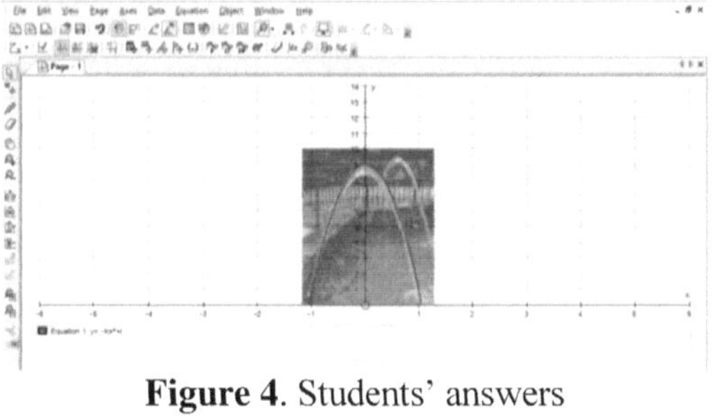

Figure 4. Students' answers

The function of the image is $y = -8.2x^2 + 9$. Moreover, the highest value of the function is 0.9. The graph of the quadratic function is $y = -8.2x^2 + 9$ that is closed when $a < 0$.

According to Figure 4, we conclude that students were able to find the quadratic function of the image, and determine the highest point of the graph. This result is consistent with [11] that reveal that Autograph software is effectively used in teaching quadratic functions in high school.

3.2. Students' perception of the lessons

At the end of the lessons, students were given the questionnaires. The results of the questionnaire show that 90% of the students had positive responses to learning with Autograph. They also stated that the language used in Autograph guideline, student's worksheets and student tasks are clear and interesting.

These positive responses are ranging from comments on the process of learning activities to comments on learning media that have been utilized. Students were happy with the learning that had taken place. Students also commented on the use of the software and the motivation in learning.

Examples of students' responses are presented in Figure 5 and Figure 6. Based on the students' comments, it shows that learning by using technology can increase students' interest and motivation. These results are consistent with [14] which states that technology as one of the mathematics learning media can bring new changes that greatly assist students in bridging the abstract world of mathematics and real-world students.

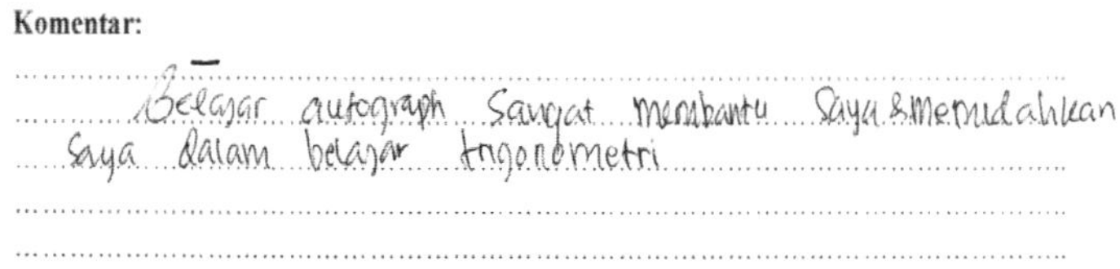

"Autograph helps me in learning of trigonometry."

Figure 5. Student's response to the use of autograph software

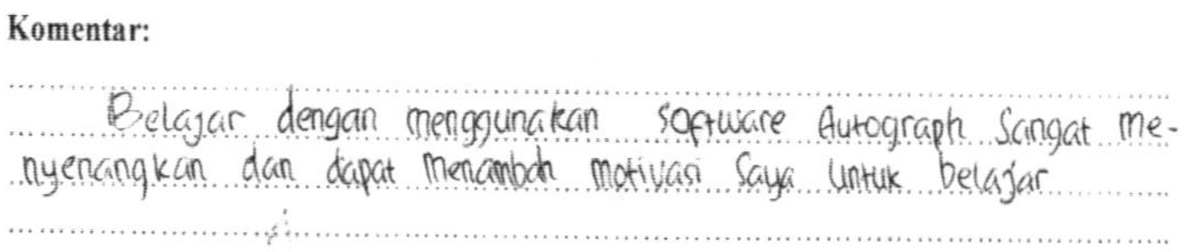

"Learning with Autograph software makes me happy and increases my motivation."

Figure 6. Student's response to the use of autograph software

4. Conclusion

In concluding this paper, based on the findings of this study, we suggest that Autograph software is an effective tool to assist students in learning of the graphics functions through problem-based learning. Moreover, students had positive responses toward the learning. Therefore, it indicates that this tool is potential to become learning media in teaching and learning of graph functions. However, in order to be more effective, we suggest that students need to be supplied with adequate ICT infrastructures such as computers or laptops. Finally, it needs further study to investigate the use of this tool in teaching and learning other topics of mathematics through problem-based learning.

References

[1] Nuralita D and Masduki 2016 Kesulitan Siswa dalam Menyelesaikan Masalah Fungsi *Prosiding KNPMP*

[2] Ramdani Y 2012 Pengembangan Instrumen dan Bahan Ajar untuk Meningkatkan Kemampuan Komunikasi, Penalaran, dan Koneksi Matematis dalam Konsep Integral *Jurnal Penelitian Pendidikan* **13** 44

[3] Hajar S 2017 *Pengembangan Perangkat Pembelajaran Integral Melalui ELPSA FRAMEWORK Berbantuan Software Matematika Untuk Meningkatkan Pemahaman Siswa SMA* (Banda Aceh: Universitas Syiah Kuala)

[4] Jannah M R 2016 *Pengembangan Perangkat Pembelajaran Program Linear Dengan Menggunakan Model Problem Based Learning Berbantuan Software Autograph* (Banda Aceh: Universitas Syiah Kuala)

[5] Irwansyah B 2015 Pengembangan Bahan Ajar Kalkulus Berbantuan Autograph *Jurnal Ilmu-ilmu Pendidikan dan Sains* **3** 45

[6] Trianto 2007 *Model-model Pembelajaran Inovatif Berorientasi Konstruktivistik* (Jakarta: Prestasi Pustaka)

[7] Dorchy F, Sergers M, Bossche P N and Gijbels D 2003 Effect of problem-based learning A meta-analysis *Learning and Instruction* **13** 533

[8] Arends R I 2008 *Learning To Teach Belajar untuk Mengajar Buku Dua Edisi Ketujuh* (Yogyakarta: Pustaka Pelajar)

[9] Savery J R 2006 Overview of Problem-based Learning Definitions and Distinctions Interdisciplinary *Journal of Problem-Based Learning* **1** 9

[10] Seragih S and Afriati V 2012 Peningkatan Pemahaman Konsep Grafik Fungsi Trigonometri Siswa SMK Melalui Penemuan Terbimbing Berbantuan Software Autograph *Jurnal Pendidikan Matematika* **18** 367

[11] Karnasih I and Rahman A 2014 Integrasi software Dinamis Autograph dalam Pembelajaran Matematika Menggunakan Pendekatan Penemuan Terbimbing *Ad Math Edu* 143

[12] Farhan M and Retnawari H 2014 Keefektifan PBL dan IBL Ditinjau dari Prestasi Belajar, Kemampuan Representasi Matematis, dan Motivasi Belajar *Jurnal Riset Pendidikan Matematika* **1** 227
[13] Sanjaya W 2007 *Strategi Pembelajaran Berorientasi Standar Proses Pendidik* (Jakarta: Kencana)
[14] Nurhasanah F 2012 Abstraksi dan Alat Peraga Maya dalam Pembelajaran Matematika *Prosiding Seminar Nasional Pendidikan Matematika UNIMED*

Dilemmatic situations for learning mathematics-related beliefs

T Y E Siswono, A W Kohar, S Hartono, A H Rosyidi, P Wijayanti and R Ekawati
Universitas Negeri Surabaya, Kampus Unesa Ketintang, Surabaya, Indonesia.

E-mail: tatagsiswono@unesa.ac.id

Abstract. This study aims to investigate how dilemmatic situations were used as context in a set of instructional materials for learning teachers' mathematics-related beliefs. The instructional materials were developed through formative evaluation steps: self-evaluation, expert review, one-to-one, small group, and field test. The formative evaluation step involved three experts and 31 pre-service teachers in examining the materials' validity, practicality, and potential effects. The dilemmatic situations included in the instructional materials provoke pre-service teachers to select one of the three types of teaching sequences corresponding to three hierarchical beliefs about the nature of mathematics (instrumentalist, Platonist, and problem-solving) in accordance with beliefs about teaching and learning. Results showed that the instructional materials were valid and practical, and have a potential effect on improving pre-service teachers' mathematics-related beliefs. In particular, the dilemmatic situations could encourage pre-service teachers to consider which beliefs they should follow regarding problem-solving teaching. On the one hand, the pre-service teachers agree to the problem-solving view, on the other hand, they are confronted with the immediate situations found in their instructional practice which makes them hesitant in having such view. However, this instructional practice lead teachers show a good understanding of the borderline between those three beliefs. Besides, the instructional practice is indicated to potentially effect on changing the pre-service teachers' mathematics-related belief to be more constructivist.

1. Introduction
Mathematics-related beliefs are known to play a great role in teaching and learning mathematics. This is because the teacher mathematics-related beliefs influence what teachers will teach, how they teach, and what they learn in the classroom [1]. In addition, Ernest [2] asserts that beliefs also affect student learning outcomes. However, some reports show that teachers, as well as pre-service teachers, are reported to have traditional beliefs about mathematics. For instance, Siswono et al. [3] reported that there were no participants in their case study having a problem-solving view. Instead, the participants considered mathematics as a discipline with a static body of knowledge with regards to Platonist view [2], or science which merely study about number and formula with regards to Instrumentalist view [2]. Therefore, change of beliefs from traditional to constructivist are considered very important for changes in teaching practice.

Considerable concern has been given to the issue of changing teacher beliefs as one of the current reform of topics in mathematics teacher education. Research provides attention on improving teacher beliefs concerning nature of mathematics [4], mathematical and pedagogical view [5], students' problem-solving [6], and students' engagement in mathematics instruction [7]. A variety of programme supporting the successfulness of reaching such attention has also been examined. Guberman & Leikin [5], for example, explored the development of teacher beliefs through a programme using systematic utilize of a multiple-solution task. Blömeke et al. [8], on the other hand, found that teachers' beliefs develop when teachers are provided with a strong climate of trust from the school where they work. Within a professional learning applying principles: *fidelity, engagement,* and *change*, Lomas & Clarke [4] reported the improvement of teachers' beliefs about mathematics and their teaching and learning. All such effort was carried out to confront teachers' initial beliefs that make them learn mathematic-related beliefs. This is because an individual learns when he/she attempts acquiring new or modifying existing knowledge, behaviors, skills, values/beliefs, or preferences [9].

Learning mathematics-related beliefs through confronting situations can also be undertaken by providing dilemmatic situations within an instructional practice. A dilemma is defined as situations in which two values, commitments, or obligations conflict and there seems to be an incorrect thing to do [10]. The conflicts cause multiple, equally possible alternatives, each of which has advantages and disadvantages [11]. By employing dilemmatic situations as contexts of learning, the authors argue that teachers would learn new schemes of knowledge and beliefs which may confront their initial beliefs to be maintained or even changed.

One aspect of dilemmas experienced by teachers is pedagogical beliefs [12]. The dilemmas occur when teachers are confronted with at least two distinct pedagogical views: traditional and constructivist view. In the traditional view, the teacher is described as the transmitter of knowledge and the director of learning, whereas the students are expected to be passive recipients of knowledge, whose responsibility is to follow the director's lead [13]. In the constructivist view, a teacher is described as a facilitator of student learning, in which students are actively engaged in constructing knowledge through collaborative learning [13]. When choosing one of those two views, teachers are faced with some variables that need to be considered occurring in their daily teaching such as immediate classroom situations (students' abilities, attitudes, and behavior; the mathematics topic at hand; time constraints), social norms (demand of curriculum; learning resources), or their previous school experience.

This study makes an effort to develop a set of instructional materials (lesson plan and worksheet) for a teacher development programme to encourage pre-service teachers to learn mathematics-related beliefs through dilemmatic situations. Therefore, the authors developed a set of instructional materials which support pre-service teachers' understanding of mathematics-related beliefs as well as improve their mathematics-related beliefs to be more constructivist. In this study, the mathematics-related beliefs learned by the pre-service teachers are those proposed by Ernest [2] and then summarised by Beswick [14]. Ernest describes three types of epistemological beliefs, namely Platonist views, instrumentalist views, and problem-solving views. In the Platonist view, mathematics is seen as a static yet integrated knowledge, whose structures and interconnections between various topics are essentially important. The instrumentalist view states that mathematics is a collection of facts, rules, and skills that are useful and essentially unrelated. Meanwhile, the problem-solving view sees mathematics as a dynamic and creative human discovery. Beswick's framework [14] present a theoretical consideration which put beliefs about mathematics [2] with their corresponding beliefs about mathematics teaching: content-focused with an emphasis on performance, content-focused with an emphasis on understanding, learner-focused [15] and learning: skill mastery, passive reception of knowledge, active construction of understanding, and autonomous exploration of learner's interests [2].

Thus, the research questions of this present study are: 1) how to develop valid and practical instructional materials supporting pre-service teachers' understanding of mathematics-related beliefs? 2) how are dilemmatic situations employed as context for learning mathematics-related beliefs? and 3) what is the potential effect of the developed instructional materials?

2. Research method

This is a design research with the type of development study design employing formative evaluation [16]. The stages of the research are preliminary stage and prototyping stage (formative evaluation) which includes self-evaluation, expert reviews and one-to-one, small group, and field test [17, 18].

IOP Conf. Series: Journal of Physics: Conf. Series **1088** (2018) 012026 doi:10.1088/1742-6596/1088/1/012026

Tessmer [17] defines formative evaluation as a judgment of the strengths and weaknesses of the instruction in its development stages for the purposes of revising the instruction to improve its effectiveness and practicality. The components involved in this development process were the designer of the instructional materials (in this case we as the authors), the experts, and the learners (in this case the pre-service teachers). The subject of this research is 31 pre-service teachers from primary school at Surabaya city and experts in mathematics education. In detail, the participants involved are five preservice teachers (one-to-one), three experts (expert review), six pre-service teachers (small group), and 20 pre-service teachers (field test). The participants of field test consist of 13 female and seven male pre-service teachers studying at the master program of primary education at Universitas Negeri Surabaya; all of them have not joined any mathematics-related course during their master program. The stages of one-to-one, expert review, small group, and field test are described to answer the first research question by examining the validity and practicality of the instructional materials.

To answer the second research question, in particular, a mathematics-related beliefs task was developed. The task encourages pre-service teachers to select one of the three teaching sequences which may arouse teachers' dilemmas on the teaching of area measurement which portrays the three philosophical beliefs summarised by Beswick [14]. Figure 1 presents one of the three options used in field test which have been revised through the development process.

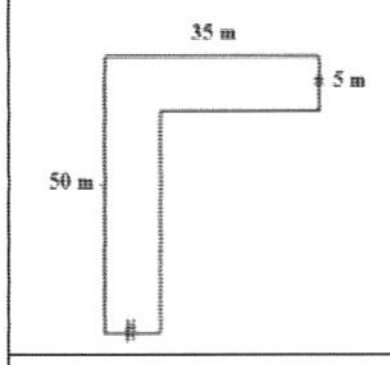

Some participants auditioned to become a singer in a queue area outside a television station building. All the participants are standing in line. With the queue area shown in the right picture, determine the best estimate of the number of participants of the audition.

Option 1 (Instrumentalist)

1. Providing an explanation on how to find the area of combined geometrical figures by firstly giving some simple examples;
2. Asking students to practice how to find the area of the combined geometrical figure based on the teacher's method;
3. Introducing the application of the area of the combined geometrical figure in real life, such as finding the number of tiles required in a floor with a particular area;
4. Ensuring that students understand how to solve such 'Tiling' problem so that it can be applied in solving the 'Queue of Singer Audition' problem;
5. Starting to present the 'Queue of Singer Audition' problem which is considered as a more difficult problem than the 'Tiling' problem;
6. Telling that there is important missing information that needed to be found, i.e., the space that a participant might occupy while standing in the queue area;
7. Directing students to a specific strategy, i.e., dividing the area occupied by participants with the size of space of a person while standing;
8. By teachers' guidance, students finally obtain the solution to the problem; and
9. Correcting students' answers based on the procedures and the final solution they performed.

Figure 1. Mathematics-related beliefs task.

At field test, the pre-service teachers followed two learning activities: 1) mathematics problem-solving knowledge for teaching developed from the Chapman's framework (see more detail at [19]), in which this topic also become another concern of supporting teachers' improvement on beliefs, and 2) mathematics-related beliefs. The dilemmatic situations were provided in the latter activities. Therefore, this does not aim to describe the first learning activities. The pre-service teachers were asked to select individually one of the options at the beginning of the lesson, give their arguments, discuss with other colleagues, confirm the results of their discussion with theoretical perspectives guided by the lecturer, and reflect on whether they change their choices at the end of the lesson.

The third research question, on the other hand, was answered by describing the change of beliefs related to teaching problem-solving task after joining the learning activities. The change was indicated by the number of choices for each of the options: Instrumentalist, Platonist, and Problem-solving selected by pre-service teachers in the field test, which was counted at the beginning of the learning activities. At the end of the activities, they were asked to consider their choices about whether they would keep or change their types of teaching sequences. Thus, the author counts the number of the pre-service teachers' choices at the end of learning activities.

3. Results and discussion

3.1. Development process of the instructional materials

At the prototyping stage, the instructional materials were developed within self-evaluation, expert review, one-to-one, small group, and field test. At the self-evaluation stage, the authors re-examined the initial prototype design while preparing the research instrument related to the development process. The study was conducted by checking the suitability of the lesson plan and learning activities in the worksheet regarding its content, constructs, and language. The results from the self-evaluation were then tested in one-to-one and expert review simultaneously. One-to-one was conducted to see the practicality of the developed materials. Information obtained from this activity is the students' comments about the clarity of the teaching materials intent and the model of pre-service teachers' responses to the instructional materials. Meanwhile, three experts in expert review validated the instructional materials concerning content, constructs, and languages.

The results of one-to-one indicates some pre-service teachers' comments to the quality of the instructional materials which relates to (1) unclear images, (2) some less familiar terms, (3) too long sentences, (4) grammatical errors, and (5) the emergence of alternative responses to problems examined in the worksheet. All these comments were then used as a basis to revise the instructional materials to meets practicality criteria. Based on the expert review, the researcher revised the instructional materials regarding some issues, i.e., correction of the typos, the correct use of capital letters, the use of more effective sentences, and the use of more legible drawings. Furthermore, the revision also concerned to the addition of questions about the selection of teaching sequences based on the type of beliefs on teaching mathematics, the refinement of mathematics-related task regarding the sequences of teaching problem-solving task that would be discussed by pre-service teachers which looks not represent the Instrumentalist, Platonist, and problem-solving view. The revised instructional materials were then used in a small group and field test. In these two stages, the learning activities designed in the instructional materials were firstly trialled and evaluated in actual classroom teaching setting although with the small number of participants.

3.2. Dilemmatic situations: instrumentalist, Platonist, or problem-solving?

Based on the results of the analysis of the subject response, four subjects chose the Instrumental option, seven subjects chose the Platonic option, and nine subjects chose the problem-solving option before the discussion took place. Figure 2 shows examples of the subject's response to the choice of the learning sequence.

Figure 2a shows the reason for the subject choosing option 1. In the subject's opinion, option 1 is the best choice because students need to learn from something simpler to something more complex gradually. Meanwhile, Figure 2b indicates the reason for the subject choosing option 2. According to him, option 2 will provide space for students to be more independent in solving mathematical problems because teachers only guide as necessary at each stage of Polya. Figure 2c shows the subject's lack of confidence in choosing option 1 and 2, due to too much teacher intervention in helping the student to solve the problem, and the timeliness of the initial problem, which should be suitable only for students who like challenges, not students that passive in learning.

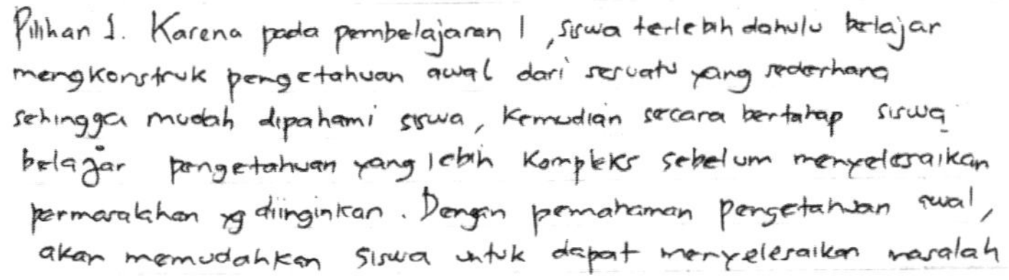
Pilihan 1. Karena pada pembelajaran I, siswa terlebih dahulu belajar mengkonstruk pengetahuan awal dari sesuatu yang sederhana sehingga mudah dipahami siswa, kemudian secara bertahap siswa belajar pengetahuan yang lebih kompleks sebelum menyelesaikan permasalahan yg diinginkan. Dengan pemahaman pengetahuan awal, akan memudahkan siswa untuk dapat menyelesaikan masalah

Translation:
Option 1. In this teaching sequence, students learn how to construct initial knowledge from simple things so that it can be easily understood, and then, they learn a more complex knowledge gradually before learning to solve a complex problem. Thus, it will make students solve such a problem.

a) Option 1. Instrumentalist view

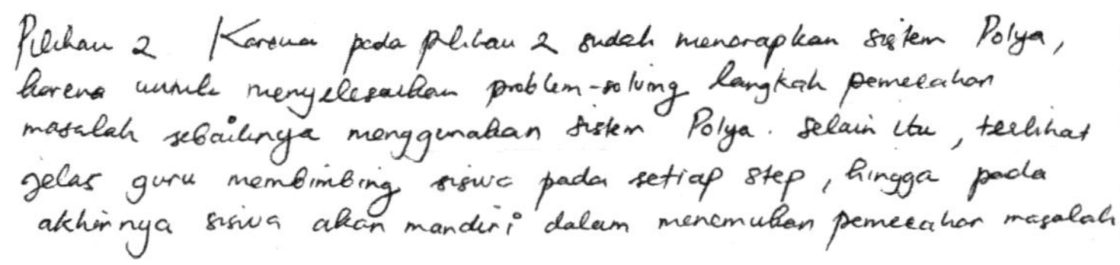
Pilihan 2 Karena pada pilihan 2 sudah menerapkan sistem Polya, karena untuk menyelesaikan problem-solving langkah pemecahan masalah sebaiknya menggunakan sistem Polya. Selain itu, terlihat jelas guru membimbing siswa pada setiap step, hingga pada akhirnya siswa akan mandiri dalam menemukan pemecahan masalah

Translation:
Option 2. This option follows Polya's model of problem-solving because to solve a problem, Polya's stages should be used. Besides, teachers guide their students at each stage so that the students will be independent in solving a problem-solving task.

b) Option 2. Platonist view

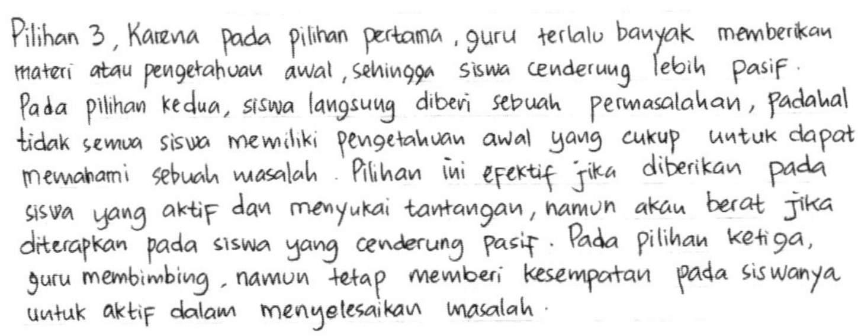
Pilihan 3, Karena pada pilihan pertama, guru terlalu banyak memberikan materi atau pengetahuan awal, sehingga siswa cenderung lebih pasif. Pada pilihan kedua, siswa langsung diberi sebuah permasalahan, padahal tidak semua siswa memiliki pengetahuan awal yang cukup untuk dapat memahami sebuah masalah. Pilihan ini efektif jika diberikan pada siswa yang aktif dan menyukai tantangan, namun akan berat jika diterapkan pada siswa yang cenderung pasif. Pada pilihan ketiga, guru membimbing, namun tetap memberi kesempatan pada siswanya untuk aktif dalam menyelesaikan masalah.

Translation:
Option 3, because option 1 indicates that the teacher provide too much initial knowledge or topics being taught and students tend to be more passive. Option 2 indicates that students are directly provided with a problem, in which not all students have sufficient initial knowledge to understand the problem. This option will be effective for those who are active and challenged. In option 3, the teacher not only keeps guiding students but also providing opportunities to be actively engaged in solving a problem.

c) Option 3. Problem-solving view

Figure 2. Pre-service teachers' responses to mathematics-related beliefs task.

Pre-service teachers encountered conflicting ideologies regarding choosing the best instructional sequences provided at the field test. Following the findings of Ding & Wang [20] regarding decision making patterns of teachers when facing dilemmatic situations, the authors categorized pre-service teachers' response within the lesson which used the developed instructional materials as follows.

3.2.1 Maintain personal beliefs and practices

In this pattern, participants who reported that they maintained their own beliefs and practices mainly justified their decisions by comparing the opposing beliefs and practices represented by other two options to highlight the beliefs on evidence found in his field teaching such as their immediate classroom situation, past school experiences. The following excerpt indicates their view.

PsT1 : I will keep my beliefs that option 1 (the instrumentalist view) is the best although many other of my friends choose another one. I believe that students need to understand a certain mathematical concept first. I have tried to give such kind of problem to my students in the previous lesson, and they got stuck to understand the problem. My teachers in the past often give my friends and I many exercises after we understand the particular basic rules of mathematics. That makes us more flexible to solve any exercises about mathematics problem. Furthermore, I think my students will be more interested in learning concepts before a real-world problem like this.

3.2.2 Change and adapt to the opposing beliefs and practices

The second pattern was to change and adapt to the opposing beliefs and practices. Participants who changed their beliefs justified the changes by viewing the potential benefits of the opposing beliefs emerging during class discussion. This gives rise them to reconceptualize the opposing beliefs by modifying the situations provided in the teaching sequences with the context where they teach. This causes them to reconceptualize the opposing beliefs by adjusting the situations provided in the teaching sequences with the context where they teach. The following excerpt indicates this pattern.

PsT2 : Yes, you [another PST] are correct. When guiding students to find the strategies, the teacher in option 1 (Instrumentalist) seems to give too much direction, like when he exemplifies some methods of finding the area. This will reduce students' opportunity to present their own strategies. Previously I select this options [option 1] since it best represents my daily teaching, but when you say it should better to give this problem at the beginning of a lesson, instead of at the end of a lesson, I start thinking to find a more suitable problem which fits students' prior knowledge. So, I agree that it is not about when we teach problem-solving task, but it is about how we select and modify an appropriate problem, and I think this problem is ok for my students.

Lecturer : Why did you change your options into option 2 (problem-solving) instead of option 3 (Platonist)?

PsT2 : Basically, I agree with the teacher in option 3 because he involves students to find the solution actively. However, the worksheet completed with a set of questions as guidance will not make students find their own methods. Thus, I change my option to option 2 which does not provide such kind of guidance.

3.2.3 Avoid taking action

In this pattern, participants show a decision-making pattern by choosing not to confront their initial beliefs or opposing beliefs. This is indicated by their view about a trust given to the curriculum designer, in this case, is the government. In other words, regardless of whether their initial beliefs or the opposing beliefs are correct, they will rely on the mandate requested by the current curriculum. That being said, these prospective teachers experienced dilemmatic tensions because of their identities as teachers as curriculum implementer. This, as Ding & Wang [20] said, is called a self-identification which became the way teachers justified their decisions to avoid taking action in response to the dilemmas. The following excerpt exemplifies one of pre-service teachers' pattern of decision making.

PsT3 : I previously select option 1, and now it is still option 1. This is not because I have a good experience in handling such kind of learning activities during my daily teaching. But, it is about how to complete all the targeted topics in the curriculum in the expected time. That is why I can only do that. However, if you give me extra time for teaching, I will choose option 2. That is good. Teachers in this case are a good facilitator.

3.3. Potential effect of the instructional materials on improving pre-service teachers' mathematics-related beliefs

Results of field test show that there were changes in the selection of belief types on the selection of learning sequences. Of the 20 participants, 12 participants changed the choice, and eight participants remain on the choice. From the 12 participants, four participants changed the choice from the instrumental option to the problem-solving option, 5 participants changed the choice from the Platonic option to the problem-solving option, and interestingly, there were 3 participants who changed the choice from problem-solving to platonic option. The latter result seems surprising since it means the beliefs of those three subjects become more traditional after joining the course. When analyzing their written work about the reasons why they changed their belief, we found that the reasons are around their students' mathematical ability background in which students at their class seems difficult to solve a real-life problem which is introduced at the beginning of the lesson. Meanwhile, from the 8 participants with unchanged options, there are six that remain on the problem-solving option, whereas only 2 participants remain on the Platonist option. In general, these results indicate that after following the course, the majority of subjects (15 out of 20 subjects) choose a learning sequence that describes the problem-solving. These results indicate that there is an influence of learning using the instructional materials developed in this study on the change of belief type choices. In other words, these teaching materials have a potential effect on the change in teachers' beliefs in mathematics and learning. Nevertheless, these results seem to need further clarified through the practice of learning which the subject will do while teaching in the classroom. This can be clarified with the findings of Siswono et al. [21] demonstrating that teacher's teaching practices of their selected participants, even with the problem-solving approach, have not fully met the requirements for problem-solving learning. However, this seems to be incompatible with the studies of Siswono, Kohar, Kurniasari, and Astuti [22] who reported that high school teachers in their studies tend to be convinced to apply problem-solving ideas as dynamic approaches while teaching math.

Accordingly, the gap between teacher beliefs and teacher teaching practices needs to be the focus of further concern, especially by professional developers of teachers to develop their knowledge and skills in problem-solving teaching practice. Minimizing such a gap have been actually carried out by a professional teacher programme designer such as by Siswono et al. [23,24]. Derived from the findings that there is a relationship between teachers' beliefs and knowledge regarding problem-solving [25], two interrelated programs have been designed with different concern, i.e. programs for improving teachers' mathematics problem solving knowledge for teaching [23], and program for improving teachers' knowledge in posing context-based problem solving task [24]. These two programs employ problem-solving and problem-posing approach as learning tools for teachers participant at those programs. Since the results of these programs were experimentally able to improve teachers' problem-solving-related knowledge, future programs may be designed by considering problem-solving and problem-posing approach as tools for improving teacher beliefs, knowledge, and practice more comprehensively. For instances, the programs may consider to include real-world situation in mathematics learning [26], support teachers to create context-based task emphasizing on the authenticity of context [27] as well as solve authentic context-based problem, instead of only camouflage context-based problem [28], and improve mathematics content knowledge which often leads teachers to have instrumentalist beliefs about mathematics [29].

4. Conclusion

The dilemmatic situations were used as context for learning mathematics-related beliefs. This context was refined along with the development process of the instructional materials within a developmental research following formative evaluation stages of Tessmer [17]. The dilemmatic situations could encourage pre-service teachers to consider which values they should follow since such values conflict, and there seems to be no right thing to do. On the one hand, the pre-service teachers agree to the problem-solving view, on the other hand, they are confronted with the immediate situations found in their instructional practice which makes them hesitant in having the problem-solving view. In addition, the instructional materials were proved to potentially effect on improving pre-service teachers' mathematics-related beliefs and attract pre-service teachers' interest in learning mathematics-related beliefs through dilemmatic situations.

Acknowledgment

The following investigation was made thanks to the contributions for the research scheme of Strategi Nasional (Stranas) year 2018. We would like to thank Ministry of Higher Education and Research, Indonesia for the grant, the rector of Universitas Negeri Surabaya, the dean of Faculty of Mathematics and Natural Science, and all the participants involved in the teacher professional development within this programme.

References

[1] Andrews P and Hatch G 2000 *Educational Studies in Mathematics* **43** 3164

[2] Ernest P 1989 The impact of beliefs on the teaching of mathematics ed P Ernest *Mathematics teaching: The state of the art* p 249
[3] Siswono T Y E, Kohar A W and Hartono S 2017 *Journal of Physics: Conference series* **812** 012
[4] Lomas L and Clarke D 2016 Opening up mathematics education research eds White B, Chinnappan M and Trenholm S *Proceedings of the 39th annual conference of the Mathematics Education Research Group of Australasia* p 391
[5] Guberman R and Leikin R 2013 *Journal of Mathematics Teacher Education* **16** 33
[6] Koichu B, Berman A and Moore M 2003 Changing teachers' beliefs about students' heuristics in problem solving In *Electronic proceedings of the 3rd Conference of the European Society for Research in Mathematics Education*
[7] Bobis J, Way J, Anderson J and Martin A J 2016 *Journal of Mathematics Teacher Education* **19** 33
[8] Blömeke S, Hoth J, Döhrmann M, Busse A, Kaiser G and König J 2015 *International Journal of Science and Mathematics Education* **13** 287
[9] Gross R 2010 *Psychology: The science of mind and behaviour 6th edition* (Hodder Education)
[10] Honig B 1996 *Social Research* **61** 563
[11] Enyedy N, Goldberg J and Welsh K M 2006 *Science Education* **90** 68
[12] Wang D 2011 *Teaching and Teacher Education* **27** 157
[13] Lim C P and Chai C S 2008 *British Journal of Educational Technology* **39** 807
[14] Beswick K 2005 *Math. Educ. Research Journal* **17** 39
[15] Van Zoest L R, Jones G A and Thornton C A 1994 *Math. Educ. Research Journal* **6** 37
[16] Plomp T and Nieveen N 2007 An introduction to educational design research *Proceedings of the Seminar Conducted at the East China Normal University* (Shanghai: SLO-Netherlands Institute for Curriculum Development)
[17] Tessmer M 1993 *Planning and Conducting Formative Evaluations: Improving the Quality of Education and Training* (London: Kogan Page)
[18] Zulkardi 2002 *Developing a Learning Environment on Realistic Mathematics Education for Indonesian Student Teachers Dissertation University of Twente Enschede* (The Netherlands Published Dissertation)
[19] Chapman O 2015 *LUMAT (2013-2015 Issues)* **3** 19
[20] Ing A C and Wang H H 2018 *Asia-Pacific Journal of Teacher Education* **46** (3) 221
[21] Siswono T Y E, Kohar A W, Kurniasari I and Hartono S 2017 *Southeast Asian Mathematics Education Journal* **7** 27
[22] Siswono T Y E, Kohar A W, Kurniasari I and Astuti Y P 2016 *Journal of Physics: Conference Series* **693** 012
[23] Siswono T Y E, Kohar A W, Hartono S, Rosyidi and Masriyah 2018 An innovative training model for supporting in-service teachers' understanding on problem-solving knowledge for teaching *Proceeding of EARCOME* **8** 321
[24] Siswono T Y E, Kohar A W, Rosyidi A H and Hartono S 2018 *Journal of Physics: Conference Series* **953** 012197
[25] Siswono T Y E, Kohar A W and Hartono S 2017 *Journal of Physics: Conference Series* **812** 012046
[26] Ekawati R and Kohar A W 2017 *International Journal of Innovation in Science and Mathematics Education (formerly CAL-laborate International)* **24** 1
[27] Ekawati R, Kohar A W, and Hartono S 2017 *Journal of Engineering Science and Technology (JESTEC)* 53
[28] Siswono T Y E, Kohar A W, Savitri D and Hartono, S 2017 *World Trans. on Engng. and Technol. Educ.* **15** 355
[29] Siswono T Y E, Kohar A W, Rosyidi A H, and Hartono S 2017 *World Trans. on Engng. and Technol. Educ.* **15** 126

Students' collaborative ability in learning geometry transformation using a scientific approach based on learning community

Hobri, J Safitri, S Romlah, E Nazareth and Susanto
University of Jember, Jl. Borneo No. 37, Campus Tegalboto, Jember, East Java, 68121, Indonesia

E-mail: hobri.fkip@unej.ac.id

Abstract. The central concept of the 2013 curriculum is the scientific approach based on the logical explanation of the fact; students are expected to be able to solve problems through a structured scientific process. However, students often make mistakes in addressing the questions about geometry transformation. Therefore, learning involving multi parties, such as friends and parents, is required. In this paper, we proposed learning based on the learning community. This research aimed to explore the improvements in student's collaborative ability by applying Learning Community-Based Scientific Approach on the topic of geometry transformation. The study was conducted in three cycles, with three steps in each cycle (plan-do-see*)*, based on the learning cycle in the lesson plan. This research was done in one of the high schools in the East Java, Indonesia, in the academic year of 2017/2018. The research subjects were Year 11 science students. The data were analysed using a qualitative descriptive method consisting of data reduction, data presentation, and conclusion. The data were gathered using observation, test, and documentation. The result shows that there is an improvement in their ability to learn mathematics collaboratively. They could work together and appreciate the opinions of their group members and another group.

1. Introduction

21st-century learning demands the students acquire the characters, often addressed as 4C, communication, collaboration, critical thinking and problem solving, creativity and Innovation [1]. In the scientific approach, learning is conducted based on facts which can be logically explained. Therefore, the students can find the answers that may not be making sense but are achieved through the structured scientific process. The success of a curriculum depends on its implementation at schools. Hence, schools are demanded to facilitate the establishment of a learning community in which students and teachers are learning from each other [2]. One of the ways to implement a learning community based scientific approach is through collaborative learning.

Learning community requires collaboration in the classrooms and encourages teachers' collegiality to work together to achieve the goal [3]. Collaborative learning is learning in a group; however, the purpose is not achieving unity through group works. The students in the group are encouraged to find various arguments or thoughts of everyone in the group [4]. The vision of the learning community is that no student can be left on their own or 'no students are to be neglected.' The teachers are demanded to know, care and educate the troublesome students by facilitating them to work collaboratively [5]. There are three important elements for success full learning community: focus on learning, collaborative culture, and result oriented. One of them is creativity, which oriented to National Council of mathematics teacher [6,8]. The established learning community helps the students to communicate with each other and to the teachers, which encourages students to be more responsible. It may also motivate the students that they find their learning process exciting. The students will often express their opinion and thoughts in transformation geometry learning through the problems related to their environment and will interpret them in answer to their misunderstanding of the forms of transformation geometry [9]. Moreover, the student may often respond to the statements or problems in the form of convincing arguments. Therefore, the research questions in this study are 'is there any significant improvement in the students' collaborative ability in learning transformation geometry in mathematics lessons in the class of XI science 1 through learning community-based scientific approach learning? Improving the students' collaboration ability can be done by applying learning community-based scientific approach to the transformation geometry material.

2. Method

This research adopted a qualitative approach with classroom action research design. The research involved three cycles of plan-do-see, following the learning cycle in the lesson. This study was conducted in one of the high schools in the east of Java during the even semester of the academic year 2017/2018. The research subject was 32 Year 11 students (18 male 12 female). The learning materials were the introduction into the types of transformation geometry in cycle 1; understanding translation and reflection in cycle 2, and were rotation and dilatation in the cycle 3. The data generation methods were observation, test, and documentation.

The observation was conducted directly on students' teaching and learning activity especially in the phases of the plan, do and see. The activities in the plan phase were creating the lesson plan and students' worksheet together with the experienced mathematics teacher. In the do phase, the teaching and learning activities were conducted based on the plan in the plan phase. Meanwhile in see phase, researchers reflected on the teaching and learning activity which has been conducted concerning students' activities to collaborate with other students and the outcome of the collaboration. Test data was used to understand students' learning outcome, while, documentation was implemented to record the teaching and learning process. The instruments used in this research were the observation guide and the achievement test questions. Data validity was obtained from sources and method triangulation. Data analysis was conducted in a descriptive qualitative method with the following procedure: data reduction, data presentation and conclusion or verification [7].

3. Results

The research is conducted for 32 Year 11 students. The summary of the observations is presented in Figure 1.

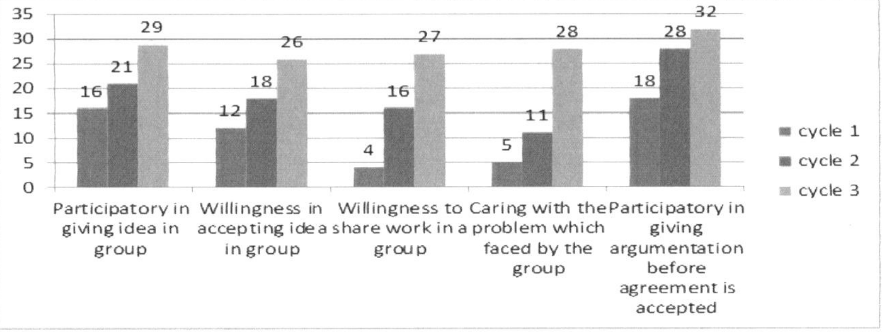

Figure 1. Diagram of the learning process based on learning community.

The student's participation in giving an idea in a group, cycle 1: 16 Students, cycle 2: 21 students, increasing by 31.25%, and in cycle 3: 29 students, increasing by 38.09%. The willingness in accepting the idea in a group on cycle 1 is 12, and in cycle 2 is 18 students, increasing by 50%, and in cycle 3:

26 students raising by 44.44%. Willingness to share work in a group on cycle 1 is 4, in cycle 2: 16 increasing four times from cycle 1, in cycle 3: 11 raised by 68.75%. Caring about the problems faced by the group in cycle 1 is 5, in the cycle 2 is 11 increasing by 120%, in cycle 3 is 28 students increasing by 154.5%. The participation in giving arguments before accepting the arguments is 18 in cycle 1, 28 in cycle 2 raising by 55.56%, and 32 students in cycle 3, increased by 14.28%. Meanwhile, students' learning outcome is as follows:

Table 1. Students' achievement.

Range score	Students		
	Cycle 1	Cycle	Cycle 3
< 50	4	0	0
50 – 59	3	1	0
60 – 69	9	4	0
70 – 79	14	18	9
80 – 89	2	8	15
90 – 100	0	1	8
Total	32	32	32

On the first cycle, 16 students achieved 70 or higher, and 16 students achieved lower than 70. Half of the students reached the score below the minimum mastery level. The average of students score in the cycle 1 is 65.1. On the cycle 2, 27 students achieved 70 or higher, and five students scored lower than 70. The average test score on the cycle 2 is 75.1. On the cycle 3, 32 students scored 70 or higher, and the average test score is 83.5. There is an increasing percentage of cycle 1 to cycle 2 by 15.36%, and cycle 2 to cycle 3 by 11.19%.

4. Discussion

4.1. Cycle 1

Cycle 1 began by applying the plan phase conducted by the researcher and the other mathematics teacher, EP and TP. The plan phase started with creating the lesson plan, students' worksheet and a test on transformation geometry material. The activity on this phase was designing the steps of learning community-based scientific approach to the transformation geometry material following the 2013 curriculum, preparing supporting media, and determining how many meeting would be required [10]. After the plan phase finished, the do phase of cycle 1 was conducted on 24 January 2018, at 06.30-08.00 WIB in the class of XI Science 1.

The teacher prepared the class and conducted the pre-activity. The students were divided into eight groups. Each group consisted of 4 students arranged by designing the students' desk at the beginning of the lesson and U-shaped in group discussion as presented in Figure 2.

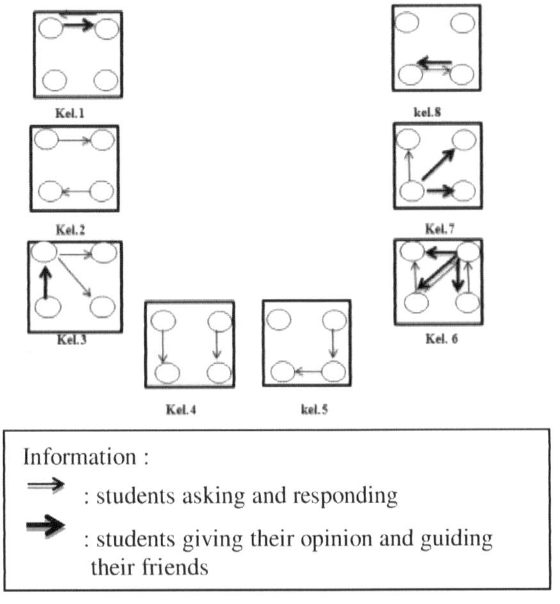

Figure 2. The illustration of teaching and learning process in cycle 1.

The teacher gave stimulus materials by providing the kind of geometry transformation in the form of pictures and another thing which needs to be filled dealing with learning community-based learning geometry transformation, in dealing with group discussion, between the group, and consultation with the teacher. Geometry provides an opportunity for students to stimulate their ways of thinking, critical thinking, and problem-solving, finding differences, generalizing and summarizing the development of student skills [11]. After the do phase, then continued to the see phase. A few things obtained from the discussion in this cycle: the learning did not run well based on the plan, for example in learning community-based the students were still accustomed to working by themselves and did not get used to ask and give opinion to his friend because they assume that the correct answers derived from the teachers. However, there was a group which participated actively in the discussion, i.e., group 6 as can be seen in the figure 3..

One question from the student in group 6 "why there are three pictures of different sizes?". One student named Rdt seemed to understand the purpose of the different sizes of the pictures and he explained to his friend that the real map is minimized twice from the original size of the map and the other is maximized 2 times so that there was a change from the original picture after being maximized and minimized 2 times. However, the others group were still passive, and there was a student who was unable to accept the opinion of a friend from a different group. In this case group 2, 4 and 5 were unable to give theirs, and they intended to be passive. A learning community (LC) requires collaborative learning in the classroom that encourages collaboration between teachers to achieve the goals [3]. However, in this first cycle, there is no communication between the groups. Hence, in the cycle 1, there was no significant improvement in the learning community. The student was not used to conducting group discussions, the students did not optimally show their courage, and they still got used to having the lecturing method from the teacher, and their thinking maturity was still low.

As the results of observation and test, sense of care against the problem faced by the group still did not emerge. Also, the willingness to share the work in a group, either asking or giving an opinion was still way beyond caring and collaborative which are the implementation of the learning community. This resulted in the still low understanding and mastery of transformation geometry of the students, and many (16 students out of 32) achieved scores below the minimum criterion, 70. The average score from the test was 65.1, still below the 70 (minimum score). Due to the weaknesses found in cycle 1, there was a need to conduct a remedial action in cycle 2.

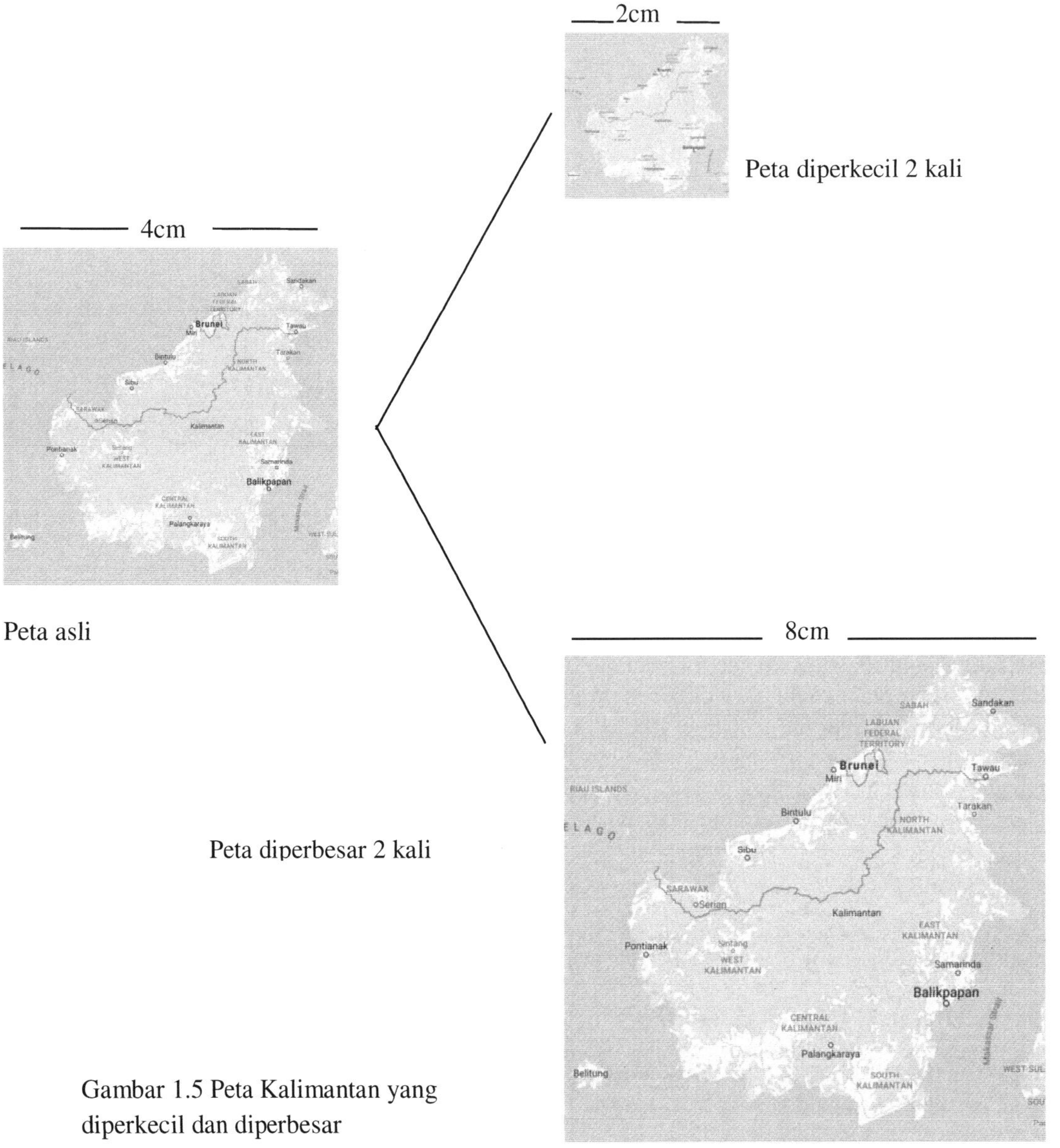

Figure 3. The example of dilatation question

4.2. Cycle 2

Cycle 2 began with doing the do phase by creating a remedial plan on the cycle 1. Phase plan started with revising the lesson plan and adjusting with the results of Cycle 1, and created the lesson plan and student worksheet on translation and reflection material, and designed the steps of learning community-based scientific approach. The do phase of cycle 2 was conducted on 25 January 2018. The implementation of the do phase of cycle 2 was better than of in cycle 1. In cycle 2, the teacher was able to design the students' U-shaped seats on the beginning of teaching and learning process in Figure 4.

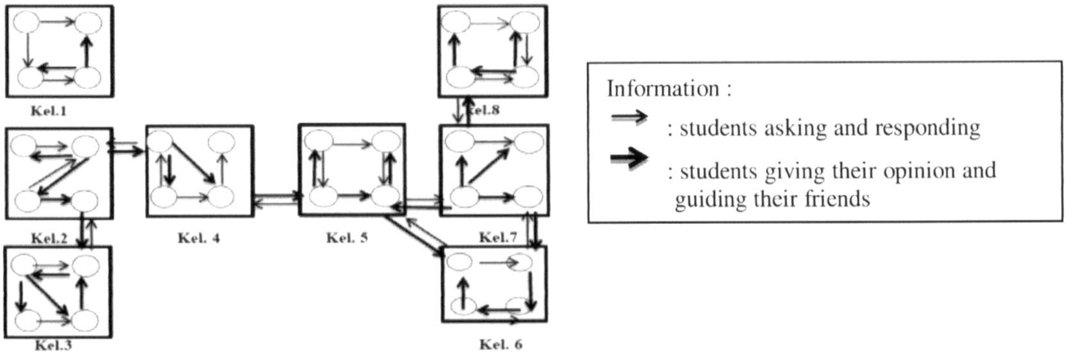

Figure 4. Illustration chart of the learning process in cycle 2.

From the illustration above, we can see that students were participating in giving their ideas in a group and giving their opinion before an agreement was increasing. The students started to get used to delivering their idea. In the cycle 2, almost half of the students got used to asking, give their idea in a group, work together with the members of their group and of the other groups. Many interactions happened in Cycle 2 even though group 1 still discussed with their own member. In cycle 2, there was an exciting event when members of group 8 actively expressed their arguments to solve this following problem:

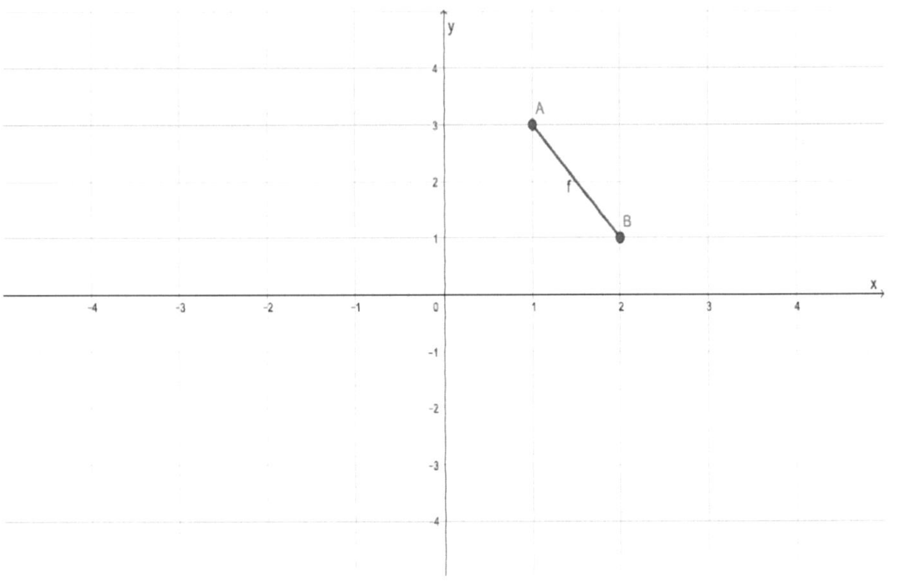

Figure 5. The example of dilatation question.

The student was asked to find the shadow of point A and B. One member of the group, called Prmd, stated that the reflection for point A (1,-3) and B is (2,-5). This made his fellow group member, named Vndn, argued that his answer was wrong, and the correct shadow point for point B (2,-1). But Prmd still insisted that his response was right, and then he asked the opinion from another group, which was group 7, and he also stated that Pond's answer was incorrect. Apparently, Prmd still did not understand reflection, which was elaborated again by his group. Then Prmd started to identify the characteristic of reflection.

The do phase was followed by the see phase. A few things obtained from the discussion on reflection i.e., the teaching and learning process was already conducted by the teacher chronologically; the teacher managed the time well and conditioned the students in the group and the class well. All students in the class participated actively in listening to the teacher's explanation, were willing to do the job assigned to each group. No student kept talking to this friend, no students who were not listening to the teacher and no student went out of the class without any permission. This shows that every student was engaged in the teaching and learning process and did as what the teacher said. Every student was concentrating, taking note and doing their work. Therefore, no student disturbed his friend, made noises and left the class due to any reasons.

The remedial plan on cycle 1 for cycle 2 had already been conducted. The teaching and learning process was better than the previous. The three most essential elements for a thriving learning community: focus on learning (ensuring that students learn), collaborative culture, and results-oriented [12]. However, the student achievement was still not optimum because there were5 students who achieved the score below the minimum criterion. Many students who actively asked questions cycle 2 is 18, increasing from 12 in Cycle 1. While, the student who gave their ideas and guiding their friends were 18, dramatically improved from cycle 1 with only four students. This affected the students' understanding and mastery of translation and reflection material as shown by the fact that 27 students achieved 70 (the minimum criterion score). The average score in cycle 2 was 75.1, which was better than cycle 1 65.1. It means that in cycle 2 there was an increase in the caring community-based learning for about 15.36% from cycle 1. However, there were still weaknesses in cycle 1 and 2, and then remedial action on cycle 3 was necessary.

4.3. Cycle 3

Just like in the previous cycle, Cycle 3 begins with plan phase conducted by the researcher in collaboration with mathematics teachers by looking at the remedial plan of cycle 2. Plan phase began with revising the lesson plan and student worksheet on translation and reflection material, designing the steps of learning community-based scientific approach. Do phase of cycle 3 conducted on 31 January 2018 and attended by the whole student. Doing the do phase of cycle 3the students were divided into eight groups and on cycle 2 the teacher was able to design the student U-shaped seat on the beginning of teaching and learning process. However, the student Ddt faster than usual, which can be seen below.

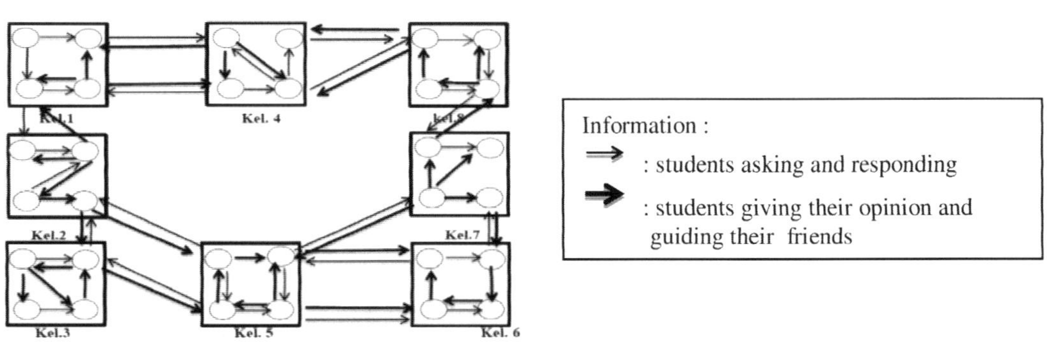

Figure 6. The teaching and learning process illustration in cycle 3.

From the figure above, we can see that the discussion in every group was running well, each student delivered their idea and those who understood guided their friends in their group. Likewise, when another group did not understand, the other group would help them. In cycle 3, as we can see from the picture above that group 5 and 6 had more opportunities than another group to collaborate with another group because of their position in the middle of the four groups. This suggests that collaborative learning is a learning that is carried out in groups, but the goal is not to achieve unity gained through group activities; however, students in groups are encouraged to find the diverse opinions or thoughts expressed by everyone in the group [4]. The results of observation, caring to the problem which faced by the group is shown through the activity. Furthermore, the willingness to share their work on the group, asking and giving their idea is fit caring and collaborative which are the implementation of the learning community. Inbuilt learning community, we can see the interaction and care among the student is rising.

5. Conclusion

It can be concluded that learning geometry transformation through learning community-based scientific approach can enhance students' collaboration ability and their learning achievement. It also increases their confidence to communicate. The result of the observation on teaching and learning process shows that there is an improvement from cycle 1 to cycle 2, and cycle 2 to cycle 3. Students' achievement also increases from cycle 1 to cycle 2 by15.36%, and in cycle 2 to cycle 3 by11.19%.

Using appropriate learning models may create a better learning environment to improve students' learning outcome. The application of Learning community-based scientific approach in teaching geometry transformation is a shred of evidence that learning also demands students' and teacher activity to develop students' potential.

Acknowledgment

The authors gratefully acknowledge the Faculty of Teacher Training and Education of Jember University for all the support provided for the completion of our project and the writing of this paper.

References

[1] Wijaya E Y, Sudjimat D A and Nyoto A 2016 *21st Proceedings of The National Seminar on Mathematics Education 2016-Universitas Kanjuruhan Malang* **I** 263

[2] Saito E and Atencio M 2015 Lesson Study for Learning Community (LSLC): Conceptualising teachers' practices within a social justice perspective *Discourse: Studies in the Cultural Politics of Education* **36** 795

[3] Widjajanti and Djamilah B 2008 *Semnas Mathematics and Mathematics Education: FMIPA UNY*

[4] Hobri 2016 *Proceedings of The National Seminar on Mathematics Education, Theme: The Role of Mathematics and Learning in Developing Local Cultural Wisdom to Support Nation Character Education*

[5] Sudirtha I 2017 Building learning community and increasing competence through lesson study *Journal of Indonesia education* **6** 30

[6] Hobri et al 2018 *J. Phys.: Conf. Ser.* **1008** 012065

[7] Wang F and Burton J 2010 Collaborative learning problems and identity salience: A mixed methods study *Journal of Educational Technology Development and Exchange (JETDE)* **3** 2

[8] Tohir M et al 2018 *J. Phys.: Conf. Ser* **1008** 012072

[9] Yılmaz G K and Koparan T 2016 The effect of designed geometry teaching lesson to the candidate teachers' van hiele geometric thinking level *J. Educ. Train. Stud.* **4** 129

[10] Hobri and Susanto 2016 *Proceedings of the National Seminar*

[11] Avci M, Yavuz A and Aidin B 2016 The van hiele levels at czech secondary schools *Eur. J. Educ. Stud.* 59

[12] Battistich V, Solomon, Watson and Schaps E 1994 Students and teachers in caring classroom and school communities *American Educational Research Association (AERA)*

Realistic mathematics learning based on the ethnomathematics in Bengkulu to improve students' cognitive level

W Widada, D Herawaty and A N M T Lubis
Universitas Bengkulu, Jl. W. R. Supratman, Kandang Limun, Muara Bangka Hulu, Bengkulu 38371

E-mail: w.widada@unib.ac.id

Abstract. Mathematical learning based on ethnomathematics applying local culture can change the perception of the society about the applications of mathematics in everyday life. The purpose of this study was to describe the increase in high school students' cognitive level through the implementation of realistic mathematics learning model based on ethnomathematics in Bengkulu. This study was an assessment phase of development research employing a 2×2 factorial experimental design with students' initial performance being controlled. There were three main findings of this study. First, mathematical understanding of students taught using realistic mathematics learning was higher than those who taught using the conventional method (the learning materials in both groups was non-ethnomathematics oriented). Second, mathematical understanding of students learned the ethnomathematics oriented materials was higher than those learned non-ethnomathematics oriented materials (realistic mathematics learning applied in both groups). Third, mathematical understanding of students who learned the ethnomathematics oriented materials was lower compared to the students learned the non-ethnomathematics materials (the conventional learning method applied in both groups).

1. Introduction

To improve the understanding of mathematical concepts, learning must begin with the real objects in daily life [2,3], which were culturally oriented concerning horizontal mathematics known as the ethnomathematics [4]. It builds a system of knowledge offering the possibility of a more harmonious and good relationship in human behavior and between human and nature [5,6,7,8,9]. Ethnomathematics is the representation of a daily mathematical practice or the culture of a particular group in a socio-cultural environment [10,11,12]. It can build mathematical reasoning and mathematics as a cultural construct [5] [11]. According to Freudenthal [13,14], mathematics is a human activity, and thus the objects of mathematics in mathematics learning should begin with real problems in everyday life or those close to the student's mind. Mathematical learning based on the ethnomathematics applying local culture can change the perception of the society concerning the lack of connection between mathematics and daily life [15,16].

Mathematics learning emphasizes more on students to understand facts, concepts, principles, and operations. Understanding of the mathematics concepts is the result of construction or reconstruction of mathematical objects. The construction or reconstruction is done through activities of mathematical actions, processes, objects organized in a *scheme* to solve questions or problems [18,19]. To understand the concepts of high school mathematics, students are expected to perform an intelligent and responsible action based on knowledge, skills, and values integrated through the process of abstraction, idealization, and generalization used to carry out tasks or solve problems associated with high school mathematics.

However, the results of preliminary research conducted in several high schools in the province of Bengkulu showed that students made mistakes of the concept and principles of understanding mathematics. The errors of concepts and principles committed by the students included the limit of a function. When the students asked to solve the problem of the limit of a function: $\lim_{x\to} \frac{\sqrt{x}-1}{x-1}$, 76% of students made mistakes concerning the concepts/principles in solving the problem. 57% of students' answer was 0/0, 14% was 1, 5% was 0, and the remaining 24% provided the correct answer of ½. Since the quadratic equation plays an important role in solving the problem related to the limit, the students' understanding of the quadratic equation was traced by asking them to solve the problem of $x^2 - 3x + 2 = 12$. Evidently, 38% of students solved the problem by presenting $(x - 2)(x - 1) = 12$, followed by $x - 2 = 12$ or $x - 1 = 12$ and concluded that the solution was $x = 14$ or $x = 13$. This result shows that 38% of the students committed errors concerning the concepts and principles in solving the quadratic equations. These errors are due to a low abstraction process leading to students having difficulty in solving mathematical problems.

Based on students' difficulties in performing the abstractions related to the concept and the principle of the limit of a function, it is implied that the mathematics learning in high school still mostly focus on theory and is not related to daily life. The students learn mathematics mechanistically without understanding the application of the theory learned. As a result, students experience difficulty in understanding the mathematics learning materials and tend only to memorize the concepts/principles learned. Most students of senior high school study mathematics merely to pass the national exam. Consequently, they find it difficult to solve everyday life problems even when they are closely related to the mathematics content being studied. Besides, mathematics learning does not connect with and explore the phenomenon around students despite mathematics being a human activity [2,3]. According to [21], to overcome the cognitive conflict and students' mistakes as experienced by the students mentioned previously, realistic mathematics learning model is required. To achieve the vertical mathematical process well, [6] [22] [24] suggested applying ethnomathematics-based learning. Mathematics learning based on the ethnomathematics enables the process of the abstraction, idealization, and generalization of concepts and principles of mathematics. When it is viewed based on the characteristics of mathematics, mathematics is a human activity which was the utilization of the reality and environment understood by the learners to expedite the process of learning mathematics to achieve a better mathematics education [2-3,13].

Based on the review discussed earlier, a mathematics learning based on the ethnomathematics is needed as a starting point. In short, based on the ethnomathematics, mathematics considers the quantitative, relational and cultural aspects of the society that are integrated with the concrete things that can be observed or understood by learners through the process of mathematization as mathematics is a reflection of the real world through an empirical abstraction process [2,3].

Improving students' mathematical ability can be conducted through meaningful learning. Learning mathematics is meaningful if it is close to the students' mind [2,3] [13] [25] so that students can be a problem solver in everyday life, have an excellent mathematical communication and provide a correct mathematical representation [31]. In learning mathematics, students perform the process of mathematical communication either orally or in writing. The improvement of mathematical communication and representation of high school students is increased when the starting point is the down-to-earth problems [31]. Mathematics learning exposes students to the process of interaction and multi-direction communication between students, teachers and learning instrument allowing a growing dynamic during the learning in the classroom.

Mathematics is the miniature of democracy and our society. Therefore, mathematics learning should be implemented using the ethnomathematics, and/or local culture as the starting point [6-7, 9, 26-27]. The results of previous studies [24] [11] showed that learning with ethnomathematics approach is more effective compared to the conventional approach. One reason is that ethnomathematics plays an essential role in building a civilization rejecting the gaps, arrogance and fanatical attitudes [8]. The ethnomathematics-oriented learning can increase student activities as well as enhance students' motivation, interest, and confidence. Ethnomathematics can build mathematical reasoning and mathematics as a cultural construct.

In everyday life, ethnomathematics forms a system of knowledge offering the possibility of a more harmonious and good relationship in human behavior as well as between humans and nature [8]. Ethnomathematics is a representation of a daily mathematics practice or a culture of a particular group in a socio-cultural environment [8,9]. However, mathematics education is not political [16]. Thus, mathematics education must be free from the influence of power. So that according to [15] [28] by The application of ethnomathematics, in the form of local culture, in mathematics learning can reduce the perception of the society concerning the lack of connection between mathematics and daily life.

By applying the surrounding culture as the starting point of mathematics learning, ethnomathematics can be the integrated component of mathematics education. Therefore, ethnomathematics can be integrated into the mathematics curriculum intensively [12]. This is rational as ethnomathematics is a sub-field of mathematics history and mathematics education with a natural connection to anthropology and science [9].

To implement mathematics learning to achieve the mastery of mathematical representation, mathematical communication, and problem-solving skills, a conceptual framework of learning model is required [29]. The learning model using problems in everyday life or that is close to the students' mind [2,3] [13], the culture of the local community [6] [8,9] [13], as the starting points. The model is a model of realistic mathematics learning based on ethnomathematics [22].

Realistic mathematics learning model based on ethnomathematics highly benefits the students in the process of abstraction, idealization, and generalization [30] [18] [20] [32], by utilizing off and for the model [13]. The abstraction ability is an ability to pay attention and determine the nature, attributes, or special characteristics that were important by putting aside different unimportant things. Based on the results of Widada's research [32] [30] [18] [20], there were six levels of abstraction performed by students, i.e., Level 0: Concrete Objects; Level 1: Semi-concrete models; Level 2: Theoretical Models; Level 3: Language in Domain; Level 4: Geometry Language; and Level 5: Inference Model. Widada's Research (2016b) found that in term of the cognitive structure of learners that was reviewed based on the abstraction ability in the formation of mathematical concepts, 74% of students had the constructive functional element and 26% of students whose constructive elements were predictive. Thus, most cognitive structures of learners in performing abstraction is functional.

Research conducted by Widada [18] [30] [32] produced a description of the learners' abstraction ability in understanding mathematical concepts of Level 0 (Concrete Objects), Level 1 (Semi-concrete), Level 2 (Theoretical Models), Level 3 (Languages Domain in example), Level 4 (Mathematics Language), and Level 5 (Inference Model). While, in the process of understanding the concept of mathematics, Widada [32] have conducted a *long-term study* concerning the students' cognitive structure in understanding mathematics. The results of Widada's study [18] reinforced and complemented the theory of Cognitive Development (scheme) of students (Extended Level Triad ++) in learning mathematics [17].

Based on the progression scheme of Extended Level Triad ++ [32], the learning model based on Extended Level Triad ++, learning model close to the student's mind is developed. The learning model is outlined in the learning syntax consisting of three cycles: Problem Launch, Exploration, and Conclusion. Based on the three main cycles, more detailed steps can be derived: (1) Preliminary phase; (2) Thinking Phase, (3) Coupled Phase, (4) Exploration Phase, (5) Discussion of Exploration Phase, (6) Conclusion Phase; and (7) Closing phase [20].

To achieve the highest level of mathematization process [14], in the horizontal mathematization, students with their knowledge can organize and solve the existing problems in real-world situations. In other words, horizontal mathematics moves from the real-world to the world of symbols. While vertical mathematization is the process of reorganizing using mathematics itself or "real-world" as the source of mathematization and a place to reapply the mathematical concepts. Thus, the vertical mathematization moves from the world of symbols. Gravemeijer [2-3] suggested that there are three main principles of realistic mathematics learning, namely: rediscovering and progressive mathematization, didactic phenomena, self-building models. Treffers [13] revealed the characteristics of realistic mathematics learning as the use of context, use of models, bridging by vertical instrument and students' contribution. By the production and construction, students were encouraged to reflect on parts they find important in their learning [23].

A study [31] showed that the mathematics problem-based learning on Earth Raflesia Bengkulu (ethnomathematics in Bengkulu) could improve students' ability to solve the problems, increase their cognitive process through a higher level of thinking, creative and critical thinking resulting in students being able to perform mathematical representation and communication correctly.

Thus, mathematics learning in high school should be taught by implementing the mathematics learning model employing a realistic mathematics education approach based on ethnomathematics in Bengkulu. The implementation of the model will enhance the understanding of mathematics concepts and the abstraction ability of high school students. Finally, the cognitive level of the student is hypothesized to reach the highest level (the extended-trans level) [32].

2. Methods

This study was an assessment phase of Plomp's developmental research model [1]. In this phase, we conducted the test, evaluation, and revision of the teaching materials, the realistic mathematics learning model based on ethnomathematics in Bengkulu. Once these materials met all the required standard, we implemented them in the classroom. In the implementation, we employed a 2×2 factorial experimental design. The preceding stage was the preliminary research and prototype phase. The learning approach included realistic and expository mathematics, while the approach of mathematics materials included ethnomathematics and conventional. The population of the study was all students in one of senior high school in Bengkulu, with a sample of 80 students. The sample was selected by the intact group technique. The realistic mathematics learning approach based on ethnomathematics in Bengkulu was implemented in the experiment class, and the conventional approach was implemented in the control class. Data collection was carried out using the instrument of mathematics understanding ability. It was used to measure students' cognitive level of mathematics understanding. Data were analyzed by Analysis Covariance (ANCOVA) test.

3. Result and discussion

Based on the data of realistic mathematics learning experiment with ethnomathematics in Bengkulu, the analysis of covariance was conducted and the results are presented in Table 1.

Table 1 shows that Fo (A) = 21.290, df = (1.79) and p-value = 0.00<0.05, Ho was rejected. It means that there was a significant difference in mathematics understanding ability between students taught by realistic and conventional mathematics learning approach after controlling the students' initial ability. Table 1 also displays that Fo (B) = 29.466, df (1.75) and and p-value = 0.00<0.05, Ho was rejected. This means there was a significant difference in mathematics understanding ability between students who were given the ethnomathematics-oriented the non-ethnomathematics learning materials after controlling students' initial ability. Besides, Table 1 also presents Fo (AB) = 40.960, df(1.75) and p-value = 0.00<0.05, Ho was rejected indicating that there was an interaction effect of the learning approach and mathematics materials orientation to the ability of mathematics understanding ability after controlling student's initial ability. It is also shown that Fo (X) = 278.460, df(1.75) and p-value = 0.00<0.05, Ho was rejected meaning that there was the linear covariate effect of student's initial ability and mathematics understanding ability. As for the corrected model, Fo = 246.049 with df (4,75) and p-value = 0,00<0,05 Ho was rejected. Thus, the students' initial ability, learning approach and mathematics material orientation together influence the mathematics understanding ability.

Table 1. Tests of Between-Subjects Effects

Source	Type III Sum of Squares	df	Mean Square	F	Sig.
Corrected Model	13518.576	4	3379644	246,049	.000
Intercept	447.232	1	447.232	32560	.000
A	292.435	1	292.435	21.290	.000
B	404.733	1	404.733	29.466	.000
A * B	562.616	1	562.616	40.960	.000
X	3824.826	1	3824,826	278.460	.000

Table 2. Parameter Estimates (1)

Parameter	B	Std. Error	t	Sig.
Intercept	-66.633	6.452	-10.327	.000
[A = 1.00]	3,599	1.159	3.106	.003

Based on Table 2, t = 3.106 and p-value = 0.003<0.05 means Ho was rejected. This means that the average ability of students' mathematical comprehension to be taught with realistic mathematics learning approaches was higher than students who were taught using conventional learning approach after controlling students' initial ability.

Table 3. Parameter Estimates (2)

Parameter	B	Std. Error	t	Sig.
Intercept	-69.954	6.175	-11.329	.000
[B = 1.00]	4,871	1.098	4,436	.000

Table 3 presents t = 4.436 and p-value = 0,000<0.05 meaning Ho was rejected. This means that the average ability of students' mathematics understanding learned using the ethnomathematics-oriented material was higher than those given non-ethnomathematics oriented material after controlling the initial ability.

Table 4. Parameter Estimates (3)

Parameter	B	Std. Error	T	Sig.
Intercept	-37.636	6.979	-5.393	.000
X	1.748	.105	16688	.000
[A = 1.00] * [B = 1.00]	12001	1.391	8.628	.000

Based on the results shown in Table 4, it is identified that t = 8.628 and p-value = 0,000<0.05, meaning Ho was rejected. It indicates that there was an interaction effect between the learning approach factor and mathematics materials orientation to the mathematics understanding ability after controlling the student's initial ability.

Table 5. Estimates Parameters (4)

Parameter	B	Std. Error	t	Sig.
Intercept	-37.636	6.979	-5.393	.000
X	1.748	.105	16688	.000
[A = 1.00] * [B = 1.00]	11.316	1.466	7.721	.000
[A = 1.00] * [B = 2.00]	-3.652	1.402	-2.604	.011

Table 5 describes the results of t-test (column A1B1) showing t = 7,721 and p-value = 0,000<0,05 indicating that Ho was rejected. Thus, mathematics understanding ability of students who were taught with a realistic mathematics learning approach was higher than those taught by the conventional learning approach (both groups of students were given ethnomathematics-oriented materials, and students' initial ability was controlled). Furthermore, the t-test result on column A1B2 shows that t = -2.604 and p-value = 0.011<0.05 meaning Ho was rejected. Therefore, the mathematics understanding ability of students learned using realistic mathematics learning approach was lower compared to the students learned by the conventional learning approach (both groups of students were given mathematics materials of non-ethnomathematics after controlling the students' initial ability).

The results of a study [31] showed that, during the learning process of realistic mathematics based on ethnomathematics in Bengkulu, the students could dominate the learning activities. Horizontal mathematical activity conducted by the students in solving the problem about "*ikan sejerek bere secupak madar.*" This suggests that mathematical models based on ethnomathematics and local culture can reduce the perception of society about the lack of connection between mathematics and to daily life as revealed by Septianawati and Puspita [15] and Franēois and Karen [16]. By applying realistic mathematics learning approach based on ethnomathematics in Bengkulu, students' cognitive level can reach the highest level as described by Widada [32].

4. Conclusion

Based on the results of the ANCOVA, it can be concluded that: after the students' initial ability was controlled, the average mathematical understanding ability of students taught in the classroom implementing the realistic mathematics learning approach was higher compared to those taught by implementing the conventional learning. Also, the average ability of students' mathematical understanding who learned the ethnomathematics-oriented materials was higher than students who were given non-ethnomathematics materials (their initial ability was also controlled).

As for the students who were given ethnomathematics-oriented materials, mathematics understanding ability of those learned by implementing the realistic mathematics learning approach was higher than students those used the conventional learning approach after controlling the students' initial ability. Also, for students learned the non-ethnomathematics oriented materials, mathematics understanding ability in the classroom utilizing the realistic mathematics learning approach was lower than their counterpart after the students' initial ability being controlled.

The results also indicate that for the students studied using the realistic mathematics learning approach, the mathematics understanding ability of students given the ethnomathematics-oriented materials was higher than the students learning with the non-ethnomathematics materials after controlling the students' initial ability. Finally, when students were taught using a conventional learning approach, the mathematics understanding of students obtained the ethnomathematics-oriented materials was lower compared to the students with non-ethnomathematics materials after controlling the student's initial ability.

References

[1] Plomp T 2013 *Educational Design Research: An Introduction* (Netherlands: Netherlands Institute for Curriculum Development (SLO))

[2] Gravemeijer K 2004 *Local Instruction Theories as Means of Support for Teacher in Reform Mathematics Education* (Utrecht: Freudenthal Institute & Department of Educational Research)

[3] Gravemeijer K 2008 RME Theory and Mathematics Teacher Education *International handbook of mathematics teacher education* **1** 283

[4] Widada W 2002 Teori APOS sebagai Suatu Alat Analisis Dekomposisi Genetik terhadap Perkembangan Konsep Matematika Seseorang *Journal of Indonesian Mathematical Society (MIHMI)* **8** 89

[5] Rubio and Jennifer S 2016 The Ethnomathematics of the Kabihug Tribe in Jose Panganiban, Camarines Norte, Philippines Malaysian *Journal of Mathematical Sciences* **10** 211

[6] D'Ambrosio U 2016 The Ethnomathematics Program and a Culture of Peace *Journal of Mathematics and Culture* **10** 1

[7] D'Ambrosio U 2001 What was ethnomathematics, and how can it help children in schools? *Teaching Children Mathematics* **7** 308

[8] D'Ambrosio U 1998 In Focus ... Mathematics, History, Ethnomathematics, and Education: A comprehensive program *The Mathematics Educator* **9** 2

[9] D'Ambrosio U 1985 Ethnomathematics and its Place in the History and Pedagogy of Mathematics *For the Learning of Mathematics* **5** 44

[10] Abiam P O, Okechukwu S A, Ugama and Gabriel O 2016 Effects of Ethnomathematics-based Instructional Approach on Primary School Pupils' Achievement in Geometry *Journal of Scientific Research & Reports* **9** 1

[11] Unodiaku S S 2013 Effect of Ethno-Mathematics Teaching Materials on Students' Achievement in Mathematics in Enugu State *Journal of Education and Practice* **4** 70

[12] Rosa M & Orey D C 2011 Ethnomathematics: the cultural aspects of mathematics. *Revista Latinoamericana de Etnomatemática* **4** 32

[13] Treffers 1991 Realistic mathematics Education in The Netherlands 1980-1990 *Realistic Mathematics Education in Primary School* (Netherlands: Freudenthal Institute)

[14] De Lange J 1987 *Mathematics, Insight and Meaning* (Utrecht: OW & OC Rijksuniversiteit Utrecht)

[15] Septianawati T, and Puspita E 2017 Ethnomathematics study: uncovering units of length, area, and volume in Kampung Naga Society *Journal of Physics: Conf. Series* **812** 012021

[16] Achor E E, Benjamin I and Emmanuel U 2009 Effect of ethnomathematics teaching approach on senior secondary students' achievement and retention in locus *Educational Research and Review* **4** 385

[17] Franēois and Karen 2010 The Role of Ethnomathematics within Mathematics Education *Proceedings of CERME* vol 6 (France: Lyon France © INRP) p 1518

[18] Widada W and Herawaty D 2017 Dekomposisi Genetik tentang Hambatan Mahasiswa dalam Menerapkan Sifat-sifat Turunan *Jurnal Didaktik Matematika* **4** 136

[19] Widada W 2016 Profile of Cognitive Structure of Students in Understanding the Concept of Real Analysis *Infinity* **5** 83

[20] Widada W 2013 The Existence of Students in Trans Extended Cognitive Development on Learning Graph Theory *Journal Math Educator Nusantara* **1** 1

[21] Widada W 2016 Sintaks Model Pembelajaran Matematika Berdasarkan Perkembangan Kognitif Peserta Didik *Jurnal Pendidikan Matematika Raflesia* **1** 163

[22] Herawaty D, Rusdi and Effie A M 2016 Pengaruh Penerapan Model Pembelajaran Matematika Realistik berbasis Konflik Kognitif Siswa terhadap Kemampuan Pemahaman konsep dan Kemampuan Pemecahan Masalah *Proc. Semirata MIPA BKS-PTN Barat* (Riau: Universitas Riau)

[23] Mega T B 2011 Etnomatematika sebagai Batu Pijakan untuk Matematika Formal *Proc. Seminar Nasional Pendidikan Matematika* (Bengkulu: Universitas Bengkulu)

[24] Widada W and Herawaty D 2017 Realistic Mathematics Learning Based on Bengkulu Ethnomathematics to Increase Cognitive Level *Bengkulu International Conference on Science and Education*

[25] Iluno C and Taylor J 2013 Ethnomathematics: The Key to Optimizing Learning and Teaching of Mathematics *IOSR Journal of Research and Method in Education* **3** 1

[26] Treffers A 1987 *Three Dimensions A Model of Goal and Theory Description in Mathematics Instruction-The Wiskobas Project* (Dordrecht: Reidel Publishing Company)

[27] Rosa M and Daniel C O 2011 Ethnomathematics: the cultural aspects of mathematics *Revista Latinoamericana de Etnomatemática* **4** 32

[28] Katsap A and Silverman F L 2008 A Case Study of the Role of Ethnomathematics among Teacher Education Students from Highly Diverse Cultural Backgrounds *Journal of Mathematics and Culture* **3** 66

[29] Joyce B and Weil M 2010 *Models of Teaching* (New Jersey: Prentice Hall)

[30] Widada W 2015 Proses Pencapaian Konsep dengan Memanfaatkan Media Pembelajaran Kontekstual *Jurnal Penelitian Pendidikan Matematika dan Sains* **22** 31

[31] Widada W and Herawaty D 2017 Realistic Mathematics Learning Based on Bengkulu Ethnomathematics to Increase Cognitive Level *Bengkulu International Conference on Science and Education*

[32] Widada W 2015 The Existence of Students in Trans Extended Cognitive Level *Journal Math Educator Nusantara* **1** 1

IOP Conf. Series: Journal of Physics: Conf. Series **1088** (2018) 012029 doi:10.1088/1742-6596/1088/1/012029

Designing student worksheet for rich mathematical tasks

F Fitriati and R Novita
STKIP Bina Bangsa Getsempena, Jl. Tanggul Krueng Aceh No.34 Rukoh, Darussalam, Banda Aceh, Indonesia.

E-mail: fitriati@stkipgetsempena.ac.id

Abstract. Implementing rich tasks in daily mathematics classroom requires a supporting document for teachers who want to use an approach. This study developed a student worksheet that would act as cognitive scaffolding to support the students' immediate construction of knowledge and provides the basis for future independent learning of the students in solving rich mathematical tasks. The study was conducted within research and development design which consisted of three phases: preliminary research, designing and experimenting phase. This paper is primarily concerned with the first and second phase of development study by focusing on the design of rich task based student worksheet and implication for pedagogy. Mathematics teachers and their students from one of the junior high schools in Banda Aceh were involved in this study. The study found that the student worksheet could be used by teachers to improve their students' ability in solving rich mathematical tasks in a daily mathematics classroom. The document also aids the students to solve rich tasks independently through tasks' scaffold questions provided within the student worksheet. However, some revisions are still needed before it will use extensively in schools.

1. Introduction

Providing an opportunity for students to develop a high level of mathematical competences as documented in TIMSS and PISA such as problem-solving, modeling, and reflective thinking become the center of current reform in mathematics education. Teachers are demanded to be able to arrange mathematics instructions and to provide teaching materials which could support students learning to achieve such competencies. This indicates that teachers are required to be more creative and innovative in designing learning approach and learning sources to their students.

It is argued that rich task is one of the approaches that can be used to promote students higher order thinking skills [1,2,3,4] because it could provide student learning opportunity to solve real-world problems that required higher order thinking skills including problem-solving, reflective thinking and reasoning [1,5]. Rich tasks are defined by Mould [6] as "A task that can engage students in the learning process, make contents meaningful and foster connections among ideas and disciplines. According to Fergusson [7], tasks in which the whole class can engage and are easily adjusted to students' level of ability. Teaching mathematics using rich tasks allows entrance into the activity at multiple levels, with multiple paths possible to obtain the resulting solution [8]. Regarding these potentials of a rich task, many education standards and curriculum guidelines throughout the world mandate mathematics teachers to create more complex tasks to prepare students to live in the challenging world. However, designing rich tasks is not an easy task for teachers. Suardja, Fitriati and Novita [9] found that mathematics teachers in Indonesia rarely developed their own mathematical tasks. Most of them use learning materials such as lesson plan, textbooks and student worksheet provided by government because they lack knowledge and skill to design mathematical tasks. In addition, there is now a mounting body of literature pointing to the fact that rich tasks are still not implemented in mathematics classrooms, or if implemented then routine mathematical tasks are being adopted.

This research has identified several issues in the literature in the international context as well as in local context that need to be addressed to facilitate the implementation of rich tasks in schools. To aid in the implementation of rich task in schools, this study aimed to design rich task based students worksheet that can be used as cognitive scaffolding in solving rich mathematical task and help mathematics teachers to provide learning resources in implementing rich tasks in daily mathematics classrooms.

In designing the student worksheet, this study used rich task design principles developed by Goos, Geiger and Dole [10] and each principle are described as follow:

1. A rich task should require application of mathematical knowledge. This knowledge includes not only fluency with accessing concepts and skills, but also problem-solving strategies and the ability to make reasonable estimations.
2. A rich task should promote positive dispositions such as confidence, initiative, and a willingness to apply mathematical knowledge flexibly and adaptively. Affective issues have long been held to play a central role in mathematics learning and teaching, and the importance of developing positive attitudes toward mathematics is emphasized in national and international curriculum documents.
3. A rich task should involve using tools. These tools may be representational (symbol system, graphs maps, diagram, drawing, tables), physical (model, measuring instruments), and digital (computers, software, calculators, internet).
4. A rich task also should be embedded in a range of contexts. These contexts may be drawn from real life or curriculum areas other than mathematics.
5. A rich task should develop a critical orientation (reflective thinking) in students since people not only know and use methods, they also evaluate the reasonableness of the results obtained and are aware of appropriate and inappropriate uses of mathematical thinking.
6. A rich task also needs to adopt investigative pedagogies which are defined as "contextualized problem-solving tasks through which students can speculate, test ideas and argue with others to defend their solutions" (p.591).

To be able to implement rich tasks, this study conjecture that the tasks for students should include a supporting document that would act as a cognitive scaffold for students in the initial stages of solving rich mathematical tasks process before they could internalize the metacognitive strategies and automatically use these strategies when faced with new rich tasks. This is important since the students not familiar with rich tasks yet. Furthermore, rich tasks related learning sources that could be accessed by students still limited. Therefore, to support them learning with rich tasks, they need a student worksheet as guidance that could be used independently. This is in line with the view that any scaffold should be gradually withdrawn as the learner becomes more competence [11].

Thus, this study was directed towards the development of this important document that is known as the student worksheet. The student worksheet is developed by incorporating the idea of cognitive scaffolding as used by Dindyal et al. [11] in developing a practical worksheet for a problem-solving task which included the use of questions, modeling, assisting with making a plan, drawing diagrams and encouraging the students to work in group. One of the reasons why it is important to develop rich task based students worksheet is because the way a task is imagined and intended by the teacher may be quite different from the way it is understood and carried out by the students. The intended learning by students may not happen if they misunderstand the tasks. Also, if teachers give too many directions, the solution process may become too trivial for students and the solution process may be reduced to a sequence of steps. In contrast, if teachers give too few directions, the students may focus on different things and the intended learning may not happen. Therefore, the implementation of rich tasks in the classroom ultimately needs to involve the teacher. This study, especially in the designing process, have paid careful attention to teacher preparation for implementing the rich task, including the use of student worksheet. As suggested by Slavit and Nelson [5], teachers were introduced to and thought about rich tasks and its characteristics as well as its principles as described before. So, they know how to help students to solve rich tasks.

2. Method

The development of student worksheet for rich tasks was conducted in three phases within research and development design [12] which have been simplified by Sukmadinata [13]. This includes: (1) preliminary study phase (literature review and survey); (2) the development phase which consist of three activities: designing student worksheet, operational field testing and main field testing; and (3) final phase is dissemination or implementation phase. This paper is primarily concerned with (1) and (2) phase of development study by focusing on the design of rich task based student worksheet and implication for pedagogy. Quantitative and qualitative data gathered from teachers and their students through questioner and interview were used in this study to develop rich tasks based student worksheet with descriptive approach.

The first stage of design research began in 2015 where the research team conducted preliminary research. In this phase, they studied the literature about rich task, analyzed the demand of solving mathematical tasks in Indonesian school curriculum and conducted a survey in four high ability junior high schools in Banda Aceh. The survey was carried out to find the information related to common teaching practices implemented by teachers in mathematics classroom, how students experience learning mathematics, what are learning sources used during instruction process.

The second stage of student worksheet development was designing process. The findings from literature and survey were used to design student worksheet. In this phase, this research paid more attention to the structure of student worksheet especially to ensure the design could work as cognitive scaffolding for students to solve rich tasks. Scrutinizing rich tasks principles was also conducted to satisfy the demand of the stated parameters. The first prototype of the student worksheet was developed in this stage. Before the designed student worksheet used in limited field test phase which is the final stage, the first prototype or prototype 1 was validated by several mathematics education experts and practitioners. This was done to check the appearance and content validity of the developed student worksheet. The prototype 1 was revised then by taking into account all suggestions from validators. This activity produced the second prototype of student worksheet (prototype 2).

The second prototype of student worksheet then used in field test through experimental design. The field test administered within two stages: (1) operational field test where the experiment conducted only in one school; and (2) main field test where experiment conducted in three schools. The operational field test was carried out in 2016 where the research team experimented on the prototype 2 of student worksheet. This study adopted a design experiment approach to produce a workable "design" (an initiative, artifact or intervention, for instance) that could be adapted by the schools. Teachers were involved in this stage where they have to use the student worksheet designed by the research team. It appeared necessary that the teachers make the proposed instructional approach a routine sufficiently familiar to them that the approach become classroom practice. To reach this aim, it seemed essential for teachers to adapt the researchers' ideas to make them their own, in the sense of their beliefs of mathematics and of solving rich mathematical tasks are transformed. Such a process would pass through a stage where the teachers negotiate and change the lesson. Therefore, the research team conducted a short workshop and discussion with teachers about the rich task and how to implement rich task based student worksheet in their mathematics classroom.

The experiment of the second prototype was conducted in one of high ability secondary schools in Banda Aceh where one teacher and twenty students of the school were involved. The observation was also conducted by the research team during the experiment session to monitor teacher teaching practice using rich task based student worksheet and to gather information related to the weaknesses of instruction process using the designed instrument. The student worksheet used by students during the teaching process and field notes during classroom observation were gathered by the research team to be used in revising the second prototype so that the third prototype will be produced.

3. Results and discussion

3.1. Choosing the right game

In solving the task in Figure 1, this study emphasized the student learning and reinforcing of rich tasks model by Goos, Geiger and Dole [9]. An audit of Indonesian 2013 curriculum identified distinctive rich task demands for mathematics subject taught in Grade 7. For example, the mathematics curriculum for Grade 7 semester 1 is organized into four topics: number, set, algebra form, linear equation and inequality. The rich task audit found that all these topic contain *mathematical knowledge* for designing project. Students must combine knowledge from a variety of previous math topic [14]. *Contexts* for rich task development included personal experience, the study of the social, economic and ecological system. Students were expected to develop *dispositions* enabling them to be active in solving real-life problems. The use of *tools* such as symbol, diagram, tables, graph and models was vital to learning in this subject. The goal of enabling students to solve real-life problems requires the development of *critical orientation* to viewing information and evaluating all possible solutions. Presenting the audit findings to teachers was intended to raise awareness of their subject' rich task demands.

Task 8

Febi and her family are in Trans Studio Bandung. They plan to enjoy the game in the playgrounds. The beside picture is one of the leaflets informing some games and the requirements.

If the information of Febi and her family's weight and height are shown in the table below:

No	Subject	Weight	Height
1	Febi	55kg	149cm
2	Brother	24kg	126cm
3	Father	70kg	173cm
4	Mother	51kg	162cm

Task 8

Febi dan keluarga sedang berada di Trans Studi Bandung. Mereka berencana ingin menikmati permainan pada wahana bermain tersebut. Gambar di samping adalah salah satu selebaran yang memuat informasi beberapa permainan dan persyaratannya.

Jika informasi tinggi dan berat badan Febi dan keluarga seperti yang ditunjukkan pada Tabel di bawah

No	Subjek	Berat Badan	Tinggi Badan
1	Febi	55 kg	149 cm
2	Adik	24 kg	126 cm
3	Ayah	70 kg	173 cm
4	Ibu	51 kg	162 cm

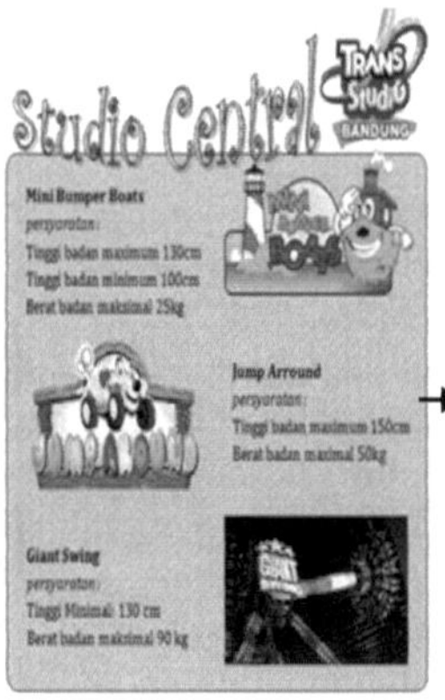

Maka, dapatkah kamu membantu Febi dan keluarganya untuk menentukan permainan apa saja yang dapat dimainkan mereka?

Then, can you assist Feby and her family in determining the games that they can play?

Mini Bumper Boats
Requirements:
The maximum height is 130cm
The minimum height is 100cm
The maximum weight is 25kg

Jump Around
Requirements:
The maximum height is 150cm
The minimum weight is 50kg

Giant Swing
Requirements:
The maximum height is 130cm
The minimum weight is 90kg

Figure 1. Example of rich task in student worksheet for inequality equation topic

The cognitive scaffolding used to help students solve the given rich task which meets also rich task' principles is provided in Table 1. This study helps students by asking several questions through follow up activities to give task specific hints, which essentially is a throwback to the usual help afforded by mathematics teachers.

Table 1. The task' scaffolding.

Rich Task Principles	Task Features	Example based on choosing the right game task
Application of mathematical knowledge	Applying inequality concept in solving the task	What are the mathematical concepts you have to apply to solve the given tasks?
Positive disposition (attitude)	Taking the initiative and a willingness to solve real-life problems	What exactly are you doing to solve the task? Why are you doing that?
Using tools	Using symbol, graphs maps, diagram, drawing, tables and modeling	What representations can you use to get a sense of what this is about?, How can the problem be modeled?
Embedded a range of context	Using student personal context	Where is the problem happen?
Critical Orientation	Viewing and evaluating the answer	Have you found all the possibilities?, do you think we have found the best solution?
Contextualized Problem Solving	Solving the problems	What objectives are required? Have you met anything like this before?, What is your plan of action? How reasonable is your solution?

The objective of student worksheet is for students to internalize the step in solving the rich task, and to ensure that they are in rich task learning trajectory through cognitive scaffolding. Assessment of rich task is certainly another issue that guides students in solving rich tasks. To this end, an accompanying assessment rubric was developed to focus students on what is valued in solving rich tasks process. At the same time, the rubric gives them feedback on their strengths and weaknesses.

3.2. The student worksheet

Based on the audit of mathematics curriculum for junior high school, the research team decided to design student worksheet for inequalities linear equation topic which is learned by Year 7 students in the first semester. As mention before that the student worksheet would act as cognitive scaffolding for students to solve rich task; therefore the structure of the designed student worksheet was arranged to include a number of activities that could help students to work independently in solving rich tasks.

The student worksheet consist of five parts: (1) basic competencies that should be achieved through the learning process, (2) what students will learn, (3) instruction how to use the student worksheet, (4) one rich mathematical tasks following by four activities to allow students understanding of the given mathematics topic (inequality equation), and (5) summary activity. The student worksheet also provides several questions related to rich tasks to guide students to solve it. The questions follow the learning trajectory expected from mathematics instruction using rich tasks especially the task' scaffolding as shown in Table 1. The questions on the worksheet include context, mathematical concepts, problem-solving, and critical thinking and all of these are arranged in the last page of student worksheet with space for students to write their answer, solution and/or reason. Example of student worksheet based on rich task is shown in Figure 2.

LKPD 3 / Pertemuan 4	Student Worksheet 3 / The Fourth Meeting
LKPD 3 **Pertemuan 4**	**Student Worksheet 3** **The Fourth Meeting**
➢ Jadi, menurutmu, apa yang dimaksud dengan pertidaksamaan linear satu variabel? Pertidak samaan linear satu variabel adalah kalimat terbuka yang memuat notasi <, >, ≤, ≥	➢ Could you please explain, what is the meaning of linear inequality of one variable? *Linear inequality of one variable is an open statement consisting of* <, >, ≤, ≥
Kamu dan kelompokmu sudah mendapatkan banyak informasi dari aktivitas di atas, dengan mengunakan informasi tersebut, kerjakanlah Task 8 dengan mengidentifikasi beberapa informasi berikut!	You and your group have find out much information from the above activities, by using the information, do task 8 by identifying some below information!
a. Dimanakah situasi di soal terjadi? studio Central / Trans Studio Bandung	a. Where do the situation happen? Studio Central / Trans Studio Bandung
b. Informasi apa saja yang kamu dapatkan dari soal? 1. tinggi badan ≤ 130 cm; tinggi badan < 100 cm; Berat badan < 25 kg 2. Tinggi badan ≤ 150 cm; Berat badan ≤ 50 kg 3. Tinggi badan > 130 cm; berat badan < 90 kg	b. What information can you get from the questions? 1. The maximum height is ≤, 130 cm; The minimum height is < 100 cm; The maximum weight is < 25 kg 2. The maximum height is ≤ 150 cm; The maximum weight is ≤ 50 kg 3. The minimum height is > 130 cm; The maximum weight is < 90 kg
c. Apa yang menjadi permasalahan dari soal tersebut? permainan apa yang bisa febi naikki bersama keluarga.	c. What is the problem of that question? What games can be played by Febi and her family?
d. Tuliskanlah cara penyelesaian masalah tersebut! yang bisa menaikki permainan 1 : adik yang bisa menaikki permainan 2 : ayah, ibu yang bisa menaikki permainan 3, ayah, ibu, Febi	d. Write the strategies to solve that problem! The first game can be played by brother The second game can be played by father and mother The third game can be played by father, mother, Febi

Figure 2. The layout of student worksheet with student answer on it.

4. Conclusion

The student worksheet developed through this study has implications for teachers, researcher, and curriculum developers. It could be used by teachers who want to incorporate rich task based instruction in a daily mathematics classroom. The documents work as cognitive scaffolding that supports the immediate construction of knowledge by students and as well as provides the basis for future independent learning of the student. However, some revisions are still needed before it will use extensively in schools. Taken altogether, the rich task design, which includes student worksheet and rich tasks based lesson plan have shown great potential in developing students' ability in solving the rich mathematical task.

Acknowledgment

We would like to thank many people who have contributed to this research and supported us through the process. We are very grateful to the Ministry of Research, Technology and Higher Education and STKIP Bina Bangsa Getsempena that supported and funded us to conduct this research.

References

[1] Watson M and Anne W no date What's in task? Generating mathematically rich activity. 1

[2] Edward E S 1997 Fostering creativity through instruction rich in mathematical problem solving and problem posing. *Int. J. Math. Educ.* **29** 75

[3] Peter A, Paul P B, Sandra S, Mathew K and Frischknecht B 2014 Teachers choosing rich tasks: The moderating impact of technology on student learning, enjoyment, and preparation *Educ. Res.* **432** 19

[4] Marjorie H and Kaye M S 1997 Mathematical tasks and student cognition: Classroom-based factors that support and inhibit high-level mathematical thinking and reasoning *J. Res. Math. Educ.* **28** 524

[5] David S and Tamara H N 2010 Collaborative teacher inquiry as a tool for building theory on the development and use of mathematical task *Journal Mathematics Teacher Education* **13** 201

[6] Philip M 2002 Rich tasks: Developing student learning around important tasks *Aust. Sci. Teach. J.* **48** 6

[7] Sarah F 2009 Teachers' Use of mathematics tasks: The impact on the mathematics learning and affective responses of low-attaining upper primary students *Crossing Divid. Proc. 32nd Annu. Conf. Math. Educ. Res. Gr. Australas.* **1** 185

[8] Piggot J 2012 *Rich task and contexts* from: http://nrich.maths.org/5662

[9] Zainal A S, Fitriati F and Rita N 2016 Developing rich task based lesson plan and student worksheet to improve teaching ability of secondary school mathematics teachers *Journal of Maju* **3** 25

[10] Merrilyn G, Vince G and Shelley D 2013 Designing rich numeracy tasks *Proc. of ICMI Study 22 on Task Design in Mathematics Education (UK)* **1** 591

[11] Jaguthsing D, Tay E G, Quek K S, Leong Y H, Toh T H, Toh P C and Ho F H 2013 Designing practical student worksheet for problem solving tasks *Proc. of ICMI Study 22 on Task Design in Mathematics Education (UK)* **1** 315

[12] Walter R G and Meredith D G 1983 *Educational Research: An Introduction* (New York: Longman)

[13] Sukmadinata 2006 *Metode Penelitian Kualitatif* (Bandung: Graha Aksara)

[14] Anna G 2016 Exploring misconception with rich tasks *Ohio Journal of School Mathematics* **73** 43

IOP Conf. Series: Journal of Physics: Conf. Series **1088** (2018) 012030 doi:10.1088/1742-6596/1088/1/012030

The development of learning instruments through the problem-based learning model to enhance students' creativity

M Iqbal[1], Yusrizal[1], and Z Abidin[2]
[1] Syiah Kuala University, Jl. Teuku Nyak Arief Darussalam, Banda Aceh 23111, Indonesia
[2] National Islamic University Ar-Raniry, Jl. Syeikh Abdul Rauf, Banda Aceh 23111, Indonesia

E-mail: yusrizal_fkip@unsyiah.ac.id

Abstract. Lack of opportunities for students to express ideas during the learning process makes students having difficulty in solving their daily life problems. Realizing the importance of mathematical creative thinking ability, it required the development of learning tool through Problem Based Learning (PBL) model. This study aimed to develop the valid learning tools through the PBL model to improve the creativity of students. Learning tools were developed based on the 4D model, that refers to define, design, develop, and disseminate. However, this research only focussed on the first three stages. The results showed that learning tools through the PBL model in improving student creativity meet the criteria of validity, in which the score was 4.13 (highly valid). Furthermore, the learning tools developed through the PBL model in this study can be used for further analysis, including in practicality and effectiveness phases.

1. Introduction

Developing students' creative thinking skill is crucial in mathematics learning. When someone applies creative thinking in solving the problem, diverging thinking will produce many useful ideas to solve the problem [2]. In addition, creative thinking contributes to finding ideas in problem-solving to produce individual solutions for students themselves [3,4]. In the Indonesian context, the national curriculum of 2013 aims to make students think creatively [1]. The curriculum requires students to think widely and convergently to develop innovations in all areas of mathematics. Therefore, mathematics teachers need to develop the learning instruments that could foster students' creative thinking skill.

Based on the preliminary research of this study, the test of students' mathematical creative thinking ability at one of the high schools in Banda Aceh showed that only 22 students (73.33%) were in middle level, 8 students (26,67%) were in the low level, and no one was at high level. This data showed that students' creative thinking skill did not show maximum results. Thus, it is necessary to develop learning instruments that can train students' mathematical creative thinking skills by embedding meaningful learning model, such as Problem Based Learning (PBL) model. Learning instruments that were developed in this study included lesson plans, student worksheets, and test. The lesson plans were developed based on the steps in the PBL model, while the student worksheets and test were developed based on appropriate indicators of mathematics creative thinking skill.

In the PBL learning process, students need to deal with real-life problems, wherein students could develop their knowledge that contributes to their learning experiences [5]. In addition, in the learning model, students were trained to model the problems into mathematical concepts. Each student could think about various solutions which could raise the level of students creativity.

The general step in the PBL model is to propose the problem. Problem-based teaching is not only to lead to certain academic principles or skills but also to direct the learning issues that would be meaningful to students [6]. The given problems were based on authentic real-life situations that allowed various solutions to one problem. Next, the students conducted authentic inquiries to find solutions to a given problem. Students were required to express various ideas and interpret the problems, develop hypotheses, collect and experiment when needed. These activities could accommodate the creative thinking aspects of fluency and flexibility. In addition, students could add original ideas to solve the problems. The final stages of problem-based learning were planning and preparing reports in the form of works that explain the solution they had found. This activity could develop ideas to enrich existing ideas, thereby developing elaboration aspects.

Based on the description of the introduction, it is generally formulated the principal issue of research as follows: How to develop learning tools through PBL model in improving the creativity of students who meet the valid criteria?

2. Methods

This study was a development research. I was used the 4D model consisting of 4 stages of development, namely define, design, develop and disseminate [7]. This study only carried out until the third stage. The define stage involved literature review to determine the characteristics of learning instruments that could improve students' creative thinking ability. Also, this stage included an initial curriculum analysis with the help of mathematics teachers to identify the mathematics topics that needed development of learning instruments. The design stage aims to design an instrument that can be used in learning mathematics. The design stage involved four steps, including the preparation of tests, the selection of media that was suitable to the characteristics of learning materials, the format selection and make the initial design according to the selected format.

The development stage involved the expert validation that followed by twice revisions. The initial learning instruments were validated by three validators, including two lecturers as the experts of teaching and learning mathematics and a high school mathematics teacher as a practitioner of mathematics learning. The validation process was conducted by using a questionnaire for each instrument, whose scale from 1 to 5 (invalid to highly valid). The criteria used for assessing the learning instruments were based on Plomp and Nieveen [8] as shown in Table 1.

Table 1. Tabel validitas score

Score			Criteria
$4 \leq$	VS	≤ 5	Highly valid
$3 \leq$	VS	≤ 4	Valid
$2 \leq$	VS	≤ 3	Less valid
$1 \leq$	VS	≤ 2	Not valid

3. Result and Discussion

3. 1 Define stage

At this stage, the researcher discussed the challenges faced by teachers when they taught the absolute value concept. The discussion aimed to obtain the information about students' condition when learning mathematics. Teachers admitted that the topics of absolute values had been one of the challenging topics to develop the proper learning tools that could improve students' creative skills. One of the challenges mostly found by students was the topics of linear equality and linear inequality. Whereas, students' comprehension on these topics was pre-requisite before students learning the topics of absolute value [9]. Students who have the poor understanding of the basic concepts would have difficulties in learning new topics, which might result in a low quality of students' creative thinking skill. Therefore, the learning tools that were developed in this study consisted of the concept of absolute values, including illustrating the graph of the problem of absolute value and solving problems related to absolute value in equations and linear inequalities.

3.2 Design stage

At this stage, the learning instruments, including lesson plans, student worksheets, and test questions, were developed based on the results of the analysis at the define stage. The results of the learning instruments at the design stage were called draft 1. The lesson plan was designed by using problem-based learning model, that was indicated by including student-centered problem, encouraging students to learn, guiding individual and group investigations, developing and presenting the work, and analyzing and evaluating the problem-solving process. The lesson lasted for four meetings, 2×45 minutes for each lesson. In general, the developed lesson plan had the similar aspects to the typical lesson plans. Only the motivation part had differences, which aimed to help students to understand the concept of absolute value by posing the daily life problems. The following question was one of the examples that was posed in the motivation section of the lesson plan.

Andi plans to travel home from Banda Aceh to Lhokseumawe through Sigli. The distance between Banda Aceh and Sigli is 110 km, while the distance from Sigli to Lhokseumawe is 166 km. Thus, the distance between Banda Aceh to Lhokseumawe is 276 km. The route of these cities could be seen in Figure 1.

Figure 1. The route of the three cities

If Sigli city is supposed to be a 0 point on the number line, Banda Aceh city is on the positive side of the number line, and Lhokseumawe city is on the negative direction of the number line. Then:

- What is the distance from Lhokseumawe to Sigli?
- Do you think the distance between Banda Aceh and Lhokseumawe is equal to the distance between Lhokseumawe to Banda Aceh? If so, could you explain why?
- If we write the distances on the number line, is it possible the distances to be negative?

In addition to motivation section on the lesson plan, the researcher also revised the at the "observing phase" on the scientific learning approach in student worksheets. The worksheets question was written in word problems, to help students to improve their ability to model mathematics question to the mathematical equation. Moreover, student worksheets were designed to encourage students to learn

IOP Conf. Series: Journal of Physics: Conf. Series **1088** (2018) 012030 doi:10.1088/1742-6596/1088/1/012030

actively and to develop students' creative thinking skills. The student worksheets were designed to measure students' creative thinking ability on equality and inequality of absolute values. Figure 2 shows one of the parts of the developed worksheets that included the "observing phase".

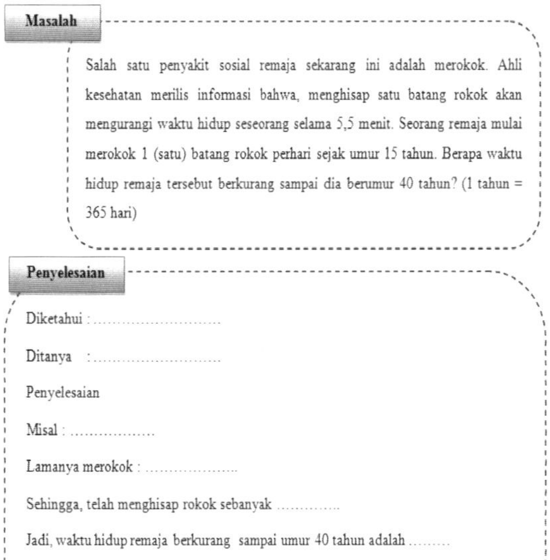
Masalah

Salah satu penyakit sosial remaja sekarang ini adalah merokok. Ahli kesehatan merilis informasi bahwa, menghisap satu batang rokok akan mengurangi waktu hidup seseorang selama 5,5 menit. Seorang remaja mulai merokok 1 (satu) batang rokok perhari sejak umur 15 tahun. Berapa waktu hidup remaja tersebut berkurang sampai dia berumur 40 tahun? (1 tahun = 365 hari)

Penyelesaian

Diketahui : ....

Ditanya : ....

Penyelesaian

Misal : ....

Lamanya merokok : ....

Sehingga, telah menghisap rokok sebanyak ....

Jadi, waktu hidup remaja berkurang sampai umur 40 tahun adalah ....

Problem

Smoking is a social disease of teenagers nowadays. Medical experts stated that smoking one cigarette could reduce life expectancy by 5.5 minutes. A teenager has started smoking one cigarette daily since he was 15 years old, how long the life expectancy of the teenager that might be reduced if he is 40 years old now? (Hint: 1 year = 365 days)

Solution:

What is known: ....
What is asked: ....
How long he's been smoking:.......
How many cigarettes that have been smoked:......
Hence, the life expectancy that has been reduced:...

Figure 2. The sample of the developed student worksheets

The design of the test questions also involved a rubric that matched the indicator on the ability of mathematical creative thinking in the topics of absolute value. Figure 3 shows the sample of the developed test questions.

Jumlah tegangan arus listrik normal yang didistribusikan PLN ke rumah-rumah di formulasikan $|x - 220| \leq 11$. Berapakah kisaran tegangan yang masih bisa ditoleransi oleh PLN?

The standard electricity current from the national electricity company that flows to houses is formulated as $|x - 220| \leq 11$. What is the approximate tolerable current that could be distributed by the company?

Figure 3. The sample of the developed test

The students who have the development of creativity have the awareness to arouse curiosity and the desire to always practice in solving math problems. This is inseparable from the scope and context of learning which is an important point in the contribution of developing creativity in the cognitive domain of students [11]. Several factors can improve the creativity of students, including 1) a suitable environment, so that students feel comfortable and not afraid to express ideas, 2) teacher personality traits that motivate students to develop ideas, 3) substantial study groups so that students exchange mind in solving problems [12]

3.3 Develop stage

At the develop stage, the learning instruments developed at the design stage were then validated by validators. The results of evaluation of the learning instruments by the validators are shown in Table 2.

Table 2. The result of the validator's assessment of the learning device in draft 1

Validated instrument	Average	Criteria
Lesson plan	4.2	Highly valid
Student Worksheets	3.4	Valid
Test	4.8	Highly valid
Average	4.13	Highly valid

Based on the validator's assessment of the learning instruments, generally, the average validation level for the developed learning instruments was at the value of 4.13, suggested that averagely the instruments was very valid [9]. Both the lesson plan and the test of mathematical creative thinking ability met the highly valid criteria, whose the average scores of 4.2 and 4.8 respectively. Meanwhile, the student worksheets had an average score of 3.5, which indicated that the instrument was at valid level. Nevertheless, these scores have not yet fullfilled the maximum targeted score, which was 5, since the instruments was revised twice only as suggested by the validators. The validity level might have been higher if it involved more revision and validation processes. Also, the number of the validators involved in this study were only three people. Whereas, the more number of validators involved might have resulted in the higher level of validity scores, as it would have provided more constructive advice from different perspectives for the improvement of the instruments.

Furthermore, a summary of revisions before and after revisions of learning tools is shown in the following table.

Table 3. The results of validation of research instruments

No	Before validation	After validation
1	Lesson plans	
	• The perception used was not compatible with the topics • The sentence used was not logical • The student worksheets were not included in the lesson plan	• The perception used has been relevant to the material • Sentences are logical • The student worksheets were included in the lesson plan
2	Student worksheets	
	• The instructions for working on student worksheets were too long • There was no image to draw the attention of learners in working on student worksheets • To provide the students with the steps of solving the questions on the student worksheets	• The instructions used have been adjusted • There are already some pictures that match the question • Only provided students with the empty space to solve the question by themselves to increase students' creative thinking skill
3	Test	
	• The third question needed to be transformed into a matter of everyday life	• The third question was adapted to the problems of everyday life

4. Conclusion

It can be concluded that the learning instruments through the PBL model in improving students' creative thinking skill met the criteria of validity. While the lesson plan and test questions had highly valid scores (4. 2 and 4.8 respectively), the student worksheets were at a valid level (3. 4). Nevertheless, on average, the developed instruments were highly valid, with an average score of 4.13. Moreover, the developed learning instruments of the PBL model to improve students' creative thinking skill can be tested for future teaching and learning practices, as well as to the further research.

References

[1] Kemendikbud 2013 *Peraturan Pemerintah tentang Pengelolaan dan Penyelenggaraan Pendidikan* (Jakarta: Badan Pengembangan Sumber Daya Manusia Pendidikan dan Kebudayaan dan Penjamin Mutu Pendidikan)

[2] Yanti N S, Yusrizal, and Gani A 2016 Penerapan Model Pembelajaran Inkuiri Terbimbing Untuk Meningkatkan Keterampilan Berpikir Kreatif dan Motivasi Siswa Ditinjau dari Jenis Kelamin Pada Materi Kalor Kelas X SMAN 11 Banda Aceh *Jurnal Pendidikan Sains Indonesia* **4** 1

[3] Mossing S 2013 The Importance of Creative Thinking and the Arts in Education *Honors Projects* **37** 1

[4] Aldig E and Arseven A 2017 The Contribution of Learning Outcomes for Listening to Creative Thinking Skills *Journal of Education and Learning* **6** 41

[5] Birgili B 2015 Creative and Critical Thinking Skills in Problem Based Learning Environments *Journal of Gifted Education and Creativity* **2** 71

[6] Rusman 2016 *Model-model Pembelajaran Mengembangkan Profesionalisme Guru* (Jakarta: PT RajaGrafindo Persada)

[7] Thiagarajan S, Semmel D S and Semmel M I 1974 *Instructional Development for Training Teachers of Expectional Children* (Minnesota: Leadership Training Institute/Special Education)

[8] Plomp T and Nieveen N 2013 *Educational Design Research* (Enschede: SLO)

[9] Septian A 2014 Pengaruh Kemampuan Prasyarat terhadap Kemampuan Penalaran Matematis Mahasiswa dalam Matakuliah Analisis Real *ATIKAN* **4**

[10] Chesimet M C, Githua B N and Ng'eno J K 2016 Effects of Experiential Learning Approach on Students' Mathematical Creativity among Secondary School Students of Kericho East Sub-County *Journal of Education and Practice* **7** 51

[11] Trnova E 2014 IBSE and Creativity Development *Science Education International* **25** 8

[12] Tan L S, Lee S S, Ponnusamy L D, Koh E R and Tan K C K 2016 Fostering Creativity in the Classroom for High Ability Students: Context Does Matter *Education Sciences* **6** 1

IOP Conf. Series: Journal of Physics: Conf. Series **1088** (2018) 012031 doi:10.1088/1742-6596/1088/1/012031

Analysis of students' errors in responding to TIMSS domain algebra problem

B A Saputro[1,2], D Suryadi[1], R Rosjanuardi[1] and B G Kartasasmita[1]
[1]Universitas Pendidikan Indonesia, Jl. Dr. Sediabudi No. 229, Bandung 40154, Indonesia
[2]Universitas PGRI Semarang, Jl. Sidodadi Timur No. 24, Semarang 50125, Indonesia

E-mail: bagusardisaputro@student.upi.edu

Abstract. The TIMSS results were referred by the Indonesian Ministry of Education and Culture to develop curriculum and textbooks of school mathematics. The purpose of this study was to understand the errors of students learning with the 2006 curriculum and students learning with the 2013 curriculum in answering the TIMSS algebra content domain. A test consisting of fifteen questions of content domain algebra and application was administered to the students. The wrong answer of the test was analysed to determine the weakness and strength of algebra learning conducted. The results showed that students studying with the 2013 curriculum were outperformed their counterpart in solving the problem of algebra content domain concerning expression of application. Misconceptions about algebra and arithmetic were also found in students' answers. The misconceptions found in this study can be further developed into an instrument to evaluate pedagogical content knowledge.

1. Introduction

The TIMSS results from three consecutive years namely 1999, 2003, and 2007 show that Grade 8 Indonesian students are still problematic in solving the TIMSS problem of algebraic content domains [1]. To overcome this problem, the Ministry of Education developed a new curriculum which is called *Kurikulum* 2013 (curriculum 2013) to replace the 2006 curriculum. The new curriculum implies substantial changes in the mathematics textbooks. However, the impact cannot be seen directly as Grade 8 Indonesian students did not participate in TIMSS 2015. So, we have not obtained a picture of the students' learning ability with the 2013 curriculum in solving the TIMSS algebra domain problem. Therefore, this article is expected to fully support the recommendations aiming at improving the curriculum of junior mathematics subjects, learning process, assessment methods, and instruments learning textbooks, and teacher training (both pre-service and in-service training). The implication of this study is to use misconception found in students to be developed as an instrument to evaluate pedagogical content knowledge.

2. Method

The study was conducted in one junior high schools which use the 2013 curriculum for the 7th grade and 2006 curriculum for the 8th grade. The students of both grades already studied the topic algebraic forms & their operation in the previous semester. To obtain student performance in algebra domain, 15 problem items from the set of TIMSS 2011 problem ware translated to Indonesian and given to the students. The problem contained ten multiple choice questions & 5 essay questions. Problems 2, 4, 13, and 15 have the main topic of equation/formulas and functions, while others are algebraic expressions. Problems 1, 2, 10, 14, and 15 have cognitive types of applying domains, while others have cognitive knowing. The answer ware analysed based on the type of the problems. The proportion of answer for multiple choice question between the two group of students was compared based on the cognitive domain. The answer to essay problems was analysed qualitatively. Firstly, the answers one grouped base on the type of errors. Secondly, the grouped answers were analysed to investigate why the errors occurred. Some students were interviewed to confirm the analysis results.

3. Result and discussion

3.1. Results of student answers analysis on multiple choice questions

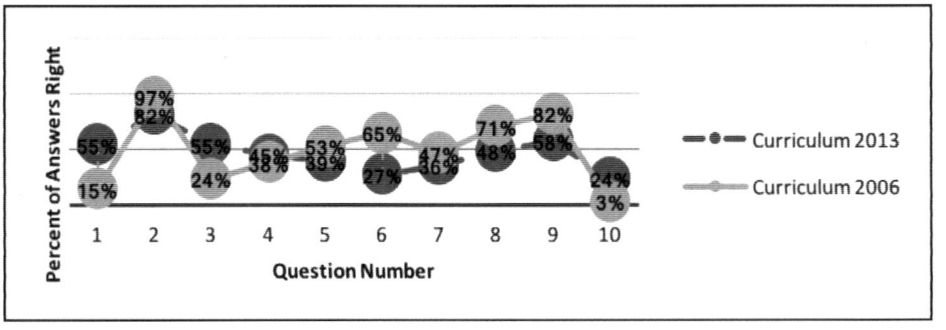

Figure 1. A comparison chart of 10 multiple choice question.

The graph of the comparison of students' answers to 10 multiple choice questions between students learning to use the 2006 curriculum and students studying with the 2013 curriculum is shown in Figure 1. The graph shows that students studying with the 2006 curriculum are superior in working on the TIMSS topic of algebraic expression that measures the domain knowledge and topic of equations/formulas and functions that measure the domain of application. While students who study with the curriculum 2013 are superior in working on the topic of algebraic expressions that measure the domain of application. This result is in line with the excellence of the 2013 textbook that has introduced a geometry representation of algebraic expressions and implementation problems involving the representation of algebraic expressions [2] so that students who have studied with the 2013 curriculum will easily answer questions 1 and 10. Besides, this may also occur because the 2013 student curriculum book contains more domain implementation questions than on knowledge domain issues [4, 5]. However, the percentage errors of students studying with the 2013 curriculum in solving domain knowledge problems outweigh the percentage of student errors in resolving domain implementation issues [6]. Therefore, the materials and questions presented in the 2013 student curriculum should also consider the domain of knowledge of algebraic expressions.

3.2. Results of Analysis on Student Answers on Essay Questions

The problem 11 and 12 presented in Table 1 is to discuss the topic of equations/formulas and functions of the knowledge domain. The students that occurred in table 2 indicate that the student made a mistake in the process of solving the problem [2]. Students who answered 10, made an error in arithmetic calculations. Students who $\frac{responded\ 0}{10}$, have a misunderstanding of integer execution operations by fractional numbers. The problem 12 discusses the topic of algebra on the domain expression of knowledge. The problem measures the skill of simplifying the expression of fractional algebraic expressions. Students who answered $\frac{3x^3}{14}$ have misconceptions on fractions. So, the students do it also on algebraic fractions. The students also have the misconception in summing algebraic expressions. Students who answered $\frac{9x^3}{8}$, have only misconceptions in summing algebraic expressions.

Table 1. Errors students make on the number 11 and number 12.

Problem	Error found	Student work
No 11. Use the formula $y = 100 - \frac{100}{1+t}$ to find the value of y when $t = 9$. Answer: ________	Simplify the terms of algebraic expressions that only contain variables.	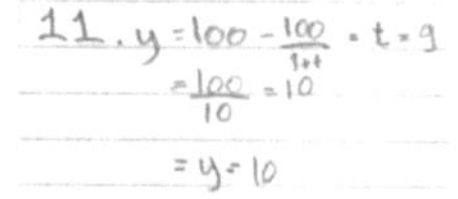
	Do not use fractional reduction rules with integers correctly.	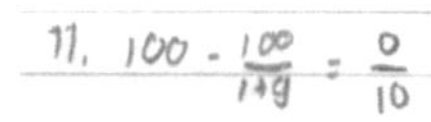
No 12. Simplify the expression $\frac{3x}{8} + \frac{x}{4} + \frac{x}{2}$. Show your work Answer: ________	Equates the denominator by summing the denominator of the algebraic fractional term and directly sums the numerator of the algebraic fractional term.	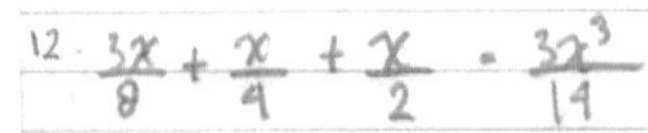
	Directly sums the numerator of the algebraic fractional term.	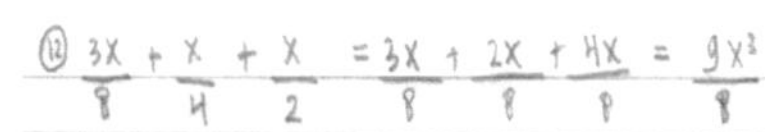

The problem 13 and 14 presented in Table 2 discusses the topic of equations/ formulas and functions application domain. The students who answered 74 made a mistake in the planning phase of problem-solving [7]. Students who received $25 + 8$ answers had misconceptions about the distributive nature of factoring. Students who $8ab$ answered have a misconception in simplifying the algebraic expression. The problem 14 discuss the topic of equations/ formulas and functions of the knowledge domain. Students who answered $\frac{51}{5}$ and $12b$ have misconceptions in the distributive nature. Students who answered 4 had misconception in understanding algebraic expressions, whereas students who answered 742 handled the numbers as mere entities rather than quantities [8].

Table 2. Errors students make on the number 13 and number 14.

Problem	Error found	Student work
13. $a + b = 25$ What is the value of $2a + 2b + 4$? Answer: __________	Specify the specific values on variables a and variables b.	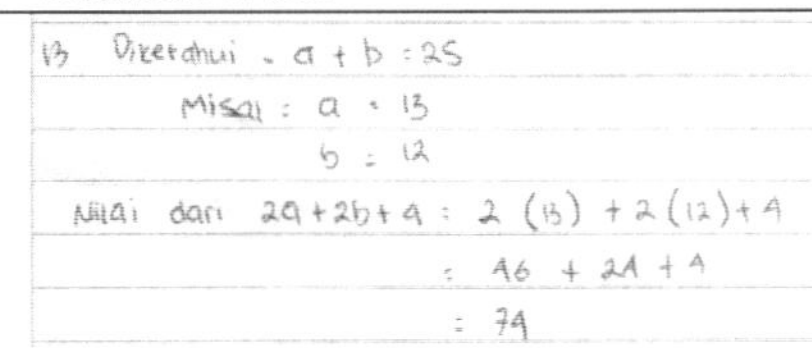
	Error in factoring	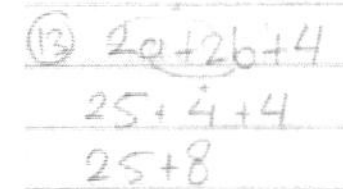
	Simplify two unlike terms into one term.	
14. $k = 7$ and $l = 10$ What is the value of P when $P = \frac{3kl}{5}$? Answer: __________	Meaning 3kl as $3k + 3l$	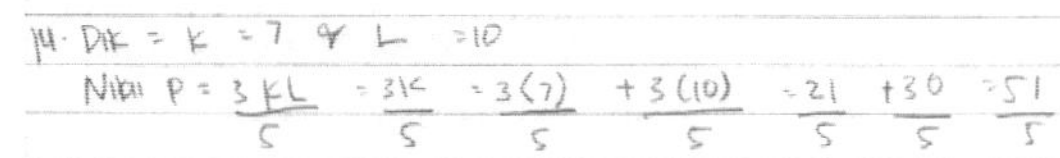
	Meaning 3kl as $3(k\times l) = (3\times k)\times(3\times l)$	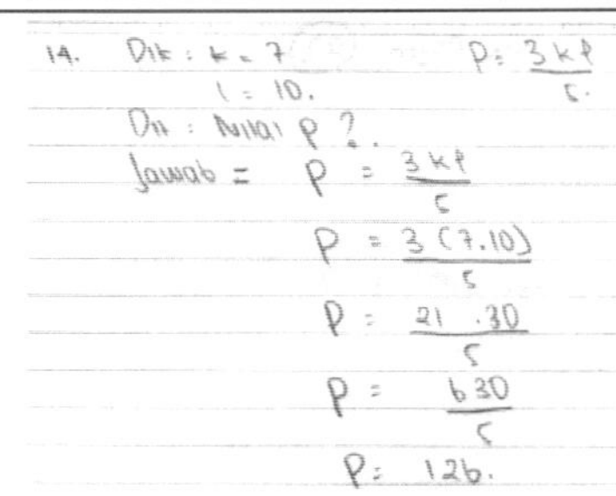
	Meaning 3kl as $3 + k + l$	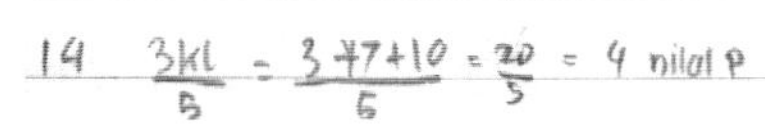
	Meaning 3kl combined number 3, k and l as a new number.	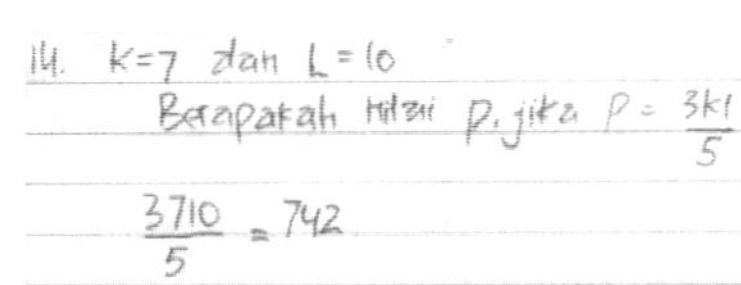

Problem 15 presented in Table 3 discusses the topic of equation/ formulas and function domain applying. All student responses show that students have started from the planning phase of problem-solving. Therefore, column error done is a classification strategy used by students. All students cannot model the situation into suitable algebraic equations [1], which is the most significant percentage of error when students solve the problem [7]. Student responses on the first line indicate that students have misconceptions about simplifying algebraic expressions. The student's answer on the third line shows that the students made a mistake in the process skill phase [7].

IOP Conf. Series: Journal of Physics: Conf. Series **1088** (2018) 012031 doi:10.1088/1742-6596/1088/1/012031

Table 3. Errors students make on the number 15.

Problem	Error found	Student work
No. 15 A piece of wood was 40 cm long. It was cut into three pieces. The lengths in cm are $2x - 5$ $x + 7$ $x + 6$ A. What is the length of the longest piece? Answer: ______ cm B. Show your work. If you use a calculator, you still must describe all the steps you used to obtain your answer.	Simplify every available algebraic expression on the problem.	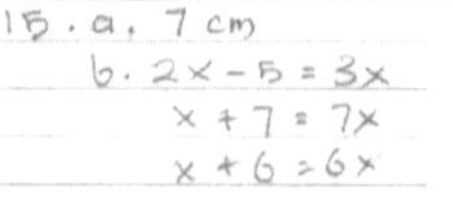
	Substituting a particular value in variable x, to determine the answer.	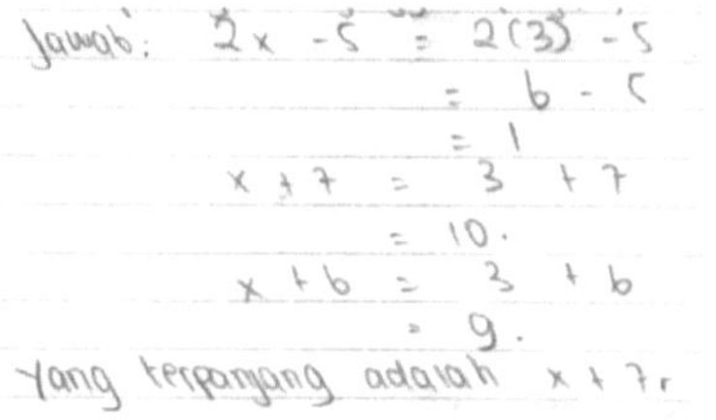
	Create three equations separately to determine the value of each algebraic expression.	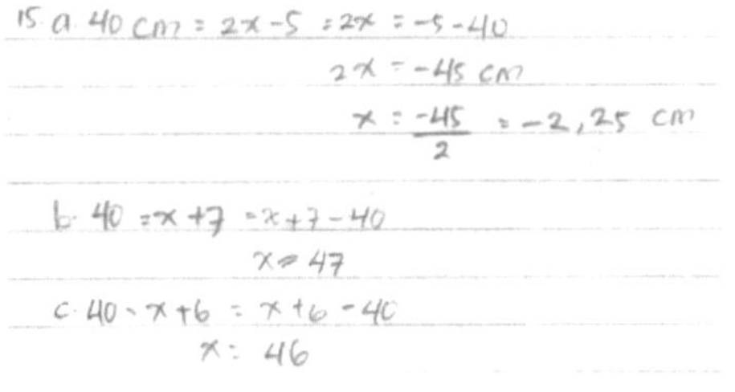

The way students in answer question number twelve were in line with Dhlamini and Kibirige's research [9]. The student answers the question by multiplying the numerators and adding the denominator respectively. This case indicates that students failed to use algorithms for fractions using least common division (LCD) which shows misconceptions on numerators (multiplying instead of adding) and errors adding different denominators leading to incorrect LCD (error leading to misunderstanding) [9]. The case also indicates that the students have generalized the multiplication rule against the addition rule [10]. Student's answer number thirteen shows that students seem to understand only that two different letters always have different numerical values, but have not realized that two different letters can have the same numerical value [11]. So that students answer about number fourteen by specifying two numbers specific for two different variables, 12 for a and 13 for b. This case can be regarded as the effect of a limited conceptual understanding of the meaning of the letter as a variable [12]. When students simplify the algebraic expressions given in Problem number fifteen, it shows that students assume that $x + 7$ is a process which interpreted as "adding x and seven." Students have not assumed $x + 7$ as an object because it contains operation symbol. Students consider that the symbol "+ or -" as an invitation to do something [13]. They also consider not to be part of the answer. Students also consider x as the unit of measurement to replace cm.

3.3. Implication

The instrument development was conducted in three stages. Firstly, presenting the student's questions and answers; secondly, submitting questions asked to teachers predicting students' thinking, what students have comprehended and what students have not comprehended; and thirdly, asking how the teacher helps the student [14]. Examples of such questions are available in Table 4. The teacher's response to student misconceptions that have been identified is to re-explain, provide situations that cause cognitive conflict, and guide students to see their mistakes [15], while the types of questions that teachers have identified are instructional questions, investigative questions, and inadequate/incompetent questions [16]

Table 4. The instrument for Evaluation Pedagogical Content Knowledge.

1. Simplify the expression $\frac{3x}{8}+\frac{x}{4}+\frac{x}{2}$. Show your work.
   Ani's solution: $\frac{3x}{8}+\frac{x}{4}+\frac{x}{2}=\frac{3x^3}{14}$
   Budi's solution: $\frac{3x}{8}+\frac{x}{4}+\frac{x}{2}=\frac{9x^3}{8}$
   a) What does/ does not the student comprehend?
   b) How could you show/ explain to the student that this answer is incorrect?
   c) How could you help them?

4. Conclusion

Results of the analysis show that students who studied under the 2013 curriculum are superior in solving problems of algebra application expression domain. However, they are lacking in solving algebraic expression problems of knowledge domains and problems of equations/formulas and functions that measure application domains. The 2013 student curriculum textbook should be designed by considering the strength of the 2006 curriculum textbook, which is proven to facilitate students well to solve problems of algebraic expression domain and problems of equations/formulas and functions measuring application domains. Moreover, there are misconceptions found when students solved TIMSS problems of algebraic content domain namely misconceptions about fractional meanings and fractional operations, misconceptions of algebraic expressions and in simplifying algebraic expressions, as well as misconceptions about distributive properties. Besides, we also found a variety of incorrect student strategies in solving the problem. It is possible that various parties can utilize this knowledge, such as being used to develop a mathematics curriculum for students, mathematics curriculum of student-teacher, mathematics textbook design, as well as to create an instrument of the national exam and improve teacher training material.

Acknowledgments

The authors would like to thank the teachers and students who have participated, and The Directorate of Research and Community Service, Directorate General of Research Enhancement and Development. Ministry of Research and Higher Education that have provided financial support to this research.

References

[1] Gunawan H 2010 *Analisis Konten dan Capaian Siswa Indonesia dalam TIMSS (Trends in International Mathematics and Science Study) Tahun 1999, 2003, dan 2007* (Jakarta: Pusat Penilaian Pendidikan Badan Penelitian dan Pengembangan Kementrian Pendidikan Nasional)
[2] As'ari A R 2014 *Matematika untuk SMP/ MTs Kelas VIII Semester 1* (Jakarta: Kementrian Pendidikan dan Kebudayaan)
[3] International Association for the Evaluation of Educational Achievement (IEA) *TIMSS 2011 Assessment* (Amsterdam, the Netherlands: Publisher: TIMSS & PIRLS International Study Center, Lynch School of Education, Boston College, Chestnut Hill, MA and International Association for the Evaluation of Educational Achievement (IEA))
[4] Cahyono B and Adilah N 2016 Analisis Soal dalam Buku Siswa Matematika Kurikulum 2013 Kelas VIII Semester I Berdasarkan Dimensi Kognitif dari TIMSS *Jurnal Review Pembelajaran Matematika* **1** 86
[5] Muklis Y M and Oktora S R 2015 Analisis Deskriptif Soal-Soal dalam Buku Siswa Kurikulum

2013 (Edisi Revisi) dan BSE Pelajaran Matematika SMP Kelas VII Ditinjau dari Domain Kognitif TIMSS 2011 *Proc. Seminar Nasional Pendidikan Matematika UMS 2015 (Surakarta, Indonesia, 10 May 2015) (Program Studi Pendidikan Matematika, Universitas Muhammadiyah Surakarta)* p 71

[6] Mahanani L G and Murtiyasa B 2016 Kemampuan Pemecahan Masalah Matematika Aljabar Berbasis TIMSS pada Siswa SMP Kelas VIII *Proc. Sempoa: Seminar Nasional, Pameran Alat Peraga, dan Olimpiade Matematika (Surakarta, Indonesia, 3 August 2016) (HMP Program Studi Pendidikan Matematika, Universitas Muhammadiyah Surakarta)* p 149

[7] Azis I I J and Fuad Y 2016 Analisis Kesalahan Siswa dalam Menyelesaikan Soal Aljabar Model TIMSS *MATHEdunesa* **3** 176

[8] Kaur B 1991 Some Common Misconceptions in Algebra *Teaching and Learning* **11** 33

[9] Dhlamini Z B and Kibirige I 2014 Grage 9 Learners' and Misconceptions in Addition of Fractions *Mediterranean Journal of Social Sciences* **5** 8

[10] Naseer MS 2015 Analysis of Students' Error and Misconceptions in pre-University Mathematics Courses *Proc. 1st International Conference on Teaching & Learning (ICTL)* Vol. 14 *(Lengkawi, Malaysia, 14-15 September 2015)* p 34

[11] Akhtar Z and Steinle V 2013 Probing Students' Numerical Misconceptions in School Algebra *Proc. of the 36th Annual Conference of the Mathematics Education Research Group of Australasia (Melbourne, Victoria, 2013) (Mathematics Education Research Group of Australasia Inc.)* p 36

[12] Steinle V, Gvozdenko E, Price B, Stacey K and Pierce R 2009 Investigating Students' Numerical Misconceptions in Algebra *Proc. of the 32nd Annual Conference of the Mathematics Education Research Group of Australasia* Vol 2 *(Palmerston North, NZ, 2009) (Mathematics Education Research Group of Australasia Inc.)* p 491

[13] Mulungye M M, O'Connor M and Ndethiu S 2016 Sources of Student Errors and Misconceptions in Algebra and Effectiveness of Classroom Practice Remediation in Machakos County-Kenya *Journal of Education and Practice* **7** 10

[14] Zuya and Elisha H 2014 Mathematics Teachers' Responses to Students' Misconceptions in Algebra *International Journal of Research in Education Methodology* **6** 22

[15] Chick H L and Baker M K 2005 Investigating Teachers' Responses to Student Misconceptions. *Proceedings of the 29th Conference of the International Group for the Psychology of Mathematics Education (Vol. 2)* 249-256

[16] Tanisli D and Kose N Y 2013 Preservice Mathematics Teachers' Knowledge of Students about the Algebraic Concepts *Australian Journal of Teacher Education* **38** 2

Teaching and learning statistics in college: how learning materials should be designed

K Idris
State Institute for Islamic Studies of Lhokseumawe, Jln Medan-Banda Aceh Buket Rata, Lhokseumawe, Indonesia

E-mail: khairiani@iainlhokseumawe.ac.id

Abstract. The evolving nature of statistics implies that the skill of reading statistical information to be essential for college students. To encourage the skill, learning materials need to be appropriately developed. This article aimed to synthesize the issues and suggestions from the current empirical research in statistics education relevant to designing statistics learning materials for college students. The data were collected from from111 research articles published from the last 15 years, which were related to teaching and learning statistics in college. Cognitive and social cognitive learning perspectives were used in reviewing and synthesizing the research, which may reveal the different concerns for a specific situation. Three key suggestions for designing statistics learning materials for college students were identified: using real-life data; familiar contexts; and interactive materials. The reviews also suggested the urgent need for research focusing on designing learning materials for statistics.

1. Introduction

The concept of lifelong learning in a broader context requires college students to interact with academic texts. Particularly, the fact that statistics is an evolving science that implies the ability to learn through reading to be prominent for college students. For example, when a student teacher intends to conduct educational research in the future, he/she would need to read the information on the current investigative method in the educational research field. In this respect, the ability to learn statistics using any available learning resources is also an essential skill for this student teacher. Moreover, some studies focusing on college statistics courses suggested that reading comprehension is significantly related to achievement [1], as well as motivation and perception of the course [2].

Reading statistics texts, as reading in other content areas, involves the interaction or transaction between a reader and a text situated in a particular context. Particularly, the density of statistical text [3], lexical ambiguity terms and multi-words phrases [4,5] may be problematic for student reading. Graphical displays, as one type of statistical texts, also have specific characteristics which might be difficult to comprehend [6,7].

Considering the obstacles students might face in reading statistical information, suggestions and recommendations are required related to how learning materials for college students should be designed. Moreover, since one of the learning principles in statistics derived by Garfield and Ben-Zvi [8] was that students learn better through their active involvement in constructing their knowledge, it is necessary to provide opportunities for college students to learn statistics through reading textbooks or didactical materials as one kind of active involvements [9,10]. Thus, how learning materials should be designed becomes an issue when reading statistical information is taken as one approach in college statistics classes. Accordingly, the goal of this article is to explore and synthesize issues and suggestions for designing learning materials for college statistics based on a review of the current empirical literature on college statistics.

2. The goals of learning statistics at college level

There was an increasingly strong call for statistics education to focus more on statistical literacy, reasoning and thinking in the late of 1990s, which arose by the concern that most statistics courses failed to produce statistically educated students [11]. Since then, the reform movement has been noted in statistics education, which was described by Garfield and Ben-Zvi [12] through changes in content (more data and analysis, less probability), pedagogy (fewer lectures, more active learning), and technology (for data analysis and simulation). More specifically, Garfield [13] classified four goals for introductory statistics course: concepts, skills, types of thinking and attitudes.

Statisticians have discussed the concepts should be taught in introductory statistics course started in the 1990s. A list of topics was proposed by Hogg [14] based on a discussion with statisticians, resulted in four levels of prioritized topics for the course. The highest priority emphasized on understanding basic knowledge, following by methods for investigation of second priority, probability and inference in the third priority, and more advanced methods for investigating data involved in the fourth level. In the more recent year, Burrill and Biehler [15] identified seven statistical ideas that seem to be fundamental for understanding and being able to use statistics in the workplace, in personal lives, and as citizens. The seven ideas are data, variation, distribution, representation, association and modeling relations between two variables, probability models for data-generating processes, sampling, and inference.

There have also been six recommendations provided by the college group of the GAISE (Guidelines for Assessment and Instruction in Statistics Education) for teaching the college statistics course [16]. The six recommendations are as follows:

- Emphasize statistical literacy and develop statistical thinking.
- Use real data.
- Stress conceptual understanding rather than mere knowledge of procedures.
- Foster active learning in the classroom.
- Use technology for developing concepts and analyzing data.
- Use assessments to improve and evaluate students' learning.

The recommendations may be related to the ways a statistics learning material should be designed. For example, the first three and the fifth recommendations may be relevant to how statistical concepts should be addressed in textbooks and how the knowledge and procedures are connected to underlying concepts. The fourth and sixth recommendations might be included as an activity or part in the exercises or homework.

3. Method

This study was a preliminary stage of a study that aimed to design and validate statistics learning materials that emphasize statistical literacy for Indonesian college students. Based on this purpose, the study may be categorized as a developmental study, which is a type of the educational design research [17]. More particularly, since this preliminary study focused on the exploration of issues and suggestions from empirical literature as the basis in the design, it was considered as the first two phases of the design processes: problem and analysis.

In exploring the issues and suggestions related to learning materials for college statistics, research articles published from 2002 to 2017 in various journal articles were searched and synthesized. However, since research focusing on reading to learn statistics texts and designing statistics learning materials has been hardly found in the current literature, the issues were synthesized from the current literature on teaching and learning college statistics. The type of literature reviewed in this article was the research articles, while the topics of the review were limited to how the content of learning materials should be designed, excluded the exercises and assessments. Using various keyword combinations, which meet the criteria described above, in two electronic search engines (*Google Scholar* and *ERIC* database), resulted in 111 journal articles considered relevant to the focus of the review. In this article, however, only several representative articles for each key suggestion were referred. In addition, by considering the recent trend in statistics education research, two learning perspectives were used in synthesizing the issues from those research articles: cognitive and social cognitive perspectives. Learning materials might be regarded as one type of learning tools in the two learning perspectives. From the cognitive perspective, tools are used to facilitate the connection of new information to prior knowledge, while from the social cognitive perspective, learning tools used in teaching need to motivate learning and decrease anxiety.

4. Results and Discussions

There were three key suggestions identified related to the design of learning materials for college statistics: *real-life data*, *familiar contexts*, and *interactive materials*. The three suggestions were elaborated in the followings with regard to the cognitive and social cognitive perspectives.

4.1. Real life data

From the perspective of cognitive learning, using real-world data and application is intended to connect students' contextual knowledge or prior experience about data set to the statistical concepts as their new knowledge. This is an effective way to organize and structure the new information to tap the students' prior experiences. For example, a qualitative study conducted by Neumann et al. [18] showed that using real-life data can promote students' engagement, enhance understanding and learning, remembering of concepts and theory. In this study, 38 psychology students were interviewed about how the use of real-life data was related to their engagement and learning. The datasets used in the course were those that have relevance to the psychology and general interest which were collected from public sources or World Wide Web data repositories. Students also reported getting a broader perspective of the relevance of statistics and more interest. Other studies [19,10] also showed that using real-world application exercises during the course may help students apply their statistical knowledge into other different contexts, the ability to which the author referred to as *far transfer*. This study using quasi-experimental design method which involved psychology student with a variety of ethnicity (Asian, Hispanic, African American and more than half were Caucasian). The materials for real-world exercises were articles from outside the academic environment which require students to summarize and critic the statistical concepts used in the articles.

Social learning theory, on the other hand, views real-world application data as the means to motivate learning, because learning and behavior are affected by students' thoughts and beliefs. Some research in this group assumed that by using real-world application data, students would appreciate the usefulness of course material and beliefs the relevance of statistics to their professions and increase attitudes. Hiedemann and Jones [20] compared two types of approaches in introductory statistics course for business students, i.e., Academic Service Learning (ASL) which require students to work with a real organization to conduct research, and Case Studies (CS) in which the task for students was to imagine themselves as business professionals or statistical consultants charged to make a recommendation or provide information to a client using available data set. Although no difference was found in examination performance, students from ASL group were more likely to agree that statistics have the application to their profession. Wiberg [21], on the other hand, revised statistics course for psychology students by using research problem relevance to psychology throughout the course. The results showed that students performed better and showed positive attitudes in the revised course.

4.2. Familiar contexts

The characteristics of statistical data, which cannot be separated from contexts, make kind of contexts as a factor plausible in student learning. Similar to that of real-life data, the familiar context from the cognitive perspective serves as the connection of students' prior experience to the statistical concepts. From the social cognitive perspective, it is suggested as a means for encouraging students' learning motivations.

The familiar contexts are those relevant to students' field of study. The studies by Hiedemann and Jones [20] and Wiberg [21] discussed above showed the positive impacts of using contexts that relevant to students' field of study. Another study by Abdelbasit [22] suggested that unfamiliar contexts might be problematic for non-native English students learning by using textbooks from English speaking country. In addition, these students tended to focus on technical parts when working with problems from the textbook and ignoring the contexts of the problems.

4.3. Interactive materials
Research suggests that students learn better if they are engaged in their learning. In addition to reading to learn the statistical concepts in textbooks [9,10] using interactive, such as simulations or interactive animations [23,24,25] can be regarded as one kind of the active involvement in learning statistics. The interactivity has been acknowledged as one advantageous feature of using technology as a learning tool.

There has been a different trend of using types of technology in teaching statistics, following the more current development of technology. From the cognitive perspective, technology is used to make the information in an organized or structured form. Several studies showed that the interactive (simulation or interactive animation) and dynamic features of technology, visualization, and constructive representations benefit student learning, which allows them to learn the concepts by their own [24], besides its capability to facilitate the numerical and graphical parts.

Social cognitive perspective, on the other hand, assumes that using technology may stimulate attention and making learning more attractive and fun for students. Some studies have shown that, due to the availability of the practice materials with immediate and targeted feedback, the interactive materials made students more willing to work extra beyond the course requirement [26] or spent more time to the course [27].

5. Conclusion
The three key suggestions identified in this study are in line with the recommendations from GAISE [17] for teaching college statistics, although some of the recommendations were not the focused on this review, such as those related to assessments. Numerous empirical studies have suggested using real-life data and familiar contexts in college statistics. Also, the contexts that are relevant to students' field of study might also be appropriate to use in statistics learning materials, since the familiar contexts might be used as the connection between students' experiences with statistical concepts and their learning motivation. Furthermore, using interactive materials can be related to the recommendation of using technology from [17]. The interactive materials, which have the dynamic visualization and constructive representations, allowed students to learn statistical concepts on their own. In addition, due to immediate feedback for the practice materials provided in the interactive materials, students may be motivated in learning.

This review also showed the urgent need for further studies focusing on designing learning materials for statistics. Furthermore, when reading to learn is intended in college statistics classes, criteria for measuring the appropriateness of reading materials for the different reading levels of students can be another important issue for future research.

References
[1] Collins K M and Onwuegbuzie A J 2002 Reading ability and the performance of African American graduate students in research methodology courses *J College Literacy & Learning* **31** 39
[2] Harwell M R, Herrick M L, Curtis D, Mundfrom D and Gold K 1996 Evaluating statistics texts used in education *J Edu and Behav Stat* **21** 3

[3] Cobb G W 1987 Introductory textbooks: a framework for evaluation: a comparison of 16 books *J American Stat Assoc* **82** 321
[4] Kaplan J J, Rogness N T, and Fisher D G 2012 Lexical ambiguity: making a case against spread. *Teach Stat* **34** 56
[5] Lesser L M, Wagler A, Esquinca A, and Valenzuela M G 2013 Survey of native English speakers and Spanish-speaking English language learners in tertiary introductory statistics *Stat Edu Res J* **12** 6
[6] Friel S N, Curcio F R, and Bright G W 2001 Making sense of graphs: Critical factors influencing comprehension and instructional implications *J for Res in Math Edu* **32** 124
[7] Shah P and Hoeffner J 2002 Review of graph comprehension research: Implications for instruction *Edu Psych Rev* **14** 47
[8] Garfield J B and Ben Zvi D 2007 How students learn statistics revisited: A current review of research on teaching and learning statistics *Inter Stat Rev* **75** 372
[9] Carlson K A and Winquist J R 2011 Evaluating an active learning approach to teaching introductory statistics: A classroom workbook approach *JSE* **19** 1
[10] Lawson T J, Schwiers M, Doellman M, Grady G, and Kelnhofer R 2003 Enhancing students' ability to use statistical reasoning with everyday problems *Teach Psych* **30** 107
[11] Garfield J B and Ben-Zvi D 2008 *Developing Students' Statistical Reasoning* (Netherlands: Springer)
[12] Moore D S 1998 Statistics among the liberal arts *J American Stat Ass* **93** 1253
[13] Garfield J 1995 How students learn statistics *Inter Statistical Rev* **63** 25
[14] Hogg R V 1990 Statisticians gather to discuss statistical education *Amstat News* **169** 19
[15] Burrill G and Biehler R 2011 Fundamental statistical ideas in the school curriculum and in training teachers *Teaching Statistics in School Mathematics-Challenges for Teaching and Teacher Education* (Springer) 57
[16] Franklin C A and Garfield J B 2006 The GAISE project: Developing statistics education guidelines for grades pre-K-12 and college courses In *Thinking and reasoning with data and chance: 2006 NCTM yearbook* ed G F Burrill and P C Elliott (Reston, VA: National Council of Teachers of Mathematics) 345
[17] Plomp T 2013 Educational design research: an introduction in *Educational Design Research* pp 11-50
[18] Neumann D L, Hood M and Neumann M 2013 Using real-life data when teaching statistics: Student perceptions of this strategy in an introductory statistics course *SERJ* **12**
[19] Daniel F & Braasch J L 2013 Application exercises improve transfer of statistical knowledge in real-world situations *Teaching of Psychology* **40** 200
[20] Hiedemann B and Jones S M 2010 Learning statistics at the farmers market? A comparison of academic service learning and case studies in an introductory statistics course *JSE* **18**
[21] Wiberg M 2009 Teaching statistics in integration with psychology *JSE* **17** 1
[22] Abdelbasit K M 2011 Learning Statistics in a foreign language In *International Encyclopedia of Statistical Science* 729 (Gainesville: Springer)
[23] Lane D M and Tang Z 2000 Effectiveness of simulation training on transfer of statistical concepts *J Edu Comp Res* **22** 383
[24] Meletiou-Mavrotheris M 2003 Technological tools in the introductory statistics classroom: effects on student understanding of inferential statistics *Int J Comp for Math Learn* **8** 265
[25] Wang P-Y, Vaughn B K, and Liu M 2011 The impact of animation interactivity on novices' learning of introductory statistics *Comp & Edu* **56** 300
[26] Ward B 2004 The best of both worlds: A hybrid statistics course *JSE* **12** 74
[27] Lovett M, Meyer O and Thille C 2008 The open learning initiative: measuring the effectiveness of the OLI statistics course in accelerating student learning *J Inter Media in Edu* **1** 13

The development of mathematics student worksheet for school literacy movement

K N S Effendi[1], Zulkardi[2], R I I Putri[2] and P Yaniawati[3]

[1]Universitas Singaperbangsa Karawang, Jl. HS.Ronggo Waluyo, Karawang, Indonesia
[2]Universitas Sriwijaya, Jl. Srijaya Negara, Bukit Besar, Palembang, Indonesia
[3]Universitas Pasundan, Jl. Sumatra, Bandung, Indonesia

E-mail: kiki.niasania@staff.unsika.ac.id

Abstract. This paper discusses the development research that produces the student worksheet of school literacy movement in junior mathematics learning. The developed worksheet is organized based on the Realistic Mathematics Education approach. This research uses a design research method with the type of development studies and the evaluation flow used is formative study, including self-evaluation, expert review, one-to-one, small group, and field test. However, this paper only would be discussed about the expert review and one-to-one phases. The research subject is the students of grade VII SMP Karawang. Data collected using walkthrough and documentation. Based on the analysis, it concluded that this research has resulted in a valid product in the form of the reading worksheet of school literacy movement in mathematics.

1. Introduction

School literacy movement, at this time, starts to be implemented by schools in Kabupaten Karawang as an effort to increase the learners' reading interest. School literacy movement is a comprehensive effort to make schools as learning organizations that make citizens literate through public engagement in their life [1]. The purpose of this movement is to cultivate the character of the learner through the culture of the school literacy ecosystem embodied in school literacy movement so that they become lifelong learners [1]. School literacy movement consists of three stages, namely: the stage of habituation, development stage, and learning phase. The activities at each stage take 15 minutes to read by *Permendikbud* No. 23 Years 2015. This reading activity is expected to foster learners' interest in reading.

School literacy movement is generally still at the habituation stage. At this stage, learners are familiarized to read, so that the learners' interest in reading will be increased. This activity is done during the first lesson when students were reading fiction books such as novels that they bring from home or books provided in the school library for 15 minutes and then determine the essence and the synopsis of the books they read. The activity is done in the first hour so every lesson using a fiction book will interest the learners without encouraging learners to look for the communication between the books read with the material to be studied. This shows that the development stage has not been implemented. Furthermore, the learning phase has not been implemented. It is because a non-lesson book, that accordance with the mathematical material to be studied, is hard to find by the teachers.

Mathematics is a lesson that is closely related to daily life. Human activity in daily life can not be separated from the utilization and application of concepts that exist in mathematics [2]. The cases those relate to the daily life of learners can be a starting point to learn a matter in mathematics school. The context or experience of students is used as the starting point of the mathematics learning process [3]. The reading text in the school literacy movement at the learning stage can serve as a starting point for learners to learn math materials. Based on the principle of school literacy movement in the learning phase, the reading book is a book about general knowledge, or passion, and can also be associated with specific subjects (not just language) and there are bills related to subjects [4]. Reading is an interactive activity to pick and understand the meaning or message contained in written materials [5]. Reading activities in school literacy movements using the reading texts with contexts favored by learners and related to the subject will develop students learning skills and develop learners' understanding of mathematics learning. One of the goals of the school literacy movement is the stages of learning can develop the ability to understand the text and relate it to personal experience so it will form a lifelong personal learner [4]. One mathematical learning theory that is suitable to be used in teaching mathematics relating to the context or situation is the Indonesian Realistic Mathematics Education. Learning with RME starts from a context that raises the informal reasoning of learners, with an accompanying set of statements used to build relationships between formal, pre-formal, and formal mathematical representations [6]. According to these principles, mathematics learning should give students the opportunity to understand and process the rediscovery of mathematics itself [7]. In this case, the used context should be meaningful and real to the students' thinking [3]. The futsal game, at the moment, became one of the sports that are very popular among various circles, both parents and children, both women and men, so that it can be used as a context for mathematics learning.

According to the International Council of Sport, sports education is a physical activity that contains the nature of the game and contains struggles with self or others and confrontations with the elements of life [8]. Some topics about the interaction between mathematics and sports are related to statistics such as how to complete relationships between tables league or sports games that are individual or grouped like chess and football [9]. Futsal is a ball game played by two teams, each consisting of five people who aim to make a goal with the ball into the opposing goal by manipulating the ball with the foot [10]. The futsal game includes a collection of players who can play futsal. In a mathematical set of objects that have properties with clear definitions is defined as a set. The set is the basic concept of all branches of mathematics [11]. The description shows the futsal game is a sport that can be used as a context in the learning of mathematics because futsal game attracted many learners and had a relationship with mathematics. Learners who learn to use sports contexts will find that math is more fun because it is different from repetitive math exercises [12]. The study used sports context successfully helps students in learning mathematics, especially in swimming. The swimming context could stimulate students' informal knowledge about the meaning of fractions which it can be used in the additional learning either the same denominator or different denominator [13].

2. Method

This research uses the method of design research with the type of research development or development studies [14]. This study was conducted in three stages, that is a preliminary or preparatory stage, the prototyping stage, and an assessment stage [15]. In the prototyping stage, the evaluation uses formative evaluation with the phases includes self-evaluation, expert review and one-to-one and small group, as well as field tests. Meanwhile, to analyze the two previous stages, the researchers use the assessment stage [16]. In this article, the discussion of research is the result of expert review phase and one-to-one. Expert review and one-to-one are phases at the prototyping stage. Prototyping phase with grooves formative evaluation includes self-evaluation, expert reviews and one-to-one, small group and field test.

In the self-evaluation stage, the researchers evaluate the text-reading design used in the school literacy movement in mathematics learning activities. The design process uses three characteristics that are content, constructs and language. The results are analyzed to form the first prototype can be seen in Table 1.

Table 1. The characteristics of the prototype

No.	Characteristics	Examined Aspects
1.	Content	Text according to RME principles (Realistic Mathematics Education). Text according to futsal game theory. The text corresponds to some mathematics material of grade VII junior high school. Text according to School Literacy Movement concept.
2.	Construct	The text constructs of learner's knowledge. The text contains various mathematical concepts. The text corresponds to the level of thinking of grade VII junior high school.
3.	Language	Text in accordance with Indonesian standard language. The flow on the text is not convoluted. Sentences on text are not ambiguous.

The text is developed based on the stages of the school literacy movement using the futsal game context. At the stage of habituation, the text tells about the history of futsal in the world and in Indonesia along with the achievements of the national team futsal's son and daughter. At this stage, the text has not been given a question to know the relevance to mathematics because at this stage the purpose is to make learners have an interesting and accustomed to reading. At the development stage, the text tells the rules in futsal game which includes field, ball, sanction, game system, and basic technique and strategy in futsal game. The questions at this stage begin to develop the learner's thinking to know the interrelation between futsal and mathematical material, such as a coaching strategy that sets up groups for players in a team to implement game patterns according to the player's ability to relate to the material sets. At the learning stage, the text presents a short story titled *Harumkan Nama Kampus Dengan Modal Kebersamaan*. The text at this stage is arranged based on the set material that is the concept of set, the universal set, the empty set, subsets and the presentation of the set, and the operation on the set. The questions in the text correspond to the material being studied because the questions are based on the purpose of the learning stage. The text is organized on the basis of the set material, so the questions at this stage can help learners understand the material sets.

At the expert review stage, the expert validated the prototype. Expert validation uses analysis in terms of content, constructs, and language. Experts who are referred in this study are mathematics education experts, and futsal game experts. The first prototype's validation process of the expert review stage is done through three ways face-to-face review, mail/review (mails review), and panel review [16]. Based on results, the product is revised. Along with the expert review, researchers test the students individually (one-to-one). Results from one-to-one are used to revise the product to be created.

3. Results and Discussions

The result of the expert review and one-to-one phases is experts' validation related to the first pre-designed prototype and it is tested on three students as a one-to-one tester. The first prototype was tested on three students of class VII K, initially: BRS, RR, and ENS. The purpose of the implementation of this trial is to determine the responses and difficulties that will be faced by learners at the time of reading and answering questions on the text. The observed responses and difficulties focus on the legibility and conspicuousness of the text along with the question on the text. The reading text in this study was prepared for the School Literacy Movement consisting of the stages of habituation, development stage, and the learning stage.

At the stage of habituation, it consists of texts on the history of futsal's world, futsal history in Indonesia, and the achievements of Indonesian futsal national team. At this stage, learners are asked to read and resolve questions related to the text and in accordance with the objectives at the stage of habituation such as increasing the love of reading outside the lesson, improving the ability to understand reading, improve self-confidence as a good reader, source of reading [4]. Based on the results of the students' answers, it can be indicated that they understand the text and there is no significant difficulty for them when answering questions from the text. The questions on the text of the readings are about like or dislike to the text by the learners, the words that are difficult for them, and retell the text they have read using their language. Also, they like the text they read because the text provides information which they did not already know.

The development stage consists on text futsal field and ball game, the basic of futsal game technique, futsal game strategy, sanction in futsal game, and futsal game system. At this stage, after the learners read they also answer questions related to the text that has been read and the question which accordance with the purpose of development stage, namely students are able to find the relationship between the books that was read by themselves and the surround environment [4], one of that is the mathematical material they learned in school.

In the text entitled futsal game strategies, learners have an opinion about the mathematical material associated with the text, so they can construct their knowledge of the mathematical material of the text. This text material presented the coach's strategy in dealing with the rivalry by creating groups of players in his team to play the game pattern. In addition, the material of attacking and defensive game patterns accompanied by images shows the quadrilateral formed of the patterns. Based on the answers of learners, they are able to answer the mathematical material they can learn from the text. Mathematical material that can be learned from the text titled futsal game strategy according to RR, it is the set; whereas according to ENS, it is a quadrilateral; and according to BRS there are rectangles and kites. Figure 1 and 2 are some pictures of the futsal game attack pattern.

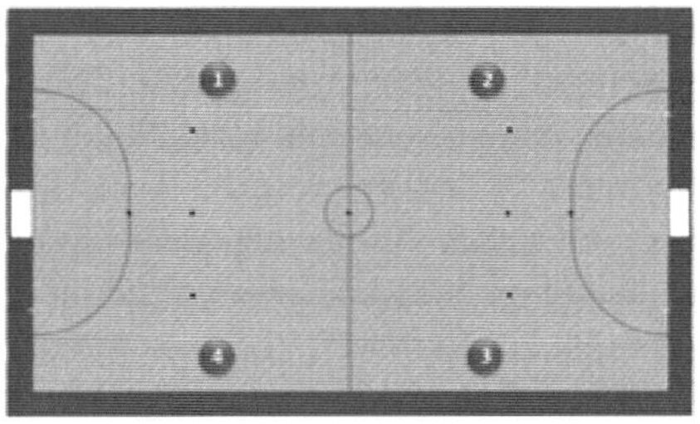

Figure 1. Attack Pattern 2 - 2

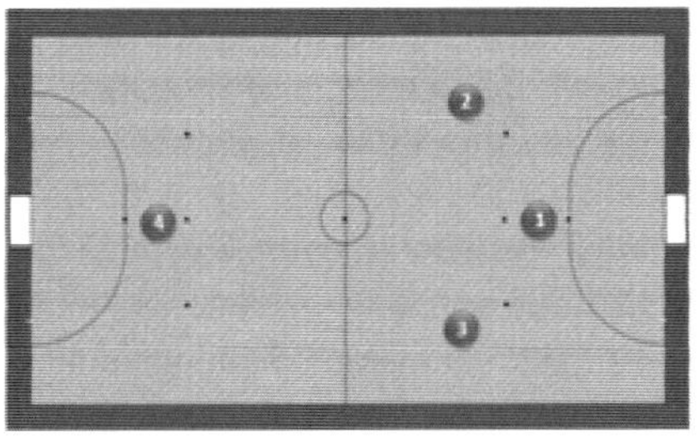

Figure 2. Attack Pattern 3 - 1

The reading text in the learning stage of school literacy movement with futsal game context can be a starting point in mathematics learning, so the learners can construct their knowledge to understand a mathematical concept. The use of context is to make a line with the characteristics of Indonesia's realistic mathematics education which aims to help learners understand a mathematical concept. Thus, the learners play an active role to understand the relevance of the context with the math material being studied, so the learners can understand concepts and solve mathematical problems. The realistic mathematical approach is suitable for mathematics learning because RME has characteristics and principles that students enable to develop optimally, such as the freedom of students to express their opinions, the existence of contextual problems that can relate mathematical concepts to real life, and it can facilitate students in solving problems [17].

Based on the learner's answer to the text relating to the subset material, it shows that they can mention the player who brought the cones is part of the team in futsal A and the reason they have pointed has already led to the subset material. Also, they can give other examples of the subsets they encounter in the text. They can see the relevance of the document they read with the mathematical material so that the text material could build their understanding into a formal concept of subsets. Through the activities of school literacy movement in the learning stage, it can enrich the knowledge of learners in learning mathematics so the obtained results will be by learning objectives. Literacy mathematics is introduced explicitly as an intervention to improve the ability of learners in mathematics learning [18]. RME strongly influences mathematical literacy because RME stresses the importance of solving mathematical problems in real-world settings so that it can be used as a fundamental theory for developing literacy-based learning math [19]. Figure 3 is the answer from one of the student.

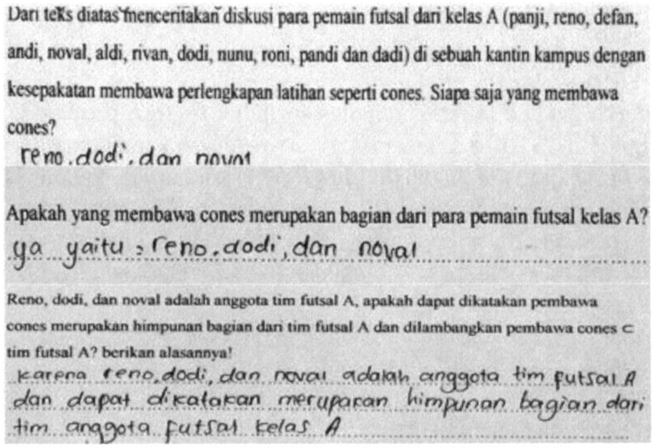
Dari teks diatas menceritakan diskusi para pemain futsal dari kelas A (panji, reno, defan, andi, noval, aldi, rivan, dodi, nunu, roni, pandi dan dadi) di sebuah kantin kampus dengan kesepakatan membawa perlengkapan latihan seperti cones. Siapa saja yang membawa cones?

reno, dodi, dan noval

Apakah yang membawa cones merupakan bagian dari para pemain futsal kelas A?

ya yaitu : reno, dodi, dan noval

Reno, dodi, dan noval adalah anggota tim futsal A, apakah dapat dikatakan pembawa cones merupakan himpunan bagian dari tim futsal A dan dilambangkan pembawa cones ⊂ tim futsal A? berikan alasannya!

karena reno, dodi, dan noval adalah anggota tim futsal A dan dapat dikatakan merupakan himpunan bagian dari tim anggota futsal kelas A

Figure 3. Student answer

After the implementation of the test, the researcher provides the students the opportunity to give comments and suggestions about the reading text of the school literacy movement. The students' comments show that they like the text because it adds to their knowledge of futsal, and they know of a connection between mathematics and futsal. Also, the given advice is to add people's photos to futsal history text. The majority of students in Indonesia is anxious and dislikes the mathematics learning, so their mathematical achievement is low [20]. The students' comments indicate an interest in reading the text. From that interest, mathematics learning will be preferred to students. It is because the learning of mathematics begins with the context that is commonly encountered learners. RME is a learning that uses reality as a starting point in the learning process that aims to support students in building and rediscovering mathematics through contextual problems [21-24].

In the expert review the researcher asks the opinion of experts who have experienced in mathematics education, futsal games, and school literacy movement. The responses and suggestions of experts and peers about the designs that have been made are written on the validation sheet as material for revising and declaring the valid text. The literature text of the developed school literacy movement is validated through one to one and expert review. This is in line with triangulation, that are the process of using conflicts of data from different sources, confirming observations from different observers, and confirming information with different data collection methods [25]. Triangulation is a data validation technique by utilizing something else beyond that (the expert and the work of learners) as a benchmark/revision of the assessment instrument. The valid aspect associated with the developed text is based on a strong theoretical rational and internal consistency [26]. The reading text for the school literacy movement on mathematics learning based on the one to one stage and expert review is valid with the revision of complete information with photos and pictures –the presented image must be clear between the player, the ball, and the direction of the ball and the direction of the player and accompanied by a description. Also, sentence writing should be in accordance with Indonesian standard language and the final text for the two parts. Validity test is conducted so that the instrument made to produce information that is appropriate, meaningful, precise, and useful in the withdrawal of conclusions made by researchers [27-28].

4. Conclusion

The result of this research is the Reading Text Materials for School Literacy Movement in Junior High School Mathematics Learning. The first prototype of the developed text material is valid based on the Expert appraisal and one-to-one test. The prototype is matching regarding content, constructs, and language. In total, three text materials for the habituation stage, five text materials for the development stage, and four text materials for the learning stage.

References

[1] Kemendikbud 2016 *Desain Induk Gerakan Literasi Sekolah* retrieved at http://dikdas.kemdikbud.go.id/index.php/desain-induk-gls-kemendikbud/
[2] Revina S and Leung F K S 2018 Educational borrowing and mathematics curriculum: Realistic Mathematics Education in the Dutch and Indonesian primary curriculum *International Journal on Emerging Mathematics Education* **2** 1
[3] Zulkardi and Putri R I I 2006 Mendesain Sendiri Soal Kontekstual Matematika *Prosiding Konferensi Nasional Matematika ke 13* (Semarang: Universitas Negeri Semarang)
[4] Kemendikbud 2016 *Panduan Gerakan Literasi di Sekolah Menengah Pertama* (Jakarta: Kemendikbud)
[5] Sumadyo S 2011 *Strategi dan Teknik Pembelajaran Membaca* (Yogyakarta: Graha Ilmu)
[6] Webb D C, *et al* 2011 Design research in the Netherlands: Introducing logarithms using Realistic Mathematics Education *Journal of Mathematics Education at Teachers College* **2** 47
[7] Fauziah A, *et al* 2018 Primary school student teachers' perception to Pendidikan Matematika Realistik Indonesia (PMRI) instruction *Journal of Physics: Conference Series* **943** 012044
[8] Lutan R S 2000 *Filsafat Olahraga* (Jakarta: Depdiknas)
[9] Haigh J 2009 Uses and limitations of mathematics in sport *IMA Journal of Management Mathematics* **20** 97
[10] Lhaksana J 2011 *Taktik dan Strategi Futsal Modern* (Jakarta: Be Champion)
[11] Murniasih T R, *et al* 2016 Media smart diagram venn untuk meningkatkan pemahaman konsep siswa pada materi himpunan *Math Didactic: Jurnal Pendidikan Matematika* **2** 70
[12] Reys, *et al* 2013 Sport courts and fields: A context for estimation and tessellation *Mathematics Teaching in the Middle School* **18** 566
[13] Putri R I I, *et al* 2018 Addition of fraction in swimming context *Journal of Physics: Conference Series* **943** 012035
[14] Akker J, *et al* 2006 *Educational Design Research* (London and New York: Routledge)
[15] Plomp T and Nieveen N 2007 An introduction to educational design research *Proceedings of the seminar conducted at the East China Normal University, Shanghai (PR China)* p 23
[16] Tessmer M 1993 *Planning and conducting formative evaluations: Improving the quality of education and training* (London: Kogan Page)
[17] Marhamah, *et al* 2011 Pengembangan materi ajar pecahan dengan pendekatan PMRI di SD Negeri 21 Palembang *Jurnal Pendidikan Matematika* **5** 171
[18] Spangenberg E D 2012 Thinking styles of mathematics and mathematical literacy learners: Implications for subject choice *Pythagoras* **33** 1
[19] Zulkardi and Kohar A W 2018 Designing PISA-Like mathematics tasks in Indonesia: Experiences and challenges *Journal of Physics: Conference Series* **947** 012015
[20] Arsaythamby V and Zubainur C M 2014 How a realistic mathematics educational approach affect students' activities in primary schools? *Procedia - Social and Behavioral Sciences* **159** 309
[21] Gravemeijer 2010 Realistic mathematics education theory as a guideline for problem-centered, interactive mathematics education *In R. Sembiring, K Hoogland & M. Dolk (Eds.) A decade of PMRI in Indonesia* (Bandung, Utrecht: APS International)
[22] Ryandi R B, *et al* 2018 Learning combinations through "handshake" *International Journal on Emerging Mathematics Education* **2** 105
[23] Risdiyanti I and Prahmana R C I 2018 *J. Phys.: Conf. Ser.* **943** 012032

[24] Wahyu K, *et al* 2017 Motivation cards to support students' understanding on fraction division *International Journal on Emerging Mathematics Education* **1** 99
[25] Sumarto S N, *et al* 2014 Proportional reasoning: How do the 4th graders use their intuitive understanding? *International Education Studies* **7** 69
[26] Krathwohl 1997 *Methods of Educational and Social Science Research, Second Edition* (New York: Longman, Inc)
[27] Akker J v d 1999 *Design Approaches and Tools in Education and Training* (Dordrecht: Kluwer Academic Publishers)
[28] Fraenkel J R and Wallen N E 2010 *How to Design and Evaluate Research in Education* (Singapore: McGraw-Hill)

Application of Microsoft Excel as an interactive learning media of acid-base titration

I Khaldun, M Hasan, and Nilawati
Syiah Kuala University, Jalan Tgk. Hasan Krueng Kalee, Darussalam, Banda Aceh Indonesia, 23111

E-mail: ibnukhdn@unsyiah.ac.id

Abstract. Senior high school students in Banda Aceh find Acid-base titration is challenging to learn as it requires knowledge and many concepts such as acid-base reaction, acid-base strength, pH solution, buffer solution, and salt hydrolysis. This study aimed at developing a Microsoft Excel-based interactive learning media program on the topic of acid-base titration. It employed ADDIE model of development design. The study was conducted at one of senior high school in Banda Aceh, Indonesia. The participants of this research were year 11 students. Interview, questionnaire, and test were employed to collect the data in this study. Data were analyzed using a descriptive quantitative technique. The results of validation from two learning media experts revealed that 79.41 were categorized into the very good category. Furthermore, two chemistry experts revealed that the 77.94 of the instrument met the very good category. In addition, in the implementation phase, the result showed the skill of science process for the students who were in the experimental class was better than students in the control class.

1. Introduction

Titration is a common laboratory method of quantitative chemical analysis that can be used to determine the concentration of an identified analytic. Acid-base titration is one of challenging topics to be learned by senior high school students. It is required knowledge or such supportive concepts as stoichiometry solution, pH of acid solution, pH of base solution, pH of buffer solution, and pH of hydrolysis solution [1]. Apart from these ones, it is also needed a laboratory practice to prove the concept [2]. It can also be done by using computer-based interactive program media other than in the laboratory. From the research finding obviously showed the use of virtual media was better than the traditional method [3,4]. Nowadays, the use of computers in learning is highly recommended by the government since it is accordance with the Strategic's Plans of Indonesia Government which hopes there is the use of technology, information, and communication within a learning activity.

This study aimed at developing an interactive learning media of acid-base titration-based Microsoft Excel-based. Microsoft Excel is not solely used in the field of statistics [5] as well as business [6], but it is also used as a medium of learning in related chemical matters to data processing from the result of chromatographic analysis [7], titration [8,9], spectroscopy [10], pharmaceutical or medical field [11].

It was a R*esearch & Development.* It purposed to make and produce a particular product then test the advisability of produced product. Advisability test is given for both media and material experts, teachers, and students as the user [12].

2. Method

The development of titration media used ADDIE model which stands for Analysis, Design, Development, Implementation and Evaluation. This research was carried out at one of Senior High School in Banda Aceh on grade 11 with the number of samples was 70 students. The following steps of research were:

- Analysis, on this step there were some matters observed such as the school condition particularly the availability of computer facility in school, curriculum analysis, difficulties analysis encountered by students on the learning material of acid-base titration.
- Design, arraying of learning cycle and media content outline, designing a media that supported the understanding of titration concept, and making a media display design were some things done on this step.
- Development, this step focused on making a Microsoft Excel-based interactive titration media and students' work sheets as the complement of the guide of media use.
- Implementation, meanwhile on this step media validation was conducted to look at the media advisability by distributing the questionnaires for both material and media experts. Afterward, it was done a trial test for senior high school students to know the skill of science process on the learning material of titration. Titration media validation was done by administering the questionnaires that contained media assessment criteria and optional responses i.e. very good, good, enough, less, very less with weight score respectively 5, 4, 3, 2, and 1. The calculation of score percentage of media validation used the equation (1) [11]. Moreover, validity criteria of media are presented in Table 1 and criteria of students' responses are presented in Table 2.

$$P = \frac{\sum_i^n x_i}{n.k} x100\% \qquad (1)$$

Note:
P : assessment percentage
ΣX : number of assessment point
N : number of subjects
K : highest assessment score

Table 1. Media validity criteria [11]

Percentage (%)	Validity Level	Description
76 - 100	very high	proper/no revision
50 - 75	high enough	proper enough/partial revision
26 - 50	less	pess proper/partial revision
<26	Invalid	improper/total revision

Table 2. Assessment percentage category of students' responses [11]

Categories	Percentage
Disagree	0% - 25%
Hesitate	26% - 50%
Agree	51% - 75%
Strongly agree	76% - 100%

- An evaluation was conducted by giving a number of questionnaires to know their responses having used titration media. The questionnaires were descriptively analyzed with the following of some categories was Strongly Agree (SA), Agree (A), Hesitate (H), and Disagree (D) weight score respectively 5, 4, 3, 2, and 1.

IOP Conf. Series: Journal of Physics: Conf. Series **1088** (2018) 012034 doi:10.1088/1742-6596/1088/1/012034

3. Result and Discussion

3.1 Analysis

The result of observation revealed that Senior High School in Banda Aceh has had computer laboratory facility which can be used for 40 students. The practice related to acid-base titration was conducted by the chemistry teachers; however, it was only restricted on HCl – NaOH titration. Moreover, it solely used one indicator type namely phenolphthalein. While, other chemical compounds were not done because there was a chemical substance. From the above finding, the researcher made an acid-base titration program which was more complete by using such solution as HCl, CH_3COOH, NaOH and NH_4OH; whereas, as its indicator used phenolphthalein, methyl red and bromothymol blue solution.

3.2 Design

Media design which was made in line with learning objectives, basic competency and the needed indicator from the learning material of acid-base titration. Design a burette picture with the volume was 25 mL and Erlenmeyer flask picture was completed with a magnetic stirrer. Burette bar picture was made from the chart graph, while; other pictures were taken ChemDraw program Ultra 8 (Figure 1).

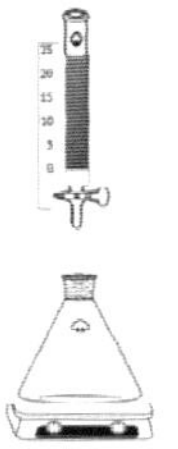

Figure 1. Burette 25 mL, erlenmeyer flask, and magnetic stirrer

Design a titration program chart that was completed with titration type, indicator type, pH range, and table of information. The information table functioned to show solution volume (cell D11 and D20), solution concentration (cell D12 and D21), K_w water equilibrium constant (cell D13), and the weak acid or the weak base equilibrium constant (cell D22).

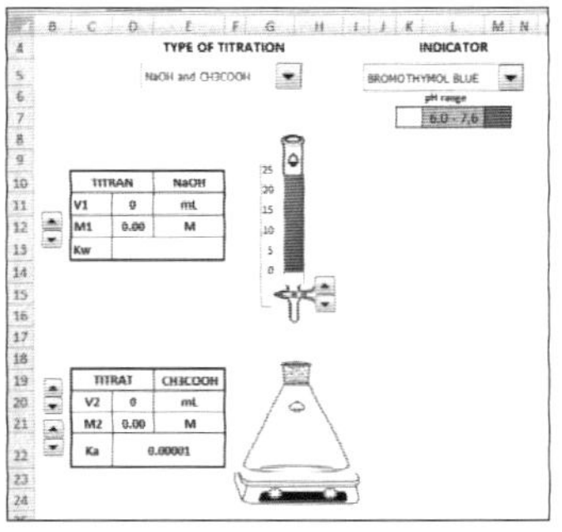

Figure 2. Design acid-based titration program

3.3 Development

Activating combo box which can set up both titration and indicator type. The titration that was used consisted of 4 types namely HCl – NaOH, HCl – NH_4OH, NaOH – HCl, and NaOH – CH_3COOH. Meanwhile, the indicator contained 3 types namely phenolphthalein, bromothymol blue, and methyl red. The change of titration type was contolled by the cell F5 and; whereas, the change of indicator type was controlled by the cell K5 which its data included into data combo box (Figure 3).

IOP Conf. Series: Journal of Physics: Conf. Series **1088** (2018) 012034 doi:10.1088/1742-6596/1088/1/012034

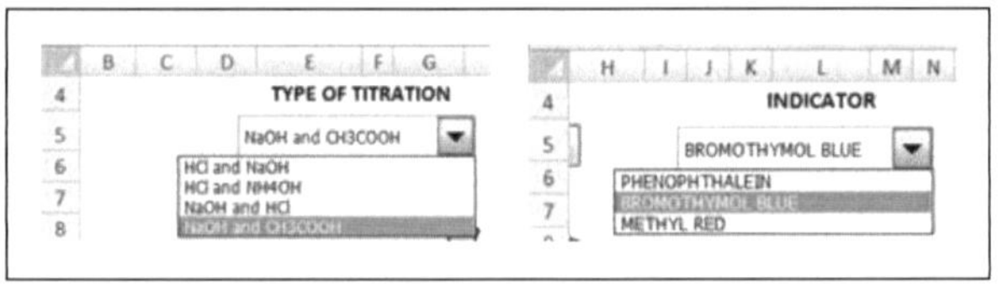

Figure 3. Titration and indicator type

Making a stoichiometric reaction chart is functioned to calculate the concentration. The reaction equation that appeared on this program is in accordance with the selected titration. The following example is the type of (Figure 4) HCl – NH_4OH titration.

	R	S	T	U	V	W	X	Y	Z
3	STOICHIOMETRIC REACTION								
4									
5			HCl(aq)	+	NH4OH(aq)	→	NH4Cl(aq)	+	H2O(l)
6	[Initial]	:	0.047	M	0.053	M			
7	[Reaction	:	0.047	M	0.047	M			
8	[Result]	:	0.000	M	0.006	M	0.047	M	

Figure 4. A reaction between 22 mL of HCl 0.1M and 25 mL of NH_4OH 0.1M

Inserting the logic formula (IF) into Stoichiometric Reaction Chart on the following cells respectively are T6, T7, T8, V6, V7, V8, and X8:

=IF(D20*D21>0,(D11*D12)/(D11+D20),"") (2)

=IF(D20*D21>0,MIN(T6,V6),"") (3)

=IF(D20*D21>0,T6-T7,"") (4)

=IF(D20*D21>0,(D20*D21)/(D11+D20),"") (5)

=IF(D20*D21>0,MIN(T6,V6),"") (6)

=IF(D20*D21>0,V6-V7,"") (7)

=IF(D20*D21>0,MIN(T6,V6),"") (8)

The next step was inserting the logic formula (IF) into the cell H25 to calculate pH of solution. The logic formula was used for each titration type can be written severally within the equation (9 – 12).

Titration of HCl – NaOH

$$= if(F5 = 1, if\left(V8 > T8, 14 + \log(V8), if\left(V8 = T8, -\log(D13^{0.5}), if(V8 < T8, -\log(T8))\right)\right)) \quad (9)$$

Titration of HCl – NH_4OH

$$= if(F5 = 2, if\left(and\left(V8 > t8, or(D11 = 0, D12 = 0)\right), 14 + \log((D21 * D22)^{0.5}), if\left(and(D11 > 0, V8 > T8), 14 + \log\left(D22 * \left(\frac{V8}{X8}\right)\right), if\left(and(D11 > 0, V8 = T8), -\log\left(\left(\left(\left(\frac{D13}{D22}\right)\right) * X8\right)0.5\right), if(V8 < T8, -\log(T8))\right)\right)\right) \quad (10)$$

Titration of NaOH – HCl

$$= if(F5 = 3, if(\left(V8 > T8, -\log(V8), if\left(V8 = T8, -\log(D13^{0.5}), if(V8 < T9, 14 + \log(T8))\right)\right)) \quad (11)$$

Titration of NaOH – CH_3COOH

$$= if(F5 = 4, if(and\left(V8 > T8, or(D11 = 0, D12 = 0)\right), -\log\left((D21 * D22)^{0.5}, if\left(and(D11 > 0, V8 > T8), -\log\left(D22 * \left(\frac{V8}{X8}\right)\right), if\left(V8 = T8, 14 + \log\left(\left(\left(\left(\frac{D13}{D22}\right)\right) * X8\right)^{0.5}\right), if(V8 < TT8, 14 + \log(T8))\right)\right)\right) \quad (12)$$

The next step is inserting the logic formula (IF) into the cell G21 change solution color based on indicator type (cell K5) and pH of solution (cell H25). The logic formula was used as on the following equation.

=IF(D20*D21>0,IF(AND(K5=1,H25<8),"A",IF(AND(K5=1,H25<9.6),"B",IF(AND(K5=1,H25>9.6),"C",IF(AND(K5=2,H25<6),"D",IF(AND(K5=2,H25<7.6),"E",IF(AND(K5=2,H25>7.6),"F",IF(AND(K5=3,H25<4.2),"G",IF(AND(K5=3,H25<6.2),"H",IF(AND(K5=3,H25>6.2),"I"))))))))),"") (13)

Based on the equation (13), for each change of indicator type that occurred (cell K5) and pH of solution (cell H25); as a result; within cell G21 will emerge the following letters are A, B, C, D, E, F, G, H, or I. After that those letters changed became the color by using "conditional formatting". The color of the letters of A, B, C, D, E, F, G, H, and I from left to right as presented respectively on figure 5.

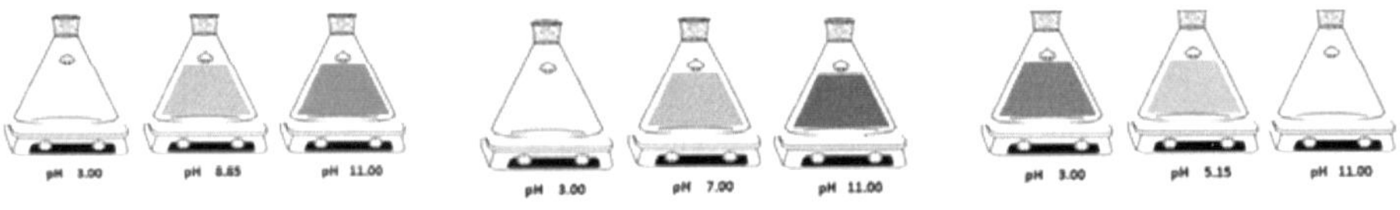

(a) Phenolpthalein (b) Bromthymol Blue (c) Methyl Red

Figure 5. Effect of the change of pH toward indicator color

Furthermore, using interactive titration media which was completed students' worksheet to identify the increase of students' science process skill. One of the experiments caried out was solution titration of NH_4OH 0.1 M with standard solution of HCl 0.1M used the methyl red indicator. When the equivalance point was achieved, the NH_4OH solution that was yellow previously changed became orange (pH 5.15).

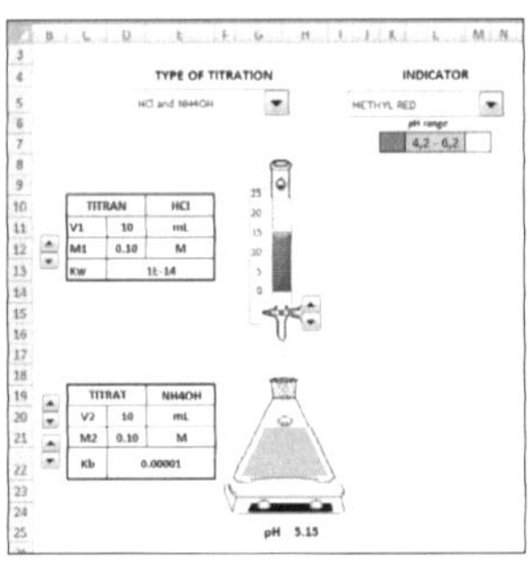

Figure 6. Titration display of HCl-NH_4OH on equivalence point using methyl red indicator, pH = 5.15.

3.4 Implementation
From the result of validation employed by both media and material experts it was respectively gained its percentage was 79.41% and 77.94%. According to the criteria presented in Table 1, it can be concluded the validity of titration media can be categorized very high and it is proper to be used in the teaching and learning process.

3.5 Evaluation
Based on the result of program trial test toward the students of Senior High School in Banda Aceh attained N-gain value was 0.38 on the experimental class was higher than the control class with N-gain value was 0.12. Generally, the students' science process skill who was in the experimental class increased having followed the learning process through Microsoft Excel-based titration learning media. The result of this study was pertinent to the previous research findings undertaken by Tatly and Ayas [2] as well as Tuyuz [3]. From the result of statistic test, both of N-gain value were significantly different. It indicated that the students who used the interactive titration media better than the other

group. Thus, the use of titration learning media by using Microsoft Excel is effective to advance the students' science process skill on the learning material of acid-base titration.

4. Conclusion

Microsoft Excel-based interactive learning media on the learning material of acid-base titration which is developed it is appropriate to be used because of its validity, practicality, and effectiveness. In addition, it can enhance the skill of students' science process at Senior High School in Banda Aceh. It suggests that this learning media is suitable as an alternative to replace laboratory-based experiment for teaching and learning of acid-base titration.

References

[1] Harvey D 2000 *Modern Analytical Chemistry* (New York: McGraw Hill)
[2] Tatly Z and Ayas A 2013 *Educ. Tech Soc.* **16** 159
[3] Tuysuz C 2010 *Int. Online J Educ. Sci.* **2** 37
[4] Morgil I, Yavuz S, Oskay O O, and Arda S 2005 *Chem. Educ. Res. and Pract.* **6** 52
[5] Spellman R 2011 *J. Access Serv.* **8** 150
[6] Hargreaves B R and McWilliams T P 2010 *Comp. Statis. and Data Anal.* **54** 1190
[7] Fasoula S, Nikita P, and Pappa-Lousi 2017 J. Chem. Educ. **94(8)** 1167-1173
[8] Kadjo A and Dasgupta K 2013 *Anal. Chim. Acta* **773** 1
[9] Burnett J and Burns W A 2006 *J. Chem. Educ.* **83** 1190
[10] Forgacs A, Balkay L, Tron L and Raics P 2014 *Appl. Rad. and Isotop* **94** 77
[11] Zhang Y, Huo M, Zhou J, and Xie S 2010 *Comp. Met. and Prog. and Biomedic.* **99** 306
[12] Sugiyono 2011 *Metode Penelitian Pendidikan: Pendekatan Kuantitatif, Kualitatif, dan R&D* (Bandung: Alfabeta)

Teachers' use of learning resources in spatial learning

C Khairunnisak, E Elizar, R Johar and T P Utami
Syiah Kuala University, Jl. Teuku Nyak Arief, Darussalam, Banda Aceh 23111, Indonesia

E-mail: cut.khairunnisak@unsyiah.ac.id

Abstract. Pursuing capabilities in the STEM (Science, Technology, Mathematics, and Engineering) needs spatial skills. However, previous studies indicated that students seem to have obstacles in learning mathematics because of low spatial skills. One of the strategies to enhance students' spatial skills is providing various learning resources. This study aimed at developing HLT using various learning resources to support students' learning of geometry. This study used design research method consisting of three phases namely preparation, teaching experiment, and retrospective analysis. Two Year 8 mathematics teachers from two public schools in Banda Aceh participated as the subjects of this study. This study focused on describing the differences and similarities of the intended and enacted learning performed by the teachers in the teaching experiment phase. Data collected from the video recording, field observation, and interviews were analysed descriptively. The result revealed that the two teachers mostly used the learning resources as intended in the HLT, even though there were some differences related to the time allocation, the flow of teachers' instruction, the way of engaging students, and the use of the online game. It is indicated that the HLT developed could be implemented with some adjustment taken into account the differences.

1. Introduction

As one of the topics in mathematics, considerable attention has been put into geometry, in particular, spatial visualization [1]. Spatial skill is also called spatial reasoning, spatial ability, spatial intelligence or spatiality. This is defined as "the ability to recognize and (mentally) manipulate the spatial properties of objects and the spatial relations among objects" [2]. Mulligan [2] also argued that the skill may be established in the early years of students and continued to be developed over time, however, it will be deteriorated without enough support and challenges. Spatial skills, as part of geometry, should not be taught as being constraint to naming shapes, determining the number of sides and faces and the basic static 2D and 3D shapes, instead, it should engage students' spatial reasoning using a dynamic approach to reveal students' potential beyond the written curriculum [2,3].

Spatial thinking plays a vital role as the basis for developing geometry in the early years of learning, and spatial reasoning is often associated with mathematics achievement [4]. Research concerning spatial reasoning has also reported that such skill does not only contribute to the successful mathematics learning at schools but also mastery of the topics across disciplines as well as assisting students in their future careers associated with STEM (Science, Technology, Engineering, and Mathematics) [2,5,6]. However, this matter was considered a little importance despite its essential role in developing mathematical ability [2,7]. Some efforts have been put up for promoting the significance of spatial thinking or reasoning, yet the actual action is still inadequate [2]. It is possible to develop students' spatial thinking, and in fact, it is required to be delivered across all level at schools [8].

However, students seem to have obstacles in learning about spatial geometric as indicated by low spatial skills [1,9] and the spatial difficulty may intervene in the difficulty of other topics [10]. This indicates that new strategies and approaches are needed for enhancing students' understanding of spatial and spatial skills. One of the strategies to be applied is providing various learning resources for easing students in grasping the concepts.

Resources are an integral part of learning and studies seeking for the effective use of learning resources have been conducted for decades. Teaching and learning at schools are likely to be more effective when the resources are sufficient which in turn can enhance students' performance [11]. Learning resources refer to the sources used for learning activities conducted in the classroom including spoken, written or visual texts [12].

The establishment of the spatial skills in the learning process will be made possible through the appropriate resources and its utilisation in the classroom. The utilisation of the learning resources in the classroom is often modified in the classroom depending on the classroom situation and the teacher's decision. The intended use of a resource may be similar or different to the use of learning resources enacted in the classroom. The definition of intended and enacted learning in this study may be formulated from the definition of the enacted and implemented curriculum proposed by Kaur et al. [13]. The intended learning refers to the plan of what the learning resources to be used and how she/he plan to use it in spatial learning, while enacted learning refers to what learning resources used in the classroom and how both teacher and students use them. The enacted learning is based on teacher decision concerning the implementation of the planned resources in the lesson and is indicated by the engagement between students and teacher as well as the resources in the classroom [3].

The examination of the use of learning resources is remained to be important as the intended use of the resources are often different from its actual use in the classroom (enacted learning). Therefore, this article will examine and discuss the use of learning resources by teachers in spatial learning, in particular, examining the difference and similarities of the intended and enacted learning.

2. Method

The present study described the implementation of a Hypothetical Learning Trajectory (HLT) for the learning of cube and cuboid developed by Utami [14]. The HLT was developed based on Design Research approach [15] consisting of three stages namely 1) preparation, 2) teaching experiment, and 3) retrospective analysis. The designed HLT had been through two cycles; each cycle consisted of the three stages of design research. The HLT on the first cycle was called HLT 1 while that on the second cycle was called HLT 2.

The developed HLT consists of learning route of instructional activities along with description and rationale for each activity [16]. The learning route was designed for four sessions; each session took 120 minutes. However, this article will only discuss the third session of the teaching experiment phase of the second cycle.

The teaching experiment phase was conducted in two public schools in Banda Aceh, Aceh, Indonesia. The participants of this study were two Year 8 Mathematics teachers (T1 and T2) and their students. T1 was a teacher in a public school in West Banda Aceh while T2 worked in a public school in East Banda Aceh. Both teachers are potential female teachers from two different favourite schools with a lot of experiences in teaching mathematics. The teachers also have participated in some previous research; thus, they do not need many assistances in teaching as expected by the researchers.

As stated before, the purpose of this study is to examine teachers' use of learning resources. The teachers' enacted activities in using the learning resources were compared to the intended activities presented in the lesson plan; the lesson plan derived from the HLT 2. The data for this study were collected from the developed lesson plan designed by Utami [14], and videos were taken during the teaching experiment. The lesson plan provided instructions on how to use the learning resources along with the examples of questions and possible explanations or answers.

3. Result and discussion

The following section describes the results of our examination of the spatial learning process. We first describe the intended lesson based on the data collected. Furthermore, we describe our analysis of the enacted lesson taught by the two teachers.

3.1. Intended lesson on the HLT 2

Considering the Indonesian National Curriculum [17], a set of the lesson was designed to present opportunities for Year 8 students to learn about the surface area and volume of cubes and cuboids. Considering the recent reports on the teaching and learning geometry [4,18], the designed lesson was embedded with spatial activities and was enriched with the use of 3D objects and computer technology (see Table 1). The designed lesson plans provided the detailed scripts, included examples of questions for teachers, and possible explanations or answers of mathematical content. Furthermore, the lesson plans also include worksheets and their possible answers.

Table 1. Lesson design for the third session of teaching experiment.

Learning objectives	Activities	Learning resources
Finding the volume of a cuboid	Solving routine problems	Student worksheet 3A
Finding the relation between the surface area and volume of cuboid by comparing its layout (horizontal or vertical)	- Determine whether the volume of the cuboid formed from four cardboards placed horizontally equals to that formed vertically - Checking students' answer by filling rice into cuboids formed from bridge cards	Cardboards, cuboids from bridge cards, rice, student worksheet 3B
Solving problems related to the surface area and volume of 3D objects	- Visualizing a 3D object (formed from some unit cubes) into the 2D drawing: drawing how the 3D object seen from the front and the right side. - Visualizing a 3D object into the 2D drawing: using "game point out the view." - Finding the surface area and volume based on its point of view	Unit cubes, online game namely "*point out the view*," student worksheet 3C

The HLT 2 for the third session consisted of three main tasks, namely: 1) finding the volume of cubes and cuboids, 2) working on the four bridge cards, and 3) examining the different perspective of cubes stack. The first task was designed to bridge students' previous knowledge while the others designed to promote students' spatial skills as they learn about the surface area and volume of cubes and cuboids. A brief description of each task is presented in the following section.

Task 1: Finding the volume of cube and cuboid. With the introduction of the lesson, based on the lesson plan, the teacher was required to recall students' previous knowledge about the volume of cubes and cuboids through related routine problems. The developed lesson plan provided some example of questions should be asked by the teacher such as "*Previously, you have studied the volume of a cube. What is the formula to find the volume?*" After the class discussion, the teacher should distribute worksheet 3A of the routine problem; the worksheet is presented in Figure 1. The teacher should ask students to solve the problems in their groups. Then, through the classroom discussion, the teacher should ensure the students understand the difference between the capacity and volume.

KEGIATAN ACTIVITY

1. Jika panjang rusuk sebuah kubus adalah 12 cm^3. Tentukanlah volume kubus tersebut!

   The side of a cube is 12 cm^3. What is the volume of the cube?

2. Jika sebuah balok memiliki panjang = 8 cm, lebar = 6 cm dan volume = 336 cm^3. Tentukan tinggi balok tersebut!

   A cuboid has dimension of length = 8 cm, width = 6 cm, and volume = 336 cm^3. What is the height of the cuboid?

Figure 1. Problems for the first task.

Task 2: Working on the four bridge cards arranged vertically and horizontally. The next activity was intended to help students understand the relationship between the surface area and volume of cuboid by comparing its layout. Before giving worksheet about cuboids from bridge cards, the teacher should present a preliminary question using cardboards as media, "*Suppose these four cardboards are arranged vertically and horizontally (while showing a glimpse of the arrangement), what is the volume? Is it the same or different? Explain*!". After waiting for some responses from the students, the teacher should experiment with filling rice into four bridge cards arranged vertically and horizontally. Some question such as "*What did you find?*" and "*What can you conclude*?" were intended to help students realize that the different layout means the different surface area and different volume. Next, the students instructed to work in groups to solve problems in the worksheet 3B (Figure 2a) to prove mathematically why the volumes can be different. Figure 2b presents one of the expected answers to the second question presented in the lesson plan.

KEGIATAN ACTIVITY

1. **Perhatikanlah kedua gambar kartu remi berikut.**
   Look at the following two figures of bridge cards.

(a)

(b)

Gambar manakah yang memiliki volume lebih besar? Jelaskan alasannya!
Which picture having more volume? Explain!

2. **Perhatikanlah kedua susunan kartu remi di hadapan kalian.**
   Look at the bridge cards structures in front of you.
   Buktikanlah secara matematis bahwa kartu remi yang disusun secara mendatar volumenya lebih besar dari kartu remi yang disusun secara tegak!
   Prove, mathematically, that the bridge cards that arranged horizontally have more volume that those that arranged vertically!

(a)

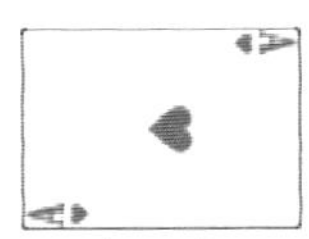

Note :
$l > w$
$l = w + a$

for bridge cards arranged vertically:
$V_1 = w^2 \times l$
$= w^2 \times (w + a)$
$= w^3 + aw^2$

for bridge cards arranged horizontally:
$V_2 = l^2 \times w$
$= (w + a)^2 \times w$
$= (w^2 + 2aw + a^2) \times w$
$= w^3 + 2aw^2 + a^3w$

Therefore, $V_1 > V_2$

(b)

Figure 2. Bridge cards arrangement problems.

IOP Conf. Series: Journal of Physics: Conf. Series **1088** (2018) 012035 doi:10.1088/1742-6596/1088/1/012035

Task 3: Examining different perspective of cubes stack. The next activity used the stack of cubes as media (Figure 3). At first, the teacher was expected to explain how to visualize the front and the right side view of the stack using the real cubes, and the students were asked to draw it on A2 paper. After some exercises using the real cubes, a more abstract media (an online game) was used in the next activity. At first, in the HLT 1, the game proposed by the researchers was game of *Isometric drawing tool*. However, based on the retrospective analysis of HLT 1, the game was changed to a game of *Point out the view* (Figure 4). This online game can be accessed from the website of [19]. The lesson plan provided instructions on how to play the game as well as how to use the game. After providing an example of how to play the game as instructed on the lesson plan, the teacher should ask some students to play the game in front of the class. The purpose of playing this game was to examine the student's understanding of spatial orientation and visualisation.

Figure 3. Example of cubes arrangement for the 3rd task.

Figure 4. Online game for the 4th task.

The last activity in session 3 was solving worksheet 3C about determining the surface area and volume of the stack of cubes. The worksheet provided the 2D picture of the front and the right side view of the cubes stack. The students were expected to solve the problems such as "*determining the minimum and the maximum number of cubes needed to form the object,*" "*finding how much paint needed to cover the object except its floor*" and finally the students were expected to conclude the relationship between the surface area and volume.

3.2. Differences between the intended and the enacted lesson
In line with previous studies [20,21], teachers could react differently toward the same textbook/task. In this study, the teachers' performances that were different from the HLT included the time allocation (by T1 and T2); the flow of teacher's instructions during the class discussion (T1 and T2); the use of contextual problem for differentiating the capacity and volume (T2); students' engagement in the class discussion (T1 and T2); and the use of online game (T2).

The lesson plan derived from HLT 2 provided the time allocation for each activity. There was a significant difference in time allocation for the activity of solving the worksheet. Students of T1 (from now on SoT1) exceeded the time allocation for working on worksheet 3A and 3B by 7 minutes and 11 minutes respectively. On the other hand, students of T2 (hereinafter SoT2) exceeded the time for working on worksheet 3A and 3B by 2 minutes and 13 minutes respectively, while the time spent for working on worksheet 3C was 2 minutes less than what intended. Related to the worksheet 3A, the time spent by SoT1 and SoT2 to work on the worksheet was not too different from what was allocated in the lesson plan. However, the teachers (T1 and T2) used more time to ask their students to present their answer on the worksheet. Therefore, the HLT 2 should be adjusted by adding an instruction to conduct group presentation for worksheet 3A.

Both teachers mainly followed the sequence of activities on the lesson plan. T1 generally acted as intended on the lesson plan, except for the introduction and group working activities on worksheet 3A as previously explained. In the lesson plan, the teacher was expected to deliver the objective of the lesson directly using PowerPoint. However, T1 presented it while referring to the learning media as cardboards, bridge cards, and cubes. On the other hand, T2 delivered the three learning objectives (see Table1) directly.

The lesson plan suggested the teachers ask about the difference between the capacity and volume before introducing the bridge cards problem. However, T1 asked the question after the problem. In contrast with T1, T2 followed the order as instructed in the lesson plan. T2 adjusted the question by relating it to students' real life: the capacity and volume of mineral water. Comparing the responses of their students, on the one hand, somehow T1's instruction was better than T2's because T1's students could directly relate the capacity and volume by relating to the rice fill in the bridge cards, which was the next resources intended to be used. However, T2 was not wrong since many studies suggested to link the lesson with the students' daily life [22,16].

Related to students' strategy to mathematically prove that different surface area means different volume, it was expected that students use the proof as presented in the lesson plan (see Figure 2b). However, both SoT1 and SoT2 performed almost similar strategy (Figure 5a presents the strategy from SoT1 while Figure 5b from SoT2). To make sure that the horizontal cuboids have a bigger volume than the vertical cuboids, the students chose numbers to substitute the length and width, and both teachers agreed with their students' strategy. This indicates the need to insert this strategy to the conjecture of student's answer on the HLT, suggesting that teachers need to help students to achieve a more formal proof, without using numbers, as presented in Figure 2b.

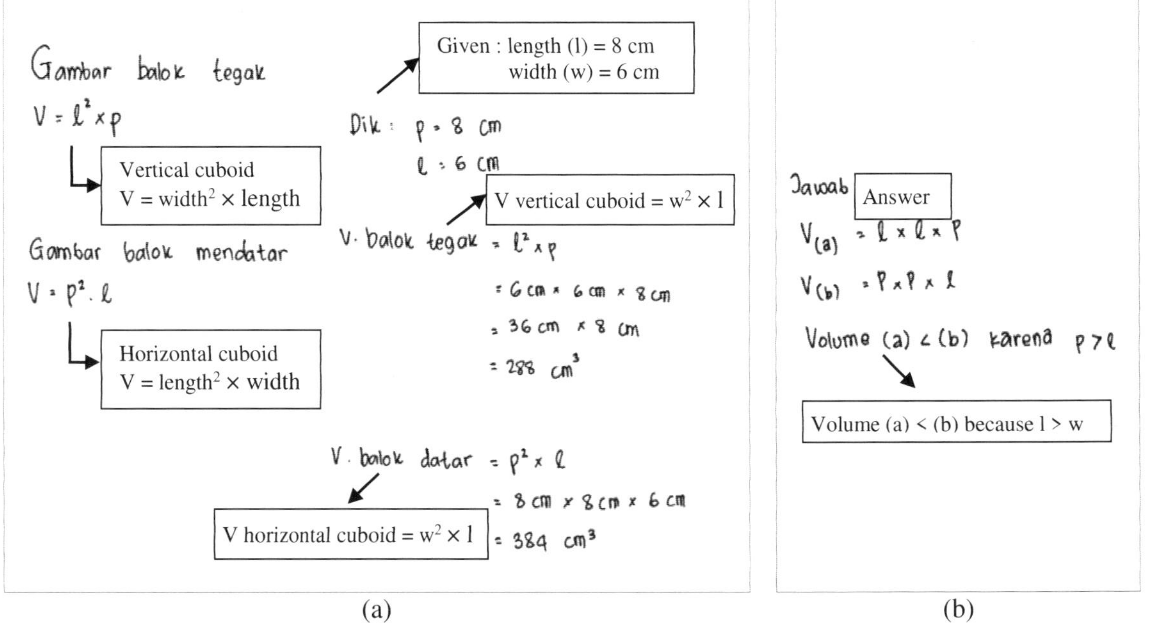

(a) (b)

Figure 5. Students' strategy to prove horizontal cuboid has a bigger volume than the vertical one.

T1 and T2 often invited students to participate in the classroom discussion actively. However, it seems that T1 did it more frequent than T2. For instance, in the experiments using the bridge cards and rice, T1 asked for students' help while T2 did it by herself as written on the lesson plan. Moreover, T1 often rechecked students' understanding of what has been explained. Based on the interview after the lesson, the teacher said that it was for exam preparation that would be held next week.

The use of game *Point out the view* was intended to help students to understand more about spatial skill. Differ from the stack of cubes, which was quite troublesome to use outside of the classroom; the online game is easier to be used anywhere. Recent studies suggested that online game promote students' learning interest [23,24]. Therefore, the HLT was designed to accommodate the online educational game. However, due to some limitation such as power failure, T2 could not utilize the game point out the view. Although the researchers had interviewed the teacher about the accessibility of the internet and other tools for playing an online game, the occurrence of the power outage was out of control.

Each activity designed in the HLT 2 had its own goal. As suggested by Gravemeijer [25], the learning mathematics should be in line with the idea of horizontal and vertical mathematization. Starting with real problems, using "real" stack of cubes (as in Figure 3), the learning activities in this study shift to more abstract problems. The real stack of cubes could be touched and moved by students. Therefore, they can easily see its view from the front or the right side. The online game was used so that the students could train their spatial orientation ability while playing it. It was expected that the students could solve problems in worksheet 3C easily.

SoT1 and SoT2 performed different work or solution while solving problems on the worksheet 3C, especially on the third problem about *determining the minimum number of cubes needed to form the figure* presented on the worksheet (see Figure 6). There were two different answers from SoT1: ten and eight cubes. On the other hand, SoT2 gave a more various answer: ten, nine, and eight cubes. The researchers conjectured that it happened because of SoT2 got fewer exercises than SoT1 because of the absence of the online game.

Figure 6. The picture on worksheet 3C.

4. Conclusion

This study developed HLT consisting of a set of learning activities to promote spatial learning through various resources. Almost all the learning resources were used as intended in the HLT 2 by the participants (T1 and T2) in this study. The differences lie in the time allocation spent, the flow of teachers' instruction, the method of engaging students, and the use of the online game. Considering the differences enacted by the teachers in this study, the HLT need some adjustments so that it accommodates more conjectures about teacher's instructions and student's reactions. Related to the online game, the HLT designers need to provide, for example, an offline game to alternate the online game when needed.

References

[1] Boakes N J 2009 Origami instruction in the middle school mathematics classroom: Its impact on spatial visualization and geometry knowledge of students *RMLE Online* **32** 1

[2] Mulligan J 2015 Looking within and beyond the geometry curriculum: connecting spatial reasoning to mathematics learning *ZDM* **47** 511

[3] Moss J, Hawes Z, Naqvi S and Caswell B 2015 Adapting Japanese Lesson Study to enhance the teaching and learning of geometry and spatial reasoning in early years classrooms: A case study *ZDM* **47** 377

[4] Olkun S, Swoboda E, Vighi P, Yuan Y and Wollring B 2017 *Proceedings of the 13th International Congress on Mathematical Education* (Switzerland: Springer Nature Switzerland AG) p 429

[5] Wai J, Lubinski D and Benbow C P 2009 Spatial ability for STEM domains: Aligning over 50 years of cumulative psychological knowledge solidifies its importance *Journal of Educational Psychology* **101** 817

[6] Titus S and Horsman E 2009 Characterizing and improving spatial visualization skills *Journal of Geoscience Education* **57** 242

[7] Hawes Z, Moss J, Caswell B, Naqvi S and MacKinnon S 2017 Enhancing children's spatial and numerical skills through a dynamic spatial approach to early geometry instruction: Effects of a 32-week intervention *Cognition and Instruction* **35** 236

[8] National Reseach Council 2006 *Learning to Think Spatially* (Washington DC: The National Academies Press)

[9] Güven B and Kosa T 2008 The effect of dynamic geometry software on student mathematics teachers' spatial visualization skills *Turkish Online Journal of Educational Technology-TOJET* **7** 100

[10] Swanson H L and Jerman O 2006 Math disabilities: A selective meta-analysis of the Literature *Review of educational Research* **76** 249

[11] Yara P O and Otieno K O 2010 Teaching/learning resources and academic performance in mathematics in secondary schools in Bondo District of Kenya *Asian Social Science* **6** 126

[12] Victoria State Government 2018 Suitable Teaching and Learning Resources *Education and Training*

[13] Kaur B, Tay E G, Toh T L, Leong Y H and Lee N H 2018 A study of school mathematics curriculum enacted by competent teachers in Singapore secondary schools *Mathematics Education Research Journal* **30** 103

[14] Utami T P *Desain Hypothetical Learning Trajectory (HLT) untuk Mengembangkan Kemampuan Spasial Siswa Melalui ELPSA (Experience, Language, Pictorial, Symbol, Application) Framework* [Thesis] Syiah Kuala University: Unpublished; 2018

[15] Gravemeijer K and Cobb P 2006 Design research from a learning design perspective. *Educational design research* (London: Routledge) p 29

[16] Gravemeijer K 2004 Local instruction theories as means of support for teachers in reform mathematics education *Mathematical thinking and learning* **6** 105

[17] Kemendikbud 2013 *Peraturan Pemerintah tentang Pengelolaan dan Penyelenggaraan Pendidikan* (Jakarta: Badan Pengembangan Sumber Daya Manusia Pendidikan dan Kebudayaan dan Penjamin Mutu Pendidikan)

[18] Bakó M 2003 *Proceedings of the Third Conference of the European society for Research in Mathematics Education*

[19] Freudenthal Institute *Rekenweb* [Internet] Utrecht, The Netherland: Freudenthal Institute; 2015 [cited 2017 April 5]. Available from: http://www.fisme.science.uu.nl/publicaties/subsets/games/

[20] Thompson D R and Senk S L 2014 The same geometry textbook does not mean the same classroom enactment *ZDM* **46** 781

[21] Patahuddin S M, Puteri I, Lowrie T, Logan T and Rika B 2018 Capturing student mathematical engagement through differently enacted classroom practices: Applying a modification of Watson's analytical tool *International Journal of Mathematical Education in Science and Technology* **49** 384

[22] Widjaja W 2013 The Use of Contextual Problems to Support Mathematical Learning *Indonesian Mathematical Society Journal on Mathematics Education* **4** 157

[23] Hwang G J, Wu P H and Chen C C 2012 An online game approach for improving students' learning performance in web-based problem-solving activities *Computers & Education* **59** 1246

[24] Sung H Y and Hwang G J 2013 A collaborative game-based learning approach to improving students' learning performance in science courses *Computers & Education* **63** 43

[25] Gravemeijer K and Cobb P 2013 *Educational Design Research Part A: an Introduction* ed T Plomp and N Nieveen (Enschede: SLO)

IOP Conf. Series: Journal of Physics: Conf. Series **1088** (2018) 012036 doi:10.1088/1742-6596/1088/1/012036

The validity of vector analysis module using *wxMaxima software*

Khairina[1], M Ikhsan[2] and Suhartati[2]
[1] *Mathematics Magister Study Program, Syiah Kuala University*
[2]*Lecturer of Mathematics Magister Study Program, Syiah Kuala University*

E-mail:ikhsan@unsyiah.ac.id

Abstract. Teachers should have the competence in applying Information and Computer Technology (ICT) in the learning process. One of the ICT application is the use of learning software such as **wx**Maxima. This research aimed to develop a valid module using the wxMaxima software. This research used the Plomp development model consisting of three phases, namely preliminary research, prototyping, and assessment. However, this research was conducted for preliminary research and prototyping phases only. The result of data analysis in the preliminary research shows that the module is feasible for further development. The validation results at the prototyping stage performed by four validators gave a score of 4.5. This score shows that the developed vector analysis module using **wx**Maxima software was highly valid. Therefore, the module as a result of the development of this research can already be tested.

1. Introduction

To train competent teachers in applying Information and Computer Technology (ICT) in the teaching and learning process, the student teacher must be familiar with the lecturing activities using ICT. Living in a technology-based community makes the learning activity applying ICT become the best method [2]. The use of ICT is not only limited to the use of powerpoint but also to the use of mathematics software which can assist the learning process. Mathematics software includes *Maple*, *Autograph*, *wxMaxima*, *Geogebra*, *Matlab,* and others. The software is mainly divided into two categories, paid and free software. One of the free mathematics software is *wxMaxima*.

ICT application in the lecture activities not only facilitates the learning process but also improves students' motivation towards the learning activities [7]. Naturally, it will affect the learning outcome, it can improve the understanding, and the lecture will be more meaningful. The student teachers' better understanding leads to the better quality of the teachers produced.

In the Indonesian National Qualification Framework (*KKNI*), it is mentioned that one of the expected competency of the bachelor degree is the capability of applying his/her competences by utilizing Science and Technology to solve the problem as well as adapting to the new situation. Meanwhile, the Mathematics Study Program of the teacher training and education faculty (FKIP) in Syiah Kuala University (Unsyiah) has not utilized ICT in the vector analysis subject. In addition, students' motivation and initiative also need to be improved.

Vector analysis subject is one of the subjects in Mathematics program study of FKIP Unsyiah provided in the 5th semester. This two credits subject has some prerequisite subjects including Calculus I, Calculus II and Solid Geometry course. However, based on the field observation, it is found that the students taking Vector Analysis Subject are still constrained by the precondition materials, even though they have passed those prerequisite subjects.

In the lecturing process, the students' independence is the most important factor. One of the factors supporting the independence is the availability of module. The module allows students to improve their learning intensity outside the classroom. In particular, for the module using software, students can conduct the exercise questions manually and then match the answers with the software output. Based on the above description, the research problem is "how can the module using wxMaxima Software in the vector analysis course be valid?"

2. Method

This research is the development research using the Plomp, development model. The Plomp development model consists of three phases, namely preliminary research, prototyping and assessment [5]. However, this paper is limited to the prototyping phase.

In the preliminary research phase, the curriculum was analyzed to examine the factors related to the research location. The university condition analysis was carried out to examine the supports, and obstacles found related to the university environment, infrastructure, and facilities. The students' characteristic analysis was carried out to investigate the basic concept and students' learning motivation. The user analysis was done to investigate the user's expectation related to the module developed. In addition, in the prototyping phase, the module was designed based on the result of the preliminary research phase to be validated.

The criteria used to assess the module developed are based on Nieveen's [8]. The module developed will be valid if it meets the following criteria:(1) the validator assessment results acknowledge that the module using **wx**Maxima software in vector analysis subject is developed based on a strong theoretical basis, and (2) validator assessment results acknowledge that the module components using **wx**Maxima software in vector analysis subject are consistently interrelated. While the criteria of validity score (VS) obtained is presented in the picture as follows:

$4 \leq$	VS	≤ 5	Highly valid
$3 \leq$	VS	≤ 4	Valid
$2 \leq$	VS	≤ 3	Less valid
$1 \leq$	VS	≤ 2	Not valid

Figure 1. Criteria of validity score

3. Result and Discussion

This section presents the development processes carried out in the following two phases.

3.1 The Preliminary Research Phase

The preliminary research phase has been started since October 2016. The activities conducted were the analyses of curriculum, university condition, student characteristic, and need assessment. The analysis processes are explained in the following section.

3.1.1 Curriculum Analysis. In this phase, the curriculum analysis was carried out based on the curriculum applied in mathematics study program of FKIP Unsyiah. The curriculum used was Indonesia National Qualification Framework (*KKNI*). The module contents were divided into two basic competencies, namely: vector differential and vector integral. The indicators of the vector analysis subject were: (1) Using differential operator; (2) Determining the scalar function gradient; (3) Determining vector Area; (4) Determining the vector function divergence; (5) Calculating line integral; (6) Calculating surface integral; (7) Calculating volume integrals.

3.1.2 Students' Characteristic Analysis. Student characteristic analysis was conducted to adapt the students' condition to the product developed. Student characteristic analyses obtained are:(1) students' attitude concerning vector analysis subject such as being enthusiastic and actively engaged; (2) students only use one guideline source; (3) students were having problems in basic mathematics concepts.

3.1.3 University Situation and Condition Analysis. University situation and condition analyses were carried out to examine the supports and obstacles found in the environment. The university situation and condition analyses showed that: (1) The location of Mathematics Study Program was quite strategic; (2) the situation and condition of Mathematics Study Program strongly supported the implementation of learning; (3) Physical condition of Mathematics Study Program was feasible for the learning activities; (4) The media supporting the courses of mathematics study program was strongly supported even though it was less maintained.

3.1.4 User Analysis. The user analysis was conducted to investigate the module requirements in vector analysis subject. The analysis results were: (1) there had been no module for vector analysis subject of Mathematics Study Program involved in the study. The students utilized the guidelines published by other universities; (2) the guideline used was not utilizing any software and the lectures had never used software in learning activities; (3)the lecturers were willing to master the module in the vector analysis subject and to apply in the learning activities, should the module was available; (4) The learning idea expected by the lecturers was that the students first learned the concept manually and followed up using the software.

```
(%i1) load(vect)$
      div([x^2*y*z,3*x*y*z^3,(x^2-z^2)]);
```
(%o2) $\mathrm{div}([x^2\,y\,z, 3\,x\,y\,z^3, x^2-z^2])$

```
(%i3) express(%);
```
(%o3) $\frac{d}{d\,y}(3\,x\,y\,z^3)+\frac{d}{d\,z}(x^2-z^2)+\frac{d}{d\,x}(x^2\,y\,z)$

```
(%i4) ev(%,diff);
```
(%o4) $3\,x\,z^3+2\,x\,y\,z-2\,z$

Figure 2. Examples of working with *wx*Maxima software

3.1.5 Material analysis. The course chosen in the learning software development was vector analysis. This material is in the 5th semester. It is the mandatory subject (not optional) consisting of 2 SKS (2 units). It is the difficult subject especially for the students who have a less basic concept.

3.2 Prototyping Phase.
The module was designed by developing module framework, consisting of (1) introduction, (2) basic competency, (3) indicator, (4) the material and exercise for each section, and (5) the answer of exercise. This framework was adopted from the module of Open University. The materials were started by manual problem solving and followed by software problem-solving. The module prototype can be summarized as follows.

3.2.1 Identifying the Learning Objectives. This identification aimed to obtain the description of students' competency after participating in the lecturing activities using the module using the software in vector analysis subject. Therefore, the module developed comprised of the exercises using both manual and software process.

3.2.2 Analyzing the Learning Objectives. The learning objective analysis aimed to examine the competency of students' understanding based on the learning objective. The learning objective analysis was started by classifying the pattern of the cognitive learning objective. Thus, the analysis results of learning objectives found were the learning process including student cognitive competency in vector analysis subject.

3.2.3 Formulating the Learning Objectives. The learning objective in this development was students' cognitive competency after participating in the learning activities of vector analysis using *wxMaxima Software.* The results of the learning were that students were able to:(1) use differential operator; (2) determine scalar function gradient; (3) Determine vector area; (4) Determine the vector function divergence; (5) Calculate line integral; (6) Calculate surface integral; (7) calculating volume integral.

3.2.4 Developing the Learning Strategy. The learning strategy developed was estimating learning time allocation and learning activities. Learning strategy applied in the modules using *wxMaxima software* in vector analysis subject included software introduction, basic competency, indicator, the material and exercises for each section, and the answer of exercises, prerequisite materials, material explanation, the example of manual problem solving, the example of the software problem solving, exercises for students, and the conclusion.

3.2.5 Developing the Module. The module development was conducted by presenting the basic competencies, indicators, and explanation. In the material explanation section, the material was divided into two subsections including the material explanation with the example of the questions both manually solved and using software, and one subsection for reviewing the exercises as the final preparation (testing the understanding test question).

Evaluasi Diri

Kerjakan soal berikut secara manual, kemudian kerjakan dengan menggunakan software wxMaxima. Cocokkan jawaban manual dengan hasil di software. Tuliskan penyelesaian manual di lembar jawaban yang disediakan.

1.Tentukan ∇f, bila diketahui:
a. $f(x,y,z) = x^2 - 3xy + 2z$
b. $f(x,y,z) = \frac{1}{2}(x^2 + y^2 + z^2)$
2.Tentukan div **F** and rot **F,** bila diketahui:
a. $\boldsymbol{F}(x,y,z) = x^2\boldsymbol{i} - 2xy\boldsymbol{j} + yz^2\boldsymbol{k}$
b. $\boldsymbol{F}(x,y,z) = x^2\boldsymbol{i} + y^2\boldsymbol{j} + z^2\boldsymbol{k}$
3.Tentukan $A.\nabla f$ dan $A \ x \ \nabla f$ di titik (1,1,-1) dengan

Self-evaluation

Solve the following questions manually on the provided sheets, and also by using *the mx*Maxima software. Match both answers.

1.Find ∇f, if:
a. $f(x,y,z) = x^2 - 3xy + 2z$
b. $f(x,y,z) = \frac{1}{2}(x^2 + y^2 + z^2)$
2.Find div **F** and rot **F,** if:
a. $\boldsymbol{F}(x,y,z) = x^2\boldsymbol{i} - 2xy\boldsymbol{j} + yz^2\boldsymbol{k}$
b. $\boldsymbol{F}(x,y,z) = x^2\boldsymbol{i} + y^2\boldsymbol{j} + z^2\boldsymbol{k}$
3.Find $A.\nabla f$ *and* $A \ x \ \nabla f$ at point (1,1,-1) with $A= 2x^2\boldsymbol{i} - y^2z\boldsymbol{j} + xz^2\boldsymbol{k}$ and $f = x^2 - y^2z$

Figure 3. Sample display practice questions on the module

3.2.6 Validity Test. In this validation phase, the validator conducted the assessment of the content, format, and language. In the comment column, the validator wrote the feedbacks that were used as the main consideration for revision before conducting the limited trial test.

3.2.7 Validity Test Analysis. The module validation results from this aspect included: 1) the validity of the content, 2) the compatibility of the basic competencies and indicators, 3) the logical grouping sections, 4) the guidance in using the software clearly formulated, 5) the compatibility with the allocation time used, and 6) the feasibility of the learning software (the validity average scores were

4.5). Meanwhile, the results of the format aspects included: 1) the validity of the distribution, 2) the clear numbering system, 3) the layout setting, and 4) the font type and size, obtaining the average validity scores of 4.5. Furthermore, the language aspects included: 1) the validity of language setting, 2) the simplicity of sentence structure, 3) the clarity of guideline and instruction, and 4) the communicative language used, obtaining the average validity scores of 4.5. The total average scores were4.5. The average value indicates that the module development criteria are classified as highly valid. Therefore, the module meets the content validity criteria. However, some minor revisions were needed to improve the module such as being less precise in the writing aspect; the use of "for" at the beginning of the sentence; in the format aspect, there were images without a caption; and the working steps without numbers. Both aspects have been revised incorporating the feedbacks, more proper sentences used, the caption provided for each image and the number in the working step.

Table 1. Module validation results

Aspect	**Criteria**	**Validator**				**Average of each criterion**	**Average of each aspect**
		1	**2**	**3**	**4**		
Content	1. The truth of the content of the vector analysis material presented in the module	5	4	4	5	4,5	4,5
	2. Compliance with basic competencies	5	5	4	5	4,75	
	3. Grouped in logical sections	5	4	5	5	4,75	
	4. Student activities are formulated clearly, so easy to be implemented by lecturers in lectures	4	4	4	5	4,25	
	5. Clarity example of a problem worked with software maxima	5	5	5	5	5	
	6. Compliance with module criteria						
	a. Self-instructional	5	4	4	5	4,5	
	b. Self-contained	4	4	4	5	4,25	
	c. Standalone	5	4	4	5	4,5	
	d. Adaptive	4	4	4	5	4,25	
	e. User-friendly	4	4	4	5	4,25	
	7. Eligibility as a learning tool	5	5	3	5	4,5	
Format	1. Clarity of material distribution	4	4	4	5	4,25	4,5
	2. The numbering system is clear	5	5	4	5	4,75	
	3. Layout settings	5	5	3	5	4,5	
	4. Type and font size	5	5	3	5	4,5	
Grammar	1. The truth of grammar	4	4	4	5	4,25	4,5
	2. The simplicity of sentence structure	5	5	4	5	4,75	
	3. Guidance clarity and direction	4	4	4	5	4,25	
	4. Communicative language used	4	4	4	5	4,25	
Average of Validity							4,5

3.3 Discussion

The product of this research should be evaluated to examine the validity level by the expert validation. As proposed by Sugiyono [8] that validate the product is done by presenting experts or experienced experts to assess the product developed The experts assessed the validity of the module developed. The experts presented in this development were the subject expert, ICT expert, mathematics lecturer, and other colleagues researching ICT product development. According to the validation process, the module developed in this research is categorized as highly valid.

The module validation results obtained from four validators, reaching the validity score of 4.5. This validity value summarizes that the module criteria is highly valid. In each learning process, the basic competencies to be achieved and the detailed indicator for each section are mentioned. Then, the explanation of the material was given together with the examples of manual and software problem-solving, in line with the expectation of the lecturer as mentioned in the analysis of users. Although students work with the software, they should also work without the software. At the end of each section, there were some exercises must be solved by the students. These exercises benefit the students in the form of reinforcement in each session. The exercises allow students to examine their difficulties. In addition, the main advantage of using the module is that the exercises accompanied by the key answers attached at the end of the module. Thus, students should be able to measure their level of understanding. The learning using module not only improved the independent learning leading to students' motivation but also allowed students to review the material independently which in turn reducing the time required in the next meeting.

4. Conclusion

The preliminary analysis of this study showed that a module of wxMaxima software in vector analysis subject needed to be developed. Meanwhile, in the prototyping phase, the module using the wxMaxima software in vector analysis subject that was developed in this study meets the validity criteria, whose a score of 4.5. This score indicates that the module is highly valid. Therefore, the module can be tested in teaching and learning process.

References

[1] Eng, Ting Seng 2005 The Impact of ICT on Learning: *A Review of Research.International Education Journal* **6** 635

[2] Kennah, MuaRodanny 2016 The Use of ICT in the Teaching and Learning Process in Secondary Schools: A Case Study of Two Cameroonian schools.Master'sThesisUniversity of Jyväskylä

[3] Kurz, T, Middleton, JA., &Yanik, H. B (2005) *A Taxonomy of Software for Mathematics Instruction*. Contemporary Issues in Technology and Teacher Education **5** 123

[4] Lavicza Z & Koren B 2015 *Integrating Technology Into Primary and Secondary School Teaching to Enhance Mathematics Education in Hungary*. Prague, Czech Republic. pp.2430- 2431, Proceedings of the Ninth Congress of the European Society for Research in Mathematics Education.<hal-01289310>

[5] Plomp T & Nieveen N (2013) *Educational Design Research.Enschede: SLO.*

[6] Tuylu, Savas (2014) *Vector Analysis Using Maxima*.Donnish Journals of education research and reviews **1** 012.

[7] Yusrizal and Hanif F (2017) Increasing of Student's motivation Learning Physics through the use of computer simulation media viewed from parent's employment background. *The International Journal of social sciences*. DOI: 10.3140/RG2.2.15473.68967

[8] Sugiyono (2014) Metode Penelitian Pendidikan (Pendekatan Kuantitatif, Kualitatif Dan R&D). Bandung: Alfabeta.

Difficulties analysis of mathematics education students on the real analysis subject

I Widiati and A Sthephani
Universitas Islam Riau, Jl. Kaharuddin Nasution No. 113, Pekanbaru Riau 28284, Indonesia

E-mail: indahwidiati@yahoo.com

Abstract. Real analysis is one of the challenging subjects for most students. The difficulties of this subject include the understanding of theorems and its proof. To solve this problem, we do study with the aim to describe the difficulties of students in Mathematics Education of FKIP UIR in Real Analysis Course. The study was conducted on the even semester for academic year 2016/2017. This research was qualitative research using a used descriptive method. The research population of this research was all students of FKIP-UIR in the 4th semester. Data collection techniques were test techniques, questionnaires, interviews, and observations. Data analysis was done by questionnaire analysis, interview sheet analysis, observation result analysis, and analysis of test result about Real Analysis. The domain results showed that the difficulties of experience by students in learning Real Analysis are: students are difficult to find the initial idea of the proof process; students still do not have the ability to think critically, logically, creatively, reasoning, and proper systematic; and students are difficult in expressing ideas into nonverbal Language. The implication of this study is to create an instrument namely book of Real Analysis, Work Sheet of Real Analysis or hand out of Real Analysis.

1. Introduction

Courses taken by students had been arranged so that the students did the learning process gradually from the lowest difficulty level to the highest difficulty level. One of all the subjects they had to go is Real Analysis course. Students obtained this course on the 4th semester, and the student must first consider the prerequisite courses that were Calculus and Number Theory. The courses are very relevance. Usually, the result obtained in the calculus and the Number Theory is directly proportional to the results obtained in the Real Analysis.

Course that is perceived to have indirect benefits is Real Analysis. Often students ask questions about what the benefits of studying this course. But before the students were asked about it, usually the lecturer gives a preliminary description of this course.

At the beginning of the meeting, the researcher attempted to give a preliminary questionnaire directly to the students taking the course of the Real Analysis and to the students who had taken this course. The results of the open-ended questionnaire show that most students do not like Real Analysis as they do not understand the material, they found the contents confusing and challenging (in term of connecting the theorems), complicated and difficult to understand. It also found that Real Analysis could train the use of sound logic, practice precision, and prudence, discover the origin of a concept, promote good reasoning skills and collaborative learning together, train patience, assist the next course, as well as improve the analysis, critical and systematic thinking skills. Furthermore, the results also reveals that students had difficulty in understanding the material, remembering the theorem and relating theorems, proving indirectly, finding the initial idea to prove, analyzing the theorem, being concentrated, providing good reasoning, being confused in choosing the theorems to be used, generating critical thinking patterns, and using logic (4). Some students who are afraid of learning Real Analysis will answer the materials of this course are difficult.

The results of these initial interviews in line with research conducted by Darmadi [1] if students began to have trouble learning Real Analysis since the beginning of the meeting. Darmadi [2] said that "almost all (99%) students find the difficulties in learning Real Analysis. The difficulties of learning Real Analysis according to the students are due to the material and the way of presentation, and the difficulties of the students start from understanding the concept that starts from the formal definition, the lack of the references or perhaps laziness of students".

Another thing was also expressed by Darmadi [3] namely "students become afraid to advance due to lack of confidence. It also causes the students to become passive in following the learning, including in learning the formal definition given".

The Real Analysis course is closely related to previous courses of Calculus and Number Theory. For the basic subjects that have been studied in the previous course, lecturers should no longer explain in detail to the students but can apply them directly. However, this can't be done because the understanding of the concept of the previous course is also not good. Therefore, students become confused, hesitant, and not confident in conducting the process of proof in this real analysis course. This also affects the low outcomes of students in the Real Analysis course.

Students had difficulties in doing the proof, though the proofing process is something that essential in this Real Analysis course. Knuth [4] says that "proof acts as: (1) to motivate that a statement is true, (2) to explain why a statement is true; (3) to communicate mathematical knowledge; (4) create new mathematics; (5) systematize the statement into an axiomatic system".

Some research results show that students are challenging to construct evidence even if the evidence is simple or routine though. The results include the introduction of proof [5], real analysis [6] and, abstract algebra [7]. Students generally do not specify what is stated in a proof [8] and the student can't determine whether a proof is valid or not.

This problem had been detected for an extended period of times by researchers; even researchers had tried to apply a learning strategy that purposes to stimulate problem-solving skills and critical thinking in the real analysis. However, the results showed that there is no influence between the applied strategy of metacognitive, the problem-solving ability and critical thinking of the students. Therefore, researchers are interested in exploring the problems or difficulties experienced by students in studying Real Analysis courses.

2. Method

This research type was qualitative research with descriptive method. The population in this study were all students of Mathematics Education Universitas Islam Riau on 4th-semester year academic 2016/2017, while the number of samples studied was 55 students of 4th semester, consist of two classes. Sample selection technique used was purposive sampling. Purposive sampling is a technique for determining samples with specific considerations. The researcher's consideration is because we teach real analysis in 2 classes so we could quickly be taking the data. Furthermore, the problem designed on the test can also be adjusted with the learning materials that have been given.

The research variables used in this study is one independent variable that is the difficulty of the student in Mathematics Education Universitas Islam Riau on Real Analysis course. Instruments and techniques of data collection used were (1) Sheet Questionnaire and questionnaire; (2) interview sheets and interviews; (3) observation and observation sheets; (4) test sheets and tests of Real Analysis courses. Data analysis techniques used were: (1) questionnaire data analysis; (2) data analysis of interview result; (3) observation data analysis; (4) data analysis of real test result.

After distributed the questionnaire to the students who were the research samples, the questionnaire data were reviewed and grouped according to the relevant answers. The result of the questionnaire was then described to be interpreted and drawn a conclusion related to the difficulties experienced by students in studying Real Analysis.

Interviews were conducted after we obtained the results of the real analysis test. This activity aimed to get in-depth information from the questionnaire data and to gather information about the causes of the difficulties and mistakes made by students. Because this interview aims to obtain detailed data about the causes of students making errors, then we selected the students who made many mistakes in the test results. After collecting data from the interview, the data was analyzed descriptively.

3. Result and discussions

3.1. The result of questionnaire analysis

The provided questionnaire is an open questionnaire to get a lot of information about student difficulties in studying Real Analysis. The questionnaires given to the students consist of 23 questions, and then we ask the students to provide reasoning based on the answers given. The result of the questionnaire of student difficulty in studying Real Analysis would be described for each question included: Based on the description of the result in the questionnaire that had been described, the following conclusions were obtained: (1) Students admitted having difficulty in finding the initial idea of the proof process; (2) Students still did not have the ability to think critically, logically, creatively, reasoning, and systematic good; (3) Students easily give up, lack of confidence, lack of care, and less ability to socialize with a heterogeneous group of peers; (4) Students felt difficulty in conducting direct and indirect evidence in determining what type of proof to use; (5) Students perceived difficulty in expressing ideas in nonverbal Language; (6) Do not remember/do not understand the previous theorem; (7) Students did not have a good understanding of the prerequisite materials; (8) Prove that it will prove or substitute the numbers into the proof of the theorem; (9) Students just memorized the theorem without understand it; (10) Difficulties in relating definitions and theorems to the given problem; (11) The lack of a source book to support the learning process.

3.2. The analysis of interview results

The analysis of the interview results is based on the interview sheets designed, but for some students, there may be addition or subtraction of questions according to the answers given by the students to the previous question. The interview was conducted with 19 students. Here are the results obtained through interview by researchers to students, namely: (1) Most of students found the difficulties to explain the existing in the book, still need additional explanation from friends or lecturers; (2) It is true that the indirect proof is difficult for the students; (3) Difficulty in composing words in writing evidence; (4) Material that is deemed difficult is the material after middle semester because theorem is not mentioned in definition or theorem to be used; (5) It is difficult to predict the direction of problem-solving; (6) Fear to try to resolve even more if there is a wrong step, it will be confused for the next step.

3.3. The analysis of observation results

The observation sheets were aimed to observe the activities undertaken by students and lecturers during the real analysis process of learning takes place. The following will be described as the results of observations when learning Real Analysis, namely: (1) The learning process takes place both in the sense of smooth, safe, comfortable, not tense, relaxed, and interspersed with humor; (2) There is a process of cooperation that occurs between group members; (3) There is also communication between lecturers and students; (4) The lecturer gives the student opportunity to discuss the proof with the group mates, to be discussed together at the end of the meeting; (5) Students are active to try to write their answers in front of the class; (6) The lecturer never demanded that the student should prove it correctly, but only asked the student to manipulate the idea. However, it appears that some students are afraid to write their answers to the front of the class; (7) each student has at least one reference book; (8) Lecturers provide keywords that can lead the students in conducting the proof; (9) there is one student who does not communicate with his or her peers but the student points on himself to present the answer in front of the class; (10) Students look enthusiastic in learning as indicated by students frequently communicating with the lecturer asking if there are constraints in the settlement; (11) There are some students who have note from senior students which is helpful in the proofing process; (12) Some students are confused during the learning process as indicated by their inactivity in writing in notebooks, discussions with peers, and not willing to write the answer in front of the class.

Based on the description of research results obtained data about the difficulties experienced by students in studying Real Analysis. The difficulties are supported by data on the results of the questionnaire, the interview sheet, and the participant's observation. Some of the core difficulties are that students have difficulties finding the initial idea of proof, difficulty in indirect proof, and difficult to express ideas nonverbally. For other difficulties can be seen from the previous explanation.

The difficulty of expressing ideas is nonverbal languages also a difficulty in learning Real Analysis. Students also have difficulty in choosing the right words to make the introduction or linking words between one process and another. For example, students are still difficult to distinguish the use of words if, suppose, select, given, and take. Students are also not accustomed to using such connecting words so, then, therefore, accordingly, by means of use, because of, and consequently. Students are accustomed to the direct process to the point. While in Real Analysis, each step should be described in detail, clearly, systematically, and there is a basis for ensuring that step. Sometimes students also have ideas, but it is difficult to transfer their ideas into written language, so it is difficult to explain them to their friends. The students should express the idea to the lecturers first, and then the lecturers help to write down the idea so that the other friends can understand. Students are also sometimes impatient with the systematic process that must be done, so some steps are passed.

The difficulties also happened to other students outside Mathematics Education Universitas Islam Riau. This is known from several previous studies. The difficulties faced are mostly the same. This suggests that to an excellent Real Analysis ability, students must be supported by excellent analytical skills so it can easily find the initial idea to be proven. The students' critical thinking ability is also low because students are unfamiliar to the given theorem, are experiencing difficulties and are confuse if conditions on the theorem are incomplete, or data is reduced, or even added. Less critical in responding to existing evidentiary evidence so as not to know the step is correct or not, is there a wrong step, not realized by the students.

The implication to develop learning instruments for the subject of Real Analysis using a specific strategy to help students in minimizing the difficulties. The instrument namely book of Real Analysis, Work Sheet of Real Analysis or hand out of Real Analysis.

4. Conclusions

The difficulties experienced by students in learning Real Analysis, namely: (1) Students find that it is challenging to find the initial idea of the proof process; (2) Students still do not have the ability to think critically, logically, creatively, reasoning, and systematic good; (3) Students are still easy to give up, lack of confidence, lack of care, and less able to socialize with a heterogeneous group of friends; (4) Difficulty in conducting indirect and indirect evidence in determining what type of proof to use; (5) Difficulty in expressing ideas into nonverbal Language; (6) Do not remember/do not understand the previous theorem; (7) Not having a good understanding of the prerequisite materials; (8) Prove that it will prove or substitute the numbers into the proof of the theorem; (9) Just memorize the theorem without understanding it; (10) Difficulties in relating definitions and theorems to the given problem; (11) The lack of a sourcebook to support the learning process.

Implication: to develop learning instruments for the subject of Real Analysis using a certain strategy to help students in minimizing the difficulties. The instrument namely book of Real Analysis, Work Sheet of Real Analysis or hand out of Real Analysis.

Acknowledge

We would like to say thank you for Universitas Islam Riau to support this research, for students in Mathematics Education Universitas Islam Riau for all of information that you give for us to support this research, and also for the 6th SEA-DR committee.

References

[1] Darmadi 2011 Imajeri mahasiswa dalam pembelajaran analisis real (Studi kasus di IKIP PGRI Madiun) *Proc. FMIPA UNY*

[2] Darmadi 2008 Spektrum hasil belajar Analisis Real mahasiswa program studi pendidikan matematika IKIP PGRI Madiun tahun akademik 2008/2009 *Seminar Nasional UNY*

[3] Darmadi 2009 Pengembangan model pembelajaran Analisis Real berbasis teori David Tall Paper presented at Seminar Nasional Universitas Negeri Surabaya; Surabaya; 2009 Aug 8

[4] Sucipto L and Mauliddin M 2016 Analisis kesulitan belajar mahasiswa dalam memahami konsep bilangan real *Beta Jurnal Tadris Matematika* **9** 197

[5] Moore R C 1994 Making the transition to formal proof *Educational Studies in Mathematics* **27** 249

[6] Bills L and Tall D 1998 Operable definitions in advanced mathematics: The case of least upper bound *Proceedings of The Conference of The International Group for* 111

[7] Weber K 2001 Student difficulty in constructing proofs: The need for strategic knowledge *Educational Studies in Mathematics* **48** 101

[8] Recio A M and Godino J D 2001 Institutional and personal meanings of proof *Educational Studies in Mathematics* **48** 83

IOP Conf. Series: Journal of Physics: Conf. Series **1088** (2018) 012038 doi:10.1088/1742-6596/1088/1/012038

The development of learning instruments based on an open-ended approach to improve students' problem-solving skill

N R Fandanu, M Ikhsan and Bahrun
Syiah Kuala University, Jl. Teungku Chik Pante Kulu No. 5, Banda Aceh 23111, Indonesia

E-mail: bahrun@fkip.unsyiah.ac.id

Abstract. Mathematics subject aims to help students to improve their problem-solving skill. One of the potential approaches that could be applied by teachers in mathematics teaching and learning process is the open-ended learning approach. This study aimed to develop the valid learning instruments on topics quadrilaterals based on open-ended approach, which included lesson plans, student worksheet, and test questions. The study focussed on Year 7 of one of the junior high schools in Banda Aceh, Indonesia. This study is a development research, which implemented a 4D model, consisting of four stages (Devine, Design, Development, and Disseminate). However, only the first three stages were conducted in this research. The validation process involved four validators specialized in mathematics education (two lecturers and two teachers). The finding showed that the average validation score of the developed instruments was 3.54, which indicated that the instruments were valid. Further studies to know the effectiveness and practicality of the instruments need to be conducted.

1. Introduction

Problem-solving skill is fundamental in mathematics, both for those studying mathematics and for those applying the skills in other fields of study and everyday life [1,2,3]. Five standards of mathematical skills must be mastered by students, namely problem-solving, communication, connection, reasoning, and representation [4]. Problem-solving is an essential part of the mathematics curriculum in relation to the learning process; students may gain experiences using their existing knowledge and skills to be applied to non-routine problem-solving. Problem-solving is an attempt to find a solution of a problem to achieve a certain goal that is not immediately achievable [5]. Therefore, problem-solving is a higher level of intellectual activity. To solve the problem, one can use certain steps that he or she controls. One of the causes of the poor quality of students' mathematics understanding in elementary and junior high schools is that teachers mostly focus on the exercises of solving procedural and calculation problems instead of understanding the mathematics concepts [6].

There are several steps of the problem-solving process as proposed by experts, which are then used to measure problem-solving skills. The steps are; Firstly, (a) reading and thinking, (b) exploring and planning, (c) selecting a strategy, (d) finding and answering, and (e) reflecting and extending [7]. Secondly, (a) understanding of the problem, (b) devising a plan, (c) carrying out the plan and, (d) looking back [8]. Thirdly, (a) recognition, (b) definition, (c) formulation, (d) test, (e) and evaluation [9]. In this study, the Polya stages were used since they are simple and implicitly cover all stages of problem-solving.

One way to improve students' mathematics problem-solving skills is by implementing the open-ended approach in teaching and learning process. Mathematics learning using open-ended questions is one of the representative methods to improve students' problem-solving skills [10, 11]. The purpose of the open-ended approach is to help develop the creative activities and students' mathematical mindset

IOP Conf. Series: Journal of Physics: Conf. Series **1088** (2018) 012038 doi:10.1088/1742-6596/1088/1/012038

through problem-solving simultaneously [12]. An open-ended approach is a learning approach presenting a problem that has more than one correct method or solution, allowing students to acquire the knowledge or experience of finding, recognizing, and solving problems using several techniques [10].

Open-ended problem is a problem that has many solutions or resolution strategies [13]. Open-ended problems are divided into three parts, namely: (a) the problem with one answer to many ways of solving; (b) the problem with one solution to many answers; (c) problem with many solutions and also many answers. Based on those three opinions above can be stated that presenting the open-ended problem can provide stimulation to the students to improve the way of thinking so that students have the freedom to express the exploration's result of reasoning power and the analysis to solve a problem actively and creatively. Thus, open-ended problems can train students' mathematical literacy skills themselves. In this study, the researcher uses the problem with one solution to many answers and problems with many solutions as well as many answers. According to Takahashi [13], there are several benefits to using open-ended problems in mathematics learning. First, students become more active in expressing their ideas. Second, students have more opportunities to use their knowledge and skills comprehensively. Third, students have abundant experience in the process of finding and receiving approval from other students for their ideas.

The open-ended learning is divided into two stages. The first stage is the teacher provide a problem, students write down ideas, summarize, and discuss the results of individual answers in the groups. The second stage: presenting of group work results [14]. Learning with open-ended approach is arranged in two stages,at stage one: students solve the problem given by the teacher at the beginning of the learning, the students are given blank paper to write their ideas, the papers are collected by teachers to generate the conclusions of the individual response, the students discuss the results of individual work in groups, students write down the results of group discussions. During the second stage: the results of each group are presented and discussed and then the learning is concluded [15]. Learning can be implemented well if it is based on appropriate learning materials and assessment instruments. Therefore, it needs to start by a valid learning instrument.

2. Method

The type of research used in this study was Research and Development. The learning instruments development model used was to modify the Four D model [16]. This model consisted of 4 stages: Define, Design, Develop, and Disseminate. This development study was limited to the Define, Design and Development stages. In the Define stage, the researcher collected information to determine the characteristics of learning instruments to improve students' problem-solving skills. At this stage also conducted an early curriculum analysis with the help of mathematics teachers to identify the mathematical material needed for the learning instruments development. The design stage was the stage of designing an instrument that would be used in mathematics learning. At this stage, there were four activities, namely the preparation of tests, the selection of formats, the selection of media in accordance with the characteristics of learning materials, and the making the early design. At this stage produced the draft 1 which was then subsequently refined at the development stage.

The development stage was the stage of validating learning instruments by experts by providing a questionnaire for each learning instruments, with a scale of 1 to 5 (invalid to valid). Learning instruments had been validated in two revisions by four validators (two lecturers and two mathematics teachers). The criteria used for assessing the learning instruments were based on Plomp and Nieveen [17] as shown in Table 1.

Table 1. The validity criteria.

Score			Criteria
$4 \leq$	VS	< 5	Highly valid
$3 \leq$	VS	< 4	Valid
$2 \leq$	VS	< 3	Less valid
$1 \leq$	VS	< 2	Not valid

3. Result and discussion

This research is a prototyping phase of the development of learning instruments based on the open-ended approach to improve students' problem-solving skills on topic quadrilaterals, which aimed to produce the valid learning instruments. The learning instruments were expected to be a reference for teachers in classroom learning, especially on the quadrilaterals topics. The research procedure used was 4D development study [16].

3. 1 Define stage

In the Define stage, the researcher discussed the obstacles that teachers found when teaching quadrilaterals materials. It was considered important to obtain information about the conditions and attitudes of students during the process of learning mathematics in class. The results of the discussion with the mathematics teachers, it was known that the quadrilaterals material was a challenging and appropriate topic to develop open-ended based learning instruments to improve students' problem-solving skills. One of the challenges that students often encountered was when deciding varieties of square and triangle planes. The learning material is a pre-requisite material before students learn topics of perimeter and area. Mathematics has a concept that is structured, logical, systematic and interrelated [18]. Students who had a poor understanding of the basic concepts would experience difficulties in understanding and learning new materials; thus resulting in low students' problem-solving skills. Therefore, the learning instruments developed in this study consisted of the materials of the perimeter and area of the quadrilaterals.

3.2 Design stage

At the design stage, the learning instruments consisted of the lesson plan, student worksheets, and test questions, which were developed based on the analysis at the define stage. The results of the learning instruments at the define stage were called draft 1. The lesson plan was designed using open-ended learning steps, where the learning provided open-ended problems centered on students, encouraged students to learn, guided individual and group investigations, summarized and discussed problems and answers in groups, presented the results of group discussions and conducted class discussions with other groups. The learning process lasted for five meetings, with each meeting 2 x 45 minutes per lesson. In general, the lesson plan developed had aspects that were similar to typical lesson plans. Where at the learning step, the researcher combined all the steps into one stage. And at the motivation stage had a difference, which aimed to understand the area and perimeter concepts by asking problems a day. Here was an example of a problem to motivate students from a lesson plan.

> "Dedi has two tiles as seen the following Figure 1, but Dedi does not know the area and how to determine the area of the tiles. Please help Dedi to determine the area of the tiles and how wide it is?"

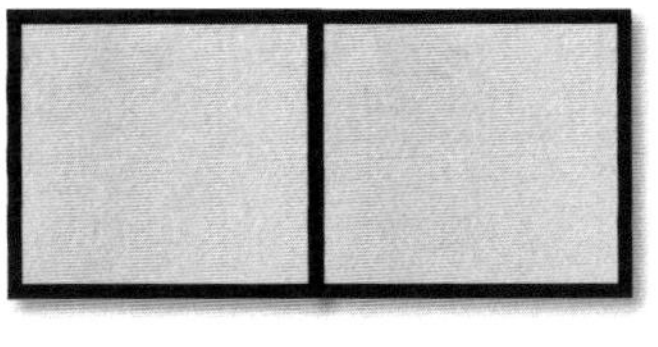

Figure 1. Two tiles that coincide with one side

Design of test question should be relevant to the form of an open-ended question on quadrilaterals material. Table 2 shows the example of test questions developed.

Tabel 2. The sample of the developed test questions.

Question	First revision	Second revision
2	A swimming pool is shaped like a shaded area below. Calculate the area and perimeter of the pool in your way! 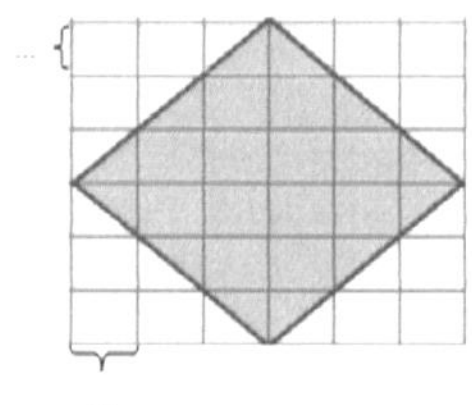	Budi wants to make ceramics on the pedestal of swimming pool shaped rhombus with the area of 160000 cm^2. Calculate how much ceramics is required for the pool according to your way! 
4	A garden is shaped rectangle with the area of 120 cm^2. If around the garden will be planted with protective trees with a distance between each tree is 2 m, how many stalks of protective trees are needed?	A garden is shaped rectangle with the area of 120 cm^2. If around the garden will be planted with areca nut trees with the distance between each tree is 2 m, how many stalks of areca nut trees are needed? 

Based on Figure 2, there were revisions of the test that would be given to the students. The revisions were based on the validator's suggestion. The first revision on questions number 2 and 4 shows that the questions have not been linked to pictures or illustrations in everyday life. Thus, it would improve students' problem-solving skills to solve the problems. Problem-solving is an attempt to find a way out of a difficulty to achieve a goal that is not immediately achievable [5]. The purpose of the open-ended approach is to help develop the creative activities and students' mathematical mindset through problem-solving simultaneously [12].

3.3 Develop stage

This stage related to the validation of product results developed. The developed learning instruments were validated in advance by experts (mathematics teachers, mathematical education experts). Table 3 shows the validation result of the learning instruments

Table 3. Learning instruments validation results.

Product	Assessment				average	category
	V1[a]	V2[b]	V3[c]	V4[d]		
lesson plan	0,87	0,86	0,86	0,80	3,39	valid
student worksheets	0,93	0,96	0,90	0,92	3,71	valid
test questions	0,80	0,97	0,97	0,80	3,54	valid
		Average			3,54	valid

[a] First Validator.
[b] Second Validator.
[c] Third Validator.
[d] Fourth Validator.

The analysis results of the learning instruments validation sheet were obtained an average value from the experts for the lesson plan of 3.39. This showed that the lesson plan with the open-ended approach was valid and could be used in the class learning process. The validation results of the students' worksheets received an average score of 3.71. This indicated that the students' worksheets with the open-ended approach were valid and could be used in the learning process in the class. The validation results of test questions got an average score of 3.54. This showed that the test questions that had been developed was valid and could be used in the learning process in the class. Overall learning instruments were obtained an average value of 3.54. This indicated that the overall learning instruments could be categorized as valid [17]. However, the score that was obtained on the questionnaire sheets had not reached the maximum value, that was 5, this was because the instrument was only revised twice in accordance with the validator's suggestion. Validity may affect the number of validators, the more validators, the more contents must be revised so that the higher level of the validity because it will provide more constructive advice from different perspectives for improving the instrument.

4. Conclusion

This study produced the valid learning instruments by implementing an open-ended learning approach on the topics quadrilaterals. The average validity score was 3. 54, which indicated that, theoretically, the instruments were valid. When designing learning instruments using an open-ended approach, teachers are supposed to consider the relevancy of the topics to the approach due to the complex process of designing open-ended questions.

Reference

[1] RuseffendiET 2006 *Pengantar kepada Membantu Guru Mengembangkan Kompetensinya dalam Pengajaran Matematika Untuk Meningkatkan CBSA (edisirevisi)* (Bandung: Tarsito)
[2] Taplin M 1998 Preservice teachers' problem-solving processes *Mathematics Education Research Journal* **10** 59
[3] Pehkonen E 1997 The state-of-art in mathematical creativityvo *ZDM* **129** 63
[4] National Council of Teacher of Mathematics (NCTM) 2000 *Principles and Standars for School Mathematics* (Reston VA NCTM)
[5] Billstein L and Lott 2010 *A Problem Solving Approach to Mathematics for Elementary School Teachers Tenth Edition* America Pearson International Edition
[6] Suherman E and Turmudi S D dkk 2003 *Strategi pembelajaran matematika kontemporer* (Bandung: JICA UPI)
[7] Krulik S, Rudnick J and Milou E 2003 *Teaching Mathematics in Middle School A Practical Guide (*Boston: Pearson Education Inc)
[8] Polya G 1973 *How to Solve It* Second Edition Princeton University Press Princeton New Jersey

[9] Sujarwo A 2012 *Proses Berpikir Siswa SMK Dengan Kecerdasan Linguistik Logika Matematika Dan Visual Spasial Dalam Memecahkan Masalah Matematika E*-Jurnal **3** Dispendik Surabaya

[10] Shimada S 1997 *The Significance of an Open-Ended Approach Dalam J P Becker & S Shimada (Ed) The Open-Ended ApproachA New Proposal for Teaching Mathematics* (Virginia: National Council of Teachers of Mathematics)

[11] Setiawan R H and Harta I 2014 Pengaruh pendekatan open-ended dan pendekatan kontekstual terhadap kemampuan pemecahan masalah dan sikap siswa terhadap matematika *JurnalRisetPendidikanMatematika* **1** 241

[12] Nohda N 2000 *Learning and Teaching Trought Open Approach Method Mathematics Education in Japan* (Tokyo: TSME)

[13] Takahashi A 2008 Communication as Process for Students to Learn Mathematical [Online] http://www.criced.tsukuba.ac.jp/math/apec/apec2008/papers/PDF/14.Akihiko_Takahashi_USA.pdf.

[14] JoharR 2016 *Strategi Belajar Mengajar* (Yokyakarta: Deepublish)

[15] Hashimoto Y 1997 *An Example of Lesson Development Shimada S dan Becker JP (Ed) The Open Ended Approach A New Proposal for Teaching Mathematics* (Reston VA NCTM)

[16] Thiagarajan S, Semmel D S and Semmel M I 1974 *Instructional Development for Training Teachers of Exeptional Children A Sorce Book. Blomington* Center of Inovation on Teaching the Handicapped Minneapolis Indiana University pp 5-9

[17] Plomp T and Nieveen N 2013 Educational Design Research (Enschede: SLO)

[18] Suhendra 2005 Senang Matematika Erlangga For Kid (Jakarta: Erlangga)

The development of learning materials using contextual teaching learning (CTL) approach oriented on the character education

R Johar, Agussalim, M Ikhsan and B Zaura
Syiah Kuala University, Jl. Teuku Nyak Arief Darussalam, Banda Aceh 23111, Indonesia

Email: rahmah.johar@unsyiah.ac.id

Abstract. The implementation of mathematics learning, especially in junior high school experiencing obstacles, one of which is the lack of learning materials linking the subject matter with the context of students' daily life. Also, there are a limited of learning materials paying attention to the aspects of students' character development. This study aimed to develop a valid and practical mathematics learning materials using a contextual teaching and learning approach oriented on the character education. The subjects of the study were Year 7 students in one of junior high school in North Aceh, Indonesia. Data were collected through the expert validation for learning materials, observation of the learning process and questionnaire of students' response. The expert validation results obtained for the lesson plan, student's book, worksheet, teacher's book, and achievement test showing highly valid criteria. The results of observation for the teachers' skill in implementing the learning materials and students' activities as a good category. The results of students' responses towards the learning materials and the implementation of learning were also positive. It is recommended that further research will test the effectiveness of the learning materials to examine the impact on the more students.

1. Introduction

The purpose of learning mathematics is to equip learners with the ability to think logically, analytical, systematic, critical, and creative, as well as the ability to work in a team. The competencies required learners having the ability to acquire, manage, and use the information to survive in an ever-changing, uncertain and competitive situation [1]. However, mathematics is a frightening subject for students it is one of the most challenging subjects to be understood by students [2]. The teaching and learning process in schools have not yielded maximum results [3]. Most of the students are not able to connect what is learned and how that knowledge is used. Therefore, it is necessary for teachers to associate the subject matter to the context of everyday life so that students can benefit from the knowledge they have acquired. One of the approaches to be applied is contextual teaching and learning (CTL).

Contextual Teaching and Learning (CTL) is a learning system based on the philosophy that students can absorb the lessons if: 1) they can grasp the meaning of the academic materials and school assignments, and 2) they can associate new information with their previous knowledge and experience. CTL is in line with the human conscience that is always eager for discovering the meaning. CTL invites students to associate school tasks in everyday life meaningfully. When students see the meaning in the tasks to be done, students will absorb the lessons and remember them [4].

One of the phenomena occurred today is the declining of characters. Various media reported the immorality involving state officials, community, teachers, and students. Cases of corruption, nepotism and immoral seem to happen commonly. Therefore, the government established a character education [5].

Hakim [6] stated that each school and the community should have the discipline and habits related to the characters to be formed. Character education is best given by indirectly influencing the students by showing examples of good behavior [7]. The students should learn on their own to understand one's character through the influencer elements provided by the teacher/educator. The formation of students' characters cultivated by the environmental factors can be done through exemplary strategies, interventions, strengthening, and habituation must be done consistently.

The law No. 18 year 2001 on Special Autonomy for the Province of Aceh, a region with strong Islamic characteristics and commitment to implement the education system based on the Islamic values. Qanun Aceh emphasizes the law regarding the provision of education asserting that the national education system organized in Aceh based on Islamic values. Based on the law, generally, it can be described that the concept of education in Aceh instructing the education concepts based on the values of Islam [8]. Furthermore, the use of the local culture can improve the quality of learning [9]. Therefore, schools in Aceh should integrate Islamic values in the learning, especially as modeling the character.

The mission of the character formation is not only the responsibility of the religious education but also other subjects. Therefore, other subjects should also foster the creation of students' character [10]. Two approaches can be taken in the character education, namely: (1) the character is as a separate subject; and (2) the character is built-in in every subject. Also, character education is essentially cultivating the character through the process of knowing the good, loving the good, and acting the good, the education process involving cognitive, emotional, and physical, so that the noble character can transform into a habit of the mind, heart, and hands [11].

However, the education practice in Aceh pays less attention to the purposes of education mandated in Qanun [12]. The education in Aceh is less integrated with the values of divinity in the process of learning and school culture, and more oriented on the development of the cognitive structure. The integration of Islamic knowledge in mathematics learning in Acehnese school is not optimal; the Islamic values carried out is still limited to the separation of seats between male and female students, praying and reading the Qur'an before the lesson and giving advice. The observation results conducted by the researcher of mathematics textbooks in junior high school showed that it is difficult to obtain the learning materials using contextual approach learning and oriented to character education oriented on the vision of Islam.

Based on the earlier review, it is necessary to develop learning materials to be applied to learning. Johar [14] has developed mathematical learning materials using a realistic Islamic nuances approach for Year 4 of primary schools consisting of the lesson plan, worksheet, and achievement test for fractions and numbers. Furthermore, Ismail, Johar, Kamarullah, and Yusmanita [15] developed a mathematical textbook of Islamic nuances for all mathematics topics in the Year 4 of primary schools such as numbers, geometry, and measurer. There has been no research developing the mathematics learning materials using CTL approach oriented on the character education for students Year 7. The purpose of this research was to create a valid and practical mathematics learning materials using CTL approach oriented on the character education.

2. Method

The development of learning materials in this study referred to the Design Research of Plomp [16] consisting of three phases namely: the preliminary research, prototyping and assessment phase. In the preliminary research phase, the identification and analysis of the mathematics curriculum, the students' condition, and the demands of the environment and community to learn mathematics were conducted. In the prototyping phase, the learning materials using CTL oriented on the character education such as the lesson plan, teacher's book, student's book, worksheets, and achievement test were designed and validated. In the assessment phase, the learning materials were implemented to Year 7 students in one of junior high school in North Aceh, Indonesia.

According to Nieveen [17], the learning materials are qualified if they meet the quality aspects of the validity, practicality, and effectiveness. The validity is attributed to two issues: (1) whether the learning materials developed based on a strong theoretical rationale, and (2) whether there is a consistency between them internally. The practicality aspect is satisfied if (1) experts and practitioners argue that the learning materials can be applied, and (2) evidence shows that the learning materials can be utilized. Meanwhile, the aspect of effectiveness concerns whether the learning materials meet the desired outcomes or not. This study focused only on assessing the validity and the practicality of the learning materials.

The research instruments in this study were: (a) validation sheet was used to determine the validity of the student's book, teacher's book, lesson plan, worksheet, and achievement test, (b) observation sheet was used to obtain data on student activities during the learning and teachers' skills to implement the lesson plans, (c) questionnaire was used to determine the students' response to the student worksheet and book, and (d) the achievement test was used to analyze the impact of the learning materials to students' learning outcome.

3. Results and discussion

3.1 The result of the preliminary research phase

In this phase, the curriculum, student, and environmental demand analyses were carried out. The results of the investigation found that the curriculum applied in Year 7 in one of junior high school North Aceh, Indonesia was the Curriculum version year 2006. The topic of fractions is one of the topics taught by utilizing contextual problems. The results of the student analysis are: (1) students only used the textbooks available in the library containing routine problems only; and (2) the context presented in the student's book did not include values to build students' character. The results of environmental demands analysis reported that: (1) teachers do not have teacher's book to help them teach in the classroom; and (2) the teaching and learning process has not accommodated the Islamic context oriented on the character education. Based on these analyses, it was concluded that it was necessary to develop learning materials with the contextual approach and oriented on the character education based on Islam.

3.2 Result of the prototyping phase

At this phase, learning materials were designed based on the contextual approach oriented on the Islamic character education, especially for Aceh as an Islamic cultural region. The learning materials consisted of the lesson plan, student worksheet, teacher's book and student's book to help students learn fractions; teacher's book to allow teachers to achieve the learning objectives by using a contextual approach for teaching fractions, test to measure students' outcome about fraction, and questionnaire to analyse students' response.

Four experts were assessing the learning materials in the validation process. The assessment included: the format, language, and content of the learning materials. Experts used grade 1 to 5 and provided comments to assess the learning materials. Based on the validation results, some revisions were made to the lesson plans, worksheets, student's books, teacher's book, and achievement test. The results from four experts showed that: (1) the lesson plan had an average of 4.21, valid criteria. Lesson plan allowed a particular time for teachers to provide guidance and character motivation for students. The lesson plan also contained good character habituation activities for students, such as religious, social care, and the environment care; (2) the student's book had an average of 4.03, highly valid criteria. Student's book contained problems that were contextually related to students' daily life, and there are stories of characters that become role models for the students in their community. Four experts concluded that the student's book could be used with minor revision; (3) teacher's book had an average of 3.7, valid and minor revision; (4) the average score of student worksheet was 4.02, also valid criteria and minor revision; and (5) the item test results had an average of 4.09, highly valid criteria.

The example of mathematics problems and character message on student worksheet is presented in Figure 1.

<table>
<tr>
<td>Problem 1
Mr. Abdurrahman has 100 million IDR savings in the Bank and gold worth 300 million IDR. All the wealth has been kept since last year. Is Mr. Abdurrahman obliged to pay zakat? How much zakat should be paid?

Note that:
The price of 1 gram of gold today 250,000.00 IDR
The formula of Zakat is:
The total amount of Zakat = 2.5% x Total wealth kept for 1 Year.</td>
<td>Character Message:
Rasulullah SAW said: "Clean your wealth with zakat, medicate your sickness with charity, and reject the disasters by prayer" (Hadith Khatib from Ibnu Mas'ud)

</td>
</tr>
<tr>
<td>Problem 2
Usman has two children, Joseph and Zainab. Both love the sweet martabak cake. One day he bought one of his favorite sweet martabak cake. Usman has three close neighbors, Mr. Sulaiman, Mr. Ibrahim, and Mr. Musa. It has become a habit of Mr. Usman's family where there is food; it should be not enjoyed alone; instead, some are distributed to the neighbors. Mr. Usman asked his daughter to share the cake equally for the family at home and the three neighbors.

How many amounts of cake are distributed to neighbors? How much cake is left at home? Express in the form of the fraction.</td>
<td>Character Message:
Rasulullah SAW said: "Whoever believes in Allah and the Last Day, let him say good or silent. And who believes in Allah and the Last Day, let him glorify his neighbor. And who believes in Allah and the Last Day, let him glorify his guest. "(Hadith Bukhari and Muslim)
Rasulullah SAW said: "O Abu Dzarr, if you cook the cuisine with broth, then multiply broth and pay attention to your neighbor." (Hadith Muslim History)
</td>
</tr>
</table>

Figure 1. Example of mathematics problems and character message on the student worksheet.

In Figure 1, the problems are presented by inserting Islamic character values based on the hadith. Acehnese students are familiar with the issue of *zakat* and fair sharing. This context leads students to be an excellent Moslem and easy to understand fraction problem.

3.3 Result of the assessment phase

In the assessment phase, a small group trial was conducted to examine the errors in writing the student's book and worksheet such as typos, missing words, and upper and lower case errors. The subjects in small group trial were 9 Year 7 students in one of junior high school in North Aceh, Indonesia including three students at each level of ability (high-, medium- and low-level ability). The information obtained from small group trial was used as a basis for improving learning materials for the field trial phase.

A field trial was conducted on 26 Year 7 students at one of junior high school in North Aceh, to examine the practicality of learning materials. The practicality of learning materials was obtained from 1) the assessment of two teachers of the learning materials, 2) the observation of student activities, 3) the observation of teacher skill during the lessons, and 4) the student responses on the learning materials and learning atmosphere.

The result of the teachers' analysis on learning materials reported the average is 88.6%, for student's book, 87.57% for teacher's book, 87.4% for student worksheet, 88,9% for the lesson plan, indicating that the learning materials satisfied the excellent criteria.

The results of observation for students' activities carried out by two observers during the learning showed that the average score of the first, second, third, fourth, fifth and sixth lesson were 86%, 93%, 96%, 98%, and 99%, and 100%. It was concluded that the students' activities in the learning process met the very good criteria.

The assessment of teachers' skill adapted from the characteristics of CTL according to Johnson [4] included: 1) presenting "models" as tools and examples in learning; 2) developing students' curiosity through questions; 3) creating an atmosphere of "learning community" by working in groups; 4) carrying out as many as possible discovery activities in the learning process; 5) developing the students' thinking of learning more meaningfully by working individually, determining, and constructing their new knowledge and skills; 6) conducting a reflection at the end of the lesson; 7) carrying out the authentic assessment; and 8) learning atmosphere. Based on the observations by an observer, teachers' skill in implementing learning materials for six lessons was found the average of 4.97 showing very good criteria.

The results of the students' response reported that more than 80% of students were happy with the learning component and stated that the learning using a CTL approach oriented on the character education was new for them. Furthermore, more than 95% of students were interested in participating in the next learning applying this learning. Regarding the language of the student's book, worksheet, and test items, more than 90% of students could understand them. Also, more than 85% of students were interested in the appearance of the student's book, worksheets, and test items. Thus, the response of students to CTL approach oriented on character education was positive.

The test results of students' outcome showed that there were 19 students (73%) who had achieved the mastery of learning (the score of 75), while seven students (27%) had not. Remedial for the topic of the fraction was conducted for the seven students. The remedial teaching aims to help students whose score below the average [18]. The validation results of the lesson plan, student worksheet, and test were 4.21, 4.02, and 4.09 respectively. Whereas, in Johar's study [14], the results of lesson plan validation, of student worksheet and test were 3.3, and 3.6. This indicates that the learning materials employing a CTL approach oriented on the character education developed to have a higher level of validity than the previous study. Besides, that students' response is positive, and 73% of students score is above 75. This is due to contextual learning allowing students to strengthen, expand and apply their academic knowledge and skills [19]. Contextual problems may be used to examine students' interest in the problem and motivating students to work on it [20]. A teacher can encourage student's thinking by providing appropriate problems and discussion in the classroom [21].

4. Conclusion

The learning materials using CTL approach oriented on the character education has met valid and practical criteria. It is suggested to the next researcher should examine the effectiveness of learning materials to find out the impact of them on more students.

Reference

[1] Badan Standar Nasional Pendidikan 2006 *Standar Isi Mata Pelajaran Kimia SMA/MA* (Jakarta: BSNP)

[2] Wahyudin 2008 *Pembelajaran dan Model-model Pembelajaran* (Bandung: UPI)
[3] Wahyudin 2007 *Peraturan Menteri Pendidikan Nasional Republik Indonesia tentang Standar Isi untuk Satuan Pendidikan Dasar dan Menengah* (Jakarta: Depdiknas)
[4] Johnson E B 2011 *Contextual Teaching and Learning (CTL) Menjadikan Kegiatan Belajar Mengajar Mengasyikkan dan Bermakna* (Bandung: Kaifa)
[5] Fauziyah L and Jailani 2014 Pengembangan Perangkat Pembelajaran Matematika yang Menunjang Pendidikan Karakter Siswa Kelas IV Sekolah Dasar *Jurnal Prima Edukasia* **2** 149
[6] Hakim R 2014 Pembentukan Karakter Peserta Didik melalui Pendidikan Berbasis Al-Quran *Jurnal Pendidikan Karakter* **2** 123
[7] Nurlina 2014 Membangun Pendidikan Berkarakter di Aceh *Jurnal Pencerahan Majelis Pendidikan Daerah Aceh* **8** 90
[8] Qanun Aceh 2008 Qanun Aceh Nomor 5 Tahun 2008 tentang Penyelenggaraan Pendidikan
[9] Saliman 2007 Penerapan Pembelajaran Berbasis Budaya Sebagai Upaya Peningkatan Kualitas Pembelajaran Pada Mata Kuliah Perencanaan Pembelajaran (Makalah: Online) http://staff.uny.ac.id/system/files/penelitian/Saliman,%20Drs.%20M.Pd./Peningkatan%20Profesionalisme%20Guru.pdf
[10] Marzuki 2012 Pembinaan Karakter Siswa Berbasis Pendidikan Agama di Sekolah Dasar dan Sekolah Menengah Pertama di Daerah (Makalah: Online) http://staff.uny.ac.id/system/files/penelitian/Marzuki, Dr. M.Ag./29.
[11] Budiyanto M and Machali I 2014 Pembentukan Karakter melalui Pendidikan Agriculture di Pondok Pesantren Islamic Studies Center Aswaja Lintang Songo Piyungan Bantul Yogyakarta *Jurnal Pendidikan Karakter* **2**108
[12] Ismail and Saminan 2013 Islamic School Culture in Aceh *Islamic Studies Journal* **1**
[13] Soraya C 2015 Pelaksanaan Pendidikan di Kota Banda Aceh: Studi terhadap Implementasi Qanun No.5 Tahun 2008 *Jurnal Ilmiah DIDAKTIK* **16** 1
[14] Johar R 2009 Pengembangan Model Pembelajaran Realistik Menuju Aceh Madani (Model PM-RAHMA) *Jurnal Pendidikan Matematika* **3** 1
[15] Ismail A, Johar R, Kamarullah and Yusmanita S 2009 *Matematika untuk kelas IV SD Pendekatan Realistik yang Islami* (Medan: Pustaka Bangsa)
[16] Plomp T 2013 *Educational Design Research: an Introduction* Ed T Plomp and Nieveen (Enschede: SLO) p 10
[17] Nieveen N 2007 Formative Evaluation in Educational Design Research *Proceeding of the Seminar Conducted at the East China Normal University* (Shanghai: PR China)
[18] Jangid N and Inda U S 2016 Effectiveness of remedial tecahing on thinking strategies of slow learners *The International Journal of Indian Psychology* ***4*** 98
[19] Washington State Consortiumfor CTL 2001 *Contextual Teaching and Learning* (Washington: STW USA)
[20] Johar R, Patahuddin S M and Widjaja W 2017 Linking pre-service teachers' questioning and students' strategies in solving contextual problems: a case study in Indonesia and the Netherlands *The Mathematics Enthusiast* **14** 101
[21] Gravemeijer K 2010 Realistic mathematics education theory as a guideline for problem-centered, interactive mathematics education eds R K Sembiring, K Hoogland and M Dolk *A decade of PMRI in Indonesia* (Utrecht: APS International) p 175

Developing learning trajectory for teaching statistics at junior high school using RME approach

A Fauzan, E Musdi and J Afriadi
Universitas Negeri Padang, Jln. Prof. Dr. Hamka, Sumatera Barat, Indonesia.

E-mail: ahmadfauzan@fmipa.unp.ac.id

Abstract. We developed a learning trajectory (LT) for teaching statistics at Year 8 junior high school using Realistic Mathematics Education (RME) approach. The research aimed at investigating the validity and practicality of the LT as well as its influence on students' statistical reasoning abilities. This was a design research that consisted of a cyclic process of preparing and conducting the experiment, and retrospective analysis. The subjects of the research were 30 Year 8 students in a junior high school in Indonesia. To collect the data, we used observations, interviews, checklist, videotaping, and analyzing the students' works. In preparing for the experiment, it was designed the LT for teaching statistics that met the criteria of validity (relevance and consistency). During one-to-one, small group, and field test evaluations we discovered that the LT could work as intended (met the criteria of practicality). We also found that the LT for teaching statistic using RME approach could help the students to reinvent the concepts about central tendency by themselves. Finally, we found the growths in the students' statistical reasoning abilities.

1. Introduction

Almost every aspect of our life is related to numbers and data. When doing activities at home, reading a newspaper, watching television, or buying something in a supermarket, we deal with numbers and data. Bailey [1] mentioned that we are drowning in data, while Steen [2] argued that the world of the twenty-first century is a world awash in numbers. To be able to deal with the data appropriately, people need to be statistically literate and have good statistical reasoning. Watson [3] said that the ability to provide reasonable evidence based arguments and to evaluate data based claims critically are important skills that all citizens should have.

As the need to improve statistical literacy and statistical reasoning grow, they become part of the mainstream school curriculum in many countries [4]. We can even find the topics of statistics in the primary school curriculum in many countries, including Indonesia. According to Garfield and Ben-Zvi [5], this condition is not surprising because statistics instruction at all educational levels is gaining more students and drawing more attention.

Although the topics of statistics have been introduced since early grades, many studies showed that the students still find difficulties when they learn similar concepts in the higher school levels [6-9]. There were some causes of this condition. Firstly, the focus of teaching statistics was more on applying formulas and computations [10,11]. Secondly, statistical concepts presented in mathematics textbooks were not sequenced in a meaningful way [12,13]. These causes were crucial because it also influenced the way the teachers teach in the classrooms that tend to be mechanistic [14-16]. To overcome these problems, we developed a learning trajectory (LT) for teaching Statistics in junior s school using realistic mathematics education (RME) approach.

A learning trajectory is the sequences of activities and tasks that designed by mathematics teachers or researchers to guide students to achieve a specific instructional goal [17]. Gravemeijer [18] argued that neither teachers nor researchers could rely on fixed teaching sequences, because they continuously have to adapt to the actual thinking and learning of her students. Therefore, in the beginning, an LT is designed in the form of *a hypothetical learning trajectory* (HLT). An HLT consists of the learning goal, the learning activities, and the hypothetical learning processes; a prediction and anticipation of how the students' thinking and understanding will evolve in the context of the learning activities [17,19]. After the HLT is tested through a cyclic process in the classroom, it becomes an LT; a theory that can be used to teach the topic by other teachers. The theory is called a local instructional theory (LIT) [20,21] or a domain specific theory [22] that explain about both the process of learning a particular topic and the means to support that learning [18]. In this research, we developed a local instructional theory to teach statistics for the students at the Year 8 junior high school.

Many researchers and math educators have developed learning trajectories [23-27] that covered various topics such as statistics, social arithmetic, optimization, and logarithm. These studies showed that the LT's could work not only for teaching mathematics in the primary schools but also in the colleges. The results of the research mostly indicated that the LT's could help the students to learn mathematical concepts in many meaningful ways [24,28].

The idea of learning mathematics in an LT is in line with the idea of horizontal and vertical mathematization in RME approach, as can be seen in Figure 1. Therefore, the LT that we designed in this research was based RME approach.

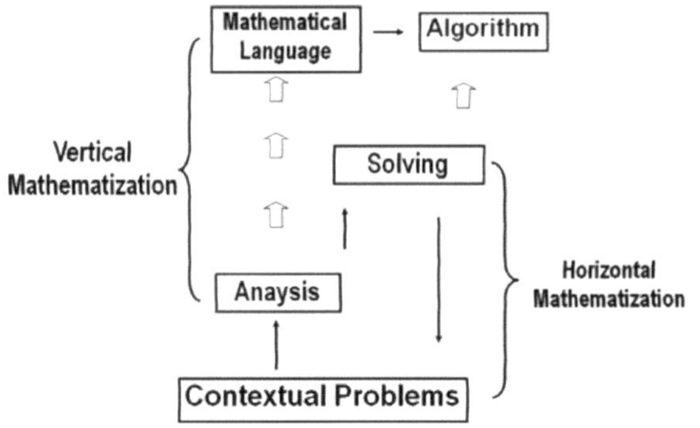

Figure 1. Processes of horizontal and vertical mathematization [19]

As we can see in Figure 1, learning mathematics using RME approach is started by giving contextual problems that can stimulate students to use their informal knowledge and their strategies to solve the problems. Gravemeijer [19] called this process as horizontal mathematization. Step by step, and after experiencing similar processes, students will use more formal knowledge, symbols, or strategies for solving the problems. It means they start experiencing vertical mathematization, until they use formal mathematical language or algorithm to solve contextual problems.

In designing the LT for teaching statistics, we also referred to the key principles of RME, namely *guided reinvention trough progressive mathematization, didactical phenomenology,* and *emerging models* [29,30]. In implementing the LT in the classrooms, we adopted the characteristics of RME for teaching and learning [31-34]. Although the RME approach has been established since the 1970's, especially in the Netherlands, it still powerful to solve current problems in mathematics education. Many researchers from different countries [35-38] recently used the approach to stimulate the students' thinking and their mathematical abilities. In this research, we investigate the effect of the RME approach on the students' statistical reasoning ability. To focus the research, we formulated a research question as follows: what are the characteristics of the learning trajectory for teaching Statistics using RME approach which are valid, practical, and effective to stimulate students' statistical reasoning ability?

2. Method

The LT for teaching statistics was developed by using design research approach mentioned by Gravemeijer and Cobb [19] because it gave more insight about the interrelatedness between teaching and learning to improve teaching [18]. The design research approach was implemented in a cyclic process of preparing for the experiment, conducting the experiment, and retrospective analysis, as illustrated in Figure 2.

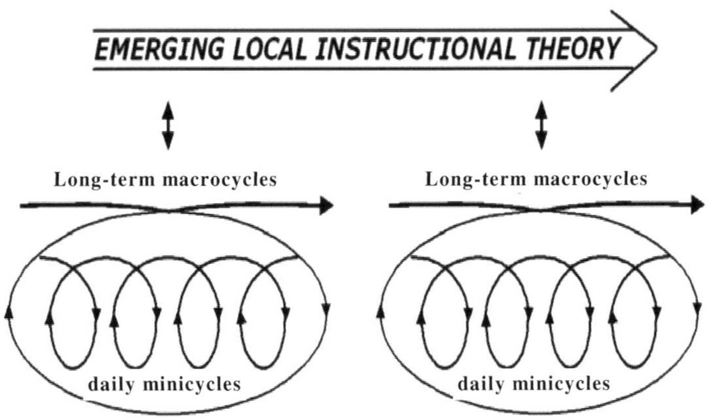

Figure 2. The cyclic process of design research [19].

In preparing for the experiment, we determined the endpoints of the instructions in which the students will reinvent the concepts about central tendencies by themselves. Therefore, we analyzed the essential concepts about statistics in the curriculum, the literature about teaching statistics, and the characteristics of the students such as their hobbies, their preferences in learning, and their preferences activities. Based on the analysis, we designed the three components of HLT [17,19] for teaching statistics using the RME approach. Then, three experts in mathematics education and one expert in instructional design validated the HLT to judge its validity (relevance and consistency) [39].

After the revision, the HLT was tried out in three cycles, namely one-to-one, small group, and field test. The first two cycles respectively involved three and six Year 8 students in a junior high school in Indonesia. To analyze the results of the tried out, we conducted the retrospective analysis and re-designed processes that involved the research team, a teacher, and an observer. The revised version of the HLT was implemented during the field test to 30 Year 8 students in the same school. To collect the data we used observations, interviews, checklist, videotaping, and analyzing the students' works. Collected data were analyzed using a descriptive technique.

3. Result and discussion

Based on the primary analysis during the preparing for experiment phase, it was designed the HLT for teaching statistics using RME approach. The learning activities in the HLT can be described as follows. The first meeting was about exploring the concept of mode in which the students collected the data about students' preference of candy flavors. Then, they were asked to present the data in various representations. By analyzing the data representations, the students can reinvent the concept of mode as the most favorite candy flavor. Besides, the students will also observe the concept through different representations. In the second meeting, the students reinvent the concept of median by exploring contextual problems about a company that wants to establish its headquarters at one of the branch offices on a highway. To reinvent the concept of mean, the students will get involved in the activities of dividing marbles, the yield of maize, and estimating the number of large quantities. We chose the activity about fair share to reinvent the concept of mean due to the opinion of Franklin and Kader [40]. They stated the mean as fair share and balance, therefore to find the mean concept informally, students can divide fairly. In these activities, the students will get an opportunity to use their informal knowledge and strategies to explore the concepts.

The relevancy of the HLT to the state of the art knowledge and the consistency of each component were validated by three experts in mathematics education and one expert in instructional design. After the revision processes, the HLT reached the criteria of validity [39] with the characteristics: 1) the HLT reflected the key principles and characteristics of RME, 2) the activities provided in the HLT could give a meaningful learning experience for the students to reinvent the concepts of central tendency, and 3) the components in the HLT consistently supported each other. Moreover, the results of one-to-one and small group evaluation revealed that, in general, the HLT could work as intended in the classrooms. Based on the results of the evaluations, the HLT was revised until it fulfilled the criteria of practicality [39].

In teaching experiment during the field test, it was observed that the HLT could help the students to reinvent the concept of mode by exploring the contextual problems. In this case, they discover by themselves the concept of mode as the most favorite candy flavor by listing the result of data collection in a table. They also could recognize the concept of mode when the data represented in various diagrams by referring to the frequency of the highest bar (in bar diagram), the biggest part (in pie diagram), or the highest point (in a line diagram). This finding indicates one of RME's principles namely *guided reinvention* [20] in which the contextual problems in the HLT could guide the students to reinvent the concept of mode. An example of a student's answer can be seen in Figure 3.

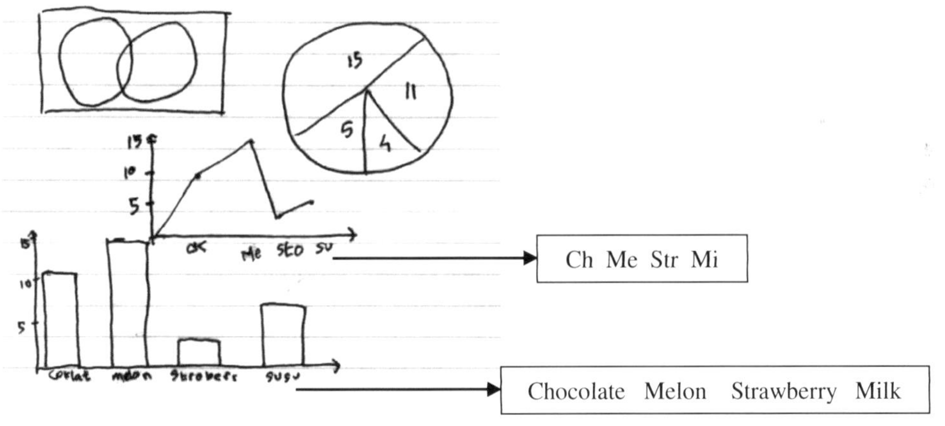

Figure 3. Student's answer when exploring the concept of mode using diagrams

To reinvent the concept of median, the students were given the next contextual problem.
A company has five branch offices that located along the main street between Padang and Painan. The first branch office is located in Padang, while the other offices are located 20 km, 28 km, 58 km, 89 km respectively from the first office. The director of the company wants to establish its headquarters at one of the branch offices so that it has the shortest distance to other branch offices. Which branch office the director must choose? Explain your answer.

In solving this problem, most students used two different strategies. In the first strategy (see Figure 4), the students tried to put each branch office in logical order, and then calculate the distance among the branch offices. By using this strategy, the students concluded that the director must choose the branch office located in the middle as its headquarters. This experience helped the students to understand the concept of median as the middle of a sorted list of data.

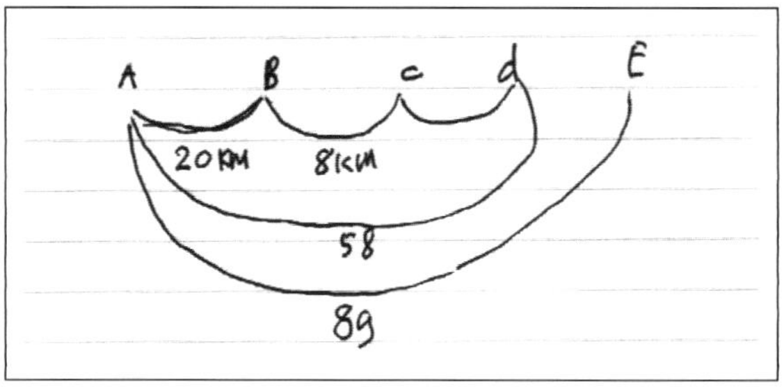

Figure 4. The students' reinvent the concept of median as the data

In the other strategy (see Figure 5), the students calculated the difference of the distance between one branch office to another. They decided to choose branch office C as the headquarters because total deviation (in distance) to other branch offices was the smallest. This strategy leads the students to discover a characteristic of median as the smallest total deviation [41].

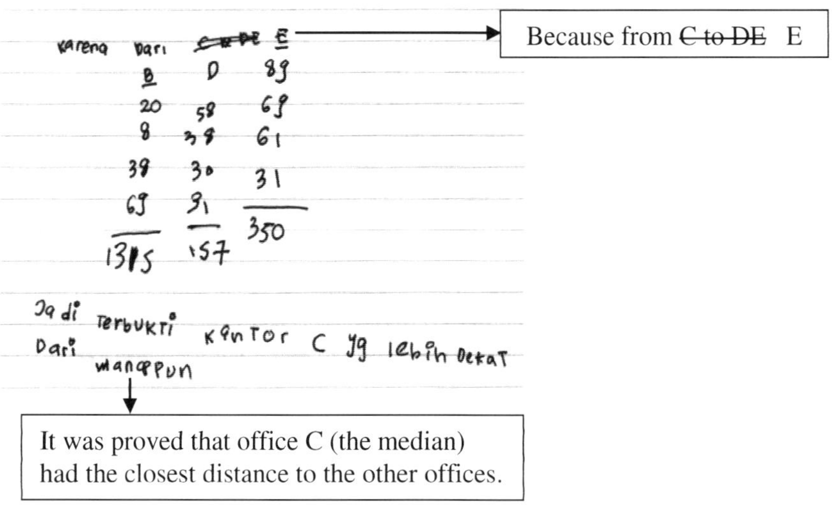

Figure 5. The students explored the median as the smallest total deviation

Based on the results of the experiments and retrospective analysis phases we can argue that the activities designed in the LT could facilitate the students to reinvent the concepts in statistics. The activities were well sequenced so that the students could learn the concepts in a meaningful way. The predictions about the students' thinking (written in the LT) mostly accurate so that the teachers could use the right probing questions (that already prepared) to guide the students to achieve the learning goals. These findings strengthen the results from previous studies that the LT was very helpful in stimulating the students to reinvent mathematical concepts [23-29]. RME approach also played an important role in building the students' understanding in this research. The contextual problems designed in the LT gave the opportunities to the students to use their informal knowledge and strategies to solve the problems. These conditions related to RME characteristics namely *students' contribution* and *students' free production* [20,29,32]. By experiencing these characteristics, the students built their confidence in learning mathematics because they can give a contribution to the learning process. Finally, during the implementation of the LT, we gradually observed the growths in students' statistical reasoning ability.

4. Conclusion

In this research, we designed a learning trajectory for teaching statistics using RME approach. The LT reached the criteria of validity, as it reflected the key principles and characteristics of RME. The activities provided in the HLT could give a meaningful learning experience for the students to reinvent the concepts of central tendency, while the components in the HLT consistently supported each other. The LT also fulfill the criteria of practicality, as it could work as intended in the classrooms. The students had more confidence to use their own strategies in solving contextual problems. The most important thing, we discovered the growth in the students' statistical reasoning.

References

[1] Bailey J 1996 *After Thought: The Computer Challenge to Human Intelligence* (New York: Basic Books)
[2] Steen L A 2001 *Mathematics and Democracy: The Case for Quantitative Literacy ed L A Steen* (Princeton, NJ: National Council on Education and the Disciplines)
[3] Watson J M 2006 *Statistical Literacy at School: Growth and Goals (Studies in Mathematical Thinking and Learning)* (New York: Routledge)
[4] NCTM 2000 *Principles and Standards for School Mathematics* (Reston, VA: NCTM)
[5] Garfield J and Ben-Zvi D 2004 *The Challenge Of Developing Statistical Literacy, Reasoning, and Thinking* eds D Ben-Zvi and J Garfield Dordrecht (The Netherlands: Kluwer Academic Publishers)
[6] Middleton J A *at al* 2008 *US Sino workshop on Mathematics and Science Education* ed L D Miller and S R Saunders (USA: Tennessee)
[7] Strauss S and Bichler E 1988 The development of children's concepts of the arithmetic average *J. Res. Math. Edu..* **19** 64
[8] Zazkis D 2013 On students' conceptions of arithmetic average: The case of inference from a fixed total *Int. J. Math. Edu. Scie. Tech.* **44** 204
[9] Shaughnessy J M and Pfannkuch M 2002 How faithful is Old Faithful? Statistical thinking: A story of variation and prediction *Mathematics Teacher* **95** 252
[10] Goldman P H 1990 Teaching Arithmetic Averaging: An Activity Approach. *The Arithmetic Teacher* **37** 38
[11] Groth R E 2006 An Exploration of Students' Statistical Thinking *Math. Educ. Res. J.* **28** 17
[12] Pfannkuch M and Wild C 2003 Statistical thinking: How can we develop it? *Proceedings of the ISI 54th* session
[13] Bremigan E G 2003 *Mathematics Teaching in the Middle School* (USA: NCTM)
[14] Fauzan A 2002 *Applying Realistic Mathematics Education (RME) in Teaching Geometry in Indonesian Primary Schools* (Enschede: PrintPartners Ipskamp)
[15] Fauzan A *et al* 2013 *Educational Design Research Part B: An Introduction* eds T Plomp and N Nieveen (Enschede: SLO)
[16] Treffers A 1991 *Three Dimensions: A Model of Goal and Theory Description in Mathematics Education* (Dordrecht: Reidel)
[17] Simon M A 1995 Reconstructing mathematics pedagogy from a constructivist perspective *J. Res. Math. Educ.* **26** 114
[18] Gravemeijer K 2015 *Proc. MADIF9 Development of Mathematics Teaching: Design, Scale, Effects* (Linköping: SMDF)
[19] Gravemeijer K and Cobb P 2013 *Educational Design Research Part A: An Introduction* ed T Plomp and N Nieveen (Enschede: SLO)
[20] Gravemeijer K 1994 *Developing Realistic Mathematics Education* (Utrecht: Freudenthal Institute)
[21] Liljekvist I *et al* 2016 *Proc. MADIF10 Development of Mathematics Teaching: Design, Scale, Effects* (Linköping: SMDF)
[22] Cobb P *et al* 2003 Design experiments in educational research *J. Ed. Researcher* **32** 9

[23] Baker A 2004 *Design Research in Statistics Education: On Symbolizing and Computer Tools* (Utrecht: Freudenthal Institute)

[24] A Fauzan *at al* 2018 A learning trajectory for teaching social arithmetic using RME approach *IOP Conf. Ser: Mater. Sci. Eng.* **33** **5**012121

[25] Prahmana R C I and Kusumah Y S 2016 The hypothetical learning trajectory on research in mathematics education using research-based learning *J Pedagogika* **123** 42

[26] Ramirez R E and Solis A H 2016 Hypothetical learning trajectories that use digital technology to tackle an optimization problem *Int. J. Techno. Math. Educ.* **24** 51

[27] Webb D C *et al* 2011 Design research in the Netherlands: Introducing logarithms using realistic mathematics education *J. Math. Edu. Technol. Coll.* **2** 47

[28] Widjaja W *et al* 2010 The role of context and teachers' questioning to enhance students' thinking *J. Scie. Math. Educ. in Southeast Asia* **33** 168

[29] Gravemeijer K 1999 How emergent models may foster the constitution of formal mathematics *J. Math. Thinking and Learning* **1** 155

[30] Gravemeijer K 2010 *A decade of PMRI in Indonesia* eds R K Sembiring, K Hoogland and M Dolk (Utrecht: Tenbrink)

[31] Streefland L 1991 *Realistic Mathematics Education in Primary Schools* (Utrecht: Freudenthal Institute)

[32] de Lange J 1987 *Mathematics, Insight, and Meaning* (Utrecht: OW & OC)

[33] Freudenthal H 1991 *Revisiting Mathematics Education* (Dordrecht: Kluwer Academic)

[34] Bahamonde A D C, Fortuny J M and Urgellés J V G 2017 Mathematical modeling and the learning trajectory: tools to support the teaching of linear algebra *International journal of mathematical education in science and technology* **48** 338

[35] Daro P *et al* 2011 *Learning Trajectories in Mathematics: a Foundation for standards, curriculum, assessment, and instruction* (Philadelphia (USA): Consortium for Policy Research in Education)

[36] Fauzan A *et al* 2017 The influence of realistic mathematics education (RME) approach on students' mathematical representation ability *Adv. Soc. Scie. Educ. Hum. Res.* **173** 9

[37] Hichendorff M 2013 The effect of presenting multi digit mathematics problems in a realistic context on sixth graders *J. Cog. Instr.* **31** 314

[38] Larson CA *et al* 2017 A hypothetical learning trajectory for conceptualizing matrices as linear transformations *Int. J. Math. Educ. Scie. Tech.* **48** 809

[39] Nieeven N and Folmer E 2013 *Educational Design Research Part A: an Introduction* ed T Plomp and N Nieveen (Enschede: SLO)

[40] Kader G and Mamer J 2007 Statistics in the middle graders: understanding center and spread *Mathematics Teaching in the Middle School,* **14** 38

[41] Abele A 1989 *Studies in mathematics education* ed R Morris (Paris: Unesco), **7** 146

The integration of science and math

Y Fitria, Y Helsa, H Nirwana and A P Zulkarnaini

Universitas Negeri Padang, Jl. Prof. Dr. Hamka, Air Tawar, Padang, Indonesia

E-mail: yantifitria@ymail.com

Abstract. The research aims to produce teaching materials for improving the quality of science learning that integrated mathematics for students in Primary Teacher Education (PTE) department of Universitas Negeri Padang. The research method used is the ADDIE model. The research stages consist of five phases. Firstly, the analysis stages conducted on the curriculum, the needs of the students and the lecturers' demands. Secondly, the product designs stage in the form of lecturing module, which is analyzed by an expert team. Thirdly, product stages consist of several chapters that integrate science and mathematics. Fourthly, the implementation stages are test stages. Its phase performed if the expert team has finished validating the used teaching materials in the review process. Lastly, the evaluation stage has been appropriately implemented such as looking back on whether all the steps. The results showed that the teaching materials developed to improve the quality of learning with the average achievement of prices of 63.66 while the average posted the value of 75.10. Then the results of the observer assessment showed that 75% of students followed the course activities well.

1. Introduction

Recently, there are so many alternatives offered to develop science, one of which is through learning development practice, i.e., integrated learning [1-4]. The essential integration of two disciplines is that of between science and mathematics [3, 4]. In line with this idea, Hurley [5] argues that science and mathematics are highly interrelated to one another and inseparable. Based on the writer's experience studying in Primary Teacher Education (PTE) department of Universitas Negeri Padang (UNP), integrated learning materials were only applied to elementary school related subject but not to the happy ones while, in fact, there are several collaborative basic concepts of natural science and mathematics. It is true that integrated learning subject has already been offered, but it still combines several disciplines for elementary school learning. During the instructional process, students involved in learning task activities without being provided with the problems to be solved in advance [6, 7]. Through the provision of these problems, students can think more creatively. As the prospective educators, students need to have the ability to integrate various disciplines of science and mathematics [8]. Through the integrative ability that they already obtain in their undergraduate learning experience, it is much easier for students to implement it and their insights will also be developed and expanded due to the familiarity with two interrelated disciplines.

In reality, many students are not able to solve the problems related to natural science subject. One of the topics integrated for the disciplines of science and mathematics is planted classification and sets as it is already given in the curriculum for the syllabus *Konsep Dasar IPA SD 1* in PTE UNP. In those learning materials, the writer will integrate science with the basic concepts of sets. The article on plant classification is highly related to basic concepts of sets. However, many elementary school teachers have not yet been able to design an assessment that develops creative thinking ability with problem-based.

Some researchers have already conducted studies related to this research [9, 10]. Science Technology Engineering Mathematics (STEM) is the currently favored interdisciplinary learning for the last few years [9]. Another research conducted by Merrit, et al. [10] reveals that PBL learning is the effective learning model to develop science and mathematics. Based on the above explanation, it is essential to conduct the research with the theme is how to developing integrated science learning problem based instructional documents on sets to improve PTE students' thinking skill.

2. Method

The research design employed is research and development. This type of study encompasses the systematic processes of planning, developing, and evaluating. This research uses ADDEI models covering analysis, design, development, implementation and evaluation stages [11]. This research performed on PTE students of UNP in the odd semester for the academic year of 2017/2018. The subjects of this research were PTE students who took *Konsep Dasar IPA SD 1* issue on the materials of plant classification and sets. The observation was conducted on the implementation of integrated mathematics and science learning problem based instructional document by students and lecturer, and it's out coming impact.

The type of data in this study is primary data, obtained from the validity result of instructional documents by the validator, practicality data from lecturers and students, and the effectiveness taken from the observation and students' evaluation [12]. The instruments of data collection in this research were practicality, validity, and the effectiveness sheet of instructional documents. The data were collected through test, process and product evaluation [13]. Besides, the questionnaire also distributed. Besides, the giving test also is done by giving a questionnaire. The type of data taken in this research is primary data from the validity result by validator and the implementation of the limited trial to the students of PTE UNP in the form of observation of the instructional documents use, and lecturer as well as students' response.

3. Results and discussions

3.1. Analysis phase

At this stage, the analysis is done on several aspects such as performance and students analysis. Performance analysis carried out by reviewing the RPS (semester lesson plan) for *Konsep Dasar IPA 2* subject. The concept of performance analysis is needed in studying material coverage, learning objectives and selection of strategies appropriate to the fundamentals of instructional documents development. It focuses on the review of the material classification of living things with the concept of sets.

Student analysis is a study of students' characteristics including age, background knowledge, cognitive development level, absorption rate and information processing as well as their talents and interests. The analysis of students' characteristics in this study focuses on the aspects of language development, intellectual, and other elements that can improve their motivation.

Analyzing the level of students' language development is necessary for considering the language of learning materials. Analyzing the intellectual development of the students is necessary for setting the level of difficulty of the problems in learning materials. Finally, learning motivation analysis is necessary for designing the presentation of the material and instructional documents appealing to students' interest and motivation [14].

3.2. Design phase

The designed RPS is that of mathematics and science integrated learning using problem-based on undergraduate students of PTE. The developed RPS shows the exposure of the learning stages on the topic of classification of living things combined with the concept of the sets. The characteristics derived from this RPS design are more detailed and easy to understand. It also helps to understand the classification of living things. The fundamental principle in RPS design is that the lesson material presented has fulfilled the scientific truth; the order of the article adjusts to the level of student development, and the preparation of the RPS adapt to the established components.

The SAP is compiled entirely and systematically regarding the developed RPS. Based on the analysis of the material, the achievement of competency indicators is implemented in 3 lectures with the allocation of time for each lecture 6 x 35 minutes. Each address is conducted in a structured way by collaboration the stages of Problem Based Learning [15].

Learning materials are designed by the curriculum and presented using the learning steps of the PBL model. Its development adapts to the needs and characteristics of students' progress [16]. These materials develop regarding the results of material analysis and formulated indicators.

3.3. Development phase

This stage is to validate the instructional documents namely RPS, SAP, and learning material validation. Validity test is conducted through confirming the learning process by experts and practitioners followed by revisions and improvement. The data collection instrument is validated by three experts that can be seen in Table 1.

Table 1. The results of instrument assessment by expert validators

No	Instrument	Total Score		Mean (%)	Category
		V1	V2		
1	RPS validation	3,86	3,71	3,79	Very Valid
2	SAP validation	3,86	3,86	3,86	Very Valid
3	Learning material validation	3,86	3,71	3,79	Very Valid
4	SAP implementation sheet	3,86	3,71	3,79	Very Valid
5	Lecturer's response sheet on material practicality	3,86	3,86	3,86	Very Valid
6	Students' response sheet on material practicality	3,86	3,57	3,72	Very Valid
7	Learning observation sheet	3,86	3,71	3,79	Very Valid
8	Students' activities observation sheet	3,86	4,00	3,93	Very Valid

There are three aspects of this phase namely validity, practicality, and effectiveness. All aspects already validate with the expert validator and revision validator. Validity test on mathematics and science integrated learning instructional documents using PBL based model consist of three parts such as RPS, SAP, and Learning material [17]. All results can be seen in Table 2.

Table 2. Validity test result

No	Validation	Results	Category
1	RPS	3.79	Very valid
2	SAP	3.82	Very valid
3	Learning Material	3.76	Very valid

The done RPS validation is on several aspects including formulation of learning objectives, presentation of learning materials, learning activities, selection of learning resources, and assessment. The overall result of RPS validation from expert validators and practitioner validator is 3.79 which belong to the category of very valid.

The done SAP validation is on several aspects including identity, objective formulation, selection of learning materials, method and detail of learning steps, selection of learning resources, and assessment. The overall result of SAP validation from expert validator and practitioner validator is 3.82 which belong to the category of very valid.

Learning material validation focuses on several aspects including content appropriation, language, presentation, and graphics. The overall result of learning material evaluation from expert validator and practitioner validator is 3.76 which belong to the category of very valid.

Practicality test of mathematics and science integrated learning instructional documents using PBL model consist of four aspects such as the observation results of SAP implementation, students' response on instructional document practicality, lecturer's response of instructional document practicality, and observation results of learning material implementation. Instructional documents considered to be valid are then tested to see the level of practicality in its use. The practicality of the developed instructional documents is found out by observing the level of SAP implementation, the response of lecturers and students to the learning materials, and the observation of students' use of the learning materials. The recapitulation of practicality result can be seen in Table 3.

Table 3. Practicality result

No	Validation	Results	Category
1	SAP implementation	3.60	Very practical
2	Students' response	3.81	Very practical
3	Lecturer's response	3.78	Very practical

The observation SAP implementation focuses on seeing if the learning process is carried out in accordance with the SAP. The results of this observation are in the category of very practical with the overall score of 3.60. Furthermore, the assessment of student responses is conducted to know students' opinions about the practicality level of the learning process. The data on students' response data are obtained from student response instrument. Lastly, the assessment of the lecturer's response is conducted to their opinions on instructional documents that have been developed. This response is obtained by using a questionnaire.

The observation focuses on the level of practicality for students in using learning materials developed. The results of this observation indicate that it is generally practical and comfortable for students to understand every concept in the learning materials. In general, students can follow every step of PBL well. In the aspect of interest and motivation to use the learning materials, students, in general, are quite interested in the display of teaching materials. The next element is students' participation and enthusiasm to perform the tasks. Overall, the students are quite active and enthusiastic in doing every work in the learning materials. Based on the above description, it can be concluded that it is easy for students to understand the concept and do various tasks in the learning materials. Thus, students do not experience significant constraints in using the substance of the classification of living things integrated with sets.

Effectiveness test on instructional documents on the topic of classification of living things integrated with sets using PBL model consist of the students' activity. Students' activity is observed by using an observation sheet instrument. The observed student activities are: observing and listening to the lecturer's explanation, paying attention to learning materials, doing the activities in learning materials, asking questions, and giving a response. The result of this observation can be seen in Table 4.

Table 4. The observation results of students' activity

No	Meeting	Observed Activity					Mean	Category
		A1	A2	A3	A4	A5		
1.	Meeting I	81,3%	87,5%	89,6%	72,9%	70,8%	80,4%	High
2.	Meeting II	89,6%	95,8%	95,8%	81,3%	75%	87,5%	Very High
3.	Meeting III	93,8%	100%	100%	87,5 %	85,4%	93.3%	Very High
		Mean					87,06%	Very High

Explanation:

A1 (Activity 1) = observing and listening to lecturer's explanation
A2 (Activity 2) = paying attention to learning materials
A3 (Activity 3) = doing the activities in learning materials
A4 (Activity 4) = asking questions
A5 (Activity 5) = giving response

3.4. Implementation stage

The implementation stage is the use of instructional documents in the broader scope. This phase in this research is conducted in another school, i.e., *SDN 12 Bukit Cangang*. The purpose is to test the effectiveness of the use of advanced products on different objects, situations, and conditions. The implementation stage was done from November 25 to December 8, 2016 at V *SDN 12 Bukit Cangang*. The effectiveness test in the implementation stage is the same as effectiveness test in product trial at *SDN 04 Bukit Apit*. Three main aspects need to be considered in this stage, i.e., students' activity.

Students' activity is observed by using observation sheet instrument. The result of this observation shows that the mean for students' activity in general 89.37% which belongs to the category of very high (Table 5).

Table 5. The observation results of students' activity

No	Meeting	Observed Activities					Score	Category
		A1	A2	A3	A4	A5		
1.	Meeting I	82,7%	86,5%	92,3%	76,9%	78,9%	83,5%	Very High
2.	Meeting II	92,3%	94,2%	96,2%	82,7%	82,7%	89,6%	Very High
3.	Meeting III	96,2%	96,2%	100%	90,4 %	92,3%	95,0%	Very High
		Mean					89,37%	Very High

Explanation:
A1 (Activity 1) = observing and listening to lecturer's explanation
A2 (Activity 2) = paying attention to learning materials
A3 (Activity 3) = doing the activities in learning materials
A4 (Activity 4) = asking questions
A5 (Activity 5) = giving response

3.5. Evaluation stages

Assessment of evaluation questions on classification of living things integrated with sets using PBL model product evaluation can be seen in Table 6. It shows that the overall mean of students' test result is 86.54 which belong to the category of very good.

Table 6. Product evaluation

No.	Students' Number	Meeting 1		Meeting 2		Meeting 3	
		Pass	Fail	Pass	Fail	Pass	Fail
1	30	26	4	28	2	26	-
	Percentage	84,6%	15,4%	91,3	6,7%	100%	0%
	Average Score	83,27		86,15		90,19	
	Overall Mean			86,54			

4. Conclusion

The effectiveness of developed instructional documents on the classification of living things integrated with sets using PBL model has already been tested. The results of the trial stage reveal the observation of student activity in a very high category (87.06%), very high grade (15%) and the average score of student test result in the good class (83.05), with the percentage of passing exceeding the minimum limit. Also, the results of implementation stage reveal the observation of student activity in a very high category (89.4%), and students' test result with the average score of (87.22) in excellent grade, with the percentage of passing, also exceeding the minimum limit.

References

[1] Birgili B 2015 Creative and critical thinking skills in problem-based learning environments *Journal of Gifted Education and Creativity* **2** 71

[2] Shahrill M, *et al* 2018 The relationships within the mathematical content of teachers' lesson sequences *J. Phys.: Conf. Ser.* **943** 012001
[3] Davison D M, *et al* 1995 What does integration of science and mathematics really mean? *School science and mathematics* **95** 226
[4] Kiray S A 2012 A new model for the integration of science and mathematics: The balance model *Energy Education Science and Technology Part B: Social and Educational Studies* **4** 1181
[5] Hurley M M 2001 Reviewing integrated science and mathematics: The search for evidence and definitions from new perspectives *School science and mathematics* **101** 259
[6] Ahmad S, *et al* 2018 The instruments of higher order thinking skills *J. Phys.: Conf. Ser.* **943** 012053
[7] Moseley C and Utley J 2006 The effect of an integrated science and mathematics content-based course on science and mathematics teaching efficacy of pre-service elementary teachers *Journal of Elementary Science Education* **18** 1
[8] De Witte K and Rogge N 2012 Problem-based learning in secondary education: Evaluation by a randomized experiment *Hub Research Papers* **121** 1
[9] El-Deghaidy H and Mansour N 2015 Science teachers' perceptions of STEM education: Possibilities and challenges *International Journal of Learning and Teaching* **1** 51
[10] Merritt J, *et al* 2017 Problem-based learning in K–8 mathematics and science education: A Literature review *Interdisciplinary Journal of Problem-Based Learning* **11** 3
[11] Branch R 2009 *Instructional design: The ADDIE approach* (New York: Springer)
[12] King K P 2012 *Technology, Science Teaching, and Literacy* (New York: Plenum Publisher)
[13] Tanujaya B, *et al* 2017 Mathematics instruction, problems, challenges, and opportunities: A case study in Manokwari regency, Indonesia *World Transactions on Engineering and Technology Education* **15** 287
[14] Thakur K R 2013 Delineation of English language teaching syllabi and its implications *The Criterion: An International Journal in English* **4** 205
[15] Chiu A, *et al* 2015 Supporting elementary and middle school system education at the whole-school level *Science Leadership Initiative* **1** 20
[16] Mumu J, *et al* 2018 Construction and reconstruction concept in mathematics instruction *J. Phys.: Conf. Ser.* **943** 012011
[17] Lonning R A and DeFranco T C 1997 Integration of science and mathematics: A theoretical model *School Science and Mathematics* **97** 212

The practicality of the quadratic function module by utilizing Autograph Software and Angry Birds game

C M Zubainur, Suhartati and Iwannitona
Syiah Kuala University, Jl. Teuku Nyak Arief Darussalam, Banda Aceh 23111, Indonesia

E-mail: cut.morina@fkip.unsyiah.ac.id

Abstract. The utilization of various learning resources makes students easier to achieve the learning objectives. The information-based and information-driven media-based learning module is a learning source that can help students understand quadratic function. This paper describes the practicality of a learning module of quadratic function that utilizes Autograph software and Angry Birds game. The data were obtained from the pilot study on the learning materials of quadratic function through a cooperative learning model, numbered heads together (NHT) with Autograph software and Angry Birds game. The pilot study was conducted to measure the practicality of the module that has been developed with the Plomp development model. The participants involved 30 students of grade 10 of MAN Model Banda Aceh, Indonesia. The research instruments used were the practicality score sheet and the module observation sheet. The result shows that the learning module of quadratic function utilizing Autograph software and Angry Birds game has fulfilled the criteria of practicality. The three experts stated that the module could be used in the classroom.

1. Introduction

Mathematics learning aims to train students to analyze and simplify the various problems encountered. Through the problems students are expected to foster logical, critical, creative, rational and dynamic thinking skills because problem-solving activities will give students the opportunity to propose ideas and strategies through cooperation [1]. The purpose of learning mathematics can be achieved maximally through the use of appropriate learning media close to students' daily life, for example, the media-based information technology [2]. Currently, students are accustomed to using various communication tools such as computers and mobile phones. Both devices are used by students to communicate with people in their environment and to obtain information promptly [3].

The proximity of students with technology can be utilized by teachers to maximize the achievement of the mathematics learning objectives. This is because mathematics uses various images and graphics ranging from the simplest to the most complex. The utilization of technology can help teachers and students create images or graphics quickly. In addition, students prefer learning in the form of illustrations instead of oral. This situation can motivate students in learning mathematics [2,4]. Furthermore, the use of technology in mathematics learning can also enhance students' creativity [5]. This can be achieved through the use of technology as a media of learning by taking into account various aspects of students. For instance, the technology used by teachers that appeals to students making them easier to understand the lessons conveyed by teachers.

Autograph is one of the mathematics software that acts as a learning media to help students do investigations so that students find new things related to the subject matter. Using the software, students can try more instances in a short time. The investigations gave the students the opportunity to discover, construct, and summarize mathematical principles, and thoroughly understand mathematical concepts [6]. The utilization of Autograph software in learning mathematics can improve students' ability [7]. Generally, the use of Autograph software as a media of learning mathematics can help teachers improve the effectiveness of learning [8].

The utilization of Autograph software can be more interesting if assisted with games that students like. Playing games could eliminate the feeling of boredom, stress, fatigue, etc. Games also can be utilized as a media to learn mathematics. The use of games as a media of learning can help high school students to learn better [9]. The learning using games makes students easier to understand concepts and create attractive learning [10].

Angry Birds is a game depicting an angry bird that tries to destroy the enemy fortress with various flying strategies. The flying bird path on Angry Birds game forms a parabola opening downward. The use of the game motivates and makes students easier to imagine motion paths of objects; so that, parabolic motion characteristics are easily understood [10]. Angry Birds game challenges students to find strategies to graph the track. It directs students to interact with projectile motion, connection, and motion apps. Also, this game plays an important role in supporting students' understanding of mathematics learning through an attractive display and the challenges of the game. Thus, Angry Birds can assist students in exploring mathematical concepts [11] and hardly gives negative impacts on students [12].

The quadratic function is a topic studied by junior high school students. In this topic, students are required to master many concepts and procedures such as multiplication, addition, substitution, and drawing function graphics well. Students often make mistakes in principle aspects, calculation, and drawing graphic [13]. The attempt to achieve the learning objectives of the quadratic function by utilizing the Autographic software and Angry Birds game can be maximized with the use of the module. Modules are part of teaching materials that are arranged systematically tailored to the needs of students who can be studied independently. The module is designed in such a way that it looks attractive and easy to understand by readers including the design, content, method, illustration, description, evaluation and reference [14,15].

The contents of the module of quadratic functions are tailored to different learning objectives and student abilities. The module needs to be adapted to the characteristics and circumstances of the students and is practically used as a learning resource in the classroom as well as at home. Students can actively learn through the module without the teachers' assistance [15,16]. The purpose of this study is to measure the practicality of quadratic function module utilizing Autographs software and Angry Birds game.

2. Method

This research is part of development research on learning materials of quadratic function by using cooperative learning model of Numbered Heads Together (NHT) with Autograph software and Angry Birds game. The learning materials were considered valid, practical and effective by the experts. The development of learning tools is carried out stages of the Plomp development model that consist of preliminary research, prototyping, and assessment [17].

The learning materials developed consist of the quadratic function module, lesson plans, student activity sheet, and tests. The learning materials applied cooperative learning model of NHT utilizing Autograph software and Angry Birds game. This paper focuses on the practicality of the module. The indicator of the practicality is that at least two of the three experts recommend that the developed module be used in the mathematics classroom, and the level of implementation of the developed materials is considered appropriate or highly appropriate [18]. The data were obtained based on the pilot study on the learning materials involving 30 students of grade 10 of MAN Model Banda Aceh.

The implementation of learning was observed by three observers. The instruments used in this study were the expert practicality score sheet and the module observation sheet.

3. Results and discussion

The quadratic function module with Autograph software and Angry Birds game was developed to be the primary learning material or supporting material that can be used and understood by teachers, general readers, and students. The module was designed to facilitate low, moderate, and high-ability students in understanding quadratic function topic.

The systematically of the module begins with introductory material, main material, model, and the connection of the topics with Angry Birds game and Autograph software. Specifically, the module consists of cover, introduction, concept maps, and sub-topics of quadratic function with Autograph software and Angry Birds game.

The cover of the module represents the title, the topic, and the description of the contents. The cover also displays images of Angry Birds to give an idea of the use of the game in the contents of the module. The Autograph media is also featured on the cover to inform that the module utilizes the software as a learning media. The cover of the module is shown in Figure 1.

Figure 1. The cover of quadratic function module with Autograph software and Angry Birds game

The module begins with an introduction that discusses the information of module development and the support received during the development process. The content of the module starts with a concept map. The concept map displays sub-topics and their sequence. It aims to provide brief information regarding sub-topics that will be discussed in the module. The concept maps listed are adapted from students' previous math module.

The sub-topics in the module include the introductory materials to determine coordinates of points in the Cartesian coordinate system, linear equation, function, and exercises. The sub-topic is continued to quadratic equation and function. The module content also describes the Angry Birds game and its relationship to quadratic functions. Then it is followed by a discussion on Autograph software, problem-solving of quadratic function using Autograph software, quadratic function solutions in Angry Birds context assisted by Autograph software, and advanced topics. The module content ends with a summary of the subtopics discussed. In detail, the module content is presented in Table 1.

Table 1. Table of content of the module.

Chapter I	:	Introduction
1.1 Determining coordinate of points in the Cartesian coordinate system 1.2 Linear equation 1.3 Function 1.4 Exercise		
Chapter II	:	Quadratic Equation
2.1 The general form of a quadratic equation 2.2 Determining the roots of quadratic equations 2.2.1 Factorization method 2.2.2 Completing the perfect squares 2.2.3 ABC formula 2.2.4 Exercise 2.3. Discrimination of quadratic equations 2.4. Determining the formula for the sum and product of the roots of quadratic equations 2.5. Solving mathematical models and interpretations of problem solutions related to quadratic equations		
Chapter III	:	Quadratic Function
3.1 Quadratic function definition 3.2 Quadratic function graph 3.3 Mathematical model related to quadratic equations 3.4 Relationship of quadratic equations and quadratic functions 3.5 Determine the equation of a quadratic function through the coordinates of points 3.5.1. The function graph that intersects the x-axis at two points 3.5.2. Graph of function tangent to the x-axis 3.5.3. Graph of function with a peak point 3.5.4. Graph of a quadratic function with three arbitrary coordinates		
Chapter IV	:	Angry Birds Game
4.1 Angry Birds game 4.2 Angry Birds in relation to quadratic functions		
Chapter V	:	Autograph Software
5.1 Autograph knowledge 5.2 Autograph advantage 5.3 Autograph disadvantage 5.4 Autograph's significance 5.5 Get to know the Autograph menu 5.6 Steps to use Autograph 5.7 Exercise		
Chapter VI	:	Solve the Quadratic Function Problems by Utilizing Autograph Software
6.1 Graphing a quadratic function by utilizing Autograph 6.2 Graphing a quadratic function with the specified intervals by utilizing Autograph 6.3 Determining and graphing a quadratic function that intersects the x-axis at two points and through a certain point		
Chapter VII	:	Quadratic Function, Autograph, and Angry Birds
7.1. Graphing quadratic functions assisted by Autograph 7.2. Problems with additional exercises with Autograph and Angry Birds game		
Chapter VIII	:	Advanced Topic
8.1. Autograph for three-dimensional (3D) objects 8.2. Autograph for other topics		
Summary		
References		

The practicality of the quadratic function module with Autograph software and Angry Birds game was measured through the data obtained from experts and practitioners and the learning implementation in the classroom. The learning was conducted simultaneously that implement the quadratic function learning materials with Autograph software and Angry Birds game using the Numbered Heads Together (NHT) co-operative learning model. The learning consists of three sessions.

Experts agreed that the module cover describes the contents and characteristics of the module well. The cover of the module also attracts students to read and understand the contents. The preface was also stated to support the existence of modules. Experts also suggested that the contents of the module can facilitate students' learning both with teacher guidance and independently. The steps displayed in the module are systematic and practical to students. The exercises also help the students develop their understanding and advanced problems can help them improve their understanding further.

Based on the experts' recommendation, the use of the Angry Birds game helps students find the context of learning. Angry Birds game is used as a context that motivates students to recognize the importance of solving math problems so that the learning becomes meaningful for the students. The experts also stated that the use of Autograph software helps the students illustrate the problems given and at the same time check the truth of their answers. In general, the experts recommended that the quadratic module can be implemented in the classroom.

Furthermore, based on the observation result, the utilization of the module helped engage students in learning. The students looked enthusiastic to do activities instructed in the module. These activities helped the students understand the material being taught. The activities involved reading and understanding the description of topics in the module. The students also worked in a group which is one of the syntaxes in the NHT learning model. The students were also motivated to solve the problems with the familiar Angry Birds context. Students could understand the problems given because they were familiar with the context and character of Angry Birds.

The observers stated that the illustrations presented in the module are obvious to students. This is shown by the activity of understanding the images independently and in a group without waiting for the teacher's explanation. The observers also stated that the illustrations in the module support students' problem-solving. The closeness between students and the illustration makes them easy to understand.

Moreover, the use of Autograph software helped students illustrate the problems on quadratic function topic. The Autograph software was also used by the students to check the answers they got. In general, the observation results show that the quadratic function module with Autograph software and Angry Birds game can be implemented in mathematics learning. The effectiveness of the module is considered very high.

This research is relevant to a study on the implementation of the Autograph-assisted linear programming module. The module of linear programming assisted by Autograph also met very high criteria. The module motivated students to interact in completing tasks [19]. The results of this present study are also relevant to other studies that conducted the implementation of the mathematical modules assisted with software. The utilization of software-assisted instructional modules helped the learning process in the classroom. Students enthusiastically performed various activities contained in the module. In addition to that, the atmosphere of fun learning and planned learning steps can be very well executed [20-22]. Another study found that the student activities in learning mathematics using Autograph were better than the student activities in the regular class without using Autograph. Thus, using Autograph are effectively used in teaching quadratic functions [23].

In general, it can be said that the utilization of Autograph-assisted module can maximize the quality of learning that teachers do in the classroom. The information technology-based media assistance improves students' math performance. The students' learning is also getting better by exploring various strategies. Students are more enthusiastic, independent and courageous in trying to find a solution to math problems [24].

The utilization of media-assisted module based on information technology on the implementation of learning helps to increase students' understanding of math concepts. Meanwhile, the use of the Angry Birds game assists students to achieve the learning objectives. This happens because students are motivated in imagining the motion picture of the object motion in the Angry Birds game [10]. Angry Birds is one of the games addictive to players without aware of the involvement of the mathematical elements in it [12].

4. Conclusion

The module of the quadratic function by utilizing Autographic software and Angry Birds game has fulfilled the criteria of practicality. That is, three experts stated that the module could be used in the mathematics classroom. Furthermore, based on the observations on the implementation of learning, it is found that the implementation of the module in the classroom was considered in the very high category.

References

[1] Kemendikbud 2016 *Regulation of the Minister of Education and Culture No. 21 of 2016 on Content Standards* (Jakarta: Kementerian Pendidikan dan Kebudayaan)
[2] Darmawan D *Inovasi Pendidikan: Pendekatan Praktik Teknologi Multimedia dan Pembelajaran Online* (Bandung: Remaja Rosdakarya)
[3] Eggen P and Don K 2012 *Strategi dan Model Pembelajaran: Mengajarkan Konten dan Keterampilan (Edisi Keenam)* (Jakarta Barat: Indeks)
[4] Hamalik O 2001 *Proses Belajar Mengajar* (Jakarta: Bumi Aksara)
[5] Munir 2012 *Multimedia, Konsep dan Aplikasi dalam Pendidikan* (Bandung: Alfabeta)
[6] Ahmadi A 2009 *Psikologi Sosial* (Jakarta: Rineka Cipta)
[7] Risdianto H, Karnasih I and Siregar H 2012 The Diffrence of Enhancement Mathematical Problem Solving Ability and Self-Efficiency SMA with MA Students IPS Program Through Guided Inquiry Learning Model Assisted Autograph Software In Langsa *Jurnal Paradikma* **6** 89
[8] Septriani D and Haron Z B 2012 Applications of Media Autograph Through Communication Skills In Mathematics Learning
[9] Kebritchi M, Hirumi A and Bai H 2010 The Effects of Modern Mathematics Computer Games on Mathematics Achievement and Class Motivation *Computers & Education* **55** 427
[10] Sari D R, Sudarmi M and Noviandini D 2013 Game Angry Birds dan Program Tracker sebagai Media Pembelajaran Fisika pada Tipok Gerak Parabola *Prosiding Seminar Nasional Sains dan Pendidikan Sains VIII* vol 4 (Salatiga: UKSW Salatiga)
[11] Lamb J H 2014 Angry Birds Mathematics Parabolas and Vektors *MatheMatics Teacher* **107** 334
[12] Nurfaidah R 2012 Anarkisme dalam Angry Birds *Jurnal Sosioteknologi* **11** 83
[13] Nuriyah F E Analisis Kesalahan Siswa dalam Menyelesaikan Soal Matematika Materi Pokok Persamaan dan Fungsi Kuadrat pada Siswa Kelas X MIA SMA Negeri 2 Wonosari Tahun Ajaran 2014/2015 [Thesis] Salatiga: UKSW Salatiga 2015
[14] Winkel 2009 *Psikologi Pengajaran* (Yogyakarta: Media Abadi)
[15] Anwar I 2010 *Pengembangan Bahan Ajar* (Bandung: Direktori UPI)
[16] Wijaya C, Djadjuri D and Rusyan A T 1991 *Upaya Pembaharuan dalam Pendidikan dan Pengajaran* (Bandung: Remadja Karya)
[17] Plomp T and Nieveen N 2007 An Introduction to Educational Design Research *Proceedings of The Seminar Conducted at the East China University, Shanghai (PR China).* (China: Shanghai University)
[18] Nieveen N 1999 *Prototype to reach product quality*, Ed J van den Akker, R M Branch, K Gustafson, N Nieveen and T Plomp (Dordrecht: Kluwer Academic Publisher) p 95

IOP Conf. Series: Journal of Physics: Conf. Series **1088** (2018) 012042 doi:10.1088/1742-6596/1088/1/012042

[19] Jannah M R Pengembangan Perangkat Pembelajaran Program Linear dengan Menggunakan Model Pembelajaran Problem Based Learning berbantuan software autograph [Thesis] Banda Aceh: Universitas Syiah Kuala 2016

[20] Dewi S 2016 Pengembangan Modul Matematika Diskrit Berbantuan Software Wxmaxima [Thesis] Banda Aceh: Universitas Syiah Kuala 2016

[21] Irwansyah B 2015 Pengembangan Bahan Ajar Kalkulus berbantuan Software Autograph di Program Studi Pendidikan Matematika STAIN Zawiyah Cot Kala Langsa *Logaritma* **3**

[22] Wasriono W, Syahputra E and Surya E 2015 Pengembangan Perangkat Pembelajaran berbantuan Autograph untuk Meningkatkan Pemahaman Konsep Matematik Siswa SMK melalui Model Penemuan Terbimbing *Jurnal Paradikma* **8**

[23] Karnasih I and Rahman A 2014 Integrasi Software Dinamis Autograph dalam Pembelajaran Matematika Menggunakan Pendekatan Penemuan Terbimbing *AdMathEdu: Jurnal Ilmiah Pendidikan Matematika, Ilmu Matematika dan Matematika Terapan* **4** 143

[24] Tarmizi R A, Ayub A F M, Bakar K A and Yunus A S M 2008 Learning Mathematics through Utilization of Technology: Use of Autograph Technology VS Handheld Graphing Calculator *Proceeding of The 7Th WSEAS International Conference on EDUCATION and EDUCATIONAL TECHNOLOGY (EDU'08)* (Venice: WSEAS Press)

Development of algebra test questions based on Bloom's taxonomy

M Husna, R Johar, Hajidin, and Mailizar
Universitas Syiah Kuala, Jl Teuku Nyak Arief, Darussalam, Banda Aceh, Indonesia, 23111

E-mail: rahmah.johar@unsyiah.ac.id

Abstract. In Indonesian mathematics textbooks, most of exercise questions for year 8 Algebra are at C1 and C4 levels of taxonomy bloom. Therefore, questions of C5 and C6 level are rarely found. This study aimed at developing Algebra test items for C5 and C6 level of the taxonomy. It employed Plomp's design research model. However, it is important to highlight that this paper only reports the validation phase of the study which was the analysis of the validity of C5 and C6 Algebra test items. Responses and feedback from four validators were analyzed descriptively and reported in this paper. The findings showed that the items test were valid and standardized at the cognitive level of C5 and C6. According to the feedbacks from validators, we revised the items. Finally, it can be concluded that the items satisfy the criteria of C5 and C6 with a validity score of 4,7. As the items satisfied the valid criteria, it is suggested that these questions can be used to further test their practicalities and effectiveness.

1. Introduction

International assessment such as TIMSS and PISA in 2015 show that Indonesian students' performances were poor. It is due to Indonesian students were not familiar with high order thinking problems that require students to analyze real-world problems in a mathematics model [1]. The students' difficulty in answering TIMSS and PISA questions indicates that students' cognitive levels do not fit with the level of questions. Therefore, they found difficulties in solving the questions.

Another reason for the problem mentioned above is that, in the teaching process, Indonesian students are hardly exposed to problems that relate to a real-world situation. It happens since students are mostly assigned to solve closed-ended questions that have only one solution, which prevents students from finding other possible solutions. Furthermore, most of the test items do not fit to measure students' competency [2,3]. In the teaching and learning mathematics in secondary schools, cognitive competence plays an important role in enhancing mathematical literacy. Therefore, the quality of test items requires to be aligned with Bloom's cognitive taxonomy.

In the Indonesian context, textbooks are required to be utilized in teaching and learning process at schools. Most of the problems and exercises in mathematics textbooks are procedural-oriented questions that do not require students' creative thinking to solve them [4]. Previous studies on textbook analysis showed that Indonesian secondary mathematics textbook mostly present low cognitive level questions, therefore, it needs improvement of the question quality [5,6,7]. For instance, compared to other Asian countries such as China and Saudi Arabia, Indonesian mathematics textbooks have lower quantities of mathematical proofs [8].

It is widely believed that students find mathematics difficult to master, including the topic of algebra. The main difficulty in learning algebra is how to use symbols of mathematics. Algebra emphasizes on the aspect of analysis as well as representation of mathematical concepts and ideas. In the curriculum 2013, Algebra is one of the main competencies that need to be mastered by students. It is essential for the students to master Algebra as the subject is applicable to daily life. For instance, NCTM reveals that Algebra helps students to connect mathematical ideas that are utilized in various fields, either mathematics of other fields [9].

Concerning the condition presented above, the researcher develops the learning instrument using the development model of Plomp and Nieveen [10]. This model consists of three phases: preliminary research, prototyping phase, and assessment phase. However, this paper only focuses on the prototyping phase to determine the validity of the learning instrument.

2. Method

This study adopts Plomp and Nieveen's model of developmental research. As previously mentioned, this paper reports the prototyping phase. The priliminary stage is divided into 2 stages, namely the analysis and problem questioning phase. The analysis phase is the first step of development research. In the analysis phase, we analyzed textbooks. In this phase, it was found that there is only one problem for the C5 level and there is no question for the C6level in exercise of algebra. Therefore we need to develop algebra item test for the C5 and C6 levels.

The second phase is prototyping. Thus, in this stage, we developed C5 and C6 questions based on indicators adapted from Bloom's taxonomy. The items were validated by a panel expert consisting of one lecturer who is an expert on high order thinking tasks as well as three mathematics teachers. It is important to remind the reader that this paper only presents the validation phase of the development of the test items.

We conducted the following steps to establish validity of the questions: (1) Conducting recapitulation of feedbacks from the validators, (2) Determining average scores of validity for each criterion, (3) Counting average scores for each aspect, (4) Counting total of average scores, and (5) Interpreting total of average scores according to tthe criteria presented in Table 1.

Table 1. Criteria of the validity

Criteria	Average score
Very Valid	4 ≤ Average Score<5
Valid	3 ≤Average Score<4
Reasonably Valid	2 ≤Average Score<3
Not Valid	1 ≤Average Score<2

If the validity scores fall into the criteria of very valid, it can conclude that the algebra test items are qualified to be utilized in the next phase which is the pilot study in the classroom. On the other hand, if the scores fall into the criteria reasonably valid and not valid, we would revise the test items.

3. Results and discussions

As mentioned earlier, this paper reports the prototyping phases of the development of algebra test items. In the phase, we developed algebra test items at C5 and C6 level of taxonomy bloom. The test items then reviewed by experts in order to determine the validity. Table 2 summarizes average scores for each criterion based on feedback from the experts.

The results showed that the test items were valid in terms of quality. Validators' reviews showed that the questions were aligned with C5 and C6 bloom's taxonomy in terms of content. The items were also valid in terms of context, competence, and indicators such as: identifying mathematical statements that were relevant to mathematics problems; developing supporting and contra arguments; using standardized language.

Table 2. Results of validity

Aspect	Criteria	Validator				Average of each criterion	Average of each aspect
		V1	V2	V3	V4		
Content	Test items align with the criteria	5	5	5	4	4.8	4.6
	Test items align with C5 and C6 bloom's cognitive level	5	4	5	4	4.5	
Format	The numbering system is clear	5	5	5	5	5	4.9
	Layout	4	5	5	5	4.8	
	Font	5	5	5	5	5	
Language	Language	4	5	4	5	4.5	4.7
	Structure of sentence	4	5	4	5	4.5	
	Simple guideline and direction	5	5	5	5	5	
	Communicative language	4	5	5	5	4.8	
	Total of average score						4.7

The next step was to revise the questions based on validators' review. We deleted unnecessary questions as well as added some important parts of the test items. Therefore, Table 3 shows the corrections of the test items.

Table 3. Validators' comments and suggestions

No.	Validator	Comment and suggestion
1.	V1	Revise item 1 and 4
2.	V2	Revise item 1 since it is not aligned with indicators of C6
3.	V3	The item is appropriate
4.	V4	The question is in accordance with the cognitive level

It shows that the validator required revising item 1 and 4. Questions' guideline and indicator were not appropriate to measure students' cognitive level, including the ambiguity of language. Therefore, based on the review from the validators, we revised the items. Table 4 shows the corrections for item 1 and 4.

In the first version of the test items, we indicated that question number 1 as a C6 item. However, after the review and correction process, based on the indicator, this question falls into the criteria of the C5 cognitive level.

Before the validation process, as presented in Table 5, we classified item number 2 into the category of C6 cognitive level. However, the validator suggested that item is suitable for C5 cognitive level. This classification is based on language and instruction of the question.

Table 4. Item 1

No.	First edition		Correction		Reasons
		Cognitive level		Cognitive level	
1.	Item number 1 Najwa chooses three positive integer numbers a, b, c. Alifah determines the value of a $+\frac{b}{c}$, and find that it is 101. Naufal determines the value of b $+\frac{a}{c}$, and find that it is 68. Dzaky determines the value of $\frac{a+b}{c}$ and find that it k .the value of k is…..	C6	Item number 1 There are three positive integer numbers a, b, and c. Alifah uses these numbers to determine the values of a $+\frac{b}{c}$ and find that it is 101. Furthermore, Naufal uses the number to determine the value ofb $+\frac{a}{c}$ and find that it is 68. In addition, Dzaky determines the value of $\frac{a+b}{c}$ dan find that it is k. What is the value of k?	C5	This item requires students to analyze their strategies and solutions as well as students' decision making in providing the best solution.

Table 5. Item 2

No.	First Edition		Correction		Reasons
		Cognitive Level		Cognitive Level	
2.	Item number 2 702, 787 and 855 when divided by m result a positive residual r, whereas 421, 722 and 815 divided by n leave a residual s, with $r \neq s$. Please determine $m+n+r+s$.	C6	Item number 2 702, 787 and 855 when divided by m result a positive residual r, whereas 421, 722 and 815 divided by n leave a residual s, with $r \neq s$. Please determine $m+n+r+s$.	C5	This item requires students to analyze their strategies and solution as well as students' decision making in providing the best solution

Furthermore, Table 6 presents results of validators' response to item 3. It shows that the validators agree that it does not require correction of this item in terms of language and cognitive level.

Another correction was for question number 4. As presented in Table 7, the first version of this question had language problems. The validators suggested that language level have to be suitable with the intended cognitive level. In the first version, language and guideline of questions for item 4 were not clear. Therefore, the item required language improvement in to develop an item that is suitable to measure the intended cognitive level.

Table 6. Item 3

No.	First Edition		Correction		Reasons
		Cognitive Level		Cognitive Level	
3.	Item number 3 When a positive integer is divided by 7, the remainder is 2. When it is divided by 8, the remainder is 3 a. Please determine a formula that satisfies the above statement b. What is the smallest positive integer that satisfies the above statement	C6	Item number 3 When a positive integer is divided by 7, the remainder is 2. When it is divided by 8, the remainder is 3 a. Please determine a formula that satisfies the above statement. b. What is the smallest positive integer that satisfies the above statement	C6	This item provides a problem that requires students to determine a mathematical formula based on provided information and solve the problem by using the formula.

Table 7. Item 4

No.	First Edition		Correction		Reasons
		Cognitive Level		Cognitive Level	
4.	Item number 4 Determine ten numbers that are not formed of a sum of several sequential numbers. For example, 5 = 2 + 3, 6 = 1+2+4.	C6	Item number 4 There are several numbers that are formed of sum of several sequential numbers. For example, 5 = 2 + 3, 6=1+2+3. Determine ten numbers that are not form of the sum of several sequential numbers.	C6	This item provides a problem that requires students to determine a mathematical formula based on the provided information and to solve the question by using the formula.

This study was conducted based on the previous study that found that there was no question for C5 and C6 cognitive level. On the other hand, according to Sudjana [11], the best proportion is that percentage of C1 and C2 is 30%, C3 and C4 is 40% and C5 and C6 is 40 %.

This paper reports the second phase of Ploom's model of design research that is prototyping phase. Hence, we only report the validation of instruments developed in this study. Several researchers have also reported findings from the prototyping phase of their studies [12,13,14]. Moreover, there are several advantages of C5 and C6 level of test items such as through high order thinking (analysis, evaluation, and creation) student would enhance their knowledge and skills [15]. In addition, students who are familiar with high order thinking questions will have high order thinking skill that supports their academic achievement [16].

However, this study has a limitation since the developed C5 and C6 level test items for algebra topic only involved validity test. Therefore, in the future study, we suggest that the test items are utilized in the classroom to examine their effectiveness.

4. Conclusion

It is concluded that this research has produced valid Algebra test items at C5 and C6 cognitive level for teaching and learning mathematics at Year 7 with validity score was 4.7. Review form expert shows that the test items are appropriate to be piloted in the classroom. Therefore, the piloting step will be the next phase of this study.

References

[1] Edo S I, Ilma R, and Hartono Y 2013 Investigating Secondary School Students' Difficulties in Modeling Problems PISA-Model Level 5 And 6 *IndoMS Journal on Mathematics Education (IndoMS-JME)* **4** 41

[2] Fan L, Zhu Y and Miao Z 2013 Textbook Research in Mathematics Education: Development Status and Directions *Journal of ZDM* **45** 633

[3] .Pulungan D A 2014 Pengembangan Instrumen Tes Literasi Matematika Model Pisa *Journal of Educational Research and Evaluation* **3** 2

[4] .Hartatiana H 2014 Pengembangan Soal Pemecahan Masalah Berbasis Argumen untuk Siswa Kelas V di SD Negeri 79 Palembang *Jurnal Pendidikan Matematika* **8** 76

[5] Johar R, Yusniarti S, and Saminan S 2018 The Analysis of Proportional Reasoning Problem in The Indonesian Mathematics Textbook for The Junior High School *Journal on Mathematics Education* **9** 55

[6] Wijaya A, Van H M and Doorman M 2015 Opportunity to Learn Context Based Tasks Provided by Mathematics Textbooks *Journal ofEduc Stud Math* **8** 41

[7] Mailizar M and Fan L 2014 Assessing A New Indonesian Secondary Mathematics Textbook: How Does it Promote Authentic Learning? *Conference on Mathematics Textbook Research and Development* **3** 19

[8] Fan L, Mailizar M, Alafaleq M and Wang Y 2018 A Comparative Study on the Presentation of Geometric Proof in Secondary Mathematics Textbooks in China, Indonesia, and Saudi Arabia In *Research on Mathematics Text books and Teachers' Resources* 53

[9] National Council of Teacher Mathematics 2000 *Principles and Standards for School Mathematics* (USA: NCTM)

[10] Plomp T and Nieveen N 2013 *Educational Design Research* (Enschede: SLO)

[11] Sudjana N 2004 *Penilaian Hasil proses Belajar Mengajar* (Jakarta: Rosdakarya)

[12] .Dündar S, Temel H and Gündüz N 2016 Development of A Mathematical Ability Test: A Validity and Reliability Study *International Journal of Mathematical Education in Science and Technology* **47** 7

[13] .Nugroho A A, Juniati D and Siswono T Y E 2018 An Instrument Measuring Prospective Mathematics Teacher Self-Regulated Learning: Validity and reliability *Journal of Physics: Conference Series* **983** 1

[14] Adnyani L P W, Kurniawan I and Pinahayu E A R 2018 Development of Creative-Thinking Instrument in Mathematics Problem Solving Based on Logical Mathematics Intelligence *Journal of Research and Advances in Mathematics Education* **3** 1

[15] Widodo T and Kadarwati S 2013 Higher Order Thinking Berbasis Pemecahan Masalah untuk Meningkatkan Hasil Belajar Berorientasi Pembentukan Karakter Siswa *Cakrawala Pendidikan* **5** 161

[16] Conklin W 2011 *Higher-Order Thinking Skills to Develop 21st Century Learners* (Huntington: Shell Education Publishing) Inc

IOP Conf. Series: Journal of Physics: Conf. Series **1088** (2018) 012044 doi:10.1088/1742-6596/1088/1/012044

Developing LTBI for addition and multiplication rules in probability theory with realistic mathematics education

H Julie
Sanata Dharma University, Jl. Affandi Tromol Pos 29, Mrican, Daerah Istimewa Yogyakarta 55002 Indonesia

E-mail: hongkijulie@yahoo.co.id

Abstract. There were three components on a hypothetical learning trajectory (HLT) developed in this paper, namely: (1) the learning goal, (2) the learning activities, and (3) the way of students' thinking and learning. A learning trajectory based instruction (LTBI) was defined as a teaching and learning trajectory using HLT for instructional decisions. In this paper, the researcher will present a LTBI using Realistic Mathematics Education approach which helps mathematics education students following the Probabilistic Theory course to construct (1) the addition rule, and (2) the multiplication rule. The type of the research was the design research developed by Gravemeijer and Cobb. There were three phases in the research development, namely (1) the design preparation, (2) the design trial, and (3) the retrospective analysis. The researcher exposure in this paper was limited to the first stage of the design research developed by Gravemeijer and Cobb. As the products developed in this paper was limited to LTBI, so to construct a local instructional theory (LIT) for the sum and multiplication rules, the researcher needs to implement this LTBI in the classroom learning process.

1. Introduction

One of the mathematics courses that must be taken by students in Mathematics Education in the third semester is the probability course. The goal of this course is to help students to reinvent the basic probability concepts, and to apply these concepts to solve probabilistic problems. One of the importance of this subject was as it contained a provision for them to develop teaching and learning process about probability theory for senior high school students and/or vocational school students. From the lecture's experience of previous years, the most difficulties experienced by students were to perform a horizontal and vertical mathematizing process [1].

According to Sztajn et al. [2], Simon was the first researcher to use the term hypothetical learning trajectory (HLT) to represent the student learning process [2]. Simon (1995) said three components on a HLT, namely: (1) the learning goal, (2) the learning activities, and (3) how students think and learn. Simon named a trajectory as a hypothetical learning trajectory because the student's learning trajectory was unknowable [3].

The learning and the teaching were often looked at as two sides of the same phenomenon, but often the studies carried out in these two areas were not connected to one another [2]. One attempt to link the research in both fields appeared in 2012 conducted by Sztajn et al [2]. Their effort to combine study in both areas was to construct what they referred to as learning trajectory based instruction (LTBI). A LTBI was defined as a teaching and learning trajectory that HLT used for instructional decisions [2].

The students' achievement taught by primary teachers were related directly and positively with mathematical and pedagogy abilities of their teachers [4,5]. There was a significantly connected between the knowledge achieved by the students and (1) the paradigm of teachers of mathematics teaching and learning process, and (2) teachers' care to the tendency of students' mathematics skills [4,5]. The mathematical knowledge of teachers related to the teachers' care of the tendency of students' mathematical skills [4,5]. The teacher mastery of the mathematics knowledge and the pedagogy would support the teachers' paradigm on the settlement of mathematical models and learning organization [4,5]. So, one of the determinants of student success in developing the probabilistic knowledge was the teacher' ability in managing the mathematics teaching and learning process and solving mathematical problems.

There were two types of understanding that students had in the process of learning mathematics, namely: (1) the instrumental understanding and (2) the relational understanding. The instrumental understanding means knowing how to use a rule or know how to use a formula to solve a problem, without understanding how the formula is derived, and why it can be used to solve the problem. Understanding in the relational understanding means knowing about (1) the relationship between concepts in mathematics (2) how to use a rule, (3) how to use a formula to solve a problem, (4) how the formula is derived, and (5) how to use a formula why the formula can be used to solve the problem [6].

From several studies reported that in the 21st century that humans not only required a content knowledge, but they also required skills that called as 21st-century skills. The 21st skills include the critical thinking and problem solving, creativity and innovation, communication and collaboration, flexibility and adaptability, initiative and self-direction, social and cross-cultural, productivity and accountability, leadership and responsibility, and information literacy [7,8]. In this LTBI, the developing of the critical thinking and problem solving were to become the focus of the researcher.

The philosophy of RME was mathematics as a human activity. It meant learning mathematics should be able to make the students thought that there was mathematics in human activities, and mathematics was used by them in real life [9,10,11]. There were five main characteristics in the RME [9,10,11], namely: (1) phenomenological exploration, (2) bridging by vertical instruments, (3) student contributions, (4) interactivity, and (5) intertwining.

According to Gravemeijer and Cobb [12], design research can be characterized as:

- Interventionist: the research leading to the design of an intervention in the real world.
- Iterative: the research incorporates a cyclic approach to the design, evaluation, and revision.
- Process-oriented: a model of research that avoids the measurement of inputs and outputs, focus on understanding and improving interventions.
- Oriented to usability: the benefits of design were measured by looking at the practicality of the design for the user in reality.
- Oriented to the theory: design (at least partially) made by theories that already exist, and field testing of the design contribute to the development of the theory.

According to Gravemeijer and Cobb [12], there were three phases in the design research, namely: (1) preparation of trial design, (2) trial design, and (3) a retrospective analysis [1]. In this paper, researchers will only discuss the results of research in stage 1.

The research question that would be answered by the researcher in this paper, namely: how to develop LTBI that could be used to build student knowledge about addition and multiplication rule using the RME approach?

2. Method

This research was classified as the design research. In this study, the researcher developed a LTBI to be used to build student knowledge concerning the addition and multiplication rule using RME approach. The research will only discuss the results at stage 1.

The subjects of this research were 38 students (7 male and 31 female) taking the Introduction Probability Theory course in one of the class at Sanata Dharma University. The research instrument used to obtain data about the students' reflection examined using the open questionnaire administered to students once the course completed. A teacher's journal was used to get data concerning teacher's reflection.

The study consisted of three cycles. Each cycle consisted of five main elements, namely context, experience, action, reflection, and evaluation. The learning process in the first cycle discussed about the counting principle, permutation and combination, experiment, sample space and events as well as the notion of the probability of the event and the definition of the axiomatic probability, the probability properties, independent and conditional events, and the conditional probability. The learning process in the second cycle discussed about random variables. The learning process in the third cycle concerned about the binomial, the Poisson, and the normal distribution. This paper is limited only at a LTBI that will help students to construct knowledge concerning addition and multiplication rules in the first cycle.

3. Result and discussion

In this paper, the researcher will develop a LTBI that will help students to construct knowledge about addition and multiplication rules. The learning goals of this LTBI were (1) students could reinvent the addition rule, and (2) students could reinvent the multiplication rule. The teaching and learning trajectory which content the learning activities and the way of students thinking and learning in the researcher's LTBI as followed:

- Students are required to solve the first problem in a way that they understand as follows: there are four green balls, and five red balls in a bag. From inside the bag, two balls will be taken at once. Please, determine how many possible outcomes from the ball-taking experiment!

Problem 1 is intended to provide an opportunity for students to explore a phenomenon that will encourage students to construct knowledge of the rules of addition and multiplication. This activity is planned by the researcher to bring up a phenomenological exploration characteristic.

The first possibility of a student answer for the problem 1 is:

Suppose, the green balls in the bag are $\mathrm{H_1}$, $\mathrm{H_2}$, $\mathrm{H_3}, \mathrm{H_4}$, and the red balls in the bag are $\mathrm{M_1}$, $\mathrm{M_2}$, $\mathrm{M_3}$, $\mathrm{M_4}, \mathrm{M_5}$; then

- Students will make the following list to state the number of events for two green balls as follows: H_1H_2, H_1H_3, H_1H_4, H_2H_3, H_2H_4, H_3H_4. So, the number of events for two green balls is six.
- Students make the following list to state the number of events of one green ball and one red ball as follows:
  H_1M_1, H_1M_2, H_1M_3, H_1M_4, H_1M_5, H_2M_1, H_2M_2, H_2M_3, H_2M_4, H_2M_5, H_3M_1, H_3M_2, $H_3M_3, H_3M_4, H_3M_5, H_4M_1, H_4M_2, H_4M_3, H_4M_4, H_4M_5$. So, the number of events for one green ball and one red ball is 20.
- Students make the following list to state the number of events for two red balls: M_1M_2, M_1M_3, M_1M_4, M_1M_5, M_2M_3, M_2M_4, M_2M_5, M_3M_4, M_3M_5, M_4M_5. So, the number of events for two red balls is 10.

Thus, the number of events in this experiment is 6 + 20 + 10 = 36.

The second possibility of a student answer for the problem 1 is:

Suppose, the green balls in the bag are $\mathrm{H_1}$, $\mathrm{H_2}$, $\mathrm{H_3}, \mathrm{H_4}$, and the red balls in the bag are $\mathrm{M_1}$, $\mathrm{M_2}$, $\mathrm{M_3}$, $\mathrm{M_4}, \mathrm{M_5}$; then

- The number of events for two green balls can be described with the diagram as follows (look at figure 1):

IOP Conf. Series: Journal of Physics: Conf. Series **1088** (2018) 012044 doi:10.1088/1742-6596/1088/1/012044

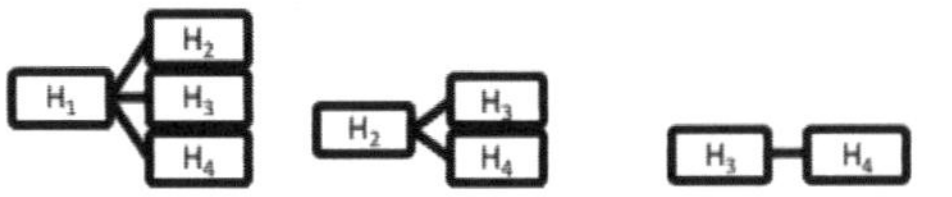

Figure 1. The possibility for two green balls.

So, the number of events for two green balls is 3 + 2 + 1 = 6

- The number of events for one green ball and one red ball can be drawn with the diagram as follows (look at figure 2):

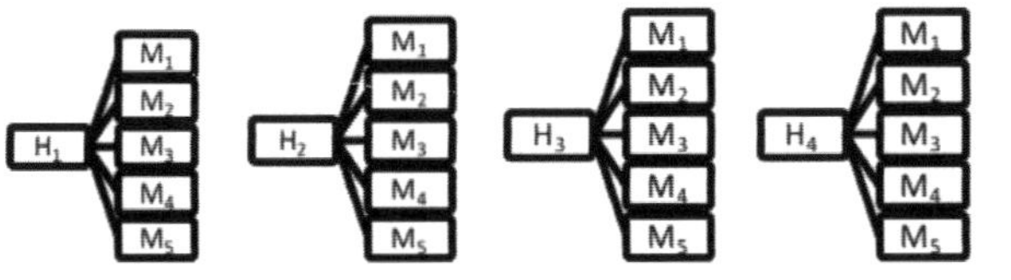

Figure 2. The possibility for one green ball and one red ball.

So, the number of events for one green ball and one red ball is$(1\times5) + (1\times5) + (1\times5) + (1\times5) = 4\times(1\times5) = 4\times5 = 20$.

- Figure 3 describes the number of events for two red balls.

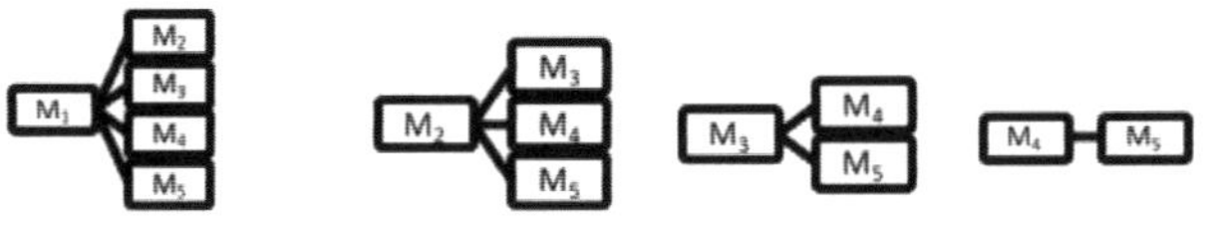

Figure 3. The possibility for two red balls.

So, the number of events for two green balls is $4 + 3 + 2 + 1 = 10$.

So, the number of possibilities in this experiment is $6 + 20 + 10 = 36$.

The third possibility of a student answer for the problem 1 is:

Suppose, the green balls in the bag areH_1, H_2, H_3, H_4, and the red balls in the bag are M_1, M_2, M_3, M_4, M_5; then

- Students will create the following chart to record the number of events for two green balls (look at table 1):

Table 1. The possibility for two green balls.

	H_1	H_2	H_3	H_4
$\boldsymbol{H_1}$	-	H_1H_2	H_1H_3	H_1H_4
$\boldsymbol{H_2}$	-	-	H_2H_3	H_2H_4
$\boldsymbol{H_3}$	-	-	-	H_3H_4
$\boldsymbol{H_4}$	-	-	-	-

So, the number of events for two green balls is 1 + 2 + 3 = 6.

- Students will create the following table to record the number of events for one green ball and one red ball (look at table 2):

Table 2. The possibility for one green ball and one red ball.

	M_1	M_2	M_3	M_4	M_5
$\boldsymbol{H_1}$	H_1M_1	H_1M_2,	H_1M_3	H_1M_4	H_1M_5
$\boldsymbol{H_2}$	H_2M_1	H_2M_2	H_2M_3	H_2M_3	H_2M_5
$\boldsymbol{H_3}$	H_2M_5	H_3M_2	H_3M_3	H_3M_4	H_3M_5
$\boldsymbol{H_4}$	H_4M_1	H_4M_2	H_4M_3	H_4M_4	H_4M_5

So, the number of events for one green ball and one red ball is $4 \times 5 = 20$.

- Students will create the following table to record the number of events for two red balls (look at table 3):

Table 3. The possibility for two red balls.

	M_1	M_2	M_3	M_4	M_5
M_1	-	M_1M_2,	M_1M_3	M_1M_4	M_1M_5
M_2	-	-	M_2M_3	M_2M_3	M_2M_5
M_3	-	-	-	M_3M_4	M_3M_5
M_4	-	-	-	-	M_4M_5
M_5	-	-	-	-	-

So, the number of events for two red balls is $4 + 3 + 2 = 1 = 10$.
So, the number of possibilities in this experiment is $6 + 20 + 10 = 36$.
The researcher provides the opportunity for students to solve the problem individually so that researchers can bring up a student contribution characteristic in the LTBI design.

- Once the student completed this problem, they are asked to discuss their solution with a friend's seat. The researcher gave the opportunity for students to present their solution to their colleague so that the researcher brings up an interactivity characteristic in the LTBI design.
- The lecturer asks the student who answers like the first, second, and third possibility for the problem 1 to write his/her answer on the board. This step is designed so that the researcher can generate a bridging by vertical instrument characteristics in the LTBI design.
- Other students are asked to observe and criticize their friends' answers. The researcher can create a student contribution characteristic in the LTBI design with this activity.
- Students are required to solve the second problem in a way that they understand individually: in SMP Maju Jaya, there is an election of OSIS management. There are four candidates. From the four people selected three people to occupy the position of chairman, secretary, and treasurer. The rule of the election is a person should not obtain multiple positions. How many forms of organizational structure can be made?

The problem 2 is intended to provide an opportunity for students to explore a phenomenon that will encourage students to construct knowledge of the rules of multiplication. This activity is planned by researchers to create a phenomenological exploration characteristic.

The first possibility of a student answer for the second problem is:
Suppose the 4 candidates are A, B, C, and D. Students write the following list as possible of OSIS stewardship arrangement: ABC, ABD, ACB, ACD, ADB, ADC, BAC, BAD, BCA, BCD, BDA, BDC, CAB, CAD, CBA, CBD, CDA, CDB, DAB, DAC, DBA, DBC, DCA, and DCB. Thus, there are 24 possible arrangements of OSIS management in Maju Jaya Junior High School.

The second possibility of a student answer for the second problem is:
Suppose the four candidates are A, B, C, and D. Students describe the following diagram to illustrate the possibility of OSIS stewardship (look at figure 4):

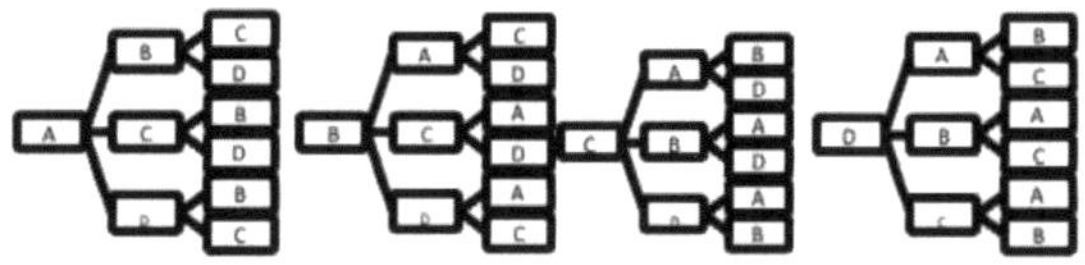

Figure 4. The possibility of OSIS stewardship.

Thus, the number of possible arrangements of OSIS management in SMP Maju Jaya is $(3\times2) + (3\times2) + (3\times2) + (3\times2) = 4\times(3\times2) = 4\times3\times2 = 24$.

The third possibility of a student answer for the second problem is:

Suppose the four candidates are A, B, C, and D. Students describe the following table to illustrate the possibility of OSIS stewardship (refertoTable 4):

Table 4. The possibility of OSIS stewardship.

Chairman	Secretary	Treasurer
Four possibility	Three possibility	Two possibility

So, the possibilities of OSIS stewardship in Maju Jaya Junior High School are $4 \times 3 \times 2 = 24$. The researcher provides the opportunity for students to solve the problem individually so that researchers can bring up a student contribution characteristic in the LTBI design.

- Once the student completed this problem, they are asked to discuss their solution with a friend's seat. The researcher gave the opportunity for students to present their solution to their colleague so that the researcher brings up an interactivity characteristic in the LTBI design.
- The lecturer asks the student who answers like the first, second, and third possibility for the problem 2 to write his/her answer on the board. This step is designed so that the researcher can generate a bridging by vertical instrument characteristics in the LTBI design.
- Other students are asked to observe and criticize their friends' answers. The researcher can create a student contribution characteristic in the LTBI design with this activity.
- Students are invited to discuss about (1) the connecting between the concepts that have been built by them when studying set theory and number theory and problem 1 and 2 and (2) concluding about the addition and multiplication rules. The expected conclusions are as follows:
    - The principle of addition: if the object A_1can be selected byn_1 ways, the object A_2can be selected byn_2 ways, until to object $A_k$can be selected by n_k ways, then the number of ways to choose objects$A_1, A_2, \ldots,$ or A_k is $n_1 + n_2 + \ldots + n_k$.
    - The Multiplication Principle: if a process can be formed from n_1 different ways followed by the next process insiden_2 different ways, and followed by the next process insiden_3 different ways, up to the k-th procedure formed from n_k different ways, then the number of ways to form the procedure is $n_1 \times n_2 \times \ldots \times n_k$.
- The lecturer gives the exercise to be done by the students to strengthen the student' understanding about the rule of addition and multiplication. Here are two practical exercises that students can use to reinforce their understanding of the sum and multiplication rules:
    - Two judges for the mathematics Olympiad were chosen from four people. Three judges for the Olympiad physics contest were selected from six people. "How many ways can you choose to choose a jury of a mathematics Olympiad or a physics Olympiad jury?" and "How many ways can you choose to choose a jury of a mathematics Olympiad and a physics Olympiad jury?"
    - The teacher will form a learning group consisting of four students from eight male students and four female students. How many study groups may be created by the teacher if in each group there are at least two male students?
- Lecturers hold class discussions to discuss student completion outcomes.

4. Conclusion

Three conclusions can be drawn from the above explanation, namely:

- To develop a LTBI to help students constructing the addition and multiplication rules using the RME approach, the researcher needs to conduct the several steps. First, providing the phenomenon to be explored by the student. Second, providing the opportunity for students to discuss with their friends. Third, asking the students to present the answers according to the formalization degree of the answers. Fourth, inviting students in class discussions to connect new material with the concepts they have learned and drawn conclusions.

- The contexts of ball-taking experiments and OSIS selection are used by researchers to help students construct their knowledge about the addition and multiplication rules.
- The RME approach can be used to assist students in constructing knowledge about the rules of addition and multiplication.

References

[1] Julie H 2017 *AIP Conference Proceedings* vol 1868 (US: AIP Publishing)
[2] Sztajn P, Confrey J, Wilson P H and Edgington C 2012 Learning trajectory based instruction: toward a theory of teaching *Educational Researcher* **41** 147
[3] Simon M A 1995 Reconstructing mathematics pedagogy from a constructivist perspective *Journal for Research in Mathematics Education* **26** 114
[4] Campbell P F, Nishio M, Smith T M, Clark L M, Conant D L, Rust A H, ... and Choi Y 2014 The Relationship Between Teachers' Mathematical Content and Pedagogical Knowledge, Teachers' Perceptions, and Student Achievement *Journal for Research in Mathematics Education* **45** 419
[5] Julie H 2017 The elementary school teachers' ability in adding and subtracting fraction, and interpreting and computing *International Journal of Science and Applied Science: Conference Series* **1** 55
[6] Skemp R 2009 *The Psychology of Learning Mathematics* (New York: Routledge)
[7] Ariyadi W 2016 Students' Information Literacy: A Perspective from Mathematical Literacy *IndoMS Journal Mathematics Education* **7** 73
[8] Stacey K 2011 The PISA View of Mathematical Literacy in Indonesia *Journal Mathematics Education* **2** 95
[9] Gravemeijer K P G 1994 *Developing Realistic Mathematics Education* (Utrecht: Freudenthal Institute)
[10] Julie H 2015 Student learning materials on the multiplication and division of fractions for grade five with realistic mathematics education *Proceedings of the 3th SEA–DR conference* (Palembang: Universitas Sriwijaya)
[11] Julie H 2016 Developing student learning materials on the multiplication fractions for grade five with realistic mathematics education *Proceedings of the 4th SEA–DR conference* (Padang: Universitas Negeri Padang)
[12] Van Den Akker J, Gravemeijer K, McKenney S and Nieveen N 2006 *Educational Design Research* (New York: Routledge)

IOP Conf. Series: Journal of Physics: Conf. Series **1088** (2018) 012045 doi:10.1088/1742-6596/1088/1/012045

Developing a physics module based on the local wisdom of Hulu Sungai Tengah regency to train the *murakata* character

S Hartini, M F Isnanda, M Wati, M Misbah, S An'nur and S Mahtari
Universitas Lambung Mangkurat, Jl. Brigjen H. Hasan Basri No.3, RW.02, Pangeran, Banjarmasin Utara, Kota Banjarmasin, Kalimantan Selatan 70124, Indonesia

E-mail: srihartini_pfis@ulm.ac.id

Abstract. This research was conducted due to the unavailability of the physics module containing the local wisdom as well as a lack of comprehension to students on the cultural diversity in South Kalimantan. Therefore, this study aimed at producing the physics module based on the local wisdom of Hulu Sungai Tengah (HST) Regency in South Kalimantan province which was feasible on the basis of validity, practicality, the effectiveness of the physics module and the achievement of murakata character. This study was a research and development using the ADDIE model. The subjects of the study were 30 students of XI MIA class at one of the high schools in Barabai, HST regency. The instruments used were a module validation sheet, a questionnaire, an achievement test and a student characteristic observation sheet. The results showed that the validity of the module was categorized as highly valid, the practicality of the module was categorized as practical, and the effectiveness of the module was categorized as high, and the attainment of murakata character was categorized as good. It could be concluded that the physics module based on the local wisdom of HST regency to train murakata character is feasible to use in the teaching and learning.

1. Introduction

Education is one of the primary needs of human life. Education is a systematic process to improve human dignity as a whole [1]. Educational institutions are not only a place to transfer knowledge, but also a place to shape youths' attitude, character, habit, and leadership [2]. One of the subjects taught in educational institutions in Indonesia is Physics. Physics is an essential subject that must exist in the curriculum school [3] because physics contribute to the development of science and technology in society. Physic learning will be more meaningful if learners can develop experience to understand better the real world through physics learning [4]. Significant physics learning can be obtained when the physics learning process is associated with the daily life of learners, especially the local wisdom of a region.

The local wisdom is local ideas that are wise, full of wisdom, good value, embedded and followed by members of the community [5]. Therefore, the implementation of local wisdom values of a nation in learning becomes very important so that it can form a personality following the noble values of the nation's culture. Local forms of wisdom in society can be cultural (norms, values, ethics, beliefs, customary laws, customs and special rules) [6]. The integration of local culture is a crucial thing to fill the cultural element in education, to make culture stronger and to be able to survive in the globalization era [7]. Based on the result of the interview with physics teachers in one of the high schools in Barabai, it shows that physics learning in this school has not yet linked with local wisdom. Teaching material used in this school is the teaching material from the publisher and has not been inserted with the local wisdom. Based on the results of questionnaires distributed to learners in the school, it shows that 63.6% of the learners are not familiar with local wisdom in South Kalimantan Local wisdom in the area of South Kalimantan, especially Hulu Sungai Tengah (HST), is the culture of *murakata* and the regional culinary of Barabai cake. Apam is a common culinary region of Hulu Sungai Tengah [8]. The process of learning physics, particularly on the material of temperature and heat, can load the examples of everyday life such as the process of making the Apam which is the application of the heat topic. The process of making the Apam can be integrated with Murakata values. Murakata which is an acronym of *mufakat* (agreement), *rakat* (society), and *seiya-sekata* (understanding each other) is the slogan of Banjar people in HST District. *Murakata* means consensus or deliberation in social life and honest behavior in everyday life [9]. Values of *murakata* character with the indicators like cooperation and tolerance can be instilled, trained, and developed through learning activities. The character is that of which distinguishes one individual from others [10]. Children's character can be formed through the integration of learning resources into learning [11]. The research by Oktaviana et al. [12] shows that the module integrated with local wisdom can increase the achievement of hard-work character and persistence. It shows that the learning process can be implemented by incorporating local wisdom through teaching materials.

One type of instructional material is a module. The module has its explanatory power and contains complete learning content so it can make students learn independently [13]. The module has the benefit that students can utilize the module without the teacher's presence [14]. According to Prastowo [15] module is a teaching material that is arranged systematically by using language that is easy to understand by learners with the level of knowledge and age learners so that they can learn independently with minimal help from an educator. Kurniasih and Berlin [16] stated that the module is a set of teaching materials that are systematically presented so that readers can learn independently. Based on those opinions, it can be concluded that the module is one form of instructional materials arranged in such a way that allows students to learn independently.

The research finding Hartini et al. [17] shows that physics-based modules of local wisdom can trace the character with excellent achievement. In line with this, the result of Cahyaningtyas et al. research [18] shows that the science teaching material integrated with the local wisdom is able to improve critical thinking skills significantly. Based on the described problems above, the researchers aim to conduct research and development of physics module on temperature and heat integrated with the local wisdom of HST to trace the character of *murakata*.

2. Method

This was a research and development using the ADDIE model [19]. The phases of the ADDIE model are Analyze, Design, Develop, Implement, and Evaluate. These phases are sequential; each depended upon the successful completion of the preceding phase. Moreover, the ADDIE model is an iterative feedback model, which means the results of the evaluation phase are returned to the feedback, closing the loop and facilitating further refinement of the learning product [19]. The steps of ADDIE development can be seen in Table 1.

Table 1. Steps for developing ADDIE model.

Step	Sample Task	Sample Output
Analysis	Needs analysis, task analysis, and instructional analysis	Learner profile and problem statement, description of instructional content and establish what must be learned.
Design	Identification of objectives, Plan instruction and Identify resources	Measurable objectives, instructional strategy, and specification.
Development	Developing learning materials, expert validation and small group testing	Module prototype and validator comments
Implementation	Tryout in real condition	Student comments and data
Evaluation	Formative and summative evaluation	Revised module prototype and the package report

The product developed was the module. The module was validated by two validators using module validation sheet. The subjects of this study were 30 students of XI MIA class at one of the high schools in Barabai in the academic year 2017/2018. The object of this research was the appropriateness of the physics module based on the local wisdom of HST regency to trace the character of *murakata*. Data collection instruments used in this study can be seen in Table 2.

Table 2. Data collection instruments.

Aspect	Instrument
Validity	Module Validation Sheet
Practicality	Questionnaire
Effectiveness	Test of the learning achievement
Murakata Character	Questionnaire

Module validation consisted of content and displays validation with the criteria based on [20]. Module practicality could be seen through three aspects: the ease of module use, the module benefit and the efficiency of learning time [21]. Module practicality criteria were based on [20]. The module effectiveness value can be analyzed by using the following equation average normalized gain $\langle g \rangle$ [22]. Module effectiveness criteria were based on [22]. Criteria for students' *murakata* character achievement was based on [20].

3. Result and discussion

The product developed in this research was the Physics module on Temperature and Heat Topic based on Local Wisdom of HST Regency to train the Character of *murakata*. The module was developed according to the criteria [1]. The module consisted of cover, table of contents, introduction (including module guideline, core and basic competencies, learning indicators and learning objectives), introduction to cooking Apam, a brief history of Kab. HST, mindmaps, chapter titles, keywords, material descriptions, a corner of *murakata* characters, summaries, exercise questions, and competency tests, bibliography and glossary and key answer of the formative tests. The product consisted of 4 sections: thermal expansion, heat and change shape, the principle of black and heat transfer. The Module was compiled based on local wisdom by linking the process of making an Apam. The difference between the developed module and the existing modules was the local wisdom of HST regency by adding an illustration of the process of making Apam in explaining the matter of physics and train murakata characters. One example was the use of firewood in the procedure of manufacture which there was heat emission through convection and radiation process. The word character could be derived from the philosophy of the heat as well as the process of making Apam that uphold the behavior of cooperation and tolerance.

Figure 1. The process of making Apam used as illustrations in the learning process heat material.

3.1. Module validity

Validation result of physic module based on local wisdom on making Apam cake to trace the *murakata* character can be seen in Tables 3 and 4.

Table 3. The calculation result of content validation of the module

Scoring Aspects	Mean	Category
Content Quality	3.45	Highly Valid
Organization	3.50	Highly Valid
Language	3.25	Valid
Evaluation	3.33	Valid
Validity	3.40	Highly Valid

Table 4. The calculation result of appearance validation of the module

Scoring Aspects	Average	Category
Consistency	3.33	Valid
Format	3.78	Highly valid
Attraction	3.58	Highly valid
Font and Font Size	3.89	Highly valid
Language	3.67	Highly valid
Validity	3.64	Highly valid

Validation result of the module content validated by the experts reached 3.40 average score which was in a highly valid category. The highly valid category indicated that the contents of the module could be used in the learning. Module appearance validation had an average score of 3.64 which was categorized as highly valid. The module appearance aspects such as consistency, format, attraction, font and font size, and language are good and correct.

The result of content validation and module appearance was valid based on the validity level. According to [23], aspects of the teaching material validity could be associated with two things. They were material developed was based on a strong theoretical rationale and internal consistency. Depdiknas [1] adds that the indicators of feasibility of a teaching material included the feasibility of content, language, presentation, and graphics. The validation result showed that the developed module had a strong theoretical basis and had internal consistency.

3.2. Module practicality

The result of students' response questionnaire to measure the practicality of physics modules is shown in Table 5.

Table 5. The analysis result of students' response questionnaire

Indicators	Average	Criteria
Easy to use	3.14	Practical
Benefit	3.07	Practical
Time efficiency	3.03	Practical
Average	3.08	Practical

The result of the practical analysis was obtained by measuring students' response questionnaire. It is shown in table 5 that the average score is 3.08 which is in the practical category. Wati et al. [24] state that an easy-to-use product is practical. Meanwhile, Sumbada's research [25] showed that the integration of local wisdom in the learning process made students active in the process of reconstructing their own knowledge, this would have an impact on the increase of students' cognitive abilities.

3.3. Module effectiveness

The Effectiveness of the developed modules could be identified from the learner's test result after learning. The test was done through two activities namely pre-test and post-test. The test was about ten essays and was calculated using N-gain. Learning achievement of learners can be seen in Table 6.

Table 6. Learning achievement of pre-test and post-test.

Pretest average score	Posttest average score	N-gain	Category
4.3	70.7	0.77	High

The calculations of students' cognitive learning achievement using the n-gain test to 30 learners can be seen in table 6. It is shown that the n-gain score obtained <g> 0.77 which is categorized as high. The N-gain value of 0.77 with the high category indicates that the developed module is effective for learning. Improvement of learning outcomes was tested using the significance test obtained values of significance 0.000 smaller than 0.05 which means there was an increase in student learning outcomes through developed modules. Daryanto [21] revealed that the use of the module was effective if an expected learning objective could be achieved. Akker et al. [23] also argued that one aspect that must be met by a teaching material to be effective was that the teaching material was operationally able to deliver the expected result.

3.4. Murakata character achievement

The *Murakata* assessed character aspect is the character of cooperation and tolerance. The result of character achievement analysis of learners can be seen in Table 7.

Table 7. Character achievement analysis result.

Character criteria	Average	Category
Cooperation	2.99	Good
Tolerance	3.14	Good
Average	3.06	Good

The assessed cooperation criteria consist of six aspects: learners' willingness to accept the division of group tasks, willingness to use time discussion, participation in giving opinions during the discussion, willingness to provide information to help to complete group tasks, willingness to solve problems during group discussions and willingness to accept decisions made by the group.

Aspects of learners' cooperation character were obtained an overall average of 2.99 which was in a good category. Based on the data, it could be interpreted that overall learners were getting used to familiarize with the cooperation character in learning that included aspects of the division of tasks, utilization of time, providing opinions and information, complete reports and problem-solving.

The assessed tolerance criteria consisted of four aspects: learners' willingness to get along with other group members, willingness to create a familiar atmosphere within the group, willingness to accept the decisions made by the group, and willingness to accept friends' opinion. The tolerance aspect obtained average score 3.14 which was categorized as good. This showed that learners had been able to appreciate differences and had not imposed personal opinions.

Overall *murakata* character obtained average score was 3.06 which was in the good category. Based on these data, the students had been able to accommodate tolerance in learning by showing a good attitude in maintaining cohesiveness and creating a familiar atmosphere within the group, and free to accept friends' opinions and decisions that had been agreed upon one another. These results are in line with Kamaruddin's opinion [26] that says that education can instill good habits in students. This is also in line with the idea of Agung [27] suggesting that the value of a character can be developed through local wisdom.

4. Conclusion

Based on the results of development and test, it is concluded that the physics module on temperature and heat materials based on the local wisdom of HST regency is feasible to use to trace the character of *murakata*. This is supported by the validity which was categorized as highly valid, the practicality which was categorized as practical, the effectiveness which was categorized as a highly effective module. The *murakata* character achievement is categorized as good.

References

[1] Depdiknas 2005 *Rencana Strategis Departemen Pendidikan Nasional 2005-2009* (Jakarta: Pusat Informasi dan Humas Depdiknas)
[2] Rokhman F and Yuliati A S 2014 *Procedia-Social and Behavioral Sciences* **141** 1161
[3] Eraikhuemen L and Ogumogu A E 2014 *Academic Research International* **5** 253
[4] Novak J D 2002 *Science Education* **86** 548
[5] Sartini 2004 *Jurnal Filsafat* **37** 111
[6] Asriati N 2012 *Jurnal Pendidikan Sosiologi dan Humaniora* **2** 106
[7] Saputra G A S 2013 *Proceeding of the Global Summit on Education* (Kuala Lumpur: World Conferences.net)
[8] Gusti A 2018 *SCIENTIFICS* **1** 21
[9] Najibuddin M 2014 *Jurnal Pendidikan Kewarganegaraan* **4** 545
[10] Ikhwanuddin 2012 *Jurnal Pendidikan Karakter* **2** 153
[11] Almerico G M 2014 *Research in Higher Education Journal* **26** 1
[12] Oktaviana D, Hartini S and Misbah 2017 *Jurnal Berkala Ilmiah Pendidikan Fisika* **5** 272
[13] Telaumbanua Y N and Surya B S M E 2017 *Journal of Education and Practice* **8** 73
[14] Resita I and Ertikanto C 2018 *J. Phys. Conf. Ser* **1022** 012
[15] Prastowo A 2015 *Panduan Kreatif Membuat Bahan Ajar Inovatif* (Jogjakarta: Diva Press)
[16] Kurniasih I and Berlin S 2014 *Panduan Membuat Bahan Ajar Buku Teks Pelajaran sesuai dengan Kurikulum 2013* (Surabaya: Katapena)
[17] Hartini S, Misbah, Helda and Dewantara D 2017 *AIP Proc. 4TH ICRIEMS Research and Education for Developing Scientific Attitude in Sciences and Mathematics* (New York: AIP Publishing) **1868** 070006
[18] Cahyaningtyas R N, Wilujeng I and Suryadarma I G P 2017 *Unnes Science Education Journal* **6** 1601
[19] Welty G 2007 *Journal of GXP Compliance* **11** 40
[20] Widoyoko E P 2016 *Evaluasi Program Pembelajaran Panduan Praktis Bagi Pendidik dan Calon Pendidik* (Yogyakarta: Pustaka Pelajar)
[21] Daryanto 2013*Menyusun Modul* (Yogyakarta: Gava Media)
[22] Hake R R 1998 *American Journal of Physics* **66** 1
[23] Akker J V, Bannan B, Kelly A E, Nieveen N and Plomp T 2007 *Proc. Sem. Conducted at the East China Normal University Shanghai* (Netherlands: SLO/Netherlands Institute For Curriculum Development) 37
[24] Wati M et al. 2018 *J. Phys. Conf. Ser* **997** 012
[25] Sambada D 2012 *Jurnal Penelitian Fisika dan Aplikasinya* **2** 37
[26] Kamaruddin S A 2012 *Journal Education and Learning* **6** 223
[27] Lukito A 2015 *American Journal of Social Science* **4** 51

IOP Conf. Series: Journal of Physics: Conf. Series **1088** (2018) 012046 doi:10.1088/1742-6596/1088/1/012046

The development of learning resources through Benthic species study in mangrove ecosystem Reuleung Leupung for invertebrate zoology learning

M Ali S, Supriyatno, Asiah M D, M Saputri, A Mursawal and Zulfikar
Syiah Kuala University, Jl. Teungku Chik Pante Kulu No. 5, Banda Aceh 23111, Indonesia

E-mail: ali_sarong@unsyiah.ac.id

Abstract. This study aimed at developing learning resources through the use of bottom water biota aquatic Mangrove Ecosystem Reuleung, Leupung, and to analyse the composition of biota bottom waters as a source of learning of invertebrate zoology of the students of Undergraduate Program in Biology Education . The research was conducted from July to September 2017 in mangrove ecosystem area of Reuleung Leupung, Indonesia. Two lecturers and three laboratory assistants from Syiah Kuala University and Ar-Raniry Islamic University participated in this study. Data on aquatic biota species as a learning resource was analysed by using percentage formula, while the dominant level of marine biota was analysed by using dominance formula. The results showed that the amount of aquatic biota used ranged from 3.13% to 93.75% and the composition of the use of each species ranged 0.03 to 0.94. It can be concluded that bottom biota which has the highest percentage as learning resource used in study is Moluska Phylum from Class Gastropoda, while Phylum Crustacea from class is widely used in learning is the Malacostraca Class. The composition bottom biota as a learning resource in Invertebrate Zoology learning is low to high.

1. Introduction

Zoology Courses is a course in the Department of Biology Education at FKIP Syiah Kuala University and in FTK UIN Ar-Raniry Darussalam Banda Aceh. This course studies various animals that do not have vertebrae on their bodies, including Porifera Phylum, Coelenterata, Platyhelminthes, Nemathelminthes, Annelids, Molluscs, Crustaceans, Mandibulates, Rotifers, Small Phylum, and Echinodermata Phylum [1].

Invertebrate Zoological Learning conducted in accordance with the curriculum of Biology Education Studies Program, implemented in theory and practice. The implementation activities are theoretically conducted in the lecture hall, which has been arranged according to space, time and lecturers who take care of the course. While the practice is done in the laboratory by using various preparations and animal specimens that belong to Invertebrates so that the learning process that is done through the approach of theory and approach of the lab work well and get the results in accordance with the curriculum expectations.

Specimens and animal preparations used in learning as learning media in learning activities, consist of terrestrial fauna and aquatic fauna obtained from the various habitats of each species of the animal.

One group of animals as aquatic fauna used as a learning resource for learning in the lecture hall and in the practicum room is the bottom biota of the waters.

Benthic is a waters fauna that occupies the bottom of the waters, including biota crawling, sticking or immersing in freshwater waters, brackish, and salt waters, especially at the bottom of waters and in water bodies [2]. Freshwater, brackish and saltwater (marine) waters are widely used as aquatic biota habitat in various regions of the world [3], especially in the tropics and one of them is the waters of Payau Reuleung Leupung River of Aceh Besar Aceh Province.

Phylum studied in the study of Invertebrate Zoology courses [4], generally using the example of species either in the form of drawings or biota as other learning sources derived from other place. Even examples of species used in the learning process in each class, order, family, genus, species, and various subspecies emerge only in the form of images originating from the literature or from other sources. This condition may emerge as obfuscation to students as learners to recognize the species of the animal, because the animals shown as examples are rarely found in their environment.

Invertebrate Zoology Learning Courses at Syiah Kuala University and at Ar-Raniry Darussalam State Islamic University Banda Aceh still use animal samples from various kinds of literature as learning resources. Many animals are used as a source of learning so far in each phylum based on existing literature books, with animal samples from other places and examples of animals shown differently from examples of animals that exist in the student environment. This can lead to an example of a learning resource for each phylum or concept conveyed in the lesson still far from the curriculum expectations.

The learning process in each course according to the curriculum provision is expected that the animal samples presented in the learning of each phylum, are sourced from the learner's environment. The use of learning resources derived from the environment of these learners is expected that learners can more easily recognize animals systematically presented in the learning of each concept of learning.

Mangrove ecosystem Reuleung District Leupung Aceh Besar district is one of the living areas of various animals belonging to Invertebrates and Vertebrata. Invertebrates found in this area according to Sarong et. al [4] including mangrove shells *Geloina erosa*, *Faunus ater*, *Turritella terebralia*, and *Saccrostrea Filum* sp from Moluska, *Macrobranchium* sp., *Ucha* sp., And *Scylla* sp. member of the Phylum Crustacea [5]. Besides, [6] research conducted found 12 species of 6 phyla, and research [10] which found 10 species of 7 phyla. While Vertebrate contained in the mangrove ecosystem Reuleung Leupung is *Valamungil* sp., *Lutjanus* sp., *Chanos chanos*, and *Varanus* sp.

Mangrove ecosystem area Releung Leupung Aceh Besar district has been widely used as a research area by students [7]. Reuleung Leupung mangrove area is easy to reach by students because the location can be reached by public vehicles, private vehicles and not far with the campus of Syiah Kuala University and the Campus of the Ar-Raniry State Islamic University of Darussalam, Banda Aceh.

Observing the condition of mangrove ecosystem area of Reuleung Leupung Regency of Aceh Besar with various species of Invertebrate and Vertebrata, of course, this area can be used as a source of learning various species of biota base of waters or various species of another aquatic biota. Species from the bottom biota of waters contained in the mangroves Reuleug Leupung this can be used as a source of learning to study Invertebrate Zoology in universities, especially at Syiah Kuala University and at the State Islamic University Ar-Raniry Darussalam Banda Aceh.

The objectives of the research are to develop learning resources through the study of the bottom biota of marine ecosystem Mangrove Reuleung Leupung in Invertebrate Zoology study as well as to analyze the composition of the bottom biota of water as a learning resource in Invertebrate Zoology study in Biology Education Study Program. The benefits of this study can provide information about the species of aquatic biota that can be used as a learning resource in Invertebrate Zoology learning in Higher Education.

2. Research methods

2.1. Place and time research
Research conducted in the area of Mangrove Ecosystems Reuleng Leupung River, Aceh Besar District, Aceh Province, to lecturers and lab assistants of Invertebrate Zoology in Biology Education Program of Syiah Kuala University and Biology Education Program of Ar-Raniry Islamic State University, Indonesia. Baseline biota use using *Ekman Grap* and quadrant using survey method, and analysis of species and species composition of marine biota is conducted in Biology Education Laboratory, Syiah Kuala University, in July 2017. Data collection using Biota Bottom waters used in the concept of learning is done on the students of these two courses are traced through a book of practicum activity assistance, and a search through a practicum book used by the student at each practicum.

2.2. Determination of observation station, sampling plot and source of
2.2.1 Animals sampling and data source learning concept of
Mangrove Ecosystem Reuleung Leupung of Aceh Besar Regency as a research area, divided into 3 stations: Station 1 mangrove ecosystem of the river estuary, Station 2 mangrove ecosystem in flow and station 3 mangrove area in the early area of the brackish waters of the Reuleung Leupung River of Aceh Besar District (Figure 1). Each station is set 5 sampling plots measuring 1 m x 1 m, conducted by purposive sampling that takes into account the presence of species of biota base waters in the mangrove ecosystem of Reuleung Leupung River, Aceh Besar District. The aquatic biota species was taken by damaging the sampling plots and then cleansing, after which each sampling plot was identified at each station. The species is recorded in number and the number of individuals tabulated in the observation table will be used as a learning resource in the learning process.

Each species and individual aquatic biota that has been obtained in the field as a source of learning, then tabulated according to concepts, phyla, and classes are taught in the Invertebrate Zoology Course. Each concept that uses aquatic biota is derived from a practicum guide used by students at the time of practicum and from an assistant who handles students doing the practicum. Meanwhile, the use of aquatic biota in learning in the lecture hall, obtained through a lecturer who teaches Invertebrate Zoology Courses. All of these data are recorded in the observation table, which has been prepared in advance.

2.2.2. Learning concept
The concept of learning using the bottom biota species of mangrove ecosystem waters of the Reuleung Leupung River of Aceh Besar District as a learning resource was obtained through the study of the Invertebrate Zoology practice manual of each Study Program [4]. To complete the data about the concept of learning that uses basal biota as a source of learning, we also conducted interviews with Lecturers and three laboratory assistants of Invertebrate Zoology Courses in each study program.

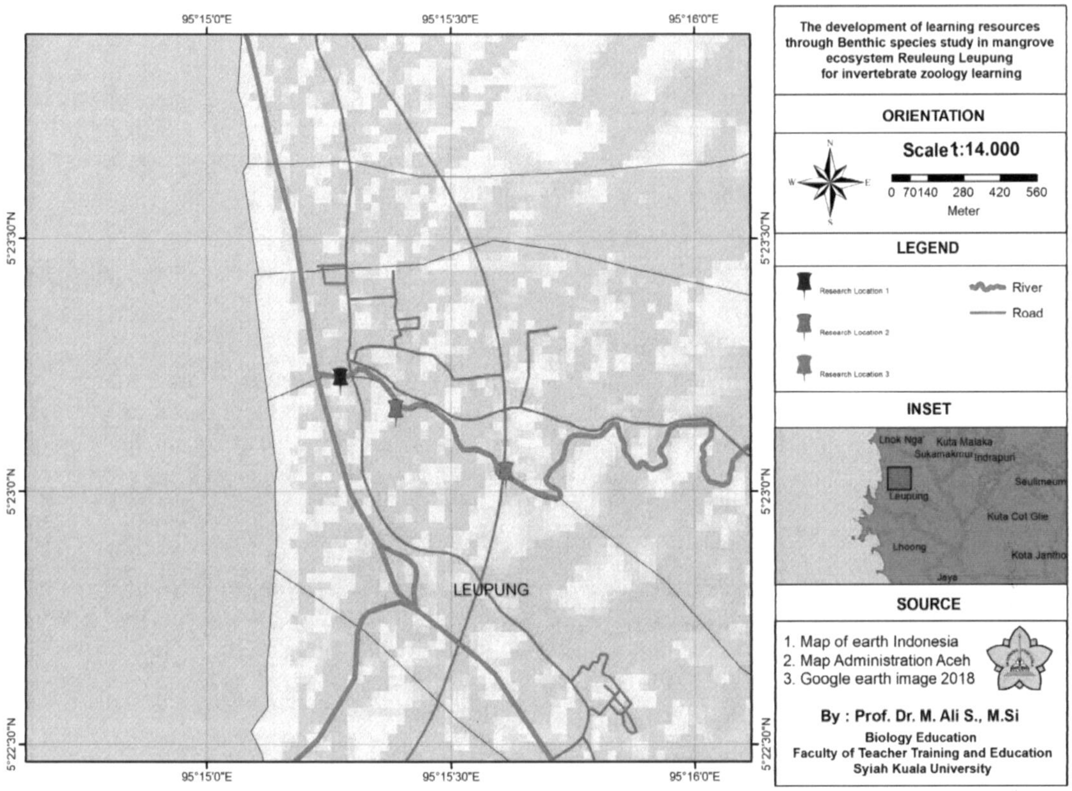

Figure 1. Mangrove ecosystem area where the data were collected

3. Data analysis

3.1. Percentage of each phylum and class as a source of learning

Percentage of each species of phyla and class which is used as a learning resource is done through a descriptive approach, after being studied through simple statistics in the form of modified percentages [5]. The formula used in this calculation is shown as follows.

$$\text{Percentage of Use} = \frac{\text{Number of species each phylum / class}}{\text{Total species of each phylum/grade}} \times 100\%$$

If a percentage of 0-20% means very low usage, 21-40% means the low use, if 41-60% means Medium usage, if 61-80% means high usage, and if the percentage of 81-100% means very high usage.

3.2. Composition

Analysis of the composition of the use of marine aquatic species in the development of learning resources conducted descriptively after the study through simple modified composition formula. The calculation formulation of the composition of the use of marine aquatic species as a source of learning with the formulation as follows.

$$\text{Composition} = \frac{\text{Number of species used}}{\text{Total species}} \times 100$$

Composition (K) use ranges from 0 -1, if K = 0,00-0,50 means low use composition, if K = 0,50-0,75 medium use composition, and if C = 0,75-1,00 high usage composition [6].

4. Results and discussion

4.1. Percentage of species each phylum and class as a learning source

Percentage of each species of phylum used as a learning resource in the Invertebrate Zoology Course ranges from 3.13% to 87.50%. Meanwhile, the percentage used by basin biota based on class as the source of learning ranging from 3.13% to 62.50%. The percentage of species of aquatic biota used as a learning resource in the Invertebrate Zoology study based on phyla and class is presented in Table 1.

Table 1. Percentage of species used learning resources learning zoology invertebrate

No	Phylum	Number of species	Percentage species phylum	Species used	Classes	Number of species	Percentage species class	Species used
1	Porifera	0	0	-	Calcarea	0	0	-
2	Coelenterata	0	0	-	Hydrozoa	0	0	-
3	Platyhelminthes	0	0	-	Trematoda	0	0	-
4	Nemathelminthes	0	0	-	Nematodes	0	0	-
5	Annelida	1	3.13	Low	Polychaeta	1	3.13	Low
					Oligochaeta	0	0	
6	Mollusk	28	87.50		Gastropod	20	62.50	Medium
					Bivalves	8	24.24	Low
					Polyplacophora	0		
7	Crustaceans	3	9.38	Low	Malacostraca	3	9.09	Low
8	Echinodermata	0	0	-	Holothuroidea	0		
9	Small Phylum	0	0	-	Acanthocephala	0	0	-
	Total	32	100	-	-	32	100	-

The results of the analysis listed in Table 1 show that there are three phyla from the bottom biota of Mangrove Ecosystem Reuleung Leupung ecosystem, which is used as a learning resource in the process of Invertebrate Zoology in Syiah Kuala University and at Ar-Raniry State Islamic University Banda Aceh. Species from Phylum Mollusca had the highest percentage of use as a learning resource of 87.50%, while the species of Phylum Annelida and Phylum Crustacea had the lowest usage rate of 3.13% and 9.38%, respectively.

The phylum molluscs learning process that uses various species as a learning resource in learning Invertebrate Zoology is necessary [11], because in addition to using the body as an animal study also uses the shell as a protective animal body. Students as learners can study more closely about the body of the molluscs concerning the body structure of each species such as the structure of the gonads and the structure of the body system it possesses, in addition to using the shells of each species to study morphology, and the anatomy of each shell [12]. Sarong et al. [13] stated that the female gonadal structure of *Geloina erosa (Bivalve)* as one of the members of the class of Phylum Mollusc is different from the female gonadal structure. This can be observed from the morphology and anatomy of each gonad studied in accordance with the objectives of learning, using the bottom biota of the water directly as a source of learning so as to obtain maximum results.

When examined for the presence of species in these three phyla, it shows that Mollusca Phylum has the largest species, used as a learning resource in the study of Invertebrate Zoology. This indicates that the species of the Mollusca Phylum are found in the waters of the Reuleung Leupung municipal ecosystem of Aceh Besar Regency. The existence of species from this phylum in the bottom waters of mangrove ecosystem Reuleung Leupung, because this mangrove ecosystem has salinity with various litter that has been able to be exploited by basin biota including Mollusca Phylum. This is in accordance with the opinion of Sarong [14], [15] which states that the basal area of mangrove ecosystem of Reuleung Leupung municipality of Aceh Besar Regency, many have mud with various leaf litter detritus of mangrove. The mangrove litter is utilized by a variety of aquatic fauna [16], especially the aquatic biota consisting of infauna and epifauna, as a source of food in life at the bottom of the waters. Nasution [17] stated that the Moluska Phylum that lives in the waters bottom of the waters of Bintan Riau, utilizing algae, plankton and detritus as the food source.

At the grade level, the species of the Gastropoda Class are widely used in the Invertebrate Zoology study process, to examine the taxonomic hierarchy ranging from the phyla, class, order, family, and at the genus level. The use of species in each of this taxonomic hierarchy, as the main specimens, assess and know more fully the existence of these aquatic biotas in each hierarchy. The class of Gastropods as the bottom biota of waters living in the waters of the mangrove ecosystem of Reuleung Leupung, has the most widely used species in the study of Invertebrate Zoology. Many species of this class that live at the bottom of the waters of this mangrove ecosystem, Konsep an existence of food components available in the bottom waters of this mangrove ecosystem. Many mangrove leaves become litter decomposed by various microbial base waters to become detritus and become a source of energy for aquatic biota including Gastropoda [18]. [13] Meanwhile, Bivalvia, Polychaeta and Malacostraca Classes are not so many species found in this region, so there is not much that can be used in Invertebrate Zoology.

4.2. The composition of species biota bottom waters as a source of learning invertebrate zoology

A composition of species Biota Bottom Waters of Mangrove Ecosystem Reuleung Leupung Aceh Besar District used as a learning resource in Invertebrate Zoology, ranging from 0.03 to 0.87. This suggests that the composition of the use of aquatic biota species as a learning resource in the Invertebrate Zoology Initiative is low to high, and can be considered in Table 2.

The concept of learning Invertebrate Zoology that has the composition of the use of Biota bottom waters Mangrove Reuleung Leupung high is the concept of animal classification with the composition 0.94 and the concept of Phylum Mollusc with a composition of 0.87. Both of these concepts comprise the use of species as a source of learning is high because the availability of species required in both these concepts is very much found in the bottom waters of mangrove ecosystem Reuleung Leupung Aceh Besar District. [5] which acquired 15 bottom biota species of waters of the Reuleung Leupung mangrove ecosystem, as aquatic biota belonging to the Gastropoda Class, Bivalves, and Malacostraca Classes. Many of the results of the research resulting from the benthos group found in the waters of the mangrove ecosystem of Reuleung Leupung of Aceh Besar Regency can be used as learning resources in Invertebrate Zoology.

Studying the concept of learning in Invertebrate Zoology more closely and producing maximum learning outcomes, can be done by using various components required in the concept [19]. The concept of classification of animals and the concept of Phylum Mollusks requires the object of direct study derived from the environment such as various species of biota base waters scattered in various bottom biota waters including the mangrove waters of Reuleung Leupung Aceh Besar [20].

Table 2. The concept of learning using species biota bottom waters as a source of learning

No	Concept	Condition species		Composition
		Used	Index composition	
1	Classification of animals	30	0.94	Height
2	Nomenclature and naming rules	20	0.62	Medium
3	Phylum Porifera	0	0	-
4	Filum Coelenterata	0	0	-
5	Phylum Platyhelminthes	0	0	-
6	Phylum Nemathelminthes	0	0	-
7	Filum Annelida	1	0.03	Low
8	Phylum Mollusca	28	0.87	Height
9	Phylum Crustacea	3	0.09	Low
10	Phylum Mandibulata	0	0	-
11	Phylum Echinodermata	0	0	-
12	Small Phylum	0	0	-

5. Conclusions

Conclusions obtained from the results of data analysis in this paper, among others, can be described as follows. The bottom water biota that has the highest percentage as a learning resource is the Moluska Phylum from Gastropoda Class, while the Phylum Crustacea from the Malacostraca Class is the least utilized in Invertebrate Zoology study used in Invertebrate Zoology study at Syiah Kuala University and UIN Ar-Raniry. Moreover, the composition of bottom water biota of mangrove ecosystems of Reuleung Leupung municipality of Aceh Besar Regency as a learning resource in low to high Zoology Invertebrate learning.

References

[1] FKIP Unsyiah 2017 *FKIP Unsyiah Curiculum* (Banda Aceh: FKIP Unsyiah)
[2] Barus T J, Arjuna A, Zuliyanti and J H Martopo 2001 *Aquatic Organisms (Benthos and Plankton)* (Medan: Unit Loser and University of North North Sumatra)
[3] Romimohtarto K and S Juwana 2007 *Marine Biology* (Jakarta: Jambatan)
[4] Sarong M A, Boer M, Dahuri R, Wardiatno Y and M Kamal 2010. Ecosystem Geloina erosa Capture Environmentally Friendly In Leupung Community of Aceh Besar District *Journal of Indonesian Molluscs* **1** 59
[5] Afkar, Djufri and Sarong M A 2015 Macrozoobenthos Association with Mangrove Flora in Mangrove Ecosytem Reuleung Leupung Aceh Besar District *Jurnal Edu Bio Tropika*
[6] Ali SM, Supriyatno, Asiah and A Mursawal 2017 *Mangrove Ecosystem of Aceh Besar District* (Banda Aceh: Natural Aceh)
[7] Ali S M 2017 *RPS Invertebrate Zoology* (Banda Aceh: FKIP Unsyiah)
[8] Ali S M 2016 *Invertebrate Zoology Practicum Guide* (Banda Aceh: FKIP Unsyiah)
[9] Rappe R A 2010 Community Structure of Different Seagrass Graves in Pulau Barrang Lompo *Journal of Tropical Marine Sciences* **2** 63
[10] Riduan 2003 *Statistic Basic* (Bandung: Alfa Beta)
[11] Sarong M A, S M I Huda and S Putra 2016 *Analysis of Ecosystem Use of Leupung Coastal Ecosystem as a Study Area of Ecology and Biology of Education* (Banda Aceh: Syiah Kuala University).
[12] Sarong M A, I Huda, S Putra, A Mursawal and Zulfikar 2017 Mangrove Ecosystem in Leupung Costal Aceh Besar District. Banda Aceh:

[13] Sarong M A, M Boer, R Dahuri, Y Wardiatno and M Kamal 2013 Morphological Structure Gonad Male Geloina erosa On Various Sizes Shell in Mangrove Ecosystem Area Reuleung Leupung River Aceh Besar District *Journal of Veterinary Medicine* **7** 105

[14] Sarong M A, M Boer, R Dahuri, Y Wardiatno and M Kamal 2010 *Management of mangrove shell of Geloina erosa (Solander 1879) based on Biological aspect in West Coast Area of Aceh Besar Regency* (Bogor: Graduate School of Bogor Agricultural University)

[15] Sarong M A, Asiah, Wardiah I, Dewiyanti and Z A Muclisin 2015 Gonad Histological Characteristic of Mud Clam (Geloina erosa) in the Estuary of Reuleung River Aceh Besar District Indonesia *Jurnal AAC Bioflux* **8** 708

[16] Yulianti P, Y Wardiatno and A M Samosir 2013 Mangrove Ecosystem Resilience to sea Level rise: A case Study of Blanakan Bay Subang Regency West Java Indonesia *Jurnal: Aquatic Science & Management* **1** 63

[17] Nasution S and M Siska 2011 The content of Heavy Metal Lead (Pb) In Sediment and Snail Strombus canarium in Bintan Island Waters *Journal of Environmental Science* **5** 82

[18] Noor Y R, M Khazali and I N N Suryadipura 2006 Guide of Introduction of Mangrove in Indonesia (Bogor: Wetlands International Indonesia Program)

[19] Agustina E, N Mahdi, and Herdanawati 2013 Development of Metamorphosis Fruit Flies (*Drosophila melanogaster*) in Natural Breeding Media as a Learning Reference in Animals Development Program *Jurnal Biotik* **1** 17

[20] Putra S, I Huda and MA Sarong 2018 Diversity, Perception of Teachers, and Spreading Pattern of Gastropoda in Mangrove Ecosystem Area Reuleung Leupung River Aceh Besar District (Banda Aceh: Syiah Kuala University)

PISA-like mathematics problems using the context of athletics in Asian Games 2018

I Pratiwi, R I I Putri and Zulkardi

Universitas Sriwijaya, Bukit Besar, Palembang, Indonesia

E-mail: ratu.ilma@yahoo.com

Abstract. The PISA result of Indonesian students on uncertainty and data content were still lacking. It is important to giving a problems with PISA characteristics in students' learning. This study aimed to produce a set of valid and practical PISA-like mathematics problems using the context of athletics in Asian Games. It used a design research method with the type of development studies consisting of two stages namely preliminary and formative evaluation. The data collection techniques used were walk through, observation of the video recording of students' discussion and interview during the learning process, and students' answer. The result of this research was valid and practical of six items PISA-like mathematics problems for the content of uncertainty and data using the context of athletics in Asian Games at level 3, 4, 5, and 6. Valid criteria can be identified based on the content, construct, and language. While practical criteria consist of easy to understand, contexts that students recognize, and can be used in learning. Also, the result of this study showed that athletics contexts used made students more active in the discussion and could to express the various reason.

1. Introduction

In the development of the 2013 curriculum, learning and assessment in Indonesia have been adapted to the material tested in international issues such as PISA (Program for International Student Assessment) as one of the government's efforts to improve the results of PISA Indonesian students In the 2013 curriculum, teachers are required to be able to design learning that begins from real problems to the abstract problem and requires students to think algorithmically and critically in solving the problems [1]. One approach consistent with the development of the 2013 curriculum is the *Pendekatan Matematika Realistik Indonesia* (PMRI) [2, 3]. Since it was developed from 2001, PMRI which emphasizes the context as the starting point of learning for the formation of concepts and motivation toward mathematics has been widely used in improving students' interest, attitude, and learning outcomes [4, 5]. The use of real problems context that students recognize and real in the students' minds can build explicit relationships between context and mathematical ideas that support student development for mathematical thinking [6,7].

One of interesting context to use is the sport at the Asian Games, especially athletics. The socialization of the Asian Games that will be held in Indonesia on 2018 is done by using the Asian Games as a learning situation as the presidential instruction [8]. Also, previous researchers who had designed a learning activities using Asian Games [9, 10], found that the Asian Games context used can help students to understand the material and make the learning process more meaningful and fun. One of the subjects that related to sports is uncertainty and data so that sports can be used as a situation in the learning of the material [11, 12].

Uncertainty and data content is one of the PISA content in the mathematical domain. In uncertainty and data content, Indonesia is only ranked 63 out of 65 countries with a score of 384, while the OECD average score for uncertainty and data is 493 in PISA 2012 [13]. This is because of Indonesian students are not accustomed to solving the problem of PISA especially high level both in learning and evaluation so that the results of PISA Indonesian students are always at the lower level of other countries [14-16]. Also, the cause of low PISA results is the limited availability of context-based questions that have characteristics of PISA problems to improve the students' literacy skills, even in books that have passed BNSP [17, 18].

Whereas the students' understanding and literacy skills on the uncertainty and data content are very important, so that students can use and solve problems related to uncertainty and data content in various fields of life especially in sports, politics, demography, business, etc. [11, 12, 19]. To that end, Indonesian teachers and students are required to improve their ability in designing mathematics problems using a context that closes to students' life and according to PISA characteristics for the process and evaluation of PISA based learning [4, 20]. This to improve the literacy skills of Indonesian students are in line with the goals of the 2013 curriculum development [1]. Therefore, the researchers are interested to develop a set of PISA-like mathematics problems on uncertainty and data content using the context of athletics in Asian Games. Based on the description before, the objective of this research is to produce a set of valid and practical PISA-like mathematics problems using the context of athletics in Asian Games.

2. Method

This study involved the tenth-grade students of Senior High School. This study used a design research method type of development studies consisting of two stages, preliminary and formative evaluation. Preliminary consisting of analysis and design, and formative evaluation consisting of self-evaluation, expert reviews, one-to-one, small group, and field test [21, 22]. However, this study only showed the result at the preliminary to the small group stage to get prototype III which were valid and practical

In the preliminary stages, the researchers analyzed the characteristics of PISA problems, the curriculum, and the students. Then, the researchers designed the initial prototype including the question grid, question cards, and assessment rubric to be used. Also, the researchers designed the lesson plans that will be used in the learning process of small groups.

In the formative evaluation, self evaluation was the first stage where the researchers evaluated the initial prototype by themselves. The result of the revision from this stage was called prototype I. Then, prototype I was evaluated by the validators at the expert reviews stage based on the content, construct, and language. Simultaneously, the one-to-one was undertaken where prototype I was tested to three students with high, medium and low ability at this stage. The result of the expert reviews stage and one-to-one stage was used to revise prototype I, and it obtained prototype II. Then, prototype II that has been valid was tested to six students with high, medium and low ability in small group stage. The video recording while students' discussion and interview during the learning process and the student's answer were analyzed to know the practicality of prototype II. The result of small group stage was used to revise prototype II, and it obtained prototype III.

Prototype III was the final prototype which were valid and practical. The data collection techniques used were walk through, observation of the video recording of students' discussion and interview during the learning process, and students' answer. It obtained the qualitative data. Then the data are analyzed by qualitative descriptive method to describe the result of each step of development.

3. Result and discussion

In this study, there were six items of PISA-like mathematics problems on uncertainty and data content using the context of athletics in Asian Games. The result of this study on each stages of the preliminary evaluation and formative evaluation are shown below.

3.1. Preliminary
In the first stage of preliminary stage, from the analysis of the characteristics of PISA problems in PISA 2015 Assessment Analytical Framework and PISA Results, it was found that PISA problems were identified in three mathematical processes (including formulating, employing, and interpreting), four mathematical contents (including change and relationship, space and shape, quantity, uncertainty and data), and four contexts (including personal, occupational, societal, and scientific). The main characteristics of PISA problems that the problems were designed using the context of real-life problems to enhance the ability of the students' mathematical literacy to solving problems in variety situations [23]. Also, from the analysis of PISA Released Items, the researchers were interested on some of PISA problems content uncertainty and data as the basic of development problems.

From the analysis of the 2013 curriculum used in Senior High School, it was found that uncertainty and data content was studied by students in class X on the material of probability and statistics. While the results of student analysis with the help of teachers in each school obtained each of the three students, who will be involved in the stage one-to-one and six students at the small group's stage.

Then, the researchers did the design phase by designing the question grid, question cards, and assessment rubric to be used. Also, the researchers also designed the lesson plans that will be used in small groups. One of the problems in this stage was Indonesian medal achievement at the Asian Games unit. It designed from M179 PISA items that using context of robberies. In this units, the researchers asked students to give some arguments of the given interpretation of the bar graph.

3.2. Formative evaluation
In this study, the formative evaluation phase will only be discussed until small group stage. At self-evaluation stage, the researchers decided to revise the whole problems. The revision at this stage was obtained prototype I. Prototype I contained three units consisting of six items on uncertainty and data content using the context of athletics in Asian Games. At the prediction level, three problems are predicted at level 3, and each one is at a level of 4.5 and 6. While in the processing ability, one problem required the ability of formulation, two problems required the ability of employ, and three problems in the ability of interpretation.

One of the problems in prototype I was marathon unit. In this unit, the researchers decided to change the context of Indonesian medal achievement at the Asian Games into the context of time record data from two Bahraini marathon athletes. Also, the researchers made it into one question from two question before. This was done so that the problem fits the athletic context.

Hereafter prototype I was evaluated in parallel at expert reviews and one-to-one to see the validity of the problem. In expert review stage, prototype I was consulted to the experts through email validation with an PISA expert from Brunei, panel validation with the lecturer of mathematics education of Sriwijaya University and colleagues whose research on the development of PISA-like mathematics problems , and face-to-face validation with a mathematics teacher.

From the results of expert reviews found that the whole question was accepted, but with some revisions according to the comments and suggestions that were given. The comments and suggestions contained in adding information of the marathon distance information, providing interesting information about the reason for choosing the Bahrain results as the data in context, adding the word of 'country' to Bahrain because Bahrain was a country that is not well known to the students, and improving the statement on the problems to be more focused for the change of graph or the drastically of the changes.

While in the one-to-one stage, the researchers asked each student to solve prototype I. From the results of student answers analysis, it found that most students only focused to assess the truth of the statement by looking at the decline alone. Though the focus of the problem was the understanding and arguments of students in declaring a change of graphs can be said drastically or not.

Based on the results of expert reviews and one-to-one, the researchers decided to change the information about Mubarak Hassan Shami's time-lapse record which could win in all of his participation in the two Asian Games events. The researchers also decided to add marathon information and change the problem statements based on the new data. The revised result of prototype I produced a valid prototype II.

After obtaining a valid prototype in content, construct, and language, prototype II was tested at the small group stage of each of the six students to see the practicality of the problem. The small group stage was done in two lessons and one final test. In the learning process, it found that students were very enthusiastic in solving problems and active during the discussion. Also, the context used had been known to students and could help students to think critically to see the difference of regular time charts that have different meanings with time charts in marathon competition. This was in line with previous research [2, 24, 25] which found that the sports contexts in Asian Games could help students to enlarge students' thinking in understanding the material and make students more motivated in a fun and meaningful learning process.

From the analysis of student, answers found that most students had been able to understand the problem even though there were some students who still focused on the reduction of time records alone without seeing the drastically of the changes. Also, there were several reasons that could be expressed by students. The Figure 1 showed one of the student's arguments. It was no drastic reduction in Mubarak Hassan Shami's achievements from 2006 to 2010.

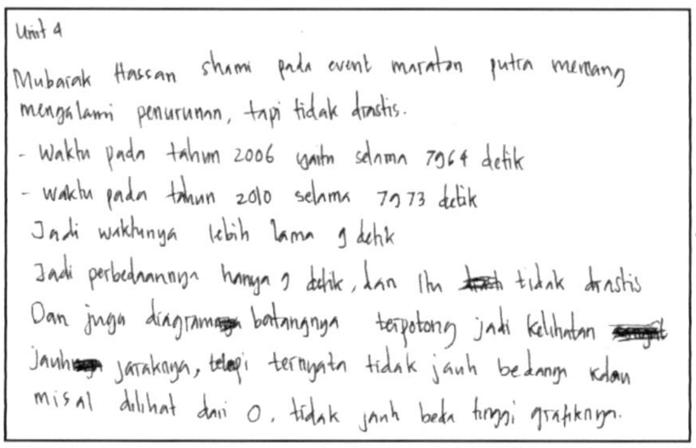
Unit 4
Mubarak Hassan shami pada event maraton putra memang
mengalami penurunan, tapi tidak drastis.
- Waktu pada tahun 2006 yaitu selama 7964 detik
- waktu pada tahun 2010 selama 7973 detik
Jadi waktunya lebih lama 9 detik
Jadi perbedaannya hanya 9 detik, dan itu tidak drastis
Dan juga diagramnya batangnya terpotong jadi kelihatan
jaraknya, tetapi ternyata tidak jauh bedanya kalau
misal dilihat dari 0, tidak jauh beda tinggi grafiknya.

→

Unit 4
Mubarak Hassan Shami on male marathon event did decrease, but it was not drastic.
- Time on 2006 was 7964 seconds
- Time in 2010 was 7973 seconds

So, the time was 9 seconds slower, and it was not drastic.
And the bar graph was truncated, so the distance was looked further. However, it did not, if it looked from 0, the difference of the bar graph was not too far.

Figure 1. One of student's argument on the problem of prototype II.

As for the analysis of time difference, the student did not give a complete reason although it was true that Mubarak Hassan Shami's record time decline was not drastic because it was only 9 seconds apart. In addition to the difference in time records, the student was also able to examine the drastically of the change from the chart correctly.

From the results of student answers analysis, it was found that students were difficult to give reasons and arguments that supported the answer clearly and completely, especially about this marathon unit was a PISA-like mathematics problem on level 6. Only one out of six students that could describe the truncated bar graph as the argument for the interpretation graph, whereas it was the main argument for this level 6 item [26]. Incomplete reasons indicated that most students were only able to solve the problem of partial credit. It was in line with previous research which stated that most students did not write an explanation of the solution they made because students were not accustomed to solving PISA type math problems that required students to communicated their answers in writing along with the reasons, especially on level 6 [27-30].

Based on the results of small groups, researchers decided to thicken the word "drastically" on prototype II revision. The revised result of prototype II was prototype III. Figure 2 showed the PISA-like mathematics problem on uncertainty and data content using the context of athletics in Asian Games that had been valid and practical.

Marathon

Mubarak Hassan Shami is a marathon athlete from Qatar. At the 42,4 km Men's marathon event in the Asian Games, Mubarak Hassan Shami won consecutive the medals in 2006 and 2010. The time records achieved by Mubarak Hassan Shami can be illustrated in the following chart:

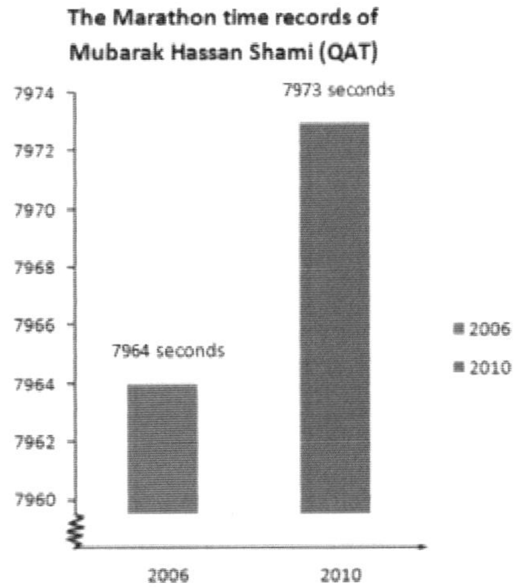

(Source: *en.wikipedia.org*)

Based on the chart above, is the following statement true?

"The time achievement of Mubarak Hassan Shami at the marathon event has decreased **drastically** from 2006 to 2010."

Give a reason to support your answer.

Figure 2. The valid and practical of PISA-like mathematics problem on uncertainty and data content using the context of athletics in Asian Games

At the marathon unit in Figure 2, it could be seen that it had the characteristics of a PISA-like problems. In mathematical process, it was included in interpreting processes ability because students should be able to reflect upon mathematical solutions, results, or conclusions and interpret it in the context of real-life problems [31, 32]. In mathematical content, it was included in uncertainty and data content because it contained a data representation problems. In context, it was included in personal context because the context of sports that used was in personal context [23]. Also, it had the characteristics of level 6 problems. It could be seen from the problems that required students to use their knowledge in relatively non standard context and be capable of advanced mathematical thinking and reasoning [33]. The context of marathon unit that contained interpretation of a truncated bar graph was non-standard context [23, 26].

4. Conclusion

The PISA mathematics problem on uncertainty and data contents using the athletic context had fulfilled the valid and practical criteria. Valid criteria can be seen in terms of content that matched the PISA mathematical literacy domain for content, constructs, and process capabilities, in terms of constructs that matched the characteristics of the PISA level and the ability of the tenth grade students, and in terms of the language corresponding to the enhanced spelling and could be understood by students. Practical criteria were seen from the context of questions that students recognized and could be used in learning. Also, the athletic context in the Asian Games used could make students to express the different reasons and be more active in discussions during the learning process. Therefore, researchers suggest teachers and other researchers to develop and use PISA-type math problems using a context that closed to students, especially a context that relates to Indonesia. It is to make the learning process to be more interesting, fun and meaningful.

Acknowledgments
The researchers would like to express gratitude to the Directorate General of Higher Education, Indonesia who has funded 'Hibah Pasca' research in 2017-2018, and to Mrs. Idda Mawaddah and her students for participating in this research. As well as those who have helped the researchers in developing the research and writing this article.

References

[1] Kemendikbud 2014 *Implementasi kurikulum 2013* (Jakarta: Kementrian Pendidikan dan Kebudayaan)
[2] Nasution M F, Putri R I I and Zulkardi 2017 Dayung context in fraction *Proceeding in the Fifth South East Asia Design/Development Research (SEA-DR) International Conference 2017* (Banjarmasin: Universitas Lambung Mangkurat) p 1
[3] Putri R I I and Zulkardi 2018 Higher-order thingking skill problem on data representation in primary school: a case study *J. Phys.: Conf. Ser.* **948** 012056
[4] Zulkardi and Putri R I I 2006 Mendesain sendiri soal kontekstual matematika *Konferensi Nasional Matematika ke-13 (KNM13)* 2006 (Semarang: Indonesia) p 1
[5] Zulkardi 2009 The "P" in PMRI: progress and problems *Proceeding of ICMA 2009 Mathematics Education* (Yogyakarta: IndoMS) p 773
[6] Putri R I I 2014 Evaluasi program pelatihan Pendidikan Matematika Realistik Indonesia (PMRI) bagi guru matematika sumatera selatan *Prosiding Seminar Nasional Evaluasi Pendidikan 2014* (Jakarta: Universitas Negeri Jakarta) p 522
[7] Widjaya W 2013 The use of contextual problems to support mathematical learning *Journal on Mathematics Education (IndoMS-JME)* **4** 151
[8] Instruksi Presiden 2016 Instruksi presiden republik Indonesia nomor 2 tahun 2016 tentang dukungan penyelenggaraan asian games XVIII tahun 2018 (Jakarta: Sekretariat Negara)
[9] Roni A, Zulkardi and Putri R I I 2017 Sprint context of asian games in the division of fractions *Proceeding in the Fifth South East Asia Design/Development Research (SEA-DR) International Conference 2017* (Banjarmasin: Universitas Lambung Mangkurat) p 22
[10] Rahayu C, Putri R I I and Zulkardi 2017 Multiplication of fraction with natural number by using hurdles *Proceeding in the Fifth South East Asia Design/Development Research (SEA-DR) International Conference 2017* (Banjarmasin: Universitas Lambung Mangkurat) p 43
[11] Hanah R, Mushetyo G and Siswor 2016 Penggunaan bahan manipulatif untuk memahamkan materi peluang pada siswa SMP dengan pendekatan pendidikan matematika Rrealistik *Jurnal Pendidikan: Teori, Penelitian, dan Pengembangan* **1** 927
[12] Yanti W, Nusantara T and Qohar A 2016 Analisis kesalahan dalam menyelesaikan soal pada materi permutasi dan kombinasi *Prosiding Seminar Nasional Pendidikan Matematika 2016* **1** p 97
[13] OECD 2014 PISA 2012 results: what students know and can do – student performance in mathematics, reading and science (Paris: OECD Publishing)
[14] Novita R, Zulkardi and Hartono Y 2012 Exploring primary student's problem-solving abillity *Journal on Mathematics Education (IndoMS-JME)* **3** 133
[15] Ahyan S, Zulkardi and Darmawijoyo 2014 Developing mathematics problems based on PISA level *IndoMS-JME* **5** 47
[16] Stacey K 2011 The PISA view of mathematical literacy in indonesia *IndoMS-JME* **2** 95
[17] Wardhani S and Rumiati 2011 *Instrumen penilaian hasil belajar matematika SMP: belajar dari PISA dan TIMSS* (Yogyakarta: Depdiknas P4TK)

[18] Fatmawati D and Ekawati R 2016 Pengembangan soal matematika pisa like pada konten change and relationship untuk siswa sekolah menengah pertama *Jurnal Ilmiah Pendidikan Matematika,* **2** 29
[19] Johar R 2012 Domain soal PISA untuk literasi matematika *Jurnal Peluang* **1** 30
[20] Kohar A W, Zulkardi and Darmawijoyo 2014 Developing PISA-like mathematics task to promote student's mathematical literacy *Proceeding in The Second South East Asia Design/Development Research (SEA-DR) Conference 2014* (Palembang: Sriwijaya University) p 14
[21] Zulkardi 2002 Developing a learning environment on realistic mathematics education for Indonesian student teachers *Ph.D Thesis University of Twente* (Enschede, the Netherlands)
[22] Tessmer M 1999 Planning and conducting formative evaluation:improving the quality of education and training (London, Philadelphia: Kogan Page)
[23] OECD 2016 *PISA 2015 Assessment and analytical framework: science, reading, mathematics, and financial literacy* (Paris: OECD)
[24] Gunawan MS, Putri R I I and Zulkardi 2017 Learning fractions through swimming context for elementary school students *Proceeding in the Fifth South East Asia Design/Development Research (SEA-DR) International Conference 2017* (Banjarmasin: Universitas Lambung Mangkurat) p 61
[25] Nasution M F, Putri R I I and Zulkardi 2018 Rowing sport in learning fractions of the fourth grade students *Journal on Mathematics Education* **9** 69
[26] OECD 2004 *Learning for tomorrow's world: first result from PISA 2003* (Paris: OECD)
[27] Edo S I, Hartono Y and Putri R I I 2013 Investigating secondary school student's difficulties in modeling problems PISA-model level 5 and 6 *IndoMS. J.M.E.* **4** 41
[28] Mardhiyanti D, Putri R I I and Kesumawati N 2013 Pengembangan soal matematika model PISA untuk mengukur kemampuan komunikasi matematis siswa sekolah dasar *Jurnal Pendidikan Matematika* **5** 16
[29] Kamaliyah, Zulkardi and Darmawijoyo 2013 Developing the sixth level of PISA-like mathematics problems for secondary school students *IndoMS JME* **4** 9
[30] Hendroanto A, *et al* 2018 Mathematics education students solving PISA's mathematics problems: An overview of students' mathematical literacy *International Journal on Emerging Mathematics Education* **2** 129
[31] Risdiyanti I and Prahmana R C I 2018 Ethnomathematics: Exploration in Javanese culture *J. Phys.: Conf. Ser.* **943** 012032
[32] Hendroanto A, *et al* 2018 Photography activities for developing students' spatial orientation and spatial visualization *J. Phys.: Conf. Ser.* **943** 012029
[33] Ahmad S, *et al* 2018 The instruments of higher order thinking skills *J. Phys.: Conf. Ser.* **943** 012053

IOP Conf. Series: Journal of Physics: Conf. Series **1088** (2018) 012048 doi:10.1088/1742-6596/1088/1/012048

Learning direct proportion by using the context of *timpan* recipes

S F Zuhra, C M Zubainur and T F Abidin
Syiah Kuala University, Jl. Teuku Nyak Arief Darussalam, Banda Aceh 23111, Indonesia

E-mail: cutmorinazubainur@unsyiah.ac.id

Abstract. Successful learning is not only based on students' final grades but also on the learning process itself. The process is crucial to assess students' understanding of mathematics concepts. This study aims to describe the learning activities of direct proportion using the realistic mathematics education (RME) approach implemented in the hypothetical learning trajectory (HLT) trial. The HLT was developed in the context of *timpan* recipes which is appropriate to the characteristics of students in Aceh. This study used design research method that involves two stages, preparation and learning trial. This HLT trial involved 20 seventh graders in a junior high school in Banda Aceh, Indonesia. Data were obtained through the observation sheet, field notes, and interview. The results showed that the students were actively involved during the HLT trial, thoughtfully analyzed the problems, positively responded to the learning, appropriately chose and used procedures or operations in solving the issues related to direct proportion. Also, the class performed effective interaction among students and the teacher.

1. Introduction

The learning trajectory was designed by teachers to help students to learn meaningfully through stages of activities involving students' previous experiences and existing knowledge [1]. All learning materials require a learning trajectory to lead students through the stages of learning from a real situation to the abstract form of mathematics. One of the topics studied by junior high school students is direct proportion. This topic is important as these concepts are applied to various daily life problems. Students' understanding of direct proportion is also required for learning other disciplines as well as being applied in the workplace.

Learning becomes meaningful if the teacher uses an appropriate learning approach to provide students with real experience and meaningful mathematics knowledge [2]. The realistic mathematics education (RME) is one of the approaches providing students the opportunity to construct concepts and formulas in mathematics [3]. Students need to be given sufficient opportunities to build their own knowledge and understanding through a set of specific activities using a context as a source of learning [4]. It is because mathematics learning focuses on not only the outcome but also the way to obtain the solution [5]. Based on the first characteristics of RME, the use of context, students carefully pay attention to the problems given by the teacher's explanation. The context used can also help the emergence of the fifth characteristic, the intertwinement with other mathematics topics or other aspects beyond mathematics, such as culture. Besides, the context used is not only in the form of real-world problems but also problems that can be imagined in the students' mind [6]. The choice of context greatly influences the students' responses to the learning process because the context is the key for students to understand the topic taught meaningfully. The teacher's chosen context needs to be closely related to the students' experience.

The context of the *timpan* recipe as a typical cake of Acehnese encountered by students in their daily life can be utilized to help students to understand mathematics especially the direct proportion topic. Although the context of food recipes has been widely used in teaching proportion, the context used is reserved for students living in Aceh. Therefore, the researchers used this *timpan* recipe context for the students in Aceh. The students will find this direct proportion concept as a ratio of two measurements and represent it in graphs and equations. This study aims to describe the learning activities of direct proportion using realistic mathematics education (RME) approach implemented in the hypothetical learning trajectory (HLT) trial. A research question "what are the students' activities occurred during the HLT trial of direct proportion using the context of the *timpan* recipe?" will be answered through this study.

2. Method

This paper describes the learning activities of direct proportion with RME approach using the context of the *timpan* recipe. The learning obtained is the result of the development of HLT trial of direct proportion appropriate to the characteristics of students in Aceh. Local instruction theory (LIT) has been developed through design research steps: preparatory steps, learning trials, and retrospective analysis [7]. At the preparation stage, a needs assessment has been made with the decision that HLT has not been formed and needs to be developed. At this stage, HLT of direct proportion has also been developed using the RME approach with the *timpan* recipe context. Then, a trial of learning with HLT was conducted at the second stage. This paper focuses on the results of the HLT trial of learning activities analyzed in the classroom.

The HLT trial of direct proportion with the context of the *timpan* recipe involved 20 Year 7 students in a junior high school in Banda Aceh, Indonesia. Data were obtained through the observation sheet, field notes, and interview. The data was then analyzed based on the observation and the field note on the student activities during the learning process. The data was analyzed and re-confirmed to the students through the interview.

3. Result and discussion

The data in this article was the results of a lesson trial. The learning objective is that the students are expected to find the direct proportion concept by using graph and equation. Students' learning activities can be observed through students' activity in observing the problems given during the learning, students' responses to the problems, students' strategies in solving the problems, and the interaction between students and the teacher.

Learning activity began with selecting flavor variants of *timpan* based on individual preference. The students worked on the given problem and chose one of the three flavors of *timpan*, namely pumpkin, jackfruit, and banana *timpan*. This activity relates to the students' experiences that have made or eaten their favorite flavor of *timpan*. Their experiences can help students solve the problems by choosing *timpan* flavor as a benchmark. Activity and student answers' of the selection of *timpan* flavor can be seen in Figure 1.

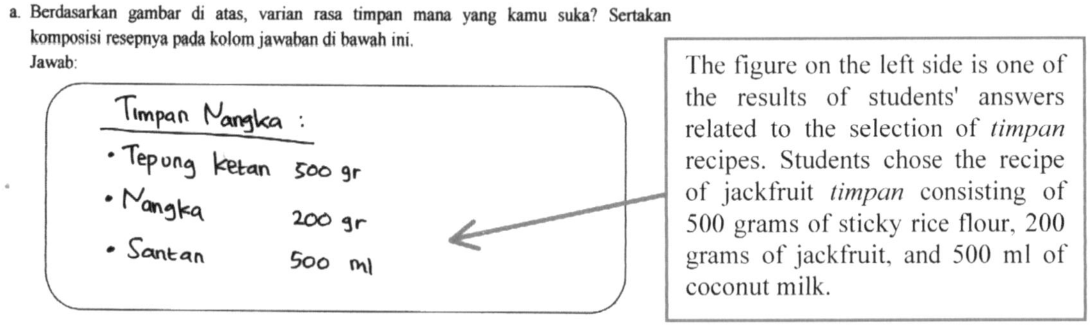

Figure 1. The results of student answers in choosing a *timpan* recipe based on their preferences.

Based on Figure 1, the selection of the recipes varied. Two groups chose the pumpkin *timpan*, two groups chose the jackfruit *timpan*, and one group chose the banana *timpan*. The group choosing the banana *timpan* questioned about the units of the banana used in the *timpan* recipe as in the following excerpt.

Teacher	*: Which timpan flavor do you choose?*
Students	*: We choose the banana one.*
Teacher	*: Good, any problem with this recipe?*
Students	*: Yes, there is. How many grams are 10 bananas?*
Teacher	*: Okay, 10 bananas are less than ½ kilograms*
Student	*: ½ kg means 500 grams, isn't it?*
Teacher	*: Yes, correct.*

Based on the excerpt above, students sought additional information to match the units of each ingredient to their chosen recipe by linking the students' prior knowledge of the units and quantities they had studied in primary school.

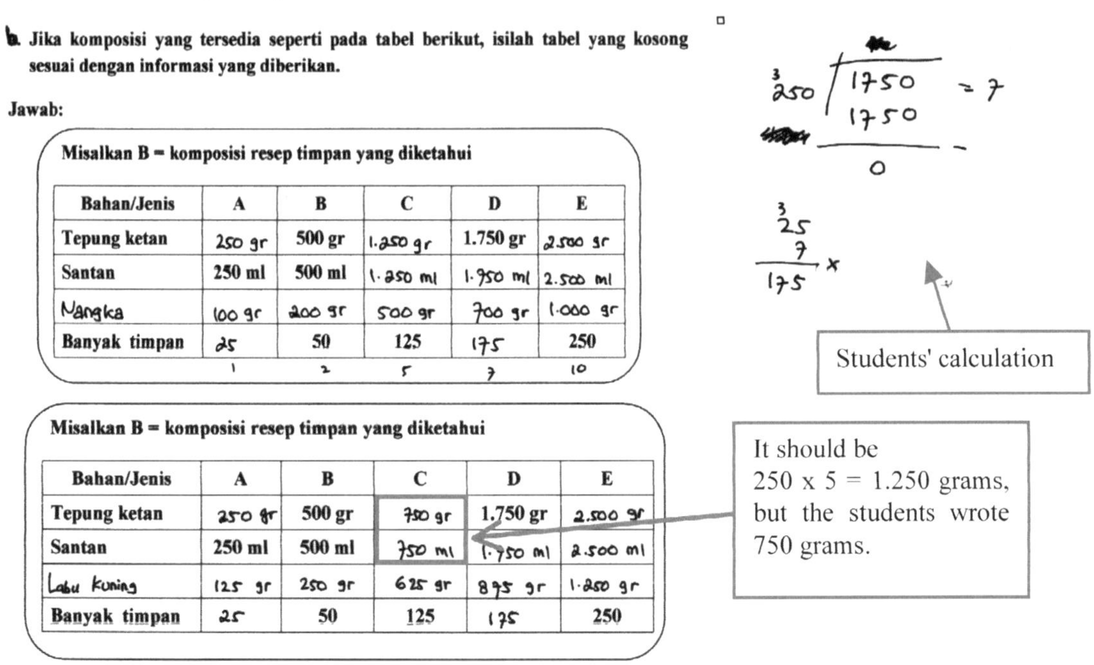

Bahan/Jenis	A	B	C	D	E
Tepung ketan	250 gr	500 gr	1.250 gr	1.750 gr	2.500 gr
Santan	250 ml	500 ml	1.250 ml	1.750 ml	2.500 ml
Nangka	100 gr	200 gr	500 gr	700 gr	1.000 gr
Banyak timpan	25	50	125	175	250
	1	2	5	7	10

Bahan/Jenis	A	B	C	D	E
Tepung ketan	250 gr	500 gr	750 gr	1.750 gr	2.500 gr
Santan	250 ml	500 ml	750 ml	1.750 ml	2.500 ml
Labu kuning	125 gr	250 gr	625 gr	875 gr	1.250 gr
Banyak timpan	25	50	125	175	250

Figure 2. The results of the students' answers to determine the A, C, D, and E recipes.

The second activity was to determine the composition of the ingredients in other recipes with the same flavor if some compositions of the ingredients in a recipe are known. The students used the recipes of *timpan* (i.e., pumpkin, jackfruit, and banana *timpan*) they chose as a benchmark in determining the composition of other ingredients by completing recipe B. After completing the composition in column B, the students were asked to find the composition in the column A, C, D, and E as presented in Figure 2.

In this activity, the students determined the composition of the ingredients in other recipes using the recipe A as a benchmark. The students knew that the composition of the ingredients in the recipe B was two parts of that of the ingredients in the recipe A. They used the recipe A to determine the recipe C, D, and E. The students used multiplication and division rules in determining the composition in the recipes. One out of five groups made mistakes in calculating multiplication. Students wrote 750 grams and 750 ml for the composition of glutinous flour and coconut milk on the recipe C, respectively. Whereas, the students had already known that the composition in the recipe C was one-fifth of that in the recipe A (i.e., A : C = 1 : 5). The composition in the recipe A should be multiplied by 5 so that the answer would be 250 x 5 = 1.250 grams of flour and 1.250 ml of coconut milk. Based on the interview results, the students admitted that they made a mistake in multiplying 25x5. They did not do the calculation on the paper, but instead, they spontaneously mentioned that 25x5 was 75.

The next step was to determine the ratio of the results of the composition of the five recipes that have been previously obtained. Figure 3 presents one of the students' answers.

Bahan/Jenis	A	B	C	D	E
Tepung ketan	1	2	5	7	10
Santan	1	2	5	7	10
Nangka	1	2	5	7	10
Banyak timpan	1	2	5	7	10

Figure 3. The results of students' answers in determining the ratio

In this activity, the students determined the ratio based on one component in a *timpan* recipe. For example, by looking at the comparison of the number of *timpan*, the students then split each component of the ingredients with the smallest number. These students' strategies were also applied to other components. Thus, the students got the recipe B is twice of the recipe A, the recipe C is five times of the recipe A, the recipe D is 7 times of the recipe A, and the recipe E is ten times of the recipe A. In another word, the ratio of the recipe A, B, C, D, and E is 1:2:5:7:10.

In the final stage of this learning, the students represented the ratio of the *timpan* recipe composition in graphs and equations. The students worked in their groups to find out the direct proportion concepts and wrote the results on the activity sheets as presented in Figure 4.

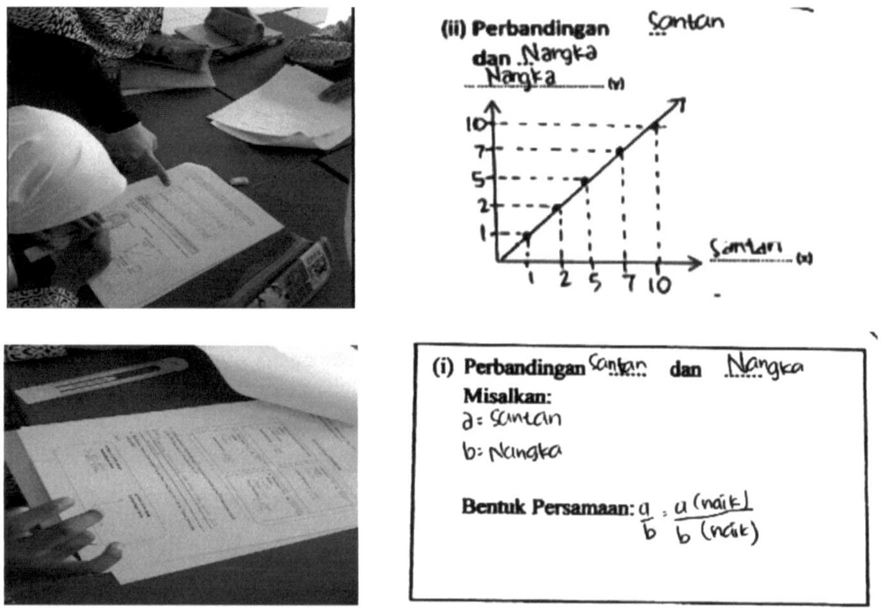

Figure 4. Students' answer to the representation of ratios in graphs and equations.

In the graph presented in Figure 4, the students randomly selected two different components as coordinates and chose ratio as scales. Although no one chose the original measurement as a scale, the teacher kept reminding the students that the original measurement of a *timpan* recipe could be used as a scale in drawing a graph. In general, the curve of direct proportion graph obtained by students was a linear graph, and the students were managed to write the equation of direct proportion correctly.

Based on the four activities previously discussed, it is indicated that the teaching and learning process went well. The students presented a direct proportion concept in graphs and equations using the *timpan* recipe context. This is aligned with other research findings that the use of context in mathematics learning leads to more meaningful learning for students. The context presents abstract concepts of mathematics in representations that can be easily understood by the students. Therefore, the student learning activities using the realistic mathematics education (RME) approach are improved [8] and [1].

4. Conclusion

The results show that the students were actively involved during the direct proportion learning using the context of the *timpan* recipe. This finding is indicated by the students' activities, students' positive responses to the learning, the utilization and the selection of certain procedures or specific operations appropriately in direct proportion problem-solving. Besides, there was an effective interaction between the students and the teacher. Also, the context of this *timpan* recipe can be used as a starting point in teaching direct proportion. For further research, it is recommended to select other cultural contexts closed to students' life as the key element in learning mathematics.

References

[1] Wijaya A 2012 *Pendidikan Matematika realistic Suatu Alternatif Pendekatan Pembelajaran Matematika* (GrahaIlmu: Yogyakarta)

[2] Zubainur C M and Johar R 2016 Teacher ability in teaching of finding π value at primary school *Desain Research for Creativity in Education Proc The Fourth South East Asia Design/Development Research International Conference* (Padang: West Sumatera Indonesia/ UniversitasNegeri Padang) pp 167-173

[3] van den Heuvel-Panhuizen M 2003 The Didactical Use Of Models In Realistic Mathematics Education: An Example From a Longitudinal Trajectory On Percentage *J Educational Studies in Mathematics* **54** 9

[4] Gravemeijer K 2004 Local instruction theories as means of support for teachers in reform mathematics education *J Mathematical Thinking And Learning* **6** 105

[5] Johar R 2006 Pendidikan matematika realistik: pendekatan baru dalam pembelajaran matematika *J Wacana FKIP Usyiah* **5**

[6] Gravemeijer K 1994 *Developing realistic mathematics education* (Ultrech: Freudenthal institute)

[7] Gravemeijer K and Van Eerde D 2009 Research as a means for building a knowledge base for teaching in mathematics education *J. Elementary School* **109** 510

[8] Anugrahana A 2015 Pengaruh pembelajaran matematika dengan menggunakan pendekatan realistik terhadap aktivitas siswa dan hasil belajar siswa di sekolah dasar *J DIDAKTIKA* **3** 226

IOP Conf. Series: Journal of Physics: Conf. Series **1088** (2018) 012049 doi:10.1088/1742-6596/1088/1/012049

Developing mathematics teaching tool using ELPSA

E Gradini and F Bahri
STAIN Gajah Putih, Jl. Aman Dimot No.10, Aceh Tengah, Aceh. Indonesia, 24519

E-mail: egagradini@staingp.ac.id

Abstract
This study aimed at measuring the validity, practicality, and effectiveness of teaching tools designed to address students' mathematical understanding in the classroom. This paper discusses validity and effectiveness only. Five components of ELPSA (Experience, Language, Picture, Symbol, and Application) are proposed to give teachers and students a more tangible way to construct the understanding and learning experience. Four-D model (Define, Design, Develop, and Disseminate) by Thiagarajan, Semmel, and Semmel was used as the developmental approach in this study. The finding showed that teaching tools designed are valid and effective to be used in the mathematics classroom.

1. Introduction

The paradigm of mathematics learning has changed from a transfer of knowledge to the construction of knowledge that assigned students as the centre of learning activities. Despite the change of curriculum in Indonesia, from *KTSP* to *Kurikulum 2013*, most Indonesian teachers still use teacher-centred instructions as opposed to student-centred classroom learning. The teacher-centered class has been embedded into Indonesian school culture for a very long time [1]. Such method leaves a little room for students' creativity and expressiveness, a subject of criticism by academic society in Indonesia [2]. In contrary of the significant role of teachers in the classroom, many Indonesian teachers have lack of teaching competence [1]. Researchers believe this is because of the long tradition of teacher-centred teaching and rote learning in the Indonesian classroom [3].

Construction of knowledge was difficult to pursue if the classroom is dominated by a teacher. Teacher-centred instruction is regarded to be undemocratic because it fails to consider students' learning dynamics and ignores students' contributions in the classroom [3]. It also prevents students to construct the mathematical understanding by the learning experience. According to an interview with one of junior high school mathematics teacher in Aceh Tengah, students learn mathematics from the teacher's explanation only. The teacher had difficulties in providing learning activities, media, and material that promote students to construct mathematical understanding [4]. Therefore, it is necessary to develop teaching tools to promote students' understanding.

ELPSA as a learning framework offers opportunities for teachers to better understand the way in which pedagogical practices and learning experiences can be effectively presented to students in ways that reflect students' concept development [5]. The learning framework provides opportunities for teachers to enhance the content pedagogical knowledge as well as provides learning experiences for children that could be classified as culturally appropriate and personalized. In ELPSA framework, learning is an active process where students construct their understanding through individual thinking and social interaction.

ELPSA consists of five components; Experience, Language, Picture, Symbol, and Application. Experience means the association students previous knowledge with new knowledge that will be studied [5]. The first component is Experience (E). ELPSA assumes that experiences are the foundations for the introduction of new learning opportunities. Learning on this phase occurs through participation and encourage high levels of engagement and interaction. By experiencing the learning, mathematical ideas/content can be introduced in meaningful ways. The foundation is embedded with the language used in promoting learning.

The second component is Language (L). It is important to connect students experience with mathematical terminology to promote students' understanding. Culture and daily language are believed to influence both students' perception and understanding. ELPSA believes that language is important for teacher and students. Explaining mathematical concept with students' language could help students to clarify and strengthen their understanding.

The third component of ELPSA is Pictures (P). This component helps student to construct their understanding of abstract. This component is built using a visual representation to represent mathematical ideas. The images used to promote students' understanding and stimulus to solve mathematical tasks before they used symbols to represent the abstract concept. The fourth component is Symbol (S). This component involves students' capacity to represent, construct, and manipulate analytic information. Symbols are best utilized when students understand a particular concept and it is necessary for students to practice using symbolic operation [5]. Kinach suggested that symbol introduced problem-solving level [6]. In common mathematics instruction, a symbol is often used at first level before students able to perform concept understanding. Students placed on the problem-solving level before they have concept and content understanding. The research concludes that if a symbol is introduced at the early phase, students are likely to use them inappropriately [7]. The last component is Application (A). At this phase, students apply mathematics idea to a situation that is considered to enhance students' mathematical understanding. This is the biggest problem of mathematics instruction in Indonesia, where students believed that mathematics is irrelevant to their daily life. Although problem-solving activities are challenging to be solved by students, they are not connected to real-life experiences [8]. The application component also provides opportunities for students to see how mathematics can be used in and out of school contexts [5]. ELPSA views this component as activities that encourage students to apply mathematics knowledge to new and related situations, both help to reinforce concepts and provide opportunities to scaffold mathematical understanding.

There were many previous studies on ELPSA framework in a mathematics classroom. Johar and Hajar implement ELPSA using technology in learning integral [9]. Syahdan applied ELPSA framework in Bruner Theory based-mathematics classroom [10]. Whereas this study discussed designing teaching tools using ELPSA framework by focusing on E and A component of ELPSA to build the connection between mathematical concept and students' daily life. The objectives of this study are to develop learning tools using ELPSA framework.

2. Research method

This study employed Thiagarajan, Semmel, and Semmel's model of developmental approach [11]. This model has four phases; (1) Define, (2) Design, (3) Develop, and (4) Disseminate [11]. This study measures the validity, practicality, and effectiveness of teaching tools. The validity of teaching tool is measured by the relevance and consistency. The expert agreement index for content validity is a comparison of the number of items from two experts with strong relevance category of overall items [12]. While the results of the relevancy tabulation (contingency tables) are presented in Table 1, the validity coefficient is presented in Formula 1.

Table 1. The relevance category scoring with two validators [12]

		Validator 1	
		Weak	Strong
Validator 2	Weak	A	B
	Strong	C	D

$$\text{Content validity coefficient} = \frac{D}{A+B+C+D)} \quad (1)$$

Teaching tools is effective if it fulfills 3 of 4 criteria; (1) more than 50% student response to material and worksheet positively, (2) more than 70% students' activities at good-level, (3) all teacher's activities at good-level, and (4) more than 75% students passed the Minimum Criteria of Masterly Learning (KKM) [13].

3. Result and finding

3.1 The validity of teaching tools

Content validity is determined using expert agreement index suggested by Gregory [12]. The index ranges from 0 to 1. It is conducted by making contingency tables on two experts, with the first category that is not relevant and less relevant become the weak relevancy category, and the second category which is for quite relevant and very relevant that is created in a new strong relevant category. The expert agreement index for content validity is a comparison of the number of items of the two experts with strong relevance category of overall items. The expert agreement index for content validity is a comparison of the numbers of items from two experts as validators with strong relevance to the overall items category.

3.1.1 Content validity coefficient of the lesson plan

Aspects assessed in validating Lesson Plan are curriculum, material, learning, language, and time allocation. Table 2 shows the relevance category scoring of two validators.

Table 2. Relevance category scoring of the lesson plan

		Validator 1	
		Weak	Strong
Validator 2	Weak	0	2
	Strong	2	17

Content validity coefficient of lesson plan is 0.809. It means that the lesson plan has high relevancy with their indicators.

3.1.2 Content validity coefficient of teaching material

Teaching material designed are teachers' book and students' book. Aspects assessed in validating students' book and teachers' books are curriculum, material, illustration and example, problem, layout, and language. While Table 3 shows the relevance category scoring of students' book, the relevance category scoring of teacher's book presented in Table 4.

Table 3. Relevance category scoring of student's book

		Validator 1	
		Weak	Strong
Validator 2	Weak	0	0
	Strong	5	17

Table 4. Relevance category scoring of teacher's book

		Validator 1	
		Weak	Strong
Validator 2	Weak	0	0
	Strong	4	18

Content validity coefficient of student's book is 0.773 and teacher's book is 0.818. It means the validator agreement index about both teacher's book and students' book relevance with their indicators is high. Although the content validity coefficient is high, validators gave some suggestions as presented follow:

- Make activities more interesting to students by inserting the activity on student worksheets.
- Give answer key for every problem and task at the end of chapter
- Add reflection part in order to make students able to do self-assessment. This part also helps the teacher to find out what students feel and know after learning.

3.1.3 Content validity coefficient of students' worksheet

Table 5 presents aspects assessed in validating students' worksheet are curriculum, material, illustration and example, problem, layout, time allocation and language.

Table 5. Relevance category scoring of students' worksheet

		Validator 1	
		Weak	Strong
Validator 2	Weak	0	0
	Strong	1	13

Content validity coefficient of students' worksheet is 0.928. It means the validator agreement index about students' worksheet relevance with their indicators is high. Although the content validity coefficient is high, validators gave some suggestions as presented follow:

- Illustration need to be placed for every problem to encourage students to construct their understanding through the picture
- Replace Avocado in problem 3 (Presented in Figure 1) with another thing because each avocado has a different weight.

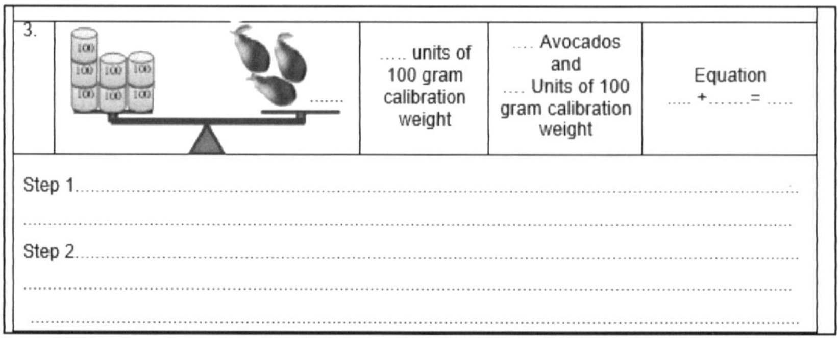

Figure 1. Problem 3 "avocado" on students' worksheet

3.2. The effectiveness of teaching tools

3.2.1. Students' response to students' book and worksheet

Students asked to assess teaching material (students' book) and worksheet. All items responded positively by 72% of students, hence teaching tools are effective to implement in a mathematics classroom.

3.2.2. Students' activities in the classroom

Six students are observed during teaching and learning process. The observation of students' activities is presented in Table 6.

Table 6. Students' activities in the classroom

No	Observation items	Lesson					Score	Category
		1	2	3	4	5		
1	Students' interest during instruction	3.00	3.17	3.17	3.08	3.17	3.12	Good
2	Students active in the utilization of teaching media	2.50	3.17	3.17	3.33	3.33	3.10	Good
3	Students active in a group discussion	3.17	3.28	3.17	3.33	3.33	3.26	Good
4	Students active in sharing idea during group discussion	2.83	3.00	3.00	3.08	3.00	2.98	fair
5	Students active in responding to teacher questions	3.00	3.32	3.32	3.17	3.17	3.20	Good
6	Students have a responsibility in their groups	3.17	3.33	3.33	3.33	3.33	3.30	Good
7	Students participation in problem-solving	3.17	3.33	3.33	3.42	3.42	3.33	Good
8	Students active in helping their friends to understand the material	2.67	2.83	2.83	3.08	3.00	2.88	Fair
9	Student responsible to his/her own task in the group	3.17	3.00	3.00	3.17	3.17	3.10	Good
10	Students give correct and relevance answer	3.50	3.53	3.53	3.58	3.53	3.53	Very Good

IOP Conf. Series: Journal of Physics: Conf. Series **1088** (2018) 012049 doi:10.1088/1742-6596/1088/1/012049

3.2.3. Teacher's competency in instruction

Table 7. Teacher's competency in instruction

No	Observed activities	Lesson 1	Lesson 2	Lesson 3	Lesson 4	Lesson 5	Average	Category
Phase 1. Clarify goals and establish set								
1	Teachers motivate students to learn	3	4	3.5	4	4	3.70	Very good
2	Teachers explain goals of learning	3	3.5	4.5	4	5	4.00	Very good
3	Teachers establish a set of learning	3.5	4	4	4	4	3.90	Very good
4	Teachers provide daily life problem	4	4	4	3.5	3	3.70	Very good
	Average score						3.83	Very good
Phase 2. Providing information								
1	Teachers provide information to students	5	4.5	4	4	4	4.30	Very good
2	Teachers demonstrate an example, problems, or skill	4	4	4	3.5	3	3.70	Very good
3	Teachers give an opportunity for students to ask difficult material	4	4	4	4	3	3.80	Very good
	Average score						3.93	Very good
Phase 3. Organize students in group								
1	Teachers organize students into groups according to their achievement	4	4	4	3.5	3	3.70	Very good
2	Teachers distribute worksheet to students	4	4	4	4	4	4.00	Very good
	Average score						3.85	Very good
Phase 4. Assist teamwork and study								
1	Teachers make sure that every student understand their task and responsibility in group	4.5	4	4	3	3	3.70	Very good
2	Teachers ask students to discuss and solve problems in the worksheet.	4	4	4	4	3.5	3.90	Very good
3	Teachers assist students who have difficulties in group discussion	4.5	4	4	3	3	3.70	Very good
4	Teachers rephrase the explanation if the student does not understand the problem/question	4	4	3.5	3	3	3.50	Very good
	Average score						3.70	Very good
Phase 5. Evaluation								
1	Teachers evaluate students individually	4	4	4	3.5	3.5	3.80	Very good
2	Teachers ask every group to present their work	4	4	4	3.5	3	3.70	Very good
3	Teacher give feedback to students' work	4	4	3.5	4	4	3.90	Very good
	Average score						3.80	Very good
Phase 6. Providing recognition								
1	Teachers recognize students' achievement	4	3.5	3	3	3	3.30	Very good
2	Teachers ask student to resume the lesson/material	4	4	3.5	3	3	3.50	Very good
	Average score						3.40	Very good
	Total score						3.75	Very good

During the teaching and learning process, the teacher uses the scientific approach and cooperative learning as the teaching method. The teacher observed for five meeting in order to see the teacher's ability in implementing teaching tools. Teacher's competency in instruction is presented in Table 7.

4. Conclusion

This study showed that students able to understand effectively when they experience the learning by themselves. In this study, content-level understanding conducted on experience (E) phase and problem-solving level on application (A) phase. By experiencing the learning process, students allowed to interact with other students and develop their understanding. This finding supports theory stated that in Experience component, students learn through participation and develop their mathematical ideas. Students able to solve the given problem accurately when the problem is routine and familiar with their experience, when they face non-routine problems, most students were not able to. Unlike other studies carried out on ELPSA, this study found that applying the concept to a new daily life situation still challenging for some students. Teaching tools were valid and effective to be implemented in the mathematics classroom. The effectiveness of teaching tools based on (1) the ability of teachers in managing to learn is well, (2) students' activity during the learning process was active, and (3) student responses to learning were positive. Thus, it recommended using the developed teaching tools in the mathematics classroom.

References

[1] Azra A 2002 *Paradigma Baru Pendidikan Nasional: Rekonstruksi dan Demokrasi* (Jakarta: Kompas)
[2] Zulfikar T 2010 The Making of Indonesian Education: an Overview on Empowering Indonesian Teachers *J.Ind. Soc. Sci.Hum.* **2** 13
[3] Wolk S 1998 *A Democratic Classroom* (Portsmouth: Heinemann)
[4] Bahri F 2017 *Efektivitas ELPSA Framework dalam mengkonstruksi Pemahaman Siswa pada Pembelajaran Matematika* (Aceh: STAIN Gajah Putih Press)
[5] Lowrie T and Patahuddin S M 2015 ELPSA as a Lesson Design Framework *J. Math. Edu* **6** 2
[6] Kinach B M 2002 Understanding and Learning-To-Explain by Representing Mathematics: Epistemological Dilemmas Facing Teacher Educators in the Secondary Mathematics "Methods" Course *J. Research on Mathematics Teacher Education* **5**153
[7] Ball D L 1992 The Mathematical Understandings that Prospective Teachers Bring to Teacher Education *J. Advances in Research on Teaching* **2** 1
[8] Lowrie T and Patahuddin SMELPSA-Kerangka Kerja Merancang Pembelajaran Matematika *J. Didaktik Matematika* **2** 94
[9] Johar R and Hajar S 2017 Implementation of ELPSA Framework in Teaching Integral Using Technology *Int J. Sci. Technol* **1** 1
[10] Syahdan S 2016 *Efektivitas Penerapan Experience, Language, Pictorial, Symbol and Application (ELPSA) Pada Pembelajaran Matematika Berbasis Teori Bruner Pada Siswa Kelas VII SMP NEGERI 29 Makassar* (Makassar: Universitas Negeri Makassar Press)
[11] Perkins DN and Simmons R 1988 Patterns of Misunderstanding: an Integrative Model for Science, Math, and Programming *J. Rev. Edu Research* **58** 303
[12] Gregory R J 2007 *Psychological Testing: History, Principle, and Applications* (Boston, MA: Pearson)
[13] Nurhuda, Lukito A, and Masriyah 2018 Effectiveness of Cooperative Learning Instructional Tool With Predit –Observe –Explain Strategy on The Topic of Cuboid and Cube *Volume J. Physics. Conf Series.* **947** 1

IOP Conf. Series: Journal of Physics: Conf. Series **1088** (2018) 012050 doi:10.1088/1742-6596/1088/1/012050

The effectiveness of learning instruments on the topic of the set using problem-based learning model at Islamic junior high school in Pekanbaru

Sakur, A Murni and R D Anggraini
Universitas Riau, Jl. Kampus Bina Widya, Km. 12.5 Simpang Baru, Riau 28293, Indonesia

E-mail: sakur@lecturer.unri.ac.id

Abstract. This study is a continuation of the research and development of learning instruments on the topic of set. Researchers through Research and Development have produced prototypes of learning instruments, namely: lesson plan and worksheets. The prototype of the learning instruments has been analyzed for validity and practicality. The learning instruments then needed to be tested for its effectiveness through experimental research in two classes, the experiment, and control class. The purpose of the study was to examine the difference of learning outcomes between classes using learning instruments on the topic of the set and the class using the conventional learning model in Year 7 at one of the Islamic Junior High School in Pekanbaru. This research used a pre-experimental research design from a static group comparison. The results showed that the students' learning outcome using learning instruments of Problem-Based Learning (PBL) model was better than their counterpart. The learning instruments using the problem-based learning model was effective to be used in Islamic Junior High School in Pekanbaru.

1. Introduction

The student-centered learning is important to create a fun and enjoyable learning. To implement student-centered learning, learning instruments, including lesson plan and worksheet employing student-centered approach should be developed [1]. One of the approaches focusing on students is problem-based learning (PBL). Learning is a mental process that occurs within a person, thus causing the change in behavior [2]. The mental activity occurs because of the individual interaction with the environment. Trianto [3] stated that learning, in general, is a process marked by the change in one's self. Changes as a result of the learning process can be indicated in various forms such as changing knowledge, understanding, attitude, and behavior, skills, and abilities, as well as changes in other aspects in the individual learning. Learning is a process of change in a person towards a better and relatively permanent as a result of his/her own experience with the environment. These changes can be knowledge, understanding, attitude and behavior, skills and abilities. The process of change is an individual's interaction with the environment. Oemar Hamalik [4] stated that learning outcomes might be indicated by the behavior of learners, which can be observed and measured in the form of changes in knowledge, attitudes, and skills.

The PBL is learning aiming to develop the ability to think and solve problems. This learning challenges learners to develop critical thinking skills, examine problems, find and actively and creatively apply their existing ability. Implementation of PBL encompasses five steps [5] as outlined in Table 1.

Table 1. Steps of problem-based learning.

Step/stage	Teacher activity
The orientation of the learners on the problem	Presenting phenomena/issues, exposing logistics to learning, motivating learners to engage in defined real-life problem-solving activities
Organize learners to learn	Help learners understand the problem of starting with the identification of learning tasks to do problem-solving planning
Guiding individual or collective investigations	Guide or motivate learners to do research (experiment if necessary) and implement problem-solving
Develop and present the work	Helps learners provide ease in providing the necessary tools and sharing tasks in presenting products or works
Evaluate the learning process	Together with the learners to summarize the contents of the lesson, reflect on the problem-solving process. And evaluation.

The role of teachers in the application of PBL is explaining the problem, facilitating the learners' investigation and dialogue or discussion and guiding learners. Besides, teachers also provide scaffolding to help learners to solve problems. Scaffolding is an effort to improve the ability of inquiry and the cognitive development of learners by guiding and motivating the learners to find or build their knowledge of what is learned.

The PBL model is learning that allows learners to share ideas and considers the most appropriate answer. This learning also encourages learners to improve their collaborative skills to solve the contextual problem mutually. Learners are expected to present their work in front of the class so that the learning process that followed by all learners are required to achieve the learning objectives. The implementation of this PBL enables learners to be engaged in learning as this learning focuses on the activities of learners in searching, processing and reporting information from various sources. There are several aspects that are analyzed to test the effectiveness of learning instruments with PBL toward the students' mathematics learning outcome. One of these is the Mathematics Initial Ability (MIA). The students' MIA is categorized into high, medium, and low, as the other variable in this research.

This problems statement of this research are: (1) Is there a significant difference in mathematics learning outcome between students in the experimental and control class?; (2) Is there a significant difference of mathematics learning outcome of learners with high MIA between the experimental and control class; (3) Is there a significant difference in mathematics learning outcome of learners with medium MIA between the experimental and control class; and (4) Is there a significant difference of mathematics learning outcome of learners with low MIA between the experimental and control class. This research aims to test the effectiveness of learning instruments on the topic of set. The learning instruments (called prototype II) are tested for the validity and reliability. The specific purpose of this research is to examine the difference of the Year 7 learners' learning outcome in Islamic Junior High School Pekanbaru. The difference in the learners' learning outcome can be interpreted to examine the effectiveness of learning instruments.

2. Method

Through research and development, the researcher has generated learning device products invalid prototype and practicality. The prototype products must then be tested for its effectiveness through experimental research. The selected school was divided into the experimental and control class determined by the normality and homogeneity test on the data of the Mathematical Initial Ability (MIA). In the experimental class, learning was conducted by using the learning instruments developed, while the control class was run in the conventional way. This research used a quasi-experimental research method that was performed in the existing environment. The MIA of the Year 7 learners were divided into high, medium, and low groups. The impact researched as a result of the treatment was the mathematics learning outcome as an achievement of basic competence. Population in this research were all students of

Year 7 Islamic Junior High School Pekanbaru on the academic year 2017/2018, consisting of four classes. The sample classes were two class, one experiment class implementing PBL (28 students) model and control class (31 students) with the conventional learning model.

The research involved two classes with designs presented in Table 2.

Table 2. Research design.

Class	Treatment	Post-test
Experiment	X_1^a	O^c
Controls	X_2^b	O^c

[a]: Problem Based learning model
[b]: Conventional learning using a scientific approach
[c]: Posttest

The data collection instruments that were used are test and non-test. The test was a daily test for basic competency about the subject of the set. The non-test instrument was the observation sheet during the learning process and interview guidelines. The test was developed by the researchers that used to measure the mathematics learning outcome of learners. The data of MIA and learning outcome of the learners were tested for normality, homogeneity and mean difference. Analysis data was conducted using the SPSS software version 20.

The data of MIA was derived from the value of daily tests on concerning number. Table 3 shows the result of the normality test with Kolmogorov Smirnov test not normally distributed.

Table 3. The test results of the normality of MIA data.

Class	N	μ	ϑ	Sig.	Description
VII-3	28	61.96	31.250	0.000	Not normal
VII-4	31	54.19	28.899	0.000	Not normal

Furthermore, hypothesis testing using nonparametric statistical Mann Whitney tests. The verbal hypothesis formula for Mann Whitney test on posttest score data of experimental class and control class is as follows.

H_0: The average MIA of grade VII-3 is lower or equal than grade VII-4
H_1: The average MIA of grade VII-3 is higher than grade VII-4

Mann-Whitney test results are presented in table 4.

Table 4. Mann-Whitney test data of MIA students.

Class	N	μ	Z	Sig. (1-tailed)	H_0
VII-3	28	61.96	-0.594	0.552	Accepted
VII-4	31	54.19			

Table 4 shows that the significance value (sig.) is more than $\alpha = 0.05$ which means H_0 is accepted, and H_1 is rejected. This means that at 95% confidence level, there is no average difference between grade VII-3 and grade VII-4.

The test of mathematics learning outcome in this research was in the form of a description problem and a test device aiming to measure the achievement of basic ability as stated in the regulation of the ministry of education (Permendikbud) No. 24, Attachment 1 [6]. Interview guides were prepared to interview teachers and some learners. The interview was used to explore teacher and student comments' related to the learning process using learning instruments developed and conducted to improve the mathematics learning outcomes.

Learners were grouped into three groups of height, medium and low. Grouping criteria were based on the mean (MIA), and standard deviations (SD) referring to the criteria used by Ratnaningsih [7] is; high group(s-MIA $\geq \bar{x}$ + SD), the medium group ($\bar{x}$ − SD $\leq$ s-MIA < $\bar{x}$ + SD), and low group (s-MIA < $\bar{x}$ − SD)

To obtain optimal results, researchers conducted observation activities on the implementation of the learning process, especially in the experimental group. Observation sheets for learners were used to observe the activities of learners in the ongoing learning using the developed student worksheets. Teacher observation sheets are used to observe teacher activities in implementing the learning. Observations were made by observers using the observation sheet.

Tests to determine the effectiveness of actions performed on the experimental class of mathematics learning outcomes obtained by learners were measured by calculating the "*effect-size*" (*d)* using Cohen's formula [8] as follows.

$$d = \frac{\bar{x}_t - \bar{x}_c}{S_{pooled}} \quad \text{and} \quad S_{pooled} = \sqrt{\frac{(n_t - 1)S_t^2 + (n_c - 1)S_c^2}{n_t + n_c}}$$

The interpretation of the result of the calculation of the value of effect size is presented is; High ($d \geq 0.80$), medium ($0.50 \leq d < 0.80$), low ($0.20 \leq d < 0.50$), and very low ($d < 0.20$).

3. Result and discussion

The results with Kolmogorov Smirnov and Lavene test shows that the score of the students' posttest of both classes was normally distributed as showed table 5.

Table 5. Test normality of posttest.

Class	N	μ	ϑ	Sig.	Description
VII-3	28	92.55	10.540	0.175	Normal
VII-4	31	48.18	24.968	0.191	Normal

The hypotheses for testing the homogeneity of posttest data were:
H_0: There is no difference in variance from posttest data of both classes
H_1: There is the difference in variance from posttest data of both classes

The result of the homogeneity test, Lavene-test, is presented in Table 6.

Table 6. Test homogeneity data posttest.

Class	N	μ	Varian's	Sig.	Description
VII-3	28	92.55	111.100	0.000	Homogeny
VII-4	31	48.18	623.404		

Based on table 6, it can be seen that the significance value (sig.) is less than $\alpha = 0.05$ which means H_0 is accepted. This means that at 95% confidence level, there is no difference between the experiment and control class. The hypothesis formula for t-test of posttest data of learners of experimental class and control class is:
H_0: the average of posttest of experiment class is lower or equal to the average of the control class
H_1: the average of posttest of the experiment class is higher than the average of the control class

The results of the independent sample t-test of students' posttest data are presented in Table 7.

Table 7. Independent sample t-test test for posttest data.

Class	N	Df	T	$t_{0.95(56)}$	H_0
VII-3	28	56	6.797	2.003241	Accepted
VII-4	31				

Table 7 shows that at the 5% significance level the average achievement of experimental class ability is better than the control class average.

The treatment of this research took place for six lessons in both the experimental and control class. At the end of the study, a posttest was administered for both classes to examine the results achieved after the treatment. Implementation of learning in both sample classes with the same subject which is the set. Based on observations and interviews it is stated that the implementation of learning in both sample classes has been verified and convinced according to the planning. Profile of the MIA group of the learners is presented in Table 8.

Table 8. Profile of learners groups.

Group	Experiment			Controls		
category	N	rMIA[a]	rHB[b]	N	rMIA[a]	rHB[b]
G-height	3	100	90.8	6	86.7	55.7
G-medium	15	81	88.2	20	54	42.6
G-low	10	22	81.1	5	16	32.8

[a] : the average of MIA
[b] : the average learning outcome

Table 8 shows the average increase in learning outcomes in the experimental class, especially in groups of students in the low and medium category. The average outcome of the controlled class learning in the MIA group was high and medium. The difference in learning result of MIA group high in experimental class decreased 9.2 points, while in control class decreased 31 points. The results of the study were medium, in the experimental class, there was an increasing while in the control class there was a decreasing. The learning result of the low MIA group in the experimental class showed an increase of 59.1 points, while in the controlled class is 16.8 points.

The changing of the average of learning outcome indicates that the learning process is effective. After the hypothesis test in the posttest scores of each class using the Independent Sample T-Test with the help of SPSS, it was found that the average experimental class was better than the average post-control class as supported by the presence of Table 6. So it can be said that there is a significant difference to the learning outcomes between the classes applied Problem-based Learning model and the conventional learning model of Year 7 students at Islamic Junior High School Pekanbaru. Based on table 4 and table 7, can be said that the average achievement of experimental class competencies increases. The average achievement of controlled class competencies decrease.

4. Conclusion

There is a significant difference in the mathematics learning outcomes of learners between the class applying Problem-based Learning model and the class applying the conventional learning model at grade VII at Islamic Junior High School Pekanbaru. There is a difference in the mathematics learning outcomes of learners with high Mathematical initial ability in experimental and controlled classes. There is a difference in the mathematics learning outcomes of learners with medium Mathematical initial ability in experimental and controlled classes. There is a difference in the mathematics learning outcomes of learners with low Mathematical initial ability in experimental and controlled classes. Thus it can be interpreting that the application of learning instruments that designed in prototype II is effective and can enhance the achievement of the learner's ability.

References

[1] Kemendikbud 2016 *Peraturan Pemerintah No. 22 tahun 2016 tentang Standar Proses Pendidikan Dasar dan Menengah* (Jakarta: Kementerian Pendidikan dan Kebudayaan)
[2] Sanjaya W 2008 *Strategi Pembelajaran Berorientasi Standar Proses Pendidikan* (Jakarta: Kencana Prenada Media)
[3] Trianto 2007 *Mendesain Model Pembelajaran Inovatif-Regresif* (Jakarta: Kencana Prenada Media Group)
[4] Hamalik O 2003 *Proses Belajar Mengajar* (Jakarta: Bumi Aksara)

[5] Ibrahim M and Nur M 2002 *Pembelajaran Berasaskan Masalah* (Surabaya: Universiti Press)
[6] Kemendikbud 2006 *Peraturan Pemerintah No. 24 Tahun 2016 Lampiran 15 tentang Kompetensi Dasar dan Struktur Kurikulum SMP/MTs* (Jakarta: Kementerian Pendidikan dan Kebudayaan)
[7] Ratnaningsih N Pengaruh Pembelajaran Kontekstual terhadap Kemampuan Berpikir Kritis dan Kreatif Matematik serta Kemandirian Belajar Siswa Sekolah Menengah Atas [Dissertation] Bandung; Universitas Pendidikan Indonesia; 2007
[8] Cohen J 1992 Quantitative methods in psychology: A power primer *Psychological Bulletin* **112** 155

The teacher's mathematical literacy for the change and relationship problems on the PISA adaptation test

A Y Anggoro, H Julie, F Sanjaya and M A Rudhito
Sanata Dharma University, Jl. Affandi Tromol Pos 29, Sleman, Yogyakarta 55002 Indonesia

E-mail: yudhianggoro@usd.ac.id

Abstract. There is evidence suggesting that teachers are one of the factors influencing student's performance in solving mathematical problems. However, the study on Indonesian teacher's ability to solve PISA's mathematical problems is inadequate. Therefore, this study aimed to examine Indonesian teachers' ability in solving PISA-adapted mathematics problems for the topic of change and relationship. This study employed a case study research design involving seven mathematics teachers from various schools in Yogyakarta. They were required to solve PISA-adapted mathematics problems classified into four areas, namely Change and relationship, Space and Shape, Uncertainty, and Quantity. The results showed that: (1) all teachers could solve the first level problem, (2) all teachers could solve the first problem at the third level, (3) five teachers (71.43 %) could solve the second problem at the third level, (4) two teachers (28.57 %) could solve a third problem at the third level, (5) six teachers (85.71 %) could solve the fourth problem at the third level, (6) four teachers (57.14 %) could solve the first problem at the fourth level, and (7) Six teachers (85.71 %) could solve the second problem at the fourth level.

1. Introduction

Mathematical and pedagogy abilities of primary teachers were directly and positively related to their students' achievement [1,2]. Teachers' perception also significantly correlates with the students' knowledge [1,2]. The teachers' perception in this study was defined as (1) the paradigm of teachers of mathematics teaching and learning process, and (2) teachers' concern of students' mathematics achievement is closely related to teachers' mathematical knowledge [1,2]. Teachers' paradigm on the settlement of mathematical models and learning organization supported teacher mastery of the mathematics knowledge and the pedagogy [1,2]. Thus, one of the determinants of students' success in solving the PISA test is teachers' ability in managing the mathematics teaching and learning process and solving mathematical problems.

In 2015, Indonesia followed the PISA test for the fifth time and the ranking Indonesia for PISA tests were 63 for mathematics, 62 for science, and 64 for reading from 70 countries. These results generally improved, especially for mathematics, and scientific literacy. In the PISA test at 2012, ranking literacy in mathematics and science was 65 and 64, while the areas of reading literacy in 61 of 65 countries. The average score on the PISA tests at 2015 were as follows 386 for math, 403 for science, and 397 for reading. The average score on the PISA tests at 2012 were as follows 375 for math, 382 for science, and 396 for reading [3,4]. PISA test involved four content namely (1) the quantity, (2) space and shape, (3) change and relationship, and (4) uncertainty and data [3, 4, 5]. There were six levels in the PISA questions related to mathematical literacy of students [1].

In 2017, Hongki et al. [6] revealed that 55.56 % of junior high school students achieved level 5 for change and relationship problem. Furthermore, 77.78 % of junior high school students achieved level 2 for change and relationship problem. These results lead us to the question: how is the teacher's mathematical literacy ability for the change and relationship problems?

Currently, studies that related to teacher's abilities in solving PISA-adapted mathematics problems are considered limited. Therefore, the objective of this paper was to describe the junior high school teacher ability for the PISA adaptation test in the quantity, space and shape, change and relationship, and uncertainty domain. This paper only describes the teachers' ability to solve the adaptation PISA test in the change and relationship domain.

If a student is equipped with mathematical literacy skills, he/she would understand the role of mathematics in his/her life [2,7,8]. Mathematical literacy is an individual's ability to identify and understand the role of mathematics in the world, to make an accurate assessment, use and involves mathematics in various ways to fulfill the individual needs as a reflective, constructive and filial citizen [5,8]. The following competencies would form the mathematical literacy skills, namely: (1) the mathematical thinking and reasoning competence, (2) the logical argument competence, (3) the mathematical communicating competence, (4) the problem model competence, (5) the proposing and solving problem competence, (6) the representing idea competence, and (7) the using of symbol and formal language competence [5,8].

Mathematical literacy was important because in the 21st century humans not only required a content knowledge, but also required skills that called as 21st century skills. The 21st skills include critical thinking and problem solving, creativity and innovation, communication and collaboration, flexibility and adaptability, initiative and self-direction, social and cross-cultural, productivity and accountability, leadership and responsibility, and information literacy [8,9]. Mathematical literacy became one of the components necessary to build 21st century skills [5].

2. Method

This research was design research. The goal of this study was to describe teachers' mathematical literacy for the change and relationship problem on PISA adaptation test. This goal was achieved by the following procedures:

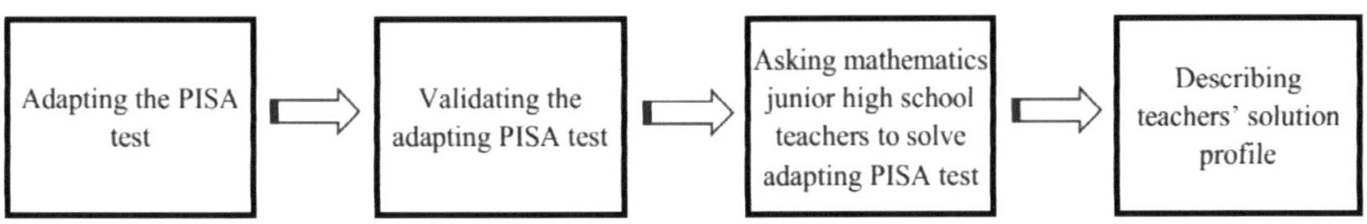

Figure 1. Research procedures.

This procedures followed research steps introduced by Akker [10]. The PISA adaptation test consisted of three change and relationship problems, four space and shape problems, two uncertainty problems, and four quantity problems. The subjects of this research were seven junior high school teachers in Yogyakarta and surrounding areas. The school was selected randomly, and each subject was the best teacher from the selected schools.

3. Results and discussion

This graph shows how the speed of a racing car varies along a flat 3 kilometer track during its second lap.

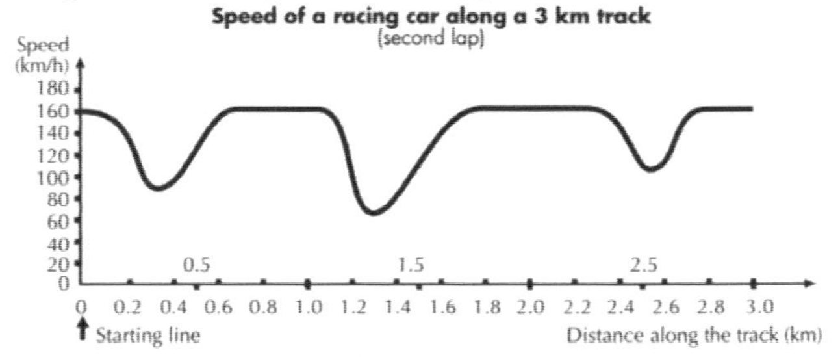

a. Where was the lowest speed recorded during the second lap?
   A. at the starting line. C. at about 1.3 km.
   B. at about 0.8 km. D. halfway around the track.

b. What can you say about the speed of the car between the 2.6 km and 2.8 km marks?
   A. The speed of the car remains constant. C. The speed of the car is decreasing.
   B. The speed of the car is increasing. D. The speed of the car cannot be determined from the graph.

Figure 2. Problem 1.

3.1. The teacher's answer for problem 1a

According to the graph presented in Figure 1, horizontal axis represented the distance along the track, and vertical axes represent cars' speed. Therefore, the lowest speed is represented as the deepest valley on the graph. Thus, the answer to the question is C. All teachers answered the question correctly. They can answer questions involving familiar contexts where all relevant information is presented, and the questions are clearly defined. Thus, we can conclude that all or 100% of teachers were in the first level.

3.2. The teacher's answer for problem 1b

At the interval (2.6, 2.8), the graph was increasing monotone. Thus, the answer is B, because the speed of the car was increasing. All teachers answered the question correctly. They can answer questions involving familiar contexts where all relevant information is presented, and the questions are clearly defined. Thus, all or 100% teacher are in the first level. The following was the example of the teacher's answer for problem 1a and 1b.

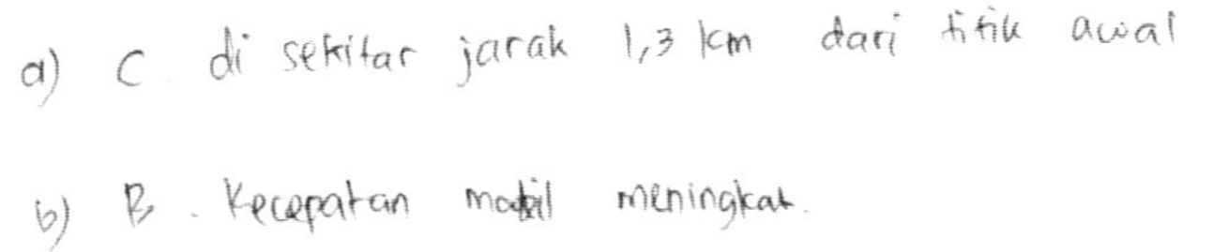

Figure 3. The example of the teacher's answer for problem 1a and 1b.

People living in an apartment building decide to buy the building. They will put their money together in such a way that each will pay an amount that is proportional to the size of their apartment. For example, a man living in an apartment that occupies one fifth of the floor area of all apartments will pay one fifth of the total price of the building.

a. There were three apartments in the building. The largest, apartment 1, has a total area of 95 m^2. Apartments 2 and 3 have areas of 85 m^2 and 70 m^2 respectively. The selling price for the building is 30 billion rupiah. How much should the owner of apartment 2 pay? Show your work.
b. Circle Correct or Incorrect for each of the following statements:

Statement	**Correct / Incorrect**
A person living in the largest apartment will pay more money for each square meter of his apartment than the person living in the smallest apartment.	Correct / Incorrect
If we know the areas of two apartments and the price of one of them we can calculate the price of the second.	Correct / Incorrect
If we know the price of the building and how much each owner will pay, then the total area of all apartments can be calculated.	Correct / Incorrect
If the total price of the building were reduced by 10%, each of the owners would pay 10% less.	Correct / Incorrect

Figure 4. Problem 2.

3.3. Teacher's answer for problem 2a

Generally, all teachers used the same technique to solve this problem. First, they counted the total area of the apartment, that is 95 m^2 + 85 m^2 + 70 m^2 = 250 m^2. After that, by using worth comparison, they counted the price of apartment 2. The result was IDR 10.200.000.000 or some teacher simply wrote 10.2 M. From this solution, we knew that teachers can interpret agreement between apartment buyer and seller and represent it by using comparison. Thus, we can conclude that all or 100% teacher was in the level 3.

Table 1. Teacher's answer of problem 2a.

Statement	Answer
A person living in the largest apartment will pay more money for each square meter of his apartment than the person living in the smallest apartment.	Incorrect
If we know the areas of two apartments and the price of one of them we can calculate the price of the second.	Correct
If we know the price of the building and how much each owner will pay, then the total area of all apartments can be calculated.	Incorrect
If the total price of the building were reduced by 10%, each of the owners would pay 10% less.	Correct

3.3.1. Teacher's answer for problem 2b.1. Four teachers answered "incorrect" for question 2b.1 and three teachers answered "correct" for this problem. Thus 57.14% teachers were in the level 4 because they can construct and communicate the reasons why they answer "incorrect" based on their interpretation of proportional understanding. About 42.86% teacher could not be leveled.

3.3.2. Teacher's answer to problem 2b.2. Five teachers answered "correct" for question 2b.2, one teacher answered "incorrect" for this problem and one teacher did not answer the question. Thus 71.43% of teachers were in the level 3 because they can communicate their interpretation of given information in the problem, the result of their thinking, and the reason for their answer. About 14.29 % teacher could not be leveled, and 14.29% teacher did not answer the question.

3.3.3. Teacher's answer for problem 2b.3. Two teachers answered "incorrect" for question 2b.3, four teachers answered "correct" for this problem, and one teacher did not answer the question. Thus 28.57% of teachers were in the level 3 because they can communicate their interpretation of given information in the problem, the result of their thinking, and the reason for their answer. About 57.14% teacher could not be leveled, and 14.29% teacher did not answer the question.

3.3.4. Teacher's answer for problem 2b.4. Six teachers answered "correct" for question 2b.4, and one teacher answered "incorrect" for this problem. Thus 85.71% teachers were in the level 4 because they can construct and communicate the reasons why they answer "correct" based on their interpretation of proportional understanding. About 14.29% teacher did not answer the question.

In 1998 the average height of both young males and young females in the Netherlands is represented in this graph. According to this graph, on average, during which period in their life are females taller than males of the same age?

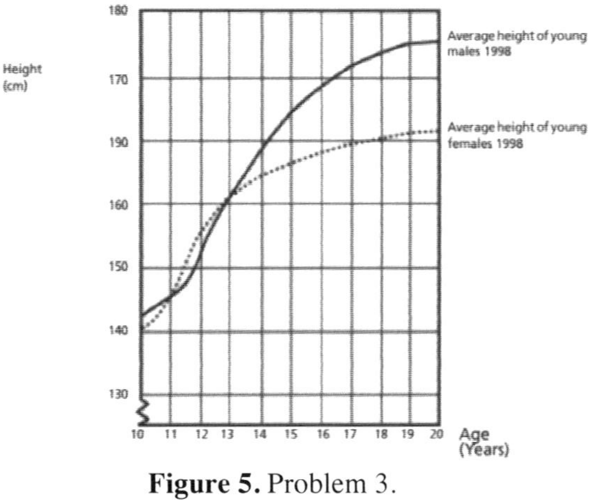

Figure 5. Problem 3.

3.4. Teacher's answer to problem 3

From Figure 5, we knew that the graph of the average height of young males in 1998 is above the graph of the average height of young females 1998 before 11 years old or after 13 years old. It means that at that period, males are taller than females of the same age. During period 11 – 13 years old, the graph of the average height of young males in 1998 is under the graph of the average height of young females in 1998. It means that at that period females are taller than males of the same age. Five teachers answer the question correctly. They claim that females are taller than male in the period 11 – 13 years old. Therefore, those five teachers were in the third level or 85.71% of the teachers are in the third level because they can communicate their interpretation of given information in the problem, the result of their thinking, and the reason of their answer. One teacher gave a different answer. She stated that females are taller than male at the age of 11 and 14 years old. Unfortunately, she didn't give any reason for her statement. Therefore, we can't level this teacher. The following is the example of the teacher's answer:

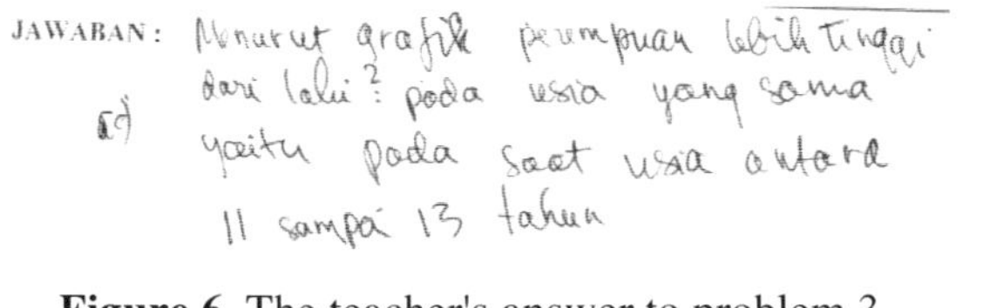

Figure 6. The teacher's answer to problem 3.

The result obtained by the teachers could be summarized in Table 2:

Table 2. The teachers' ability for the change and relationship problems for the PISA adaptation test.

Problem	Teacher's Achievement Level	Reason	The number of teachers	Percentage
1a	Level 1	Teachers can answer questions involving familiar contexts where all relevant information is present, and the questions are clearly defined.	7	100 %
1b	Level 1	Teachers can answer questions involving familiar contexts where all relevant information is present, and the questions are clearly defined.	7	100 %
2a	Level 3	The teacher can interpret the agreement between apartment buyer and seller and represent it by using comparison.	7	100 %
2b.1	Level 4	Teachers can construct and communicate the reasons why they answer "incorrect" based on their interpretation of proportional understanding.	4	57.14 %
	Could not be leveled	-	3	42.86 %
2b.2	Level 3	Teachers can communicate their interpretation of given information in the problem, the result of their thinking, and the reason for their answer.	5	71.43 %
	Didn't answer the question	-	1	14.29 %
	Could not be leveled	-	1	14.29 %
2b.3	Level 3	Teachers can communicate their interpretation of given information in the problem, the result of their thinking, and the reason for their answer.	2	28.57 %
	Didn't answer the question	-	1	14.29 %
	Could not be leveled	-	4	57.14 %
2b.4	Level 4	Teachers can construct and communicate the reasons why they answer "correct" based on their interpretation of proportional understanding.	6	85.71 %
	Didn't answer the question	-	1	14.29 %
3	Level 3	Teachers can communicate their interpretation of given information in the problem, the result of their thinking, and the reason for their answer.	6	85.71 %
	Could not be leveled	-	1	14.29 %

From the table above, we knew that teacher's ability in solving problem level 1 - 4 is good. Furthermore, from [1,2] we knew that teacher's mathematical abilities were related to student's achievement. These facts were chance for us to upgrade student's literacy ability by choose appropriate pedagogy to teach them.

4. Conclusions

Based on the results, it can be concluded that: (1) the first level one problem could be solved by all teachers; (2) the first level three problem could be solved by all teachers; (3) the second level three problem could be solved by five or 71.43 % teachers; (4) the third level three problem could be solved by two or 28.57 % teachers; (5) the fourth level three problem could be solved by six or 85.71 % teachers; (6) the first level four problem could be solved by four or 57.14 % teachers; (7) the second level four problem could be solved by six or 85.71 % teachers.

Acknowledgments

Our thanks to the Ministry of the Research, Technology and Higher Education has been funding this research so we could finish this research and present this publication through Penelitian Strategis Nasional Institusi Tahun 2018 grant distributed through the Sanata University Dharma.

References

[1] Campbell, Patricia F, et. all. 2014 *Journal for Research in Mathematics Education* **45** 419
[2] Julie H 2017 *International Journal of Science and Applied Science: Conference Series* **1** 55
[3] OECD 2013 *PISA 2012 Results: What students know and can do. Student Performance in mathematics, reading, and science* (Paris: OECD)
[4] OECD 2012 *Assessment Framework. Key Competencies in Reading, Mathematics and Science* (Paris: OECD)
[5] Julie H and Marpaung Y 2012 *Widya Dharma* **23**
[6] Hongki J et. al. 2017 *J. Phys.: Conf. Ser.* **890** 012089
[7] Christiansen I B 2006 *Pythagoras* **64** 6
[8] Stacey K 2011 *Journal Mathematics Education* **2** 95
[9] Wijaya A 2016 *IndoMS Journal Mathematics Education* **7** 73
[10] Akker, Den J V, Gravemeijer K, McKenney S and Nieveen N 2006 *Educational Design Research* (New York: Taylor and Francis Group)

IOP Conf. Series: Journal of Physics: Conf. Series **1088** (2018) 012052 doi:10.1088/1742-6596/1088/1/012052

Junior high school mathematics teachers' pedagogical content knowledge in teaching of polyhedra

Ma'rufi, M Ilyas and Salwah
UniversitasCokroaminotoPalopo, Jl. Latammacelling No.19, Tompotika, Wara, Palopo91911, Indonesia

E-mail: marufi.ilyas@gmail.com

Abstract. A teacher plays an important role in developing students' knowledge and skills in learning mathematics. This article focuses on the development of mathematics teachers' Pedagogical Content Knowledge (PCK) in the teaching of polyhedra, particularly on cube and cuboid for Year 8 students. PCK is a special knowledge for teacher integrating content knowledge, pedagogical knowledge, and knowledge of students. Content knowledge is defined as knowledge on facts, concepts, relations among concepts, and procedures about mathematics, particularly the concept of polyhedra. Pedagogical knowledge refers to teacher's knowledge in organizing teaching and learning process of polyhedra, while knowledge of students is defined as teacher's knowledge on students' difficulties and mistakes on polyhedra. Data were gathered by recording the teaching and learning process and interviewing the students. The data were analyzed by reduction, categorization, and interpretation. The result showed that the teacher tended to combine the conceptual and procedural explanation to correct the mistakes and address the difficulties of students on polyhedra by providing scaffolding and employing peer tutoring.

1. Introduction

The law on teachers and lecturers states that teachers in carrying out their duties must have four competencies: pedagogic, professional, personality, and social. The competencies indicate that the teacher should; (1) knowing the students well (2) master the subject and the teaching materials, (3) be able to organize educational learning, which includes the plan and implementation of the learning, the evaluation of learning process and outcomes, and follow-up for the improvement of learning, 4) develop the personality and professionalism in a sustainable manner. This is in line with the meaning of Pedagogical Content Knowledge (PCK) mentioning that teacher is one of the critical factors of students' achievement. Teachers are required to master the content subject and an effective way to deliver it, as well as interpret the students' thinking so that they understand the subject matter well [1]. The development of teachers' PCK can be done through training as professional development and training are important to improve teacher's PCK in mathematics learning [2,3].

"PCK is a teacher's method for understanding student cognition so that students can reduce misunderstandings, difficulties, and misconceptions" (p.9) [4]. Students' difficulties in learning mathematics at schools are closely related to the mathematical and pedagogical knowledge of the teachers. "Pedagogical content knowledge was defined as teacher's ability has to use their knowledge of mathematics to "unwrap" the mathematical topics and present the content in ways for students to successfully learn mathematics"[5]. A teacher should be able to use his or her mathematical knowledge to examine mathematics topics, using various means and representations to strengthen students' understanding, thus reducing student difficulties, errors, and misconceptions [6].

2. Method

This research aimed at developing a Mathematics Mentoring model based on Pedagogical Content Knowledge (MGM-PCK). The model developed in this research referred to the Plomp's model [7] consisting of five phases, namely: (a) initial assessment phase, (b) design phase, (c) construction phase, (d) testing, evaluation, and revision phase, and (e) implementation phase. This article focuses on the development of PCK of mathematics teachers in the learning of polyhedra in Year 8 during the pilot study.

Data were collected through the recording of learning activities and interviews. Two lessons were recorded in this research. Researchers and five lecturers from the mathematics education program were directly involved in data collection. The instructional videos were repeatedly observed and then transcribed. Data were analyzed using the qualitative analysis including the reduction, categorization, interpretation and drawing conclusion. Components of PCK in this research were: (1) teacher's knowledge about polyhedra, (2) teacher's pedagogical of knowledge polyhedra learning (3) teacher's knowledge in overcoming student's difficulties and mistakes on polyhedra.

3. Result

PCK of Year 8 mathematics teachers on polyhedra includes knowledge of polyhedra, pedagogy, and knowledge of students. The topic for lesson 1 is to discover the formula of the surface area of the cube and rectangular, while lesson 2 is to find the formula of the surface area of the prism. The learning outcomes of each session are described in the following section.

3.1. Lesson 1

3.1.1 Knowledge of subject matter

The teacher did not directly explain the facts related to cubes and rectangular at the beginning of the lesson. Instead, they were delivered through examples and questions that guided students to construct definitions of terms in cubes and cuboid, such as edge and face. Teachers also always related the concept of the cube/cuboid with the prerequisite material, i.e., plane geometry (square and rectangle). Besides, the link between concepts was explained by analogizing the geometric shape of a cube/cuboid with the form of concrete objects that are familiar to the students and can easily be found around them such as cabinet and shoe boxes. Procedurally, the teacher also asked about the formula of area of a square ($A = face^2$) and the area of a rectangle ($A = l \times w$) and then wrote it on the board correctly. Nevertheless, there was one term that was still difficult to understand by the students, the term of congruency, since it had not yet taught in Year 8. Therefore, the teacher used another congruent replacement term that is, two plane geometry of the same shape and size. On the other hand, the mathematical procedure was used to find out the formula of the surface area of the cube and the cuboid that had been written in the student worksheets but cannot be appropriately done by the students.

3.1.2 Knowledge of pedagogy

The steps to achieve the learning objectives include: the teacher conveying apperception on the area of plane geometry; presenting the lesson objectives; showing the cube and cuboid through video; distributing worksheets and media to each group of students; guiding each group to work on the student worksheet; asking a group to explain the work, and confirming the conclusion together.

The broad concept of cubes and cuboid was represented through visual media in the form of animated video consisting of the formation of cubes and cuboid from their nets and concrete media in the form of cubes and cuboid nets made of cardboard. The use of this two media aimed to show the students the process of opening a cube/cuboid into a net, and by calculating the total area of the net, it is expected that the student can understand that it is the process of discovery of the concept area of polyhedra. In the implementation of learning, visual media was used at the stage of information delivery. Concrete media was used in the core learning activities, by sharing media with each group of students to encourage them to be active learners. Students started to be engaged when they hold the cube and cuboid, cut, and measured them. During the learning activities, the teacher did not explain the material/instructions on the board, but she/he visited each group. The teacher gave direction to each student to investigate the model of cube nets. However, not all students did so. Therefore, the teacher provided limited assistance with scaffolding techniques on difficult parts. Teacher's role was still dominant so that the planned guided discovery did not work effectively since the students were not accustomed to working independently and waited for instructions from the teacher. This process took longer than expected and as a result, individual and group assessments could not be conducted due to the insufficient time.

3.1.3 Knowledge of students

The teacher explored students' background knowledge by asking several questions related to the characteristics of cubes and cuboid including the number of edges, vertices, faces, the area of squares and rectangles, asking them to analogize objects resembling cubes/cuboid in their surroundings so that they can distinguish the characteristics of both. When a student had difficulty in answering the point mentioned previously, the teacher guided him/her by showing the cube and cuboid models through a concrete media and demonstrated how to open the cube/cuboid nets in each group. Besides, misconceptions also occurred when students referred to edges as faces. To overcome this, the teacher re-explained the difference of square as plane geometry and the cube as a polyhedron. For a square, the boundaries are called sides, while for a cube, the face is a plane bounded by four edges. Another challenging concept for students concerned the formula of each face of cuboid and cube into the formula of surface area. Students could only calculate the size of each edge and then multiplied it, but they were not successful in writing it into an algebraic form of $A = 2\,lw + 2\,lh + 2\,wh$. Besides these challenges, there was also a misconception of students mentioned that the kite was a square and a triangle was a polyhedron. This occurred during the apperception. The teacher then gave a personal explanation to each student having this misconception. The teacher also always motivated students who seemed to be lethargic and less enthusiastic.

3.2. Lesson 2

3.2.1 Knowledge of subject matter

The facts, definitions, and symbols used in the lesson can be seen when the teacher drew and showed some plane geometry like square and triangle, but she/he were not consistent in naming the sides. Each side can be called with the letter/name of the corner facing the side. However, the teacher did not write it down. For the relations between concepts, the teacher associated the shape of the prism base with particular plane geometry. If the prism base is square, then it is called a rectangular/square prism, similar case for a triangular base. The teacher also wrote the general formula and the circumference of some plane geometry that can be the base of the prism such as the area and circumference of the square and the area and circumference of the triangle. To relate between concepts and formulas, the teacher provided explanation for some mathematical procedures associated with the invention of the surface area formula of the prism; the student first determined the base area and the area of the prism cover (top base) and then discovered the area of the vertical faces of prisms and summed the three.

3.2.2 Knowledge of pedagogy

Teachers' learning steps were in line with the lesson design, starting with the apperception of several forms of plane geometry, area and circumference, conveying the learning objectives, briefly explaining the surface area of the prism, distributing student worksheets to each group, working with the students, asking one of the groups to present their work, and confirming the conclusion. In explaining the materials, the teacher tended to explain the contents and relevant information of the student worksheet directly.

The surface area of the prism was explained through cardboard of the net of triangular prisms. The teacher communicated the representation of a prism in the form of the net, determined the area of each net, and then summed the area of each face to obtain the total surface area of the prism. The teacher also gave some examples of plane geometry at the beginning of learning such as triangle, square, and how to determine the square area, determine the height of the triangle, and the triangle area contextually. The examples were written on the board, and then the teacher explained them to the students. In addition to the examples, teachers also assigned individual tasks for the students to solve two problems. To engage students in group work, the students were also encouraged to ask questions. In this case, teachers selected the student randomly to obtain some questions from them. Some students were also asked to help their friends who were struggling to understand the worksheet. In term of assessment, assessment instruments were not used, however, overall, teachers always appreciated the students by praising and giving applause for those who actively proposed questions and provided correct answers.

3.2.3 Knowledge of students
Teachers may find the initial knowledge of students through the apperception of the square area, the height of the triangle, and triangle area by asking. Not many students remembered them, so the teacher re-explained how to determine the square area, the height of the triangle, and the triangle area. When there were students facing difficulties, the teacher guided them by giving simple examples (scaffolding) and slowly reduced the amount of assistance until the students can independently solve the problem. Any possible difficulties, errors, and misconceptions described by the teacher during the planning, just as students would have difficulty composing algebraic forms from the surface area of the prism, the error in determining the sum of each faces in the prism net, and misconceptions about the prism base on the position of the prism. To overcome these issues, the teacher explained the correct concepts in rotation from one group to another. Because of the high enthusiasm and curiosity of students, the teacher needed to perform this process repeatedly, explaining the same thing to each group.

4. Discussion
4.1. Lesson 2
The knowledge of Year 8 high school teachers of polyhedra are divided into three categories: factual, conceptual and procedural knowledge. Factual knowledge is shown by facts and mathematical symbols associated with polyhedral presented accurately in the lesson. The knowledge of concepts and their relationships between mathematical concepts was explained accurately; the teacher properly connected the concept of plane geometry and polyhedra, mathematical formulas such as the area of a square, rectangle, and triangle were well defined; mathematical procedures related to polyhedra was explained in systematic steps. Knowledge of subject matter is referred to knowledge of facts and concepts of mathematics and their relationships [8]. Likewise, knowledge of subject matter is classified into two categories, namely conceptual knowledge and procedural knowledge [9]. Knowledge of teaching materials may include relating certain mathematical concepts to others and explaining the underlying reasons for certain mathematical procedures. Teachers should understand the rules, procedures, and facts of mathematics and the reasons underlying the procedures to be able to enhance students' conceptual understanding.

4.2. Lesson 2
The pedagogical knowledge of Year 8 teachers in learning polyhedra is divided into two categories: learning organization and assessment. Knowledge of mathematics learning strategies includes the ability of teachers to use appropriate activities in mathematics learning, real-life examples and analogies in mathematical learning as well as different learning strategies in presentation, and the ability to use different representations in mathematics learning [10]. Mastering the teaching material is

not enough to achieve the desired learning objective [11]. This indicates that teachers should be able to teach the material in a way to be easily understood by the students.

The learning organization of teachers was shown by the use of discovery method including the use manipulative and worksheets to guide students to discover the formula of the surface area of cubes, cuboid, and prisms; to explore students' prerequisite knowledge by relating the lesson to the previous one. Furthermore, teachers involved students in learning by giving them the opportunity to find their formula of the surface area of cubes, cuboid, and prisms. Students' active learning was created through teacher's explorative questions. The strategy used by teachers was to provide scaffolding to students during the learning. The teacher assessed the learning process through performance appraisal. Teachers need to have different knowledge and skills to plan and implement learning strategies that encourage students' understanding and learning.

4.3 Knowledge of students

The teacher's knowledge of the students in learning polyhedra is divided into two categories: teacher's knowledge about student's difficulties and how to overcome the difficulties. The teacher obtained the information about the student's difficulties based on the same questions from each group. The teacher explored the students' initial knowledge by asking questions related to the topic to be studied. The teacher overcame the errors, difficulties, and misconceptions of students by giving a simple analogy and applying the scaffolding technique. It is essential for the teachers to recognize certain aspects of the topic that may cause problems for students, teachers should also have the ability to address the issues faced by the students regarding learning materials [12]. Teachers should not only confirm that a solution is correct but also explain why it is correct [13]. Teachers who have content knowledge of polyhedra and good pedagogical knowledge but are lack of the knowledge of students will have difficulties in formulating learning strategies that meet the characteristics of students as they do not have a clear picture of possible errors and student misconceptions.

5. Conclusion

This research shows that PCK is important and necessary in a learning, including learning polyhedra in Year 8. The PCK of Year 8 teachers of in polyhedra is the teacher's knowledge of the polyhedra integrated with the knowledge of students which have an impact on teacher pedagogical knowledge in the learning organization. Knowledge of students requires both the knowledge of teaching materials as well as the pedagogical knowledge.

References

[1] Ma'rufi 2016*The 1st International Conference on Mathematics, Science, and Computer Science (ICMSC)* (Balikpapan: American Institute of Physics) p 1

[2] Lee J 2010 Exploring kindergarten teachers' pedagogical content knowledge of mathematics: Department of Curriculum and Instruction-EC4, College of Education and Health Profesional*International Journal of Early Childhood***42**27

[3] Tanisli D and Kose N Y 2013Preservice Mathematics Teachers' Knowledge of Students about the Algebraic Concepts*Australian Journal of Teacher Education***38**

[4] Shulman L S 1986 Those who understand: Knowledge growth in teaching *Educational Researcher***15** 4

[5] Ma L 1999*Knowing and teaching elementary mathematics: Teachers' understanding of fundamental mathematics in China and the United States*(New York: Routledge)

[6] Ma'rufi 2017*International Conference on Natural and Social Sciences* (Makassar: Universitas Cokroaminoto Palopo) p 264

[7] Plomp T 1997 *Educational and training system design.* (The Netherlands: University of Twente)

[8] Kilic H 2011*Turkish Online Journal of Qualitative Inquiry***2**

[9] Maat S M and Zakaria E 2014Analyzing pedagogical content knowledge of algebra using confirmatory factor analysis *Indian Journal of Science and Technology***7** 249

[10] Guzel E B 2010 *Journal Scientific Research and Essay***5** 1872

[11] Isiskal M and Cakirogu E 2011*Jurnal Math Teacher Education***14**213

[12] Hawkins W J 2012An investigation of primary teachers' pedagogical content knowledge when teaching measurement three and four*12th International on Mathematics Education*

[13] Kleve B 2009Aspects of a teacher's mathematical knowledge on a lesson on fractions*Proceeding of The British Society for Research into Learning Mathematics*vol29 p 67

The effect of creative problem-solving learning model using geometry transformation book based on Al-Qur'an on students' van Hiele thinking level and learning outcome

F Kristanti, C Ainy and S Shoffa
Universitas Muhammadiyah Surabaya, Jl. Sutorejo No. 59, Surabaya 60113, Indonesia

E-mail: febriana.pendmat@fkip.um-surabaya.ac.id

Abstract. This study aimed at analyzing the differences between Creative Problem Solving and conventional learning model to students' van Hiele geometric thinking level and learning outcome. This study used quasi-experimental design, and this paper focused on the implementation phase of design research. Research subjects were determined by random sampling technique, i.e., an experimental class of 12 students and a control class of 11 students. The Geometry Transformation book was developed based on Al-Quran and includes a discussion on the relationship between geometry concepts and Al-Qur'an. The results of this study showed that the level of geometric thinking ability of 33.33% of the students in the experiment class is at level 4, while only 9.09% of the control class is at the same level. The ANAKOVA test showed that there was a significant difference ($P = 0.041 < \alpha = 0.0050$) in cognitive learning result between the experimental class (63.08) and the control class (59.27).

1. Introduction

The development of education and teaching is closely related to the progress of Science, Technology, and Arts. Furthermore, from the Islamic perspective, the survival of humankind depends on how human beings develop their scholarship in receiving education and teaching as life regulators. The presence of the ASEAN Economic Community (AEC) with its free market system constitutes a form of the impact of the progress in the fields of science and technology on economics. On the other hand, the development of science and technology can negatively impact and even endanger human life and dignity. Currently, ethics and modesty are often forgotten by society [1] and this can be a threat to the next generation. Thus, university as a higher education institution is tasked with organizing education and teaching based on cognitive abilities. [2]. The Department of Mathematics Education of Muhammadiyah University of Surabaya has developed a learning process using Al-Quran-based Geometry Transformation. Therefore, the role of religion as a guide of life becomes the main lesson to be taught based on Islamic aqidah.

CPS model is a variation of learning by problem-solving through systematic techniques in organizing creative ideas so that the learning can take place engagingly, thus generating motivation and encouraging students to construct their knowledge [3]. In this case, the basic skills students need in CPS learning are the abilities to (1) declare the sequence of problem-solving steps; (2) find possible problem-solving strategies; (3) evaluate critically; (4) choose an optimal solution option; and (5) develop a plan for implementing problem-solving strategies. The goals of CPS instruction are: (1) students are able to state the sequence of problem-solving steps in CPS; (2) students are able to find possibilities of CPS strategy; (3) students are able to evaluate and select those possibilities in relation to the existing criteria; (4) students are able to choose an optimal solution option; (5) students are able to develop a plan in implementing CPS strategy; and (6) students are able to articulate how CPS can be used in various fields or situations. Furthermore, as a combination of Von Oech and Osborn procedures, CPS steps in mathematics learning are namely (1) problem orientation; (2) finding facts; (3) finding problems; (4) finding ideas; (5) finding solutions; (6) finding acceptance [5]. CPS learning model emphasizes understanding a concept and the importance of direct experience involved in the learning process in building meaningful experiences on the students. Meaningful learning process means that the learning can make the students understand and realize that a mathematics concept has meaning and usefulness [6]. Therefore, CPS learning is designed based on Al-Quran by connecting mathematical concepts with the Qur'an, especially in Transformation Geometry topic. The learning design using Al-Qur'an approach connects mathematics and Qur'an, especially in geometry. The relationship between geometric concept and the Qur'an is included in a textbook that serves as an additional reference book for students. Each chapter of the textbook will be extracted from the essence or outline of the problems associated with Islamic values to build meaningfulness in the process of Islamic geometry learning.

Geometry is learned with a concept that can be used to motivate and attract students' attention and stimulate their imagination. Activities in geometry can be used to introduce new ideas by starting with something tangible, an experience that manipulates to produce new insights [7]. The geometrical ability can be measured using van Hiele theory of five thinking levels. van Hiele geometry thinking levels are (1) level 0 (visualization); when students are on shape recognition level; (2) level 1 (analysis); when students' thinking ability develops by describing a shape using their language according to the previous level, (3) level 2 (abstraction); when students use a language to know the difference of each shape according to the previous level; (4) Level 3 (informal deduction), when students' deductive thinking has begun to develop and deductive reasoning as a way to build geometric structures in an axiomatic system has been understood; (5) Level 4 (formal deduction), when students can work in various axiomatic deduction structures [8]. The improvements of geometric thinking ability in each van Hiele level is used by a lecturer to help students improve their geometric thinking ability as indicated in students' learning outcome.

Learning outcomes can be assessed and measured that are used as evaluations to determine the success of a learning process. There are three domains of learning outcomes to be studied as the indicators of student learning outcomes, namely cognitive, affective, and psychomotor domains. Cognitive domain is measured through tests. Affective domain assessed is the Islamic attitude that supports students' understanding of mathematical concepts. Islamic attitudes students can obtain through CPS learning model on Quran-based books include: (1) Honesty, as seen in the need for absolute and unchangeable consistency of truth based on facts in the process of solving mathematical problems; (2) Patience, as seen when the students cope with a complicated and long mathematical completion process that requires a spirit of unyielding; and (3) Carefulness and thoroughness, which can be seen during the mathematics learning process that demands high level of carefulness and analysis. These three traits are closely related to the values contained in the Qur'an [9].

The CPS learning model can improve the mastery of concepts and skills [10]. In addition, CPS learning model also affects the ability of mathematical problem solving [11]. Therefore, the purpose of this study is to find out the influence of the complementary Quran-based CPS learning model textbook on student's van Hiele thinking levels and learning outcomes in Transformation Geometry course. Therefore, this study aimed at analyzing the differences between Creative Problem-Solving and conventional learning models in students' van Hiele geometric thinking levels and learning outcome.

2. Method

This research used quasi-experimental research method which was applied to two groups: experimental group (E), treated with Creative Problem-Solving learning model using Geometry Transformation's book based on Al-Quran, and control group (K), treated with conventional learning model. The research design used Pre-test-Post-test Control Group Design; both the experiment and control groups took a pre-test and a post-test before and after learning. The research involved 23 students who were divided into two classes, and the Department of Mathematics Education of Muhammadiyah University of Surabaya were chosen as the research population. Sample determination was done randomly through cluster sampling, with an experimental group of 12 students and a control group of 11 students. The two instruments used were treatment and measurement. Syllabus treatment instrument was in the form of syllabus, lesson plan, and student worksheet. Measurement instruments used in this study consisted of learning assessment instrument (cognitive, affective, and psychomotor) and van Hiele geometry rating instrument of thinking levels. Content validation by three experts was conducted before the instruments were used. Empirical validity test was conducted on 40 mathematics education students in State Islamic University of Sunan Ampel Surabaya, assisted by a set of validity test using SPSS program. The test showed that 5 out of 6 test questions on cognitive learning outcomes were valid. Only 1 test question was not valid, while the other 5 test questions of geometric ability made based on the indicators of van Hiele's levels of thinking were all valid. The test on the questions dealing with cognitive learning outcome yielded the coefficients of 0.761, while the test on geometry based on van Hiele's thinking levels yielded the coefficients of 0.784. The prerequisite test resulted in homogeneous and normal data. Descriptive and covariance analyses (ANAKOVA) were used in data analysis. Descriptive analysis was used to describe the learning process of students' psychomotor and affective assessment. Covariance analysis was used to test the research hypothesis.

The affective learning outcome of Creative Problem-Solving learning model using Geometry Transformation's book based on Al-Quran was obtained by observing the students' Islamic attitudes during the learning process, which include: (1) Honesty; Mathematics is a science that is closely related to honesty as can be observed in how the students solve math problems. A mathematical problem is a matter of consistent, absolute and unchangeable truth. This is in accordance with the meaning of honesty; saying something with facts. (2) Patience; Learning mathematics teaches students to be patient in dealing with everything in life. Solving difficult mathematical problems requires patience and persistence to arrive at the correct result. (3) Carefulness and thoroughness; learning mathematics trains students to be more careful and thorough in their action to avoid misunderstanding the graphs or miscalculations.

3. Results and discussion

In this study, students' learning outcomes are in the form of cognitive, affective, and psychomotor learning outcomes. The cognitive learning outcomes were measured through a written test. Description of the cognitive test learning outcomes is shown in Table 1. Table 1 indicates that the mean values of the control class pretests are greater than those of the experimental class, while the mean score of post-test of the experimental class is greater than that of the control class. The mean value of cognitive learning outcome of experimental class (63.08) is higher than that of the control class (59.27). This implies that the treatment given to the experiment class was able to produce greater post-test value than the untreated control class. The result of the analysis of the influence of CPS learning model based on Al-Qur'an approach to the cognitive learning outcomes showed that there are significant differences in post-test values of cognitive learning outcomes between experimental and control classes ($p = 0.000 < \alpha = 0.050$). Furthermore, there is a difference of post-test value in terms of cognitive learning outcomes between experimental class and control class as indicated by the significant pre-test data value ($p = 0.041 < \alpha = 0.050$)

IOP Conf. Series: Journal of Physics: Conf. Series **1088** (2018) 012053 doi:10.1088/1742-6596/1088/1/012053

Table 1. The average value of cognitive learning outcome.

	Pretest	Post-test		
	Mean	Mean	Std. Deviation	N
Experiment	52.6	63.08	1.17	12
Control	56.0	59.27	1.22	11
Total	108.6	122.36	2.39	23

Psychomotor and affective learning outcomes were obtained through observation during the learning process. Psychomotor assessment includes assessment of work performance by observing students' activities in completing Student Worksheet. The assessment aspects include (1) Completeness, (2) Accuracy of calculation and (3) Accuracy of writing formulas and units. Description of the average value of psychomotor learning outcome can be seen in Table 2.

Table 2. The average value of psychomotor learning outcome.

	Psychomotor learning outcome			Average
	Completeness	Accuracy of calculation	The accuracy of writing formulas	
Experiment	87.5	78.90	83.00	83
Control	81.0	75.40	78.00	78
Average	84.3	77.15	80.50	81

Affective learning outcome was obtained by observing the students' Islamic attitudes during the learning process, which include: (1) Honesty, (2) Patience, and (3) Carefulness and thoroughness. The overall description of the affective learning outcome can be seen in Table 3. Based on Table 3, general the average value of affective learning outcome of the experimental class is higher than that of the control class.

Table 3. The average value of affective learning outcome.

	Affective learning outcome			Average
	Honesty	Patience	Careful and thorough	
Experiment	82.3	81.80	85.80	83
Control	81.0	77.10	73.50	77
Average	81.7	79.45	79.65	80

The data on students' geometric ability can be measured using 5 description questions. The data on the van Hiele thinking rate is determined by categorizing each item. Category 1 is at level 0, category 2 at level 1, category 3 at level 2, category 4 at level 3, and category 5 at level 4. The description of van Hiele's levels of thinking can be seen in Table 4. Table 4 shows that the average pretest value of van Hiele thinking levels in the control class is greater than that of the experimental class, while the average of the experiment class post-test is greater than that of the control class.

Table 4. Geometry ability based on van Hiele's thinking level.

	Pretest	Post-test		
	Mean	Mean	Std. Deviation	N
Experiment	51.9	65.83	16.33	12
Control	53.5	59.91	10.89	11
Total	105.5	125.74	27.22	23

The van Hiele's geometric levels of thinking can be seen in Table 5. Table 5 shows that after learning Transformation using CPS model Qur'an-based textbook, the students were no longer at level 0-2. After the post-test, the percentage of Year 3 students in the control class was 90.91%, as compared to 66.67% in the experimental class. This suggests that the van Hiele's thinking level of most students in the control class had not reached a formal deduction stage; they still had difficulties in geometric axiomatic system reasoning. However, while 33.33% of the students in the experimental class were at level 4, only 9.09% of those in the control class was at level 4. This means that the CPS model of learning can develop students' creativity in solving the problem.

Table 5. Van Hiele's Geometry thinking level.

No Test	Van Hiele's level	Pre-test				Post-test			
		Experiment		Control		Experiment		Control	
		Frequency	%	Frequency	%	Frequency	%	Frequency	%
1	0	0	0	0	0	0	0	0	0
2	1	0	0	0	0	0	0	0	0
3	2	7	58.33	6	54.55	0	0	0	0
4	3	5	41.67	5	45.45	8	66.67	10	90.91
5	4	0	0	0	0	4	33.33	1	9.09
Total		12	100	11	100	12	100	11	100

The prerequisite analysis test was performed before the research hypothesis test. The prerequisite analysis test included normality and homogeneity tests. The normality test used the Simonov Kolmogorov method to determine whether the data used was normally distributed or not. The decision-making basis of this test was calculated by using the p-value. The significance level of test results greater than the alpha of 5% indicated that the data used was in normal distribution. The Kolmogorov-Smirnov test value (p-value) on cognitive value and van Hiele levels of thinking in the experimental class and control class were more than 0.05, meaning that each data was normally distributed. The homogeneity test was conducted by using the Levene method in order to determine whether the data used had the same variation among treatments or not. The basis for the decision making of this test was determined by using the p-value. The significance value of the test results greater than the alpha of 5% indicated that the data used had the same range among treatments. The significance value (p-value) in the determination of cognitive value and the value of van Hiele thinking is more than 0.05, meaning that the type between the experimental class and the control class is the same or homogenous.

The content of Geometry Transformation's book based on Al-Quran includes the discussion on how geometry concepts and Al-Qur'an are related. The book is the product of design research using questionnaires and small-scale trials validated by experts. As a result of the development of this book, the average results of the validation and testing showed that it was a good fit for use in accordance with the table eligibility criteria and product revision level. This book is a product designed with average validation results, and the tests are good matches to use in accordance with table eligibility criteria. Also, its product revision level as a whole is valid. The textbook review process was used to produce Quran-based geometry transformation book that conforms to the Islamic context. The quantitative data analysis indicators of the textbook validation and its integration into Qur'an include the integration of materials on transformation geometry with Al-Quran, the integration of Al-Quran with each material on transformation geometry, the suitability of reading letters to Al-Quran with translation, the stance of the Islamic terms, and the conformity of Islamic terms.

The results of this study are in line with the results of several other studies on CPS and the level of thinking of van Hiele geometry. Research conducted by Helie (2010) on the theory of CPS is one form of understanding produced by integrating new theories. Therefore, CPS is a development of process-based creativity theory [12]. The findings of a study by Yingsu and Vincent (2010) are used to view cognitive abilities based on problem-solving. Problem-solving is a cognitive process, so this study is closely related to CPS as a mathematical problem-solving process [13]. Research conducted by Fatih and Murat (2015) shows that van Hiele's ability level can be observed in the activities of the students who are trained in geometry-based activities [14]. Research conducted by Sasha and Margaret (2014) indicated that van Hiele's thinking recalls involving complex reasoning abilities, so it is not easy to conduct a process of inquiry [15].

The implications of this research can be used to improve lecturers' creativity in developing learning models that can provide students with thinking motivation through the degree of problem orientation, finding the facts, finding problems, finding ideas, finding solutions, and finding acceptance. As a result, students will possess geometry skills in solving problems based on van Hiele thinking levels.

4. Conclusion

Based on the results of the research and the data analysis of the hypothesis test results, it can be concluded that: after participating in the geometry learning with CPS model, the level of thinking ability of van Hiele geometry among students increased. 4 out of 12 students (33.33%) in experiment class are currently at level 4 (formal deduction), i.e., they begin to understand the axioms as statements that can be used to prove theorems in geometry formally, and realize how important the precision of the basic principles underlying a proof is, so as to create Islamic traits of honesty, patience, and carefulness. In the control class, however, only 90.91% or 10 out of 11 students are currently at level 3; they cannot fully understand transformation geometry by formal deduction. There are three domains in the assessment of learning outcomes, cognitive, psychomotor, and affective domains. The result of cognitive learning in the experimental class was 63.08 while that of the control class was 59.27. Psychomotor learning result in experiment class was 83 while that in control class was 78. The result of the experimental class' affective learning was 83 while that of control class was 77. Because CPS learning model can be applied with various approaches, it is advisable for other researchers to examine CPS learning model by using technology-based media, so that time constraints in constructing concepts do not become obstacles.

Based on the results of research and hypothesis test results, it can be concluded that there is a difference between van Hiele thinking levels between students taught by using CPS model and those taught using conventional model. There are also differences in learning outcomes between students who learn using CPS learning model and those who learn using conventional model. Although this study shows that the mean value of van Hiele thinking level in the experimental group is higher than that in the control group, the average score is still unsatisfactory, with van Hiele thinking level being 65.83 and cognitive learning outcomes being 63.08. This figure is still below the Minimum Passing Level (MPL). In order to effectively implement CPS learning model, especially in managing time, lecturers are advised to accumulate time in Semester Lesson Plan appropriately. Due to the fact that CPS learning model can be applied using various approaches, it is suggested that other researchers examine CPS learning model using technology-based media, so that time constraints in constructing the concepts do not become obstacles.

Acknowledgments

Alhamdulillahirabbil'alamin. "All the praises and thanks be to Allah, who is the Lord of the universes." May prayers and peace be always upon the respected Prophet Muhammad PBUH and his family, friends and followers until the end of the world. We would also like to express our sincere thanks to KEMENRISTEK DIKTI who has given us a chance to work in the field of educational research that, if God wills, provides benefits for the next generation of the nation.

References

[1] Tartre L A 1990 Spatial orientation skill and mathematical problem solving *Journal for Research in Mathematics Education* **21** 216

[2] Cheng Y and Mix K S 2014 Spatial training improves children's mathematics ability *Journal of Cognition and Development* **15** 2

[3] Pittalis M and Christou C 2010 Types of reasoning in 3D geometry thinking and their relation to spatial ability *Educ. Stud. Math.* **75** 191

[4] Holmes J Adams J W and Hamilton C J 2008 The relationship between visuospatial sketchpad capacity and children's mathematical skills *European Journal of Cognitive Psychology* **20** 272

[5] Rasmussen C and Bisanz J 2005 Representation and working memory in early arithmetic *Journal of experimental child psychology* **91** 137

[6] Cakmak S Isiksal M and Koc Y 2014 Investigating the effect of origami-based instruction on elementary students' spatial skills and perceptions *The Journal of Educational Research* **107** 59

[7] National Academy of Science 2012 *Learning to Think Spatially: GIS as a Support System in the K-12 Curriculum* (Washington DC: National Academy of Science)

[8] Foundation N S 2010 *Preparing the Next Generation of STEM Innovators: Identifying and Developing Our Nation's Human Capital* (National Science Foundation)

[9] Australian Curriculum, Asessment and Reporting Authority 2010 *Australian F-10 Curriculum* Retrieved from https://www.australiancurriculum.edu.au/f-10-curriculum/general-capabilities/numeracy/

[10] Miller D I Newcombe N S and Uttal D H 2013 Exploring and Enhancing Spatial Thinking Links to Achievement in Science Technology Engineering and Mathematics *Current Directions in Psychological Science* **22** 367

[11] Olkun S 2003 Making connections: Improving engineering drawing activities *International Journal of Mathematics Teaching and Learning* **3** 1

[12] Stieff M and Uttal D 2015 How much can spatial training improve STEM achievement? *Educational Psychology Review* **27** 607

[13] Lowrie T, Logan T and Ramful A 2017 Visuospatial training improves elementary students' mathematics performance. *British Journal of Educational Psychology*

[14] Battista M T 2007 The development of geometric and spatial thinking *Second Handbook of Research on Mathematics Teaching and Learning* **2** 843

[15] National Council of Teachers of Mathematics 2000 *Principles and Standards for School Mathematics* (Reston: NCTM)

[16] Clements D H and Stephan M 2004 Measurement in Pre-K to grade 2 mathematics *Engaging young children in mathematics: Standards for early childhood mathematics education* 299-317

IOP Conf. Series: Journal of Physics: Conf. Series **1088** (2018) 012054 doi:10.1088/1742-6596/1088/1/012054

Gender differences of mathematical critical thinking skills of secondary school students

Mawaddah[1], A Ahmad[1] and M Duskri[2]
[1] Syiah Kuala University, Jl. Teuku Nyak Arief Darussalam, Banda Aceh 23111, Indonesia
[2]Ar-Raniry State Islamic University, Jl. Syeikh Abdul Rauf Kopelma Darussalam, Banda Aceh, Indonesia

E-mail: anizarahmad.aud@fkip.unsyiah.ac.id

Abstract. One of the competencies in the curriculum of 2013 is to develop creativity, curiosity, the ability to formulate questions to form critical thinking that is necessary for modern life and lifelong learning. It is therefore essential to create an instrument to measure students' mathematical critical thinking skills. This study aimed to describe of mathematical critical thinking skills of secondary school students in solving mathematical problems concerning gender based on developed instruments validated by four experts. Four students selected from 30 students of Year 9 in one of junior high school in Banda Aceh. The instruments used in this study are critical thinking skills and interview guidance. The results showed that the critical thinking skills of female students were slightly better than that of male students. The findings of this study imply that there is a gender difference in mathematical critical thinking skills and therefore it is required to develop instruments that enhance the critical thinking of both male and female students.

1. Introduction
Critical thinking is needed by students to solve everyday and mathematics problems. Critical thinking as the ability to analyze and evaluate information (the ability to analyze and evaluate data or information) [1]. Critical thinking skills include reasoning skills and reflective thinking focusing on deciding what to believed and do [2]. One of the objectives of mathematics learning in the 2013 curriculum is to develop students' skills in understanding mathematical concepts, explaining interconnectedness between concepts, and using concepts or algorithms flexibly, accurately, efficiently and appropriately in problem-solving. Based on the demands of the curriculum, the current learning process developed in Indonesia demands students to be actively involved in the process of teaching and learning activities to develop problem-solving ability. Associated with the aspects of problem-solving skills in mathematics, a student is required to have higher order thinking skills. Ministry of Education and Culture Republic of Indonesia Press Workshop stipulated that "Implementation Curriculum 2013" in 2014 said that learning mathematics is designed so that the students have to think critically to solve the problem. Critical thinking skills should be incorporated into the mathematics curriculum so that students can learn critical thinking skills and apply them to improve understanding of logical connections between ideas, create and evaluate arguments, and solve problems systematically [3]. Therefore, mathematics learning should emphasize the thinking process of students.

Many factors must be considered in studying mathematics, such as willingness, ability, particular intelligence, readiness of teachers, readiness of students, curriculum, and presentation methods. One factor that is not less important is the gender factor of students. Gender differences certainly cause physiological differences as well as affect the psychological differences in learning, so that male and female students have differences in learning mathematics.

Psychologically males and females are different. Females are more interested in real life issues, whereas males are more interested in abstract aspects [4]. The difference between males and females in learning mathematics, that the males are superior in reasoning, whereas females are superior in accuracy, precision, carefulness, and thoughtfulness [5].

Since childhood, male students are known to be more natural at recognizing problems. However, their concern in solving the problem is lower than of female students who tend to give more effort to problem-solving, commonly found male students lazy around in the classroom during the learning process [6].

Based on the above description, it is necessary to develop critical thinking instruments to analyze the critical thinking skills of mathematically male and female students using these instruments. This paper only examines the mathematical critical thinking skills of male and female students using the developed instrument, then the problem formulation in this research is how the mathematical critical thinking skills of secondary school students in solving mathematical problems reviewed of sex?

2. Method

This study aims to explore information about students' mathematical critical thinking skills. This research was conducted on Year 9 students of one secondary school in Banda Aceh. The subjects were selected based on test result about critical thinking skills. According to the test data of critical thinking skills, the researcher chose four students who were communicative and had high and medium mathematical skills because the researcher wants the students who could give or convey the ideas, ideas, and reason so that the researcher could explore deeper to the research subject. To choose a communicative student then the researcher discussed with the teacher of mathematics teacher in the class, this is because the teacher knows better the characteristics of the students. Based on the results of the discussion, then selected four subjects consisting of two male students and two female students.

Instruments used in this research consist of critical thinking skills and interview guidelines. The question instrument of critical thinking skills in this research is a question of mathematical story for the material system of two linear equations, the number pattern and set. This instrument was developed based on validation process by four experts, namely two lecturers of mathematics as an expert of learning evaluation and mathematics subject matter, one mathematics teacher as a practitioner of mathematics learning, and one colleague who has attended mathematics learning evaluation workshops and have done publication of paper scientifically like this research. Based on the assessment of the four validators it was concluded that the problem used met the criteria "worthy using with a little revision." The problem is given to the subject of research to be done for approximately 1 hour and continued with the interview. The interview guide is intended to guide researchers in uncovering the process of critical thinking of students in solving mathematical problems. The questions in the interview guide are structured according to the stages of the critical thinking indicator. Interview guidelines are used to keep the interview focused on the issues to be expressed.

The data collected in this study there are two types, namely the data of the work (answer) the subject of solving mathematical problems and subject interview data. The process of taking data is also carried out in two stages, the first stage is given the test of critical thinking skills-I followed the interview of each subject after the subjects did the test, and the second stage was given the test of critical thinking skills-2 followed the interview of each subject after the subject did the test. The critical thinking indicator used in this study is the critical thinking indicator according to Facione [7] which consists of interpretation, analysis, evaluation, inference, and explanation.

3. Result and discussion

Based on data analysis and test results, the following describes students' critical thinking skills for each indicator: interpretation, analysis, evaluation, inference, and explanatory.

3.1. Interpretation

According to the results of students' critical thinking skills test for interpretation indicator, male and female students could interpret the problem, that is to understand and explain the problem, and could interpret the meaning of data from a problem, meaning that the interpretation is easy for the students. This is in line with Anderson [8] stating that the interpretation is part of the level of understanding and belongs to the low-level thinking, so most students are able to meet this indicator. Here is one example of student answers in interpreting the problem.

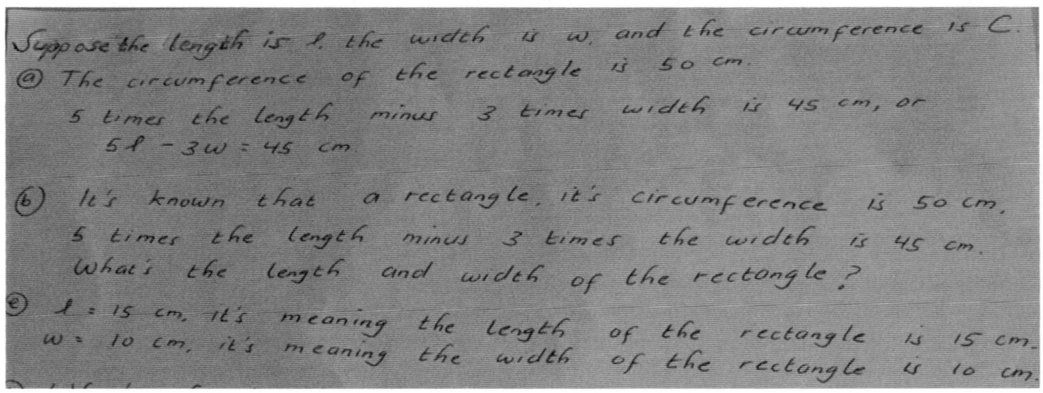

Figure 1. Student's answer in interpreting problems.

3.2. Analysis

According to the results of students' critical thinking skills test for analysis indicators, male and female students could determine the questions correctly and examine the outcome of the problem, but both male subjects are unable to use the correct strategy to solve the problem. This indicates that female students could solve the problem well. According to research Robbins [9], in females, the smaller the hippocampus, the better it works. Experts investigate how the hippocampus works and say that the female brain works more effective and efficient because the neurons that arrange the female brain communicate better with each other than the neurons in the male brain. Therefore, females have the ability to complete the tasks assigned without having to involve large numbers of neurons in the process, and their small size may represent more intense neuronal packing of cells or more active signals in females so they can work more effectively and efficiently in many ways. Here is one example of student answers in analyzing the problem.

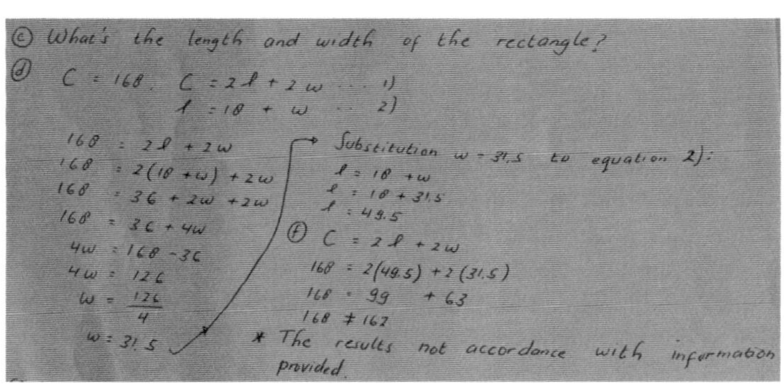

Figure 2. Student's answer in analyzing problems.

IOP Conf. Series: Journal of Physics: Conf. Series **1088** (2018) 012054 doi:10.1088/1742-6596/1088/1/012054

3.3. Evaluation

According to the results of students' critical thinking skills test for evaluation indicators, male and female students access connection between statement correctly and reject a statement based on predetermined criteria, but both male subjects could not use the right strategy to solve the problem. Lack of exercise in evaluating resulted in low students' critical thinking skills [10,11]. This indicates that the female students are more rigorous in problems solving than male students. This is also supported by the results of research Krutetskii [5] which states that there are differences in character between males and females. Generally, males are better at reasoning while females are better regarding accuracy, thoroughness, and thoughtfulness of thinking. Here is one example of student answers in evaluating the problem.

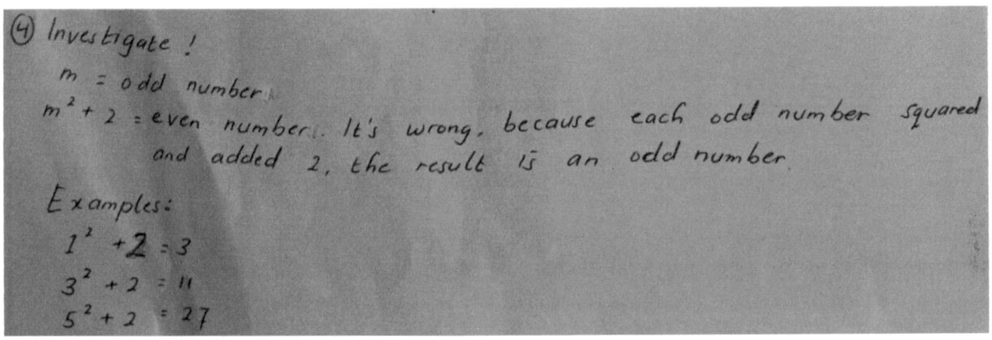

Figure 3. Student's answer in evaluation problems.

3.4. Inference

According to the results of students' critical thinking skills test for inference indicators, male and female students could identify the elements needed to conclude. This is following the results of the research Arif [12] which said that male and female students could solve problems in the inference indicators. Here is one example of student answers in the problem.

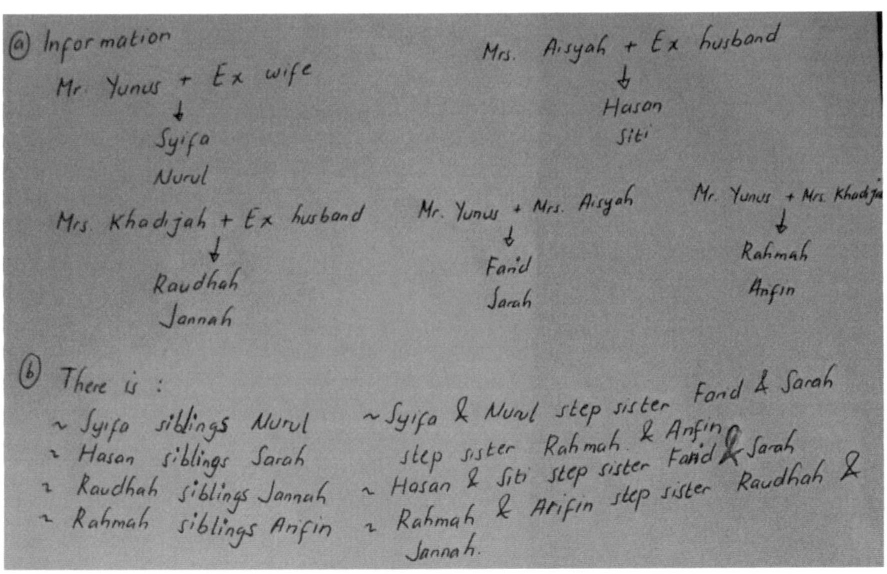

Figure 4. Student's answer in inference problems.

3.5. Explanatory

According to the results of students' critical thinking skills test for explanatory indicators, male and female students could explain or state the thought based on evidence, methodology, and provide logical reasons based on the results obtained. This is in accordance with the results of the research Arif [12] which said that male and female students could meet the explanatory indicators. Here is one example of student answers in explaining the problem.

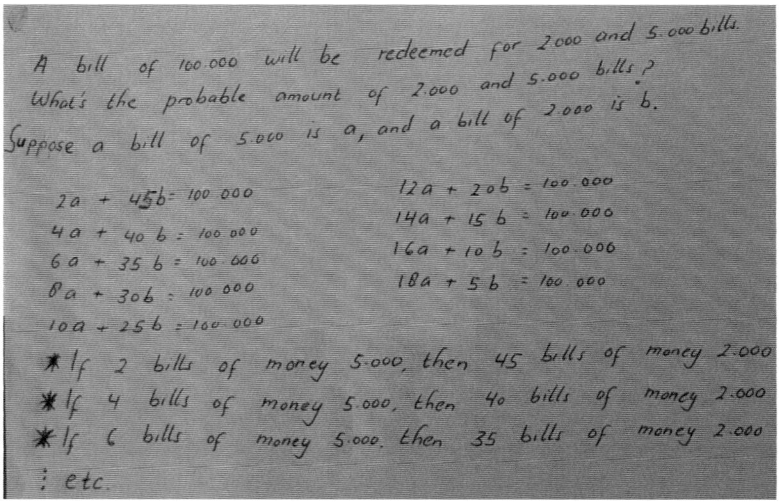

Figure 5. Student's Answer in explanatory problems.

4. Conclusions

From the analysis and discussion that has been described, in interpreting the problem, female and male students could write down the information from a given problem, write down the information from the problem with their language, and could give the data meaning of the problem. At the analyzing the problem, male and female students could determine the question of the problem, re-examine the outcome of the problem, but the male students could not use the correct calculation. At the evaluating the problem, male and female students could access connection between statement correctly, could to rejected a statement based on predetermined criteria, but the male students could not use the correct strategy to solve the problem. At the inference, the problem, male and female students could use relevant information in the matter, could to explain the connection of each information available, and could to make conclusions based on the information provided in the matter. Concerning explaining the problem, male and female students could explain the completion that has been done and gave reasons based on the conclusions taken. Based on the above description, female students meet all aspects of critical thinking indicators, while male students meet only 12 aspects of critical thinking indicators, so it can be concluded that the critical thinking skills of female students are slightly better than male students solve math problems.

References

[1] Duron R, Limbach B and Waugh W 2006 Critical Thinking Framework for Any Discipline *International Journal of Teaching and Learning in Higher Education 2006* **17** 160

[2] Ennis R H 1985 *A Logical Basis for Measuring Critical Thinking Skills* (Educational Leadership)

[3] Chukwuyenum and Nelson A 2013 Impact of Critical thinking on Performance in Mathematics among Senior Secondary School Students in Lagos State. *IOSR Journal of Research & Method in Education (IOSR-JRME)* **3** 18

[4] Ekawati A and Wulandari S 2011 Perbedaan Jenis Kelamin Terhadap Kemampuan Siswa dalam Mata Pelajaran Matematika (Studi Kasus Sekolah Dasar) *Jurnal Ilmu-Ilmu Sosial* **3** 19

[5] Krutetskii V A 1976 *The Psychology of Mathematics Abilities in School Children* (Chicago: The University of Chicago Press)

[6] D'Zurilla J Maydeu-Olivares A and Kant G L 1998 Age and Gender Differences In Social Problem-Solving Ability *Journal Personality and Individual Differences* **25** 241

[7] Facione 1990 *Critical Thinking: A Statement of Expert Consensus for Purposes of Educational Assessment and Instruction "The Delphi Report" Executive Summary* (California: The California Academic Press)

[8] Anderson L W and Krathwohl D R 1999 *A Taxonomy for Learning, Teaching, and Assessing: A Revision of Bloom's Taxonomy of Educational Objectives* (New York: Longman)

[9] Robbins S P 1998 *Perilaku Organisasi* (Jakarta: PT. Prenhallindo)

[10] Snyder L G and Snyder M J 2008 Teaching Critical Thinking and Problem Solving Skills *Spring/Summer* L(2)

[11] Peter E E 2012 Critical Thinking: Essence for Teaching Mathematics and Mathematics Problem Solving Skill. *African Journal of Mathematics and Computer Science Research* **5** 39

[12] Arif M and Hayudiyani M 2017 Identifikasi Kemampuan Berpikir Kritis Siswa Kelas X TKJ Ditinjau dari Kemampuan Awal dan Jenis Kelamin Siswa di SMKN 1 Kamal *Edutic-Scientific Journal of Informatics Education* **4** 1

IOP Conf. Series: Journal of Physics: Conf. Series **1088** (2018) 012055 doi:10.1088/1742-6596/1088/1/012055

Ethnomathematics analysis on Jambi plait art as the mathematics learning resources

Kamid, A Wandari and Rohati
Universitas Jambi, Jl. Raden Mattaher No. 16, Jambi, Indonesia

E-mail: kamid.fkip@unja.ac.id

Abstract. This research was based on the fact that mathematics is considered a difficult subject and most students have less understanding of mathematics. Meanwhile, culture is believed to be a good learning resource that can help teachers to teach effectively even for mathematics. Art and culture such as plaited mats and hat could be used as a mathematics learning resource that might make mathematics easier and interesting. This qualitative research aimed to understand and analysis the mathematics aspects of the Jambi plait art for mathematics learning as well as the connection of ethnomathematics and Jambi plait art that could be used as mathematics learning resource. Research data were collected by observation, interview, and documentation technique of Jambi plait art products and artists. The result showed that three aspects of Jambi plait art could be used as mathematics learning resource: including counting, measuring, and explaining on some topics such as arithmetic, linear comparison, the surface area of 2-dimensional figures, line pattern and sequences, geometric patterns, and linear programming.

1. Introduction

Mathematics is knowledge of logical about shape, arrangement, scale, and concepts that are classified into three fields, such as algebra, analysis and geometric. Many view mathematics as a difficult subject that cannot be applied much in daily life and is not related to the cultures at all. Most people do not understand how to apply mathematics to solve any problems in daily life, wherein the advantages of mathematics are not significant and could lead to mathematics-phobia.

Based on the observation of some senior high school students, it showed that mathematics was believed as a difficult subject, even scary for the students. Students believed that when learning mathematics, they were only required to recognize many formulas, to understand counting concepts. Students also felt that mathematics is an abstract and absurd subject. The view of mathematics as a difficult subject was caused by the formal mathematics learning environment and often beyond what found in daily life [1]. Mathematics is widely used in daily life, also as a tool and language to solve problems either in mathematics subject itself or other daily life problems.

Education is a part of a culture and has many aspects that influence each other. One of the most important elements of education in the interconnection to the particular culture is the integrity of cultural values into the curriculum content [2]. Since mathematics is applied to daily life, it is needed to figure out the alternatives to make mathematics easier and culturally involved.

One of the alternatives which is growing nowadays is cultural based-mathematics. Involving culture context in teaching and learning process can give and create meaningful learning to achieve the understanding of the integrated scientific information, as well as the understanding of the implementation of the scientific information on cultural problems context [3]. The cultural community has many mathematics ideas which can be studied and researched.

All formal mathematics was a cultural interaction process, which caused every student experienced the cultural conflicts in the learning process [4]. Education cannot be separated from the culture. There is a strong relationship between education and culture, they complete and support each other. Therefore, education has an important role in the process of cultural heritage; thus cultural values needed to implement in the learning process. Education as the cultural scaffolding, as well as the culture will develop education [5].

Ethnomathematics is a study of the differences in how society solves mathematics problems and simple algorithm according to their mathematics perspectives, which refer to various mathematical shapes as the consequences of the cultural events [6]. In relation to the previous point of view [5], it could be said that ethnomathematics is used as the tool to act in the world. Therefore, ethnomathematics contributes to the social role in an academic context. Cultural based-mathematics learning is one of the ways to create the meaningful and contextual mathematics learning that is related to cultural community, in where mathematics will be studied and implemented.

According to the meaning of ethnomathematics, a study could be conducted on the plait art as part of cultures in its relation to mathematics concept and learning. Some of the plait arts that are commonly used are plait mats and hat. These arts apply patterns which are related to mathematics. Plait is a psychomotor art which involves the common senses and awareness of the symbol with a particular meaning as a traditional art in a community. Plait art still exists until now.

Apart from many usages of the plait arts, the arts are also familiar because of the simplicity. The plait arts are developed on various shapes and motifs that make the shape and motif dynamics. Thus, plait is a society psychomotor activity to create useful objects by overlapping and folding between loom and woof that strengthen each other. This plait art could be considered as the mathematics learning resource that help students to recognize mathematics concept without starting by learning formal mathematics concepts. Jambi is one the provinces in Indonesia that still maintains the development of the plait arts. Thus, the Jambi plait art can be used as a mathematics learning resource that could make mathematics learning easier, unique, and joyful. It is also believed that the plait arts could increase motivation and self-confidence in learning mathematics.

In relation to the descriptions above, it is essential to conduct a further study of mathematical concept and learning that is related to the Jambi plait arts. This study is the preliminary stage to develop learning materials based on the cultural context. This study aims to analyze mathematics aspects of Jambi plait art and its implementation in classroom mathematics learning. Further, it is a study about the connection of ethnomathematics of Jambi plait art to the mathematics curriculum as the mathematics learning resource.

2. Research method

This research was a descriptive research by the qualitative approach. In terms of research purpose, the qualitative research means something related to quality aspect, value or meaning which exist in reality [8]. The subject of the study is the ethnomathematics analysis of plait art related to mathematics learning. The description process involved an observation of the mathematical values of plait arts. The researchers also interviewed the research informants, the artists of plait arts in one of the villages in Jambi, to dig information deeper about plait art and its relationships to the mathematics learning.

This research was conducted on the Jambi plait art workshop, *Rumah Anyaman Pandan*, *Koperasi Usaha Bersama (KUB) Radesta* at one of the villages in Jambi province. The selection of the research informants was by purposive sampling [9]. The research instruments were the observation sheet and an interview. The researchers conducted the research by observing and noting the important information and documentation, such as pictures and videotape recording. The interview was an unstructured interview. Interview result was recorded by tape recording and pictures documentation for supporting data. The data were collected from a mathematics teacher, an artist, and a culture expert. Data validity of the research was done by the credibility test using time triangulation, which was the repetition of the process of plait art information at the different time and the same informant. The data were analyzed by categorizing the data systematically.

3. Result and discussion

The qualitative data were in the form of words and action. The words data were gained from interviews of the research informants, and the action data were acquired by observing the activities of the artist during the production process. Documentation data were gained from references and Jambi museum archives. The picture data were obtained from direct documentation on *Rumah Anyaman Pandan*, *Koperasi Usaha Bersama (KUB) Radesta* in Jambi province, as well as from the pictures archives.

3.1 Mathematics aspects in the Jambi plait art.

According to identification and coding results that were conducted by the researcher, the findings showed that there were three aspects of the plait arts that could be used as learning materials, including counting, measuring, and explaining aspects. These results are in line with Bishop [10]. The analysis result from ethnomathematics point of view revealed that culture influenced the impact of activities on a particular environment. By employing ethnomathematics, someone could see the existence of mathematics as the knowledge that can be learnt not only in the classroom but also outside the classroom. While mathematics in the school was recognized as academic mathematics, ethnomathematics is mathematics that is implemented on the cultural community and is identified as the ethnic society, the labor community, a certain age of children, professional class, etc. [6]. Mathematics aspects of the plait art included:

3.1.1 Counting. In the Jambi plait art, counting aspect is found in the raw material collection phase. This activity can be related to a cultural aspect which was done by answering the question 'how many', with counting tools and numbering [10,11,15]. When the process of taking pandan leaves, the plait artist usually counts how many leaves have been taken, and then selects them on the plait art. The unit of pandan leaves using "*ikat*" or "*ikat*an". The words of the counting or numbering process can be noted as a natural number, even number and odd number as the numbering concept.

3.1.2 Measuring. Generally, measuring is related to the questions of "how long, how wide, or how many" [15], [19]. The measurement that is related to the number of leaves was used on the plaiting process. Measuring is an activity of selecting materials of the plait. Measuring aspect is also the activity of marketing process, measuring length, width, height, area, volume, speed, etc. This measuring aspect occurred in the selecting the raw materials and in creating the plait mats and hat. When selecting pandan leaves, the artist measured the length and width of the leaves that were needed. During the creating process, the artist used the measuring aspect to measure the length and width of the mat in order to make a proportional mat.

3.1.3 Explaining. In relation to ethnomathematics, mathematics can be learnt by connecting it to local culture. This connecting process is the core of teaching and learning method [12]. Based on that theory, there is the explaining aspect of the daily life activity in the plaiting process. Explaining aspect is related to the understanding of why a certain number pattern occurs, why geometrics shapes have the same patterns, why one result refers to other results, why several relations of nature are related to mathematics law. In the process of creating plait mats and hat, re some steps could be used to represent the explaining aspect. The process of creating the plait would also include an explanation of the creating process.

3.2 The relationships between the mathematics concepts with plait mats and hat as a mathematics learning resource.

Based on the identification of cultural elements that were embedded to mathematics lesson, several cultural values were related to mathematics concepts as could be seen in the following descriptions.

3.2.1 Multiply two (arithmetic) and linear comparison. Based on counting and explaining aspects, the artists have implemented mathematics concepts indirectly, which were the multiplication of two and the linear comparison when selecting the raw materials. The unit that was used on the counting the number of the leaves was "*ikat*". There were 50 leaves in 1 *ikat*. Thus, if one mat needed 2 *ikats*, then the artist would need 100 leaves. In relation to the linear comparison concept, the students were asked to compare the number of *ikats* of pandan leaves that would be needed to make a certain number of mats.

2 *ikat* pandan leaves = 1 mat
4 *ikat* pandan leaves = 2 mats
6 *ikat* pandan leaves = 3 mats
8 *ikat* pandan leaves = 4 mats
2n *ikat* pandan leaves = n mat

Thus, the linear comparison concept would be achieved. Meanwhile, regarding the explaining aspect, almost all mathematics learning use explaining aspect, and also on the plaiting process [15]. Plait process also needs an explanation from the plait artist.

3.2.2 Surface area and perimeter of 2-dimensional figures. According to the measuring aspect, people have used the mathematics concept such as surface area and perimeter of 2-dimensional figures when selecting the raw materials. In this case, the artist measured the length and width of pandan leaves to plait mats and hats in centimeter (cm) or meter (m). 1 pandan leave had the width of 1 cm and the length of 2 m (200 cm), and 1 mat needed 2 *ikat* pandan leaves or equal to 100 pandan leaves. Plaiting a mat required two *ikats* of the pandan leaves, 1 *ikat* to cover all area of the mat and 1 *ikat* for the plait layer. Hence, the surface area and perimeter of the mat were 10,000 cm^2 and 500 cm respectively. Those results were obtained because each pandan leave had a length of 200 cm and width of 50 cm. Thus, this problem has taught the concept of surface area and perimeter, which also involved the explaining aspect.

3.2.3 Number patterns and sequences.

- *Number pattern.* The understanding of what aspects the number pattern and sequences involved could be acquired by answering the question "how many", that is included in counting aspect. Number pattern is the rule of a group of numbers with a certain sequence rule. On the plait art, especially on plait mat, the artist made the pattern as the natural number pattern before plaiting. The basic sequence when the artist plaited the mat was 1, 2, 3, 4, 5,... and put them on a sequence. The artist then inserted other pandan leaves to the odd number of the basic sequence of 1,3, 5, 7, 9, ..., and in turn, to the even number 2, 4, 6, 8, 10, ... Thus, mathematics concept suddenly used on this process (counting process).
- *Sequences.* Likewise the number pattern, sequences also involved the counting aspect [3,10,14,15]. This counting aspect was conducted by answering the question "how many", by using counting tools and numbering. The plait art also implemented line and sequences, which were used to arrange the distance of inserted leaves to produce the proportional mat.

3.2.4 Geometry. Tesselation principle is widely used in daily life, such as floor tile, cloth motif design, wallpaper pattern, etc.[11]. According to a theory, there is explaining aspect on plait mats and hat directly or indirectly[13].

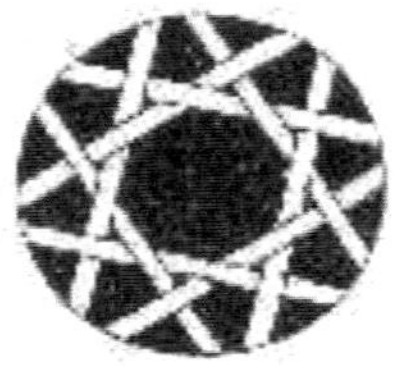

Figure 1. Jambi plait pattern

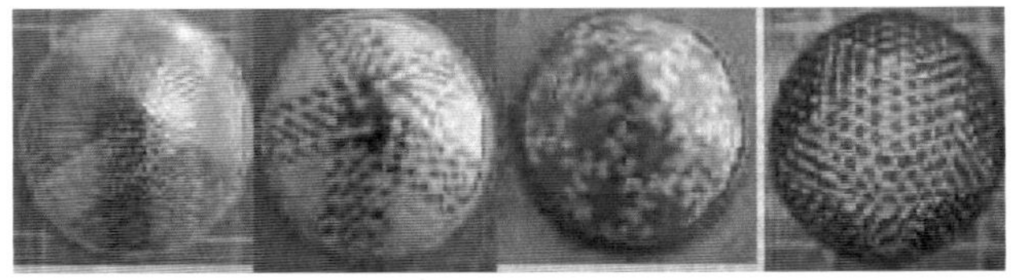

Figure 2. Jambi plait pattern on hat production

When plaiting a hat, the artist used the reflection concept in geometry, which says that a reflection to a line s is a function M to the P point on a side. In this case, if the plait hat were divided by 2, there would be 2 same sides in the same shape, same length, and same pattern (Figure 1 and Figure 2). This idea was in line with the findings of other ethnomathematics research [13].

3.2.5 Linear programming. Linear programming also consists of measuring aspect. In this research, there were measuring aspect in relation to the length of pandan leaves which was used in the plait art. Linear program concept was used indirectly, such as determining the patterns, the number of the materials, the timeline that needed to create 1 mat or hat [3,11,15]. Besides, a linear program could be used to minimize the production cost, and to determine material and the number of the plait artist.

4. Conclusion

Several aspects of mathematics could be explained by the use of cultural materials and arts such as plait and mat made from pandan leaves in Jambi culture. This study could give more insights for the researchers to develop meaningful and helpful teaching and learning materials of mathematics. This cultural mathematics learning material can be a good solution to improve the quality of teaching mathematics for the students. Thus, by using such learning materials, it is expected that both teachers and students will find mathematics fun, easy, and interesting to be learned, especially on topics such as arithmetic, linear comparison, the surface area of 2-dimensional figures, line pattern and sequences, geometric patterns, and linear programming.

References

[1] Hiebert J and Carpenter TP 1992 *Learning With Understanding*. In Grouws D A (ed) Handbook of Research on Mathematics Teaching and Learning A Project of NCTM (New York: Macmillan)
[2] Sirate S F 2012 *Jurnal Lentera Pendidikan* **14** 123
[3] D'Ambrosio U 1985 *For the Learning of Mathematics* **51** 44
[4] Bishop A J 1990 *Western Mathematics* **32**2 51
[5] Orey D C and Selin, H (ed) 2000 *Mathematics across culture: The History of non-Western mathematics* 239
[6] D'Ambrosio U 1999 *Literacy, Matheracy, and Technoracy* **2** 131
[7] Richey R C, Klein J D and Tracey M W 2011 *The Instructional Design Knowledge base: theory, research, and practice*. (New York. Routledge)
[8] Gunawan I 2013 *Metode Penelitian Kualitatif: Teori dan Praktik* (Jakarta: BumiAksara)
[9] Moleong, L J 2015*Metodologi Penelitian Kualitatif; Edisi Revisi* (Bandung: PT. Remaja Rosdakarya)
[10] Bishop A J 1997 *Journal of Teacher Education* **42** 17
[11] Rahayu K P and I Gusti Nila Putra 2014 *Jurnal Matematika* **4** 2 80
[12] Shirley L 2008 *Journal International Congress of Mathematics Education* 6
[13] Herlina R, Somakim and Hartono Y 2017 *Prosiding Seminar Nasional Etnomatnesia* **6** 849
[14] Kamid, Resmita A, and Rohati 2016 *Aksioma: Jurnal Pendidikan Matematika.* **5** 3 26
[15] Hartoyo A 2012 *Jurnal Pendidikan Matematika* **13** 114

The feasibility of an Android-based pocketbook as mathematics learning media in senior high school

M Saputra, T F Abidin, B I Ansari and M Hidayat
Syiah Kuala University, Jl. Teuku Nyak Arief Darussalam, Banda Aceh 23111, Indonesia

E-mail: taufik.abidin@unsyiah.ac.id

Abstract. Learning media is a tool to deliver the materials to students during teaching and learning process and used as a source of learning like textbooks. However, textbooks are sometimes considered less practical and boring. The innovation that follows the technology advancement is undeniably needed. Nowadays, students are very creative using smartphones. The students' creativity should be supported for better learning. One of the efforts is making use of a smartphone or mobile technology as a learning media. This study aims to develop an Android-based learning media called Trigonometry app and evaluate the validity and practicality of the learning media. This learning media is expected to help students learn mathematics. This study used the ADDIE development model. The learning media was validated by six experts and tested by 95 respondents. The results show that the learning media met the valid criteria with kappa value ranged from 0.562 to 0.843. The learning media was also considered practical based on the responses from the users. 86.32% of the users stated that the learning media was excellent and best imaginable. Therefore, the learning media is declared valid and practical to use in the classroom. It is necessary to conduct further research to investigate the use of the media in the classroom and its impact on students' learning and teaching practices.

1. Introduction

Teachers should have the ability to plan, choose and use multiple sources and learning media to make the learning more effective [1]. Choosing a learning media that can support and facilitate students' learning is a teacher's responsibility. The role of the teacher in the learning process is crucial, especially in selecting instructional and design media to motivate students to learn. Learning media is a tool to deliver the materials in the learning process [2]. Utilizing an exciting, innovative, and creative learning media is an essential aspect of achieving successful learning in schools. Through the media, the expected learning will be easy to implement.

Learning media is also used as learning resources. Learning resources commonly used in the learning process are textbooks and worksheet [3]. Excellent textbooks make teachers easier to teach and provide learning materials that fit students' interests. However, textbooks also have some shortcomings, especially in this technological era. Some students rarely use textbooks because they are lazy to carry large package books [4]. Textbooks sometimes cannot be used all the time and are considered boring by some students [5]. Textbooks should be interestingly designed so that students will feel comfortable and interested in reading the textbooks. Therefore, we need innovation for textbooks to follow the technological progress and students will be easier in learning. One of the learning media that are often used to increase students' interests in textbooks is the media in the form of a pocketbook. Pocketbook is a small printed book that contains information and can be stored in a pocket so easy to carry anywhere. The use of pocketbooks can encourage students to learn independently and improve students' learning outcomes [5]. However, printed instructional media like a book sometimes make students challenging to open and receive course material quickly [6].

Pocketbooks can be generated through technology and information device in the form of mobile technology. Mobile technology is used not only for verbal communication among people but also for other things, such as recording and accessing sources of knowledge and information. Nowadays, students are very creative using smartphones. Some students even no longer take notes; instead, use a smartphone to photograph what written on the board. Since more students have smartphones, it is an excellent opportunity to use the phone for the learning process. Learning media using mobile technology is called mobile learning (m-learning). M-learning is a learning device with personal digital (PDA), tab, iPhone, or HP as the primary device [7,8]. M-learning is used as a complement to learning that can be used by students anytime and anywhere and gives students the opportunity to learn more about the topics they feel uncomfortable. One of the advantages of m-learning is that it can be used as an alternative to substitute a textbook, called mobile e-book [9]. Pocketbooks digital-based android application is one of the developments of mobile learning (m-learning) that can access the application of pocketbooks, and user-friendly. M-learning helps increase students' interests, independence, participation, motivation, and achievements [10-12]. Therefore, the development of Android-based learning media designed effectively and practically for students is essential. The additional book is needed to facilitate students' learning. Therefore, this present study aims to develop Android-based learning media to help students' learning in mathematics.

2. Method

This study is development research that used the ADDIE model (Analysis, Design, Development, Implementation, and Evaluation). At the analysis stage, the students' needs and characteristics were identified. After that, the standard and basic competencies were analyzed to match the curriculum. At the design stage, the researchers designed a media, set topics, composed questions and answers, and collected background and image. At the development stage, the product was created using the Android Studio app. Once the product finished, then it was validated by experts who master the content of the topic chosen (called subject experts) and who understand the media developed in this study (called media experts). At the implementation stage, trials were conducted to test the practicality of the product that had been developed. At the evaluation stage, the practicality of the products was analyzed and administered based on the evaluation in the trials. However, this study only focuses on the validity and practicality criteria of the product.

This study used a questionnaire to obtain data from three media experts, three subject experts, and respondents. The validation of the product was done by the experts. After the experts validated the product, the analysis of the validation is also strengthened by inter-rater test to assess the agreement among the validators. The data were analyzed by calculating the average score given by the experts. Then, the average score was categorized into the criteria guideline presented in Table 1 [13].

Table 1. The guideline for assessment criteria of learning materials.

Score Interval	Criteria
$\bar{x} > 4,2$	Very good
$3,4 < \bar{x} \leq 4,2$	Good
$2,6 < \bar{x} \leq 3,4$	Sufficient
$1,8 < \bar{x} \leq 2,6$	Poor
$\bar{x} \leq 1,8$	Very bad

Moreover, data about the practicality were obtained from the opinions of the respondents on the product. To get the data, two stages of trials were conducted. The first stage is the small-scale trial that involved 10 senior high school students as the respondents. The second stage is the large-scale trial conducted using an online questionnaire from Google Forms. In this trial, respondents could try the application more conveniently at any time. The number of respondents in this large-scale trial was 95 respondents that consist of senior high school students, teachers, and college students.

The questionnaire used for the respondents was designed based on System Usability User (SUS) developed by Brooke [14]. SUS is a survey tool that can be used to test the usability of various kinds of products [15]. SUS's questionnaire is highly appropriate for a small scale of respondents [16]. SUS consists of 10 general statements regarding a respondent's response after using a new application or product. This questionnaire used the Likert scale with five options: strongly disagree, disagree, neutral, agree, and strongly agree. The responses were then converted to a scale with interval 0 to 4 [14]. After that, the sum of the conversion results was calculated, and then multiplied by 2.5 to get the usability score. The score was then categorized based on the assessment criteria in Table 2.

Table 2. Assessment criteria for the usability.

Score Range	Qualification
$V_{sus} > 85.00$	Best imaginable
$73.50 < V_{sus} \leq 85.00$	Excellent
$52.50 < V_{sus} \leq 73.50$	Good
$38.00 < V_{sus} \leq 52.50$	Ok
$25.00 < V_{sus} \leq 38.00$	Poor
$V_{sus} \leq 25.00$	Worst imaginable

3. Results and discussion

The product of this study is an Android-based learning media called Trigonometry app. It contains trigonometric materials, examples of trigonometric problems in the form of videos and exercises. This Trigonometry app was developed using Android Studio software assisted with Adobe Dreamweaver software and the videos in the application were made using Video Scribe software. In the Trigonometry app, there are nine menus, and each has different functions. The menu are: (1) Introduction; (2) The size of the angle (degrees and radians); (3) Comparison of trigonometry in a right triangle; (4) Special Angles; (5) Relation angle; (6) Identity of trigonometry; (7) Rules of sine and cosine; (8) Exercise 1; and (9) Exercise 2. The screenshot of the application can be seen in Figure 1 below.

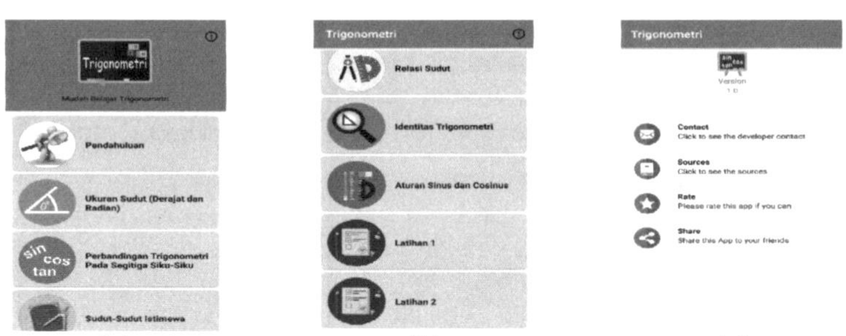

(a) (b) (c)

Figure 1. The appearance of Trigonometry app: (a) Main menu, (b) Main menu (continued), (c) Menu information

The validation of Trigonometry app is obtained through a questionnaire given to the subject and media experts. The results of the validation can be seen in Figure 2.

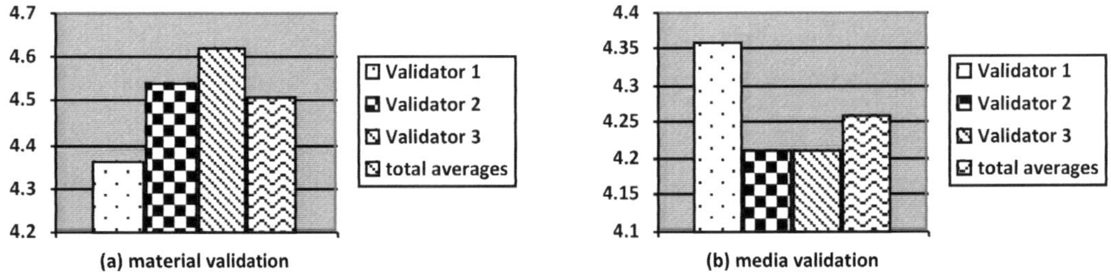

Figure 2. The validation results on subject and media.

Based on Figure 2a, the average score of 4.38 was obtained from Validator 1, 4.54 from Validator 2 and 4.62 from Validator 3. The total average for the subject validation is 4.51which indicates the validity of the content was considered very good criteria. Based on Figure 2b, the average score of 4.36 was obtained from Validator 1, 4.21 from Validator 2 and 3. The total average for media validation reached 4.26, categorized under very good criteria.

Based on the inter-rater test results for the subject validation, Validator 1 and 2 obtained kappa's coefficient of 0.698, Validator 1 and 3 got kappa's coefficient of 0.562, and Validator 2 and 3 obtained kappa's coefficient of 0.843. All of those coefficients were considered in the criteria of a very good agreement. Based on the results, it can be concluded that the perceptions or assessments from the three validators were categorized well.

Regarding the inter-rater reliability test for the media, the results found that Validator 1 and 2 obtained the kappa's coefficient of 0.705. Similarly, Validator 1 and 3 gained the kappa's coefficient of 0.705 while kappa's coefficient obtained from Validator 2 and 3 is 0.714. Similar to the results of the subject validation, those coefficients indicated good agreement among the validators. Therefore, it can be concluded that the perceptions or assessments from the three validators were categorized well. After the experts validated the application, then the researchers conducted trials to evaluate the practicality of the application. Trials were conducted twice: a small-scale trial and a large-scale trial. The results of the small-scale and large-scale trials are shown in Figure 3.

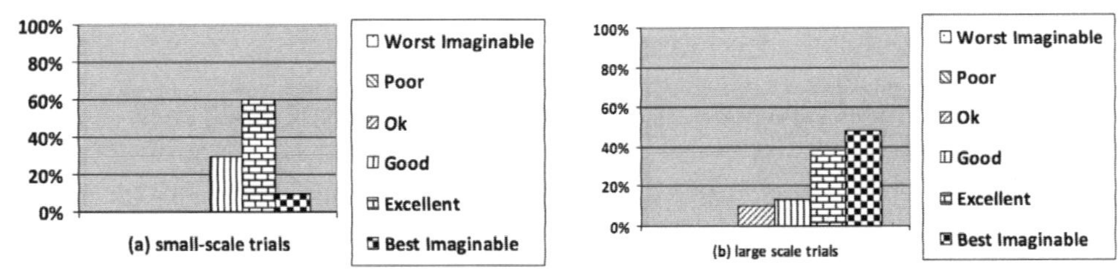

Figure 3. The trial results on the Trigonometry app.

The results of a small-scale trial (see Figure 3a) show that 10% of respondents considered the usability of the application in the best imaginable criteria and 60% of respondents gave excellent criteria. Meanwhile, the rest of the respondents found the application was good. Based on the assessment from the respondents in a small-scale trial, it can be concluded that in general, the Trigonometry app has a good quality.

Furthermore, based on the results of the large-scale trial as seen in Figure 3b, 48.42% of the respondents stated that the usability of the application is best imaginable. Then, 37.89% of the respondents gave excellent criteria, 12.63% of the respondents provided good criteria, and the rest, 1.05% of the respondents, considered poor criteria. From those findings, it can be concluded that generally, the Trigonometry app has a good quality.

Trigonometry app, the final product of this study, is in the form of a pocketbook application that can be installed in Android-based smartphones. This application is a learning media that contains materials, video and exercise questions. The materials in this application cannot be added directly. If users want to add some materials, they have to unpack this application and re-edit the program. It is different from the Google Classroom app developed by Google LLC that can add contents without editing the program. The Google Classroom app is more like an online class with all the materials provided by the teacher. While the application that the researchers developed contains materials, examples of problems and exercises already. Thus, teachers do not need to prepare the materials again.

In her study, Purbasari also developed an Android-based application on the topic of three-dimensional objects [17]. In her application, there is an evaluation menu used to assess students' understanding after using the application. Similarly, in the Trigonometry app that the researchers developed, it can be found in the evaluation menu in the form of exercises with multiple-choice questions. However, the application developed by Purbasari is considered less interactive because there is no animation used in it. Thus, the researchers added videos in the Trigonometry app to make it more interesting for the users. Another study that developed an android application is a study by Astra [18]. That is, an application for ideal gas simulation and the target users are senior high school students. To install the application, users should install Adobe AIR on the smartphone first, which is different from the application developed in this study. The Trigonometry app can be directly installed without requiring additional applications beforehand.

4. Conclusion

The learning media, The Trigonometry app, developed in this study has a very good quality based on the validation from the experts. The subject validation obtained 4.51, and the media validation obtained 4.26 for the total average. These findings indicate that the validity of the learning media was considered very good. Furthermore, the results of the practicality of the learning media show that 86.32% of respondents believed the learning media is excellent and best imaginable. Therefore, the learning media in the form of the Android-based pocketbook is declared valid and practical. It is recommended for further research to apply the Trigonometry app in the learning process and investigate its effects on students' learning and teachers' teaching practices.

References

[1] Sanjaya W 2006 *Strategi Pembelajaran Berorientasi Standar Proses Pendidikan* (Jakarta: Prenada Media Group)

[2] Sadiman A S, Rahardjo R, Haryono A, & Hajito 2014 *Media Pendidikan: Pengertian Pengembangan, dan Pemanfaatannya* (Jakarta: PT Raja Grafindo Persada)

[3] Nur M F 2012 Pemanfaatan Sumber Belajar dalam Pembelajaran Sains Kelas V SD Pada Pokok Bahasan Makhluk Hidup dan Proses Kehidupan *JESBIO.* **1** 14

[4] Nuzula E F & As'ari A R Pengembangan Buku Saku Volume Kubus, Balok, dan Limas sebagai Media Pembelajaran untuk Siswa SMP [Skripsi] Malang: Universitas Malang; 2013

[5] Sulistyani N H D, Jamzuri & Rahardjo D T 2013 Perbedaan Hasil Belajar Siswa Antara Menggunakan Media Pocket Book Dan Tanpa Pocket Book Pada Materi Kinematika Gerak Melingkar Kelas X *Jurnal Pendidikan Fisika* **1** 164

[6] Marlinda L 2015 Pengaruh CD Education dan Animasi Berbasis Android Sebagai Alat Bantu Pembelajaran Untuk Meningkatkan Minat Belajar Siswa SMU IPA *Prosiding Simposium Nasional Inovasi dan Pembelajaran Sains 2015 (SNIPS 2015)*

[7] Sulastri Y L & Hakim L L 2014 Pembelajaran Berbasis Mobile *Jurnal Pengajaran MIPA* **19** 173

[8] Wirawan I M A & Ratnaya I G 2011 Pengembangan Desain Pembelajaran Mobile Learning Management System pada Materi Pengenalan Komponen Jaringan *Jurnal Penelitian dan Pengembangan Pendidikan* **5** 314

[9] Tamimuddin M 2014 Pemanfaatan Mathematics Mobile Learning Dalam Pembelajaran Matematika *P4TK Matematika*

[10] Handayani T S & Suharyanto 2016 Pengembangan Mobile Learning Berbasis Android sebagai Media Pembelajaran pada Materi Fluida Statis untuk Meningkatkan Minat dan Hasil Belajar Ranah Kognitif Peserta Didik *Jurnal Pendidikan Fisika FMIPA UNY* **1** 1

[11] Handayani F 2015 *Pengembangan Multimedia Pembelajaran Biologi Berbasis Android Materi Interaksi dalam Ekosistem untuk Meningkatkan Kemandirian Belajar dan Hasil Belajar Kognitif Siswa Kelas X SMA* (Yogyakarta: Universitas Negeri Yogyakarta)

[12] Irawan P 2015 *Pengembangan Media Pembelajaran Berbasis Mobile Learning pada Smartphone dengan Platform Android sebagai Sumber Belajar untuk Meningkatkan Motivasi dan Hasil Belajar Siswa pada Materi Listrik* (Yogyakarta: Universitas Negeri Yogyakarta)

[13] Widoyoko E P 2009 *Teknik Penyusunan Instrumen Penelitian* (Yogyakarta: Pustaka Pelajar)

[14] Brooke J 2013 SUS : A Retrospective *Journal of Usability Studies* **8** 29

[15] Gatsou C, Politis A & Zevgolis D 2014 An Exploration to User Experience of A Mobile Tablet Application Through Prototyping *International Journal of Computer Science and Applications* **11** 56

[16] Bangor A, Kortum P & Miller J 2009 Determining what individual SUS scores mean: Adding an adjective rating scale *Journal of Usability Studies* **4** 114

[17] Purbasari R J, Kahfi M S & Yunus M 2013 Pengembangan Aplikasi Android sebagai Media Pembelajaran Matematika pada Materi Dimensi Tiga untuk Siswa SMA Kelas X *Jurnal Online Universitas Negeri Malang* **1**

[18] Astra I M, Nasbey H & Nugraha A 2015 Development of an Android Application in the Form of a Simulation Lab as Learning Media for Senior High School Students *Eurasia Journal of Mathematics, Science & Technology Education* **11** 1081

IOP Conf. Series: Journal of Physics: Conf. Series **1088** (2018) 012057 doi:10.1088/1742-6596/1088/1/012057

Integrated model in science for elementary school

R Amini, Usmeldi and Y Helsa

Universitas Negeri Padang, Jl. Prof. Dr. Hamka, Air Tawar, Padang, Indonesia

E-mail: risdamini@yahoo.co.id

Abstract. The interaction of the integrated learning model used in science lessons in elementary school did not necessarily have an impact on the outcomes of elementary students. It means that learning with the integrated model was not focused on students who had high motivation alone; learners who had low motivation could also obtain high learning results. The purpose of this research is to reveal the effect of the integrated model used by students. This research methodology used experimental research. The amount of contribution of motivation in improving learners' learning outcomes could be seen on the average learning outcomes obtained by learners who had high motivation in the experimental class was 86.2, while the learning outcomes of low motivated learners were 78.9. In the control class who had high motivation was 74.8 and average learning outcomes of learners who had low motivation were 63.1.

1. Introduction

Education is one of the decisive factors in efforts to improve human resources [1]. The government always tries to improve the quality and management of knowledge in various ways such as developing curriculum, facilities and infrastructure, learning resources, improving the quality of educators, as well as conducive learning climate [2, 3]. Beginning in 2013 the government implemented a new curriculum as a refinement of the KTSP curriculum in 2006. The 2013 curriculum used an integrated thematic approach. Integrated thematic learning was a learning approach that integrates the various competencies from various subjects into themes [4].

Primarily integrated thematic learning is developed to create education in which learners themselves are mentally active to build knowledge; based on the experience they have [5]. The successful use of this integrated thematic approach demands the ability of educators to apply it in the classroom learning process. A teacher has a significant role in determining the quality and quantity of teaching carried out in the learning process [6]. One of the most practical and realistic efforts in improving the quality of learning processes and learning outcomes of learners is the improvement and refinement of the learning system.

The fact in the field, thematic learning has been implemented because it refers to the book of teacher and student's book, but the learning process conducted by some educators still use conventional education that is the delivery of the material by the lecture and question and answer method, then ended by giving exercise in the form of questions. The role of learners who should be directly involved in the learning process only acts as a good listener. It makes learning less meaningful and impacts on less than optimal learning outcomes. The learner's motivation also influences less optimal learning outcomes for learners. Educators mostly answer the tasks contained in the book students in question and answer activities. Learners quiet more and only hear the explanation from educators; consequently they will lose the experience of direct and natural learning (direct experiences). Sensory experiences that form the basis of the general learning abilities of learners become untouchable, and it is a significant characteristic of the development of elementary school age children [7].

In line with this fact, Mudiono's [8] studied on the development of thematic learning models states that although educators had implemented the 2013 curriculum, most educators had no understanding of curriculum implementation and even the learning strategy itself. Also, the results of his research showed that students were less active in learning so that learning became less meaningful.

Educators are required to be able to choose, design and implement learning model by the characteristics of learners in the learning process, and foster learners motivation learn to improve the process and optimal learning outcomes. Learning models used by educators are not focused on thematic integrated only. Three models of integrated learning can be used in the elementary school such as connected model, webbed (theme) model, and an integrated model [9].

Researchers provide a solution to the above problems by applying the integrated learning model. Blended learning integrated model is learning that combines several subjects by prioritizing the concepts, skills, and attitudes that can be connected with each subject [10]. Through the integrated learning model, learners can gain hands-on experience, so it can add strength to receive, store, and apply the concepts they have learned. Thus learners are trained to be able to find themselves a whole range of ideas that are studied holistically, meaningfully, authentically and actively that improve their learning processes and outcomes. It is following the stage of cognitive development of elementary students.

The implementation of integrated learning using a blended learning model indicated that there was an increased understanding of learning for students [11]. To improve the quality of learning in elementary school to obtain optimal and meaningful learning outcomes, applied learning model centered on the involvement of learners and pay attention to the way of thinking. Learners themselves actively build knowledge based on the owned cognitive structure.

2. Method

The type of research was experimental research. This research includes quasi-experiment because the variables cannot be fully controlled as real experiments [12]. The kind of analysis used was Random Research Group Control. In this study, the learners grouped into two classes, the experimental class, and the control class. The experimental level is the class treated by the integrated model, while the control class is a class that uses the conventional approach to learning at the same time. Furthermore, at the end of the learning, the two groups were given the identical posttest that was the result of a learning test. The population in this study was all students of V MIN 1 Padang in the year of academic 2017/2018 consisting of 3 classes. The researchers took two sample classes in this study. The sampling technique used to determine the sample in this study was Simple Random Sampling (randomly). The first class selected was used as the experimental class, the V-A class and the second class selected to be the control class, the V-B class. Sources of data in this study were primary data sources obtained directly by researchers and sources of seconder data sourced from others. This study used three instruments: questionnaire learning motivation, test and observation sheet (assessment) for the attitude aspect of learners.

3. Result and Discussion

The learning outcomes of "animal and human circulatory organ" learners MIN 1 Padang City studied showed that learners who were taught by the integrated model both whole and groups of learners who had high motivation and groups of learners who had low motivation were higher than the learner who was taught by conventional learning model. Based on the prerequisite test the analysis obtained normally distributed data and had a homogeneous variance so that it was qualified to continue the hypothesis test. Hypothesis testing aims to see the treatment given to the sample research. Hypotheses 1, 2 and 3 used two path Anava test or use the F test. The hypothesis test results showed in Table 1.

Table 1. Results of variance analysis

Varians Sources	SS	df	MS	F_{count}	F_{table}	Annotation
Between treatment	1.198,8	16		F (1,60)=34.14		H_a accepted
Factor (A)	698.94	1	698,94	F (1.60)=24.51	4,00	H_a accepted
Factor (B)	481,27	1	481.27	F (1.60)=0.91		H_a rejected
Factor AXB	18,62	1	18.62			
Within Treatment	1.228,23	60	20.47			
Total	2.427,06	63				

Based on the results of calculations in Table 1 by using two-way ANOVA test, the hypothesis could be described such as:

- Hypothesis 1. Based on the analysis results obtained Fcount of 34.14 and Ftabel of 4.00 because Fcount greater than Ftabel, which is 34.14 > 4.00 then Ho is rejected, and Ha was accepted. It means that there was a significant difference between the learning outcomes of learners taught with an integrated model with the learning outcomes of learners taught with conventional learning model. Learning outcomes of learners taught with an integrated model were better than the learning outcomes of learners taught with traditional models.
- Hypothesis 2. obtained Fcount of 24.51 and Ftabel 4.00 because Fcount was bigger than Ftabel, that was 24.51 > 4.00 then H0 was rejected, and Ha was accepted, this means there was a significant difference between learning result of learners who had high motivation with learners had low motivation. The learning outcomes of highly motivated learners in the two sample classes were better than the low-motivated learners' outcomes in both samples.
- Hypothesis 3. Based on the calculation results, Fcount was 0.91, and Ftabel was 4.00, so Fcount was smaller than Ftabel (Ftabel < Ftabel), that was 0.91 > 4.00 this means H0 accepted and Ha rejected, meaning there is no significant interaction between the integrated model and motivation to learners' learning outcomes on animal circulation animal and human organ.

3.1. Learning outcomes of students of circulatory organ material of animals and humans in experiment class and control class

The data of students' learning outcomes in this study were obtained through the final test of a written test in the form of multiple choice questions given to the experimental and control classes at the end of the learning process meeting, the results of the attitude assessment through self-assessment and assessment of partners and skills assessment through performance appraisal and products implemented in the learning process of each meeting. Description of data about students' learning outcomes aspects of knowledge, attitude, and skills can be seen in Figure 1.

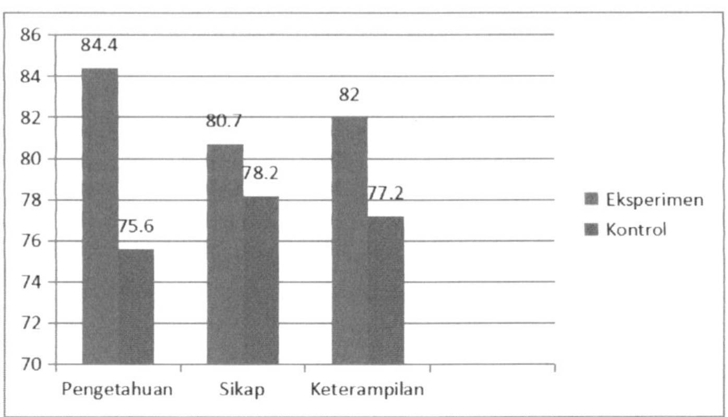

Figure 1. Average learners' learning outcomes overall experiment class and control class

Based on known final test results given in both classes showed that the average value obtained in the experimental class was higher than the average grade of the control class. The experiment class maximum value was higher than the maximum amount of the control class. While the minimum value of experimental courses was also higher than the control class. The learning outcomes of the overall learners consisting of final test scores and attitude and skill assessment assessments can be seen in Figure 2.

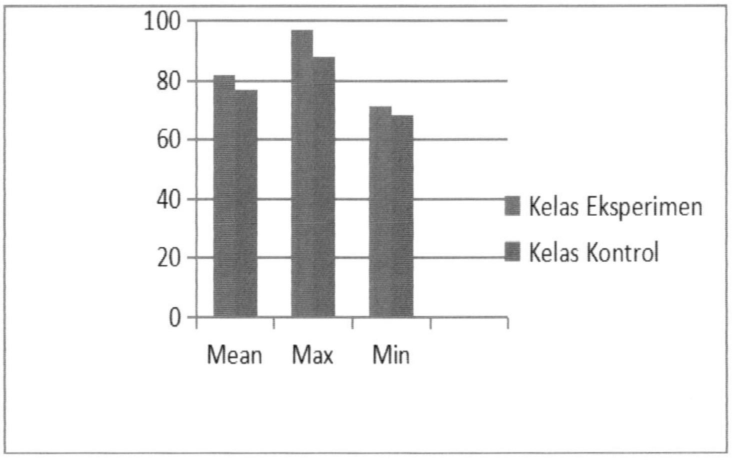

Figure 2. Average student learning outcomes overall experiment class and control class

Based on Figure 2, the results of the study consisting of the final test, attitude assessment and students' skill assessments as a whole of the experimental class obtained the average, maximum and minimum values higher than the control class. Learning outcomes of learners based on high and low motivation in the experimental class and control class can be seen in Table 2.

Table 2. Learning results of classroom experiment and control class students

Classes	X	X max	X min	S	S^2
Experimental	82	96.5	73.5	6.24	38.93
Control	76.9	87.5	67.5	4.93	24.29

Table 3 shows that the average learning outcomes of experimental class students based on high motivation were better than average learning outcomes control class that had high motivation. Both high-motivated experimental and control levels had a higher average learning outcome than the low-motivated average learning outcomes of learners.

Table 3. Learning outcomes of learners based on motivation in experiment class and control class

Classes	Motivation	X	X max	X min	S	S^2
Experimental	High	86.28	96.5	84.5	5,8	33.49
	Low	78.59	84.5	73.5	3.4	11.53
Control	High	80.97	87.5	76.5	6.95	48.3
	Low	72.94	77.5	67.5	3.93	15.43

4. Conclusion

The learning outcomes of learners taught in the integrated learning model is higher than the learning outcomes of learners taught with conventional learning models. Learning outcomes of learners is who have high motivation higher than the learning outcomes of learners who have low motivation [13]. There is no significant interaction between the learning models Integrated with the motivation to learn to learners' learning outcomes. The amount of contribution of motivation in improving learners' learning outcomes could be seen on the collective learning outcomes obtained by learners who had high motivation in the experimental class was 86.2, while the learning outcomes of low motivated learners were 78.9. In the control class who had high motivation was 74.8 and average learning outcomes of learners who had low motivation were 63.1.

References

[1] Lawson-Martin R and Normore A H 2006 Effects of cooperative and individual integrated learning system on attitudes and achievement in mathematics *Proceedings of the Fifth Annual College of Education Research Conference: Urban and International Education Section* (Miami: Florida International University) p 64
[2] Mumu J, *et al* 2018 *J. Phys.: Conf. Ser.* **943** 012011
[3] Tanujaya B, *et al* 2017 *World Transactions on Engineering and Technology Education* **15** 287
[4] Majid and Rochman 2014 *Pendekatan Ilmiah Dalam Implementasi Kurikulum 2013* (Bandung: Remaja Rosdakarya)
[5] Muhtadi D, *et al* 2018 *J. Phys.: Conf. Ser.* **943** 012020
[6] Usman 2010 *Mendesain Model Pembelajaran Inovatif Progresif* (Jakarta: Kencana Prenada Media)
[7] Hernawan A H and Resmini N 2008 Konsep dasar dan model-model pembelajaran terpadu *Modul Pembelajaran Terpadu di SD* (Jakarta: Penerbit Universitas Terbuka)
[8] Mudiono A 2016 *LINGUA: Journal of Language, Literature and Teaching* **11** 95
[9] Daryanto 2005 *Evaluasi Pendidikan* (Jakarta: Rineka Cipta)
[10] Fogarty R 1991 *The Mindful School How to Integrate the Curricula* (Illinois: Skylight Training and Publishing Inc.)
[11] Amini R 2017 *Unnes Science Education Journal* **6** 1586
[12] Gravetter F J and Wallnau L B 2016 *Statistics for the Behavioral Sciences* (New York: The College at Brockport)
[13] Tella A 2007 *Eurasia Journal of Mathematics, Science and Technology Education* **3** 149

IOP Conf. Series: Journal of Physics: Conf. Series **1088** (2018) 012058 doi:10.1088/1742-6596/1088/1/012058

Students' perceptions on the implementation of e-learning: Helpful or unhelpful?

L Vitoria, M Mislinawati and N Nurmasyitah
Syiah Kuala University, Jln. Teuku Nyak Arief, Darussalam, Banda Aceh, Aceh, 23111, Indonesia

E-mail: lindav@unsyiah.ac.id

Abstract. E-learning is an essential part of today's education. This article reports a design research conducted to develop an e-learning web-based module at the Primary Education Department of Syiah Kuala University in Indonesia. The study employed the Four-D Model of instructional development which consisted of Define, Design, Develop, and Disseminate. This article focuses on the Develop stage where the users' perception was assessed regarding the implementation of the module. A number of 19 third semester students participated in the study. The data was collected using a questionnaire and interview. The questionnaire was constructed based on the Technology Acceptance Model (TAM) which suggested that two factors influencing someone's acceptance of technology were perceived usefulness and perceived ease of use. The result showed that the students perceived the e-learning web-based module to be useful in improving their understanding, independence, self-discipline, motivation to learn, and interactions with each other and with the teacher. The students also agreed that the e-learning web-based module was easy to use. This study implies that the inclusion of technology in education at the university is beneficial.

1. Introduction

The implementation of information technology is proliferating in many aspects of our lives today, including education. Consequently, Syiah Kuala University as the leading university in Aceh Province of Indonesia has been promoting the implementation of ICT in the teaching and learning process in the form of blended learning. Blended learning is a combination of e-learning and face-to-face lectures. E-learning is "a system that uses internet technology to deliver information to students with interactions through computer interfaces" [1].

At the Primary Education Department of Syiah Kuala University, a limited number of courses have started implementing e-learning. This article reports an investigation on students' perception of the implementation of an e-learning web-based module at the Primary Education Department. The investigation was part of a design research conducted to develop the module for the ICT-Based Learning Course. We consider it is essential to assess students' perception because the way students view a teaching approach greatly affects their interestedness in learning and class participation [2]. Lorenzi and Riley [3] state that behavioral factors, such as perceptions, play more important role than technical factors in the success of an approach.

Many studies have been conducted to investigate students' perceptions on the inclusion of ICT in their learning. Generally, students respond positively towards the implementation of e-learning [4-5], but there are also some students who prefer traditional on-campus learning because they consider e-learning can cause uncertainty [6]. These studies show varied students' perceptions towards e-learning.

At the Primary Education Department of Syiah Kuala University, most of the students are active users of the internet, emails, and social media, but they are not yet familiar with e-learning. E-learning is not yet widely implemented in universities in the Aceh Province. Therefore, there is a very limited number of studies on this topic that have been published. This article is intended to give insight into how students perceive the implementation of e-learning in Aceh.

Being an active user of ICT does not guarantee one's acceptance of e-learning because learning using technology is very different from daily experience with ICT [7]. Popovici and Mironov [7] suggest that the main factors that influence someone's acceptance of new technology are needs and demands. If a student perceives e-learning to be useful and helpful for their study, then they are likely willing to accept it. On the other hand, hindrance such as lack of ICT skills and lack of access can prompt negative perceptions towards e-learning [8].

For this study, we employed Davis' [9] technology acceptance model (TAM) which posited that someone's acceptance of technology was determined by two factors: 1) perceived usefulness, and 2) perceived ease of use. Perceived usefulness refers to the degree of user's acceptance that the technology brings benefits for them in that it can enhance their performance, and perceived ease of use refers to the degree of user's acceptance that the technology can be accessed effortlessly [10]. The research question addressed in this study is: What are the students' perceptions on the implementation of the e-learning web-based module in the ICT-Based Learning Course at the Primary Education Department of Syiah Kuala University Indonesia?

2. Method

This study employed Research and Development (R&D). We used the Four-D Model of instructional development proposed by Thiagarajan, Semmel, and Semmel [11]. The Four-D Model consisted of Define, Design, Develop, and Disseminate. The Define stage is aimed at defining objectives and constraints; the Design stage is to design initial product; the Develop stage is where modification takes place in accordance with feedback from experts and users; and the Disseminate stage is to distribute the final product [11].

The product that we developed was an e-learning web-based module for the ICT-Based Learning Course at the Primary Education Department of Syiah Kuala University, Indonesia. The module contained course objective, course outline, lessons, links to reading materials and videos, discussion forum, and assessment. The module was available online for the students who were enrolled in the course at www.elearning.unsyiah.ac.id. Figure 1 is a screenshot of some of the module's contents.

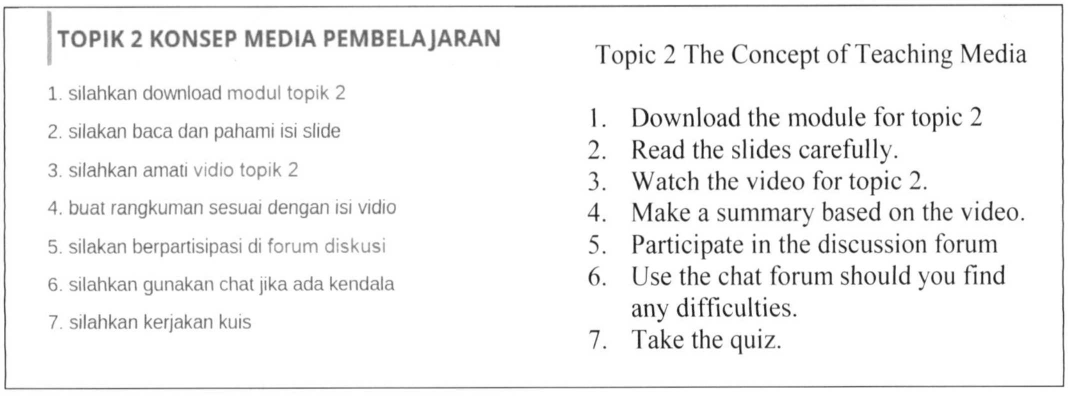

Figure 1. Teacher's instructions for a particular topic in the module.

In this article, we focus on the Develop stage where we report the practicality of the e-learning web-based module. Van den Akker [12] explains that practicality refers to the degree that users perceive a product as "appealing and usable". Therefore, in this study, practicality refers to students' perceived usefulness and ease of use of the e-learning web-based module. The investigation of students' perceptions as the end-users of the module is an important part of the module development because as Thiagarajan, Semmel, and Semmel [11] state that during the Develop stage, users' responses and comments serve as a basis for revision of the instructional development.

A number of 19 third-semester (1 male and 18 female) students who were enrolled in the ICT-Based Learning Course participated in the study. Their average age was 20 years old. Regarding the students' awareness of technology, all of the students were familiar with the internet and word processor. A preliminary survey showed that over 84% of the students owned a personal computer; 94.74% owned a smartphone; 89.47% of the students used the internet and social media on a daily basis while 10.53% spent at least 10 hours per week on the internet. Therefore, it could be established that none of the participants were ICT-illiterate.

For data collection, a questionnaire and interview were utilized to assess the students' perception on the implementation of the e-learning web-based module. The data collection was conducted at the end of the odd semester in 2017. The questionnaire was developed based on Davis' Technology Acceptance Model (TAM) which suggested that two factors influencing someone's acceptance of technology were: 1) perceived usefulness, and 2) perceived ease of use [9]. The questionnaire contained 20 questions using a 4-point Likert scale ranging from strongly disagree (SD) = 1, disagree (D) = 2, agree (A) = 3, and strongly agree (SA) = 4. The data obtained from the questionnaire were analyzed statistically by calculating the mean score for each question.

In order to gain in-depth insight into the students' perception, we conducted a semi-structured interview with two students. One student earned A on their final mark for the ICT-Based Learning Course, and the other student earned B. No one in the class earned less than B. The interview was intended to clarify the students' perception on whether the module was useful and easy to use. Each interview lasted for approximately 30 minutes. In the following section, the interviewees are referred to as Student A and Student B.

3. Results and discussion

This section is reported in two subsections: students' perceived usefulness and students' perceived ease of use.

3.1. Students' perceived usefulness

Table 1 shows the students' perceived usefulness of the e-learning web-based module.

Table 1. Students' perceived usefulness.

Items No	Score				Mean
	SA	A	D	SD	
1. The e-learning web-based module provides me with broad sources of learning materials.	6	13			3.3158
2. E-learning gives me more time to explore learning sources.	6	13			3.3158
3. The module helps my understanding of a topic.	11	8			3.5789
4. E-learning improves my motivation to learn.	10	9			3.5263
5. E-learning helps me to be more independent in my learning.	13	6			3.6842
6. On-line tasks improve my self-discipline.	9	10			3.4737
7. E-learning improves my self-confidence in expressing ideas and opinions.	7	12			3.3684
8. The e-learning web-based module improves my ICT skills.	11	8			3.5789
9. E-learning makes learning more enjoyable.	9	10			3.4737
10. The e-learning web-based module improves my interactions with friends and the teacher.	1	15	3		2.8947

Table 1 shows that the students' perception on the usefulness of the module was positive. None of the students strongly disagreed with the statements in the questionnaire. All of the students agreed or strongly agreed with items 1 to 9, and 84% of students agreed or strongly agreed with item 10. The item that most of the students strongly agreed with was item no.5 (mean = 3.6842), and the least agreed with was item no.10 (mean = 2.8947).

As evidence in table 1, the students agreed that e-learning provided them with wide sources of learning materials. During the interview, the students explained that e-learning gave them more time to read the materials so that they could understand well about a particular topic prior to class meeting.

Student A: We have access to the sources before face-to-face meeting in the classroom. It is convenient because I can read the source materials at my own time. I usually read through the materials to have a good understanding of them so that I feel more confident to participate in the classroom when we do the face-to-face meeting.

Student B: The source materials are there, I don't have to look anywhere else so I can save time. There are articles, powerpoint slides, and videos. That is very helpful.

The students also declared that e-learning improved their motivation to learn (item no.4, mean = 3.5263). The interviewees explained,

Student A: With blended learning like this, I think it improves my motivation to learn because there is a variety so I don't feel bored. Usually, we only do face-to-face meetings in the classroom. Now we have online meetings as well. The teacher has provided various sources online, so I am motivated to read them because I want to be ready for the face-to-face meetings and online discussions.

Student B: The teacher always reminds us that we need to read through the materials in order to understand better. I am motivated to learn because I want to understand.

The students agreed that e-learning helped improve their independence (item no.5, mean = 3.6842) and self-discipline (item no.6, mean = 3.4737). With the availability of sources and clarity of what to expect at every stage throughout the course, the students were motivated to manage their time well.

Regarding students' interactions, 15.79% of students disagreed that the e-learning web-based module helped improve their interactions. Student B explained that there was no difference between e-learning and traditional learning in terms of students' interactions with each other or with the teacher. However, 5.26% of students strongly agreed and 78.95% agreed that e-learning improved their interactions. Student A described as follow.

Student A: I think e-learning improves my interactions. I mean, when I am having difficulties in understanding an article or in doing my task at home, I can use the forum to ask for help. It is different from traditional classroom in that I usually ask only when we are having a class meeting. With e-learning, there is more time to communicate with everyone.

All the data above showed that the students perceived the e-learning web-based module to be useful. First of all the availability of various sources online which the students could access at their convenience was perceived by the students to be helpful in their effort to understand a topic. This statement is in line with Al-Dosari's [5] finding that online course users considered accessibility to be the greatest advantage of e-learning.

Second, the students claimed that e-learning improved their learning motivation. E-learning offered an alternative mode of learning, various forms of teaching materials, and more time to communicate especially with the teacher. All these contributed to the students' preparedness for face-to-face lectures, which the students considered to be highly important. This finding is in line with a study conducted by Aristovnik, Kerzic, Tomazevic and Umek [13] that students' perceived usefulness of e-learning was significantly influenced by the consistency of e-learning with traditional classroom meetings and teacher's responsiveness.

Third, the students viewed that the availability of discussion forums improved their interactions because they could continue exchanging information outside the classroom through online communication. This result is in line with Al-Dosari's [5] finding that the online courses improved students' participation in discussions that even the shy ones tended to be more conversational in online communication. Improvement in students' interactions is beneficial for students' learning because as Vygotsky states that learning happens when students interact and communicate with each other [14].

Overall, the implementation of e-learning, as a supplement to traditional face-to-face lectures, enhanced students' learning experience [1]. From the data above, it can be concluded that the students perceived the e-learning web-based module to be useful.

3.2. Students' perceived ease of use

Table 2 displays the students' perception on the ease of use of the e-learning web-based module.

Table 2. Students' perceived ease of use.

Items No	Score				Mean
	SA	A	D	SD	
1. I have the technology to access the module.	10	9			3.5263
2. I have time to access the online materials.	7	12			3.3684
3. I know how to access the materials provided online by the teacher.	13	6			3.6842
4. I know how to navigate through the e-learning web-based module.	14	5			3.7368
5. The language of the module is easy to understand.	7	12			3.3684
6. The teacher's online instructions are easy to understand.	9	10			3.4737
7. It is easy to do my tasks with the e-learning web-based module.	11	8			3.5789
8. Internet connection is not a problem for me.	9	10			3.4737
9. The online forums ease my interactions with classmates and the teacher.	8	10	1		3.3684
10. I often encounter technical problems with the e-learning web-based module.		3	15	1	2.1053

Table 2 shows that the students have a positive perception regarding the module ease of use. The most favorable item was item no.4 (mean = 3.7368), and the least favorable item was item no.10 (mean = 2.1053). This is not surprising given that the students were already familiar with ICT. First of all, the students had ready access to the module (item no.1, mean = 3.5263). They owned a personal computer or a smartphone.

Being familiar with the internet, the students claimed that they knew how to access the source materials online (item no.2, mean = 3.3684), and how to navigate the e-learning web-based module (item no.4, mean = 3.7368). The students considered the language used in the module was easy to understand (item no.5, mean = 3.3684). So were the teacher's instructions (item no.6, mean = 3.4737)This is as explained by the interviewees as follow.

Student A: I find it quite easy to work my way through the website and the forums. They are quite similar to other websites that I am used to accessing. As for the forums, they were new to me at first, but they were not difficult to understand.

Student B: The teacher's instructions are quite clear and easy to understand, so it is not difficult to follow the steps to navigate the web and the forums.

The students also claimed that internet connection was not a difficulty for them (item no.8, mean = 3.4737). Student A stated that the students usually used Wi-Fi available at the campus and while outside campus they used a modem.

The students also agreed that it was easy to do tasks with e-learning (item no.7, mean = 3.5789). Student A elaborated,

Student A: For the task, it is basically the same as a traditional task, but easier. For example, one time we were asked to make a summary on a specific topic. I do it the way I always do. I type my summary on Word Processor and I go online to retrieve some articles that I need. With e-learning, it is easier because the sources are already available on the website, and I don't have to print out my assignment, I submit it online.

The survey shows that more than 80% of the students did not often encounter technical problems. However, 15.79% of students agreed that they often encounter a technical problem. Student B was one of them. She explained that the problem was a slow connection. In regards to other aspects of e-learning ease of use, she agreed that the web-based module was easy to use.

Nawaz and Qureshi [15] stated that the most crucial factor that determined someone's comfort in accessing e-learning was their ICT literacy. This is as Popovici and Mironov [7] found that students' expertise in technology affected their acceptance of e-learning. Basic ICT skills are required to be able to participate in e-learning without stress. As we mentioned earlier, the students who participated in the study were ICT literate. Therefore it was comprehensible that they did not encounter serious technical problems with the e-learning web-based module, except for a few students who had difficulties with a slow internet connection. Overall, the students agreed that the module was easy to navigate, the sources were easy to access, and the teacher's instructions were easy to understand. Therefore it can be concluded that the students agreed that the e-learning web-based module was easy to use.

4. Conclusion

The students who participated in the study perceived the e-learning web-based module to be useful and easy to use. They agreed that the module was helpful in improving their understanding of course materials, time discipline, interactions with each other and with the teacher, and that e-learning was enjoyable. The students also claimed that it was easy to access the module. Presumably, the students' familiarity with ICT in their everyday life played an important role in their familiarizing with e-learning. This finding mentions that the development of e-learning should be continuously updated with the latest trends.

In this study, we addressed the issue of students' perceived usefulness and ease of use of the e-learning web-based module used at the Primary Education Department of Syiah Kuala University. For future research, we suggest an investigation on students' preferences of an e-learning system to gain information about what students, as the end users of e-learning, wish from e-learning.

References

[1] Al-Adwan A, Al-Adwan A and Smedley J 2013 Exploring students acceptance of e-learning using Technology Acceptance Model in Jordanian universities *International Journal of Education and Development using ICT* **9** 4-18
[2] Suprihatiningrum J 2016 *Strategi Pembelajaran: Teori dan Aplikasi* (Yogyakarta: Ar-Ruz Media)
[3] Lorenzi N M and Riley R T 2000 Managing change: An overview *Journal of the American Medical Informatics Association* **7** 116-24
[4] Almarabeh T 2014 Students' perceptions of e-learning at the University of Jordan *iJET* **9** 31-5
[5] Al-Dosari H 2011 Faculty members and students perceptions of e-learning in the English department: A project evaluation *Journal of Social Sciences* **7** 291
[6] El-Gamal S and Aziz A E 2011 The perception of students regarding e-learning implementation in Egyptian universities: The case of Arab Academy for science and technology *Proc. 3rdInt. Conf. on Mobile, Hybrid, and On-line Learning (France, 23 – 28 February 2011)* pp 1-5
[7] Popovici A and Mironov C 2015 Students' perception on using elearning technologies *Procedia – Social and Behavioral Sciences* **180** 1514-19
[8] Silviyanti T M and Yusuf Y Q 2015 EFL teachers' perceptions on using ICT in their teaching: To use or to reject? *Teaching English with Technology* **15** 29-43
[9] Davis F D 1989 Perceived usefulness, perceived ease of use, and user acceptance of information technology *MIS Quarterly* **13** 319-40
[10] Lederer A L, Maupin D J, Sena M P and Zhuang Y 2000 The technology acceptance model and the world wide web *Decision Support System* **29** 269-82
[11] Thiagarajan S, Semmel D S and Semmel M I 1974 *Instructional Development for Training Teachers of Exceptional Children: A Sourcebook* (Indiana: Indiana University)
[12] Van den Akker J 1999 Principles and methods of development research *Design Approaches and Tools in Education and Training* (Netherlands: Springer) pp 1–14
[13] Aristovnik A, Kerzic D, Tomazevic N and Umek L 2016 Determining factors of students' perceived usefulness of e-learning in higher education *Proc. of the Int. Conf. e-learning (Portugal 1–4 July 2016)* ed M B Nunes and M McPherson pp 3-10
[14] Darmadi H 2017 *Pengembangan Model dan Metode Pembelajaran dalam Dinamika Belajar Siswa* (Yogyakarta: Budi Utama)
[15] Nawaz A and Qureshi Q A 2010 Sustained technical support: Issue & prospects for e-learning in Heis *Global Journal of Management and Business Research* **10** 32-9

IOP Conf. Series: Journal of Physics: Conf. Series **1088** (2018) 012059 doi:10.1088/1742-6596/1088/1/012059

The development of an instrument to explore non-routine problem solving strategies among mathematics education students

R M Bambang S, Salasi R, M Hasbi and Mardhiah MZ
Syiah Kuala University, JL. Tgk. Chik Pante Kulu No 5, Banda Aceh, Indonesia, 23111

E-mail: bambang_radenmas@unsyiah.ac.id

Abstract. This research aimed at describing the validation results of instrument development to explore the non-routine problem solving strategies, especially the non-routine strategy assignment and interview guidelines. The validity of the assignment instrument is viewed from the aspect of accuracy, strategic direction, cognitive level, scope of the knowledge, and legibility of the problems. The interview guidelines instrument is evaluated by the ability to disclose the strategy, the nature of the question by being exploring and not guiding, and the clarity of the students' response. This study is part of the broader development research. The subjects were Mathematics Education students of Syiah Kuala University. The research instrument was a validation sheet about non-routine problem solving strategy and interview guidelines. Data were collected by obtaining expert's opinion on the content and construct validity of the non-routine problem solving strategies assignment instrument, while the data collection for the trial was conducted through the task-based, semi-structured interview. Expert validation data then was analyzed by reviewing thoroughly the experts' suggestion and opinion on the instrument validation, while the result of the trial is analyzed by describing the emerging strategies from student's response to the given problems. The validation result of instrument development indicates that the instrument and interview guidelines are applicable in order to explore strategies used by students in non-routine problem solving.

1. Introduction

Problem solving is one of the cognitive activity procedures used in everyday life [1]. Gagne [10] asserts that solution finding is high-level thinking process. Thus, problem solving is the process of thinking that requires the implementation of procedures or strategies in mathematical problem-solving activities. The success of an individual in obtaining the solution of a problem is influenced by the knowledge, skills, and the ability to link the knowledge structure in the mind [10], and the process of using those strategies in problem-solving [3].

Findings from a study [4] in non-routine problem-solving strategies at first-year students show that students are generally successful to apply strategies in problem solving. The success of a student in carrying out non-routine problem solving activities is also influenced by the problems encountered. One question may become a problem for student A, which is not necessarily a problem for student B, or vice versa. Therefore, developing non-routine problem-solving instruments and interview guidelines are considered important.

2. Methods

This research is a development research using descriptive approach [7]. The sample was students of Mathematics Education of Syiah Kuala University. The data were collected by (1) distributing validation sheet to assess the content and construct of the non-routine problem-solving strategy, and validation sheet to assess the content and construct of the interview guidelines to 2 *validators* for assessing the content and construct aspects. A *validator* is an expert and Doctor of Mathematics education, (2) enrolling the test of the non-routine problem-solving strategy assignment instrument towards 15 students, at the same time, the interview guidelines instrument was tested towards 2 students based on the completion of the assignments. Data were collected using an expert validated sheet for non-routine problem-solving strategy assignment and an expert validated instrument of the interview guidelines. The raw data were analyzed by reviewing and revising the results from the assessment towards the problems in the problem-solving strategy task and the questionnaires in the interview guidelines; while the test results from the trial were analyzed by examining and interpreting the strategies that emerged in student's attempt to solve the given problems.

3. Results and discussion

The result of this study comprises the results from the development of non-routine problem-solving strategy instrument and the interview guide instruments. The result from the instrument development is in the form of the validated instrument and the tested instrument of problem-solving strategy task and interview guide. Results from the development of the two instruments are presented as follows:

3.1 Results of Validation and Testing of Non-Routine Problem-Solving Strategy Assignment

Experts' validation was performed to evaluate the contents validity. Expert validation results are used as a basis for the revision and refinement of the research instruments. The revised instrument then used to explore the strategies students used in solving non-routine mathematical problems [5,12]. The *validators* in this study consist of two mathematics education experts. Based on the suggestions and comments from the *validators*, several revisions and improvements were made as the first draft, which later compiled into the second draft.

Table 1. Validation result of non-routine problem solving strategies assignment

No	Submitted	Revision
1	A square root of a swarm of bees from its nest heading to the west, while the 5/7 parts of this swarm turned eastwards. Whereas 4/3 of the bees that go eastwards minus 15 multiplied by 7 yields 15. How many bees present at first?	A square root of a swarm of bees from its nest heading to the west, 5/7 of this swarm turned eastward, whereas 4/3 of the bees that go eastwards minus 15 multiplied by 7 yields 15. How many bees present at first?
2	Santi and Tini ran along a circular trajectory. Both begin to run at the same time from point P, but take the opposite direction. Santi ran 1 1/2 times faster than Tini. If PQ is the center line of the circle path and both meet for the first time at the point R, determine how much ∠RPQ?.	Santi and Tini ran along a circular trajectory. Both begin to run at the same time from point P, but take the opposite direction. Santi ran 1 1/2 times faster than Tini. If PQ is the center line of the circle path and both meet for the first time at the point R, determine how much ∠QPR?.
3	Mr. Ali has a water tank to fill in the reserve water. The water tank has an inverted cone with circular top which radius r meters and width h meters where the length of the basis radius is a half of the cone height. If water is pumped into the tank at a rate of $2m^2$/min, determine the rate of water level increase at a depth of 3 meters.	Mr. Ali has a water tank to fill in the reserve water. The water tank has an inverted cone with circular top which radius *r* meters and width *h* meters where the length of the basis radius is a half of the cone height. If water is pumped into the tank at a rate of $2m^2$/min, determine the rate of water level increase at a depth of 3 meters.

The non-routine problem solving assignment sheet consists of 3 questions. The items in the non-routine problem solving strategy sheet related to the concept of numbers, algebra and geometry, and analytic calculus in the form of illustrated questions. The preparation of this non-routine problem solving assignment begins with developing the main point needed for the assignment. This main point comprises the strategies that may be used by students in solving problems through Polya's four steps, which is conformed on how the instrument was assembled [4,11]. Furthermore, two mathematics education experts validated the non-routine problem solving strategy assignment sheet. The *validator* assesses and provides inputs on the validation sheet. The results from the assessment of the two *validators* can be seen in Table 1.

Table 1 shows the result from the revision of the assignment based on the suggestion from the *validator*. In the first item, both *validators* suggested that the word "while" to be omitted. Next, the second item, the two *validators* also provided suggestions on the question to determine how much ∠RPQ, changed in to question to determine how much ∠QPR?, in the third item, the first *validator* does not give suggestion, but the second *validator* provides suggestion on the writing of the symbol of "r "And" h ". The validation results show that the non-routine problem solving strategy assignment is accurate, divergent, liable, broad knowledge coverage, easy to understand, cognitive level appropriateness, and feasible to use after revision. Thus, the assignment instrument can be used to explore the strategies that might be used by students in order to solve the non-routine problem. [6]
The test to assess the instrument was enrolled on March 24, 2018 towards 15 third-year students of Mathematics Education Program. One of the student works of the non-routine problem solving strategy assignment as shown in Figure 1, 2 and 3.

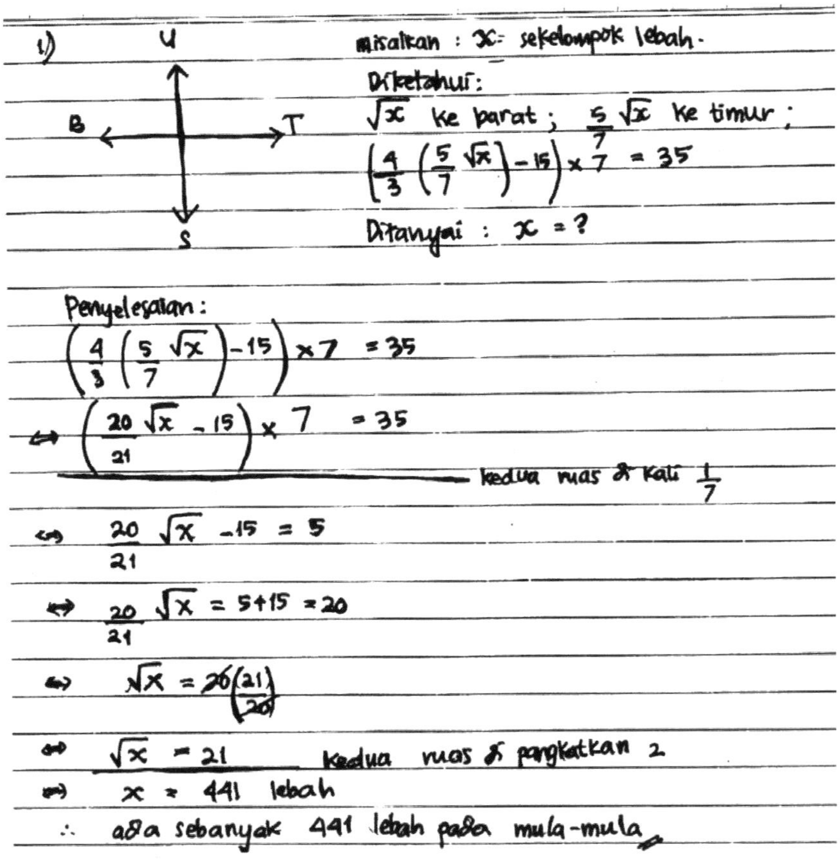

Figure 1. Student answer to question number 1

Figure 1 shows the student used a compass drawing to understand the problem [6], and then write down the known and questioned data correctly, the student then creates the model and continue to solve the model. The student completes the model by using the backward strategy by multiplying the two segments by 1/7 then summing by 5 then multiplying the two segments by 21/20 then rooted by two to obtained solutions. From this student work, it shows that problem 1 contains various strategies to gain its completion. [3], completed sequentially, with drawing illustrations [4], using rules to obtain solutions, and showed to be a correlation to the cognitive level of students. Based on the subject's answer to question 1 and the instrument assessment indicators of non-routine problem-solving tasks it can be concluded that the task of non-routine problem has various settlement strategies, corresponded to the cognitive level of the student, and the knowledge coverage is only related to the numbers. Thus, question 1 indicates a non-routine problem that has strategies in progress.

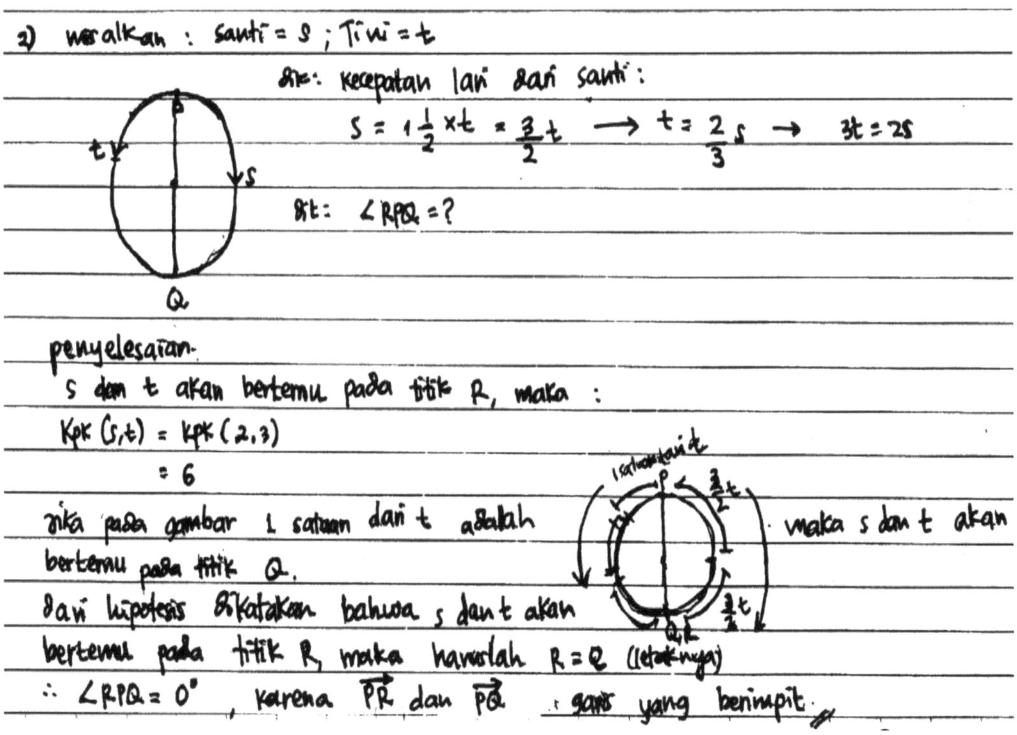

Figure 2. Student answer to question number 2

In figure 2 the student makes the illustration with complete and correct data [6], but there was an error in determining the speed equation [13]. To complete the task student used trial and error strategy, but fails to get the solution. The student work shows that problem 2 has various strategies to offer its solution. [3] Although drawing of illustrations was correct, the student failed to optimize the strategy [4] by miss calculating the speed equation. Based on the subject's answer to problem 2 and the instrument assessment indicators of non-routine problem-solving tasks, it can be concluded that the subject completed the task of non-routine problem by trial and error (only one strategy), knowledge coverage including the distance, speed (physics), geometry, algebra, and student cognitive level. Thus, problem 2 indicates a non-routine problem that has strategies emerged while solving it.

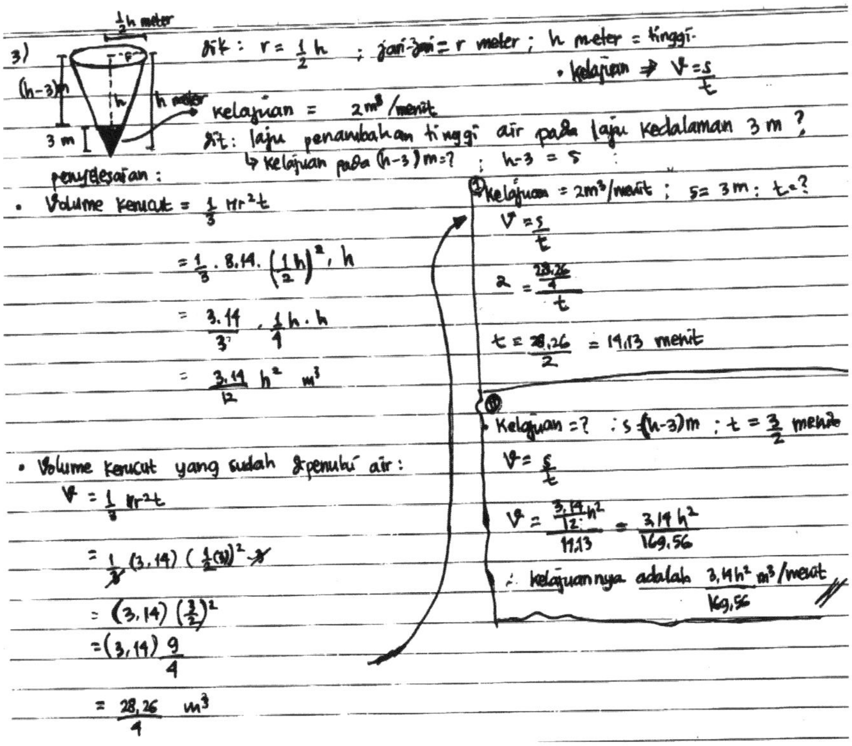

Figure 3. Student answer to question number 3

In figure 3 the student creates a cone illustration along with the data, and then creates symbols of the tube size, generates important data from the question information, then creates an equation showing the relationship between volume and height of the tube, uses a similar triangular image to create the equation relationship between the radius and tube height, models the tube volume relationship with tube height, then calculate the speed of water over time using derivative rules. From this student work, it shows that problem 3 contains various solving strategies [3], [9] completed sequentially, drawing illustrations [4,6], knowledge coverage covering geometry, algebra, derivative (differential), and appropriate with student cognitive level [5]. Based on the subject's answer to the problem 3 and the instrument assessment indicators of non-routine problem-solving tasks, it can be concluded that the task of non-routine problem has various coping strategies, broad knowledge coverage, and cognitive level conformity. Thus, problem 3 indicates a non-routine problem that has strategies in progress.

3.2 Results of Validation and Interview Testing Instrument

The interview guide instrument contains questions-statements to explore the process of thinking about the use of strategies in solving non-routine problems of students which viewed through Polya steps, namely; understanding problems, planning, executing and checking results. Questions that are compiled on the interview guide are key questions which could be developed during the interview. This is to adapt with the student answer or response to the question. The items in the interview questions are consist of (1) the stage of understanding the problem: what is understood from the problem, how to understand and why to use it, (2) to plan: what should be done to solve the problem, how, why use it? (3) to solve: what should be made to get the solution, how to get the solution, why use it that way? (4) check the results: how to check the certainty of the solution obtained, how to check it, why use it that way? [11], [12] It is in accordance with consideration [2] that semi-structured interviews are used to explore the thinking process in solving problems by reading problems, explaining what is understood, explaining what strategies or knowledge they possess, explaining how to obtain solutions, and what should be done to check the results.

The items question in the interview guidelines which have been developed is further validated by 2 mathematics education expert *validators* using a validation sheet. *Validators* assess and make suggestions in the interview validation sheets. The results of the assessment of the two *validators* are shown in Table 2.

Table 2. The result of validation of contents and constructs of interview guidelines

No	Interview Guideline Instrument	Score	
		Validator 1	Validator 2
1	Indicator completely described	Partially describe	Partially describe
2	Ability to reveal strategies in non-routine problem solving	Good	Very good
3	Open question	Open	Open enough
4	Suitable with student cognitive level	Yes	Yes
5	Triggered, not guided	Good	Good
6	Multiple interpretations	No	No

Table 2 shows the validation result of the interview guideline instrument is eligible to be used with minor revisions based on *validator*s' recommendation. Aspects of questions in the interview guide includes: 1) questions to trace the strategy in order to understand the problem, how to understand the problem, why use it, 2) the strategy of planning the settlement, that is how to plan, why use it, 3) the strategy of implementing the problem solving, that is how to solve it, why use it, 4) strategy to check results, that is how to check the solution obtained, why use it that way. The validation results of the two *validators* to the interview guidelines show that the interview questions; has no direct mention of the indicators at each stage, has the power of disclosing strategies in non-routine problem solving, in the form of open questions, conforms with the cognitive level of students, are in the nature of digging and not guiding, and does not lead to multiple interpretations. Thus, the interview guideline is applicable to explore strategies used by students in non-routine problem solving.

4. Conclusion

Based on the results of research and discussion, it indicates that the development of the problem solving strategy assignment instrument and the interview guideline instrument to explore the strategies used in problem solving is viable to be used. This article develops task instruments and interview guides to explore non-routine problem-solving strategies related to number theory, algebra and geometry and calculus, we strongly suggest to explore other materials in mathematics. We recommend testing non-routine problems on college students in order to gain a certain cognitive level in solving problems.

Acknowledgments

We thank Syiah Kuala University for providing the senior lecturer research grant funding, contract number 288/UN11.2/PP/PNBP/SP3/2018.

References

[1] Aljaberi N M 2015 University Students' Learning Styles and Their Ability to Solve Mathematical Problem *International Journal of Business and Social Science* **6** 1
[2] Angateeah K S 2017 An Investigation of Students' Difficulties in Solving Non Routine Word Problem at Lower Secondary *International Journal of Learning and Teaching* **3** 1
[3] Aveu S & Aveu R 2010 Pre-service Elementary Mathematics Teachers' use of Strategies in Mathematical Problem Solving *Prodia Social and Biheviorical Science* **9** 1282 Available online at www.sciencedirect.com

[4] Celebioglu B, Yazgan Y, & Ezentas R 2010 Usages of Non Routine Problem Solving Strategies at first Grade Level. *Prodia Social and Biheviorical Science* **2** 2968 Available online at www.sciencedirect.com
[5] Docktor J & Heller K 2009 Rebust Assessment Instrument for Student Problem Solving *Proceeding of the NARST 2009* Annual Meeting
[6] Duru A Peker M, Bozkurt E, Akgitn L, & Bayrakdar Z 2011 Pre-service Primary School Teachers' Preference of The Problem Solving Strategies for Word Problems *Prodia Social and Biheviorical Science* **15** 3463 Available online at www.sciencedirect.com
[7] Frenkel J R 1990 How to Design and Evaluation Research Instrumen Educatioan McGrwa Hill Publishing Coy
[8] Isoda M dan Katagiri S 2012 *Mathematical Thinking: How to Develop it in the Classroom* Singapure: World Scientific
[9] Nonotna et.al 2014 Problem Solving in School Mathematics Based on Heuristik Strategies *Journal on Effeciency and Responsibility in Education and Science* 7 1
[10] Orton A 1992 *Learning Mathematics: Issues, Theory and Classroom Practice*, Second Edition New York: Library of Congress Cataloging in Publication Data
[11] Sajadi M, Amiripour P, and Maikhalifeh M R 2013 The Examining Mathematics Word Problem Solving Ability under Efficient Representation Aspect *Journal International Mathematics Education Trends and Research*, Available online at www.ispacs.com/metr
[12] Samuel G & Andreas F 2013 Measuring complex problem solving an educational application of psychology theories *Journal for Education Research online* **5** 38 Um.nbn:de:0111.opus.80196
[13] Tambychik T and Meerah T S M 2010 Students' Difficulties in Mathematics Problem-Solving: What do they Say? *Prodia Social and Biheviorical Science* **8** 142 Available online at www.sciencedirect.com

IOP Conf. Series: Journal of Physics: Conf. Series **1088** (2018) 012060 doi:10.1088/1742-6596/1088/1/012060

Light Emitting Diode (LED) as an essential prop component for STEM education in the 21st century: A focus for secondary school level

Irwandi, R Oktavia, Rajibussalim, A Halim, Melvina
Syiah Kuala University, Jl. Teuku Nyak Arief Darussalam, Banda Aceh 23111, Indonesia

E-mail: irwandi@unsyiah.ac.id

Abstract. Indonesian secondary science curriculum is often late to keep up with the rapid development of technology. For instance, the widely-used LED (Light Emitting Diode) technology is not familiar as props in secondary science instructions in Indonesia. This study aims to: (1) provide professional development for teachers in developing experiments using LEDs for learning physics concepts, and (2) increase teachers' beliefs on their knowledge on using LEDs for physics experiments. The professional development only used a single group pre- and post-test research design in the form of a two-day workshop attended by 20 science teachers from secondary level schools in the city of Banda Aceh. The workshop used Investigative Science Learning Environment (ISLE)-STEM approach, an inquiry-based physics learning that incorporate STEM education approach in active learning and collaboration. The workshop used LEDs as experimental props. From the qualitative analysis of the pre- and post-questionnaires' responses, we found that there are positive improvements on participants' understanding of ISLE-based STEM approach and of the use of LEDs as props. From the questionnaire administered after the workshop, we found that most of the teachers were excited about the training using LED as props and 89.4% of them feel that their knowledge increased.

1. Introduction

Since its first participation on Programme for International Student Assessment (PISA), Indonesia consistently performed poorly in STEM subjects. In 2015, Indonesia ranked at the 63rd for Mathematics and at the 62nd for Science from 70 countries participated in the assessment. This result shows that the abilities of Indonesian students to apply and reason in Science and Mathematics are below average. One reason could be that common teaching practices in Indonesia rarely promote critical thinking skills [1]. Therefore, revision and improving curriculum to increase skills in STEM education are essential for teachers in Indonesia. However, the revision of the secondary school curriculum is often late to be conducted, and most of the time the curriculum was not able to follow the rapid development of technology. Meanwhile one of the educational goals of the 21st century is equipping students with problem-solving skills of today's world challenges. For example, in Indonesia, the use of LED technology (Light Emitting Diode) as components for toys and high-precision instruments were extensive, but the use of the technology as props is infrequent. In National Curriculum 2013 revised in 2016, the discussion of LEDs is only a part of discussions about Light Wave [1].

The lack of the use of LEDs as props in school- physics experiments instigate the following questions. The first question is "Given different examples of ISLE-based STEM experiments using LEDs, how will teachers develop their understanding of ISLE-based STEM approach and the use of LEDs as props for physics?", and the second question is "How does the professional development increase teachers' beliefs on their knowledge?" These questions directed this study to aim for the following goals: (1) to provide professional development for teachers in developing ISLE-Based STEM experiments using LEDs for learning physics concepts, such as conductivity, electrical currents, and resistances, light wavelength, etc.; and (2) to increase teachers' confidence in using LEDs for physics experiments.

1.1. ISLE-based STEM

STEM stands for Science, Technology, Engineering, and Mathematics. There are several definitions of STEM, and the most precise definition is "STEM education is an interdisciplinary approach to learning that removes traditional barriers separating the four disciplines of science, technology, engineering, and mathematics (and) integrates them into a real-world, rigorous, and relevant learning experience for students" [9]. A STEM is an approach that involves the application of skills and knowledge to solve problems or tasks and use engineering design processes to create products and solutions. Based on the sequence in the abbreviation we can say that STEM learning is learning Science by utilizing Technology in Engineering process as well as involving Mathematics calculation. The STEM approach is an approach that combines knowledge and skill in solving problems with various models including inquiry-based learning (IBL), discovery-based learning (DBL), and project-based learning (PBL). One of the models used in this study is the Investigative Science Learning Environment (ISLE). Prof. Eugenia Etkina from Rutgers University of New Jersey, USA developed ISLE models for learning physics. ISLE is a model that helps students mimic the way physicists create knowledge [3]. We have modified a simplified ISLE model [4] as shown in Figure 1.

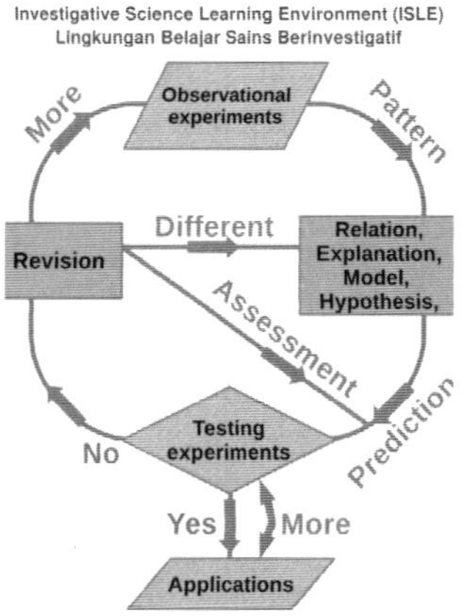

Figure 1. The simplified version of the ISLE model [5] modified from http://www.islephysics.net/isle.php.

The ISLE learning cycle in Figure 1 showed that to figure out the physics concepts underlying a real-life phenomenon, we start by observing the phenomenon under different conditions (observation experiments) until we find patterns that will lead to the development of a relation, explanation, model, or hypothesis. Assumed that the hypothesis is correct, then we predict what would happen if we conduct testing experiments. If our prediction is accurate then, we convince that the hypothesis is correct; otherwise, we need to revise it or do some more observation experiments and continue the process until we get the proved hypothesis. Then we apply the proved hypothesis to solve real-life problems or to explain related phenomena.

One of the advantages of ISLE is that it is creating an environment favorable to the 21st-century skills such as critical thinking, problem-solving, communication, and collaboration. One of the authors' experience was attending Workshop of PHYSWARE 2015: Preparing 21st Century Physics

Teacher. Trieste, Italy January 25-30, 2015 that intensively created collaboration atmosphere as shown in Figure 2 and communication atmosphere as shown in Figure 3. In Figure 2, the author's group did an experiment that shows human bodies are conductors as well as media for currents to flow. In Figure 3 the author with his group discuss the patterns that they saw on observational experiments using LEDs to develop hypotheses. In Figure 3, we can see that the use of LEDs as props in experiments could lead to meaningful learning. Why do LEDs become essential for physics experiments? The facts that (1) LED (Light Emitting Diode) technology is widely used in our daily life in this digital era, (2) their cost is inexpensive, and (3) their effectiveness is extensive to explain physics concepts, are some of the reasons why students need to be familiar with this technology.

Figure 2. Team collaboration to conduct investigations that the human body can act as a current conductor and a current flow between each other.

Figure 3. The team is communicating with each other to discuss the relationship between voltage and current that flows in the LED.

1.2 The 21st century skills and challenges
Why do we need to use STEM approach such as ISLE in learning STEM? The answer is because the approach will help students to develop skills, abilities, and learning dispositions. These skills are those that are required for success in the 21st-century society and workplaces. These skills are not the same as skills traditionally prepared in schools, which are primarily content-knowledge-based skills. There are similar concerns on changing the focus of educations in many countries around the world. The education system should shift its focus to prepare students to master analytic reasoning, complex problem solving, and teamwork that lead to deeper learning, the skills needed to be successful in our current dynamic digital society [7].

In addition to the skills mentioned above, another important skill that could be considered as the ultimate skill needed to face the rapid changes in society and to overcome the challenges that an individual face now or lie ahead is the ability to "learn how to learn." The four pillars of learning: learning to know, learning to do, learning to live together, and learning to be the fundamental principles of the 21st century's education. While mastering content-knowledge-based skills, students must also develop mastery of the essential skills such as critical thinking, problem-solving, communication, and collaboration, to be successful in this new era. A study suggested that skills like traditional reading, writing, and arithmetic are no longer the top skills demanded by U.S. Fortune 500 companies by the year of 2000 which preferred teamwork, problem-solving, and interpersonal skills [8,9].

1.3 LED technology and educational props

Russian inventor Oleg Losev invented light - emitting diode in 1927. His invention was distributed in Soviet, German and British scientific journals, but no practical use was made of the discovery for several decades [10]. Nick Holonyak at General Electric developed the first practical visible-spectrum (red) LED in 1962. Soon after that, LEDs became available as electronic components, and thus they gained the attention of the physics teaching community. The first paper on LEDs in The Physics Teacher appeared in 1974. Awareness that LEDs can be used in teaching several introductory physics topics emerged as early as in by Jewett 1991[11]. He described some activities with LEDs in the areas of electricity, optics, and modern physics. Since the LED technology developed and produce brighter, efficient, wide range of wavelengths of LED, they found many applications in the 21st century live. Etkina, E. and Planinši, G., publish four articles series about LED during 2014-2015 related the various application LED publish in The Physics Teacher [12,13,14,15].

In making the props, we always consider the ease of obtaining the materials, the assembly, and the cost. Based on the experience from the ISLE workshop [5], we have implemented the LED components to create customized props with available materials in Banda Aceh. This LED components had been used for workshop and training activities that we held in Banda Aceh [5]. Figure 4 shows a collection of props that utilizing LEDs technology. At the bottom, there are six types of LEDs with various spectra of infrared, red, yellow, green, blue, and violet. LED spectrum can be used to study the energy band gap, and if it is related to the wavelength, radiated Plank constant can be determined. Especially for infrared LED lights, red and yellow mounting on the 3V power source, they must use a resistor, and we make a resistor ladder (Figure 4a above) that can be used for various purposes. Connection terminal used for brass nails that will simplify the process of soldering. By adding a single transistor, LED lights (Figure 4b) can be used as a tool to check the conductivity, such as experiments conducted in figure 2. When using two transistors, the LEDs will result in a higher sensitivity so that it can be used to detect the presence of electrostatic charge.

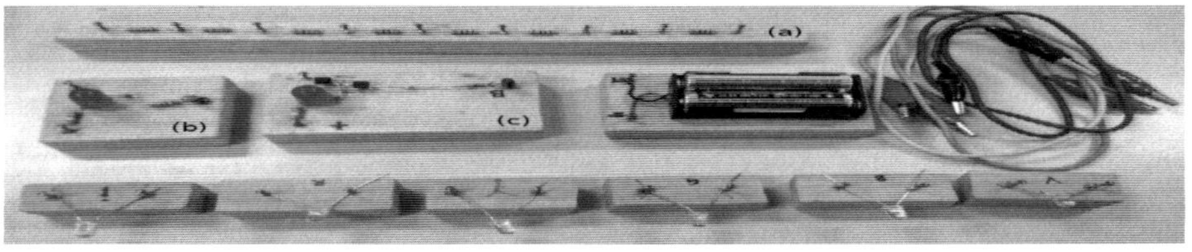

Figure 4. LEDs as physics learning props.

2. Method

The professional development conducted in this study only used a single group pre- and post-test research design. Twenty secondary school science teachers in the city of Banda Aceh participated in a two-day workshop held in Syiah Kuala University, in Banda Aceh, Indonesia. The workshop used ISLE-based STEM approach, an ISLE learning that incorporates STEM education approach in active learning and collaboration. The workshop used LEDs as experimental props. To measure the effectiveness of the workshop in developing teachers' understanding of ISLE-Based STEM approach and the use of LEDs in physics experiments, we administered a pre- and post- questionnaires consisting of nine items. At the end of the workshop, we also administered a questionnaire on participants' beliefs on the benefits of the workshop to their knowledge.

3. Results and discussion

According to the results on the pre- and post-tests, the teachers showed an increase in their enthusiasm in participating in this workshop since they had learned and mastering new skills and knowledge. Figure 5 shows the participants conducting experiments and then presented the results (Figure 6).

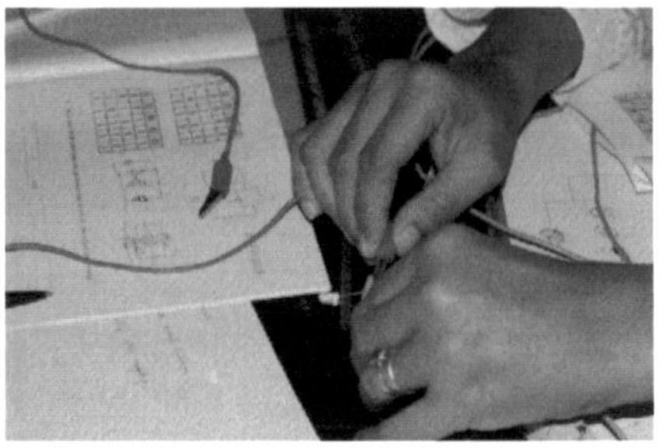

Figure 5. Participants experiment with LED.

Figure 6. Participants showed their research results on LEDs.

The results of the qualitative analysis on the pre- and post-questionnaires' responses, we found that there are positive improvements on participants' understanding of ISLE-based STEM approach and of the use of LEDs as props. For example, one participant answered in the pre-test that *hypothesis* is the temporary answer from research findings, estimating, and observations, and did not answer what *prediction* is. This participant answered on the post-test that *hypothesis* is a temporary answer developed after observations, meanwhile, the *prediction* is the temporary guessing on what will happen on the testing experiment, a correct interpretation of the two important terms in ISLE. Another example of the improved understanding of a participant is evident when one participant who cannot answer in pre-test about where in the curriculum we can use LEDs as props, answered on the post-test that we could use LEDs on electronics lesson by changing the light bulbs that were usually used with LEDs. From the questionnaire on the participants' satisfaction with the workshop, we found that 89.4% of teachers experience that their knowledge is increasing and around 10.6% of the teachers acknowledge that their knowledge increased significantly. Also, 84.2% of teachers mentioned that the ISLE was easy to understand, in comparison to 15.7% of teachers who thought that ISLE was hard to understand. No teacher has an extreme opinion that ISLE is either very easy or very hard to understand.

4. Conclusions

Utilizing LEDs as props for physics lessons helped teachers realizing many objects in our everyday life are easy to find, inexpensive, and can be assembled without hassle to teach essential concepts in physics. The skills that the teachers gain from ISLE-based STEM approach is that they could apply this method, especially with LEDs in their classroom especially in creating an environment to build students' critical thinking, problem-solving, communication, and collaboration when the learning is fun for them. Most of the teachers participated in this study agreed that ISLE is a teaching approach that is easy to be understood. Most of them also agreed that the workshop had increased their knowledge and skills. Many more such professional developments will enhance science teachers' knowledge and skills in understanding concepts in physics and help students to understand the concepts.

Acknowledgments

Authors wishing to acknowledge ICTP (International Center for Theoretical Physics) and Director of Workshop Prof. Eugenia Etkina for opportunity attending Workshop Physware 2015 Preparing 21st Century Physics Teacher. This study is initially funded by LPPM Unsyiah (IbM) Award Number No. 1711/UN11/SP/PNBP/2016 support for public service activity, entitle Training of Inquiry-Based Learning (IBL) Method Through Investigating Utilization of LED (Light Emitting Diode) To Improve Teachers' Abilities in Designing Scientific Based Experiment. This study is funded partially by the NAS and USAID under the USAID Prime Award Number AID-OAA-A-11-00012, and that any opinions, findings, conclusion, or recommendations stated in the article are from the author only, and do not always reflect the view of USAID or NAS.

References

[1] Ilyas H P 2016 Infusing critical thinking into English course books *Journal of ELT Research* **1** 113
[2] Kemendikbud 2016 *Silabus Mata Pelajaran Fisika Sekolah Menengah Atas/Madrasah Aliyah (SMA/MA) Kurikulum 2013 Revisi 2016* (Jakarta: Kementerian Pendidikan dan Kebudayaan)
[3] Vasquez J A 2013 *STEM lesson essentials, grades 3-8: Integrating science, technology, engineering, and mathematics* (Great Britain: Heinemann)
[4] Etkina E and Heuvelen A V 2001 *Investigative Science Learning Environment: Using the processes of science and cognitive strategies to learn physics*
[5] Irwandi, Melvina, Syurki U 2017 *Training of Inquiry Based Learning (IBL) method through investigating utilization of LED (Light Emitting Diode) to improve teachers' abilities in designing scientific based experiment* (Report of Public Services: unpublished)
[6] Etkina, Eugenia, Gorazd P, Irena D and Leos D 2015 *ICTP Workshop Physware 2015, Preparing 21st Century Physics Teacher* (Italy: Trieste)
[7] Dede C 2010 Comparing frameworks for 21st century skills *21st century skills: Rethinking how students learn* **20** 51
[8] Cassel R N and Kolstad R 1998 The critical job-skills requirements for the 21st century: Living and working with people *Journal of Instructional Psychology* **25** 176
[9] P21 (Partnership for 21st Century Skills) 2011 *Framework for 21st Century Learning*
[10] Zheludev N 2007 The life and times of the LED —a 100-year history *Nature Photonics* **1** 189
[11] Jewett J W 1991 Get the LED out *The Physics Teacher* **29** 530
[12] Planinšič G and Etkina E 2014 Light-emitting diodes: A hidden treasure *The Physics Teacher* **52** 94
[13] Etkina E and Planinšič G 2014 Light-emitting diodes: Exploration of underlying physics *The Physics Teacher* **52** 212
[14] Planinšič G and Etkina E 2015 Light-emitting diodes: Learning new physics *The Physics Teacher* **53** 210
[15] Planinšič G and Etkina E 2015 Light-emitting diodes: Solving complex problems *The Physics Teacher* **53** 291

IOP Conf. Series: Journal of Physics: Conf. Series **1088** (2018) 012061 doi:10.1088/1742-6596/1088/1/012061

The ethnomathematics: exploration of Gayo tribe local wisdom related to mathematics education

B Yustinaningrum, Nurliana and E Rahmadhani
Sekolah Tinggi Agama Islam Negeri Gajah Putih, Jl.Yos Sudarso/A. Dimot No.10 Takengon, Aceh, Indonesia

E-mail: bettri_yustinaningrum@yahoo.com

Abstract. Education cannot be separated by cultural value. However, in practice, learning is still far from the daily culture of students. The study aims to investigate the sort of Gayo tribe local wisdom related to ethnomathematics, and to investigate the Islamic character found in Gayo tribe local wisdom. The results of the investigation will be usedto develop ideas of mathematics learning integrating the Gayo tribe local wisdom and the Islamic character. This studyused a semi-structured interview that focused on four questions: *where to look, how to look, what it is, and what it means*. The data were analyzed qualitatively, followed by the concept given by Miles and Huberman.The finding shows that the local wisdom of Gayo tribe relates to mathematics. Measurementconcept was found in*kal, bambu, kaleng, naleh, kunce, and padang* used to measure the dose of *zakat*. On the other hand, geometry concept was in relation to*kerawang* motives, such as *matanelo, pucuk rebung, puter tali, emun berkune, peger, emun berangkat, rante, emun beriring* and *cucuk pengong*. The value of Islamcan be foundin the motives of *kerawang* Gayo. The study implies that the Gayo tribe local wisdom and Islamic characteristics could be used for mathematics learning.

1. Introduction

Education is one of the factors that support the development of a country. Hall and Marrhewas claim that training and education aspects and knowledge are useful for the progress of the nation [1]. Whereas, National Education Law number 23 of 2003 states that education is the human effort to create a learning atmosphere and process so that students are able to actively improve their self-potential to have spiritual strength, self-control, personality, intelligence, noble character, and skill needed for themselves, their society and their nation [2].

Education has a significant role in forming the intelligence, self-confidence, and commitment to promoting nation. Furthermore, the aims of education concern not only on life goals but also on helping human development to have life and survival skills. The Islamic thinkers are more oriented toward the ideal aspect and purpose, as well as spiritual thing related to God [3].

Education cannot be separated by cultural value. Education is one of the most effective ways to preserve the culture. Bishop claims that mathematics is a form of culture [4]. Drake and Reid believe that one way to address other initiatives such as environmental education, character education and the new literacies (media, critical and technological) is integrating the curriculum [5]. Learning integrating cultural value teaches core concepts and skills by connecting multiple subject areas to a unifying theme or issue. It can be implemented in mathematics learning since mathematics learning has integrated into social life. This contradicts the "conventional" insight that sees mathematics as a subject "free of culture" and "free of value" [6]. The ethnomathematics experts state that basically, the growth of mathematics would never release from culture and social value in society.

Indonesia consists of many kinds of ethnics such as Gayonese, Acehnese, Javanese, and others. The majority are Moslems. Based on the BPS (Indonesia's Central Department of Statistics) data in 2010, the number of Moslem in Indonesia is 207,2 million people or 87,8 percent [7]. This condition makes Islamic culture indirectly affect Indonesian culture. Ethnics have their own culture, noble value, and local wisdom. According to Wagiran, local wisdom has broad coverage: (a) idea/thought, attitude and language action, art, and literature that has a philosophy nuance (b) idea/thought, attitude and the work related to artifacts such as decoration, paintings, etc. (c) idea/thought, attitude and social behavior [8].

Gayonese is one of the ethnics who lived in Takengon, Aceh. They have some local wisdom such as dance, literature, and decoration. The Islamic Shariais applied in their daily life.The Islamic sharia affects their culture. For example, it is found on the Kerawang carve in the *umah pitu ruang* (Gayo traditional house) ornaments. Kerawang carving contains the philosophy of life. The number of*pitu ruang* (seven rooms) symbolizes the Islamic faith associated with human and his God (*Allah SWT*). Culture and education are crucial as it is a unity and complete entity applied in society and education and become a fundamental need for a person.

Culture and education can be linked to cultivating characters in mathematics learning through ethnomathematics. Ethnomathematics is used to represent a field of mathematics that studies different types of mathematics arising from different cultures. Ethnomathematics includes mathematics that is practiced, used or simply incorporated in the cultural practices or activities of different groups in society [9]. Through ethnomathematics, it is expected that students cat on understand mathematics in culture. Thus, the teacher will easily cultivate the cultural value that is part of the nation's character in the students [10]. Therefore, this study aims to excavate the Gayonese local wisdom related to ethnomathematics accompanied by Islamic character.

2. Method

This paper investigates the cultural wisdom of the Gayo tribe and the Islamic character related to ethnomathematics. The results of this study will then be used to develop learning materials integrating the Gayo tribe local wisdom and the Islamic character. This paper describes the investigation result on preliminary research, a phase from Plomp development model. This study used a snowball sampling technique. The result investigation on the cultural wisdom of the Gayo tribe and the Islamic character obtained is discussed through focus group discussion (FGD). The research instrument used a semi-structured interview that refers to four questions: where to look, how to look, what it is, and what it means. Those four general questions become the essence to utilize *ethnography* principle as shown in Table 1.

Table 1. The guideline of the research question.

Generic Question	Initial Answer
Where to look?	It is seen in the activity of Gayonese society.
How to look?	The investigation of the aspects related to the Gayonese activity.
What is it?	Alternative thinking in the prior process
What does it mean?	It is essential for culture and mathematics.

3. Result and discussion

Based on FGD, the information about the cultural wisdom of the Gayo tribe and the Islamic character of the Gayonese was obtained. That information relates to mathematics topics taught at school, specifically measurement and geometry. The following will describe four important findings of the investigation onthe cultural wisdom of the Gayo tribe and the Islamic character on the mathematics topics.

The first finding relates to the answer to the question "where to look?",the activity of Gayonese society. Measurement topic is found in the livelihood of the Gayonese. Because of the geographical condition, the livelihood of the Gayonese society is centered in agriculture. The traditional tools they used to measure the agriculture products, such as rice and coffee is *bamboo* as seen in Figure 1.

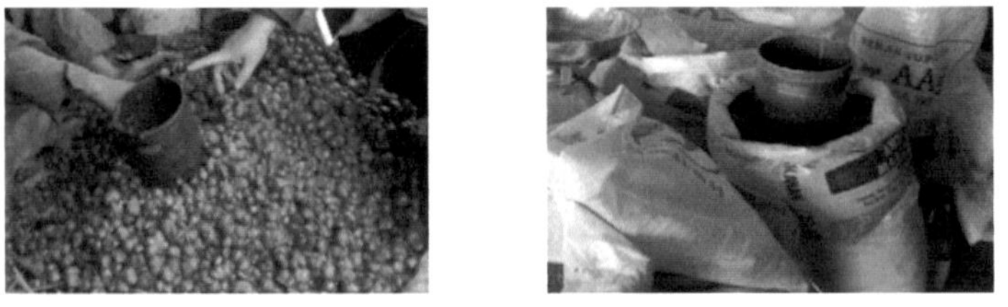

Figure 1. *Bamboo,* the traditional measuring tool used by Gayonese.

The finding shows that mathematics in an ethnic group is influenced by their cultural background because they do mathematics based on what they see and feel. This is in line with Pinxten stating that mathematics is a symbolic technology that grows on cultural skills or activities of a cultural nature [12].

On the other hand, geometry topic is found in the Gayo traditional house called *umah pitu ruang*. *Umah pitu ruang* is a big house that has seven rooms (*panaek)* with 14 connecting roofs. *Umah pitu ruang* was made by the king of Linge kingdom for his seven children. At first, there was no carve in it. Then, the king asked some thinkers to make the home more beautiful by adding some decorations such as sculpture, carvings and woven. To actualize the idea, the thinkers and diviners discussed it together. The imagination is the result of *ker* (idea from the thinker) that means a sudden feeling and *rawang* (the idea from the diviner) that means visualization. The concepts of *kerawang* Gayo basic motive *tali puter tiga* comes from *Merhum* King. *Matanelo* (the sun) and *emun berkune* is the idea from the Queen, *pucuk meluhor pucuk rebung* from the Imam (religious leader), pagar (fence) and *cucuk pengong* from the elders, and *embun* from the Crown Prince. At that time, the carve is made of *kertan, bedem, cike, benyet, bengkuang* and etc. which are usually called *lintem*. Kerawang Gayo motive can be found on *Umah Pitu Ruang's* walls, roofs, ladders and foundation in Toweren. Those motives resemble with two-dimensional figure in mathematics and every motive has a meaning and character. Kerawang Gayo motive can be seen in Figure 2.

Figure 2. *Kerawang* Gayo motive.

The second finding relates to the answer to the question "how to look?"This part discusses the investigation of qualitative aspects, relational, spatial on activities of Gayonese related to mathematics. Based on the observation and the interview conducted with the head of the Cultural Section of

Education and Teaching of Central Aceh Regency, traders, and farmers, it is found that the measuring tools used for those three activities have some similarities and differences. For market and coffee refining, the Gayonese use *bamboo* and *kaleng*. Meanwhile,in the rice field, the Gayonese use*kal, bambu* or *are, kaleng* or *tem, naleh, kunce, padang, gateng*. The Gayonese is still using those traditional measuring tools until present. This is one way to preserve their local wisdom. The measurements used are 1 *kal*= ¼ kg, 1 *bambu/are* = 4 *kal*, 1 *kaleng* = 10 *bambu*, 1 *naleh* = 16 *bambu*, 1 *gateng* = 5 *kaleng*, 1 *kunce* = 5 *kaleng* = 50 *bambu*, 1 *padang* = 5 *bambu*. Those measurements are used by the Gayonese in their daily life. According to Bishop, the broader conception of mathematical calculations illustrates that mathematics based on culture will generate mathematical knowledge to acrhieve community goals that are expected to solve problems and to define rules in life [13]. Mathematics is the social construction within the context of the community, where the notion is negotiated, and the conversion is approved.

Moreover, based on the interview and the observation, it is found that there are some *kerawang* gayo motives such as *matanelo, pucuk rebung, puter tali, emun berkune, peger, emun berangkat, rante, emun beriring,* and *cucuk pengong*. Those motives can be found on *Pitu Ruang's* wall, roof, foundation, and ladder. Those motives relate to geometry topic.

The third finding is concerning the answer to the question "what is it?" that refers to alternative thinking in the prior process. This part discusses the value of the Islamic character integrated into mathematics and Gayo culture. Based on the interview conducted with the custom figure and the head of the Cultural Section of Education and Teaching of Central Aceh Regency, it is found that the values of Islamic character are integrated into the Gayonese' everyday life through the kerawang Gayo motives that relate to geometry topic. The brief description of the Islamic values in the Gayo motives is presented in Table 2.

Table 2. The description of motive and character value of *kerawang* Gayo.

Motive	Meaning
Matanelo	The life blood of every creature and as gratitude for every favors and patient to God The vertical and horizontal relationship, either the relationship of a person to God and a person to others for the salvation of the hereafter.
Pucuk rebung	*Tut wuri handayani* which means following from behind and giving moral encouragement. *Taring merai* (if a person is left behind, hewill be picked up), *pantas beruru* (if a personis too fast, he will be pulled back).
Puter tali	Support each other for every good deed
Emun berkune	The same thought, speech, and deed
Peger	Gayonese society always live under Gayo local wisdom and Islamic Sharia; otherwise they are not protected.
Emun berangkat	Symbolize the loyalty of Gayonese. Wherever they go, they will be together
Tapak seleman	Every problem should be handled wisely
Rante	Symbolize unity and togetherness
Emun beriring	Symbolize the substantial unity in the social life especially in putting ourselves wherever we are
Cucuk pengong	The same faith, perception, and destination.

Furthermore, in their activity, the Gayonese society uses measuring tools, e.g., *bambu* not only to measure the agriculture, forestry and selling product but also to measure the dose of *zakat*. *Zakat* is the required payment under Islamic law on certain kinds of property for charitable and religious purposes. For example, if the agriculture product, such as rice and grain, reach two *kunce*, the Gayonese must pay *zakat* because the amount of the product meets the *nisab* (the minimum amount required to pay the *zakat*). Two *kunce*is equal to 100 *kaleng* or 1000 *bambu* or 2000 liters in international measurement unit.

The fourth finding relates to the answer to the question of "what does it means?". This section discusses the relationship that occurs between mathematics and culture. Based on the interview and the observation results, it is found that the size of the bamboo is the same as 2 liters in the international measurement unit. To convert the traditional measuring unit into the international measurement unit, we can use the following formula.

$$\rho = \frac{m}{v} \text{ or } m = \rho . v$$

Where:
ρ = density (Kg/m^3)
v = volume (m^3)
m = mass (Kg)

Moreover, regarding *kerawang* Gayo motives, it is found that its shape is similar to two-dimensional objects. For instance, *matanelo* (sun) has a shape that resembles a circle in geometry. *Matanelo* symbolizes the lifeblood of every creature, gratitude for every favors, and patience to God. It also represents the vertical and horizontal relationship, the relationship between human and his God and the relationship between human and others humans for the salvation of the hereafter. *Pucuk rebung* (the top of the bamboo shoots) resembles the triangles in geometry. The meaning of *pucuk rebung* relates to education.That is *tut wuri handayani* that means following from behind and giving moral encouragement. *Pucuk rebung* also has a meaning of *taring merai* (if a person is left behind, he will be picked up), *pantas beruru* (if a person is too fast, he will be pulled back). *Peger* (fence) represents a rectangle. It symbolizes that the closeness of the Gayonese society with Islamic Sharia; they live under Gayo local wisdom and Islamic Sharia. They will not be protected if they break the Islamic law. *Tapak seleman* resembles the two-stacked triangles. *Tapak seleman* has a meaning that every problem has to be solved wisely. *Rante* (chain) resembles small circles arranged vertically. *Rante* represents unity and togetherness within the Gayonese community. *Cucuk pengong* resembles a large circle with small circles surrounding it. *Cucuk pengong* represents the same faith, perception, and destination. This is in line with the constructivist understanding that a person's knowledge is built through existing experience in his environment [14]. Thus, if the teacher makes a connection between mathematics and students' experience and objects in their environment, the students can understand the meaning of the lesson better.

4. Conclusion

Based on the research findings, there is some potential ethnomathematics in Gayo society that can be developed in mathematics learning.The traditional measuring tools used in activities of the Gayo community can be developed for teaching measurement topic in the second and third grade. Gayo ethnic art artifacts such as Gayo *kerawang* motives and a traditional house *Umah Pitu Ruang* can be developed for teaching geometry topics in elementary and junior high school levels, such as angle, line, plane geometry, and solid geometry.

References

[1] Hall J and Matthews E 2008 The Measurement of Progress and The Role of Education *European Journal of education* **43**
[2] Direktorat jenderal Pendidikan Islam Departemen Agama 2006 *Undang-undang dan Peraturan Pemerintah RI Tentang Pendidikan Indonesian Government Laws and Regulations about Education* (Jakarta: Departemen Agama) p 5
[3] Hasbiyallah and Sulhan M 2015 *Hadis Tarbawi* (Bandung: Remaja Rosdakarya) p 11
[4] Bishop J A 1994 *Cultural Conflicts in the Mathematics Education of Indigenous people. Clyton* (Viktoria: Monash University)

[5] Drake M S and Reid J 2010 *J What Works? Research Into Practice* **28**
[6] Tandililing E 2013 *Proceeding Penguatan Peran Matematika dan Pendidikan Matematika untuk Indonesia yang Lebih Baik* (FMIPA UNY) (Yogyakarta)
[7] BPS 2011 *Kewarganegaraan Suku Bangsa, Agama, dan Bahasa Sehari-hari Penduduk Indonesia Hasil sensus Penduduk 2010* (Jakarta: Badan Pusat Statistik) p 10
[8] Wagiran 2012 Pengembangan Karakter Berbasis Kearifan Lokal Hamemyu Hayuni Bawan (Identifikasi Nilai-nilai Karakter Berbasis Budaya) **2** 3
[9] Mosimege M and Ismel A 2015 *Proceedings of the 10th International Congress of Mathematics Education* Ethnomathematical Studies on Indigenous Games: Examples from Southern Africa (Copenhagen) (Denmark)
[10] Wahyuni A 2013 *Proceeding Penguatan Peran Matematika dan Pendidikan Matematika untuk Indonesia yang* (FMIPA UNY) (Yogyakarta)
[11] Miles B M and Huberman A M 1994 *Qualitative Data Analysis* (London: International Educational and Professional Publisher) p **12**
[12] Pixten R 1994 *Journal For the Learning of Mathematics* **23**
[13] Bishop J A 1991 *The Symbolic Technology Calet Mathematics its Role in Education* (De Belgique: Bullatin De La Societe Mathematique)
[14] SutrisnoL 1995 *Implementasi Tradisi Constroctivis Dalam Pendidikan di Indonesia* (Cirasua Bogor: Forum Komukasi Bidang Pendidikan)

The development of teaching material: Rigorous mathematical thinking in a geometry classroom

M Meilantifa[1] and M T Budiarto[2]
[1] Universitas Wijaya Kusuma, Jl. Dukuh Kupang XXV/54 Surabaya, Dukuh Kupang, Dukuh Pakis, Surabaya, Jawa Timur 60225, Indonesia
[2] Universitas Negeri Surabaya, Jl. Raya Kampus Unesa, Lidah Wetan, Lakarsantri, Surabaya, Jawa Timur 60213, Indonesia

E-mail: meilantifa_fbs@uwks.ac.id

Abstract. This study aimed to develop a teaching material, a textbook, using a *Rigorous Mathematical Thinking* (RMT) approach. It was constructed by considering some activities involving problem-solving, problem posing, open-ended problem, connection problem, conjecture, creativity task, mini lab, and drawing skill. The subject in this study was the undergraduate mathematics students who enrolled in the geometry subject in the University of Wijaya Kusuma Surabaya. A textbook was developed because of the necessity. It was developed using the Mediated Learning Experience based on the Rigorous Mathematical Thinking. This study employed Plomp's Research Design model consisting of three phases: the preliminary research, prototyping phase, and assessment phase. However, this paper focused only on the validation of the textbook involving five experts. Students have been mediated to construct their understanding by using mental operation. One important task about creative thinking including in the textbook dealing with the trapezium definition. Students were asked to create a new definition of trapezium which is different from the preceding definition they learned at school. The results showed that the development of student book using Rigorous Mathematical Thinking approach was valid.

1. Introduction

Geometry contains symbols which are challenging to be understood by students without guidance, directions, and mediation from teachers or geometry experts. The symbols can be used to help students to solve geometry problems. However, students with low capabilities in geometry still experience some difficulties in solving them. Therefore, it is necessary to provide a geometry expert to mediate the learning process of geometry concepts. Mediator stimulates students to optimally utilize their mathematical psychological tools for a good understanding of the concepts so that they will not encounter any difficulties in solving the problems. The textbook is one of the teaching materials used to assist the learning process to achieve the objectives. It is categorized as a handbook written by the relevant experts satisfying the rules of textbooks, being officially published and disseminated [8].

A geometry textbook is designed to utilize the mathematical psychological tool as one of the aspects emphasized on teaching processes involving Rigorous Mathematical Thinking (RMT) interventions [1]. Vygotsky stated that the development of the higher mental process of students depends on the mediation in the interaction with the environment. Vygotsky emphasized that the symbolic mediator is tailored to the students' specific socio-cultural context. Based on the personal experience, there are some issues found in a geometry classroom, including: (1) the delay of students in following the lecturers' explanations, (2) students who cannot draw the cross line ask the other students who also cannot draw the line intercept resulting in the student was not successful in drawing (3) students do not try to search other resources instead they only rely on the lecturers, and (4) students' ability to look at the spatial perspective using a textbook is lacking, resulting in the classroom learning being organized, especially in designing the learning activities playing the mediator role between students [3] and the lecturer's role is changed from being an information provider to the source of the mediated learning experience [4].

The Rigorous Mathematical Thinking (RMT) paradigm enables to produce a textbook which culturally describes the abstract representations to build an understanding, transforming, generalizing, generating the conceptualization and understanding of coherent and logical ideas as well as planning how to use ideas to facilitate problem solving, to reduce the new knowledge in writing and systematic in the form of textbooks. Some rules of writing textbooks are using scientific references, writing clearly, correctly, nor abstractly or out of the field of knowledge; using language effectively; presenting an attractive style of the layout; well organized in term of both the content and the writing standards including the design of the cover, colour, and graphics [5]. Thus, the textbook contents must satisfy the scientific truth and in line the description of the geometry course. Also, the language used in the textbook must align with the rules of excellent language and legibility. The textbook must be arranged in a systematic and coherent hierarchy of science. The layout and the design are tailored to the students as the users.

The textbook discussed in this paper used the Plomp development model consisting of the preliminary investigation phase, the design phase, the realization/construction phase, and the phase of test, evaluation, and revision, and implementation.

2. Method

The process of developing the geometry textbook used the Mediated Learning Experience based on the Rigorous Mathematical Thinking. It is developed by referring to the Plomp's Research Design model involving three phases: the preliminary research, prototyping phase, and assessment phase. Preliminary research phase was conducted by doing the prerequisite analysis, the context analysis, literature study, and create the development of concepts. In the prerequisite analysis phase, we collected some information as the requirement from books or the internet. Then, we selected and grouped the materials according to the descriptions of the geometry courses aligned with the curriculum. In the phase of the context analysis, the researcher performed student and material analysis. The literature study was conducted to obtain a learning theory suitable for developing the geometry textbook, in this study the theory of Rigorous Mathematical Thinking (RMT) was adopted. The illustration of the textbook development process can be seen in Figure 1.

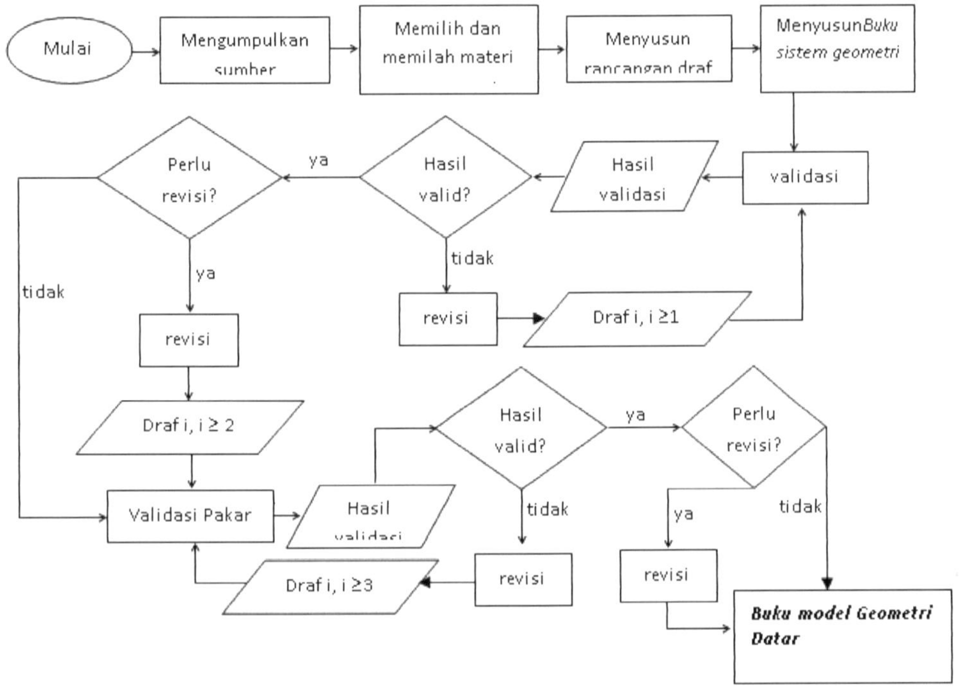

Figure 1. Geometry textbook development process.

The next activity in this study was gathering some information related to what, why, who, where, which, and how. According to Figure 1, the final result is a draft of the geometry textbook for students. Thus, the next step was to determine the geometry materials to be compiled. Based on the theory referred, the RMT-based geometry textbook should include the cognitive development phase, the content as a development process, and the conceptual cognitive construction practice. The implementation activities undertaken were the preparation of the early prototype of geometry material draft. Next, the draft contents were then compiled into a geometry book draft.

Following the previous step, we developed an instrument to validate the textbooks including the content feasibility components such as the dimensions of knowledge and skill dimensions. The instruments also had the feasibility of presentation consisting of the presentation techniques, supporting the materials and learning presentation. The feasibility of the content in the knowledge dimension included the scope of the materials consisting of the coverage, breadth, and depth. The accuracy of the material dealt with the accuracy of the facts/notation/symbols, concepts/definitions, principles (theorems, axioms, propositions, properties, rules, laws), as well as procedures/algorithms, samples, and questions. Based on the law and legislation, the contents of the textbook is a subject of the intellectual property that should be free from racism, pornography as well as gender, area, and occupation bias. The skill dimensions criteria included the coverage and activity accuracy. The feasibility component of the content in the presentation technique involved the systematic consistency criteria presented in the chapter as well as the presentation of logic and demands. To support the material, the criteria of the suitability and accuracy of illustrations was needed, including the organization and the concept maps at the beginning of the chapter, summary and exercise problems at the end of each chapter, and the precision of numbering and naming of the tables, figures, and appendix. The presentation of learning consisted of the criteria of students' active engagement and focused on the students. The final presentation criteria comprise of the preliminary criteria, table of contents, glossary, bibliography, and index.

3. Result and discussion

Table 1 presents the result of the validation process conducted by five selected validators. Overall, the total average validity is 2.545. It means that the textbook is valid and it can be used as a geometry teaching material.

Table 1. Validation Result.

Aspect	Criteria	Validator					Mean Criteria	Mean Aspect
		1	2	3	4	5		
Feasibility of Content	**Knowledge Dimension**							
	1. Completeness	2.33	2.33	2.33	2.67	2.67	2.5	
	1. Breadth-ness	3	2.5	2,5	3	3	2.8	
	2. Depth-ness	3	3	2	3	2	2.6	
	3. Fact/Notation/Symbol accuracy	2	3	2	2	2	2.2	
	4. Definition/Concept accuracy	3	3	2	3	3	2.8	
	5. Principle accuracy (theorem, axiom, propositions, properties, rules, laws)	3	3	2	3	3	2.8	
	6. Procedure/Algorithm accuracy	2	3	2	3	2	2.4	2.63
	7. Example accuracy	3	3	1	3	2	2.4	
	8. Exam accuracy	3	3	3	3	3	3	
	9. Patent adherence	2.5	3	3	2.5	1.5	2.5	
	10. Free of racism, pornography, and BIAS (gender, area, and occupation)	3	3	3	3	3	3	
	Skills Dimension							
	11. Skills scope	2	3	2	3	2	2.4	
	12. Activity accuracy	2.33	3	2.67	3	2.67	2.73	
Feasibility of Presentation	**Presentation technique**							
	1. Consistency	3	3	2	2	2	2.4	
	1. Logic	3	3	3	3	3	3	
	2. Demands	2	3	3	3	3	2.8	
	Presentation Support							
	3. Compatibility and precision of illustrations	2.5	3	2	2	3	2.5	
	4. Advance Organizer	3	3	3	3	2	2.8	
	5. Concept Mapping	2	2.5	2.5	1	1	1.8	
	6. Exercise	2.5	3	2.5	2.75	2.25	2.6	
	7. Reference	1	3	2	2	2	2	
	8. The precision of numbering and naming tables, drawings, and attachments.	2	3	3	2	3	2.6	2.46
	Learning Presentation							
	9. Active involvement of learners and centered on learners.	2.5	3	2	3	2	2.5	
	10. Interactive Communicative	3	2	3	2	1	2.2	
	11. Variation	2	3	2	2	1	2	
	Completeness of Presentation							
	12. Introduction	2	2.33	2.67	2.33	1.67	2.2	
	13. Table of Content	3	3	3	2	2	2.6	
	14. Glossary	3	3	2	1	1	2	
	15. Reference	3	3	3	3	3	3	
	16. Index	2	3	3	3	3	2.8	
The total average of validity								2.545

The average score provided by the validators was 2 for the reference/ source, text, table, figure, and appendix. The follow-up effort involved a thorough examination of the font, the table, the source, the number of images, the appendix as well as index and glossary. The average score of the feasibility of content in knowledge dimension with material coverage criterion was 2.5. To improve the scores, the researcher added some illustrations, new information, relevant concepts and challenging activities. The variations in the presentation obtained the score of 2.5, some illustrations, examples, and non-examples were added to improve the variations. Also, fixing the order of summaries, clarifying the phrase, and re-examining the items that have been and have not been summarized were some of the efforts to improve the summary sections.

In addition, based on the expert discussion conducted in Aceh on 26-28 June 2018, some revision was required, namely: 1) the layout should be revised based on the standard of textbook writing proposed by the ministry of higher education (DIKTI), (2) there should be an accurate naming, scale and image cropping, (3) the order of image number should be revised and (4) it is necessary to add glossaries.

4. Conclusion

Based on the results of the field analysis of the Geometry textbook, it is concluded that the textbook is valid to be used as literature on geometry course, especially for students of Wijaya Kusuma Surabaya University. Based on the validation conducted by five validators and discussion results, the average score of validation is 2.545 indicating that the textbook is valid and can be used as a geometry textbook with some revision. The validity of the textbook is assessed regarding knowledge and skills. The instruments are assessed for the materials, presentation, legibility and the overall illustrations of the contents of the Geometry book model.

Reference

[1] Kinard J T and Kozulin A 2005 Rigorous mathematical thinking: Mediated learning and psychological tools *Focus on Learning Problems in Mathematics* **27** 1

[2] Kozulin A 2002 Sociocultural theory and the mediated learning experience. *School Psychology International* **23** 7

[3] Kozulin A and Presseisen B Z 1995 Mediated Learning Experience and Psychological Tools: Vygotsky's and Feuerstein's Perspectives in a Study of Student Learning. *Educational Psychological* **30** 67

[4] Kozulin A, Gindis B, Ageyev V S and Miller S M 2003 *Vygotsky's Educational Theory in Cultural Context* (New York: Cambridge University Press)

[5] Kinard J T and Kozulin A 2008 *Rigorous Mathematical Thinking: Conceptual Formation in the Mathematics Classroom*. (New York: Cambridge University)

[6] Kinard J T and Kozulin A 1998 *Psychological Tools: A Sociocultural Approach to education*. (London: Harvard University Press)

[7] Kinard J T and Kozulin A 2001 *Creating Rigorous Mathematical Thinking: A Dynamic that Drives Mathematical and Science Conceptual Development*. Retrieved on October 21, 2009 from www.umanitoba.ca/unevoc/conference/ papers/ kinard .pdf

[8] Kepmendiknas no 36 2001 from http://www.unm.ac.id/files/surat/Kepmen36-D-O-2001PenilaianKredit.pdf access 22th May 2018

PISA-like mathematics problem with karate context in Asian Games

H Nizar, R I I Putri and Zulkardi

Universitas Sriwijaya, Jl. Srijaya Negara, Bukit Besar, Palembang, Indonesia

E-mail: ratu.ilma@yahoo.com

Abstract. The background of this research was based on the low achievement of Indonesian students in working on the PISA problems. The research objective produced valid and practical PISA-like mathematics problem with karate context in Asian Games. The subject of this research was tenth-grade students of Senior High School. The research method used is design research with the type of development studies. This research had two stages that were the preliminary and formative evaluation. In this research, only presented the result on self-evaluation, expert reviews and one-to-one, and small group phase. Data collection techniques used were walkthrough, documentation, observation, and interview. The research had produced valid and practical PISA-like mathematics problem with karate context in Asian Games. Valid could be seen from the assessment of validator (expert) regarding content, construct, and language used and artistry of students in the phase one-to-one. Practical could be seen from the skill of students in small group phase where the student could understand the intent or language of the problem. Also, karate context used could make students actively engaged in the discussion, and it could help students to solve the problem and give the argument.

1. Introduction

Programme for International Student Assessment (PISA) was an international study to measure skills and capabilities of students aged 15 years and was held every three years [1]. The skills and capabilities which were assessed in PISA include four types of literacy those were scientific literacy, reading literacy, mathematical literacy, and financial literacy [1]. The PISA problem on the mathematical literacy was developed based on four contents: change and relationship, space and shape, quantity, and uncertainty and data [1]. PISA results were one of the external factors which revised KTSP (Kurikulum Tingkat Satuan Pendidikan) to the 2013 curriculum in Indonesia [2]. So it could be said that PISA was very important, because it became a reference of curriculum development in Indonesia. Indonesia was one of the countries that had been participating in PISA since 2000 and was expected to get achievements in PISA. However result of PISA study on mathematical literacy, Indonesia from 2000 until 2015 was always in lower rank. The result of PISA study on mathematical literacy in 2012, Indonesia was ranked 64 out of 65 countries [3, 4]. Then the result of PISA study on mathematical literacy in 2015, Indonesia was ranked 62 out of 70 countries [5]. The low achievement of Indonesian students in working on PISA problems were caused by the students were not familiar to do the problems that contain context, high level difficulty, and just doing the usual problems. As the result, many students made mistakes when working on PISA problems. This was the same as with the low PISA result of Indonesian students due to the lack of problem-solving capabilities of high level or non-routine, the evaluation system in Indonesia was still using low-level problems, and students were accustomed to acquiring and using formal mathematics knowledge in the classroom [6-8]. Zulkardi [9] suggested to design the PISA-like mathematics problems and use it in the learning of mathematics in the class.

Context was a situation or phenomenon/natural occurrence associated with the concept of mathematics that was being studied [10]. The use of context in mathematics had several benefits, such as: useful for the formation of concepts, access and motivation to mathematics, model formation, providing tools for thinking using procedures, notations, drawings and rules, reality as the source and application domain, and specific practical skills in specific situations. Pendidikan Matematika Realistik Indonesia (PMRI) was a learning innovation that adapts from a similar learning approach as RME (Realistic Mathematics Education) [11, 12]. In PMRI, learning was begun by presenting a contextual/realistic problem [13]. Using of context was important because it could motivate students in learning mathematics and help students to understand the topics [14-17].

Asian Games is a sporting event of Asian countries which is held every four years [18]. Indonesia is hosting the Asian Games in 2018. Asian Games in 2018 will be consisted by 40 sports and followed by 45 countries in Asia. The kind of the sport at the Asian Games are the games that most students do or watch like karate, so it's familiar to them and a real context for them. Also, previous researchers that used the context of sports branch at the Asian Games 2018 such as using sprint context [19], *dayung* context [20], shot-put context [21], swimming context [22], and hurdles context [23]. The results showed that the learning of mathematics with the context of the sports branch could help students in understanding mathematics material.

Several kinds of research already documented their research on the development of PISA-like problems [24-26]. Mardhiyanti et al. [24] developed the PISA-like problem to measure students' mathematical communication skills. Next, Silva et al. [25] already developed PISA-like problems on uncertainty content to measure students' mathematical problem-solving abilities. Lastly, Kamaliyah et al. [26] also developed PISA-like problems level 6 for junior high school students.

Based on the background, the researchers were interested to develop PISA-like mathematics problem using one of the sport at the Asian Games in 2018 that was karate. The objective of this research was to produce valid and practical PISA-like mathematics problem with karate context in Asian Games.

2. Method

The method in this research is design research with the type of development studies. This research consisted of two stages, which were preliminary and formative evaluation that includes self-evaluation, expert reviews and one-to-one, small group, and field test [27, 28]. The subject of this research was tenth-grade students of Senior High School. Data collection techniques used were walk through, document, observation, and interview.

In the preliminary stage, the researchers analyzed of research subject, curriculum, 2015 PISA framework, and make the instrument of the problem which included problem grid, problem card, and the scoring rubric. In self-evaluation phase, the researchers evaluated and reviewed the initial prototype and to form prototype 1.

In expert reviews phase prototype 1 was validated by the experts in a way observed, assessed and evaluated. Expert validation used analysis regarding content, construct, and language. In expert reviews, researchers also did the panel discussion (item panelling) of the problem. Item panelling or panel discussion was one of some essential steps in the development of high-quality test items [29]. It was a means of subjecting draft test material to the scrutiny of experts who could provide a rigorous evaluation of the quality of the material, and where appropriate could propose ways in which the material may be improved. It was an essential quality assurance process before any use of the material with students or other test candidates.

Along with the expert reviews, researchers did the test to students in individual (one-to-one). In one-to-one phase, the test was done to 3 students with high, medium and low ability. The one-to-one phase was used to view the readability of the problem. From the expert reviews and one-to-one phase would produce a valid prototype 2.

The next phase was small group. In the small group phase, prototype 2 was tested to 6 students with high, medium, and low ability. In this phase, the problem was used in the learning process. This small group phase was used to see the practicality of the problem. From the small group phase would produce a valid and practical prototype 3.

3. Result and discussion

This research was conducted in Senior High School which involving three students in one-to-one phase and six students in small group phase consisting students who have high ability, medium ability, and low ability. This research had done through 3 phases of 4 phases of design research development studies type, namely: expert reviews, one-to-one, and small group.

3.1. Preliminary

In the preliminary stage, the researchers analyzed research subjects, curriculum, 2015 PISA framework, and design an instrument that included the grid of problem, problem card, and scoring rubric. The researchers developed PISA-like mathematics problem from the original PISA mathematics problem.

On the problem the researchers developed PISA-like mathematics problem from PISA mathematics problem in 2006 that was "Spring Fair". Content of the problem was uncertainty and data. This Spring Fair problem asked how the possibility of Sue will win the prize. Solve the problem with the concept of probability with regard condition that the spinner stops on an even number. PISA-like that was developed was “Karate Kick Training”. The problem asked player most probability to win the match in the other match. Resolve the problem by using the concept of probability by calculating the score obtained and compared with the maximum score. The context of the problem used was occupational, the content used was uncertainty and data, then prediction level was level 4. From the preliminary stage was obtained initial prototype.

3.2. Formative Evaluation

In self-evaluation phase, the researchers evaluated and reviewed the initial prototype based on the three characteristics that became the focus of the prototype regarding content, constructs, and language. Three characteristics were validated by researchers, supervisors, and peers. The results obtained at this phase was prototype 1 which would be tested at expert reviews and one-to-one phase.

The expert reviewers in this research were RT and KS (from Australia), and HJ (from Indonesia). The experts reviewed and evaluated the PISA-like mathematics problem regarding content, construct, and language. In the expert reviews phase, panel discussion was also conducted by researchers with 9 masters of mathematics education of Sriwijaya University and 3 students of mathematics education of Sriwijaya University who also did a research on PISA-like mathematics problems, and also 3 lecturers of mathematics education of Sriwijaya University.

Along with the expert reviews phase, the problems were tested in one-to-one phase. In this phase, the PISA-like problem was tested to 3 students of tenth-grade class with different abilities. The three students were AR, AL, and FA. They were asked to work on the problem and then each student asked for opinions, comments and suggestions on the problem. This aimed to observe the responses and constraints of students when working on the problems and focus on the readability and clarity of the problems. The followings were the comments/suggestions from Experts and Students as well as the revision decision on problem in Table 1.

Table 1. Comments/suggestions from experts and students

Validation	Comments/Suggestions	Revision
RT	1. Not a probability, but a relative frequency for the key answer 2. Added a formula to calculate the relative frequency of the number of kicks which had been done and associated it with a score on the answer key	1. Adding words to the question becomed: "From the data above which players are most likely to win match at the upcoming 2018 Asian Games?" 2. Adding words in the problem "every karate player is given different kicking opportunities" 3. The answer key used the concept of relative frequency 4. Exclamation mark (!) In the sentence "give your explanation!" on the problem was replaced with full stop (.)
KS	1. Not a probability, but a proportion for the answer key 2. For the answer in terms of the number of kicks, there was a possibility that also Chamil may be better.	
HJ	1. Ask the criteria for winning the game 2. The answer key was not to use probability, but relative frequency	
Panel Discussion	1. Add "match at the upcoming Asian Games 2018" to the problem 2. After the words "give your explanation" use full stop (.) , not an exclamation mark (!)	
Students	1. Students could read the table of karate kick training result clearly 2. Students assumed the words "win the game" was an ongoing trainings	

After the problem was revised then the problem became prototype 2 that had been valid. The validity of the problem from the validator's assessment of the content, construct and language and from suggestions and comments one-to-one phase to the clarity/readability of the problem [30]. The following was prototype 2 in Figure 1.

Karate Kick Training

Source : http://asset.indosport.com

5 Indonesian karate players are doing joint kick training to prepare for the upcoming 2018 Asian Games . They practice against coach one by one. They practice kicks against coach one by one with every karate player is given a chance kick different. The system of karate matches that kicks on the head gets score of 3, while the kick on the body get score of 2. In the karate training, the data obtained as follows:

Number	Players Name	Number of Kicks	Head kick	Body kick
1	Iwan Bidu Sirait	13	4	3
2	Muh. Riski	11	3	2
3	Chamil Nazbudin Irfandi	15	4	4
4	Diki Hermawan	14	4	3
5	Rakha Agung Suryandaru	12	2	4

Question :
From which data are the most likely player to win the match at upcoming 2018 Asian Games? Give your reasoning.

Figure 1. Prototype 2

In small group phase, the researchers tested PISA-like mathematics problem that had been developed to 6 students of tenth-grade class with different abilities in the learning process. The six students were AN, TM, AM, MF, MHD, MRF. The following was an example of student answers in small group phase in Figure 2.

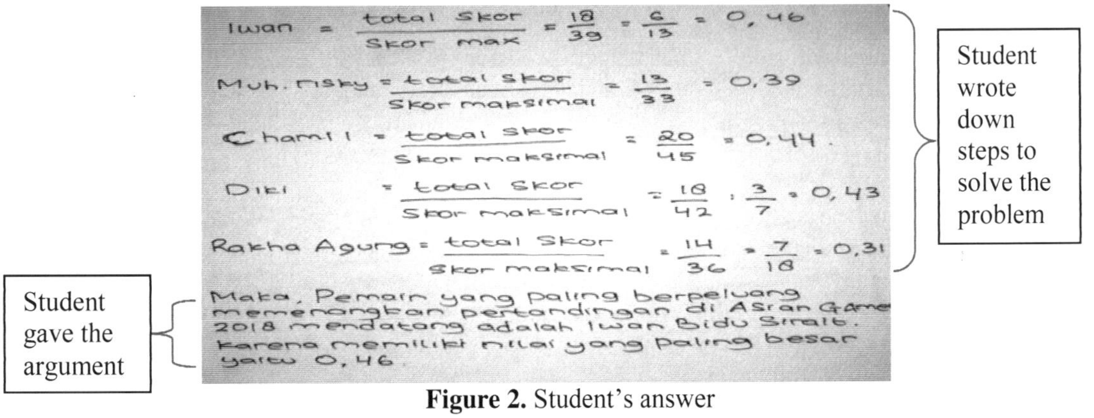

Figure 2. Student's answer

In Figure 2, it could be seen that the student could understand the problem well, the student could write down the steps to solve the problem which was calculating the relative frequency of each karate player with calculating the total score and compared with the maximum score, and the student could give an argument of their answer correctly, which was determining Iwan Bidu Sirait as a player who most likely to win the match at upcoming 2018 Asian Games. It was based on the relative frequency value, which was the highest among the other karate players that was 0.46.

Based on the results of student work from small group phase, it could be concluded that the problem was easy to understand, explanation or clue on the problem was clear, students were able to answer the problem correctly. Also, in the learning process, the students were actively engaged in discussion with karate context that familiar with them, and also it could help students to solve the problem and give the argument. It was no revision from prototype 2 to prototype 3. Based on the small group phase obtained valid and practical prototype 3. Practicality was seen from the ease of students in understanding the problem [27, 30, 31].

4. Conclusion

The research had produced valid and practical PISA-like mathematics problem with karate context in Asian Games. Validity was shown from the validator assessment results at the expert reviews phase regarding of content, construct, and language problem and also from students' suggestions/comments of one-to-one phase to the clarity/readability of the problem. For practicality in the small group phase, that students had been able to understand the problem well. Also, karate context that was used could make students actively engaged in the discussion and it could help students to solve the problem and give the argument.

Acknowledgements

The researchers would like to express gratitude to the Directorate General of Higher Education Indonesia who has funded Postgraduate Grant research in 2017-2018, and to Ms. Eriga and her students for participating in this research. As well as those who have helped the researchers in developing the research and writing this article.

References

[1] OECD 2016 PISA 2015 *Assessment and Analytical Framework: Science, Reading, Mathematics, and Financial Literacy* (Paris: OECD)

[2] Kemendikbud 2014 Permendikbud No. 59 Tahun 2014 Tentang Kurikulum 2013 Sekolah Menengah Atas/Madrasah Aliyah (Jakarta: Kemendikbud)

[3] OECD 2014 *PISA 2012 Results in Focus: What 15-year-old know and what they can do with what they know* (Paris: OECD)
[4] Putri R I I and Zulkardi Z 2018 Higher-order thinking skill problem on data representation in primary school: A case study *J. of Physics: Conf. Series* **948** 012056
[5] OECD 2016 *PISA 2015 Result (Volume 1): Excellence and Equity in Education* (Paris: OECD)
[6] Stacey K 2010 Mathematical and scientific literacy around the world *J. of Science and Mathematics* **33** 1
[7] Wu M 2012 Using PISA and TIMSS mathematics assessments to identify relative strengths of students in Western and Asian *J. of Research in Education Sciences* **1** 67
[8] Novita R, Zulkardi and Hartono Y 2012 Exploring primary student's problem-solving ability by doing tasks like PISA's question *J. on Mathematic Education* **3** 133
[9] Zulkardi 2010 *PISA, KTSP and UN Proc. KNM XV* (Manado: UNIMA) p 53
[10] Zulkardi and Putri R I I 2006 *Mendesain Sendiri Soal Kontekstual Matematika Proc. KNM13* (Semarang: UNNES) p 1
[11] Putri R I I 2011 Pembelajaran materi bangun datar melalui cerita menggunakan pendekatan Pendidikan Matematika Realistik Indonesia (PMRI) di sekolah dasar *J. Pendidikan dan Pembelajaran* **18** 234
[12] Risma D A, Putri R I I, and Hartono Y 2013 On developing students' spatial visualisation ability *J. International Education Studies* **6** 1
[13] Merpaung Y and Julie H 2011 *PMRI dan PISA: Suatu Usaha Peningkatan Mutu Pendidikan Matematika di Indonesia* Available from https://www.usd.ac.id/fakultas/pendidikan/pen_matematika/f113/PMRI%20dan%20PISA.pdf (Accessed 25th September 2017)
[14] Widjaja W 2013 The used of contextual problems support mathematical learning *J. On Mathematics Education* **4** 151
[15] Fauziah A, Putri R I I , Zulkardi, and Somakim 2017 Primary school student teachers' perception to Pendidikan Matematika Realistik Indonesia (PMRI) instruction *J. of Physics: Conf. Series* **943** 012044
[16] Arifin S, Zulkardi, Putri R I I, Hartono Y, and Susanti E 2017 Developing Ill-defined problem solving for the context of "South Sumatera" *J. of Physics: Conf. Series* **943** 012038
[17] Rejeki S and Putri R I I 2018 Models to support students' understanding of measuring area of circles *J. of Physics: Conf. Series* **948** 012058
[18] Wulandari D and Atmojo D 2014 *Fresh Update Top No. 1 Soal-Soal CPNS* (Jakarta - PT. Wahyumedia)
[19] Roni A, Zulkardi and Putri R I I 2017 *Sprint Context of Asian Games in the Division of Fractions Proc. 5th South East Asia Development Research (SEA-DR) International Conference 2017*, (Banjarmasin: ULM) p 22
[20] Nasution M F, Putri R I I and Zulkardi 2017 *Dayung Context in Fraction Proc. 5th South East Asia Development Research (SEA-DR) International Conference 2017* (Banjarmasin: ULM) p 1
[21] Putri, R I I and Zulkardi 2017 Fraction in shot-put: a learning trajectory AIP Conference Proceedings **1868** 050005
[22] Putri R I I , Gunawan M S, and Zulkardi 2017 Addition of fraction in swimming context *J. of Physics: Conf. Series* **943** 012035
[23] Rahayu C, Putri R I I and Zulkardi 2017 *Multiplication of Fraction With Natural Number by Using Hurdles Proc. 5th South East Asia Development Research (SEA-DR) International Conference 2017* (Banjarmasin: ULM) p 43
[24] Mardhiyanti D, Putri R I I and Kesumawati N 2013 Pengembangan soal matematika model PISA untuk mengukur kemampuan komunikasi matematis siswa sekolah dasar *J. Pendidikan Matematika* **5** 16

[25] Silva E Y, Zulkardi and Darmawijoyo 2013 Pengembangan soal matematika model PISA pada konten uncertainty untuk mengukur kemampuan pemecahan masalah matematika siswa sekolah menengah pertama *J. Pendidikan Matematika* **5** 48

[26] Kamaliyah, Zulkardi and Darmawijoyo 2013 Developing the sixth level of PISA-like Mathematics problems for secondary school students *J. on Mathematic Education* **4** 9

[27] Tessmer M 1993 *Planning and Conducting Formative Evaluations* (London: Kogan Page)

[28] Zulkardi 2006 *Formative Evaluation : What, why, when, and how* Available from http://www.oocities.org/zulkardi/books.html (Accessed 15th September 2017)

[29] Turner R 2000 *Item panelling , or cognitive walk-through* Available from https://works.bepress.com/ross_turner/21/ (Accessed 11th October 2017)

[30] Zulkardi 2002 *Developing A Learning Environment on Realistic Mathematics Education for Indonesian Student Teachers* Published Dissertation (Enschede: University Of Twente)

[31] Riyanto B, Zulkardi, and Putri R I I 2017 Mathematical modeling in realistic mathematics education *J. of Physics: Conf. Series* **943** 012049

Students' metacognitive ability in mathematical problem-solving learning based on lesson study for learning community (LSLC)

Hobri[1], S Romlah[1,2], A C Prihandoko[1], J Safitri[1,3] and E Nazareth[1,4]

[1] Postgraduate Program of Mathematics Education Department, University of Jember, Indonesia.
[2] Senior High School 1 Wondana, Teluk Wondana, Papua, Indonesia.
[3] Senior High School Muhammadiyah 3 Jember, East Java, Indonesia.
[4] Vocational High School of Health Analyst "Medika Farma," Jember, Indonesia.

E-mail: hobri.fkip@unej.ac.id

Abstract. This research aimed to analyze students' metacognitive skill through the application of mathematical problem solving learning based on Lesson Study for Learning Community (LSLC). This research used a mixed method, qualitative and quantitative. This research started by developing learning material collaboratively with peers, through Plan-Do-See. The data were analyzed by independent t-test. Researchers and some mathematics teachers developed learning materials. The learning materials included the syntax of mathematical problem solving on LSLC. Learning material was not only developed to make students able to resolve the problem solving and improve the metacognitive capability but also to create an atmosphere for the students to learn from each other and not to leave anybody neglected due to incapability. In addition, the developed learning materials also provided "jumping task," exercises designed for students who were more advanced to solve problems with a much higher level to increase their capacity. The research results showed that the implementation of mathematical problem solving based on LSLC significantly affected students' metacognitive capability. Metacognition ability is part of the core competence that must have by senior high school students. Metacognition can be established when students work on problem-solving, work in a small group through collaboration in a community which cares for and learns with one another in the form of LSLC.

1. Introduction

The purposes of the Indonesian national Curriculum 2013 are to help to establish students' metacognitive skills and to guide students' learning by questioning whether or not they have understood the lesson that they are dealing with. The correlation between problem-solving model and students' metacognitive skills has been stated in previous studies, stating that students were better in solving problems when they had a high altitude of metacognitive skills [1][2][3][4][5].

Metacognition is one's knowledge and awareness of his cognitive process. In this case, students were designed to be aware of what they know and what they do not. Metacognition per se often refers

[1] To whom any correspondence should be addressed.

to the process of "thinking about thinking", which expresses that "metacognition is a high-level thinking skill involving someone metacognitive awareness during the learning process" [6][7].

Problem-solving can be interpreted as a series of learning activities, that emphasize on the process of resolving the problems scientifically. One of the problem-solving models is Polya's model. There are four steps in the problem-solving learning model proposed by Polya, namely: (1) understand the problem, (2) determine the plan of problem-solving strategies, (3) complete the problem-solving strategies, and (4) re-examine the answers obtained. Learning begins by giving a problem, then the students practice to understand, strategize, and implement a strategy to derive a conclusion [8]. In addition, problem-solving ability can be improved by applying collaborative learning, when students work effectively together on projects or tasks. The brain is a social brain and learns more effectively in a learning community than in isolation (two heads are better than one) [9].

The efforts to improve metacognition ability and student learning activity begin by designing the school habit and integrating it into the learning process. Therefore, learning tools were needed to accommodate applying these characters in learning. Learning equipment also becomes guidelines for teachers in implementing learning. Preparation of learning equipment requires a good plan [10][11][12][13]. Metacognitive capability can be monitored from the learning journal, applied during a lesson study for learning community, prepared by the teacher [14]. With the LS learning, collaboration skills can be developed and trained through group work collaborative learning to create an atmosphere that allows students to learn from each other, not leaving any student feel neglected due to incapability or two. Therefore, the problem in this research was how the learning apparatus of LSLC-based learning with Problem Solving Model was applied and how was its effect on students' metacognitive capability. The expected goal was to increase the understanding of concepts, improve students' metacognitive capability, increase the sense of belonging between the members of the learning community, and minimize the mistakes when solving problems.

2. Method

This research used a mixed Method with concurrent triangulation between the qualitative and quantitative formula to address similar problems [15]. The research was conducted in the Year 7 students Junior High School in Jember, East Java, Indonesia. In quantitative research, the sample consisted of two classes that were selected by using random cluster sampling.

In qualitative research, researchers used the phase of lesson study consists of Plan-Do-See. Researchers developed learning equipment and its instruments collaboratively with peers, refer to the development of Thiagarajan, Semmel and Semmel. Furthermore, this study employed an in-depth interview to explore more about the students' activities. Validity test was carried out on the instruments before the test, and the interview was fulfilled. The data were validated using triangulation which was based on test and interview. Then, the interview result was analyzed using data analysis framework proposed by Miles and Huberman. Research procedure was presented in Figure 1.

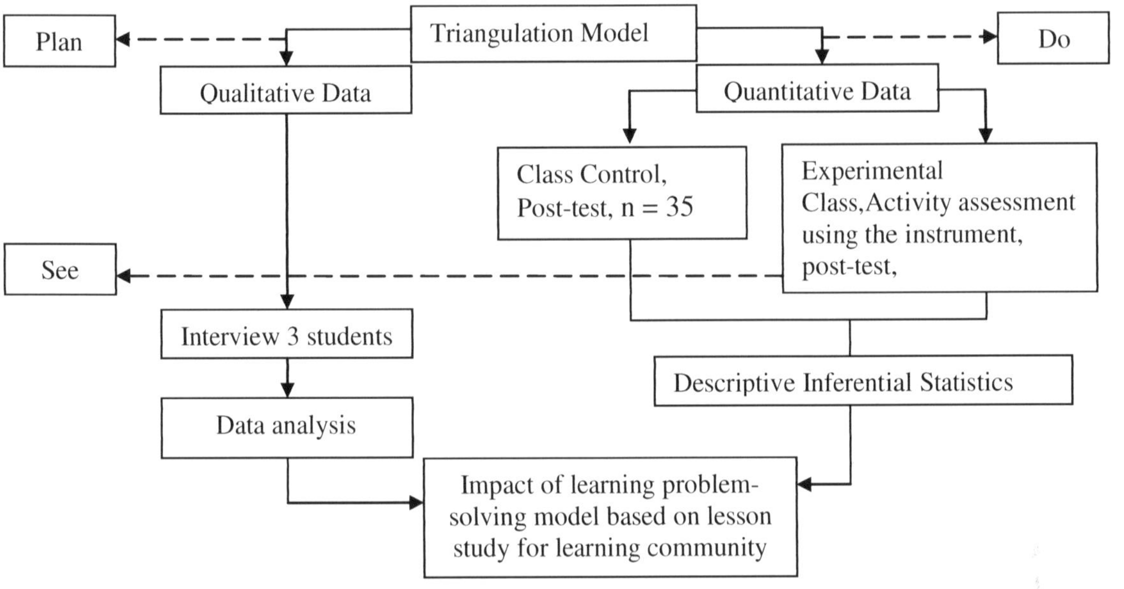

Figure 1. Research Design

3. Results and Discussion

Lesson Study for Lesson Community (LSLC) is part of the Lesson Study (LS), so its phase consists of Plan-Do-See. In the first phase, Plan, a pre-learning activity, the researchers developed learning equipment and its instruments collaboratively with peers, refer to the development of Thiagarajan, Semmel, and Semmel. According to the concept of analysis, mutual care, metacognition ability is part of the core competence that can be achieved at the senior high school level. Metacognition ability can be developed in a learning environment, which is focussing on problem-solving on the topic of matrix.

Before giving treatment to the experimental class and control class, researchers and some mathematics teachers developed learning materials according to the learning model that later was used in the experimental class. The learning materials included the syntax of mathematical problem solving based on LSLC. Learning materials were developed to allow students to be able to resolve the problem solving and improve the metacognitive capability, as well as to create an atmosphere for the students to learn from each other and not leaving anybody neglected due to incapability. In addition, the developed learning materials also provided "jumping task," exercises designed for students who were more advanced to solve problems with a much higher level to increase their capacity.

The next phase is Do, implementing the learning equipment as a development result, that was tested in the experimental class. Learning activity in experimental class and control class has been done four times meeting, the first meeting talked about matrix concept, secondly discuss the operation of matrix, thirdly discuss determinant, and fourthly discuss completion systems of linear equations with matrix and determinant.

In the experimental class has done by opening class and using the scientific approach with problem-solving learning, collaborative learning, and caring community. In control class, conventional learning has been used learning materials that were provided in the school. On experimental class, 36 students are divided into nine small groups and each group observed by one observer. Besides controlling the student activity, one observer controls the teacher activity in the learning process. One of the observer observation results in the fourth meeting to the third group discussion result illustrated in Figure 2.

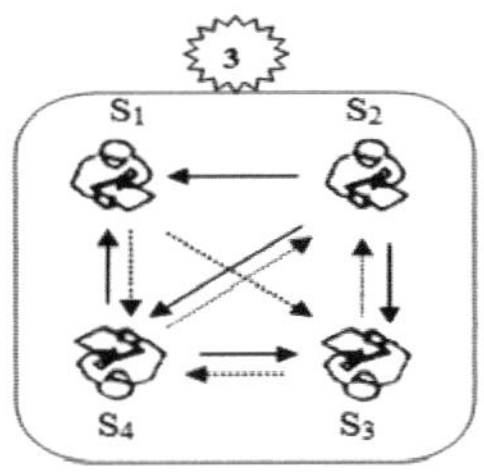

Notes: ⟶ : Students ask or give feedback.
⟶ : Students express opinions and/or guide a friend

S_1: Student 1 S_3: Student 3
S_2: Student 2 S_4: Student 4

Figure 2. Illustration of group discussions 3

During the discussion of the completion of two variables linear equation system (TVLES) with the determinant method, the results are as follows.

S_4 Friends, for TVLES completion with the determinant method, we need to distribute the task, S_2 and S_3 use Sarrus rules, S_1 and I use the Minor-Cofactor method, then we will mutually check the results of our work

S_2 S_2 asked S_3 to change the form of TVLES, whether the equation: (1) as the first column, equation (2) as the second column, and equation (3) as the third column, as has been written in the following:

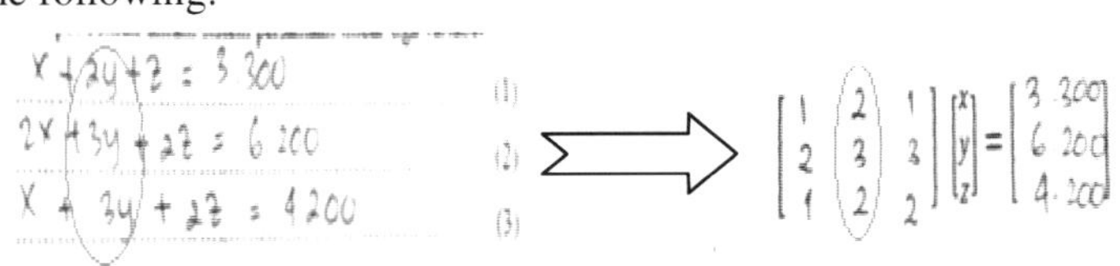

Figure 3. The result of student

S_3 S_3 explained that what S_2 was working on, was still incorrect, S_3 mentioned that the x, y, z variables needed to put into column matrix. Therefore, to change TVLES into matrix form, the equation (1) as the first line, equation (2) as the second line, and equation, (3) as the third line.

S_1 S_1 wanted to check his work and compare the results with S_4. The results of S_4's work are as the following.

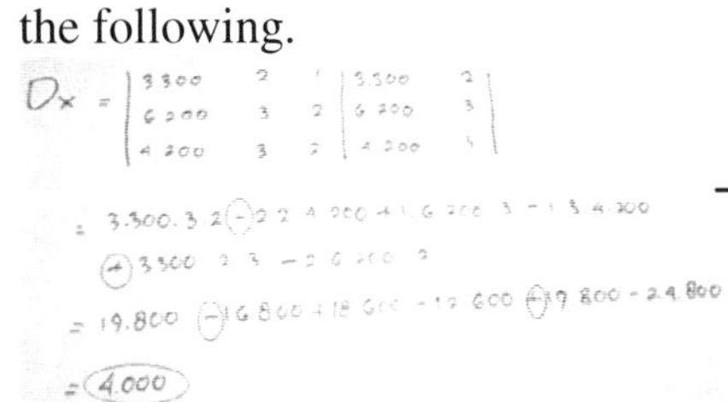

After checking on S_4's work, it was found that $S_4$was careless in completing the task. Therefore, he made a mistake in the next steps as well. S_1 told S_4 in which part he made a mistake so that it was easier for him to fix the answers.

Figure 4. Th result of Student

S_1 showed his conclusion that each portion of the meal cost 2,000 rupiahs, a piece of soybean cake cost 400 rupiahs, a glass of iced orange juice cost 500 rupiahs.

In the experimental class, in accordance with Vygotsky's Zone of Proximal Development (ZPD) theory, the researcher gave the jumping tasks developed in the student worksheet; this jumping task was for high-ability students in the experimental class aimed to get new challenges in learning and to createing possible knowledge. However, the activity of giving the question of the jumping task was not done well because there were no students who can solve the whole jumping problem from the problems that have been given even though they have been guided by the teacher [16].

In the learning process of the control class, students also worked in groups; each group consisted of five students. In the discussions, students were dominating the group from the first to the fourth meeting, so those students did the tasks while other members simply relied on the group. They were copying the work of other students. When facing difficulties in answering the questions in the worksheet, they were more likely to ask for guidance from the teacher than asking friends in their group. This was due to distrust among the students about the ability of their knowledge. Moreover, the groups that already completed their task refused to share knowledge and to teach other groups that need assistance in learning. This did not increase interpersonal relationships because smart students refused to offer help and considered the work of the others as incorrect. In addition, smart students preferred to do the work on their own and gave the answers for granted to their group members.

The process of the conventional learning activities with the group discussion in the control class can be performed as illustrated in Figure 5.

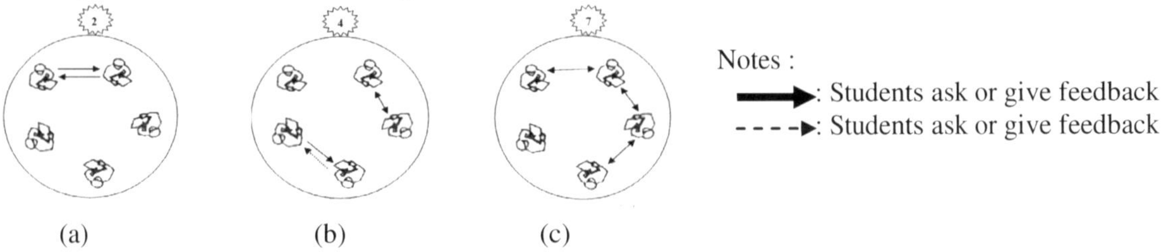

Figure 5. Students' interaction

During the discussion (a) in group 2, the two students were dominating the tasks and only two of them mutually asking or responding to one another. They refused to guide the other three group members. Other three members were passive and told merely to copy the work of the domineering students, (b) in group 4, there were only two pairs communicating because they were seatmate refused to cooperate with the other group members, (c), unlike the other groups, in group seven, four members of the group asked questions in the discussion, but there was one student who refused to join the discussion because he could not understand the subject mattered and refused to ask his friends for help.

After learning activity finished, the observers present their observation result during the learning process, which is reflection phase (see) to LS phase. Explanation of the observation result during open class will be used for remedial the next learning process. In experimental class from observation result, learning activity and students interaction are supported by learning equipment with problem-solving learning model based on lesson study for learning communicate has gotten improvement for each meeting. Students' communication skill between students in-group as well as another group in experimental class. Students activity are very good, and the students learning result is very high like in the discussion activity, understand the case, solve the problem is very well, among the students appear a sense of mutual care, and not be ignored. The students, who are more capable, will use communication ability and the students who need an explanation during work in-group can receive an argument in order to their knowledge clearly. The process of functioning communication and argumentation also cannot be separated from the principles that contain in metacognition ability component are planning, information management, monitoring, revising and evaluating.

The appearance of mutual care, mutual learning, sharing knowledge so learning activity by this model, not students stand out nor students left alone and did not ignore each other. In the learning process using the problem-solving model based on lesson study for learning community is an implementation of lesson study for the learning community. This case accorded to the philosophy of learning communities formed into two main pillars namely collaborative learning and caring community [11]. Likewise also the results of research, in designing collaborative learning the elements appearance are learning that make children become as a main role, from teaching to learning and subsequently learning, learning with dialogue and collaboration. The role of the Lesson study in collaborative learning was not prioritized to solve problem and difficulties but sharing problem, interaction and sharing and think to solve together. [13].

The results of observations in each meeting were used to reflect (See) on various findings in the learning implementation. Teachers and observers shared the findings related to students' learning activities during the learning process to make any improvement in the next meeting in the experimental class.

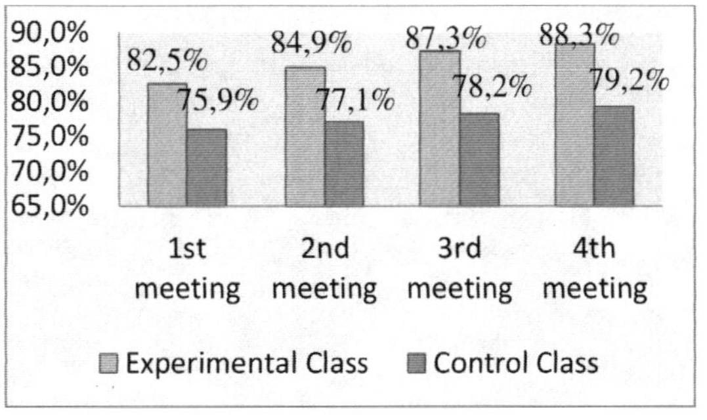

From four meetings, the average rate of students' learning activities in the experimental class was 85.75% while the average rate of students' learning activity in the control class was 77.60%. It can be concluded that the activities of students in the experimental class with the implementation of mathematical problem-solving learning based on LSLC is better than the activities of students in the control class.

Figure 6. Activities rate of the experimental class and the controlling Class

In line with Hobri's research, the application of student worksheets combined with the concept of collaborative learning, learning community, caring community, and jumping task, student activity is very good, and student-learning outcomes as a whole are very high. Student activities in discussing, understanding the subject matter "matrix," and solving the problem very well [12]. In addition to increasing students' activities, students also responded positively to the mathematical problem-solving learning based on LSLC. It can be seen from the results of students' questionnaire responses distributed to 36 students from the experimental class. Based on the questionnaire responses, students' response to all aspects was 91.28%. Since the result was above 80%, the student's response was positive.

In relating to the students' metacognition ability after applied problem-solving learning based on LSLC, has gotten through the task result of learning task result. Furthermore, the researcher interviewed to a student with who has high mathematics competence, one student with medium mathematics competence, and one student with low mathematics competence. The test was solved by Polya's steps, which used metacognition indicator ability in problem-solving. The same number and type task were given to students of the experimental and control class. The test has been analyzed by t-test. The result can be seen in Table 1.

Table 1. Independent samples t-test

		Levene's Test for Equality of Variances		t-test for Equality of Means			
		F	Sig.	t	df	Sig. (2-tailed)	Std. Error Difference
Student_ Metacognition	Equal variances assumed	3.176	.079	3.730	69	.000	5.8079
	Equal variances not assumed			3.742	66.08	.000	5.8079

Based on the independent sample t-test on the Table 3, the value of sig (2-tailed) is 3.742. This showed the implementation of problem-solving learning model based on LSLC has a significant effect on the students' metacognition ability on matrix topic at the grade 11th senior high school. It is caused during the implementation of problem-solving learning model based on LSLC, the students not only aware about the benefit of learning model but also can endeavor their metacognition ability to planning, monitoring, evaluating their result task.

This learning model could potentially increase learning activities and students' metacognition ability. The implementation of LSLC showed improving the students' communication ability and argument. It can be separated from the principles contained in metacognition ability components; those are planning, information management, monitoring, revising, and evaluating. According to Polya's theory about the steps of problem-solving, can be stated that all the steps those are proposed pointing to students' awareness and arrangement to the process has been done to get appropriate solution [17]. The step of problem-solving that Polya proposed has been a basis for developing of metacognition ability and most of the researcher has become reference especially by mathematics teacher [18][19][20]. In addition, students' metacognition skills were also improved through a learning journal developed on the student worksheet on learning by applying a lesson study [14] [21]. By applying the model of problem solving-based on LSLL can increase student activity, students' ability in solving problems that resulted in increasing of students' metacognitive ability.

4. Conclusion

The student worksheet that implemented mathematical problem-solving learning based on LSLC in this study combined the concepts of collaborative learning, learning community, caring community, and jumping task. The worksheet resulted in very good student activity and a very high students' learning outcomes. The students' activities in the experimental class with the implementation of mathematical problem-solving learning based on LSLC through utilization of students worksheet was better than the activities of students in the control class with conventional learning. The implementation of mathematical problem-solving learning that was based on LSLC through utilization of students' worksheet significantly affected students' metacognitive capability, particularly on subject matter matrix and its operation. It can be suggested that the implementation of mathematical problem-solving learning based on LSLC through utilization of students worksheet is one alternative of learning models that are beneficial for classroom use. However, its effectiveness still needs to be clarified in other classes even in other schools with a variety of different conditions.

Reference

[1] Garrett, A.J ,Mazzocco Garrett , and Mazzocco, Baker, L 2006 Development of the Metacognitive Skills of Prediction and Evaluation in Children with or without Math Disability *Learning Disabilities Research & Practice* **21**(2) pp 77–88.

[2] Keichi, S 2000 *Metacognition in Mathematics Education in Japan* (Japan: JSME)

[3] Mohamed, and Nai, Tan Ten 2005 The Use of Metacognitive Process in Learning Mathematics. *The Mathematics Education into the 21st Century Project (Universiti Teknologi Malaysia Reform, Revolution, and Paradigm Shifts in Mathematics Education Johor Bahru Nov 25th – Dec 1st*) pp 159 – 162

[4] In'am, A. A 2012. Metacognitive Approach to Solving Algebra Problems. *International Journal of Independent Research and Studies (IJIRS)* **1** (4) pp 162-173

[5] Purnomo, D., T. Nusantara , Subanji , and S. Rahardjo 2016 Metacognition Process Characteristics of The Students in Solving Mathematics Problems IOSR *Journal of Research & Method in Education (IOSR-JRME)* **6** (5) pp 26-35.

[6] Livingston, J. A 1997 Metacognition: An Overview. Retrieved from http://www.gse.buffalo.edu/fas/ shuell/cep564/Metacog.htm, 1-5. Retrieved February 15, 2017.

[7] Costa, A. L. and Kallick, B 2009 Leading and Learning with Habits of Mind: 16 Characteristics for Success (*Alexandria, VA: Association for Supervision and Curriculum Development)*

[8] Ahmadi, A. and Prasetya, J.T .1997 *Teaching and Learning Strategies* (Bandung: Pustaka Setia)

[9] Kryza, K., Duncan, A., and Stephens, S. J. 2009 *Inspiring Elementary Learners: Nurturing the Whole Child in a Differentiated Classroom* (Thousand Oaks, CA: Corwin Press) p 31

[10] Mariati, P. S. 2012 Development of Problem-Based Physics Learning Model for Improving Metacognition Skills and Student Concept Understanding *Journal of Physics Education Indonesia* **8** pp 152-160

[11] Hobri, Dafik, and Hossain 2018 A The Implementation of Learning Together in Improving Students' Mathematical Performance *International Journal of Instruction* **11**(2) pp 483-496

[12] Purwaaktari, E. 2015 The Effect of Collaborative Learning Model on Problem Solving Ability of Mathematics and Social Attitude of V Students of SD Jarakan Sewon Bantul *Journal of Educational Science Research* **8**(1) pp 95-111

[13] Rudolph, J. et al. 2017 Metacognitive Confidence Judgments and Their Link to Complex Problem Solving *Intelligence* **63** pp 1 – 8.

[14] Setiawan and Susilo 2015.Improving Metacognitive Skills of Biology Study Program Students through The Application of Journal of Learning with Jigsaw Strategy Combined PBL Based Lesson Study in General Biology Course *Proceedings of National Biology Education Seminar 2015* (Biology Education Prodi FKIP University of Muhammadiyah Malang) *March 21, 2015* pp 359-69.

[15] Sugiyono 2017 *Combined Research Methods (Mixed Methods)* (Bandung: Alfabeta) p 17

[16] Vygotsky, L.S. 1978 *Mind in Society: The Development of Higher Psychological Processes* (Harvard: Harvard University Press)

[17] Gama, C. A. 2004 *Integrating Metacognition Instruction in Interactive Learning Environment* (D. Phil Dissertation: University of Sussex)

[18] Garrett, A.J, Mazzocco Garrett, and Mazzocco, Baker, L. 2006 Development of the Metacognitive Skills of Prediction and Evaluation in Children with or without Math Disability *Learning Disabilities Research & Practice* **21**(2) pp 77–88

[19] Keiichi, Shigematsu and Yoshio, Katsumi 2000 Metacognition: The Role of the "Inner Teacher" (6): Research on The Relation Between a Transfiguration of Student's Mathematics Knowledge and "Inner Teacher" *Proceedings of the Conference of the International Group for the Psychology of Mathematics Education (PME)* (24th,Hiroshima: Japan, July 23-27, 2000) **1** pp 4-137

[20] Mohamed, Mohini and Nai, Tan Ten 2005 The Use of Metacognitive Process in Learning Mathematics *The Mathematics Education into the 21st Century Project* (Universiti Teknologi Malaysia Reform, Revolution and Paradigm Shifts in Mathematics Education: Johor Bahru) Nov 25th – Dec 1st

[21] Harlita, Riezky Maya Probosari 2010 *Use of Journal of Learning to Increase Mastery of Animal Embryology Concept of Student Education* (Study Program Biology Faculty of Teacher Training and Education UNS National Seminar on Biology Education: Semarang) 2010 pp 200-206

IOP Conf. Series: Journal of Physics: Conf. Series **1088** (2018) 012065 doi:10.1088/1742-6596/1088/1/012065

The impact of using LINE@ on the cooperative learning to improve the critical thinking skills of high school students

D Suliswor[1], S Daimah[2], M Toifur[3] and A Suryadi[4]
[1,3] Ahmad Dahlan University, Jl. Pramuka No. 42, Yogyakarta 55164, Indonesia
[2]SMA Negeri 1 Jetis Bantul (Senior High School), JL. Imogiri Barat KM. 11, Yogyakarta 55781, Indonesia
[4]University of Muhammadiyah Jakarta, Jl. KH. Ahmad Dahlan, Tangerang 15419, Indonesia

E-mail: dwi.suliswor@uad.ac.id

Abstract. The widespread use of social media enables new opportunities in various fields including in education. This study aimed to determine the influence of social media usage, in this case, LINE@, on cooperative learning environment to improve critical thinking skill. This research was conducted on Year 11 students on physics, especially for Elasticity subject. The method used in this research was the quasi-experiment with pretest-posttest controlled group design. The independent variable of this research was learning strategy. The dependent variable was critical thinking skill measured by essay. The covariates used were prior knowledge measured by multiple-choice questions. Statistical analysis technique used ANCOVA with the error margin of 5%. The results showed that students who received cooperative learning using LINE@ tend to have better critical thinking skills after learning compared to conventionally taught students. Prior knowledge can be used as a predictor to examine the level of critical thinking skills. The contribution of previous knowledge was relatively small indicating that the role of learning strategy was relatively high in influencing the improvement of critical thinking skills.

1. Introduction

Social media is now days being utilized in various fields with higher and higher intensity from time to time. The users of social media in Indonesia are increasing each year [1]. There is a tendency that societies get more comfortable with using social media in their activities. School-age children, who are digital natives, even get more comfortable in using telecommunication media rather than adults who are digital immigrants [2]. This situation also occurs in the school environment where most students use the smartphone in their daily activities. Most activities that use the smartphone are through social media [1]. It becomes opportunities for educators to develop more comfortable and flexible learning for students by utilizing social media.

The use of social media surely gives both positive and negative impacts on students. On the negative side, social media activities open the potential for a cyber-bullying, online harassment [3], negative word of mouth [4] and others. Accompaniment and teaching for normal internet usage need to be carried out by educators to mitigate those impacts. On the positive side, numbers of researchers find out the result of learning by utilizing social media on various subjects at school [5,6,7]. Review of the result show there is many positive aspects. However, among the researchers, there are only a few numbers of research that utilize social media for cooperative learning that encourage higher order thinking skills improvement. Therefore, this study is famous as the learning demand-oriented in 4C (critical thinking, creativity, communication, and collaboration) as addressed by UNESCO.

Various studies on cooperative learning (CL) showed that STAD type (Student Teams Achievement Division) is relatively more prominent compared to other models for various reasons. The implementation of this strategy in multiple fields showed the excellent result [8,9,10]. It is possibly because this strategy is relatively practical and consistent with the teaching philosophy. Therefore, this strategy has broad generalization level on the outcome. In today learning active development, learning strategy with STAD has the potential for being able to improve learning performance related to 4C.

Considering new opportunities in a better learning practice by utilizing social media and cooperative learning strategy, this research aims to determine the influence of learning strategy towards critical thinking skills if observed from prior knowledge. The researchers expect that this research will contribute to learning practice which meets current student's needs that relatively have the different focus on demand for learning outcome performance.

2. Method

2.1. Participants
Participants of this research were Year 11 Science students. The study was conducted on physics subjects on the topic of Elasticity in the academic year of 2017/2018 in Yogyakarta Special Province, Indonesia. The location of the school is in the suburban area. Generally, the parent's income level is from middle to low-class society. The school quality in this area is in average level indicated from the scoring average of students' national examination from year to year. Even though in the suburban area, teachers and students at the school have smartphone used in their daily life including for social media activities. Thus, students have understood how to use Line@. Teacher and students' understanding to use this application becomes an important part to ensure the internal validity of the research process.

2.2. Research design
This research is quasi-experiment with post test only controlled group design. Since student's randomization was technically impossible to be done, randomization was only conducted when choosing classes as a control and treatment group from 5 classes (156 students). Sampling for control and treatment group was taken by using simple random sampling technique. Control group consisted of 31 students and treatment group consisted of 32 students. To ensure internal validity, students considered in the research were only ones entirely attended all learning activities. Ones who did not fully participate in all actions were taken out from the research sample.

In control group, students were taught with the conventional method by using the textbooks at school while in the treatment group, students were taught with cooperative learning strategy, which in this research was STAD (student team achievement division) by using, by Line@. Thus, in treatment group blended learning was carried out by combining face-to-face learning at the classroom with independent activities and enrichment via Line@. The research variables were learning strategies as the independent variable, critical thinking skill as the dependent variable, and prior knowledge as covariates.

2.3. Material and learning activities
Line@ is a social media application that can be used on both computer and smartphone. This application can be downloaded via Play Store on an Android smartphone or link https://line.me/en/download. Students and teacher can independently install this application on their device to be used for learning. To run this application, teacher and students should register to have an account to use in social media activities.

In conducting learning, the teacher prepared a lesson plan which had been validated, so it met the learning objective. Learning strategy in treatment class was cooperative learning with STAD type. This strategy is relatively a simple strategy but good for encouraging teamwork. On Elasticity material, student's groups were given various information and problems which could be accessed on their Line@ account. Class activities were used to discuss among groups and confirm students' understanding by the teacher. The stages of learning implementation with STAD type by using Line@ were the modification of what had been developed by [11], as follows:

- At the first meeting, the teacher explained the learning objective and delivered a preface related to the learning process.
- Next, the teacher divided students into small groups consisting of five or six students (6 groups) which were heterogenetic concerning previous learning achievement and sex. These groups remained until the end of learning.
- During the learning process, the teacher and students interacted by using Line@. The teacher posted subject material, assignment, and quiz on Line@. During the lesson, students could have the discussion, and the result was also posted on Line@.
- At every end of lesson season, the teacher gave a quiz or question to be finished individually according to the material that had been learned in group. The result was submitted to be compared to the previous score.
- This individual score would contribute towards their group's score. At the end of the meeting, the teacher announced each group's score and gave appreciation to improve students' learning motivation.

2.4. Data collection and analysis
The data of critical thinking skill was taken by using five essay questions. The questions were scored based on the stages of solving a complex problem related to Elasticity. The questions were developed with the reference of taxonomy bloom on the level of C3 and C4. Before used, the questions had been tested on item discrimination, the degree of difficulty, validity, and reliability.

The data analysis used parametric statistic by considering data normality and homogeneity. Next, the data was analyzed using Covariance or ANCOVA to find out the significance of the learning strategy influence towards critical thinking skills and the covariate as the predictor.

3. Result and discussion

3.1. Result
Learning activities with Line@ enable higher interaction between students and the teacher without being known by other students. With this application, the teacher can give different focus portion for each student. Some screenshots of activities between student and teacher, student and another student are shown on Figure 1.

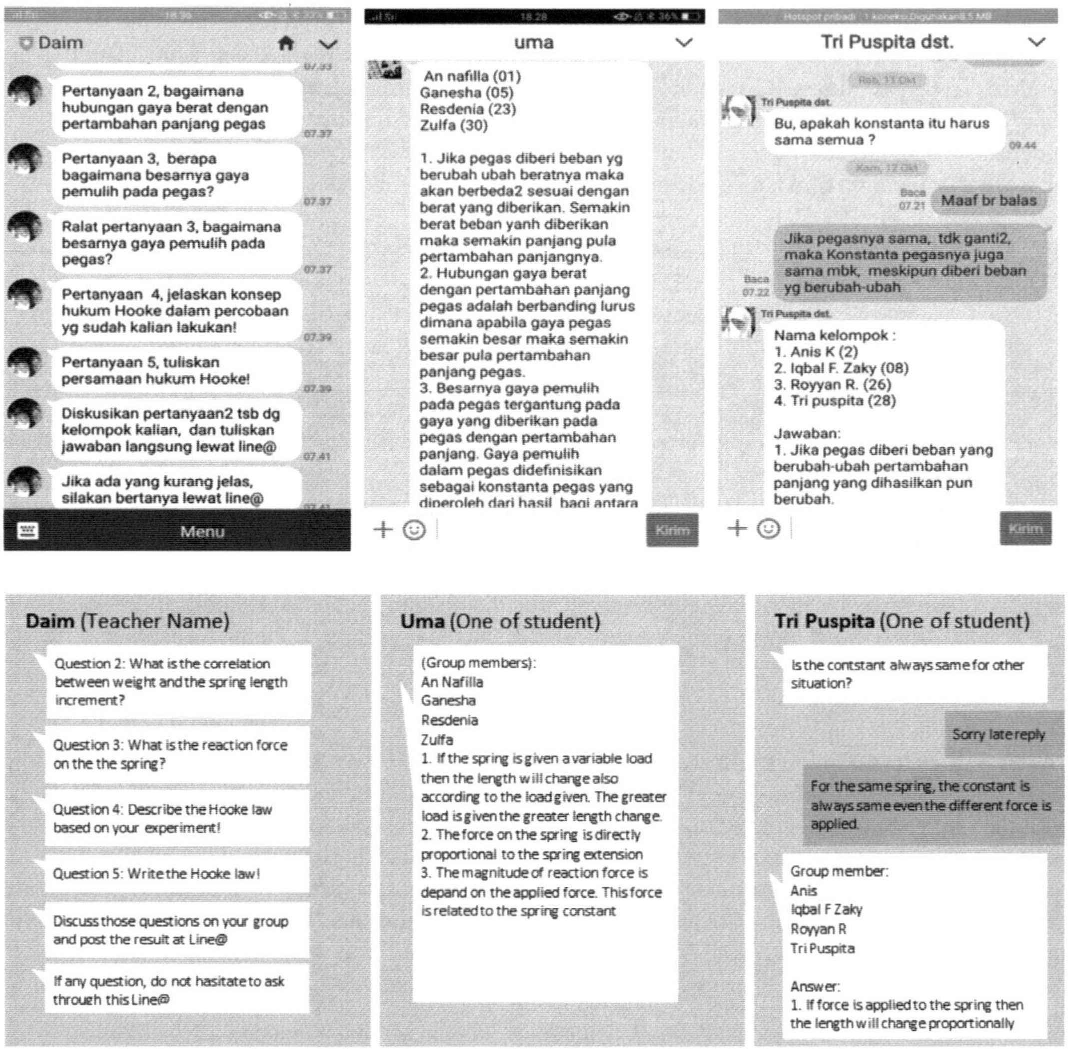

Figure 1. The screenshot of some activities on Line@ (above) and its translation in English

The measurement data of critical thinking skills in control and treatment group was then analyzed with descriptive statistic and ANCOVA. Table 1 shows the result of the descriptive statistics. It is seen that CTS average score for treatment group (66.91) is higher than the score for the control group (59.87), while the standard deviation of treatment group is narrower (6.87) than the control group (14.1). These two findings become essential to the effect of cooperative learning by using Line@ at class. There is a tendency that treatment group consisting of students who study Elasticity with cooperative learning by using Line@ tend to have better critical thinking skills after learning. Also, in the whole group, the distribution of critical thinking skills seems to be similar among students at the end of learning. This similarity can be identified from the standard deviation of treatment group which is smaller than the control group (students who studied conventionally). The difference of standard deviation also shows that at the end of learning both groups become heterogeneous groups. This heterogeneity can be identified from the result of the test of homogeneity of variance with Levene's statistic ($F= 28.91$; $p= 0.002< 0.05$).

Table 1. Descriptive statistics dependent variable: CTS after learning

Learning Strategy	Mean	Std. Deviation	N
Conventional	59.8710	14.10140	31
Using LINE@	66.9091	6.87097	33
Total	63.5000	11.45592	64

The ANCOVA test on the error margin of 5% was conducted to find out a difference of CTS score between the control and treatment group, and whether prior knowledge could be used as a predictor. The dependent variable was the CTS score, and the independent variable was the learning strategy. The test result is shown in Table 2.

Table 2. Tests of effects between-subjects, dependent variable: critical thinking score after learning

Source	Type III sum of squares	Df	Mean square	F	Sig.
Corrected Model	1435.177[a]	2	717.588	6.406	.003
Intercept	1.702	1	1.702	.015	.902
PreK	643.388	1	643.388	5.744	.020
LS	1146.381	1	1146.381	10.234	.002
Error	6832.823	61	112.013		
Total	266332.000	64			
Corrected Total	8268.000	63			

a. R Squared = .174 (Adjusted R squared = .146)

The results indicate that there is a significant difference in the average score of critical thinking skills between the control and treatment group. This difference can be identified from the significant influence of learning strategy (LS) ($p= 0.002 < 0.05$). The learning outcome (CTS) of students taught with STAD strategy by using Line@ is significantly higher than the conventional one.

3.2. Discussion

The result of statistical data processing shows that traditional learning is relatively less effective in improving critical thinking skills in physics compared to learning with STAD strategy by using Line@. This founding is parallel to the other research on physics learning using cooperative learning [12,13] and learning using social media [7,14,15]. Some factors are influencing this difference in learning outcomes. Cooperative learning, which was designed to encourage social interaction among students, tends to support information sharing among students during the learning process [16]. This information sharing occurs when students among groups discuss solving problems.

During the discussion, students will get feedback on their understanding from other students. This feedback will influence the change of perception of students on the idea they have. Higher interaction in STAD learning also motivates students in learning, and it does not happen to students in traditional learning. High motivation tends to improve student's learning outcome at the end of the learning process [17,18].

The level of student's comfort in interaction by using Line@ also becomes a factor that makes students get more natural and more comfortable in delivering the idea. In traditional learning, there is a tendency that students feel reluctant to provide their opinion because of various communication obstacles in their learning environment. It is different from when they write their idea on Line@. They tend to be more self-confident. From this process, it seems that students on learning with this strategy can show better learning outcomes in critical thinking skills. Nowadays, the technology has given high impact to the learning philosophy and practices [19,20] for better student performance.

4. Conclusion

This study shows that learning strategies influence the learning outcomes. Students who are taught with STAD learning strategy by using Line@ have higher critical thinking skills after learning compared to students taught conventionally. This research also reveals that interaction in STAD learning by using Line@ improve the level of students' critical thinking skills, although it is undeniable that their learning outcome a is also influenced by their prior knowledge. These research findings support the notion that social media has great opportunities to be used on cooperative learning to improve student's learning performance in various aspects.

References

[1] Kemp S 2018 Digital in 2018: world's internet users pass the 4 billion mark *We Are Social* https://wearesocial.com/blog/2018/01/global-digital-report-2018

[2] Ahn J and Jung Y 2016 The common sense of dependence on smartphone: A comparison between digital natives and digital immigrants *New media & society* **18** 1236

[3] O'Keeffe G S and Clarke-Pearson K 2011 The impact of social media on children, adolescents, and families *Pediatrics* **127** 800

[4] Pfeffer J, Zorbach T and Carley K M 2014 Understanding online firestorms: Negative word-of-mouth dynamics in social media networks *Journal of Marketing Communications* **20** 117

[5] Mao J 2014 Social media for learning: A mixed methods study on high school students' technology affordances and perspectives *Computers in Human Behavior* **33** 213

[6] Kimmerle J, Moskaliuk J, Oeberst A and Cress U 2015 Learning and collective knowledge construction with social media: A process-oriented perspective *Educational Psychologist* **50** 120

[7] Sulisworo D, Rahayu T and Akhsan R N 2016 The students'academic writing skill after implementing blended learning using Facebook *Information Technologies and Learning Tools* **56** 176

[8] Yeung H C 2015 Literature Review of The Cooperative Learning Strategy-Student Team Achievement Division (STAD) *International Journal of Education* **7** 29

[9] Alabekee E C, Samuel A and Osaat S D 2015 Effect of cooperative learning strategy on students learning experience and achievements in mathematics *International Journal of Education Learning and Development* **3** 67

[10] Hasanah MA and Surya E 2017 Differences in the Abilities of Creative Thinking and Problem Solving of Students in Mathematics by Using Cooperative Learning and Learning of Problem Solving *International Journal of Sciences: Basic and Applied Research (IJSBAR)* **34** 286

[11] Xu L and Liming L 2010 Probe into STAD Cooperative Learning Based on Moodle. InEducation Technology and Computer Science (ETCS) *2010 Second International Workshop* **2** 424

[12] Gambari A I, Yusuf M O and Thomas D A 2015 Effects of Computer-Assisted STAD, LTM and ICI Cooperative Learning Strategies on Nigerian Secondary School Students' Achievement, Gender and Motivation in Physics *Malaysian Online Journal of Educational Sciences* **3** 11

[13] Sulisworo D, Agustin S P and Sudarmiyati E 2016 Cooperative-blended learning using Moodle as an open source learning platform *International Journal of Technology Enhanced Learning* **8** 187

[14] Khan M L 2017 Social media engagement: What motivates user participation and consumption on YouTube? *Computers in Human Behavior* **66** 236

[15] Gregorcic B and Bodin M 2017 Algodoo: A tool for encouraging creativity in physics teaching and learning *The Physics Teacher* **55** 25

[16] Borokhovski E, Bernard R M, Tamim R M, Schmid R F and Sokolovskaya A 2016 Technology-supported student interaction in post-secondary education: A meta-analysis of designed versus contextual treatments *Computers & Education* **96** 15

[17] Ciampa K 2014 Learning in a mobile age: an investigation of student motivation *Journal of Computer Assisted Learning* **30** 82

[18] Shernoff D J, Csikszentmihalyi M, Schneider B and Shernoff E S 2014 Student engagement in high school classrooms from the perspective of flow theory. In *Applications of Flow in Human Development and Education* 475 (Springer Netherlands)

[19] Sulisworo D and Toifur M 2016 The role of mobile learning on the learning environment shifting at high school in Indonesia *International Journal of Mobile Learning and Organisation* **10** 159

[20] Sangrà A and González-Sanmamed M 2100 The role of information and communication technologies in improving teaching and learning processes in primary and secondary schools *ALT-J* **18** 207

IOP Conf. Series: Journal of Physics: Conf. Series **1088** (2018) 012066 doi:10.1088/1742-6596/1088/1/012066

The teachers' mathematics literacy ability for solving uncertainty problems on a PISA adaptation test

F Sanjaya, A Y Anggoro, H Julie and M A Rudhito
Sanata Dharma University, Jl Affandi Tromol, Yogyakarta, 55002, Indonesia

E-mail: febi@usd.ac.id

Abstract. In research related to PISA, students have earned many portions as research subjects. However, research on teachers is rarely conducted. It is widely believed that there is a link between students' ability and teachers' ability. This study aimed at describing teacher's solution on PISA adaptation test. The PISA adaptation test consists of 13 questions from 4 scopes. However, this research focuses on the scope of uncertainty consisting of 3 test questions. In this study, the subjects were 7 teachers who were randomly selected from a Junior High School. This study adopted Cobb and Koeno's model of design research. The finding showed that all teachers responded correctly in the scope of uncertainty at level 2 and level 4. Moreover, the results revealed that 42,86% of teachers achieved level 5 and 28,57% teachers achieved level 6. Therefore, to improve the quality of students it is necessary to improve the quality of teachers.

1. Introduction

Student achievement was closely and directly related to the mathematics and pedagogical skills of elementary teachers [1, 2]. There was a significant relationship between teachers' perceptions and knowledge achieved by students [1, 2]. In this study teachers' perceptions were defined as (1) teacher paradigm in the learning process of mathematics, and (2) teacher attention to students' math skills [1, 2]. The teacher's attention to the students' mathematical skills was dependent on the teacher's mathematical knowledge [1, 2]. Based on the opinion, the ability of teachers in managing the process of learning to teach mathematics and solve mathematics problems was one of the determinants of student success in completing the PISA test. In research related to PISA, students have earned many portions as research subjects. However, research on teachers is rarely done. This study aimed to describe the Junior High School teachers' solution on PISA adaptation tests in quantity, space and shape, change and relationship, and uncertainty area, but in this paper, researchers will only describe the results of the research for teachers' solution on PISA adaptation tests in the uncertainty area [3, 4, 5].

The literacy of mathematics was the ability of the individual in understanding mathematics and applying it in daily life [6]. Using mathematical literacy, students could understand and apply the role of mathematics in the context of real life. [2, 7, 8]. According to Jan de Lange, the ability of a person to identify and understand the role of mathematics in real life, to make accurate judgments, use and involve mathematics in various ways to meet the needs of the individual as a reflective, constructive and devoted citizen called the literacy of mathematics. The following matters affect mathematical literacy skills, including: (1) the thinking and reasoning mathematically ability, (2) the argument logically ability, (3) the communicating mathematically ability, (4) the problem model ability, (5) the proposing and solving problem ability, (6) the representing idea ability, and (7) the using symbol and formal language ability [7, 9].

PISA problems related to students' mathematical literacy have 6 levels [3, 10]. Based on the research of Julie [10], 83.33% students achieved level 4 and 38.89% of students achieved level 5. Furthermore, this study will also look at whether there is a correspondence between the teacher's ability and these results.

2. The Research Methodology

Two of design research characteristics were process-oriented and oriented to usability [11]. A design of a test that adapted from PISA was developed by the researcher. In addition, the teacher's solution for the test was described. Therefore, this study may be classified in the design study.

One of this study aims was to describe the teachers' solution on PISA adaptation tests. The steps that researchers did in this research can be seen in the chart below.

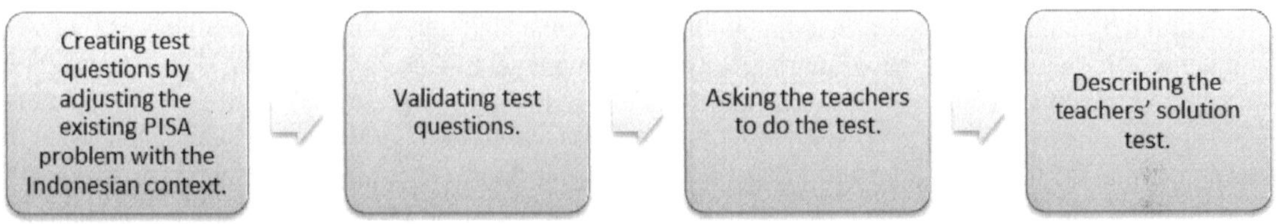

Figure 1. Research step

This test contained 13 questions, namely: (1) 3 questions for quantity, (2) 3 questions for uncertainty, (3) 3 questions for change and relationship, and (4) 4 questions for space and shape. The test was done by the teacher within 90 minutes.

This research used 7 Junior High School teachers as the subject of research in Yogyakarta and surrounding areas. The selection of schools as the subjects of this study was conducted randomly proportional, then the best teachers from each school were selected as research subjects.

3. The Results and Discussion

In this paper, the research results that would be presented were the result of the PISA adaptation test on the uncertainty area [3, 4, 5]. The research result were described as follow:

Problem 1. The charts below show information about exports from Zedland, a country that uses zeds as its currency (see Figure 2).

a. What is the total value (in million zeds) of exports from Zedland in 1998?
b. What is the value of fruit exported from Zedland in 2000?

A. 1,8 million zeds. D. 3,4 million zeds.
B. 2,3 million zeds. E. 3,8 million zeds.
C. 2,4 million zeds.

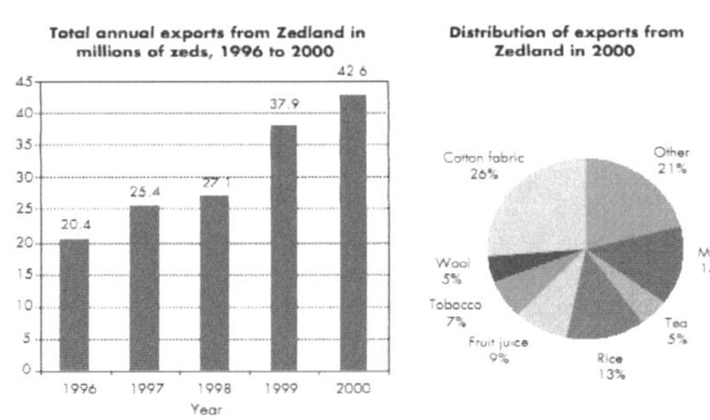

Figure 2. Distribution of exports from Zedland

Teacher's answer for problem 1a:
The total value (in million zeds) of exports from Zedland in 1998 was 27.1. There were seven teachers who made solutions like this solution. Teachers' answer as above could be categorized in level 2 math literacy skills, because they must chose the relevant information about exports in 1998 from the chart.

Teacher's answer for problem 1b:
The value of the fruit is exported from Zedland in the year 2000 $= \frac{9}{100} \times 42{,}6 = 3{,}834$ million zeds$\approx$ 3,8 million zeds. The answer was E. There were seven teachers who made solutions like this solution. Teachers' answer as above could be categorized in level 4 math literacy skills, because they must integrate different representations from chart 1 and chart 2, and then link the information that they need that were data of fruits and data export in year 2000.

Problem 2. Many scientists fear the increasing levels of CO_2gas in Earth's atmosphere caused by climate change. The following diagram shows that the level of CO_2 emissions in 1990 (see bright bars) in some countries (or territories), emissions levels in 1998 (see dark bars), and percentage changes in emission levels between 1990 and 1998 (indicated by arrows and percentage).

a. In the diagram, you can read that in the USA, the increase in CO_2emission level from 1990 to 1998 was 11%. Show the calculation to demonstrate how the 11% is obtained.
b. Mandy analysed the diagram and claimed she discovered a mistake in the percentage change in emission levels: "The percentage decrease in Germany (16%) is bigger than the percentage decrease in the whole European Union (EU total, 4%). This is not possible, since Germany is part of the EU." Do you agree with Mandy when she says this is not possible? Give an explanation to support your answer.

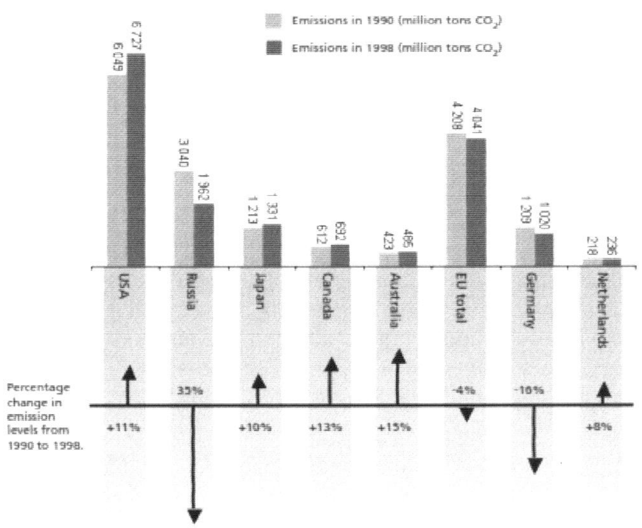

Figure 3. The level of CO2 emissions in 1990

Teachers' answer to problem 2a:
The increased CO2 emission levels from 1990 to 1998 is $= \frac{(6727-6049)}{6049} \times 100\% = \frac{678}{6045} \times 100\% = \frac{67800}{6045}\% \approx 11\%$. There were seven teachers who made solution like this solution.
Teachers' answer as above could be categorized in level 5 math literacy skills, because they must did some following action:

1) Choose data that they need that was the emission data of USA, namely: emission of CO_2 in 1990 was 6049 and emission of CO_2 in 1998 was 6727,

2) Choose the formula that satisfies real situation, and the percentage changing was $\frac{newdata-olddata}{olddata} \times 100\%$.
3) Compare and evaluate the result of the formula that they made with 11%.

Teachers' answer to problem 2b:

1. The teacher did not agree with Mandy opinion because he did not know the percentage of other EU member states. There was the possibility of other EU countries that percentage CO2 emission levels was more than4%. There were two teachers who made solution like this solution.
   Teachers' answer as above could be categorized in level 6 math literacy skills, because:
   a. They could utilize information that was the percentage decrease in Germany was 16% and the percentage decrease in EU was 4%, to be conceptualized.
   b. They also could formulate and communicate their thoughts precisely and in accordance with the original situation that the percentage decrease in EU was 4% not mean that all of the percentage decrease EU countries was 4%.
2. The teacher agrees with Mandy opinion because the percentage of all countries in the EU was 4%.
   There was one teacher who made solution like this solution.
3. There was one teacher who agrees with Mandy opinion without a reason.
4. There were three teachers who did not answer this question.

Problem 3. In a country, a number of polls are conducted to find out support for an upcoming presidential candidate. Four newspapers conducted a poll with results as presented in Table 1.

Table 1. Results of the vote for newspapers.

Newspaper name	Result of vote	Information
Newspaper 1	36.5 %	The poll was conducted on January 6 to 500 randomly selected people from voting citizens.
Newspaper 2	41.0 %	The poll was conducted on January 20 to 500 randomly selected people from voting citizens.
Newspaper 3	39.0 %	The poll was conducted on January 20 to 1000 randomly selected people from voting citizens.
Newspaper 4	44.5 %	The poll was conducted on January 20th to 1000 readers by phone

Which newspaper gave the best predictions about the number of voters, if the elections take place on 25 January? Give three reasons to support your answer!

Teachers' answer to problem 3:

1. The best prediction is Newspaper 3.
   The reason was: (a) the poll was conducted at a time that is closer to the election time, so the owner of the voting option was assumed to be more secure with his choice and ready to make his choice, (b) the samples were more than others, then the poll results will be more valid, and (3) polls conducted on citizens who have voting rights. There were two teachers who made solution like this solution.

Table 2. Teachers' ability in the uncertainty area for the PISA adaptation test.

Problem	Teacher's achievement level	Reasoning	The number of teacher	Percentage
1a	Level 2	The teachers could choose the relevant information about exports in 1998 from the chart.	7	100,00%
1b	Level 4	The teachers could integrate different representations from chart 1 and chart 2, and then link the information that they need that were data of fruits and data export in 2000.	7	100,00%
2a	Level 5	The teachers could did some following action: a. Choose data that they need that was the emission data of USA, namely: emission of CO_2 in 1990 was 6049 and emission of CO_2 in 1998 was 6727, b. Choose the formula that satisfies real situation, and the percentage changing was $\frac{newdata-olddata}{olddata} \times 100\%$, c. Compare and evaluate the result of the formula that they made with 11%.	6	85,71%
	Could not answer		1	14,29%
2b	Level 6	The teachers could did some following action: a. They could utilize information that was the percentage decrease in Germany was 16% and the percentage decrease in EU was 4%, to be conceptualized. b. They also could formulate and communicate their thoughts precisely and in accordance with the original situation that the percentage decrease in EU was 4% not mean that all of percentage decrease EU countries was 4%.	2	28,57%
	Could not be leveled		2	28,57%
	Could not answer		3	42,86%
3	Level 5	The teachers could did some following action: a. Develop models and specifying assumptions for complex situations b. Choose, compare, and evaluate problem solving strategies for complex problems related to the models.	3	42,86%
	Could not be leveled		4	57,14%

2. The best prediction is Newspaper 3.
   The reason was: (a) respondents were selected at random, (2) more respondents, and (3) respondents were selected from citizens who have voting rights. There was one teacher who made solution like this solution. Teachers' answer as above could be categorized in level 6 math literacy skills, because the teachers were able to do the following activities:
   a. Develop models and specifying assumptions for complex situations
   b. Choose, compare, and evaluate problem solving strategies for complex problems related to the models.
3. The best prediction was Newspaper 4.
   The reason was: (a) the result was the largest percentage, (b) the polling time was close to election time, and (c) the number of respondents taken more than newspapers 1 and 2. There were four teachers who made solution like this solution.

Table 2 summarizes the results obtained by the teacher. From the table, it can be seen that the teachers have been able to reach level 2 and level 4. Some teachers have been able to reach level 5 and 6, but not all teachers were able to achieve it. Based on the research of Julie [10], 83.33% students achieved level 4 and 38.89% of students achieved level 5.These results correspond to the results of this research, that were 100,00% of teachers achieved level 4 and 42,86% of teachers achieved level 5. Therefore, to improve the quality of students it is necessary to improve the quality of teachers.

4. Conclusions
From the above explanation, it can be concluded that all teachers could solve one level 2 and one level 4 uncertainty problems. One of the seven teachers has not been able to complete a first level 5 uncertainty problem. Four of the seven teachers have not been able to complete the second level 5 uncertainty problems. Five of the teachers have not been able to complete an uncertainty problem for level 6. This means the teachers were still having difficulties with the PISA problems level 5 and level 6. Therefore, the quality of teachers needs to be improved in order to be able to assist students to reach the maximum level in PISA.

Acknowledgments
Our thanks to the Ministry of the Research, Technology and Higher Education has been funding this research so we could finish this research and present this publication through Penelitian Strategis Nasional Institusi 2018 grant distributed through the Sanata University Dharma.

References
[1] Campbell, Patricia F, et. all. 2014 *Journal for Research in Mathematics Education* **45** 419
[2] Julie H 2017 *International Journal of Science and Applied Science: Conference Series* **1** 55
[3] Julie H, and Marpaung Y 2012 *Widya Dharma* **23**
[4] OECD 2012 *Assessment Framework. Key Competencies in Reading, Mathematics and Science* (Paris: OECD)
[5] OECD *2013 PISA 2012 Results: What students know and can do. Student Performance in mathematics, reading, and science* (Paris: OECD)
[6] Ojose, B 2011 *Journal of Mathematics Education* **4** 89
[7] Christiansen, I B 2006 *Pythagoras* **64** 6
[8] Stacey K 2011 *Journal Mathematics Education* **2** 95
[9] Wijaya A 2016 *IndoMS Journal Mathematics Education* **7** 73
[10] Julie H, Sanjaya F and Anggoro A Y 2017 *AIP Conf Proc* **1868** 050026
[11] Akker, Den J V, Gravemeijer K, McKenney S, and Nieveen N 2006 *Educational Design Research* (New York: Taylor and Francis Group)

Students' creative thinking ability in learning mathematics through learning model of Logan Avenue Problem Solving (LAPS) – Heuristic

U Husna, C M Zubainur and B I Ansari
Syiah Kuala University, Jl. Teuku Nyak Arief Darussalam, Banda Aceh 23111, Indonesia

Email: cutmorinazubainur@unsyiah.ac.id

Abstract. Students' mathematical creative thinking ability has not met the expectation. One of the efforts to develop creative thinking ability is implementing the learning model of LAPS-Heuristic. This paper focuses on the development of students' mathematical creative thinking ability through the LAPS-Heuristic model. The learning instruments used in the LAPS-Heuristic model have been developed through the ADDIE development model. The data were students' achievement on mathematical creative thinking ability obtained from the pilot test. This test is part of the implementation stages of the ADDIE model. The participants were 30 students in grade 8 of a junior high school in Banda Aceh, Indonesia. The results show that the development of students' mathematical creative thinking abilities during the learning with the LAPS-Heuristic model varied. The indicator of fluency and flexibility decreased in the second meeting but was followed by an increasing result in the subsequent meetings. The indicator of originality had an increasing result at the third meeting and remained the same for the following meeting. Meanwhile, the elaboration indicator showed an increasing result in each meeting. Overall, the development of students' mathematical creative thinking abilities could be developed by maximizing the heuristic questions posed to students.

1. Introduction

Mathematical creative thinking is an ability that students should develop through mathematics learning. This ability is a part of high order thinking skill which should be acquired by students to solve mathematical problems. This mathematical skill is necessary to develop the ability to think and to reason [1].

The ability to think creatively includes the aspects of fluency, flexibility, originality, and elaboration [2]. Fluency is the ability to spark different ideas, answers, ways or suggestions for solving a problem or answering a question. Flexibility is the ability to generate various ideas, alternative solutions or questions to see a problem from different perspectives, and to change the way of thinking. Originality is the ability to create a new and unique response. The originality aspect relates to the ability to construct and make an unusual combination of mathematical procedures. Elaboration is the ability to develop ideas or products. It is also concerned with the ability to find out the details of an object, idea or situation to be something more interesting to explore. Those all abilities can be developed through problem-solving activities. The activities can stimulate students to find multiple solutions or ideas [3].

Generally, students' mathematical creative thinking abilities have not met the expectation. This is because the creativity aspect still gets less attention compared to other aspects of learning mathematics. The problems presented in the class are not challenging enough and do not give the students the opportunity to use a variety of strategies. The problems given generally have a single way of completion so that students are not used to doing creative thinking. [2].

It is necessary to develop the ability of students' mathematical creative thinking through learning that can generate students' creative thinking abilities to solve mathematical problems [4]. One of the efforts that can be done is implementing the learning through the model of Logan Avenue Problem Solving (LAPS)-Heuristic. Learning with the LAPS-Heuristic model can create a sense of curiosity and motivation that makes students more creative [5]. The learning with the LAPS-Heuristic model consists of understanding the problem, planning the solution, solving the problem, and reviewing the solutions [6].

The researchers have developed learning instruments with the LAPS-Heuristic model as an effort to help foster students' mathematical creative thinking abilities. The purpose of this paper is to discuss the development of students' mathematical creative thinking abilities during the learning with the LAPS-Heuristic model using the learning instruments that have been developed.

2. Method

The learning instruments with the LAPS-Heuristic model to foster students' mathematical creative thinking abilities have been developed using the ADDIE development model [7]. The learning instruments developed have fulfilled the validity and practicality criteria based on the recommendations from experts, practitioners, and pilot test. This paper describes the development of students' mathematical creative thinking ability through the LAPS-Heuristic model. This model was implemented in five meetings of the class session on proportion topic. Students were taught the concept of proportion in the first meeting and the concept of ratio in the second meeting. Then, the students learned direct proportion in the third meeting, and inverse proportion in the fourth and fifth meetings.

Data on the development of students' mathematical creative thinking ability was obtained through a test of mathematical creative thinking ability given at the end of each class session. This study involved 30 students in grade 8 of a junior high school in Banda Aceh, Indonesia. The data was analyzed by examining student's strategies using a rubric of mathematical creative thinking ability.

3. Result and discussion

The results show that students' creative thinking ability during learning with Logan Avenue Problem Solving (LAPS) –Heuristic model developed variously. Indicators used to measure creative thinking ability include fluency, flexibility, originality, and elaboration [2]. Students meet the fluency indicator if they can provide more than one solution with the same concept. Flexibility indicator is achieved if students can provide multiple solutions involved different concepts. Students meet the originality indicator if the solution given is novel and different from the solution of the majority of students. Students attend the elaboration indicator if their solution is explained in detail.

The percentage of the number of students met the fluency indicator for mathematical creative thinking ability reached 70% in the first meeting, 60% in the second meeting, 73.3% in the third meeting, 66.66% in fourth meeting and 83.33% in the fifth meeting. This finding shows an improvement on the fluency indicator at the first, third and fifth meetings. The students could understand the concept of proportion so that the solutions given were varied and involved the same concept. On the other hand, at the second and fourth meetings, the number of students who met the indicator of fluency decreased from the previous session. The questions for the fluency indicator given at the second meeting were more difficult than the questions presented at the first meeting. Besides, in the first meeting, the students discovered the concept of direct proportion using manipulatives so that the concept they had is more profound than the concept they got in the second meeting. At the fourth

IOP Conf. Series: Journal of Physics: Conf. Series **1088** (2018) 012067 doi:10.1088/1742-6596/1088/1/012067

meeting, the students made mistakes in solving inverse proportion problems; they used the concept of direct proportion to solve those problems.

An example of the student's answer regarding the fluency indicator at the first meeting is shown in Figure 1. On the problem, the student was asked to determine some forms of direct proportion from the statement: "The number of Honda motorcycle is three times more than the number of Yamaha motorcycle with the ratio of 3: 1."

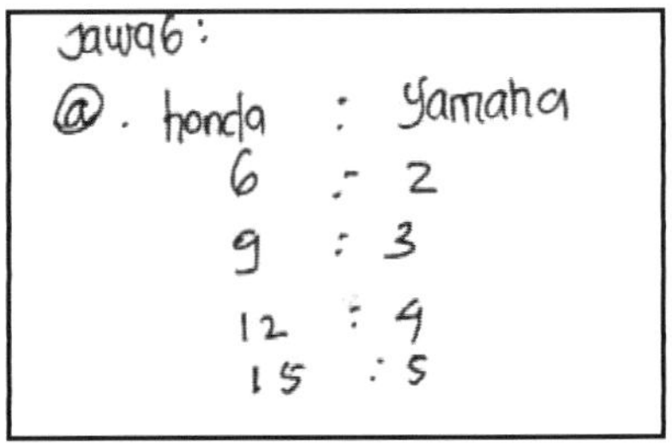

Figure 1. A student's answer that met the fluency indicator.

Moreover, the percentage of the number of students who met the flexibility indicator at the first meeting was 53.33%, at the second meeting was 50%, at the third meeting was 70%, at the fourth meeting was 73.33% and at the fifth meeting was 80%. The students' creative thinking ability for the flexibility indicator had improved. However, it decreased by 3.3% at the second meeting compared to the percentage in the first meeting. This happened because the students were not used to solving problems in different ways. To deal with this situation, teachers should sharpen the heuristic questions further in the next meeting to train students to solve problems in various ways.

Heuristic questions posed by teachers allowed students to be more explorative and creative in solving the problems. Examples of the heuristic questions are "What information do you obtain from the problem given?", "Please write the information you get with your own words," "Is the information provided has been already complete?", "How do you solve the problem?", and "Is there any other ways to solve the problem?".

Andi wants to travel by car. He spends IDR 135.000 for the gas to drive a distance of 50 km. If Andi has IDR540.000 for gas, how many km Andi could travel? Solve the problem using different strategies.

Figure 2. The problem for the flexibility indicator in the third meeting.

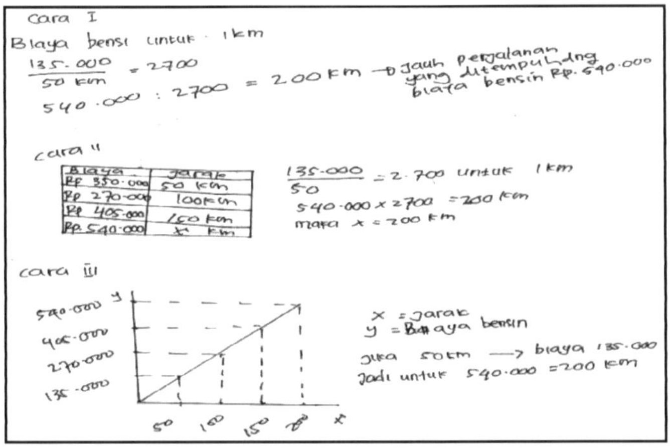

Figure 3. Example of a student's answer that met the flexibility indicator in the third meeting.

Figure 3 shows an example of a student's answer regarding the flexibility indicator. In that problem (see Figure 2), the students were required to provide more than one solution with different concepts. Based on the student's answers, the student could provide more than one solution with different representations, i.e., equations, table, and graph.

Furthermore, the students who met the originality indicator reached 13.33% in the first meeting, 10% in the second meeting, 16.66% in the third meeting, 10% in the fourth and fifth meetings. The students' creative thinking ability improved only in the third meeting. The students who met the originality indicator used various and novel ways of solving problems while the other students generally gave the same solution. The problem with the originality indicator is shown in Figure 4, and an example of a student's answer that met the originality indicator is presented in Figure 5.

> Andi wants to travel by car. He spends IDR 135.000 for the gas to drive a distance of 50 km. If Andi has IDR540.000 for gas, how many km Andi could travel? Solve the problem using a novel strategy.

Figure 4. The problem for the originality indicator in the third meeting.

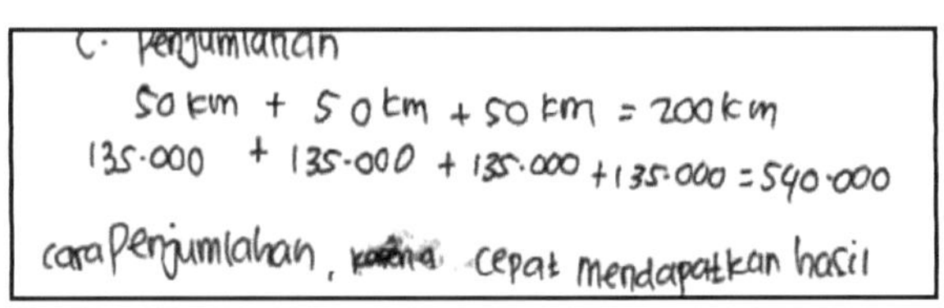

Figure 5. Example of a student's answer that met the originality indicator in the third meeting.

Generally, the students solved the problem (see Figure 4) using the form of logic, proportionality, and summation. Figure 5 is a student's answer that used summation.

Furthermore, the finding shows that the percentage of the number of students who met the elaboration indicator was 40% in the first meeting, 60% in the second meeting, 73.33% in the third meeting, 74.07% in the fourth meeting and 83.33% in the fifth meeting. This percentage shows that the students' creative thinking ability on the elaboration indicator kept increasing in every meeting. The heuristic questions helped them solve the problem in detail. The problem for the elaboration indicator in the fifth meeting is provided in Figure 6, and an example of student's answer is shown in Figure 7. The answers given by the students shows that the students were able to elaborate on their answer began with understanding the problem, planning a solution, solving the problem, and drawing a conclusion.

> Mr. Lukman is a chicken farmer. At the beginning of a month, Mr. Lukman has 480 chickens. The chickens could finish 80 kg of chicken feed within a week. Mr. Lukman buys 8 bags of the chicken feed (1 bag contains 120 kg) in one time. After he sells his chickens, there are 160 chickens left on his farm. Mr. Lukman needs your help to find how many months he could feed the 160 chickens with the amount of the feed left.

Figure 6. The problem for the elaboration indicatorin the fifth meeting.

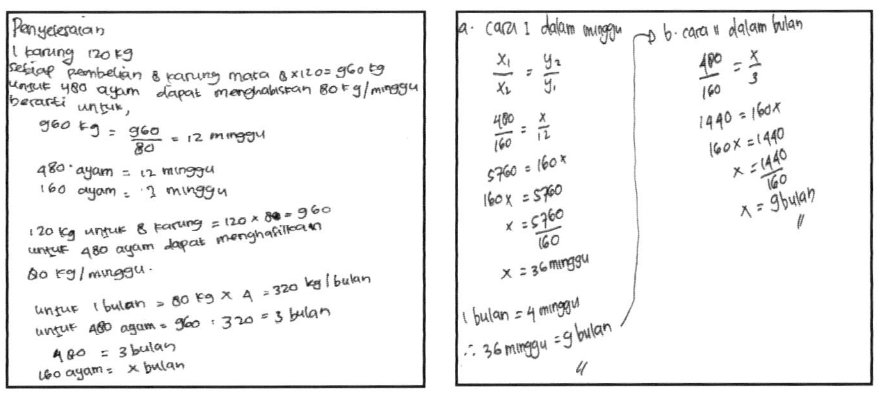

Figure 7. Example of a student's answer that met the elaboration indicator in the fifth meeting.

Overall, the development of students' creative thinking ability varied. At the beginning of the meetings, the students' ability on indicators of fluency, flexibility, and originality had not improved yet because the students were less interested in solving problem-based tasks. Teachers tried to motivate the students to explore multiple strategies to solve problems by asking heuristic questions.Heuristic questions helped students to develop their thinking process to reveal new relationships, see new things and form new combinations of two or more concepts that had been learned before.In this way, the teacherscould help develop students' creative thinking abilities [8].

However, some students still did not follow the steps of problem-solving as required. The students preferred to complete the problems quickly without thinking of other possible strategies. The students needed more time to get used to solving the problems that require multiple strategies. This is because the students were familiar with routine problems that only have a single correct answer. This condition needs to be changed to help the students develop their mathematical abilities [9,10].

Improving mathematics learningcould be done by conditioning the students not only as listeners but also as active students that eager to search for information usefulto solve problems. The students also need to learn the most crucial aspect in solving mathematical problems; that is, the process of finding the right solution involving the higher order thinking process [11] through problem-based tasks. The problem-based tasks allow the students to generate curiosity and motivation, to develop creativity, to analyze problems, to evaluate the results obtained, to get new solutions, to found and use various strategies, and tointegrate new and prior knowledge.

4. Conclusion

Based on the results of this study, it can be concluded that the creative thinking ability of junior high school students through Logan Avenue Problem Solving (LAPS) -Heuristic model varied for each indicator. The development of students' creative thinking ability during learning through the LAPS-Heuristic model for the indicator of proficiency decreased by 10% in the second meeting, but it showed a progress in the next meeting. For the indicator of flexibility, there was a decrease of 3.3% in the second meeting while in the next meeting, the students experiencedan improvement. For the indicator of originality, there was an improvement in the third meeting. The improvement also occuredin each meeting forthe elaboration indicator. The development of creative thinking abilityin solving problems is also supported by the heuristic questions. Therefore, it can be concluded that the LAPS-Heuristic model can be used to improve students' creative thinking abilities.

References

[1] Kwon O N, Park J S and Park J H 2006 Cultivating divergent thinking in mathematics through an open-ended approach *Asia Pacific Education Review* **7**51
[2] Munandar U 2004 *Pegembangan Kreativitas Anak Berbakat* (Jakarta : Rineka cipta)
[3] Silver E A 1997 Fostering creativity through instruction rich in mathematical problem solving and thinking in problem posing *ZDM* **2** 1615
[4] Coelho A and Cabrita I 2015 A creative approach to isometries integrating geogebra and italic with 'paper and pencil' environments *Journal of the European Teacher Education Network* **10** 71
[5] Anggrianto and Desi N 2016 Improving critical thinking skills using learning model Logan Avenue Problem Solving (LAPS)-Heuristic *Journal of Education and Practice* **7** 67
[6] Shoimin A 2014 *68 Model Pembelajaran Inovatif Dalam Kurikulum 2013* (Yogyakarta: Ar Ruzz Media)
[7] Branch R M 2009 *Instructional Design: The ADDIE Approach* (USA: Springer. Science Business Media)
[8] Siswono T Y E 2010 Leveling students' creative thinking in solving and posing mathematical problem *IndoMS. J.M.E* **1** 17
[9] Polya G 1985 *How to solve it : A new aspect to mathematical method* (Princeton: Princeton University Press)
[10] Švecová V, Rumanová L and Pavlovičová G 2014 Support of pupil's creative thinking in mathematical education *Procedia-Social and Behavioral Sciences* **116** 1715
[11] Bey A 2014 Developing skills resolution mathematical primary school students *International Journal of Education and Research* **2** 601

IOP Conf. Series: Journal of Physics: Conf. Series **1088** (2018) 012068 doi:10.1088/1742-6596/1088/1/012068

The improvement of mathematics teachers' Pedagogical Content Knowledge (PCK) through mentoring

T Zubaidah[1], R Johar[1] and Suparno[2]
[1]Syiah Kuala University, Jl. T Nyak Arief, Darussalam, Banda Aceh, 23111, Indonesia
[2]SMA Negeri 2 Banda Aceh

E-mail: tuti.zubaidah@fkip.unsyiah.ac.id

Abstract. This study aimed to examine the improvement of mathematics teachers' Pedagogical Content Knowledge (PCK) through mentoring as the foundation to develop a mentoring model to enhance teachers' PCK. This study employed Research and Development method consisting of three phases: the preliminary research, prototyping and assessment phase. At the preliminary research, the teachers and mentors prepared the learning. The initial observation indicated that the initial PCK of teachers was lacking and therefore mentoring was needed to improve teachers' PCK. The instrument of PCK used was validated for its language to assist the observers in assessing teachers' PCK. In the following phase, the prototyping, the mentoring model was designed and then implemented for two teachers in one of the senior high school in Banda Aceh, Indonesia. The results showed that the instrument used to assess teachers' PCK was valid to be used and the mentoring model enhanced teachers' PCK. Thus, this study should be continued in the assessment phase. In the next phase, the mentoring model should also be validated and implemented for more teachers with a more intensive observation.

1. Introduction

Many factors contribute to improving the quality of education. One of which is the role of teachers in the learning process. Teachers are demanded to create a learning environment engaging students which in turn will enhance students' understanding of the subject matter. Teachers are expected to have a pleasant personality to understand students as well as to have the skills to present the teaching materials so that they can easily understand the teacher's explanation [1].

Teachers will perform the most effective way of teaching when: they are equipped with excellent pedagogy knowledge of the learning materials; realize that they are not the primary source of knowledge, recognize that students' thinking is full, rich, and varied of knowledge so that teaching means connecting the students' existing knowledge to the current learning materials [1] as well as utilize students' perspective and experience to create a favorable classroom climate to motivate students [2].

An excellent teacher should not assume that they are the perfect learning resource as students have background knowledge that needs to be explored and utilized during the learning [3]. Teachers should also appreciate students' opinions and works so that students are more confident and acknowledged in solving the problem [1]. Thus, the teachers should be not only able to master the materials but also the pedagogical strategies to understand students to prepare the conditions enabling students to learn successfully. Shulman [1] defined this capability as the Pedagogical Content Knowledge (PCK).

Pedagogical knowledge involves methods and processes of teaching including the understanding of classroom management, tasks, lesson planning and students' learning. Pedagogical knowledge should at least consist of the understanding of the insights or educational foundations, students, curriculum development/syllabus, learning design, the implementation of educational and dialogic learning, the use of technology in learning, the evaluation of learning outcomes, and the students' development to actualize various potentials of the students [1]. When a teacher discovers the materials that students do not understand, they should be able to modify their teaching strategies including the understanding of the subject matter and its delivery or adjust the teaching activities based on students' achievement [4].

Pedagogical content knowledge (PCK) is teachers' ability to understand the concepts, problems, or emerging issues adapted based on students' talents, interests, and abilities. Teachers' mastery of concepts influences their students' mastery of concepts. Most of the concepts grasped by students are from teachers [5]. Learning will be more effective, efficient, and well-targeted if the teacher delivers it by implementing the appropriate teaching method considering both the physics and mental of the students.

Therefore, the competencies of teachers include The Content Knowledge, Pedagogical Content Knowledge (PCK), Curriculum Knowledge, General Pedagogical Knowledge, Knowledge of Learners, Knowledge of Educational Contexts, and Knowledge of Educational Ends [1]. Among those competencies, teachers' PCK play a crucial role as it reflects on teachers' special skill in teaching covering both the Pedagogical Knowledge and Content Knowledge.

One of the abilities in PCK concerns the strategies to address students' misconception indicated when students systematically use inappropriate rules or appropriate rules, yet they are not in line with the expected learning objectives. For example, the study conducted by [6] found that when students are taught how to quickly memorize multiplication of 10 by merely adding zero to the unit, it is likely that students will have a misconception when they experience multiplication of fraction or decimal in which zero is after the decimal point.

Students' misconceptions may occur during the lesson due to the basic concepts of mathematics formed in students experiencing the misconceptions or the misconception arising from delivery of teachers that contradicts the existing cognitive structure in students. Studies revealed that teachers who have excellent PCK would be able to anticipate misconceptions emerged in students to enhance students' learning outcomes [6, 7].

The results of a preliminary survey conducted by the researcher to two mathematics teachers in one high school in Banda Aceh showed that the PCK of teachers was generally inadequate. The average PCK of targeted teachers in the initial survey was 29 out of 60. This was indicated by teachers' less effort in guiding the students to discover their knowledge during the learning. The teachers delivered the lesson by following these teaching procedures: (1) conveying the concept or procedure of problem-solving, (2) explaining some examples of problems, (3) asking students together with the teacher to solve problems with similar contexts to teacher's previous example, (4) asking students to solve the practice problems independently. These procedures assisted students in solving the problems successfully; however, students found it challenging when they faced problems in different contexts. The instrument used for the initial survey was adapted from Monica and Helen [8].

PCK is both the understanding of concepts and the process of teaching including the appropriate method to teach and formulate the concepts to be understood by students [9]. To be successful in teaching, teachers should combine both Pedagogical Knowledge and Content Knowledge into PCK where the component of materials presented based on the student's previous performance. Basically, PCK can be defined as the teacher's effort to teach the subject matter to the students. Having an excellent understanding of PCK, teachers will find different methods to effectively teach the subject matter based on the characteristics of the students in the classroom [10].

A teacher cannot improve PCK without the help from others. Teachers need mentoring from experts to develop their PCK during the teaching and learning, starting from the early preparation, during and after the lesson (to obtain feedback concerning the effectiveness of the lesson conducted). The mentoring is not meant to patronize the teachers instead the mentors act as partners to provide feedback and observe the components of PCK during the lesson [11]. Mentoring is an activity to assist each other to be responsible for the development of their abilities to be assessed afterward.

There are nine fundamental principles for the effective mentoring: a) sharing knowledge, attitudes, skills, and experience are the main change expected in mentoring activities; b) the expected outcome of mentoring is the participants' positive self-change; c) the participants specifying the sections that will be the main issues to be developed and changed; d) trainers or mentors are not only the instructors but also facilitators in the learning process and self-development of participants. e) the mentoring results consists of the self-change of the participants; f) the framework of the change process facilitates the expected direction of change; g) skills obtained include the development of knowledge, self-potential and the impact of emerging changes; h) the quality of the mentors is appropriate to support the participants to change; i) the feasibility of the mentors' ability should be able to improve coaching and mentoring of participants [12].

The mentoring occurred when there is a mutual relationship between the mentor and the participants to improve the participant's potential. The core of mentoring is an open relationship that will facilitate and lead to the career development. The mentoring process occurs when the experienced individuals organize regular meetings and discussions, determine the interests of the participants and support the development of the skills perceived to be lacking [12].

There are some mentoring models include: 1) The Grow model, seeking to develop the skills and self-development goal expected by the participants. 2) The Skilled Helper, assisting the participants to identify the expected results and how to achieve them. 3) The Inner Game, motivating participants to think positively. 4) Non-directive, providing the right conditions enabling the participants to become trainers for themselves. 5) Co-Active Coaching, assisting in designing the needs of participants by listening, intuition, curiosity, trying to implement and self-management. 6) Solution-focused, finding what has not been done and improve it. 7) Team coaching, creating a mentoring team with similar interests [13]. Mentoring is expected to enhance teachers' PCK. Mentoring activities are an important element of teachers' capacity building and the development of teachers' strategies for self-reflective [14].

Based on the review previously mentioned, there has no mentoring model to improve teachers' PCK. This study aimed to describe the development of mathematics teachers' PCK through a mentoring model.

2. Method

This study employed a Research and Development method developed by Plomp [15] consisting of three phases. However, only two phases were used in this research, namely: the preliminary research and prototyping phase. The preliminary research phase involved agreeing on the mentoring model preferred by the teacher before, during and after the learning. The developed mentoring model was adapted from the stages of lesson study (i.e., plan, do, see) by involving two observers from peers in the field of mathematics. This phase also dealt with the content analysis involving the analysis of the learning materials based on the senior high school curriculum, standard competences, basic competencies, as well as the analysis of essential concepts considered to be difficult to understand by teachers or students. Besides, the literature review was also conducted to review of relevant research concerning the common misconceptions experienced by students on the topics to be taught and examine the supporting literature on teaching materials and current issues related to the workshop concerning the understanding of PCK by the mentor lecturer.

The phase prototyping is carried out following the preliminary phase. This phase involved adapting the instrument assessing the PCK of teachers presented by Monica and Helen [8] and validating the instrument to evaluate teachers' PCK validated by three mathematics education experts. It also covered the analysis of the early PCK of mathematics teachers participated in the research as an initial step to improve teachers' PCK through mentoring. Also, this phase also concerned on observing the classroom to examine the teacher's PCK improvement at the first lesson; recording and producing the first lesson video and reflecting on the first lessons conducted. Finally, this phase also repeated the

IOP Conf. Series: Journal of Physics: Conf. Series **1088** (2018) 012068 doi:10.1088/1742-6596/1088/1/012068

preliminary research and prototyping phase for the second and third lessons as well as developed a mentoring model shown in Figure 1. This model was adapted by merging the two-phase developed by Plomp [15] and the phase of lesson study.

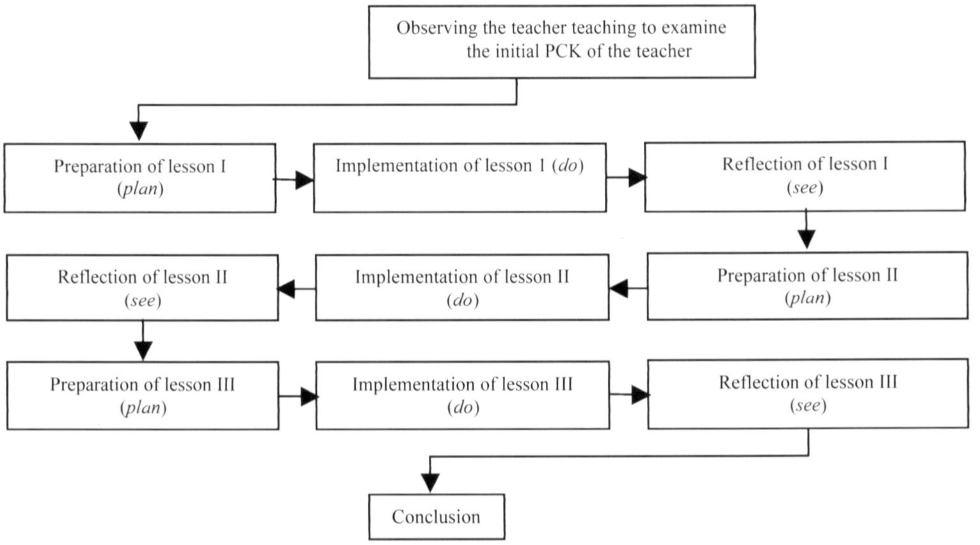

Figure 1. The flowchart of research to improve the PCK of teachers through mentoring.

The assessment instrument to evaluate the PCK of teachers used in this study adapted from Baker's research [8] consisting of three abilities. First, the PCK includes eight indicators including a) teaching strategies, b) thinking of the students, c) thinking to overcome students' misconception, d) the demands of the cognitive level of the students' task, e) the conformity of presentation with the characteristics of concept, f) knowledge of learning resources, g) knowledge of curriculum and h) the relevance of teaching materials with other relevant materials and their implementation. Second, the ability to understand the concepts related to the pedagogic ability, consisting of five indicators, includes: a) excellent understanding of the fundamentals of mathematics, b) describing the content of the lesson into the core components, c) mathematical structure and connections, d) procedural knowledge and e) methods of mathematics problem-solving. Third, pedagogical skills related to conceptual understanding ability, consisting of three indicators, includes: a) learning objectives, b) striving and maintaining the focus of students, and c). classroom management.

The teachers involved in this research were those being observed for their PCK in the preliminary research. They were two mathematics teachers in one of the senior high school in Banda Aceh, Indonesia. Both teachers have been teaching for more than 20 years and experiencing teaching at different Year level. G1, a female teacher, taught in Year 10 and G2, a male teacher, taught in Year 11. Both G1 and G2 worked at the same school, and both treated equally in the study. They participated in the workshop and were mentored by two mathematics education lecturers. The mentor lecturers assisting in the research were two lecturers from the teacher training and education faculty of Syiah Kuala University (FKIP Unsyiah) as well as the authors of this paper.

The improvement of teachers' PCK for one lesson was assessed by computing the average assessment of teachers' PCK based on two teacher observers and one mentor lecturer. Next, the average of the score obtained and the score of teachers' PCK based on watching the learning video of the first lesson. The same procedures were conducted for the second and third lesson. Once all data from the three lessons gathered, the analysis of the improvement of teachers' PCK was conducted.

3. Result and discussion

Some preparation before the research was conducted in the preliminary research stage including 1) developing the lesson plan and student worksheet together with the mentor lecturers according to the curriculum and characteristics of the students in the classroom; analyzing the teaching materials enabling students to do the task independently and minimizing misconceptions as well as discussing the possibilities of difficulties experienced by students during learning as well as how to overcome them, and preparing a PowerPoint presentation to be approved by the mentor lecturers; 2) predicting misconceptions that may occur during the learning and how to overcome them, conducting a workshop concerning the understanding of PCK by mentor lecturers for teachers involved in the research, organizing workshop about the utilization of multimedia in learning as per teacher request. Multimedia has also introduced: the use of more advanced PowerPoint, and innovative learning in teaching mathematics.

In the Prototyping phase, the research instrument was adapted to assess and observe the PCK of teachers as discussed in the research findings [8] by adapting the indicators of PCK to ease the process of the assessment conducted by the observer during the study, validating the instrument to assess the PCK of teachers. The validation results showed that the instrument was eligible to be used with a minor revision in the description of the PCK indicators (2a, 2b, 2e, and 3b) which were initially less understood leading to the observer being doubtful in assessing the PCK of teachers due to the non-operational words. The revision results are presented in Table 1:

Table 1. The observation sheet of teachers' PCK before and after the validation.

No	Before Validation	After Validation
II	The ability to understand the concepts and its relation to the pedagogical ability	
	a. Showing a high level of understanding of mathematical aspects	a. Demonstrating a deep and thorough understanding of mathematics aspects
	b. Elaborating the content of learning into fundamental components so that they are easier to be understood and implemented	b. Identifying the essential elements of mathematics concepts critically so that they are easier to be understood and implemented
	e. Using a method (a certain way) in solving mathematics problems	e. Mastering the steps of solving mathematics problems systematically and accurately (the ability to understand mathematics concepts is not taken into account)

The mathematics teachers' initial PCK was examined in the study, the PCK of G1 and G2 was 33 and 24 respectively (out of 80). The initial PCK reflected the teachers' ability in the initial observation before the study conducted. This study confirmed that teachers' PCK were improved in the first, second and third lessons. The improvement is presented in Table 2.

Table 2. The teachers' PCK (G1 and G2).

Lesson	PCK score	
	G1	G2
Initial observation	33	24
The first lesson	37	28
The second lesson	47	31
The third lesson	51	37

Even though the PCK of both teachers were improved, the PCK of G1 increased higher than the other. Some factors may contribute to the differences including the fact that G1 prepared the presentation of the concepts that were well–structured and easier to understand leading to the students learn and independently work on the tasks without asking many questions related to the task given by the teacher (indicator 2.b). On the other hand, many students of G2 waited for the teacher to come to them and ask about the parts they did not understand. GI, at the beginning of the lesson, have been explained in detail all the students' needs to solve the problems.

When G1 gave a problem from the example, students could solve the problem provided (indicator 1.b). In contrast, the students of G2 experienced obstacles when the problems are given was slightly different from the example. G1 prepared the lesson without any significant barriers (indicator I.a). G1 also tried to predict the possible misconception (indicator 1.c), while G2 were doubtful of the need to predict misconception in detail considering the lack of students' enthusiasm to learn mathematics.

In each lesson, the number of students in G1's classroom who were focused and engaged in the learning rose (indicator 3.b), and fewer students violated the rules set together (indicator 3.c). On the other hand, students in G2's classroom did not increase or remained stable. The fact that the lesson was not scheduled at the beginning of the school hours leading to students being less focus in learning may contribute to this issue.

Among other factors supporting the improvement, the teacher's PCK was that the mentor lecturers being familiar with both teachers before the research activities carried out so that the reflection can easily be accepted by teachers and communication during the mentoring was intimate and nurturing. The school involved in this study was also one of the partner school of Syiah Kuala University, and both teachers (G1and G2) were actively involved in some activities involving the university such as being the mentor teacher for the student teaching internship, and the teacher professional education program and therefore the relationship of mentor lecturers and teacher became familiar. Only the teachers who had a high commitment to make changes were selected to participate in the study. Thus, the mentoring process run well without any significant constraints and the teachers voluntarily spent time after the lesson to consult with mentor lecturers. The mentoring conducted benefited the teachers to prepare the lesson optimally. The indicators on the PCK instrument, the preparation for teaching, monitoring the implementation of the learning process, and providing feedback to the teacher after the lesson, were well implemented. Besides, the collaboration between teachers and lecturers run harmoniously without any significant pressure. Teachers and mentor lecturers agreed on a schedule of mentoring by respecting the individual schedule.

Although the PCK of both teachers were improved, the scores were not satisfactory. This is because teachers were accustomed to teaching using their old method and it was difficult to change in a short time (in three lessons and mentoring). The teachers involved in the study were rarely invited by the government to attend training for improving their teaching performance. According to Tjeerd and Nienke [15], one of the reason contributes to the low teaching performance of teachers is teachers' less participation in the training leading to their pedagogical knowledge being outdated and less motivated for self-development. A professional teacher (supported by teacher career development workshops) should be concerned with students' thinking and prepare challenging problems [15]. Students taught by professional teachers usually have an integrated, more coherent, and higher abstraction of concepts understanding [16,17].

Based on the discussion with both G1 and G2, teachers' motivation to change was lacking because of the overcrowded non-teaching and teaching scheduled leading to teachers had less time to prepare the lesson. This finding is in line with the study conducted by Patricia and Masako [18] found that teachers' motivation is the main predictor of students' interest. Highly motivated teachers will have an impact on how well the lesson prepared and the increase in students' learning outcome.

4. Conclusion

The results of the study show that the development of mathematics teachers' Pedagogical Content Knowledge (PCK) through the mentoring program is increased. However, in term of the indicators of PCK, some indicators showed no improvement or declining in each lesson (for the three lesson). The PCK scores of G1 for each indicator remained stable include: the cognitive level of the problems given by G1 in each lesson was classified as C2 and C3 level (indicator 1d), G1 always had an excellent score for the knowledge of the curriculum (indicator 1g), G1 mastered the mathematics concepts of all topics taught (indicator 2a), G1 always presented the objectives of learning at the beginning of the lesson prepared on PowerPoint slide (indicator 3a).

The PCK scores of G2 that remained stable were: the learning resources being the same in each lesson, the textbook, and PowerPoint (indicator 1.f), always linking the relevant concepts (indicator 2.c), neither oral or written teaching objective presented (indicator 3.a), no change in the strategies to engage students (indicator 3.b). One indicator of G2's PCK decreased, the complexity of the problem given to the students decreased (indicator 1d) due to time constraint and the increasing level of the subject matter difficulty towards the application of concepts in the field other than mathematics.

It can be concluded that generally the mentoring improves teachers' PCK. Thus, this research should be continued to the assessment phase in which the mentoring model will be re-validated and implemented on a broader scope of teachers following by intensive observation.

Reference

[1] Lee S 1987 Knowledge and teaching: Foundation of the new reform *Harvard Educational Review* **1** 57

[2] Bill C and Michael G 2012 *Teaching Mathematics: A guide for postgraduates and teaching assistants* (UK: The Maths, Stats & OR Network)

[3] Beth H, Randall W and Sarah N 2013 The impact of social interaction on student learning *Reading Horizons Magazine* **52** 375

[4] Pavinee S, Jari L and Kalle J 2013 Primary school teachers' interviews regarding Pedagogical Content Knowledge (PCK) and General Pedagogical Knowledge (GPK) *European Journal of Science and Mathematics Education* **1** 84

[5] Chien L S, Rohaida M S and Siow H L 2015 The knowledge of teaching Pedagogical Content Knowledge (PCK) *The Malaysian Online Journal of Educational Science* **3** 40

[6] Romelia V M, Hal A and John M 2003 *Mathematics Pedagogy and Content in a Blended Teacher Education Program* (California: Teacher Education Quarterly Magazine)

[7] Baumert J, Kunter M, Wermer B, Martin B, Thamar V, Alexander J, Uta K, Stefan K, Michael N and Yi M T 2017 Teachers' mathematical knowledge, cognitive activation in the classroom and student progress *Sage Journal* **47** 133

[8] Monica B and Helen C 2006 Pedagogical content knowledge for teaching primary mathematics: A case study of two teachers *Australian Research Council* **1** 60

[9] Maher Z H 2005 Teacher pedagogical constructions: A reconfiguration of pedagogical content knowledge *Teachers and Teaching: Theory and Practice Journal* **11** 273

[10] Sonia G 2017 *Pedagogical Knowledge and The Changing Nature of The Teaching Profession* (Paris: OECD Publishing)

[11] Gillian J, Jos F and Andrew D 2018 *Mentoring Handbook* (Scotland: Scotland's Rural College (SRUC))

[12] Mary C and Julia P 2012 *Coaching and Mentoring at Work* (England: Open University press)

[13] Jessica J, David L and Annette F T 2006 *The case for Coaching: Making Evidence Based Decisions* (London: Chartered Institute of Personnel and Development (CIPD))

[14] Martyna W P 2012 The role of mentoring in organizational *Management Knowledge and Learning International Conference* 627

[15] Tjeerd P and Nienke N 2013 *Educational Design Research: An Introduction* (Netherland: SLO)

[16] James K, Pamela S and Linda B 2014 *Effective Teaching* (UK: Education Development Trust)

[17] Sahin D and Dilek T 2017 Examination of mathematics teachers' pedagogical content knowledge of probability *Malaysian Online Journal of Educational Sciences* **5** 16

[18] Patricia F C, Masako N, Toni M S, Lawrence M C, Darcy L C, Amber H R, Anne A, Jill N D, Toya J F, Matthew J G and Youyoung C 2014 The relationship between teachers' mathematical content and pedagogical knowledge, teachers' perceptions and student achievement *Journal for Research in Mathematics Education* **45** 419

Cartography in designing digital map using Adobe Flash CS6

Y Miaz, Y Helsa, Desyandri and R Febrianto

Universitas Negeri Padang, Jl. Prof. Dr. Hamka, Air Tawar, Padang, Indonesia

E-mail: yalmiaz@gmail.com

Abstract. This study created an IT-based learning model in designing digital atlas for elementary students using Adobe Flash CS6 on social science learning. The purpose of this media was to motivate the students to study social science thoroughly, anytime and anywhere without being limited by space and time. The reasons to choose Adobe flash CS6 media in this research were because it is accessible, attractive, update and can be animated. Through digital map media, it expects that social science learning in elementary school will be more exciting and social science learning objectives can be achieved well. This research used a development research method which included: analysis, design, and evaluation. The developing procedures for IT-based learning media divided into three stages such as analyzing social science curriculum for elementary school about natural appearance, then designing digital map media using CS 6 software, and the final step was validation process. It obtained criteria that stated good product and user effectiveness equal to 83, 5. Thus, the result showed that digital maps appropriately used in social science lessons for elementary school students.

1. Introduction

Learning is the process of communication between students, teachers, and teaching materials. The conversation in the form of the learning process will not run correctly without the help of a medium to convey information or messages. The structures of stimulus used as a medium are human relationships or interactions, reality, moved or unmoved pictures, written and recorded sounds. The instructional media is one of the factors for learning success [1, 2]. The students can be motivated, actively involved physically and psychologically by using the media. It also maximizes all the senses of the students in learning and makes learning more meaningful [3].

Describing the natural features of the surrounding environment is one of the materials that the students learn in grade IV of the elementary school in social science learning. The teachers in social science lessons are expected to provide knowledge, attitudes, and skills to the students. For the achieved learning objectives, the teachers should be able to create an exciting learning process and provide a convenience lesson atmosphere for the students. One of the components that can help the achievement of social science learning objectives is the use of instructional media.

Further, the development of information technology has influenced the use of various types of media as a learning tool. One of them is the use of interactive multimedia in social science learning. Interactive multimedia is a technology that combines various media sources such as text, graphics, sound, video, and so forth that is conveyed and controlled by an interactive computer system. Additionally, it stated that multimedia increases the students' success, positively affects students' attitudes, and make lessons more enjoyable and understandable [4].

In fact, the development of technology has not been in line with its application in the field of education, especially on social science subjects in elementary school. The use of interactive multimedia in social science lessons in elementary school is still less to be available. Problems found in learning social science related to instructional media used in the material about identifying natural features in the environment based on the observations and interviews conducted by the researchers at SDN 22 Alai, Sutera sub-district, Pesisir Selatan, West Sumatera on Wednesday, 05 April 2017 until Wednesday, May 03 2017. Firstly, the teachers used unmatched instructional media like map picture, atlas, and artificial earth to help the students in understanding the lessons about the natural features. Secondly, the teachers did not follow the development of science and technology in the instructional media development. Thirdly, map media that was used by the teachers had not shown the form of natural features in the environment around the students. Lastly, the lack of interactive multimedia tools in the learning is about the map that can display the way of physical characteristics in the environment around the students. In fact, these matters affected the students: (1) it was difficult for the students to understand the lessons (2) the students seemed to be less active in learning, (3) the students were not able to express their ideas and combine the senses in learning activities, and (4) the lessons were not meaningful for the students.

The results of other studies show that the problems found in elementary schools were the lack of instructional media that facilitate the teachers in the social science learning process in grade IV. The problem faced by the schools in Indonesia today was lack of availability of social science learning software for elementary level in the form of interactive multimedia [5]. The development of interactive multimedia-based instructional media about the natural features of the surrounding environment for social science lessons of elementary level in Indonesia is still not widely developed. Some studies about interactive map multimedia have not discussed the natural features of the surrounding environment yet. The research results showed that the media still present two-dimensional map (2D) and the media have not been adequately equipped by the route of the natural features of Payakumbuh city as a whole. It was only limited to tourist attractions in the form of video [6-7].

One of the innovations that can help in overcoming the problems above is by developing an interactive multimedia-based map media. Further, alternative media that can be used to present the material of the surrounding natural environment features in the form of fact, data, image, photo, and video components by using Adobe Flash CS6 application to create an interactive multimedia-based map media are needed. The reason for using Adobe Flash CS6 application in making the interactive multimedia map in this research is because this software can combine interesting images, audio, and video. The Adobe Flash CS6 application is one of the programs devoted to designers or programmers to produce animation [8].

This study aims to produce valid, practical, and useful interactive multimedia-based map media in learning social science using Adobe Flash CS6 application. Theoretically, the benefits of this study are to provide an alternative instructional press for teaching natural features of the surrounding environment material, and practically it helps educators, learners, and readers. Based on the above explanation, the author wants to research with the theme is the development of interactive multimedia-based map media in learning social science for elementary level.

2. Method

This type of research was development research. It is a research method used to produce a particular product and test the effectiveness of the product [9]. This research used 4-D development model (Four D) which includes defines, design, develop, and disseminate. The advantages of the 4-D model (Four D) are procedures or steps in the development are more detail, more appropriately to be used as a basis for developing learning tools, the description seems to be complete and systematic, and involves the expert in its development, so that prior to the field trials the learning device had been revised based on the expert's assessment, suggestion, and input [10].

This research uses qualitative and quantitative approaches by describing the data of validity, practicality, and efficacy test results. The first data in the form of instructional media validation were given by validators who were expert in media, material, and language. The second data obtained on the trial test. In this experiment, the data collected in the form of the practicality test result (teacher and student response) and the efficacy test result (learning result).

Product trials were conducted on Tuesday, February 6, 2018, at SDN 22 Alai sub-district of Sutera Pesisir Selatan district on social science subject. The study time allocation was 4 x 35 minutes (2 x meetings). The target of this research was to assist the teacher and grade IV elementary students in social science learning. The research subjects were 20 grade IV elementary students in SDN 22 Alai, District Sutera, Pesisir Selatan Regency, West Sumatra. This research involved validators who were lecturers of media, language, material, teachers, and students.

The research data were obtained in the form of validity assessment, practicality assessment, and student learning outcomes. The data were collected using observation and test techniques. The data collection instruments were collected using validity questionnaires (media, material and language experts), practicality test questionnaires (teachers and students), evaluation tests (cognitive aspects), and observation sheets (effective and psycho-motor aspects). The data collection instruments were tools or facilities used by the researchers in collecting the data so that the work was more comfortable and the result was better to be more accurate, complete, and systematic. So that it was processed and analyzed efficiently [11]. The data analysis techniques in this study used qualitative and quantitative data analysis. Qualitative data on the quality of the interactive multimedia was obtained from the input of suggestions and criticism of the subject teachers, the media lecturers, and the students as subjects of the trial. Then, the data were collected and concluded to improve the products developed. While the quantitative data obtained from the questionnaires conducted for the subject teachers, the media lecturers, and the students. Moreover, quantitative descriptive data were presented through tables.

3. Results and discussion

Following the objectives and development procedures used in the research that had been conducted by the researchers, the data results obtained as follows.

3.1. Defining stage (Definition)

The needs analysis was done with the purpose to find out the needs, advantages, and disadvantages of instructional media that have already existed and used in learning social science for grade IV elementary level on the material about natural features. The researchers did the needs analysis by observing SDN 22 Alai Sutera sub-district on Wednesday, April 05 2017 until Wednesday, May 03 2017 and collected information from various sources such as books, journals, expert opinions, and others. The result of the needs analysis of instructional media is that the researchers found four necessary items. First, the media used by the teachers in learning social science grade IV of the elementary school on the subject matter of natural environment features were about less appropriate. Second, the existing instructional media had not yet presented the natural environment features. Third, the lack of availability of interactive multimedia-based instructional media is in learning social science in elementary level. Last, interactive multimedia-based instructional media can help to improve learning outcomes and stimulate the students' learning motivation.

The curriculum used in the development of instructional media in this research was the education unit level curriculum (KTSP). The purpose of the curriculum analysis was to know the description of the instructional media design that was suitable to be developed following KTSP. In developing this interactive multimedia-based map media, the researchers first conducted curriculum analysis on social science subject for grade IV elementary school, including competency standard (SK), basic competence (KD), indicators, learning objectives, and materials.

The competency standard of social science subjects in grade IV of the elementary school in this research was to understand history, the features of nature, and the diversity of tribe in Regency/city and province. The basic competence of social science subjects in grade IV of the elementary school in this research was to describe the features of nature in the district/city and provincial and its relationship with social and cultural diversity. The indicators for the development of interactive multimedia-based map media were identifying the facts form of the natural environment features, explaining the natural and artificial environments, defining the natural climate features and the surrounding environment, describing the values contained in the characteristics of nature, showing the attitude of preserving the natural environment, and using interactive multimedia as instructional media. The learning objectives of this research were (1) by observing the images in interactive multimedia, the students were able to identify the facts of natural features form correctly, (2) by reading the material in interactive multimedia, the students were able to explain the forms of the elements of the natural environment, (3) by observing maps in interactive multimedia, the students were able to solve the natural and artificial environment in the surrounding environment correctly, (4) by discussing with group members, the students were able to explain the values contained in the natural features well, (5) by following learning activities, the students were able to demonstrate the attitude of preserving the natural environment well, and (6) by using interactive multimedia in learning, the students were able to use interactive multimedia well. The social science teaching materials for grade IV in the media that developed in this study was the features of the surrounding natural environment.

The researchers analyzed instructional materials by referring to the syllabus of social science subject for grade IV elementary school on the material about the features of nature in the district/city and provincial, and its relationship with the diversity of society and culture. In the development of interactive multimedia-based map media, the researchers focused on the material about the features of the natural environment around Pesisir Selatan district, West Sumatra. The result of the material analysis in the instructional media that the researchers developed had been per the syllabus and the characteristics of the learning materials on education unit level curriculum (KTSP). The material presented in this instructional media contained teaching materials about the features of natural land, waters, inherent nature, artificial nature, and natural landscape map in the form of the video of Pesisir Selatan district, West Sumatera.

The researchers did media analysis by doing observations in schools, interviews, and discussions with the teachers grade IVA SDN 22 Alai Sutera district, the experts, the lecturer majoring Primary Teacher Education, Faculty of Education, Universitas Negeri Padang who was an expert in the field of social science, that was Prof. Yalvema Miaz, M.Si, Ph.D. From the results of media analysis that the researchers did through observation to the school, the researchers got information that the multimedia-based interactive media in learning social science in elementary school had not been widely used and developed. Particularly on the features of the surrounding natural environment, the teachers had not used suitable media. Based on the results of interviews and discussions that researchers did with the teachers of grade IVA on Monday, August 7, 2017, it was found that the teacher had never used an interactive multimedia-based map media in learning social science for elementary students. He strongly supported the instructional media that the researchers developed. The results of interviews and discussions that researchers did on Saturday, August 12, 2017, with the experts, it was concluded that the interactive multimedia-based map media was still not widely developed. He also advised and supported an interactive multimedia-based map media that the researcher developed.

The researchers did student analysis by observing the students in the school, discussing with IV grade elementary school teachers, and seeking information from books about the theory of the students' characteristics in elementary level. The student analysis aimed to determine the ability and interest of the students in the use of interactive multimedia-based map media. The discussion result that the researchers conducted with the teachers of grade IVA SDN 22 Alai on Tuesday, August 08 2017 were (1) interactive multimedia-based map media can improve the students' learning outcomes in learning social science because during this time the teachers had not used it in learning, (2) the students were interested and happy to learn to use electronic media, such as computers, (3) the fourth graders had been able to use the computer.

3.2. Designing stage (Design)

The researchers began to design an interactive multimedia-based map media that developed and tested to grade IV students of SDN 22 Alai sub-district of Sutera Pesisir Selatan district. The stages were included preparing the instructional material about natural landscape, waters, physical environment, and artificial natural environment, making questions for evaluation test, selecting the media, selecting the formats by designing the components of instructional media to be created, creating media design, and creating an interactive multimedia-based map media using Adobe Flash CS6 application.

3.3. Developing stage (Development)

The validity test aimed to determine the validity level of instructional media that developed. The aspects of validity to be assessed were aspects of media, material, and language. An expert validator evaluated each element in their field, i.e., a media expert, material expert, and linguist. The data of validity test for media aspect were obtained from one lecturer who was expert in the field of instructional media, that was Mr. Darmansyah, S.T, M.Pd as a lecturer of Curriculum and Education Technology of Universitas Negeri Padang and Rector of Dhamasraya University of Indonesia (UNDARI). The validity test was done by appraising the media developed using the assessment sheet in the form of a questionnaire. The variables assessed on the validity aspects of instructional media were appearance, legibility, ease of use, and test or assessment.

The data of validity test for material experts were obtained from one lecturer who was expert in social science material for elementary school, namely Mr. Drs, Arwin, M.Pd as a lecturer majoring in Primary Teacher Education, Faculty of Education, Universitas Negeri Padang. The assessment was done by assessing the interactive multimedia-based map media product using an assessment sheet in the form of the questionnaire. The validator evaluated the aspects of social science materials for elementary level on the made instructional media. The variables assessed on the validity test of the material elements of the media were the content, objectives of learning, presentation of materials, instructional, and checks or assessment.

The data of validity test for language aspect were obtained from one lecturer who was expert on language majoring Primary Teacher Education, Faculty of Education, Universitas Negeri Padang language lecturer, namely Dr. Taufina Taufik, M.Pd. The validity test was done by assessing the interactive multimedia-based map media product using an assessment sheet in the form of the questionnaire. The validator evaluated the language of the media aspect according to its use for IV graders of the elementary school in learning social science. The variables assessed on the elements of style were grammar, communicative, conformity with the students' development and compliance with social science learning steps (Table 1).

Table 1. The recapitulation of validity test results

Aspect	Score P1	Score P2	Category
Media	3.7	4.6	Very Good
Material	4.5	4.9	Very Good
Language	3.5	4.4	Very Good

Note: P1 = 1st assessment
P2 = 2nd assessment

Practicality test of the instructional media was done by the teacher and student as the user of the instructional media. The practicality test results can be seen in Table 2.

Table 2. The recapitulation of practicality test results

Aspect	Variable Criteria	Average Score	Category
Teacher	a. The ease of the users b. Benefit c. Display d. Time	4.53	Very Good
Students	a. The ease of the users b. Benefit c. Display d. Time	4.16	Very Good

The efficacy test of instructional media was obtained from the grade IVA students' learning in SDN 22 Alai seen from the cognitive, affective, and psychomotor aspects. The results of the efficacy test can be seen in Table 3.

Table 3. The recapitulation of the students' learning results

Aspect	Variable Criteria	Average Score	Category
Cognitive	a. The features of natural view b. Natural environment c. Artificial environment	81.25	Very Good
Affective	a. Active b. Environmental care c. Respect others	75.20	Good
Psychomotor	a. Using media b. Discussion	78.50	Good

3.4. Disseminating stage (Dissemination)

At this stage, the researchers disseminated the interactive multimedia-based map media to several schools located in Pesisir Selatan District, West Sumatra. It showed that the results of interactive multimedia-based map media products assessment using Adobe Flash CS6 applications on grade IV elementary school had been very good. It can be seen based on the validity test (aspects of media, material, and language), practicality (the teachers and students), efficacy (learning outcomes), and the students' learning outcomes at the dissemination stage were obtained in the good and excellent category. The results of this study indicated that the use of interactive multimedia had a positive impact in learning social science for elementary level. In the process of teaching and learning with interactive multimedia-based learning, applications provided many benefits which were efficient in delivering the material and increase the student's interest in the material of natural appearance. Besides, multimedia-based interactive media was very feasible to use in learning. The instructional media can help the students to learn at school and independently at home [12-13].

Some previous research results also positively impact on the use of interactive multimedia such as use of technology in social studies as a means to motivate students by engaging students in the learning process with the use of a familiar instructional tool that improves students' self- efficacy and self-worth [14]. All computer-based instructional strategies somehow reinforce the constructivist classroom environment. However, social studies teachers still are not comfortable with applying all or some computer-based instructional strategies. We believe that a follow-up study can be done to review the beliefs and attitudes of social studies teachers toward these strategies. The results of the study might reveal the reasons for the lack of using computer-based instructional strategies among social studies teachers [15]. The use of multimedia in education has proven its importance due to its positive impact on the teaching and learning process. For students, the better academic achievement and positive attitude is development. The study should be replicated in other disciplines as well as elementary level [16].

Furthermore, the use of multimedia is helpful to build active learning [17-19]. It helps the students develop their knowledge. In social science learning, the use of multimedia can also help the teachers to transfer abstract concepts to semi-concrete by visualizing in the form of drawings or diagrams [20]. Also, interactive multimedia is not only used in social science subjects only, but it can also be used on all items. One of the research results on mathematics subjects explained that result showed that in such math skills at this age, using programs or multimedia enhanced methods of teaching can be useful in getting students attention especially when cartoon characters are used [21].

4. Conclusion

The developing an interactive multimedia-based map media began by conducting needs analysis, curriculum analysis, material analysis, media analysis and student analysis. After all the study were done, the next steps were to compose teaching materials, evaluation tests, media selection, format selection, and initial media design (creating the flowchart, storyboard, and media creation process). Second, the validity test results of the instructional media done by the media experts, material experts, and linguists were in the excellent category. Third, the practicality test results of the instructional media done by the teachers and students of grade IVA SDN 22 Alai Sutera district of Pesisir Selatan Regency was in an outstanding category. Fourth, the efficacy test of the instructional media showed that the grade IVA students' learning result of SDN 22 Alai had increased compared to the students' learning result before using interactive multimedia-based map media with passing percentage was 85%.

References

[1] Fathurrohman M, *et al* 2017 *International Journal on Emerging Mathematics Education* **1** 11
[2] Muhtadi D, *et al* 2018 *J. Phys.: Conf. Ser.* **943** 012020
[3] Muhibuddin F 2015 *Jurnal Dimensi Pendidikan dan Pembelajaran* **3** 24
[4] Ilhan O G and Sahin O 2016 *Academic Journal Yidiz Technical University* **1** 1
[5] Fanny, A M 2013 *Journal Prima Edukasia* **I** 1
[6] Oktaviani L 2017 *Jurnal Dinamika Informatika* **8** 156
[7] Novaliendry D 2012 *Jurnal Teknologi Informasi dan Pendidikan* **5** 17
[8] Pranowo G 2011 *Kreasi Animasi Interaktif dengan Action Script 3.0 pada Flash CS6* (Yogyakarta: Graha Ilmu)
[9] Sugiyono 2012 *Metode Penelitian Pendidikan Kualitatif, Kuantitatif, dan R&D* (Bandung: Alfabeta)
[10] Trianto 2009 *Mendesain Model Pembelajaran Inovatif-Progresive* (Jakarta: Kencana Prenada Media Group)
[11] Arikunto S 2013 *Dasar-Dasar Evaluasi Pendidikan* (Jakarta: PT. Bumi Aksara)
[12] Anshary H, *et al* 2016 *eProceedings of Applied Science* **2** 1113
[13] Diyan F and Upik Y 2016 *Jurnal Biodik* **2** 1
[14] Heafiner T 2004 *Cite Journal* **4** 42
[15] Mehmet A 2005 *The Turkish Online Journal Technology* **4** 18
[16] Iqbal S and Khan M 2015 *David Publishing Journal* **5** 349
[17] Masniladevi, *et al* 2018 *J. Phys.: Conf. Ser.* **943** 012059
[18] Young J R 2017 *International Journal on Emerging Mathematics Education* **1** 71
[19] Zainil M, *et al* 2018 *J. Phys.: Conf. Ser.* **943** 012046
[20] Nugraha F F 2017 *Journal of Elementary School Primaryedu* **1** 17
[21] Sawsan N, *et al* 2012 *Journal Acta Didactica Napocensia* **5** 17

IOP Conf. Series: Journal of Physics: Conf. Series **1088** (2018) 012070 doi:10.1088/1742-6596/1088/1/012070

Differences in students' learning outcomes between discovery learning and conventional learning models

R D Anggraini[1], A Murni[1], and Sakur[1]
[1]Mathematics Education Study Program, Riau University, Pekanbaru, Indonesia

E-mail: dianrini62@yahoo.com

Abstract. The number of students' learning outcomes of Year 9 SMP Negeri 1 Pekanbaru, Indonesia, were still under minimum completion criteria (less than 82). The objective of this research is to see the difference in learning outcomes between students who follow the learning through the discovery learning model implementation and conventional learning on 3-dimensional shapes with curved surfaces. This study used a pre-experimental research design (Static Group Comparison) involved 2 sample classes, experiment, and control classes. The results showed that there were differences in learning outcomes between students who learned by discovery and conventional learning models.

1. Introduction

Education is the key to form the intelligent society. According to the Indonesian government regulation, the goals of mathematics learning are to have knowledge and skills, to foster faithful and devoted persons to the one almighty God, to be noble, to have physical and spiritual health, to have a solid and independent personality, and to have a sense of community responsibility and nationality [1]. The achievement of the learning objectives will affect the learning outcomes. Good learning will result in a good quality of learning outcomes. Based on the national 2013 curriculum, a good lesson is student-centered learning, where students will have hands-on experience in learning. Thus, the success of a learning process can be seen in the learning outcomes.

According to ECTS Users' Guide in [2], the learning outcomes refer to what a learner is expected to know, understand and/or be able to demonstrate after completion of a process of learning. Thes' learning process can determine Students'learning outcomes. Learning outcomes are the level of students' abilities in accepting, rejecting, and assessing the information gained through the teaching and learning process [3]. Thus, the learning outcomes are the results that are achieved by students in the form of scores after participating in a test at the end of each lesson.

The result of mathematics learning is defined by Gagne in [4] as the change of students' behavior after receiving mathematical learning experiences, that can be measured through their knowledge, behavior, attitude, and skills. Based on the definition, it can be concluded that the results of learning mathematics are the score obtained by students after learning mathematics. The result of learning mathematics in this research is students' scores after following discovery learning in the experimental class and conventional learning in the control class on the topic of 3-dimensional shapes with curved surfaces.

This research aimed to develop learning materials on 3-dimensional shapes with curved surfaces for Year 9. The study used 4D development model. This paper explained the development phase related effectiveness test toward learning materials that have been developed.

Several studies have shown low learning outcomes of Junior High School students [5][6]. There are several factors that affect the low learning outcomes of natural sciences especially students of Junior High School level, such as: (1) the high difficulty level of the materials; (2) the lesson presented by the teacher is still a conventional learning model; (3) students were relatively passive in accepting and following what is presented by the teacher; (4) teachers rarely used appropriate instructional media so that students' basic understanding was still low; (5) teachers had difficulty in finding the right models and methods to present innovative learning; and (6) student's learning motivation was still low, indicated by less attention by playing with their friends [5]. Besides, many teachers used the teacher-centered learning the model [6].

The results of observations by the researcher in some junior high schools in Pekanbaru showed that many schools had implemented the national 2013 curriculum on Year 9. However, the learning model that was used was still teacher-centered. Teachers used lecturing method, which caused some students inactive and unengaged. Whereas, learning activities should use learning models, learning methods, learning media, and learning resources that could adjust to the characteristics of students and subjects [7].

Based on the interview results with teachers, many students were less active and not focus during the learning process, which caused a decline in students'learning outcomes. Based on the data of the students'learning outcomes in SMP Negeri 1 Pekanbaru on the current daily test results, the score were 42% in Year 9 Hang Jebat, 100% in Year 9 Tuanku Tambusai, 72.5% in class Hang Tuah, 87.8% in Hang Nadim class, with minimum completion criteria score is 82. Based on the results of interviews with some students, it is found that students were less interested in mathematics because it only focuses on solving problems with many formulas, students had a poor understanding of what the teacher explained in the classroom.

The potential solutions to overcome these problems is to apply the discovery learning model. Discovery learning is a model for developing active student learning by discovering by themselves, self-investigating, the results obtained will be long-lasting in memory and will not be easily forgotten by students[8]. In discovery learning, the teaching materials are not presented in the final form. Students are required to perform various activities of collecting information, comparing, categorizing, analyzing, integrating, organizing materials and making conclusions [10]. Also, students are required to find a concept with the help of a given stimulus, and this will make the learning will be students oriented. The procedure of applying the discovery learning model according to Syah (in [9]) is shown in Table 1.

Table 1. The procedure of discovery learning model implementation.

Stages	Teacher Activity
Stimulation	Ask the question, recommend reading a book and other learning activity that lead to the preparation of the problem
Problem Statement	Give opportunity for students to identify and analyze problems that students face so can create a hypothesis
Data Collection	Give the opportunity for students to gather as much as possible relevant information to prove whether or not the hypothesis is correct.
Data Processing	Guide students in processing data and information that has been obtained either through interviews, observation and so on and interpreted.
Verification	Guide students to perform a careful examination to prove whether or not the hypothesis has been established and associated with the data processing result.
Generalization	With students draw a conclusion that can be the general principles and apply to all the same events or problems, pay attention to the verification result

The objective of this research was to know the difference in mathematics learning outcomes between students who learn through discovery learning model and conventional learning. The research hypothesis was "there is a significant difference in mathematics learning outcomes between students who learned through discovery learning and conventional learning."

2. Method

Validation process, small group tests, and big group tests have been conducted at the developed phase to produce mathematical learning materials that are feasible to be tested for effectiveness. This paper describes the developed phase related to the effectiveness of learning materials that have been developed. Experiment design in Tabel 2 that was used in this research is *The Static Group Comparison Design* [11].

Table 2. The static group comparison design.

Group	Treatment	Post-test
Experiment	${}^{a}X_1$	${}^{c}O$
Control	${}^{b}X_2$	${}^{c}O$

${}^{a}X_1$: Discovery learning model
${}^{b}X_2$: Conventional learning
${}^{c}O$: Post-test

The population in this research was all student Year 9 of SMP Negeri 1 Pekanbaru academic year 2017/2018 consisting of six classes. All students were given a mathematical pre-test to assess students' early mathematical abilities, which would be used as research samples. Furthermore, the purposive sampling technique was used by looking at the schedule of mathematics subjects that were not overlapped, including four classes (Year 9 Hang Jebat, Year 9 Hang Nadim, Year 9 Tuanku Tambusan, and Year 9 Hang Tuah). The random sampling technique was used to determine the sample class from classes that have not crossed the timetable by drawing a draw. Testing was then conducted to see mathematical ability. The sample was Year 9 Hang Jebat to implement the discovery learning model (experimental group) and Year 9 Tuanku Tambusai to implement conventional learning (control group).

3. Result and Discussion

3.1 Mathematical Early Ability Test Data Analysis

An early test is given in the form of pre-requisite questions to learn the main subject of curved side space. Testing was done by using SPSS software version 23 for windows. Test results with Kolmogorov Smirnov test and Leneve-test indicated that the test scores students' mathematical pre-test were not normally distributed. Since the data were not normally distributed, the average analysis was performed by non-parametric statistical tests. The nonparametric test used was the Mann Whitney test. The hypothesis formulation for the Mann Whitney test on the test scores of mathematical pre-test of students Year 9 Hang jebat and Year 9 Tuanku Tambusai are as follows.

H_0: There was no a significant difference between test scores of mathematical pre-test of students Year 9 Hang Jebat and Year 9 Tuanku Tambusai

H_1: There was a significant difference between test scores of mathematical pre-test of students Year 9 Hang Jebat and Year 9 Tuanku Tambusai

Table 3 shows students' pre-test results by using Mann Whitney test in software SPSS version 23 for Windows

Table 3. Mann-Whitney test result of students' pre-test score using SPSS.

Class	N	Average	Z	*Sig.(2-tailed)*	H_0
IX Hang Jebat	37	89.59	-1.855	0.064	Received
IX Tuanku Tambusai	36	83.94			

Based on Table 3 above shows that the significance value (sig.) is more than $\alpha = 0.05$ which means H_0 was accepted and H_1 was rejected. This value means that at the value 95% confidence level, there was no average difference between Year 9 class Hang Jebat and Year 9 Tuanku Tambusai or in other words students in both classes have the equal mathematical ability.

3.2 Post-test Data Analysis

A post-test analysis is an analysis of the students'learning outcomes in the curved side space material. The result of the normality test with Kolmogorov Smirnov test showed that the posttest score data of the student was not normally distributed. Furthermore, hypothesis testing using Mann Whitney nonparametric statistical test. The verbal hypothesis formula for Mann Whitney test on posttest score data of experimental class and control class is as follows.

H_0: There was no a significant difference in students' learning outcomes between experiment class and control class

H_1: There was a significant difference in students' learning outcomes between experiment class and control class

Table 4 shows the result of hypothesis testing by using Mann Whitney test on software SPSS version 23 for windows.

Table 4. Mann-Whitney test result of post-test score data using SPSS.

Class	N	Average	Z	*Sig.(2-tailed)*	H_0
Exsperiment	37	95.65	-6.336	0.000	Rejected
Control	36	65.89			

Based on Table 4, it could be seen that the significance value (sig.) was less than $\alpha = 0.05$, so H_0 was rejected, and H_1 was accepted. This value means that at value 95% confidence level, there was a significant difference in students' learning outcomes between experiment class and control class.

3.3 Discussion

The discovery learning model is one of the learning models that can be applied in the implementation of the curriculum 2013 [9]. The implementation of the model discovery learning could give students stimulus so that students can think critically and at the end, they can solve the assigned problems. It is expected to increase students' skills so that ultimately able to improve student learning outcomes. Table 5 shows the percentage of students' completeness in the sample class.

Table 5. Percentage of the completeness of class sample students

Class	Students' score in %	
	Complete	Incomplete
Experiment	89.2	10.8
Control	36.1	63.9

Based on Table 5, it shows that the class that has the highest learning achievement was the experimental class while the control class had lower learning achievement than the experimental class. After the hypothesis test on the post-test score of each class using Mann-Whitney test with SPSS help obtained sig = 0.000, which is smaller than the level of significance 0.05. This value shows that H_0 was rejected, which means that there was an average difference between the experimental classes treated with the discovery learning model and the control class average with conventional learning. So it could be said that there are differences in learning outcomes between experimental class students with the discovery learning model implementation and control class students with the application of conventional learning.

4. Conclusion

Based on the results of data analysis and hypothesis testing, it can be concluded that there were differences in learning outcomes between experimental class students with the discovery learning model and control class students with the conventional learning on Year 9 SMP 1 Pekanbaru. The average student learning outcomes used the discovery learning model was 95.65 and the average of student learning outcomes used conventional learning was 65.89.

Based on the research result and discussion, the researcher recommends for teachers or researcher who want to do research using the discovery learning model and using student worksheet to design the stimulus and problem statement carefully. It is because in this study many students who less able to establish a temporary hypothesis that exists in the problem statement stage.

References

[1] Depdiknas 2003 Undang-Undang Republik Indonesia No. 20 Tahun 2003 tentang SistemPendidikan Nasional (Jakarta: Kemdiknas)

[2] Declan Kennedy, Aine Hyland, and Norma Ryan 2012 Writing and Using Learning Outcomes: a Practical Guide(Online) http://www.procesbolonski.uw.edu.pl/dane/learning-outcomes.pdf

[3] Hamdu G dan Agustina L 2011 Pengaruh Motivasi Belajar Siswa terhadap Pestasi Belajar IPA di Sekolah Dasar (Studi Kasus terhadap Siswa Kelas IV SDN Tarumanagara Kecamatan Tawang Kota Tasikmalaya) *Jurnal Penelitian Pendidikan* **12** 90-6

[4] Robert Gagne 1985 The Condition of Learning (Fourth Edition) (New York: Holt, Rinehart & Winston)

[5] Mamik Suendarti 2017 The Effect of Learning Discovery Model on the Learning Outcomes of Natural Science of Junior High School Students Indonesia *International Journal of Environmental of Science Education* **12**2213-16

[6] Fitriyah 2017 Pengaruh Model Pembelajaran Discovery Learning terhadap Hasil BelajarMatematika Siswa MAN Model Kota Jambi *Jurnal Pelangi* **9** 5

[7] Depdikbud 2016 Permendikbud No. 22/2016 Standar Proses Pendidikan Dasar dan Menengah(Jakarta: Kemdikbud)

[8] Christoper Pappas 2014 Instructional Design Models and Theories: The Discovery Learning Model(Online) https://elearningindustry.com/discovery-learning-model

[9] Kemdikbud 2013 Materi Pelatihan Implementasi Kurikulum 2013 Tahun 2013 (Jakarta:Kemdikbud)

[10] Kemdikbud 2015 Materi Pelatihan Guru Implementasi Kurikulum 2013 Jenjang SMP Tahun 2015 (Jakarta: Kemdikbud)

[11] Cottrel, Randal R dan James F McKenzie 2011 Health Promotion and Education ResearchMethod (Second Edition) (USA: Jones and Barlett Publishers)

Students' abstraction in solving system of linear equations with two variables

F Pangaribuan
Nommensen HKBP University, Jl. Sutomo No. 4A, Perintis, Medan, 20235, Indonesia

Email: firmanpangribfkipuhn@gmail.com

Abstract. In teaching and learning process, the teacher usually guides students to follow the demonstrated procedure; however, the result of learning process sometimes shows that students cannot construct new knowledge. In constructing new knowledge, abstraction is important for students. The objective of this research is to describe students' abstraction in solving system of linear equations with two variables problem. This research uses the qualitative method. The subjects are two female students from seven grade of junior high school. These students have not yet studied the system of linear equations with two variables formally. The data were gained as a result of solving the contextual problem in student worksheet and from an interview with the students. The results of this study were that students used their prior knowledge of the linear equations with one variable formally. Then students could solve the system of linear equations with two variables in formal symbolic by abstraction. The students needed scaffolding while abstracting. In teaching and learning process, teachers need to facilitate students by abstraction.

1. Introduction

The system of linear equations with two variables (SLETV) is learned in second-year of junior high school. The basic competence is that students explain the solution of SLETV in a contextual problem. Students' experience of resolving contextual issues becomes the path to achieve basic competencies in understanding the concept of solving SLETV in formal symbolic.

In teaching learning process, usually, the teacher guides students to follow the demonstrated procedure. However, the result sometimes shows that students cannot construct new knowledge for themselves. This is not in accordance with the curriculum, in which teaching and learning processes are expected to increase students' thinking skills so they can demonstrate logical, critical, creative, and innovative thinking skill.

Knowledge is not given as a product that ready to consume by the students. Students must construct it for themselves [1]. By constructing, students are expected to think critically or creatively. Piaget noted that all new knowledge requires abstraction [2]; without abstraction, there will be no new knowledge.

To improve long-term conceptual learning, the framework formulated here suggests that the whole curriculum must be framed with an awareness of the abstraction process to produce thinkable concepts at every stage [3]. Understanding the concept must go through the process of abstraction step by step.

Constructing knowledge is an abstraction activity. Abstraction is the process in which the learner constructs new knowledge, by what we call vertical mathematization. It is the process of constructing new knowledge with vertical mathematization, through recognizing and gathered previously possessed knowledge, and the knowledge identified and gathered must be relevant to the new knowledge to be constructed [4]. Vertical mathematization in realistic mathematics education is a process of increasing the complexity of a concept in the mathematical system itself. Horizontal mathematization is the process of transforming understanding from real situations to mathematical situations. In this research, The abstractions observed and those that appear in students' minds are three components: 1) recognizing, 2) building-with, and 3) constructing.

Abstraction process of solving the SLETV will be focused on observation: 1) recognizable element of knowledge that is the initial knowledge possessed by student in solving SLETV, 2) the aspect of knowledge that is strung together is a combination of recognized element, for example solving routine problem (not solving problem which new for the students), and last is 3) construct the new knowledge for the student which is the answer of the problem. In problem solving process, a new knowledge or concept formed is expressed in objective learning.

Some researchers stated that abstraction plays an important role in constructing mathematical knowledge. Dooley shows that student able to construct new knowledge about decimal fraction expansion and non-terminating decimals [5]. Ozmantar and Monaghan also demonstrate that student constructed a method to sketch the graphs of $|f(|x|)|$ while $f(x)$ is a linear equation [6]. They have demonstrated how one can abstract knowledge empirically. Unfortunately, they have not shown how students abstract in solving SLETV. This research aims to describe students' construction of solution in SLETV problem with abstraction stage. Theoretically, this research adds more input to the theory of abstraction that can be used in building knowledge of SLETV material. Practically, could be the input on the learning management as alternative learning with abstraction theory on the SLETV.

2. Method

This research uses the qualitative method. The subjects of this research are two female students of seven grade of junior high school, and they have not studied formal SLETV material yet, but already understood the linear equation with one variable. Data collection was done by giving three SLETV problems in each worksheet that contains pictures. The three SLETV had been validated for students and shown that problems can be understood and can be solved by them. The results of the written answers were then being interviewed in depth, then the written and oral answers were validated by seeing the consistency of both the answers. The data collection was recorded by using the audio-visual recorder, then transcribed and analyzed. The conclusion is made by giving meaning and explanation to the valid data, which are in recognizing, building with, and constructing phases.

3. Results and discussion

The following description describes of how the abstraction takes places on solving three SLETV problems. In worksheet 1, the subjects are given a picture containing two packages that must be sold at a store. Package A includes 1 glass and 2 hats for Rp260,000 and packet B comprises two glasses and 2 hats for Rp280,000. Subjects are asked to determine the price for each item.

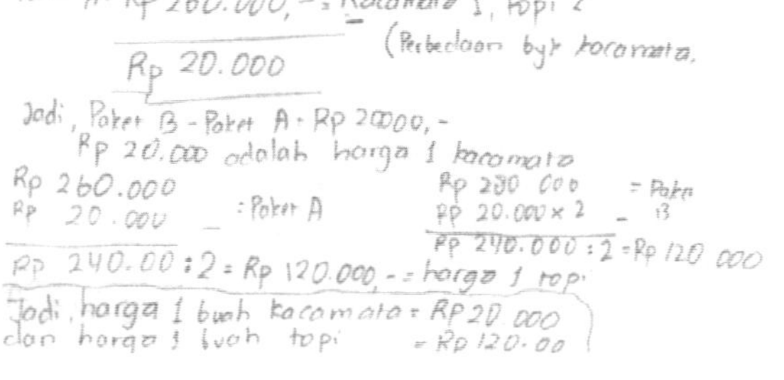

Figure 1. Student's solution before using variable.

IOP Conf. Series: Journal of Physics: Conf. Series **1088** (2018) 012071 doi:10.1088/1742-6596/1088/1/012071

The subjects recognize the solution of the SLETV in their own way after observing the given problem in the picture. Their first step is categorized as the recognition of the problem because the subjects complete it based on their experience and the subjects' written as Figure 1. This step has not requested the use of variables yet.

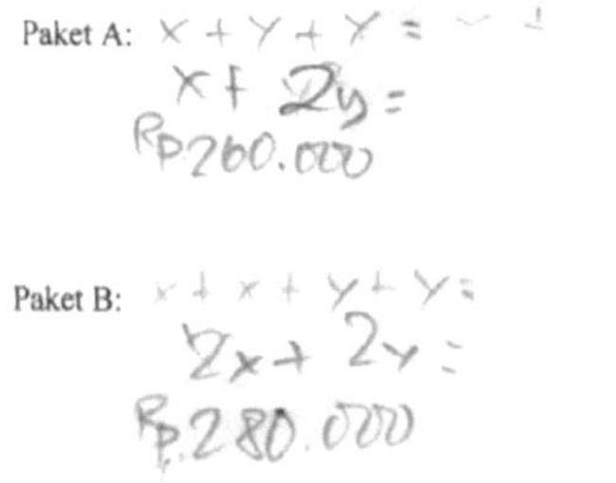

Figure 2. Students' work of linear equations with variable.

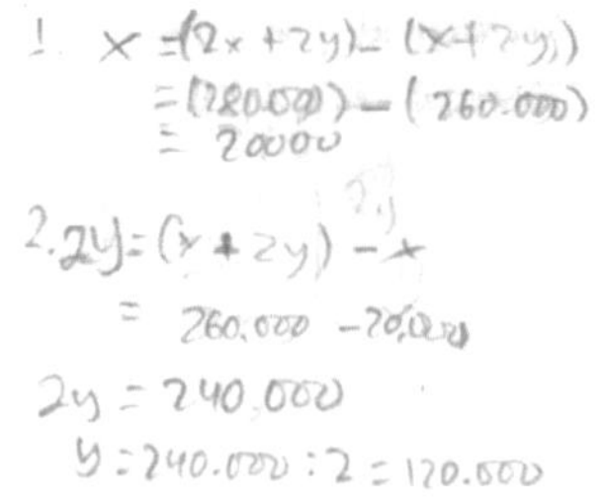

Figure 3. Students solution in formal symbolic.

Subsequently, subjects are asked to write a mathematical model by giving the price of one glass is x and the price of one hat is y. Subjects recognized the mathematical model of the given problem by changing the image or visual into a formal symbol form with the use of x and y variables. Initially, the subjects did not add the corresponding variable in the mathematical model but exposing the process in the figure. For example, package A becomes $x + y + y = x + 2y = 260.000$ and does not directly write $x + 2y = 260{,}000$. Similarly, with package B the subject writes $x + x + y + y = 2x + 2y = 280{,}000$. Likewise, the writing of equations is not written in one line, but into three lines and students write the linear equation with variables in Figure 2. In this solution, the subject had built-with the combination of using the model and the solution by himself. In this worksheet 1, the subject had not constructed a new knowledge, because there was no obstacle to the student in comprehending it. On the answers, the subject writes $x = (2x + 2y) - (x + 2y) = 280{,}000 - 260{,}000 = 20{,}000$, for package A, and the subject still writes $2y = (x + 2y) - x = 260{,}000 - 20{,}000. 2y = 240{,}000$ and this subject's text indicates that algebra reduction is already recognized and students solution as formal symbolic in Figure 3.

The more complex problems are prepared in worksheet 2. The subjects were also given a picture containing 2 packages to be sold at a store. Package A contains 1 glass and 2 hats, its total price is Rp220,000 and package B contains 2 glasses and 3 hats for Rp360,000. Subjects are asked to determine the price for each item.

Problem worksheet 2 cannot be solved in the same way as in worksheet 1, because the structures are different. The subject does so by trial and error with the price of one hat and the price of one glass, then testing the price given by substituting the figure on the problem. The subject gets the price of one glass is Rp60,000, and a one hat price is Rp80,000. Another subject answered by forming a "new package" that is the difference between package B and A; the price of one hat and one glass are Rp140,000. This new package combines with package A to get the price of a hat. The price of package A (2 hats and 1 glass) is Rp220,000 along with "new package's" price contained one hat, and one glass is Rp140,000. So the subjects get the answer of the price of one hat as the difference between the two packages. In changing the SLETV into the form of the picture and its description into the form of a formal symbol, the subject can recognize the mathematical model by writing x + y + y = x + 2y = 60,000 + 160,000 = 220,000 of package A and 2x + 3y = 120,000 + 240,000 = 360,000.

To find out the subject of how to "building-with" in solving problem 2, it is obtained through the subjects answer in section 2c "Write down your solution by using x and y." Building within this section is to combine the solution obtained by the subject through problem 2a and 2b that is the combination of the solution by the subject itself and the mathematical model of the given SLETV. Solution 1b and 2b are about writing the mathematical model in a different way. The subject does not write the price per unit while solving problem 1b since on problem 2b, the subject writes the prices x, 2y, 2x, and 3y. These prices are obtained through the answer on problem 2a, and the subject has written SLETV in formal symbol.

The subjects exposed his work by telling "*Now we see package B is 360,000 and package A is 220,000. So, package B minus package A is 140,000. One hundred and forty is still the price of 1 glass and 1 hat. It is still the price of 1 glass and 1 hat. We want to find the price of each hat and glass. While in package A there is 1 glass and 2 hats that cost 220,000. One hundred forty thousand, sorry. So packet A is worth 220,000. It is subtracted by 140,000 is equal to 80. The price of 1 hat and 1 glass is 140,000 and the price of 2 hats and 1 glass is 220,000. Two hats and one glass less than one hat and one glass with one hat. One hat is 80,000"*. The subject then determines the price of a glass with the following exposure. "*One hat is 80.000. So, if the two hats are 160,000. Two glasses and 1 glass are 220,000. So 1 glass is 220,000 minus 160,000 equals to 60,000* ". Exposure to this subject is already the construct of a new knowledge for him even though he did not finish it in a formal symbol. Particularly his ability constructs a new statement from two packets A and B, by taking the difference. The completion of this subject has also used *elimination*, i.e. eliminating the price of a glass to obtain the price of a hat. This subject will not determine the solution directly from the SLETV form of formal symbols with the reason that he has gotten it. In that solution, the subjects perceive variables on SLETV have meaning as contents of the original problem. His understanding of SLETV does not show meaningless variables.

The problem in worksheet 3 is more complex than in worksheet 2. Package A contains one glass and two hat price is Rp300,000 and Package B contains four glasses and one hat price is Rp360,000. Problem 2 can be solved by two stages, first by subtracting one package from another, then eliminating one variable, such as package A is subtracted with package B. In worksheet 3, it could not be solved similar like in worksheet 2.

The subject recognizes this SLETV problem but cannot recognize the solution yet. The subject recognizes the purpose of the problem in a formal symbol. A sample of the subject written "*A* package: x + y + y = x + 2y = 300.000 and *B* package: x + x + x + y = 3x + y = 525.000". The subject identified elements to be "built-with" in this SLETV solution process, even though the subject has not found the solution. For example, a combination of problems in the form of a picture and their descriptions and combined with a formal symbol. Also, it found elements are built-with between integer operations and meaning in the picture.

The subject experienced an obstacle while solving problem 3. Their inability to show the answer shows this obstacle. Subject who tries to determine the price of a glass and the price of a hat did not find the answer. It shows that this problem is more complex than the previous problem. Many subjects try to subtract package A from package B or vice versa but failed in interpretation.

To overcome the obstacle of the subject found in solving the problem, the interviewer scaffolded subject by giving a question as follows. "*Can it be solved if there are some similar packages?*". The subject could describe his solution by making 3 similar package A, solved it similar to the answer written in answer sheet 2, and found the price of glass is 150,000 that is $x = 150$ and the price of a hat is 75,000, that is $y = 75$.

The subject can solve the problem after the researcher made scaffolding. The interviewer did not provide direct help, but with that question, the subject became triggered to construct his knowledge by completing a new knowledge of SLETV.

The problem given to the subjects is a problem they can image or a problem in contextual form, which expected they attempt to solve the problem packet in SLETV. It is possible if the problem is not related to real life problem, or in the form of a formal symbol, the subject will not be motivated to solve it, because subjects do not see the problem for what benefits. Through a given contextual problem, subjects have informal knowledge directed toward formal knowledge. This corresponds to one of the contents of the theory of realistic mathematics education that the child's learning begins with a real situation.

On the solution of the subject using the x and y symbols, x is the price of a single glass, and y is the price of a hat. The variable x and y is a variable meaning, i.e. a value, and it is not just a meaningless letter. Although the subject often calls a glass while pointing or writing the letter x and one hat pointing the letter y. Subject' understanding of these letters accords with the first stage of understanding of variables according to Kucheman [5], that variables are a letter that will be determined by its value.

New knowledge is constructed by the subject while solving SLETV in worksheet 2. Usually, the SLETV solution is generally doubled package A so that it gets 2 glasses and 4 hats, then calculates the difference of package A duplicated with package B to obtain the price of one hat. The activity of the subject as demanded by the Curriculum 2013 is student-centered learning so that students find themselves new concepts. This finding is in line with the opinion of Glasersfeld [2] that the knowledge given is not in the finished form ready-made), but the student must construct that knowledge for himself. The activity of this subject is an abstraction of knowledge possessed formally and informally, so that constructed a new knowledge for student self. This is in accordance with Piaget [2] that all knowledge is based on abstraction.

In solving the problem SLETV number 3, subjects' experience obstacles, the procedures they use on problem number 1 or number 2 cannot be used to solve the problem in number 3. Ozmantar and Monaghan [7] said that student is not easy to construct new knowledge; the teacher needs to do scaffolding. In order for subjects to construct this new knowledge, the researcher provides minimal assistance, by asking "Can it be solved if there are some similar A package?" This question is a trigger for the subject to expose his potential to solve SLETV problems. In accordance with Vygotsky's theory that one has actual abilities and potential abilities and the area between these two abilities is expressed by the Zone of Proximal Development [8]. In order to the subject to arrive at this potential skill, a more mature person such as a peer or teacher needs to provide scaffolding.

4. Conclusion

The SLETV problem is presented in the form of a story or contextual problem with the picture, so students can recognize and even may feel challenged to solve it. Through a given contextual problem, subjects have informal knowledge directed toward formal knowledge. Students recognize the knowledge they have through their own understanding and write symbols informally. Students use their prior knowledge in a contextual and linear equation of one variable formally then directed by formal symbols, so that students can find the formal symbol (mathematical model) of the SLETV problem. Students need support in constructing new knowledge through solving SLETV problems. For further research, observations of students performing abstractions need to be done on more complex SLETV topics. In the learning process of SLETV, student-centered method with minimal guidance needs to be implemented so that learning is more meaningful.

Reference

[1] Freudenthal H 1973 *Mathematics as an educational task* (Dordrecht The Netherlands: Reidel)

[2] Glasersfeld E 1991 Abstraction, Re-Presentation, and Reflection, an Interpretation of Experience and Piaget's Approach In L Steffe (ed), *Epistemological foundations of Mathematical Experience* (New York: Springer)

[3] Hershkowitz R, Schwarz B B and Dreyfus T 2001 Abstraction in Context: Epistemic Action *Journal for Research in Mathematics Education* **32** 195

[4] Kuchemann D 1981 *Algebra* In K M Hart (Ed.), *Children's understanding of mathematics* (London: Murray) p102

[5] Dooley T 2007 *Construction Of Knowledge By Primary Pupils: The Role Of Whole-Class Interaction* In D Pitta-Pantazi & G Phillipou (Eds.) Proceedings of the fifth congress of the European society for research in mathematics education Larnaca: University of Cyprus pp 1658–1667

IOP Conf. Series: Journal of Physics: Conf. Series **1088** (2018) 012071 doi:10.1088/1742-6596/1088/1/012071

[6] Ozmantar M F and Monaghan J 2006 Abstraction, Scaffolding And Emergent Goals *Proceedings 30th Conference of the International Group for the Psychology of Mathematics Education* **4** 305
[7] Ozmantar M F and Monaghan J 2007 A Dialectical Approach to the Formation of Mathematical Abstractions *Mathematical Education Research Journal* **19** 89
[8] Johnson R and Koedinger K R 2005 Designing Knowledge Scaffolds to Support Mathematical Problem Solving Bethany Rittle-Johnson in *Cognition And Instruction* Lawrence Erlbaum Associates, Inc **23** 313

IOP Conf. Series: Journal of Physics: Conf. Series **1088** (2018) 012072 doi:10.1088/1742-6596/1088/1/012072

Students' ability to convert a situation into a mathematical model or diagram using problem solving approach

H Arianti, C M Zubainur and Hizir
Syiah Kuala University, Jl. Teuku Nyak Arief Darussalam, Banda Aceh 23111, Indonesia

E-mail: cut.morina@fkip.unsyiah.ac.id

Abstract. Students' mathematical communication skill needs to be developed through appropriate instructional strategies; one of which is the problem solving approach. One of the indicators of the mathematical communication skill is the ability to state the situation into a mathematical model/ diagram. This article aims to describe the development of students' ability to convert a situation into a mathematical model/diagram using the problem solving approach. This study used mathematical communication skill tests and a semi-structured interview. The data was obtained from the implementation of the quadrilateral learning using the problem solving approach involving six Year 7 students in a junior high school in Aceh Besar, Indonesia that consist of two high, middle, and low-achieving students. The finding shows that students' ability to convert the situation into a mathematical model/diagram using problem solving approach was varied. High-achieving students can convert the situation into the model/diagram in every lesson. Middle-achieving students can convert the situation into a mathematical model/diagram in the first lesson, but the ability decreased in the second lesson and increased in the following lesson. Low-achieving students could not convert the situation into a mathematical model/diagram in the first and second lesson. However, they were able to do that in the third lesson.

1. Introduction

Mathematical communication is one of the mathematical abilities required in learning mathematics [1,2]. Mathematical communication is the ability to write, read, listen, review, interpret and evaluate ideas, symbols, terms, and mathematical information [3]. Aspects of mathematical communication skill include the ability to express, to interpret and evaluate mathematical ideas, and the ability to use terms, symbols, and structures. The ability to declare, to interpret and evaluate mathematical ideas can be conducted orally, in writing, and by visualization [4]. The ability is essential to model the situation or mathematical problems.

Developing students' mathematical communication begins with familiarizing students with problems related to problem solving activities [5]. Problem solving approach provides an opportunity for students to develop their mathematical communication skills and mathematical ideas [5,6]. Problem solving is a learning approach exposing students to a problem and asking them to solve it. Problem solving approach trains students to understand problems, plan solutions, solve problems, and re-examine the answers [3].

A well-established plan will optimally enhance the efforts to develop students' mathematical communication using the problem solving approach. Also, developing the learning instruments using appropriate procedures is crucial in planning the learning. Good learning instruments need to satisfy the criteria of validity, practicality, and effectiveness [7]. The practicality and effectiveness criteria are evaluated based on the learning implementation in the classroom.

This article investigates the development of students' mathematical communication skill using formative tests that have been developed. This aims to answer a research question "how is the development of the students' ability to convert the situation into a mathematical model/ diagram using a problem solving approach?"

2. Method

This design research aims to develop learning instruments using the problem solving approach to improve students' mathematical communication skills. The learning instruments developed using ADDIE development model that consists of analysis, design, development, implementation, and evaluation phase. The developed products were a lesson plan, student worksheet and tests for mathematical communication skills of the topic of quadrilaterals. This article focuses on students' ability to convert the situation into a mathematical model/diagram as an indicator of mathematical communication skills.

Data of students' ability to convert the situation into the mathematical model/diagram was obtained from the effectiveness of the learning instruments, i.e., the increase of the students' skills measured [8]. The participants were six Year 7 students in one of the senior high schools in Aceh Besar, Indonesia. The ability of the students was varied, high, middle, and low. The students were selected based on the pre-test results and the teacher's recommendation.

3. Result and discussion

Quadrilateral learning instruments using problem solving approach were developed for five lessons. The first lesson discussed the area and perimeter of squares and rectangles. The second lesson discussed the area and perimeter of parallelograms. The third lesson discussed the area and perimeter of trapezoids. Meanwhile, the area and perimeter of rhombus and kites were presented in the fourth and fifth lessons.

At each end of the learning, students were given formative tests individually. Formative tests were created in the form of descriptions and developed based on the indicators of mathematical ability. The development of the students' skills to convert the situation into the mathematical model/diagram was measured by the tests and clarified by semi-structured interviews. The ability to convert situations into a mathematical model/diagram was identified when the students were able to convert the situation in a problem into a mathematical model/diagram appropriately.

The students were asked to solve a problem about the area and perimeter of the rectangle in the first formative test. The expected answer was that students could draw a rectangle with a length of 30 m and 20 m, $\frac{1}{3}$ parts for the pool, $\frac{1}{2}$ parts for the garden, the remaining for the park. Based on the students' answers, it was indicated that high- and medium-achieving students were able to convert the situation into the mathematical model/diagram. Meanwhile, the low-achieving students were not able to convert the situation into the mathematical model/diagram. The students' answers in solving the problem of mathematical communication skills based on indicators of the ability to convert the situation into the mathematical model/diagram are shown in Figure 1.

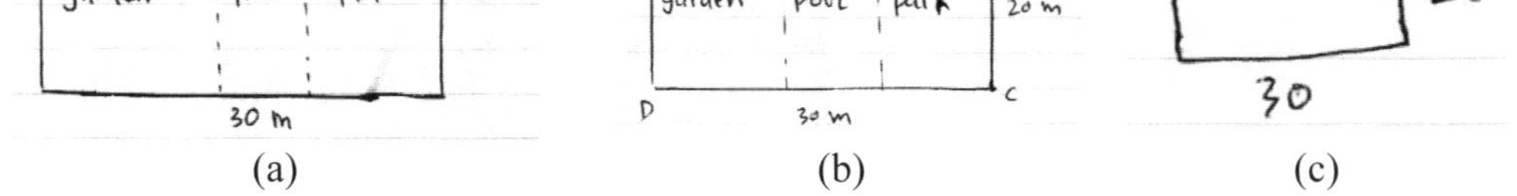

(a) (b) (c)

Figure 1. Examples of students' answers: (a) high-achieving students, (b) medium-achieving students, and (c) low-achieving students.

Figure 1 shows that high and medium-achieving students could draw a rectangle with the length of 30 m and width of 20 m as well as the location for the pool, gardens, and parks as specified in the problem. While low-achieving students only illustrated a rectangle with a length of 30 m and width of 20 m without specifying a predetermined part of the problems. This shows that high- and medium-achieving students were able to convert the situation into a mathematical model/diagram. The excerpt from the interview presented as follows.

Teacher :"What should be done to make calculation easier?"
Student :"Illustrates the situation stated in the problem
Teacher :"Yes, good! Can you convert the situation into diagram so it would be easy to understand?"
Student :"Yes, I can." (Drawing rectangles, the length is 30 and width is 20 as well as drawing sections for gardens, pools, and parks)

Based on the above explanation, it is indicated that high- and medium-achieving students could convert the situation into the mathematical model/diagram. This means that medium- and high-achieving students satisfy the indicator of the ability to convert the situation into a mathematical model/diagram. Figure 1 also shows that low-achieving students have not been able to convert the situation into the mathematical model/diagram.

In the second formative test, students were asked to solve a problem about the area and perimeter of parallelograms. The expected answer was that students could draw a parallelogram with the adjacent side lengths of 50 m and 52 m and one of the 50 m sides without trees. Based on the students' answers, it was found that high-achieving students were able to state the situation in the mathematical model/diagram, but the low- and middle-achieving students had not. The answers of low-, medium- and high-achieving students are presented in Figure 2.

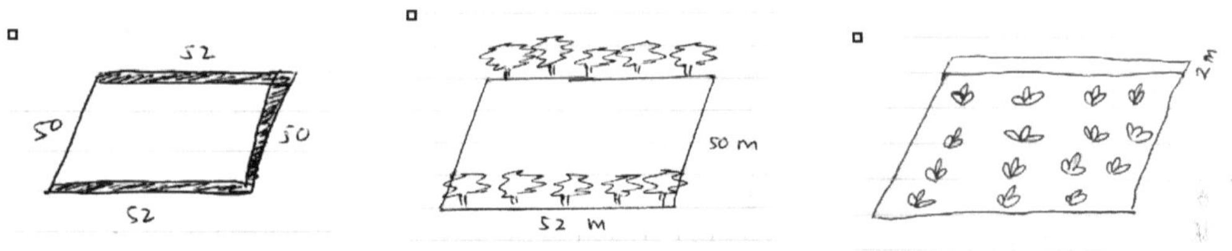

(a) (b) (c)

Figure 2. Examples of students' answers: (a) high-achieving students, (b) middle-achieving students, and (c) low-achieving students.

Figure 2 shows that high-achieving students appropriately described the land of parallelogram with adjacent side lengths of 50 m and 52 m and one of the 50 m sides was not shaded, while the middle-achieving students correctly described the land of parallelogram with adjacent side lengths of 50 m and 52 m, but they incorrectly drew the trees that should be on one side of the 50 m only (as mentioned in the problem). Low-achieving students drew a parallelogram with trees planted inside it with 2 m length without trees. This indicates that only high-achieving students were able to convert the situation in the problem into the mathematical model/diagram that was confirmed in the following interview excerpt.

Teacher :"Can you convert the situation on the problem into the diagram with one of the 50 m side planted no trees?"
Student :"Yes, I can." (Drawing a parallelogram with adjacent side lengths of 50 m and 52 m, and the 50 m side is without trees)
Teacher : "Why don't you draw the trees on the 50 m side?
Student : "Because that side is not planted with trees."

It can be said that high-achieving students were capable of expressing the situation in the problem into the mathematical model/diagram. That means that they satisfied the indicator of the ability to convert the situation into a mathematical model/diagram. However, the medium- and low-achieving students had not met the indicator yet.

In the third formative test, the students were asked to solve a problem about the area and perimeter of the trapezoid. Based on the students' answers, it was found that high-, medium-, and low-achieving students were able to convert the situation into the mathematical model/diagram. Figure 3 presents the students' answers.

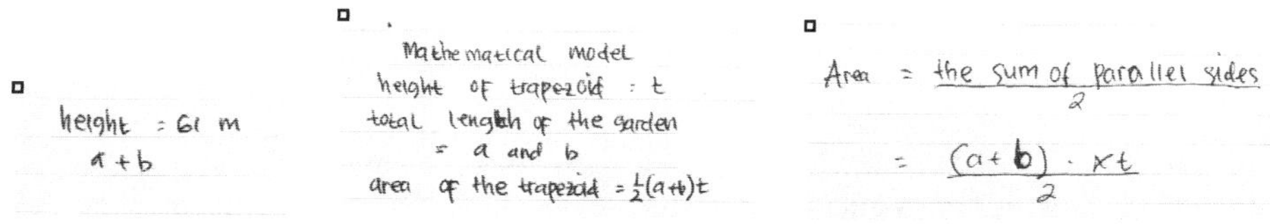

(a) (b) (c)

Figure 3. Examples of students' answers: (a) high-ability students, (b) medium-ability students, and (c) low-ability students.

Figure 3 shows that the high-achieving students expressed the mathematical model for the parallel length of the garden as a and b and the height of trapezoid as t, whereas the medium-achieving students stated the mathematical model for the height of trapezoid as t and the total length of the garden as a and b, area of the trapezoid as $\frac{1}{2}x(a+b)xt$. The low-achieving student stated that the mathematical model for the parallel sides of the garden as a and b and the height of trapezoid as t. This shows that high-, medium-, and low-achieving students were successful in transforming the information in the diagram into the mathematical ideas as it is confirmed in the following interview excerpt.

Teacher : "What is the mathematical model of this problem (showing a problem)? "
Student : "$a + b$."
Teacher : "Only that? If it is a trapezoid, what about the height?"
Student : "the height of the trapezium is t."

Based on the above explanation, it can be concluded that high-, medium-, and low-achieving students were able to convert the situation in question into the mathematical model/diagram.

In the fourth formative test, students were asked to solve a problem about the area and perimeter of a rhombus. The expected answer was that students could draw a rectangle where a rhombus is in the middle of the rectangle. The students' responses are presented in Figure 4.

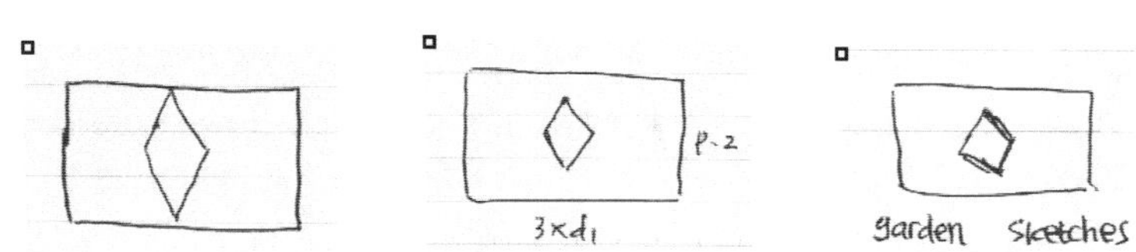

(a) (b) (c)

Figure 4. Examples of students' answers: (a) high-achieving students, (b) medium-achieving students, and (c) low-achieving students.

Figure 4 shows that high-achieving students drew a rectangle where a rhombus is in the middle of the rectangle, while medium- achieving students were able to draw the similar one with additional information of the length ($3d_1$m and *p-2* m). The low-achieving student also drew a rhombus in the

IOP Conf. Series: Journal of Physics: Conf. Series **1088** (2018) 012072 doi:10.1088/1742-6596/1088/1/012072

middle of a rectangle. This shows that high-, medium, and low- achieving students could appropriately express the information in the diagram into mathematical ideas as shown in the following interview excerpt.

Teacher : "Before calculating the area of the park outside the pool, did you sketch the park so that it would be easy to understand? Can you show me the sketch?"
Student : "Yes, I can." (drawing a rectangle in the middle of the rhombus without writing the size)
Teacher : "How to express the mathematics model?"
Student : $p = 3d_1$, $l = p-2$

Based on the above explanation, it can be concluded that the students were able to convert the situation into the mathematical model/diagram.

In the fifth formative test, the students were asked to solve a problem about the area and perimeter of the kite. The students' answers are presented in Figure 5.

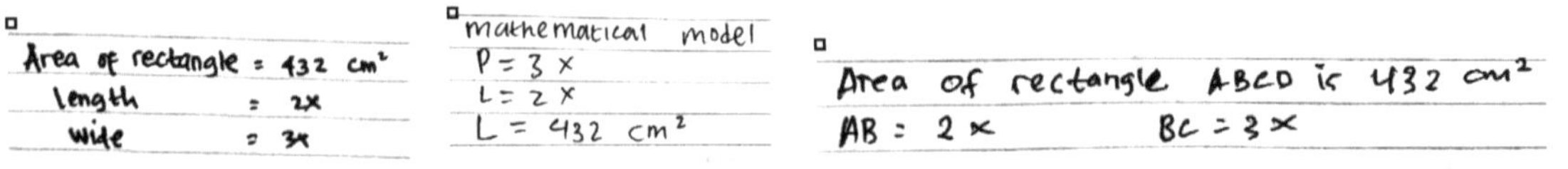

(a) (b) (c)

Figure 5. Examples of students' answers: (a) high-achieving students, (b) medium-achieving students, and (c) low-achieving students.

Figure 5 indicates that the high-achieving students stated the mathematical model for the area of the rectangle as 432 cm^2, length as 2x, and wide as 3x, while the medium-achieving students described a mathematical model for pas 3x, l as 2x, and L as 432 cm^2. The low-achieving student expressed the mathematical model for the rectangle of ABCD is 432 cm^2, AB as 2x, and BC as 3x. This finding shows that high-, medium- and low-achieving students were capable to accurately describe the information in the diagram into mathematical ideas as revealed in the following interview excerpt.

Teacher : "What is the mathematical model of this problem? "(Pointing to the problem)
Student : "The area of the rectangle is 432 cm^2, the length equals to 2x, and the width equals to 3x."

Based on the explanation mentioned earlier, it can be said that high-, medium- and low-achieving students had been able to convert the situations presented in the problems into the mathematical model/diagram. This occurred because the learning provided enough opportunities for the students to express their mathematical thinking [6]. Besides, the problem solving approach might also motivate students to be more active in discussing the problems with their group members. This discussion positively contributes to the students' ability in converting the situation into the mathematical model/diagram [9].

Based on the results of formative test analysis, students' ability to convert the situation into the mathematical model/diagram had progressed from the first to the fifth meeting. This suggests that the learning instruments developed met one of the indicators of the effective learning instruments [8].

4. Conclusion

The results show that the students' ability to convert the situation into the mathematical model/diagram improved and it was varied based on their ability. High-achieving students were able to express the situation in the question into a mathematical model/diagram in each lesson. The medium-achieving students were able to express the situation into the mathematical model/diagram in the first lesson, but it decreased in the second lesson and improved in the third lesson. Meanwhile, the low-achieving students had not been able to convert the situation into the mathematical model/diagram in the first and second lesson, but then they could do it in the third lesson.

References

[1] Jl K and Parman M S 2013 Mathematical Communication Skills of Junior Secondary School Students in Coastal Area *Sains Humanika* **63** 77
[2] Cooke B D and Buchholz D 2005 Mathematical communication in the classroom: A teacher makes a difference *Early Childhood Education Journal* **32** 365
[3] Afgani D J 2011 *Materi Pokok Analisis Kurikulum Matematika/MPMT5204/3* (Jakarta: Universitas Terbuka)
[4] National Council of Teacher of Mathematics 2000 *Principles and Standards for School Mathematics* (Reston, VA: NCTM)
[5] Mahmudi A 2009 Komunikasi dalam Pembelajaran Matematika *Jurnal MIPMIPA UNHALU* **8** 1
[6] Nayazik A 2016 Peningkatan Kemampuan Komunikasi Matematis Mahasiswa dengan Model Ideal Problem Solving Mata Kuliah Kapita Selekta Matematika 1 *Math Didactic: Jurnal Pendidikan Matematika* **2** 47
[7] Nieveen N 1999 *Prototyping to Reach Product Quality*, ed Plomp T, Nieveen N, Gustafson K, Branch R M and van den Akker J (London: Kluwer Academic Publisher)
[8] Kemp J E, Marisson G R and Ross S M 1994 *Developing evaluation instrument. Designing effective instruction* (New York, NY: macMillan college Publishing Company)
[9] Hodiyanto 2017 Kemampuan komunikasi matematis dalam pembelajaran matematika *AdMathEdu* **7** 6
[10] Cai J, Jakabcsin M S and Lane S 1996 Assessing students' mathematical communication *School Science and Mathematics* **96** 238

Integrating technology and media into mathematics learning

U Rahmi, Y Helsa and Azrul

Universitas Negeri Padang, Jl. Prof. Dr. Hamka, Air Tawar, Padang, Indonesia

E-mail: ulfia@fip.unp.ac.id

Abstract. The purpose of this study was to determine the size of developed product validation. The resulting product is interactive multimedia for learning mathematics in third class in elementary school. This development informed with the consideration that technology and learning media assist teachers in creating active, engaging, and meaningful learning. It uses in education makes learning more effective. Validator consists of media experts, linguists, and learning material experts. The instrument used is a validated questionnaire for each expert. The data were analyzed quantitatively by descriptively. The results show that the media is valid regarding media (78%), linguistic (81%), and material (80%). Indicates that the multimedia of mathematics learning for elementary school is appropriate to be used and continued to the stage of practical assessment and effectiveness.

1. Introduction

Technology and instructional media assist teachers in creating active, engaging, and meaningful learning. The emergence and development of technology can improve the effectiveness and interactivity of education [1]. However, obtaining reliable technology and instructional media is a complex task [2] for both the instructional media developers and the teachers. They need to consider various aspects of providing technology and media for students, such as students' current knowledge, learning styles, motivation, age, optimizable sensory tools, and sociocultural circumstance [2-4]. The consideration of these aspects is dynamism technology and media in the classroom involving various types of instructional media.

Instructional media can be used at all levels of education and teaching materials both formal and informal [5], including in mathematics learning for elementary school. This mathematics learning is structured on the level of material difficulty, from the easy to difficult one and from concrete to abstract one. Mathematics for elementary school also begins with the underlying mathematical principles used during the learning process, one of which is' Number Operation and Measurement. However, most elementary school teachers abstractly teach this material, which seems difficult for students. Meanwhile, it is very much beneficial to present to material concretely manipulating the visual cues, so it is easier for students to engage the material being studied [6, 7].

The problem of presenting concrete, accessible, and relevant material to students can be solved by using mathematics learning multimedia. Instructional media developers and teachers can improve the quality of learning by using various types of media [8], supported by multiple technologies [9]. Some criteria for choosing the excellent quality of technology and multimedia for learning are suitability to standards, outcomes and learning objectives; accurate and up-to-date information; age-appropriate language; student's level of interest and involvement; technical quality; ease of use; inability and user manual [10]. This multimedia development process follows these criteria to get the multimedia that fits the needs of the students. The most important thing is to use it appropriately, as it can increase students' interest in the lessons they are currently studying [11].

It is believed that the higher the students' interest in learning, the greater their chances of meaningful learning. Multimedia learning can present interesting and concrete learning by manipulating mathematical concepts into things relevant to students' lives. This manipulation not only increases students' understanding of concepts but also improves their positive attitudes such as attention and motivation. In multimedia learning, manipulation used is computer manipulation. It provides an interactive environment where students encounter problems of everyday life and their relation to teaching materials. Concrete experience in mathematics learning is not defined as a characteristic of the real world, but how to interpret the relationship between mathematical concepts and real situations [7]. This manipulation will possibly increase students' understanding of mathematical concepts [12, 13]. For example, for the material of number operations and measurements, students engage in meaningful learning by summing up the numbers and relating them to real life, such as summing up the books they have.

2. Method

The validation of mathematics learning multimedia for elementary school uses quantitative data analysis techniques. The done procedure is assessing the validity of mathematics teaching multimedia for an elementary school that has been designed based on students' needs, learning objective, and facility availability analysis [14]. The assessment is assessed by expert validators of instructional media experts, language, and determining the material from Universitas Negeri Padang. The instrument used is a validated questionnaire for each expert. Then, the analyzed data is a quantitative descriptive technique.

3. Result and discussion

This research begins by contacting the validators required to assess the validity of mathematics learning multimedia for elementary school. These validators are academicians and experts from the disciplines of mathematics, primary education, instructional media and language at Universitas Negeri Padang. Figure 1 shows a validated home media page.

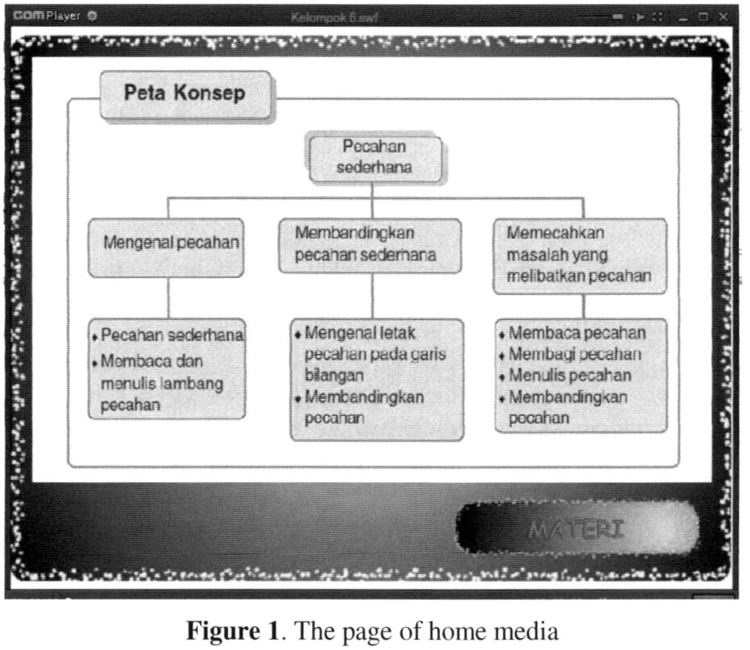

Figure 1. The page of home media

The multimedia of mathematics learning is validated by following the five steps, such as

1. Developing validation instruments
   It is the first step in validating mathematics learning multimedia for elementary school. The instrument is designed based on the principles of instructional media, mathematical concepts, and principles of learning for elementary school students.
2. Calculating instrument validity and reliability
   Before the validator uses the instruments, they estimate on the level of validity and reliability. Table 1 shows the valid and reliable test results of the tools used.

Table 1. Validity and reliability of validation instruments for mathematics learning multimedia

Instrument	Validity	Reliability
Instructional Multimedia	valid	reliable
Contents of Instructional Multimedia	valid	reliable
Language of Instructional Multimedia	valid	reliable

3. Revising instruments
   Some invalid and unreliable items on the instruments are omitted.
4. Validator assessment of products using valid and reliable instruments
   The experts of instructional media, language, and learning material s assess the feasibility of mathematics learning multimedia for the elementary school that has been developed. Each instrument is given according to the expertise of the validator. In addition to the questionnaire, the validator also provides suggestions for improvement of the developed multimedia. The recommendations are used to revise the product according to students' input and needs.
5. Analyzing validation data
   The data from the validator are collected and analyzed using the quantitative descriptive technique. The result of this analysis is provided in Table 2. It shows that the developed multimedia is valid, so it is ready to be tested for the stages of practicality and effectiveness. The validity of multimedia learning is supported by a combination of various learning media formats including text, audio, visual, video, and model [10]. The combination offers flexible and more interactive learning so that students are interested in learning math. However, such design implies that the combination of various media formats is not significant enough to give effect to the learning outcomes because it prioritizes more on students' interactivity [15].

Table 2. The result of data analysis on mathematics learning multimedia validation

Validator	Assessment	Result	Suggestion
Instructional media expert	78	Valid	Moderate Revision
Language expert	81	Very Valid	Mild Revision
Learning material expert	80	Very Valid	Mild Revision

This obtained validity implies the success in designing manipulation object to make it look concrete for students. The manipulation of multimedia content presentation is done by adjusting to learning objectives, combining various media formats, adapting to the content being taught, and merely displaying [16]. In particular mathematical content, this displayed manipulation helps students to connect real-world situations with mathematical symbols, work together to solve problems, discuss mathematical concepts and ideas, verbalize mathematical thinking, do the presentation in front of large groups, find ways to resolve the issues and solve mathematical problems based on teacher guidance [17-21].

4. Conclusion

The development of mathematics learning multimedia for elementary aimed is at meeting the students' need for their interaction with technology. It is also an effort to make the mathematical concepts concrete during the learning process. This instructional multimedia test before students widely uses it.

Based on the research results, this multimedia is valid from the aspects of instructional media, content, and language use.

References

[1] Wibowo E J 2013 *Seminar Riset Unggulan Nasional Informatika dan Komputer FTI UNSA* **2** 75
[2] Smaldino S E, *et al* 2014 *Instructional technology & media for learning: Teknologi pembelajaran dan media untuk belajar* (Jakarta: Kencana)
[3] Slavin R E and Davis N 2006 *Educational psychology: Theory and practice* (Boston: Allyn & Bacon)
[4] Pritchard A 2013 *Ways of learning: Learning theories and learning styles in the classroom* (London: Routledge)
[5] Odero J O and Chinapah V 2016 *J. Educ. Res.* **5** 107
[6] Collett P and Steyn C 2017 Using African designs in virtual manipulatives for geometrical concept development in *Bridges 2017 Conference Proceedings* (Melbourne: Tessellations Publishing) p 455
[7] Durmus S and Karakirik E 2006 *TOJET Turkish Online J. Educ. Tech.* **5** 107
[8] Leow F T and Neo M 2014 *TOJET Turkish Online J. Educ. Tech.* **13** 99
[9] Slavin R, *et al* 2013 Effects of Cooperative Learning and Embedded Multimedia on Mathematics Learning in Key Stage 2: Final Report *Project Report* (Toronto: University of York)
[10] Smaldino S E, *et al* 2008 *Instructional technology and media for learning* (New York: Pearson)
[11] Isnaini M S, *et al* 2017 *Eur. J. Spec. Educ. Res.* **2** 1
[12] Uttal D H, *et al* 1997 *J. Appl. Dev. Psychol.* **18** 37
[13] Hidayah I, *et al* 2017 *Journal of Physics: Conference Series* **824** 12042
[14] Branch R M 2009 *Instructional design: The ADDIE approach* (Berlin: Springer Science & Business Media)
[15] Rasch T and Schnotz W 2009 *Learning and Instruction* **19** 411
[16] Suydam M N and Higgins J L 1976 Review and synthesis of studies of activity-based approaches to mathematics teaching *Columbus* (Ohio: ERIC/SMEAC)
[17] Istiandaru A, *et al* 2018 *J. Phys.: Conf. Ser.* **943** 012023
[18] Masniladevi, *et al* 2018 *J. Phys.: Conf. Ser.* **943** 012059
[19] Heddens J W 1997 Improving mathematics teaching by using manipulatives (Kent: Kent State University)
[20] Ahmad S, *et al* 2018 *J. Phys.: Conf. Ser.* **943** 012053
[21] Zainil M, et al 2018 *J. Phys.: Conf. Ser.* **943** 012046

The development of student worksheet by using discovery-based approach: a case study in learning media course

E B Rahaju and P Wijayanti
Universitas Negeri Surabaya, Jl. Ketintang Surabaya, JawaTimur, 60213, Indonesia

E-mail: endahrahaju@unesa.ac.id

Abstract. With regard to the national curriculum for the primary and secondary school, the Department of Mathematics of Universitas Negeri Surabaya through Indonesian National Qualification Framework (KKNI) has tried to accommodate the demands of the National Curriculum 2013. The department proposed an innovation through Learning Media course. This study aims at producing student worksheet which had been developed by students themselves by using a discovery-based approach. It adopted Plomp's model of developmental research. Thirty-eight students participated in this study. Research instruments used were validation sheet and student worksheets. This study produced 11 student worksheets for junior high school and eight student worksheets for senior high school. The results showed that discovery activity matched learning objectives and student worksheet involved students' initial knowledge. Furthermore, it revealed that discovery stage matched student's cognitive development. The finding also showed that discovery activity guided students to make a conclusion and the answers for student worksheets were correct and matched the worksheets. However, student worksheet still had several grammatical errors; therefore, revision was needed.

1. Introduction

Learning is a process of interaction between students and teachers using some resources in one environment. It is a process to help students to learn well. The process will lead to the learning achievements. It will be effective when it is supported by students' learning motivation and teacher's creativity. The use of learning media can improve students' interest, passion, motivation and stimulation [1,2,3]. At the beginning of learning process, the media will improve the effectiveness of the learning process and deliver the messages and content of lessons. Hence, learning objectives can be achieved optimally. The learning process in primary and secondary school is regulated through the Regulation of the Minister of Education and Culture No. 22 of 2016 [4] concerning a Process Standards, which contains the basic concept of the learning process. Students are viewed as subjects who have the ability to actively search, process, construct, and use knowledge. In line with this view, learning should be concentrated upon giving the learners opportunities to construct knowledge in their cognitive processes. Furthermore, in order to truly understand and be able to apply knowledge, they need to be encouraged to work, to solve problems, to find everything for themselves, and strive to realize their ideas.

In line with the basic concepts and principles of learning, scientific approach or a scientific-based approach is used in the national curriculum. In order to strengthen the scientific approach, it is necessary to apply discovery/inquiry learning. In addition to the scientific approach, other learning models, such as discovery learning, project-based learning, problem-based learning, and inquiry learning, can be applied. in regard to the curriculum development in high school, the Department of Mathematics at one of the universities in Surabaya through KKNI Curriculum has tried to innovate and accommodate the demands of national curriculum, one of which is by setting up innovative learning course II. This course is intended to develop students' understanding of such learning models with a scientific approach as problem based learning, project-based learning, inquiry-discovery learning, and contextual learning as well as their application in task-based and IT-based learning simulations. In the course, students learn about learning theories, learning characteristics and learning syntax by using the aforementioned models and develop learning instruments that are relevant to those models. Based on the evaluation, it was found that many students in Innovative Learning II course still lacked ability to develop student worksheets for discovery learning.

As a continuation of Innovative Learning II, students can take Learning Media course. The objective of this course is that students understand the definition, type/classification, functions, basic development of learning media as well as apply them in learning process as their learning strategies. Up to now, students are only requested to develop contextual (manipulating the environment component) and ICT-based learning media. Among the media students can develop is student worksheet (specifically based on discovery). To develop various kinds of learning media is one important element a teacher candidate must possess in order to improve his/her competency as a professional teacher.

A successful use of learning media is for improving learning outcomes depends on the content of the message, how it is delivered, and the characteristics of the message recipients. Thus, in choosing and using the media, it is necessary to consider three factors. If the three factors can be delivered through learning media, optimum results will certainly be achieved. Student worksheet is a part of visual media [1]. There are several types of student worksheets: 1) one which helps students to find a concept, 2) one which assists students to apply and to integrate the various concepts that have been found, 3) one which serves as a learning guide, 4) one which serves to strengthen, 5) one which serves as a practical guide.

One important aspect that must be considered by the lecturers of Learning Media course is students' ability to develop various kinds of learning media. It is because students must have this competence when they start teaching at schools. A good teacher is one who can design a good learning, design learning instrument, implement learning with appropriate learning methods or models, and assess the implementation of learning.

2. Research methods

This research is categorized as developmental research, aiming at developing student worksheet for junior and senior high school levels. The students who took Learning Media course developed the worksheets. This research was conducted in the Department of Mathematics at a university in Surabaya. The subjects were all students in year 2015 who enrolled in Learning Media course. The research was conducted in the second semester of 2016/1017 academic year involving 38 students.

Two research instruments were employed in this research, namely validation instrument and student worksheets that had development tasks. An analysis was also conducted and used to answer the research question. In analysing the data, initially the resulted student worksheets for junior and senior high schools were grouped. Next, the results of task development were validated. This included some activities such as: 1) determining the suitability of the established indicators to the type of selected student worksheet, 2) determining the suitability of the established indicators to the tasks assigned; 3) conforming the activities in student worksheet to students' cognitive development ; 4) conforming the activities developed in the student worksheets to discovery activities, 5) conforming the discovery tasks to the possibility of eliciting alternative solutions, 6) conforming the activities in student worksheet to selected learning model, 7) matching the student worksheets with their answers, 8) checking the correctness of the answers, and 9) checking the adequacy of the language used.

3. Results

This research used developmental research model as proposed by Plomp [5]. It consists of five phases; 1) Initial Investigation Phase, 2) Design Phase, 3) Realization Phase, 4) Test, Evaluation, and Revision Phase, and 5) Implementation Phase. Unfortunately, this research was conducted only until the fourth phase. Because of limited time, the implementation phase cannot be conducted. Each phase needed to be revised until the classification was valid. It can be described in detail as follows:

3.1. Initial investigation phase

The initial investigation phase was undertaken to determine the teaching materials needed in developing student worksheet. At this stage, theoretical analysis, curriculum analysis, student analysis, and teaching material analysis of the student worksheets were conducted by collecting and analysing the supporting information in order to determine the next plan. The four stages can be explained as follows:

i. Theory supporting the development of student worksheets

To support the creation of student worksheets in accordance with discovery activities, the researcher investigated some literature to study the theories of student worksheet development and its application in learning [6,7,8,9,10].

ii. Curriculum analysis

At this stage, the curriculum used in junior and senior high schools especially in Surabaya was a reviewed. The researchers focused on the national curriculum 2013 (K-13). The curriculum would be implemented simultaneously throughout Indonesia commencing from 2018-2019 academic years. In addition, interviews were conducted with some junior and senior high school teachers in Surabaya to find out the benefits of student worksheet in the on-going math learning. The results of the interviews indicated that the student worksheets were still needed to enable students to understand mathematical concepts. The student worksheets available so far contain exercise questions.

iii. Student analysis

Students were investigated to examine their characteristics in accordance with the design of student worksheet with discovery-based approach. The students' characteristics investigated included their background knowledge of mathematics, their initial ability they had already possessed, and academic ability.

iv. Analysis of teaching materials

Analysis of teaching materials was intended to select, define, itemize and systematically compile teaching materials relevant to the development of student worksheets for junior and senior high schools. In the analysis of teaching materials, researchers mapped out K-13 materials in accordance with discovery activities. Teaching materials were selected by considering the conformity of the concept and the content of the material to the discovery activities. Based on the results of teaching material analysis, the researchers determined the appropriate junior and senior high school materials for discovery activities. There were 19 materials for junior high school and 10 materials for senior high school [11,12].

3.2. Design phase

At this stage, every student designed a prototype of student worksheet based on discovery-based approach for junior and senior high schools. Along with these activities, a research instrument, which consisted of student worksheet validation sheet and Development Task, was also developed.

The validation sheet focused on a) the suitability of the established indicators to the tasks assigned; b) the suitability of activities in student worksheet to the cognitive development of students; c) the suitability of the activities developed in student worksheets to the discovery activities, d) the suitability of the discovery tasks to the possibility of eliciting alternative solutions, e) the suitability of discovery activities in student worksheet to the selected learning model, f) the suitability of student worksheets and the answers, g) the correctness of answer, and h) the adequacy of language used.

The Development Task contains assignment that students must complete. It asked each student to develop a student worksheet with discovery-based approach. Students selected one of the mapped materials suitable for discovery activities.

3.3. Realization phase

In this phase, the college students tested the worksheets by assigning them to other college students enrol in Learning Media course. The students were from class 2015A. Each of them identified the topics that can be developed by using discovery-based approach in junior or senior high schools. They developed 19 student worksheets consisting of 11 worksheets for junior and 8 worksheets for senior high school. The student worksheets in this phase were referred to as the first prototype student worksheets.

3.4. Revision phase

Prototype student worksheet 1 was validated by three lecturers of mathematics education. The result of validation showed that a) the discovery activity in student worksheets matched the predefined indicators, b) the prototypes had integrated students' initial knowledge, c) the discovery stage was based on students' cognitive development although it had not been sequenced, d) discovery activities led students to conclude the findings, e) discovery activities that elicited alternative solutions were still less than 50%, f) Student worksheet answers were correct and suitable for the student worksheets, g) Some directions were still confusing, requiring a slight improvement.

After the validation, the students had to revise the first prototype, resulting in what was called prototype 2. Furthermore, it was found that student worksheets compiled by the students could help learners find a concept [1]. The developed student worksheets were in accordance with predetermined indicators. This shows that the students had been able to determine the materials that could be learned through the discovery activities. The student worksheet provided student-centred activities to construct knowledge. In addition to their compliance with the indicators, the worksheets were already developed by considering the characteristics of students as the recipients of the message. These characteristics included the students' initial knowledge, level of cognitive development, and language adequacy.

The developed student worksheets showed that the students were able to develop learning media in accordance with predetermined learning strategies (especially discovery-based approach). The results indicated that the worksheets could develop students' competence in developing learning media in addition to the already-existing contextual and ICT-based learning media.

4. Conclusion

Based on the data analysis, it can be summarized that initial investigation phase gave a foundation for the development of student worksheets so as to implement national curriculum 2013. There were 19 materials suitable for discovery activities. At the next stage, every student developed student worksheet with a discovery-based approach. The student selected one of the mapped materials which were suitable for the discovery activity. In the realization stage, the students developed 11 student worksheets for junior high schools. Based on the validation results, it was found that 82% of the student worksheets developed by the students were highly relevant with the indicators; 45% of the tasks given were highly relevant with the indicators; 55% of the tasks given were relevant with the indicators; 64% of the activities were highly suited to learners' cognitive development; 36% of the activities were suited to learners cognitive development; 91% of the activities developed in the student worksheets were categorized as *good* and *excellent* discovery activities; 27% of the invention tasks leading to alternative solutions were categorized as *good*, *very good*, and *excellent* tasks, while 72% of discovery activities lacked alternative solutions; 100% of student worksheet answers were appropriate and correct; 55% of the language used was very appropriate; and 45% of the language used was in suited to the students' level of cognitive development.

In the initial investigation phase of the development of student worksheets for senior high school based on Curriculum 2013, there were 10 materials suitable for discovery activities. At the next stage, every student developed student worksheets with discovery-based approach. They selected one of the mapped high school topics suitable for discovery activities. At the realization stage, the students developed 8 student worksheets. Based on the validation, 86% of student worksheets were very suitable for discovery-based approach student worksheet; 71% of the tasks were highly relevant with the indicators; and 29% of the tasks were relevant with the indicators; 86% of the activities were very suited to learners' level of cognitive development; and 14% of the activities were consistent with the level of learners' cognitive development; 91% of the activities developed in the student worksheets were discovery activities with good and very good criteria, 27% of the invention tasks leading to alternative solutions were categorized as *good* and *excellent*, while 72% of discovery activities lacked alternative solutions; 100% of student worksheet answers were appropriate and correct; 86% of the language used was very consistent with the learners' level of cognitive development; and 14% of the language used was suited to learners' level of cognitive development.

Regarding the findings, we suggest that other researchers develop students' worksheets that are suitable for implementing project-based and problem-based learning for other mathematics topics as regulated in *Kurikulum 2013*.

References

[1] Heinich R, Molend, M, Russell JD and Smaldino S 1999 *Instruction Media and Technologies for Learning* (New Jersey: Merrill, Prentice Hall)
[2] Molenda M, Russell JD and Smaldino S 1999 *Instruction Media and Technologies for Learning* (New Jersey: Merrill, Prentice Hall)
[3] Nur and Mohamad 2000 *Media Pengajaran dan Teknologi untuk Pembelajaran* Makalah tidak diterbitkan
[4] Permendikbud no 22 2016 *StandarProses Pendidikan Dasar dan Pendidikan Menengah* (Jakarta: Kemendikbud)
[5] Plomp and Tjeerd 2006 *Educational and Training System Design* (Enschede The Netherlands: University of Twente)
[6] Brodie K 2010 *Teaching Mathematical Reasoning in Secondary School Classroom* (New York: Springer)
[7] Bruner J S 1990 *Acts of Meaning* (Cambridge, MA: Harvard University Press)
[8] Bruner J S 2009 *The Process of Education* (Harvard University Press)
[9] Csapo N & Hayen R 2006 The Role of Learning Style In The Teaching/ Learning Process. *Issues in information systems* **7** 129
[10] Merril Physical Science 1995 *Teacher Resource Guide* (New York: Glencoe MacMillan/McGraw Hill)
[11] Staples M & Bartlo J 2010 Justification as a learning Practice: Its Purposes in Middle Grades Mathematics Classrooms *Center of Research in Mathematics Education*
[12] Permendikbud No 24 2016 *Kompetensi Inti dan Kompetensi Dasar Pelajaran Pada Kurikulum 2013 pada Pendidikan Dasar dan Pendidikan Menengah* (Jakarta: Kemendikbud)

Generative learning model to improve mathematics problem solving skills on polyhedron

Andriani, M Ikhsan, S Munzir and C Khairunnisak
Syiah Kuala University, Jl. Teuku Nyak Arief Darussalam Banda Aceh 23111, Indonesia

Email: smunzir@unsyiah.ac.id

Abstract. The purpose of this study was to develop valid, practical and effective generative learning instruments to improve the mathematics problem solving skills. This research was development research using Thiagarajan, Semmel & Semmel's 4D model consisting of designing, defining, developing, and disseminating stages. However, this study was limited to the validation stage. The instruments developed were lesson plan, student worksheet, and mathematics problem solving test. Data collection techniques used were questionnaires and tests. The participants are 24 year 8students in one of secondary school in Pidie, Aceh, Indonesia. The validity data of learning instruments were obtained based on expert validation while the validity data of the test were derived based on students' answers. The results of this study showed that the learning instruments developed was valid, with the score for (a) lesson plan was 4.54 (excellent category), (b) student worksheet was 4.32 (good category), and (c) the test of mathematics problem solving skills was 4.4 (good category). Furthermore, the developed test of mathematics problems solving skills showed that the average results of validity and reliability were categorized into the high criteria, difficulty index into medium criteria and differentiation index into good criteria.

1. Introduction

Mathematics is one of the subjects playing an important role in education. It is taught in educational institutions, from elementary to university levels. One of the objectives of mathematics subjects as stated in the Standard of Mathematics Subject Content of Junior High School is that students can solve mathematics problems including the ability to understand problems, design mathematics models, solve models and interpret the solutions [1]. Therefore, the mathematics learning that we do must be oriented to problem solving [2,3], and it is should be teacher's goal help students understand and resolve the problems as most problem solving in mathematics learning is about the problems in everyday life.

Problem solving is the process of applying the previously acquired knowledge to new and different situations [4]. "Problem solving is the heart of mathematics" meaning that the fundamental of mathematics is problem solving [5]. Recognizing the importance of mathematical problem solving skills, the teachers are required to deliver learning that makes students more active and free to express ideas and reasons for the given problems. Therefore, students will better understand the knowledge that formed themselves and the learning process will be more optimal.

One of the efforts that can be done is through developing learning instruments of generative learning model. Generative learning theory is developed to explain the cognitive processes that occur during learning [6]. It defines learning as a generation of knowledge of "the creation of new understanding" [7]. Generative learning is a learning model emphasizing the active integration of new knowledge by using the students' initial knowledge so that students express what they have acquired in their own words [8].

There are four stages in the generative learning model, namely the exploration stage, the focusing stage or introduction of concepts, the challenge stage, and the application stage [9]. In the first stage, the teacher guides students to explore the initial knowledge, ideas, or concepts derived from daily experiences or gained from previous levels of learning. In the second stage, students are given the opportunity to express their ideas about the concepts being studied. In the third stage, the teacher gives students the opportunity to compare their opinions with other students' and express the superiority of their opinion of the concept being studied. In the last stage, students are given the opportunity to test alternative ideas that they build to resolve the varied issues. In this learning process, students are active in the discussion because it is another way of deepening the understanding of concepts through social interaction [10].

Considering the background described above, this study aimed to develop valid, practical and effective generative learning instruments to improve the mathematical problem solving skills.

2. Method

The type of this study is a Research and Development using the 4-D model. This model consists of four development stages of Define, Design, Develop, and Disseminate [11]. However, this study was limited to the Develop stage.

In the first stage, the learning conditions should be defined. Considering the five steps of the defining phase [11] that have been performed, it was concluded that most students were having difficulty in understanding and solving problems of the polyhedron. This is because students had difficulty in determining the mathematics model and they did not understand the problems so that the significant information not being used to solve the problem. Therefore, problem solving is hard for students.

The participants in this study were 24 Year 8 students in one of secondary school in Pidie, Aceh, Indonesia. Students' ability and interest were varied. The common teaching method used in the school was lecturing, discussion and solving problems together. Concept analysis and task analysis generated an instructional objective, that is learners can solve problems related to cubes and cuboids through generative learning.

The next stage of the development is Design. There are four steps in the Design stage, namely criterion-test construction, media selection, format selection, and initial design. Through the four steps, the initial learning instruments were developed. The learning instruments developed was lesson plan, student worksheet, test questions and self-efficacy questionnaire [12]. However, this paper will only discuss the lesson plan, student worksheet, and test questions. The lesson plan was developed based on the generative learning model syntax, whereas student worksheet and test questions were developed based on the indicators appropriate to mathematics problem solving

The third stage of the 4-D model is Develop. According to Thiagarajan et al. [11], this stage consisting of two steps, expert appraisal and developmental testing. In the expert appraisal step, the initial learning instruments that have been designed were validated to determine its quality. The validators in this study consisted of two mathematicians, two mathematics teachers, and one colleague who had attended various seminars. Based on the validation results of the lesson plan and student worksheet, the instruments were revised for subsequent use in the study.

3. Result and discussion

The developed learning instruments included: (1) lesson plan, (2) student worksheet, and (3) mathematical problem solving test. The learning instruments developed needs to be validated and the validity is measured based on the validity results [13]. Therefore, the initial product of these instruments was called Draft 1, and it was ready to be validated by the validators. In addition to these learning instruments, researchers also designed the instrument for assessment of the learning instruments. The assessment instruments consisted of the validation sheets of lesson plan, student worksheet and mathematics problem solving accompanied by assessment guides.

IOP Conf. Series: Journal of Physics: Conf. Series **1088** (2018) 012075 doi:10.1088/1742-6596/1088/1/012075

The validation process of the draft I was conducted to examine the content validity of the instruments developed. Learning instruments validation results were used to determine whether the learning instruments can be used or not. Suggestions from the five experts were used as a basis for the improvement of the instruments developed. Once the validators validated them, revisions were made following the feedbacks of the validators and Draft II was obtained. The validation result of Draft I is presented in Table 1.

Table 1. Results of validators' assessment toward the draft I of learning instruments.

Validated instrument	Average	Criteria
Lesson plan	4.56	Excellent
Student worksheet	4.32	Good
Mathematical problem solving skill	4.4	Good
Average	4.43	Good

Based on the validator's assessment of the items on the mathematics problem solving skills, they fulfilled the content validity and can be used with little revision. The average score of validators' assessment for the lesson plan is 4.54 (excellent category), student worksheet is 4.32 with a good category, and the test of mathematical problem solving ability is 4.4 (good category). Overall, the average validation score for the learning instruments, including the lesson plan, student worksheet, and mathematics problem solving skills is 4.43 and falling into a good category (a scale of 5). Therefore, it was concluded that the learning instruments developed can be used with little revision.

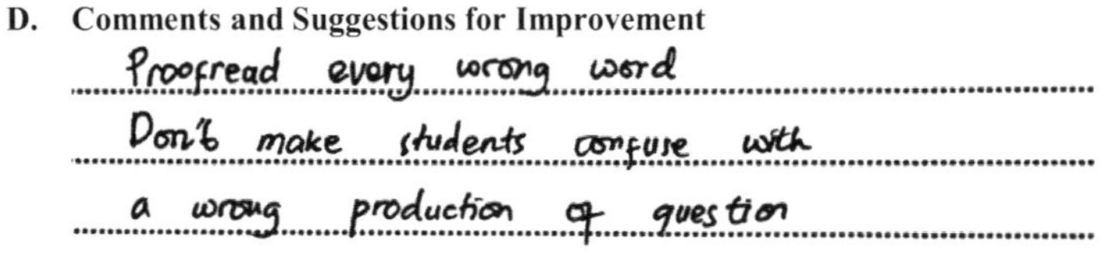
D. Comments and Suggestions for Improvement

Proofread every wrong word

Don't make students confuse with

a wrong production of question

Figure 1. Validators' comments to the test of mathematical problem solving ability

Example of validators' feedback or comment to the test of mathematical problem solving ability can be seen in Figure 1. Furthermore, the summary of validators' feedbacks toward the learning instruments is presented in Table 2.

Table 2. Revisions to the learning instruments according to validators' feedbacks

No	Before validation	After validation
I	Test of mathematical problem solving ability	
	• Students are not familiar with the words or sentences used.	• The words or sentence has been revised.
	• The information in the questions is not systematic and appropriate.	• The information given is complete and appropriate.
II	Lesson plan	
	• The sentences in the lesson plan do not clearly emphasize the activities of students and teacher.	• The sentences in the lesson plan have emphasized the activities of students and teacher.
III	Student worksheet	
	The allocated time is not in line with the content presented.	The allocated time has been revised based on the content
	Avoid using 'follow the procedure…'	The words have been removed from the lesson plan.

Learning instruments developed as previously described, namely lesson plan, student worksheet, and the test of mathematics problem solving skills. In this research, lesson plan was designed to follow generative learning model to improve mathematics problem solving skills that were the exploration or preliminary stage, the focus or introduction stage, the challenge stage, and the implementation stage. Learning took place for four meetings with a time allocation of 2 x 40 minutes for each meeting. Based on the lesson plan validity results, it was valid with a minor revision. Student worksheet in this study is designed to encourage students actively in learning activities, practice the skills of learning, and develop problem solving skills and curiosity of students. From the validation result student worksheet obtained valid criteria with a minor revision. Figure 2 presents part of the developed lesson plan and student worksheet.

Activity 2. Solving Problem Related to Volume of Cuboid

Pak Erik is building a store. The dimension of the store is 4m × 7m × 3m with the thickness of the wall is 12cm. Pak Erik has made the foundation of his store using six pillars, the dimension of each pillar is 25mx30cmx3m. What is the minimum money needed by Pak Erik to build the wall?

(The cost of a wall with dimension 1mx12cmx1m is Rp65.0000,00 and the front side of the shop entirely using a metal door).

(a)

Identify the Issues
Berdasarkan kegiatan 3 apa yang dapat kamu ketahui?

Diketahui :
Ditanya :

Devise a Plan
Langkah atau strategi apa yang dapat kamu gunakan untuk menyelesaikan kegiatan 2

Hal-hal yang perlu dilakukan :

Carry Out the Plan
Laksanakan penyelesaianmu

Look Back
Apakah yang dapat kamu simpulkan dari kegiatan 2

(b)

Figure 2. Example of the problem on the developed student worksheet

Mathematics problem solving test was designed to measure students' problem solving skills on polyhedron. The preparation of the test instruments of mathematical problem solving skills involved the preparation of test prediction and scoring rubrics. The items of the mathematics problem solving skills test were consulted to the supervisors and validators to obtain the content validity. From the results of the validation, the mathematical problem solving skills test acquired was valid with minor revision.

Table 3. Table of the values of mathematical problem solving skills test results

No	Explanation			
	Validity	Reliability	Difficulty index	Differentiation index
1	High	High	Medium	Fair
2	High	High	Medium	Good
3	High	High	Medium	Good
4	High	High	Medium	Good
5	High	High	Medium	Good
6	High	High	Medium	Good
7	High	High	Medium	Fair
8	High	High	Medium	Good

The items of mathematics problem solving skills test were tested to assess its validity, reliability, difficulty and differentiation index to examine the feasibility of the items. Therefore, the final product of the mathematics problem solving test is a set of questions meeting the valid, reliable, difficulty, and differentiation index of good or excellent. The results of test items validity, reliability, difficulty index, and differentiation index are presented in Table 3.

Based on Table 3, it can be concluded that the test of mathematics problem solving skills and reliability are classified as the high criteria, while difficulty index is classified as the medium criteria and differentiation index as the good criteria. Mathematics problem solving test items consists of pretest and posttest items, item 1, 2, 3, and 4 are the questions of pre-test while item 5,6,7, and 8 are the items of post-test. The correlation coefficient of validity test for the pretest items were 0.704, 0.853, 0.844, and 0.837 subsequently with the high and valid criteria. The result of reliability test for the items of mathematics problem solving skills test was 0.613meaningthe reliability is very high. The difficulty indexes of pretest items were 0.631, 0.623, 0.621, and 0.530, falling into for medium category. As for the differentiation index result for the pretest items were 0.310, 0.421, 0.510, and 0.405 categorized as the fair criteria for the item 1 and good criteria for items 2, 3, and 4.

The results of correlation coefficient calculation for the validity of post-test questions in the sequence were 0.794, 0.846, 0.780, and 0.864 with the high and valid criteria. The result of the reliability test for the items of the mathematics problem solving skills testis 0.623 meaning that its reliability is very high. The difficulty indexes of post-test items are 0.673, 0.644, 0.537, and 0.647 subsequently with the criteria for all questions is medium. As for the differentiation index, the results obtained the pretest items are 0.600, 0.496, 0.327, and 0.483 subsequently with fair criteria for item 3 and good criteria for item 1, 2, and 4.

Figure 3 presents one of student's answer to the posttest. As shown in figure 3, the student was able to solve the given problem by following the steps of problem solving approach: identify the issues, devise a plan, carry out the plan, and look back [2].

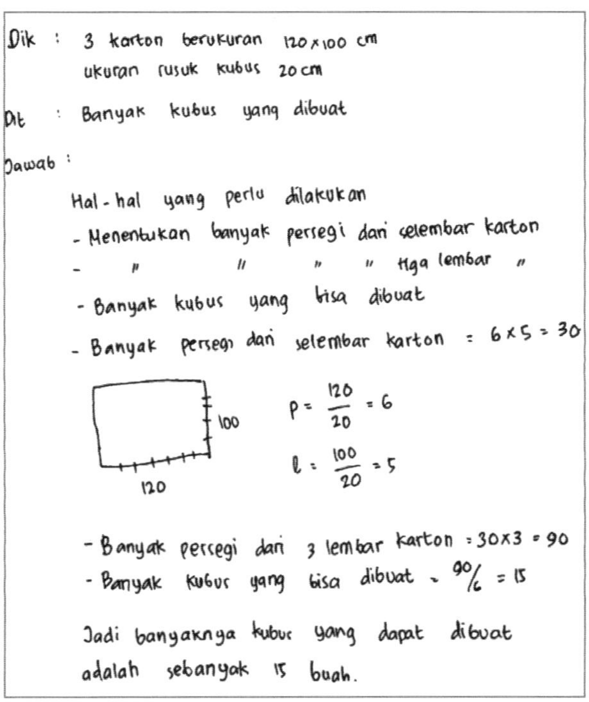
Dik : 3 karton berukuran 120x100 cm
ukuran rusuk kubus 20 cm
Dit : Banyak kubus yang dibuat
Jawab :
Hal-hal yang perlu dilakukan
- Menentukan banyak persegi dari selembar karton
- " " " " tiga lembar "
- Banyak kubus yang bisa dibuat
- Banyak persegi dari selembar karton = 6x5 = 30

100, 120

$p = \frac{120}{20} = 6$

$l = \frac{100}{20} = 5$

- Banyak persegi dari 3 lembar karton = 30x3 = 90
- Banyak kubus yang bisa dibuat = 90/6 = 15

Jadi banyaknya kubus yang dapat dibuat adalah sebanyak 15 buah.

Given : 3 cardboard with dimension 120×100cm
The length of the cube edge is 20 cm

Asked : The number of cubes can be formed [from the cardboards]

Answer :

Things to be done
- Determining the number of squares can be formed from one cardboard
- Determining the number of squares can be formed from three cardboards
- Determining the number of cubes can be formed
- Number of squares formed from a cardboard = 6×5=30

$$p = \frac{120}{20} = 6$$
$$l = \frac{100}{20} = 5$$

- Number of squares from 3 cardboards = 30×3 = 90
- Number of cubes $\frac{90}{6} = 15$

Thus, we can form 15 cubes.

Figure 3. Example of student's answer to the mathematics problem skill test.

Students' problem solving skills are excellent because the appropriate learning instruments used, one of which the learning instrument of the generative learning model. This is in accordance with a

IOP Conf. Series: Journal of Physics: Conf. Series **1088** (2018) 012075 doi:10.1088/1742-6596/1088/1/012075

study that revealed that there is an influence of the application of generative learning models on the mathematics problem solving skills [13]. In addition, the role of teachers and methods applied in learning influence students' problem solving skills [14, 15]. This can be interpreted that the selection and application of a learning instrument to enhance students' mathematics problem solving ability.

4. Conclusion

The mathematics learning instruments developed using generative learning model on the topic of polyhedron meets the criteria of validity. It shows that the average validation of experts is 4.43 of scale 5 (good category), the average score of the lesson plan is 4.54 (excellent category), student worksheet is 4.32 (good category), and mathematics problem solving skill test is 4.4 (good category). Overall, the developed learning instruments can be used with minor revision. Likewise, the test of the mathematical problem solving skill is also valid. The average results of validity and reliability are categorized into the high criteria, difficulty index into the medium criteria and differentiation into the good criteria. Therefore, the learning instruments using generative learning model on polyhedron that have been developed can be tested in the field as it has met all valid criteria.

References

[1] DEPDIKNAS Kurikulum 2006 *Kurikulum Tingkat Satuan Pendidikan (KTSP) Mata Pelajaran Matematika* (Jakarta)

[2] Baroody A J 1993 *Problem Solving Reasoning and Communicating K-8 Helping Children Think Mathematically* (New York: Macmillan Publishing Company)

[3] NCTM 1989 *Curriculum and Evaluation Standards for School Mathematics* (Reston: VA NCTM)

[4] NCTM 2000 *Principles and Standards for School Mathematics* (Virginia: NCTM)

[5] Lester F K and Kehle P E 2003 From Problem Solving to Modeling The Evolution of Thinking About Research on Complex Mathematical Activity *Beyond Constructivism Models and Modeling Perspectives on Mathematical Problem Solving Learning and Teaching* eds R Lesh & H M Doerr (Mahwah NJ Lawrence Erlbaum Associates)

[6] Wilhelm-Chapin M K and Koszalka T A 2016 *Generative Learning Theory and its Application to Learning Resources* (Syracuse University RIDLR Project)

[7] Grabowski B L 2004 Generative learning contributions to the design of instruction and learning *Handbook of Research on Educational Communication and Technology* 2 ed D H Jonassen (Penn State University)

[8] Istarani, Ridwan M, Muhammedi, Kamaliah, Lubis M S A, Yusuf H M and Dasopang 2017 *Strategi Pembelajaran Kooperatif Mengenal Tipe Strategi Model dan Teknik Pembelajaran Kooperatif* (Medan: Media Persada)

[9] Wena M 2011 *Strategi Pembelajaran Inovatif Kontemporer Suatu Tinjauan Konseptual Operasional* (Jakarta: Bumi Aksara)

[10] Kosko K W and Wilkins J L 2010 Mathematical communication and its relation to the frequency of manipulative use *International Electronic Journal of Mathematics Education* **5** 79

[11] Thiagarajan S, Semmel D S and Semmel M I 1974 *Instructional Development for Training Teachers of Exceptional Children: A Source Book 5th Edition* Bloomington Center of Innovation on Teaching the Handicapped (Minneapolis: Indiana University)

[12] Andriani *Kemampuan Pemecahan Masalah Matematis dan Self-Efficacy Siswa SMP melalui Model Pembelajaran Generatif* [Thesis]. Banda Aceh: Universitas Syiah Kuala; 2018

[13] Nieveen N 1999 *Prototyping to Reach Product Quality* eds Jan van den Akker et al. (London: Kluwer Academic Publishers)

[14] Hakim A R 2014 Pengaruh model pembelajaran generatif terhadap kemampuan pemecahan masalah matematis *Jurnal Formatif* **4** 196

[15] Temmel S 2014 the effects of problem-based learning on pre-service teachers' critical thinking dispositions and perceptions of problem-solving ability *South African Journal of Education* **34** 769

[16] Akyuz, Ibrahim H, Keser and Hafize 2015 Effect of educational agent and its form characteristics on problem solving ability perception of students in online task based learning media *Educational Sciences* **10** 265

IOP Conf. Series: Journal of Physics: Conf. Series **1088** (2018) 012076 doi:10.1088/1742-6596/1088/1/012076

The analysis of junior high school students' mathematical abstraction ability based on local cultural wisdom

I Dewi[1], N Siregar[1] and A Andriani[1]
[1]State University of Medan, Medan, Indonesia

E-mail: dewi_lubis62@yahoo.co.id

Abstract. Mathematical abstraction is widely used in everyday life. For example, *pengeret-eret* ornaments derived from Tanah Karo culture, North Sumatera, Indonesia, are composite of a parallelogram resembling the image of a lizard that has two heads. This study aimed to develop a model of mathematics learning based on local cultural wisdom to improve communication and mathematical abstraction ability of junior high school students in North Sumatra. This paper examined the profile of students' abstraction ability of year 7, that were used to develop mathematical learning models and tools. The data were obtained from observation, interview, and mathematical abstraction ability test from two regencies in North Sumatera, Tanah Karo and Mandailing Natal. The participants were 119 students in Year 7. The findings of this study showed that: (1) student could not bring up the abstraction process in solving problem, (2) students paid attention to the characteristics of the observed objects, however, they could not see the relationships among the observed objects, and (3) students noticed the characteristics of the observed objects but they have not done the actions that were supposed to be done on the objects to be manipulated. Therefore, it is necessary to develop a model of learning mathematics based on local wisdom that can improve students' abstraction ability.

1. Introduction

Conceptually, local wisdom is human wisdom that is based on traditionally institutionalization of philosophical values, traditions, and behaviors. Philosophically, local wisdom can be interpreted as indigenous knowledge systems that are empirical and pragmatic. Empirical aspect means that the local societies leave the facts that occur in their lives, while pragmatic means that all the concepts constructed as a result of thinking in the knowledge systems that aim to solve daily problems.

Local wisdom becomes important and beneficial to society because the local people who inherit the knowledge systems are willing to accept and apply the culture to be the cultural values of their lives. Thus, local wisdom can be referred to as the soul of the local culture. It can be seen from the expression of local wisdom in daily life because it has been internalized well. From all these descriptions, it can be concluded that local cultural wisdom can not only be used as an educational product but also can be used as a bridge for students to be more familiar and closer to the cultural values in their region. These ideas are relevant to the educational objectives formulated in the regulation of the Indonesian ministry education and culture No. 20 Year 2006 on content standards that mentioned that education is both a form and a cultural product.

North Sumatra is one of the provinces in Indonesia which has various tribes, such as the tribe of Melayu, Batak Karo, Batak Toba, Mandailing, Nias and so on. These are some examples of ornaments used by the tribe of Karo, Melayu, and Mandailing in daily life.

IOP Conf. Series: Journal of Physics: Conf. Series **1088** (2018) 012076 doi:10.1088/1742-6596/1088/1/012076

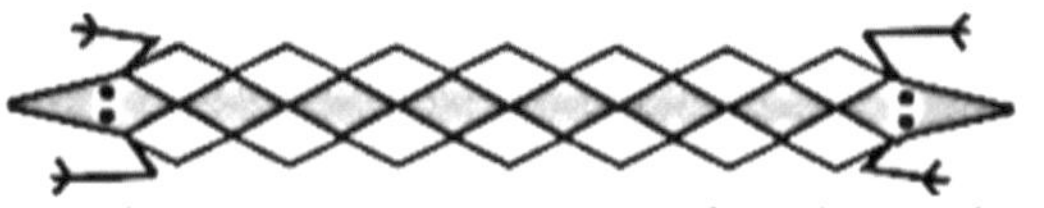

Figure 1. *Pengeret-eret*: ornament from karo culture

The *Pengeret-eret* ornament is derived from the Karo culture; it has lizard shape with two heads facing left and right. This ornament is usually installed in front of Karo custom house, according to the belief serves as a force to resist the catastrophes and threats of evil spirits that threaten homeowners [1]. Regarding the ornament shape, the shape forms a combination of geometrical shapes.

The traditional ornament used in the lid of Sopo Godang (customary court hall) and Bagas Godang (the great house of kings). Mandailaing is called Bolang. In custom, Mandailing Bolang has the values, ideas, concepts, norms, rules and laws used in living life [2].

Figure 2. Bolang: ornament from Mandailing natal culture

From these two symbols can be seen that since ancient times the process of abstraction has been attached and entrenched in daily life. When we look at the two symbols of those two regions, it is seen that although it has different shapes, both have the same philosophy that is the symbol of the force used to resist the catastrophe. It can be understood that the symbols of a region are the results and products of the mathematical abstraction process of abstract thinking, generalizing, and composing problems in daily life. Mathematical abstraction arises through a natural mechanism of the human brain in which complicated phenomena are compressed into what they call thinkable concepts [3]. Thus abstraction ability is an activity performed in daily life such that it needs to be developed in learning, especially in mathematical learning.

Based on the above descriptions, the activity of mathematical abstraction is an important thing to be developed in mathematical learning by using local cultural wisdom. In learning mathematics, abstraction is a fundamental process and have an important role in forming a concept. Through that process we can measure how far students have understood and built the concepts they have learned. On the other hand, due to the abstractness of mathematics, students often find difficult to learn mathematics. This is by the opinion which said to improve long-term conceptual learning, educators are aware of the abstraction process. The importance of abstraction ability can be enhanced in the learning process that gives students the opportunity to abstract by utilizing the local culture wisdom. Thus, it is required a learning model that can improve students' mathematical abstraction ability. The problem statement in this research is "how the profile of the abstraction of mathematics students' ability?". Those students' mathematical abstraction profile will be used to develop mathematical learning models and tools. This research is the first phase of development to develop Mean-Ends Analysis (MEA) model based on local culture wisdom to improve students' mathematical communication and abstraction ability of Junior High School (SMP) students in North Sumatra.

2. Research methods

This research was conducted on February 2018 in two regencies in North Sumatera that can represent North Sumatra culture. The subject of this study was students in Year 7 which were 60 students from junior high school 1 and 59 students from junior high school 2.

The data were obtained from 2 problems of the validated abstraction abilities test, i.e. to measure the ability to identify the manipulated object (indicator 1), and to transform the problem (idea) in symbol form (indicator 2). The development method used adapted from Plomp [4]. This paper is a research on the assessment stage of students' mathematical ability based on local wisdom.

Table 1. The rubric of scoring test

Score	Criteria
Score 4	The information/explanation given is correct and complete.
Score 3	The information/explanation given is correct but not complete at the end of the completion.
Score 2	The information/explanation given is complete but there is still an error at the end of the completion.
Score 1	The information/explanation given is correct only at the beginning of completion and cannot be developed.
Score 0	No information/explanation at all

3. Results and discussion

The results of students' mathematical abstraction ability of junior high school 1 and school 2 can be seen from Figure 3.

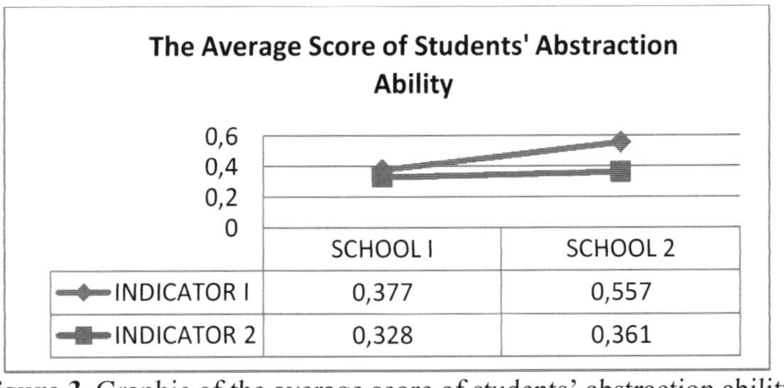

Figure 3. Graphic of the average score of students' abstraction ability

Figure 3 shows the indicator 1 (identifying the manipulated object) of the average score of students' abstraction ability that school 1 obtained 0.377, while in school 2 the average score of 0.557. The problem of indicator 2 (transforming the problem (idea) in symbol form) the average score of abstraction ability that school 1 obtained 0.328, while school 2 obtained the average score of 0.361. The data showed that the students' ability to identify the manipulated mathematical object both schools was better than the students' ability to transform the problem (idea) in symbol form. This result led to an analysis of why students have difficulty in transforming ideas into symbols (indicator 2). The problem given for viewing this indicator 2 in Figure 4.

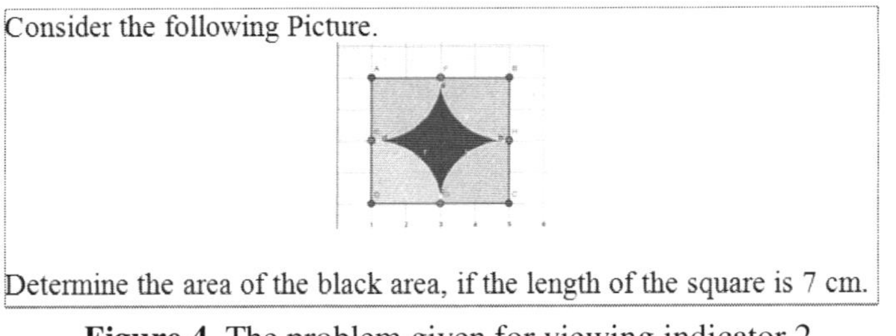

Figure 4. The problem given for viewing indicator 2

IOP Conf. Series: Journal of Physics: Conf. Series **1088** (2018) 012076 doi:10.1088/1742-6596/1088/1/012076

There were profiles of students' observable answers: (1) student did not provide the answer at all, (2) student just guessed the answer without any argument, (3) students answered by using incorrect formula. Table 2 shows the pattern of answers given by the students.

Table 2. Analysis of students' answers profile

Answers Profile	Students school 1	Students school 2
No answer	2	3
Guessing	18	38
Incorrect/Mistake formula	39	19

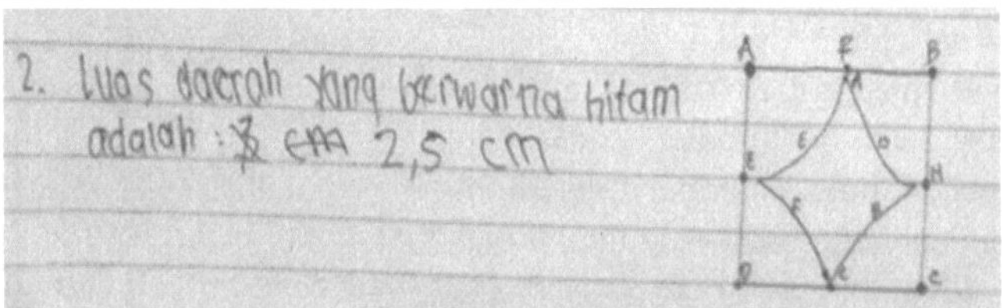

The area of the black face is 2.5 cm

Figure 5. Profile of student's answer by guessing

From Figure 5. can be seen the students just wrote "*the area of the black is 2.5 cm*" without giving any argument toward the answer given.

The next profile is students making the formula in solving the problem but the formula used does not lead to completion.

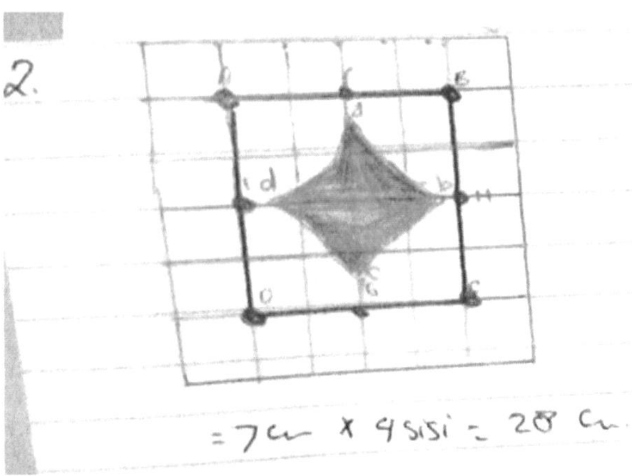

Figure 6. Profile of student's answer used formula is not correct (7 cm x 4 faces = 28 cm)

From Figure 6. student looks just "7 x 4 faces = 28 cm". Although the student has given the argument the formula used was not correct and did not lead to solving the problem. In general, no student obtained score 4 even many students who did not understand that the area of the black is the area of the square reduced by the area of a circle. The results of this data analysis suggested that when solving the problem given, the students failed to generate the process of abstraction.

According to [5], in mathematical learning, there are three circumstances that could lead to the process of abstraction, including: 1) the students' abstraction process arises from observing the observed characteristics of the object, 2) it arises when focusing attention on the actions to be performed on objects that lead to symbols and then manipulated, and 3) it arises when a student formulating a theoretical set of concepts to construct a concept that can be thought through a series of mathematical proofs.

In line with the ideas, the situation students did not give an answer means that all the abstraction process did not appear in solving a problem. The researchers did not see this issue was caused by time limitation, since the time given was sufficient and even some students submitted the answers before the due time.

In situations where the students guess only the answer, the researchers estimate that the students have considered the characteristics of the observed object (even some students measured the image using a ruler) but they did not see the relationship between the observed object. Thus, students could not solve the problem correctly. In situations students gave answers with a mistake formula, they did not lead to completion means students already pay attention to the observed characteristics, i.e. square. But to solve the problem, students did not do the actions that are supposed to be done on the object to be manipulated. As a result, students did not see that areas that were outside the black area are four of ¼ circle area, which means students could not construct the formula used in solving the problem. The material used in this research is geometry. According to [6], students' scores in geometry always decreased. Geometric objects are considered abstract by students so that they have difficulty to understand the geometry concepts. In this case, students could identify the mathematical object but they found difficulty in writing symbols or into the language of mathematics. Similarly, [7] argues that the process of student abstraction in solving geometry problems belongs to empirical abstraction which emphasizes the representing mathematical objects into mathematical languages.

With regard to the students' age at year 7, ranging from 12-13 years old, students still could not do the abstraction perfectly yet. According to [8], children learn things based on manipulation and observation of specific objects. Therefore, it might be difficult for children to apply abstraction to learning. Despite having difficulty in abstraction, this abstraction activity can be developed in schools by giving students' activities of theoretical models such as schematization activities that make up special experiences. Furthermore, a concept is the final product of abstraction [9]. Thus, abstraction abilities can be developed by giving students the abstracting activities associated with their daily lives. This resulted in a student's ability to link his own experience of mathematical ideas to a better and lasting understanding of mathematics because he can see the interrelationships between a concept in mathematics, with non-mathematical contexts, and with daily life experiences [10]

4. Conclusion

This research examined the mathematical abstraction ability of junior high school students who come from two different regions of language and local wisdom in North Sumatra. The profile was used to develop mathematical learning models and tools to improve students' mathematical abstraction ability. The different profile of students' mathematical abstraction led to learning models and materials that appropriate with the characteristics of each region. The results of this research showed that the profile of students' mathematical abstraction evidently the same. The profiles are: (1) student could not bring up the abstraction process in solving problem, (2) student paid attention to the characteristics of the observed object but they did not see the relationship between the observed objects, and (3) students noticed the characteristics of the observed objects but they have not done the actions that were supposed to be done on the objects to be manipulated.

The results of this research indicated that students' abstraction ability was not influenced by the local cultural wisdom of a region. Mathematical development occurred because of the life challenges and the ability of human beings to abstraction. Every culture and local wisdom develops mathematics according to the needs of its own society. Therefore, mathematics is a cultural product of the result abstraction of the human mind, as well as an activity done in daily life. Thus, it is necessary to develop mathematical learning models and tools that can improve students' abstraction ability based on culture and local wisdom.

References

[1] Roy G.T 2014 Inspirasi Gerhana: Ornament Tradisional Suku Karo *Retrieved on February 2017* from http://inspirasigerhana.blogspot.co.id/2014/03/ornamen-tradisional-suku-karo.html.

[2] Nasution E 2010 Bolang-Ornamen Tradisional *Mandailing APAKABARSIDEMPUAN Retrieved on February 2017* from https://www.apakabarsidimpuan.com/bolang-ornamen-tradisional-mandailing/

[3] Gray E and Tall D 2007 Abstraction as a Natural Process of Mental Compression *Mathematics Education Research Journal*, **19**(2) pp 23-40

[4] Plomp T 1997 *Educational and Training System Design.* (Enschede: University of Twentee)

[5] Tall, D. 1991. *Advanced mathematical thinking. Mathematics Education Library.* Dordrecht: Kluwer Academic Publishers Group

[6] Utami, A K D, Mardiyana, and Pramudya, I 2017 *Analysis of Junior High School Students' Difficulty in Resolving Rectangular Conceptual Problems.* AIP Conference Proceedings **1868**, 050008 (2017); Retrieved from https://doi.org/10.1063/1.4995135

[7] Nurhasanah J, Sabandar J, and Kusumah Y S 2013 *Abstraction Processes in Learning Geometry Using GSP.* ResearchGate Retrieved from https://www.researchgate.net/publication/263199291.

[8] Hong J E and Kim M K 2015 The Solving od Ill Structured Problem of Elementary School Student in Korea. *Eurasia Journal of Mathematics, Science & Technology Education,* **12**(2), 267-281

[9] Ferrari P.L 2003 *Abstraction In Mathematics* The Royal Society Publishing *retrieved in April 2018* from http://rstb.royalsocietypublishing.org/content/358/1435/1225

[10] National Council of Teachers Mathematics (NCTM) 2000. *Principle and Standards for School Mathematics.* Reston: NCTM.

IOP Conf. Series: Journal of Physics: Conf. Series **1088** (2018) 012077 doi:10.1088/1742-6596/1088/1/012077

The quality of learning materials through mathematics reaslitic to improve students' mathematical communication ability in the elementary school

Nurfadhillah, R Johar and A Ahmad
Syiah Kuala University, Jl. Teuku Nyak arief Darussalam, Banda Aceh 23111 Indonesia

E- mail: rahmah.johar@unsyiah.ac.id

Abstract. Learning mathematics aims to help students in solving daily-life problems. Therefore, students require mathematical communication ability. In fact, the mathematical communication ability of students in Indonesia is low. Mathematical communication ability can be enhanced by applying realistic mathematics in which students are engaged to be active in learning using real problems. Islamic context can be used as a real context for Acehnese student embracing Islamic culture. The learning materials incorporated with Islamic context is currently limited in mathematics learning. Thus, it is necessary to develop learning materials using realistic mathematics incorporating Islamic nuances to improve students' communication ability. This study is the research and development, only focused on the validity and practicality of the learning materials. The participant was Grade 5 students at the integrated Islamic elementary school, in Aceh Besar, Indonesia. The products of this study were a lesson plan, a student worksheet, and a communication test. The results showed the average score of the lesson plan, worksheet and test were 3.7, 3.6, and 3.8 respectively. It means learning materials is fulfill valid criteria. The average score of the implementation process is 3.5. It means learning materials was fulfill practical criteria. The further research can use the learning materials to test their effectiveness.

1. Introduction

Mathematics is all around us however we are not aware that we are performing mathematical operations. In fact, many students are trained to perform mathematical calculations rather than mathematical thinking [1]. In solving a mathematics problem, students must have the mathematical ability [2] and to communicate ideas of mathematics clearly [3]. However, the results of the assessment by the Program for International Student Assessment (PISA), a study focused on literacy reading, mathematics, and science in 2015, showed that Indonesia was ranked 65 out of 69 the participating countries. Mathematical communication is needed in solving PISA problems as a realistic problem [4]. Therefore, the teacher needs to design learning materials that links mathematics with real problems to improve student communication, one of which is by applying a Realistic Mathematics Education (RME) approach.

Freudenthal as a founder of RME said mathematics is a human activity in solving problems that are close to students' experiences [5]. Realistic in RME is about something real in the students' mind, it was not just about real word problems but also about the fantasy problems or formal problems as lons as common sense for students [6,7]. It is necessary to study mathematics about students' culture, especially the students in the province of Aceh.

As stated in Qanun no 18 of 2001 that the province of Aceh should conduct education by incorporating the Acehnese peculiarities and following the Islamic Shari'a. One of the solutions is by developing realistic mathematics learning materials that include Islamic values, a mathematics learning based on the contextual of Islamic values derived from Al-Quran and Hadith. In fact, realistic Islamic mathematics learning materials are not yet available. The implementation of Islamic nuances education in Aceh has not been implemented optimally, the integration of Islam is still limited to certain Islamic principles such as separating the seats between male and female students, praying and reciting the Qur'an before commencing lessons and giving advice [8].

Therefore, the development of mathematics learning materials using a realistic mathematical approach that is in line with the Islamic values needs to be done. The research question of this study is "how is the process and result of the development of valid and practical learning materials of Realistic mathematics with Islamic nuances to improve mathematical communication ability of students in the Elementary School?

2. Method

This research can be categorized as one type of research and development. Research and development is the research producing products. The products are the learning materials. The subjects of the study were Grade 5 students of Islamic integrated primary school in Aceh Besar, Indonesia. Development procedures following the design of [9], namely: a preliminary Investigation Stage, Prototyping Stage, and Assessment Stage.

Stages of development of learning are as follows:

- Preliminary Investigation Stage. This stage aims to obtain data about school curriculum, learning materials and learning resources, learning materials that were used, and the situation.
- Prototyping Stage. At this stage, the learning device is designed with an Islamic realistic mathematical approach by optimizing student activeness. Activities at this stage include the design of the lesson plan, the student worksheet, and the test. The design result is called prototype 1. The expert will assess prototype 1 at the assessment stage. The next stage is (1) validating and (2) practicing.
- Assessment Stage. The Quality of learning materials is based on the [10] criterion, which is valid and practical. Validity can be seen from the data based on the expert assessment and recommendations. Realistic mathematical learning materials in Islamic nuances on the lesson regarding speed is said to be valid if the average value of learning materials validity should at least be 3. Practicality is assessed by some practitioners. The data can be obtained by finding the average score of teacher's practical questionnaire and the percentage of the tool's implementation. The steps are adopted from [11], learning materials are said to be practical if at least the criteria for the implementation of learning materials and the practicality test of teachers obtained are 3. If the score is less than the specified criteria, they must be re-tested.

3. Result and discussion

The development of this learning materials follows the development model described in the previous section, which follows the steps (1) preliminary stage, (2) prototype stage, (3) assessment stage. The development results of each stage are as follows:

3.1 The result of Preliminary stage

The results of the preliminary investigation are as follows.

- A realistic approach can be used to seek out Islamic societies because of mathematical learning with a realistic approach giving students the opportunity to be actively involved physically and mentally in constructing knowledge linked to the students' real-life experiences.

- Supporting theories of learning are a constructivist theory, discovery learning theory, Piaget learning theory, and Vygotsky theory.
- Theories about Islamic values such as respecting the opinions of friends when expressing views, responsibilities, discipline, and honest.
- Islamic values derived from Al-Qur'an and Al-Hadith related to the topic of Speed such as Surah Al-Isra 'Verse 1 "Isra' Mi'raj" events about the speed of light experienced by the Prophet as a miracle and the nature of the obligatory and impossible for God.
- The preliminary condition of Grade 5 at the integrated Islamic elementary school, Aceh Besar, Indonesia as a field test has been studied in a group, but students were not used to learning mathematics with the realistic approach. The classroom setting school involved supports the realistic approach as the class consisted of 20-25 students. The Topic of speed, distance and time were taught in Grade 5 Elementry School, and usually, teachers found difficulties to teach this topic to the students. Accordingly, a preliminary idea was obtained to develop a mathematics learning tool that can increase student learning interest, that was developed by incorporating problems close to the daily life of students and inserting Islamic values.

3.2 The result of Prototyping Stage

At this stage, realistic mathematics learning materials of Islamic nuances was designed. The results obtained were designing the characteristics of realistic mathematics learning of Islamic nuances: (1) starting the learning the contextual problems in accordance with Islamic religious values, (2) giving the students an opportunity to model the problem informally with the strategies they understand, (3) Conducting the learning in groups to develop interactions among learners, (4) linking the subject matter with other materials as a unity.

While the learning materials developed were (a) Lesson Plan, (b) Student Worksheet and (c) Test. one of which is presented in Figure 1:

Student Worksheet
(1)

Group: ....
Names of group members:
1. .... 4. ....
2. .... 5. ....
3. ....

Hint:
1. Read and answer the questions of this activity, then discuss with your friends.
2. If there is a problem with the student paper below, ask your teacher!

Activity 1
Read and observe the hadith about the good of the Friday prayers below!
From Abu Huraira. Prophet Mohammed sallallaahu'alaihi wasallam said "If come Friday, then the angels stand at the doors of the mosque, then they write the people who come early as early. People who come early are assumed to sacrifice a camel. the one who comes after it is supposed to be the one who sacrificed a cow. and who afterwards sacrificed a goat. after that like sacrifice a chicken. then after that like the person who sacrificed the egg. Then, after the priest ascended to the pulpit, the angels closed their charity logbook and sat listening to the lecture. "(Ahmad in Musnad No. 10164)

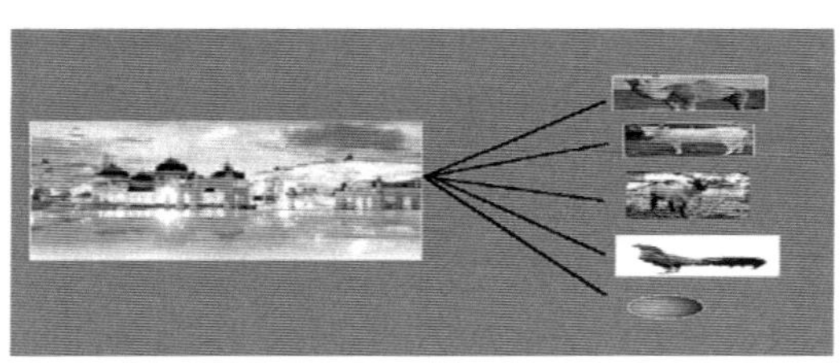

Figure 1. The Student Worksheet of Islamic Realistic Math Learning Materials.

Figure 1 is the result of the student worksheet design. The content was adapted from the hadith of the virtue of the Friday prayer and complete with pictures to make it easier for students to understand the text. The indicator to be achieved is the ability of students to understand the reading in the picture. Then, students can write more abstract mathematical symbols.

3.3 The result of Assessment Stage

3.3.1 Validation Results
In this step, experts validated prototype 1. Prototype 1 was assessed by four experts, including of two lecturers who taught mathematics education and two mathematics teachers of Grade 5. The result of the analysis is presented in Table 1.

Table 1. Results of the analysis of the validity, revision, and decision.

Component	Result analysis	Before Revision	Decision	After Revision
Lesson Plan	Average is 3.7	The learning objectives in lesson plan 3 Learners cannot solve the problem relating to the distance	Can be used with minor revision	The learning objectives in LIP 3 Learners can solve the problem relating to the units of distance (hm, dam, and dm)
Students worksheet	Average is 3.6	Students worksheet 3 activity 2 Hints about the question no 2 in the filling table is less clear.	Can be used with minor revision	Students worksheet 3 activity 2 Hints about the question no 2 in the table filling become: Table 1: Fill in the following table according to the note. Table 2: Fill in the following table by adding 2 lengths according to table 1 Table 3: Fill in the following table by subtracting 2 lengths based on with Table 1.
Test	Average is 3.8	Figure in figure 1 is too small	Can be used with minor revision	Figure in figure 1 was enlarged

The table above shows that prototype 1 can be used with minor revision. One of the revisions made to the student worksheet. The preparation of student worksheet must meet various requirements, including the didactic requirements, construction requirements and technical requirements [12]. Therefore, based on the suggestion of the expert, the student worksheet 3 was revised. Revisions are made related to the instruction of the sentence in the table in the student worksheet 3, and this was conducted so that students can easily understand the instruction of the activities. The result of the revision of prototype 1 is called prototype 2. Prototype 2 will be tested for its practicality.

3.4 Results of practicality

The practical test is done by small group and field test. Small group test was administered on five students at the integrated Islamic elementary school. Based on the results of small group test, the following information is obtained.

- Many students asked questions on the student worksheet 3 activity 1 no 2 that is "Change the meter in 300.00.000 m as you know!". The command sentence is refined to "change the unit mater at 300,000,000 m according to the other distance unit (minimum 2)!
- The figure in the student worksheet 1 number 4 and the table at the activity 2 number 3 was too small. The revision was made with the figure and table enlarged.

From a small group test, we can conclude that prototype 2 did not change much, only a few revisions were made to improve the language in students worksheet 3. Using simple and short sentences, it will make it easier for learners to capture what is hinted in student worksheet [12]. Prototype 2 that has been validated and tested for the legibility then revised following the advice of the expert and the results of a limited test. The result of the revision of prototype 2 is called prototype 3.

The prototype 3 was then tested on a field test. The test was held on 9-13 April 2018 in Grade 5 at the integrated Islamic elementary school, Aceh Besar. The test was conducted to obtain data about the implementation of learning.

Table 2. Observation results of learning materials implementation.

Learning Class	Observer 1	Observer 2
Learning on the first day	3,4	3,4
Learning on the second day	3,6	3,7
Learning on the third day	3,8	3,8
Learning on the Fourth	3,8	3,7
Learning on the fifth	4	4

Table 2 shows that observers assessed the implementation of learning materials is running well. In the first teaching and learning process, the observer gave a rating of 3.4. This result was relatively lower than the next day. According to observers, on the first day, the teacher had not mastered the class, so that the teacher-centered learning was dominant. Teachers gave many explanations about the hadiths contained. The impact was short learning time. On the second day, the average score observer 1 rated 3.6 and observer 2 rated 3.7 indicating that the learning process is better.

4. Conclusion

Realistic mathematical learning materials with Islamic nuances can be used to improve students' mathematical communication skills in the elementary schools. This is based on the learning process development process mentioned earlier, obtained by valid learning materials with the average of the validity of the lesson plan, students worksheet and test are of 3.7; 3.6; and 3.8. These learning materials are also practical; it can be seen from the implementation of learning with the average of 3.5 meaning that the implementation of learning in the good category. It is expected that these learning materials can be developed not only on the topics of measurement, distance, time and speed only but also on other topics and their effectiveness can also be tested.

References

[1] Noyes A 2007 *Rethinking School Mathematics* (London: Paul Chapman Publishing)
[2] Grootenboer P and Zevenbergen R 2008 *Proc. of the 31st Annual Conference of the Mathematics Education Research Group of Australasia Identity as a lens to understand learning mathematics: Developing a model Navigating currents and charting directions* **1** pp 243-250
[3] Government Regulations by 2016 *on the standards of the primary and secondary education processes*

[4] OECD 2016 *PISA 2015 Results in Focus* (New York: Columbia University)
[5] Freudenthal H 1991 *Revisiting Mathematics Education* (Dordrecht: Kluwer Academic Publishers)
[6] van den Heuvel-Panhuizen M and Wijers M 2005 Mathematics standards and curriculum in the Netherlands *ZDM* **37** 287
[7] Gravemiejer K 2011 How concrete is concrete *IndoMS JME* **2** 1
[8] Soraya C 2015 Pelaksanaan Pendidikan di Kota Banda Aceh: Studi terhadap Implementasi Qanun No.5 Tahun 2008 *Jurnal Ilmiah DIDAKTIK* **16** 1
[9] Plomp T 2013 *Educational Design Research* Educational Design Research: anIntroduction, Ed T Plomp and NNieveen (Enschede: SLO) p 10
[10] van den Akker J, Bannan B, Kelly A E, Nieveen N and Plomp T 2013 *Educational Design Research* Enschede (Netherlands: Netherlands Institute for Curriculum Development) (SLO) Retrieved from http://international.slo.nl/publications/edr/.
[11] Sudjana N 2005 *Penilaian Proses Hasil Belajar Dan Mengajar* (Bandung: PT Remaja Rosdakarya)
[12] Darmodjo H and Kaligis J R E 1993 *Pendidikan IPA* (Jakarta : Dirjen Dikti Depdikbud)

Designing teaching instructions catering students' needs: teaching solid geometry through problem-based learning (PBL)

K Umam and S Maulina
Syiah Kuala University, Jl. Teuku Nyak Arief Darussalam, Banda Aceh 23111, Indonesia

E-mail: khumam77@unsyiah.ac.id

Abstract. In a mathematics classroom, students are diverse. They have different learning styles, interests, readiness, paces and abilities. It is challenging to know how each student learns and what learning methods appropriate for him or her. Thus, teachers need to be assisted in developing appropriate instructions based on students' needs. This issue prompted the researchers to find learning with a model that meets the needs of students on the volume of prism topic. The researchers have developed the learning materials on the volume of prism topic using problem-based learning (PBL). The appropriateness of the learning implementation to the students' needs then needs to be analyzed. The study involved 66 eighth graders in a junior high school in Banda Aceh, Indonesia. Data in this study were obtained from students' achievement after the PBL implementation and analyzed using Fisher's discriminant analysis. The finding showed that 50 students were in a successful group and 16 students were considered unsuccessful. This finding implies that the PBL accommodated the needs of the majority of students. Further, identifying characteristics of the two groups is necessary to modify the learning that has been previously developed to meet the needs of the unsuccessful group.

1. Introduction

As teachers, making sure all students learn mathematics well is necessary. To achieve the goal, teachers need to design an instruction that meets students' needs. However, in a mathematics classroom, students are diverse. Every student has different ways of learning, abilities, paces, readiness, and interests. Different students learn well by different learning models. Students will not learn well if teaching style conflicts students' learning style [1]. A shift from teacher-centered to student-centered approach is a recognition that learning should be focusing on students' needs. In a student-centered approach, teachers need to learn how to differentiate instruction [1]. This issue prompted the researchers to find teaching instructions that meet the needs of students on the volume of prism topic and to categorize students based on their achievement after implementing the instructions to examine the group of students who experience success through the learning model.

One of the student-centered learning models that can be used to teach geometry is problem-based learning (PBL). PBL appeared as recognition that students learn by doing and by thinking through problems. Originally, PBL was developed for adults, to train doctors in how to approach and solve medical problems in the real world wherein one case it might be difficult to identify symptoms of a patient, and in another case, a patient might have multiple symptoms [2]. Through PBL, the medical students researched a situation, developed appropriate questions, and produced their own plan to solve the problem. It was found that PBL developed the students' abilities to improve their knowledge and to learn how to deal with the new illnesses. They became "self-directed learners" and used available resources to help them to learn [2,3]. This successful learning then leads PBL adapted to other fields such as mathematics education.

Some researchers have used PBL in teaching mathematics [4-6]. Schettino used PBL to teach geometry. He defines "PBL as an instructional approach where students' learning and content materials are constructed through the use, facilitation, and experience of contextual problems ... in a discussion-based classroom setting where student voice, experience, and prior knowledge are valued" [4]. He believed that PBL allows students to use skills such as communication, collaboration, and technology literacy required by the current curriculum. Through the learning with PBL, he found that students made connections and saw the relationship among the math concepts, had the opportunity to create their own strategies and expressed their ideas. Thus, implementing PBL is one of the ways to meet not only the curriculum demand but also the students' needs.

Another study examined the effect of PBL on the students' performance in learning geometry [6]. The study revealed that students experienced independent learning and developed communication and research skills through PBL. However, some students still struggled in gathering the information in problem-solving so that they ended up with poor decision-making. This study suggests future research on PBL in a more extended period.

Those studies implemented PBL and described how students learned mathematics using PBL, their responses toward the implementation, and the challenges found during the implementation. However, none of the studies [4-6] groups students based on their achievements using Fisher's discriminant analysis after the PBL implementation. Therefore, this present study aims to examine students based on their performances after implementing PBL to investigate which group of students experience success. This will lead teachers in finding out teaching instructions that fit the need of the students. In addition to that, this study will describe the PBL implementation and students' response toward the implementation.

2. Method

This present study is part of development research on mathematics learning with PBL model on the volume of prism topic. This study focuses on examining the appropriateness of the learning to the students' needs. The participants were 66 students of grade 8 in a junior high school in Banda Aceh, Indonesia. This study was conducted in two phases. In the first phase, students were given a pre-test about the volume of a prism. Then, in the next session, both classes were taught the volume of a prism using PBL by the same teacher for one meeting. Two days after the class session, they were given a post-test. Students' grades on both tests were collected to see their progress on the topic. Also, data about students' math scores on the previous grade report were collected. Based on their achievement on the topic, the students from both classes were categorized into two groups, a group who was successful and had not, with minimum completeness criteria 65. To do this process, the researchers used discriminant analysis by Fisher method. Discriminant analysis is a statistical technique used for classifying observations, for example, predicting success in academic programs [7]. "Discriminant function analysis produces functions that help define the groups; it distinguishes the first group from others, then the second group from the rest, and so on" [8]. The researchers used SPSS (version 16.0) to get the formula of discriminant analysis. After that, students from each category were interviewed to find out their responses toward the implementation of PBL.

3. Results and discussion

During the first phase, the students were taught using PBL on the topic of prism volume. With PBL, the students were expected to be able to find the volume of a prism in real-life problems. Previously, the students had learned the volume of cubes and rectangular prisms. Using their prior knowledge, the students figured out the volume of triangular prisms. To motivate the students, the teacher showed them an application of the volume of prisms in real life. For example, the teacher showed a picture of a shower tub shape that looks like a prism and asked them how to find the volume of the water in the tub. In this way, the students realized that they needed to know the volume of the prism to be able to find the volume of the water.

In a similar way, the teacher gave students a real-life problem. However, at this time, the students worked in a group to solve the problem. Since PBL focuses students as a center of learning, students need to involve actively by working collaboratively in their groups [3, 5]. Then, to launch the problem, the teacher asked the students what they understood from the problem and what they had not. Also, the students were encouraged to pose questions about the problems to help them in solving the problem and write them all. Examples of student questioning are: What is the shape of the swimming pool? What are the length and the width of the pool? What is the volume of the water? What does it mean by sloping floor in between two parts of the pool? What is the width of each part of the pool? As solving the problem, the students also did some research from the textbook and recalled their previous knowledge. Working in a group allowed them to discuss and learn from others. Interaction among the students helps them to communicate their ideas and move them forward. This is aligned with what Schettino and Ahamad et al. revealed, that students would develop communication skills through PBL [4, 6].

A swimming pool has a dimension 30 m x 10 m with two different depths 3:4. On the surface of the pool, there is a sloping floor between two parts of the pool. If the length of each part is the same, that is, 14,4 m. Find the volume of water if the pool is full.

Figure 1. The problem assigned to students.

While students were working in a group, the teacher assisted them by asking purposeful questions to assess their thinking and to help them think forward. However, the teacher did not tell the students the answers. Instead, the teacher supported students' productive struggle. For example, when a group could not get started, the teacher asked them, "What do you understand from the problem? What information do you know about the problem? Could you please rewrite them?" Also, the teacher encouraged the students to look back at the questions they had posted before. This helped them to recall the information needed for solving the problem.

Students recorded their solutions on a poster then presented their strategies in front of the class. The teacher asked them to compare the solutions found by different groups and discuss it during the whole class discussion. This activity allowed students to clarify their thinking and revealed students' misunderstanding about the volume of prisms. A group of students found the volume of a prism by making a connection to the volume of a rectangular prism. To get the volume of a triangular prism, the students divided the volume of the rectangular prism by two (see Figure 3). Then, when a student showed a misconception, the teacher brought it to the class for further discussion so that other students will not perform the same misconception in the future. For instance, a student incorrectly modeled the problem. She misinterpreted the sentence: "On the surface of the pool, there is a sloping floor in between two parts of the pool" and drew a picture as in Figure 2.

IOP Conf. Series: Journal of Physics: Conf. Series **1088** (2018) 012078 doi:10.1088/1742-6596/1088/1/012078

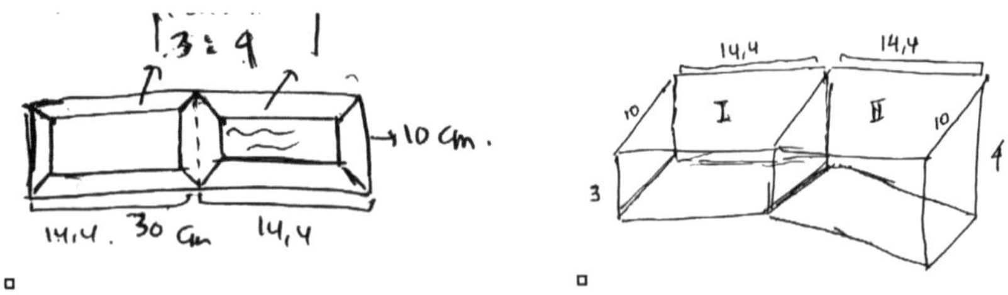

Figure 2. Student misinterpretation of the problem.

Then, the teacher asked a question to the class, "If we model the problem like this (as in Figure 2), how would you use information of the depth 3:4 in the picture? How would you put the numbers in the picture?" Some students reasoned that they could not determine the depth of another part of the pool if the surface is not flat. Some argued that there is a sloping floor between two parts of the pool; thus, the surface of both parts of the pool should be flat. Interestingly, a group of students believed that the volume of water could be varied because the depth of the parts of the pool is 3:4. That means that it could be either 3 m and 4 m, 9 m and 12 m, or 30 m and 40 m. Those will affect the volume of water in the pool. Such a discussion promoted students' reasoning and justification that helped them to understand the concept of the volume of a prism.

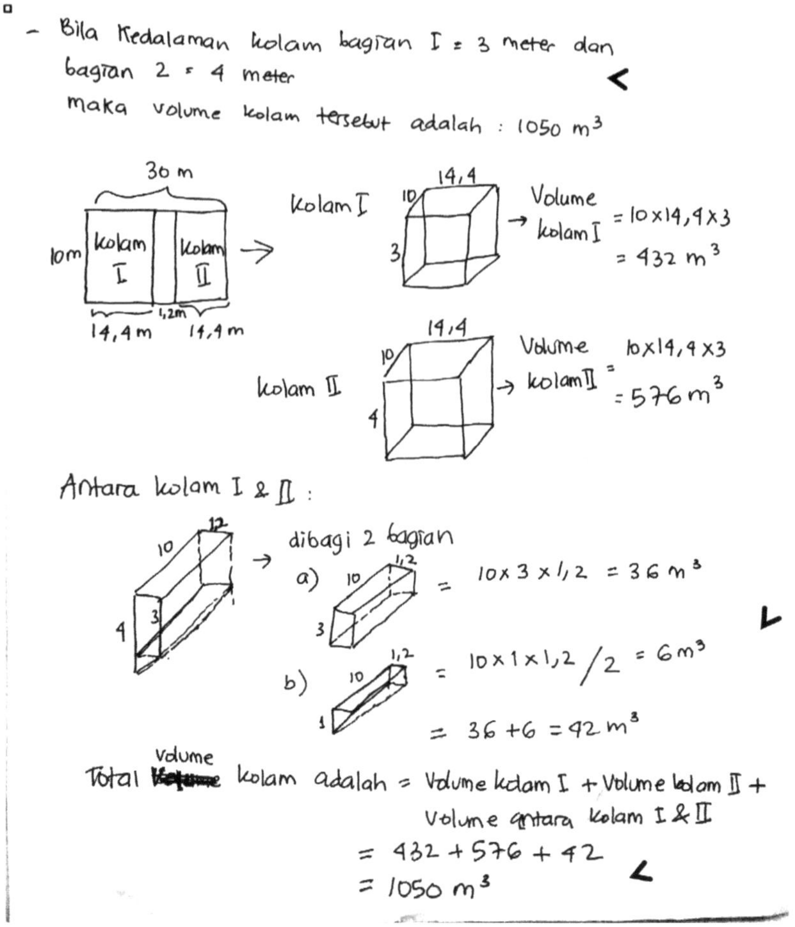

Figure 3. Example of a student's solution.

The student assumed that the depth of pool part I was 3 m and the depth of pool part II was 4 m

The student divided the volume of rectangular prism by two to get the volume of triangular prism

The student found the total volume of the pool is 1050 m^3

After learning the volume of a prism, the students were given a post-test in the following session. Then, students' achievements on both pre- and post-tests were examined. The data were normalized using the natural logarithm to minimize the error. By using SPSS software (version 16.0), the Fisher's linear discriminant functions were obtained in the following table.

Table 1. Fisher's linear discriminant functions.

	Classification function coefficients	
	Group of students	
	Successful students	Unsuccessful students
Math score	3.463	3.470
Pre-test	.015	-.011
Post-test	.618	.255

Based on the table above, the discriminant functions for the two groups were obtained as follows.

Successful group = $3.463X_1 + 0.015X_2 + 0.618X_3 = 0$
Unsuccessful group = $3.470X_1 - 0.011X_2 + 0.255X_3 = 0$
Where:
X_1 = Math score on previous grade report
X_2 = Pre-test on the volume of prism topic before the class session (phase 1)
X_3 = Post-test on the volume of prism topic after the class session (phase 1)

In SPSS (refer to Table 1), every column to which variables correspond consists of coefficients that form a function for a group. These coefficients are called coefficients of Fisher's linear discriminant functions and could be used directly for group classification. Each coefficient in each formula is used to determine a subject to which group he/she belongs based on the highest score got from the formulas [9]. By referring to the procedure, scores of each student were substituted into the formula to find to which group the student belongs. A student will be in a group that has the highest score. It was found that 50 students belong to a successful group and 16 students were considered unsuccessful. This finding indicates that learning the volume of a prism using PBL met the needs of the 50 students.

When asked about the implementation of PBL, students from the successful group responded positively. They said the learning method used was different from they usually did, and they enjoyed the learning process. A student stated that he liked a whole class discussion because he could share his ideas, which was unusual in a mathematics classroom. As in a study by Ahamad et al., some students felt that they learned better using PBL than the traditional method and they could be independent learners [6]. However, some students from the unsuccessful group did not like discussion implemented in PBL. They found it hard to express what they thought and needed more time to process their thinking. They also were not confident in sharing their thoughts. A student mentioned that she preferred solving problems by herself because she needed to contemplate the problem. When asked "what would you do if you struggle?", she answered that she will ask the teacher and go back to think and solve the problems by herself. Other students still struggled to understand the concept of the volume of a prism and could not identify a prism in the complex solid geometry. Ahamad et al. also found that some students struggled with seeking information and discovery as required in PBL [6]. Students in the unsuccessful group were also asked their preference about the instructions implemented in the classroom. Some of them prefer working in a small group without having a whole class discussion. Some of them like working individually first then discussing with partners to ensure their answers.

These findings suggest that different students have different preferences for learning, paces, and interests. In general, the implementation of PBL accommodated the needs of the majority of the students. That implies the learning materials and instructions met the characteristics of the students. However, 16 students were still considered unsuccessful after the PBL implementation. This finding recommends that the learning with the PBL model on the volume of prism need a modification appropriate to the characteristics of the unsuccessful students. Further identification on the two groups of students is also necessary to design proper instructions for them.

4. Conclusion

Teaching geometry using PBL is one of the promising ways to enhance students' learning. This suggestion was based on the finding that majority of the students were successful through the learning with PBL model. PBL enables students to explore the problem by working in a group and students will be the center of learning. This activity could actively engage students in problem-solving and promote a meaningful discussion between students and teacher. However, some students might be inconvenient to learn geometry through such instructions. Some students learn well by contemplating the problem by themselves and cannot get distracted by others. Another prefers learning by discussion. Those differences lead teachers to find the best teaching instructions for their students and decide what kind of instructions and activities teachers should use in their classroom. Therefore, identifying the needs of the students who do not experience success through PBL is needed to design modified instructions to the learning process.

References

[1] Brown K L 2003 From teacher-centered to learner-centered curriculum: Improving learning in diverse classrooms *Edu.* **124** 49
[2] Delisle R 1997 *How to Use Problem Based-Learning in the Classroom* (Alexandria: Association for Supervision & Curriculum Development)
[3] Hmelo-Silver C E 2004 Problem-based learning: What and how do students learn? *Educational Psy. Rev.* **16** 235
[4] Schettino C 2011 Teaching geometry through problem based-learning in the classroom *Math. Teacher* **5** 346
[5] Ridlon C L 2009 Learning mathematics via a problem-centered approach: A two-year study *Mathematical Thinking and Learning* **11** 188
[6] Ahamad S N S H *et al.* 2017 Implementation of problem-based learning in geometry lessons *J. Phys.: Conf. Ser.* **943** 012008
[7] Klecka W R 1980 *Discriminant Analysis* (Beverly Hills: Sage)
[8] Martinez D 2001 Predicting student outcomes using discriminant function analysis *Annual Meeting of the Research and Planning Group (CA, 2-4 May 2001)*
[9] Kathleen M and Carmen A 2002 *SPSS for Institutional Researchers* (PA: Bucknell University)

IOP Conf. Series: Journal of Physics: Conf. Series **1088** (2018) 012079 doi:10.1088/1742-6596/1088/1/012079

Calistung literacy through the application of Lectora

Masniladevi, Ritawati and Yullys Helsa

Universitas Negeri Padang, Jl. Prof. Dr. Hamka, Air Tawar, Padang, Indonesia

E-mail: masnila.devi@yahoo.co.id

Abstract. *Calistung* is an abbreviation in the Indonesian language that means reading, writing, and counting. *Calistung* literacy is basic in recognizing letter and number. Several experts state that *Calistung* literacy facilitates written communication using letter and number. This research is development research that produced animation using Lectora application. This study aims to improve students' literacy in *Calistung* through animation in the fourth grade in primary school. The learning model can be used practically, valid and useful, so the learning outcome reaches optimally. The content expert scored the content 92.22%, which means the content is excellent. The media expert scored the media 77.33%, which means the animation in this research is representative. The software expert scored the software 80%, which means the software used is useful for designing teaching material. Eighty percent of students state that the used media is right and 87% of teachers say the same.

1. Introduction

The Indonesian language is one of the studied subjects in schools, ranging from elementary school to college level [1]. Learning Bahasa (Indonesian language) is directed to improve the students' ability to communicate in the Indonesian language properly and correctly, both spoken and written, and foster the appreciation of Indonesian literary works [2]. One of the lessons in school is *Calistung* which stands for reading, writing, and arithmetic [3-6]. Appropriate learning methods for the future of children can still be processed [4-6]. All children who use SD should be familiar with the size of the individuals who support the children's favorite elementary school or famous [7]. *Calistung* literacy is the basic stage when the students can recognize letters and numbers [3]. Many experts consider the importance of *Calistung* literacy to facilitate communication in the form of written language and numbers [4-7]. *Calistung* literacy is mostly delivered informal education such as school [4-6]. By reading and writing allow the students to understand and convey any information they receive. Meanwhile, counting enables the students to be able to develop aspects of logical thinking well, especially maximizing the function of the left hemisphere. Many activities can be conducted to pursue literacy in primary schools. The teachers often use rote techniques and exercises that rely on cognitive abstractions which are not directly related to the student life. Consequently, the students' interests are constrained by scholastic tasks that given too early. Given this fact, it is necessary to hold a unique learning method about *Calistung*, by training students with appropriate models and media. One used media is the moving image media.

Learning with animated image media is expected to counter balance the development of *Calistung* literacy. In learning, the students are divided into groups of learners that consist of 4-5 people. The teacher displays animated images, the students and their group analyze the pictures, and the analysis is recorded on the paper to be read in front of the class. Individually, the *Calistung* literacy based on the animated images presented by the teacher is continued with the conclusion. The use of animated image media in learning *Calistung* can be used as one way to achieve one of the general goals of Indonesian language and literacy learning. It is used to improve the ability and skills of the students to communicate in the Indonesian language correctly and adequately, both spoken and written.

2. Method

The term model can be interpreted as an object or concept in the form of graphical display, work procedures that are regular and systematic, and contains critical thoughts following suggestions used to present a thing. The development model is based on two objectives, namely development to obtain product prototype and formulation of methodological suggestions for designing and evaluation of the prototype [8].

The development model to be planned in this study followed the path of Sivasailam et al. [9]. The main 4-D development model is Define, Design, Develop, and Disseminate or adapted into a 4-P model, defining, designing, developing, and deploying. Implementation of the main step in the research is not only tracing the original version but adapting the characteristics of the subject and place of origin of the examinee [10]. Besides, the model to be followed will be adjusted to the needs of development in the field. The research procedures would be implemented in 4 stages to make the research more focused, which can be stated in the chart of the steps of developing the fishbone (Figure 1).

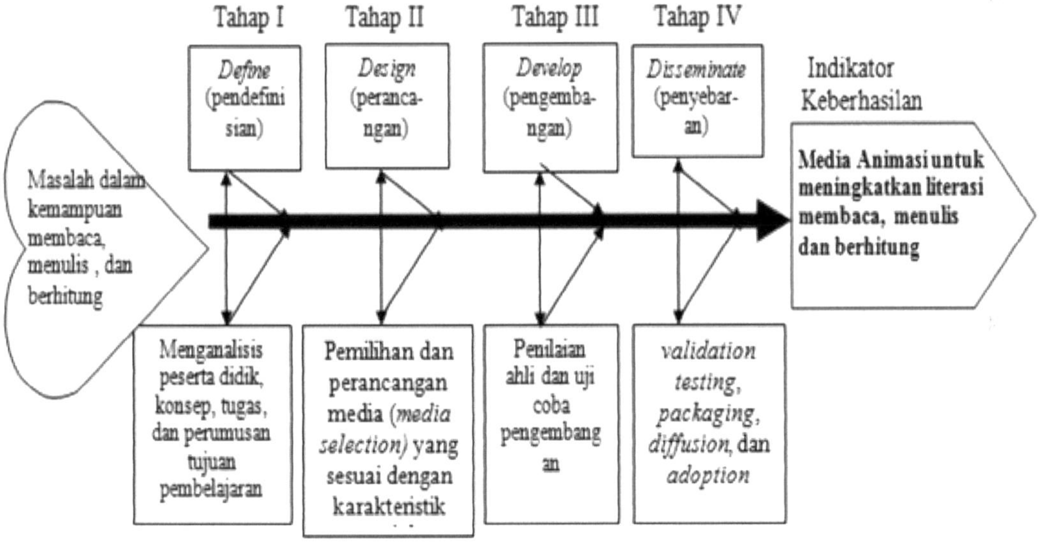

Figure 1. Fish Bone Images

3. Results and discussion

3.1. Defining stage

KTSP curriculum renewal into the 2013 curriculum requires more innovative learning in schools for the achievement of learning objectives [11]. The 2013 curriculum presents new challenges for the teachers to create education that complies with lawsuits [12]. The curriculum component is directly related to the *Calistung* animation media product. The material contained in the taken media from material in semester 1 for grade 1 of elementary school.

Student analysis concerning the condition and character of the student which include age, motivation to the subject, academic ability, psychomotor, social skill, interest and talent, and the tendency of learning style the students have. Based on the material presented in the *Calistung* animation media for grade 4 students of Primary School, it can be found that there are learning objectives that must be achieved or expected to be mastered by the students. Firstly, the students can understand the texts presented in the interactive *Calistung* animation media. Secondly, the students can solve problems in the form of calculations which is presented in interactive *Calistung* animation media.

3.2. Designing stage

The design of learning modules with Lecotora Inspire is made as a tool in the learning process. Therefore the material in making this media must be suited to the learning materials so that media that have been made can be used and can be utilized by subject teachers to make the learning process run well and the learning objectives can be achieved. Some of the pictures in learning media can be seen in Figure 2.

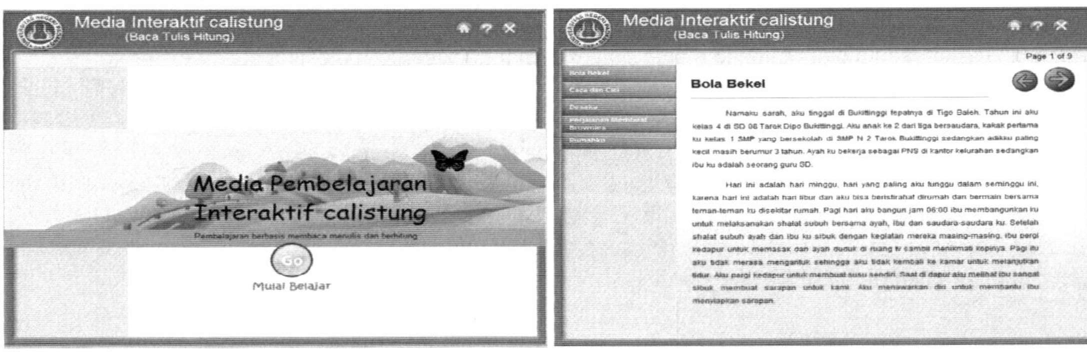

Figure 2. The first screen outlook (left) and Each menu outlook (right)

3.3. Developing stage

Developing stage is the testing phase of the finished program. If there is a program error, the error will be corrected, and if it is going well, the process will be directed to the next stage which called distribution. At this stage, some criteria will be seen such as Validity, Practicality, and Effectiveness of animation media.

Validation of interactive *Calistung* animation media was done by material experts, media experts, and linguists. Dr. Mardiah Harun, M.Ed., validated for material experts. Next, Nur Azmi Alwi, S.S., M.Pd., is the validator for the linguists. Lastly, Dra. Ritawati Mahyudin. M.Pd., and Antoni Hilman, S.Kom., validated by media experts. It was a result of validation by the experts (Figure 3).

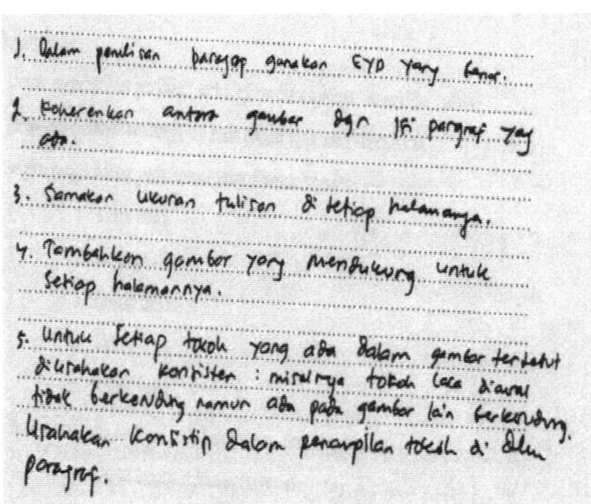

→

1. It should be using the correct sentences in writing paragraph.
2. The relationship between figure and the content of paragraph should be coherent.
3. The size of font in every sentences must be same.
4. Adding the supported figure for every pages
5. For every character including in that figure should be consistent. For example, the Caca's character in the beginning is not using veil, however in the other figure, she uses veil. Trying consistent for every character in this paragraph.

Figure 3. Teachers validity result

Based on the validation results, it could be concluded that the Interactive *Calistung* Animation Media Validity rate was 82% with the functional criteria on the Very Practical level. For the practical level, it was found that the level of Practicality of Interactive *Calistung* Animation Media was 86% with the functional criteria on the Very Practical level. The result of the effectiveness was seen in assessing the students learning outcomes in the cognitive domain using a written test. The test consisted of 8 essay questions given to the students at the 4th meeting. The students' final test scores were very ranged from 60 to 100 with an average score of 81.94, and the percentage of students' completeness was 77.27%. The result is supported by several researcher talked about the problems in production and utilization of instructional media of student teachers and the implementation of school literacy in primary school [13-18].

3.4. Distributing stage
In this process, the finished media was burned to a CD (Compact Disk) and packed so that, the product became ready to use. In this media, there were several supporting files such as:
- File.exe, which was the file that run the program.
- File.swf, like flash movie as completeness.

This media design did not have a minimum software limit that must be installed on the computer to run it. This media is also autorun, so it did not need a Lectora Inspire device to run it. This application can run on Windows XP, Vista, Seven (7) and Eight (8) operating systems.

4. Conclusion
The module of equations and linear inequalities which the authors had designated was valid and practical. Validation obtained from five validators and practicality obtained from a limited field test to the students. For a validity rate, it had 83% percent with very valid criteria. As for the practical level, it had a percentage of 86% with efficiency standards, and the rate of effectiveness was 77.27% with high criteria.

References
[1] Badan Standar Nasional Pendidikan 2006 *Standar Isi* (Jakarta: Badan Standar Nasional)
[2] Nurgiyantoro B 2001 *Penelitian dalam Pengajaran Bahasa dan Sastra* (Yogyakarta: Gajah Mada University Press)
[3] Susilo T A 2011 *Belajar Calistung itu Asyik* (Yogjakarta: Javalitera)
[4] Istiyani D 2014 *Jurnal Penelitian* **10** 1
[5] Faroqi A and Maula B 2015 *Jurnal ISTEK* **8** 229
[6] Aulina C N 2012 PEDAGOGIA: Jurnal Pendidikan **1** 131
[7] Eprilia U H and Prasetyarini A 2011 *Jurnal Penelitian Humaniora* **12** 126
[8] Van den Akker J 1999 Principles and methods of development research *Design approaches and tools in education and training* (Dordrecht: Springer) p 11
[9] Sivasailam T, *et al* 1974 *Instructional Development for Training Teachers of Exceptional Children* (Blomington: Indiana University)
[10] Arikunto S 2002 *Prosedur Penelitian Suatu Pendekatan Praktek* (Jakarta: Rineka Cipta)
[11] Mulyasa E 2013 *Pengembangan dan Implementasi Kurikulum 2013* (Jakarta: Remaja Rosdakarya)
[12] Amri S 2013 *Pengembangan dan Model Pembelajaran dalam Kurikulum 2013* (Jakarta: Prestasi Pustaka)
[13] Srichantha S 2013 A Study about the Problems in Production and Utilization of Instructional Media of Student Teachers *Proceedings of the International Conference on Information, Business and Education Technology (ICIBIT 2013)* (Paris: Atlantis Press)
[14] Kurniati S 2017 The Implementation of School Literacy in Primary School *Proceedings of the 1st International Conference on Intellectuals' Global Responsibility (ICIGR 2017)* (Paris: Atlantis Press)
[15] Aksa 2017 *Advances in Social Science, Education and Humanities Research (ASSEHR)* **158** 37
[16] Hendroanto A, *et al* 2018 Mathematics education students solving PISA's mathematics problems: An overview of students' mathematical literacy *International Journal on Emerging Mathematics Education* **2** 129
[17] Masniladevi, *et al* 2018 *J. Phys.: Conf. Ser.* **943** 012059
[18] Mumu J, *et al* 2018 *J. Phys.: Conf. Ser.* **943** 012011

The validity test of the lesson plan to reduce students' misconceptions using the cognitive conflict strategy

D Fitri, R Johar and A Ahmad
Syiah Kuala University, Jl. Teuku Nyak Arief Darussalam, Banda Aceh 23111, Indonesia

Email: rahmah.johar@unsyiah.ac.id

Abstract. Students' misconception will hinder the effectiveness of learning as well as disrupt the students' thinking in accepting the subsequent knowledge. The use of students' misconceptions is essential in improving students' conception through meaningful learning. One of the learning strategies to reduce the misconception is the cognitive conflict strategy. Therefore, it is necessary to develop a lesson plan to lessen students' misconception in learning using the cognitive conflict strategy. This study is limited to the analysis of the quality of the lesson plan on the topic of fraction by using the cognitive conflict strategy. The development method used in this research was the Plomp method focusing on the prototyping and assessment phase. The source of data for the validation test was obtained from three validators and one observer. The trial of the lesson plan was conducted for 14 Year 5 students in one of the elementary schools in Bireuen district of Aceh. Based on suggestions from the validators, the lesson plan needs to be revised. The results of the validity test showed that the lesson plan for reducing students' misconception using the cognitive conflict strategy satisfied the valid criteria and can be tested for its practicality and effectiveness.

1. Introduction

Students' misconceptions will affect the construction of the subsequent knowledge. A belief that is not in accordance with the scientific concept but is considered correct by the students is a sign of misconception [1]. The misconception is private, stable, and students who have the misconceptions do not feel the need for a coherent view [2]. The nature of misconceptions that is resistant to change, make it difficult for a misconception to be eliminated. Changing students' conceptions cannot be done by simply telling them what the misconceptions are since this strategy might only last temporarily in students' mind.

Students have learned fraction since the third grade of elementary school. However, many students' still show a poor understanding of fraction concepts. Based on the results of the National Assessment of Educational Progres (NAEP), it consistently shows that students have a poor understanding of the concept of fraction [3]. Students who have lack of the understanding of these concepts might find difficulties in obtaining meaningful learning when solving procedural questions. Based on the results of diagnostic tests on the fifth-grade students in elementary school, it is known that 40% of students have misconception related to fraction. This shows that misconception is a problem that should be immediately addressed. Thus, it is paramount to develop a lesson plan to reduce the students' misconception on fraction concept.

Several previous studies showed a wide range of misconceptions experienced by students in understanding the concept of fraction. Students compare the fraction by looking at the differences between the denominator and the numerator, and this led to students' misconceptions in comparing the value of fraction [4,5]. Students sort the fraction by looking at the least denominator of the fraction [4]. Students are surprised to learn that multiplication on fractions does not necessarily lead to greater answers [6]. This suggests that students remain on their initial knowledge of integer operations; students have not been able to integrate their initial experience with newly acquired skills.

Correcting a student's misconception can be done by utilizing a prior knowledge owned by the students [7,8]. Using students' initial conception to improve students' conceptions will be a meaningful learning experience for the students. However, this is not easy because of the misconception experienced by the students are generally resistant to change. Therefore, it is necessary to apply a strategy that can stimulate the occurrence of conceptual changes. One of the concept-changing strategies is the cognitive conflict strategy. Cognitive conflict is the individual's consciousness about the conflicting information that impacts a concept on the cognitive structure itself [9]. Cognitive conflict strategies have a general pattern of explicitly exposing students' initial conceptions, discussing and evaluating the students' initial concepts, creating conceptual conflicts on the initial concepts as well as encouraging and directing students to restructure their concepts [7]. Based on the explanation previously mentioned, the research question in this study is "How is the quality of the lesson plan to reduce student misconception using the cognitive conflict strategy."

2. Method

This study used Research & Development (R & D) method, employing Plomp model consisting of three stages: a preliminary research, prototyping phase and assessment phase [10]. In the first stage, the initial research needs analysis of the characteristics of students' misconceptions affecting the subsequent knowledge construction process was carried out. In the second stage, the prototyping phase, the development or creation of a prototype in the form of a lesson plan and validation were conducted. The third stage, the assessment phase, the assessment or evaluation developed lesson plan were done by conducting field tests. This paper only discussed the results of the validation and trial results of a limited sample.

Three competent validators did the validity of the lesson plan draft on fraction using the cognitive conflict strategy. The draft of the lesson plan was tested on five students aged 14 years old. The instruments of the data collection in this study were the validation sheet and observation sheet of implementation of the lesson plan. Data analysis techniques performed were calculating the average validity score of the three validators and analyzing the implementation of the lesson plan from an observer (a classroom teacher). The criteria for the validity of the developed lesson plan are as follows [11].

Table 1. Validity criteria of the lesson plan.

Criteria	Average
Highly valid	$4 \leq KV^{a} < 5$
Valid	$3 \leq KV < 4$
Less valid	$2 \leq KV < 3$
Invalid	$1 \leq KV < 2$

[a]Average validation results from the experts on the learning instruments.

Based on Table 1, if the validation results indicating that the data obtained is valid or highly valid, the draft of the lesson plan is eligible to be tested for field trials. However, if the validation results are less valid or invalid, the draft of the lesson plan must be revised.

3. Results and discussion

The design of a lesson plan using the cognitive conflict strategy aiming to reduce the students' misconceptions was based on the students' needs related to how misconceptions experienced by the students. The learning steps were designed by uncovering, discussing and evaluating students' initial conceptions, creating conceptual conflicts, and directing students to reconstruct their initial concepts. After designing the lesson plan using the cognitive conflict strategy, the product, a lesson plan, was validated by the experts on the lesson plan components and learning activities-aspects. The validation results of the lesson plan by the three experts are presented in Table 2.

Table 2. The results of the lesson plan validation by experts.

No	Aspect of assessment	Average	Criteria
1.	Presentation of lesson plan components	3,7	Valid
2.	Material conformity	4,3	Highly valid
3.	Activities conformity with the cognitive conflict strategy		
	a. Uncovering students' initial conceptions	3,7	Valid
	b. Discussing and evaluating students' initial conceptions	4	Highly valid
	c. Creating a conceptual conflict	4	Highly valid
	d. Directing students to restructure the initial concept	4	Highly valid
4.	Language	3,9	Valid

Table 2 indicates that the validity of the lesson plan for each aspect was highly valid or valid. The criteria of material conformity, evaluating the students' initial conception, creating conceptual conflict, directing the student to restructure the initial concept aspects were highly valid. The presentation of the lesson plan components, uncovering the students' initial conception and the language aspects were valid. Thus, the designed lesson plan can be used with minor revision.

Based on the validation results by three validators, there were some suggestions for improving the lesson plan using the cognitive conflict strategy. The validators suggested that the numbering of the learning steps should be clarified, the compilation of the learning objectives should contain the elements of the audiences, behavior, condition, and degree according to the demands of the 2013 curriculum, and enclose the teaching materials. The validators also argued that revealing the initial conception of the students should be done through a written test, followed by a question and answer session while providing opportunities to students to complete the same written test as the reflection of the change in conception after the emergence of the cognitive conflict. Some examples of revisions to the aspects of the lesson plan are presented in Table 3.

Table 3. The revised presentation of the lesson plan components.

Before revision	After revision
Numbering system using *bullets*	Numbering system using the number on each learning activity
Students can determine the pictures that represent a fraction	Through group discussion students are expected to be actively involved in the learning activities, be curious, and able to determine the pictures that represent the fractions correctly.
Without teaching materials	Attaching teaching materials

IOP Conf. Series: Journal of Physics: Conf. Series **1088** (2018) 012080 doi:10.1088/1742-6596/1088/1/012080

Revisions were made to the aspects of the presentation of the lesson plan components to meet the demands of the 2013 curriculum implemented in elementary schools. The numbering would ease the implementation of the learning steps. The attached teaching materials were presented briefly so that teachers could gain a basic understanding of it that would be taught the material.

The validators' suggestion for the initial conception of the students to be revealed through a written test, followed by a question and answer session while providing opportunities for students to complete the same written test as the reflection of the change in the conception after the emergence of the cognitive conflict, it was not just questioning. One example of the activities using a cognitive conflict strategy at the stage of uncovering students' conceptions before and after revision can be seen in Table 4.

Table 4. The results of the activity revision of uncovering students' initial conception.

Before revision	After revision
Teachers conduct a question and answer session for the students' ideas in determining the pictures representing fractions.	The teacher asks the students to choose the pictures representing a fraction individually. Look at the following pictures! Put a tick (√) for the shading picture representing $\frac{1}{4}$! 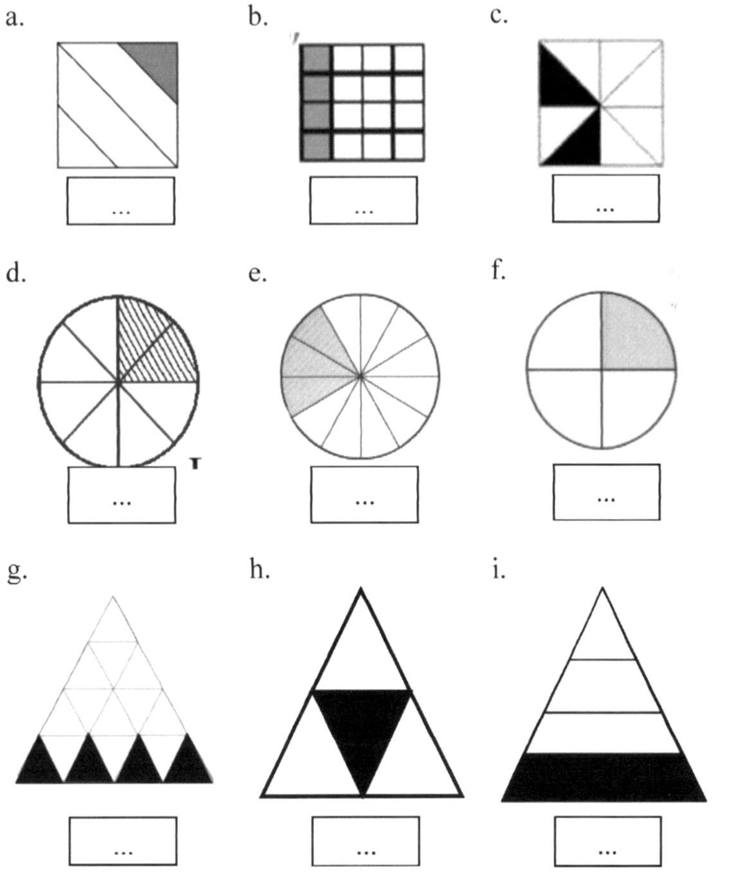 Next, check students' answers through questions and answers session and ask students to explore the reasons for choosing the picture.

The revisions of the activities for encouraging students' initial conceptions were made by providing students the opportunity to choose pictures representing a fraction. Based on these answers, the teacher dug up information of the students' conception through a question and answer session so that students can express the reasons for choosing the image. Revealing the students' initial conception is an important element in reducing misconceptions. Some attempts to reduce misconceptions do not work because teachers do not know exactly the reasons behind the misconceptions [7]. The revision results of the act of revealing the students' initial conceptions enabling the teacher to find the fault concept on fraction experienced by the students. Asking students to provide reasoning for choosing the answer was a way of knowing specifically how the misconceptions experienced by the students.

The validated and revised lesson plans were then tested for 14 of the Year 5 students. The trial results based on the observer observations are presented in Table 5.

Table 5. The trial results of the lesson plan.

No	Aspect of Assessment	Score	Criteria
1.	Introduction	5	Highly valid
2.	Core activities		
	a. Uncovering students' initial conceptions	4	Highly valid
	b. Discussing and evaluates students' initial conceptions	4	Highly valid
	c. Creating conceptual conflict	3	Valid
	d. Directing students to restructure the initial concept	3	Valid
3.	Closing	5	Highly valid

Based on Table 5, it is noted that the lesson plan tested had valid criteria. The product of designed lesson plan was able to reveal the students' initial conception through written test followed by the question and answer session; the discussion of the students' initial conception makes students aware that there was something wrong with their ideas. The raised cognitive conflicts were managed to waver students' conceptions through a demonstration of the fraction. Emerging the cognitive conflict strategies by demonstration was an attempt to provide different stimuli to students' cognition [7]. One of the activities to create cognitive conflict is to demonstrate [12]. The inconsistency between students' initial conception and the results of the demonstration stimulates students to find the appropriate concepts so that the process of reconstructing the students' conception occurs. Providing an opportunity for students to resolve the same written test was an attempt to obtain information about the changes in the students' initial conceptions.

Based on the validation results, the designed lesson plan can be considered valid. Student activities in the learning process can stimulate the emergence of the cognitive conflict and can reduce students' misconception. Therefore, the designed lesson plan can be used with minor revision.

4. Conclusion

Based on the results and discussion, the validation and the limited test previously presented the developed lesson plan using the cognitive conflict strategy is considered valid and feasible to be used. However, the lesson plan needs to be further tested for its practicality and effectiveness.

References

[1] Ormrod J E 2009 *Psikologi Pendidikan: Membantu Siswa Tumbuh dan Berkembang* (Jakarta: Erlangga)
[2] Dahar R W 2011 *Teori-Teori Belajar & Pembelajaran* (Jakarta: Erlangga)
[3] Walle J A Van De 2008 *Matematika Sekolah Dasar dan Menengah* jilid 2 (Jakarta: Erlangga)
[4] Cramer K, Post T and Delmas R 2002 Initial fraction learning by fourth- and fifth-grade students: a comparison of the effects of using commercial curricula with the effects of using the rational number project curriculum *Journal for Research in Mathematics Education* **33** 111
[5] Harvey R 2012 *Stretching student teachers' understanding of fractions* Mathematics Education Research Group of Australasia, Inc. 13394-012-0050-7
[6] Cramer K and Wyberg T 2009 *Efficacy of different concrete models for teaching the part-whole construct for fractions* Mathematical Thinking and Learning **11** 226
[7] Suparno P 2013 *Miskonsepsi & Perubahan Konsep dalam Pendidikan Fisika* (Jakarta: Grasindo)
[8] Sela H and Zaslavsky O 2007 Resolving Cognitive Conflict With Peers–Is There A Difference Between Two And Four *Proceedings of the 31st Conference of the International Group for the Psychology of Mathematics Education* **4** (Seoul PME) pp 169-176
[9] Madu B C and Oriji M 2015 *Effects of Cognitive Conflict Instructional Strategy on Students' Conceptual Change in Temperature and Heat* Sage Open 1-9
[10] Plomp T 2013 Educational Design Research: an Introduction, Ed T Plomp and N Nieveen (Enschede: SLO) p 10
[11] Khabibah S *Pengembangan Model Pembelajaran Matematika dengan Soal Terbuka untuk Meningkatkan Kreativitas Siswa SD* [Disertasi] Surabaya:UNESA 2006
[12] Ibrahim M 2012 *Konsep Miskonsepsi dan Cara Pembelajarannya* (Surabaya: Universitas Press)

The use of CIPP model for evaluation of computational algorithm learning program

S J Hartati, N Sayidah and Muhajir
Universitas Dr. Soetomo Surabaya, Jl. Semolowaru No.84, Indonesia, 60118

E-mail: sulis.janu@unitomo.ac.id

Abstract. The objective of this research was to evaluate the learning program of computational algorithm subject in two private universities in Surabaya. The evaluation was performed by using the theoretical framework of CIPP model. There were two research questions presented in this study: what is the research design to evaluate the learning program and what is the profile of the learning style-based computational algorithm learning program according to CIPP model. The results showed that the research design consists of 5 steps, which are: to determine the focus of the research, to collect the data, to test the instruments' validity and reliability, to analyse the data, and to draw a conclusion. Every research step is formulated based on 4 components, which are: context, input, process, and product. The learning program has several profiles. Firstly, the pre-required environment does not correspond with the characteristics of the population. Secondly, students' mathematical ability does not meet the qualification set in the lesson plan. Thirdly, the supporting learning facility and instruments are not sufficient. Next, the implementation of the lesson plan is adjusted according to students' condition. Lastly, the average of students' computational algorithm scores is 49. It is recommended to conduct matriculation program before the actual learning program for students that do not meet the pre-requirements stated in the lesson plan. The matriculation module consists of the basics of mathematical reasoning, which includes: arithmetic logics, algebraic logics, and mathematical logics.

1. Introduction

Context, Input, Process, and Product (CIPP) is the first evaluation model used for education and was introduced by Stufflebeam in 1965[1][2][3]. In this work, this model was used because it focuses on the system improvement instead of proving a certain point inside the system [1][3]. Moreover, CIPP is systematically designed as the guide to evaluate the learning process, starting from the beginning of the assessment (context and input evaluation), during the implementation (input and process evaluation), to the final evaluation of the learning outcomes (product evaluation) [2]. The nature of this model corresponds with the objective of this research, which is to improve the design of computational algorithm for students with below-average competency in basic arithmetic skills. Additionally, these students had the tendency of having kinaesthetic and/or tactile learning style [4][5].

The learning process has been improved since 2013 [6]. The improvement was made by adding media devices, such as software applications. The purpose of the media devices was to help the students to visualize the concept of design and iteration. The result showed an improvement in the average score. However, the dimension of the cognitive process, which was based on Bloom Taxonomy [7], remained in C1.

Another research also improved the learning model by using the theoretical framework of APOS [8]. The result showed an increase in the cognitive process dimension from C1 to C2. However, this result could only be achieved when outside stimulus, which was the logic and algorithm learning software application, was present. Based on the characteristics, this level of understanding was categorized by Asiala [9], DeVries [10], Dubinsky [11], and Tall &Vinner [12] as the understanding in the action level.

On the other hand, the learning outcome of the computational algorithm subject is to be able to create logical automation processes, which are presented in the form of flowchart and pseudo code [13]. By having this characteristic, the knowledge learned in the programming algorithm subject is considered as metacognitive knowledge [7]. As stated by Skemp [14], in order to learn metacognitive knowledge, the ability to draw a correlation between one concept to another so that a new, more complex concept is produced, is needed. This new and more complex concept is called schema. In the APOS theoretical framework [9][10][11][12], the establishment of schema corresponds to the highest level of understanding. APOS itself stands for action, process, object, and schema.

In this work, there is a very wide gap between the students' level of understanding and the learning outcome of computational algorithm subject. Because of this reason, evaluation of existing learning design is essentially needed. In this research, there are two important research questions: what is the research design for computational algorithm learning program evaluation based on the learning style, and what is the profile design of learning style based computational algorithm subject learning according to CIPP model.

2. Method

The program was evaluated by using the quantitative research paradigm, with the process approach. For this reason, CIPP was chosen as the evaluation model. The scheme of CIPP is described in Figure 1.

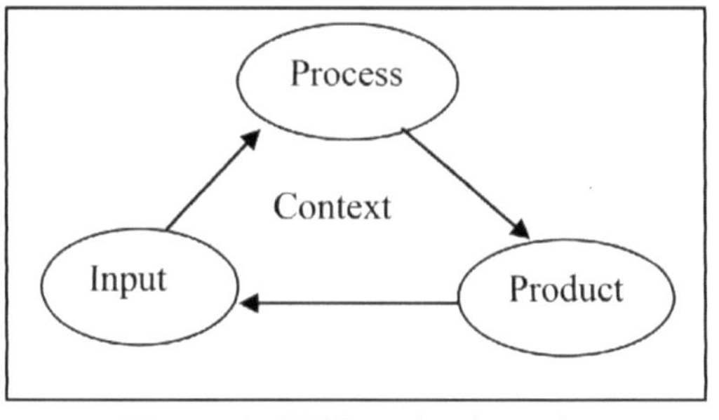

Figure 1. CIPP evaluation scheme

There are 5 steps to design the evaluation of the research program, which are: (1) to determine the focus of evaluation, (2) to determine the data collection procedure, (3) to make the instrument used for data collection and to test its reliability and validity, (4) to analyse the data by using descriptive statistics, (5) to draw a conclusion based on the 4 components of CIPP.

The focus of the evaluation was determined based on the 4 components of CIPP. Context evaluation aims to obtain the profile of the program environment, unmet needs, the characteristics of the targeted population, and the objective of the program [1][2][3]. Input evaluation aims to obtain the profile of human resources, facilities and supporting instruments, funds or budgets, procedures and regulations required to fulfil the objective [1][2][3]. Process evaluation aims to check if the prepared plan can be directly executed or requires further improvement and innovation [1][2][3]. Product evaluation stresses on the achievement of the purpose determined during the planning [1][2][3].

The subjects of this research were the students in the batch of 2015/2016 and 2016/2017. The data collection techniques were tests, document reviews, questionnaires, observations, and interviews. The tests were given in 4 topics, which were: learning style, arithmetic ability, algebra, and programming algorithm. The data analysis used was the descriptive statistic with the observed parameters of mean and proportion.

3. Result and discussion

The focus of the evaluation in this research was the 4 components of CIPP. The subject of the research was the students in the batch of 2015/2016 and 2016/2017. From the learning style test result, it was learned that 8% of the students had the tactile learning tendency, 3% had kinesthetic learning tendency, and 88% had not the tendency of visual, auditory, kinesthetic, and tactile learning style. The result of learning style test was only used as additional information, as the main focus of this research was the learning program of computational algorithm subject based on the students' learning style.

The data were collected using tests, document reviews, questionnaires, observations, and interviews. The validity of the instrument was tested by using content validation. On the other hand, the reliability of the instrument was tested in timely manner, where the subjects were tested in different occasions with three weeks intervals. The data was then analysed by using the average parameter. The conclusion was drawn based on CIPP model, by comparing the value of an average parameter with the learning achievements in the syllabus.

3.1. Context evaluation

The focus of this components is to get the profile of program environments, unmet needs, targeted population's characteristics, and the objective of the program [1][2][3].

The profile of program environments was obtained by reviewing the syllabus of programming algorithm subject in the academic year of 2015/2016 and 2016/2017. The document review results showed that the syllabus was arranged to accommodate the different learning styles of every individuals [15], especially for students with kinesthetic and tactile learning tendency. The learning media used was the software applications aimed to help students with kinesthetic and tactile learning styles to understand the abstract concepts in learning the computational algorithm subject. The abstract concepts involved variables, data, constants, operators, data processing, and flowchart [8]. The students that took the program were expected to be proficient in arithmetic operations, be it integers or fractions, and to understand the logic of basic mathematics.

In order to obtain the data of students' proficiency in arithmetic operations and basic mathematics logics, a test was conducted to the research subjects, which were the final year students in the batch of 2015/2016 and 2016/2017. The test consisted of 3 main topics, which were: (1) integer arithmetic operation problems, (2) fractional arithmetic operation problems, and (3) basic mathematics logic problems. The test results showed that: (1) the average score of integer arithmetic operation test was 76.85, (2) the average score of fractional arithmetic operation test was 43.24, and (3) the average score of basic mathematic logic test was 45.67. These results showed that the pre requirement of the program environment did not match the characteristics of the population.

The objective of the program that was written in the syllabus was that "after taking this subject, semester 1 students are able to design algorithms and evaluate their designs to solve computational problems, which are presented in the form of flowchart and pseudo codes, both individually and in team, and to possess 3 pillars of characters: honesty, responsibility, and teamwork". The data collection technique that was used to achieve the objective of the program was tested. The test instruments consisted of 2 main problems, which were to create a computational algorithm with simple variables and 1-dimensional array. The test results showed that the average score of the students was 49 and only one student scored 70. This result showed that the objective of the program was not achieved yet.

3.2. Input evaluation

The focus of the input evaluation was to obtain the profile of human resources, facilities and supporting instruments, funds or budgets, procedures, and regulations required to reach the objective [1][2][3].

The human resources that became the focus of the research were students. The profiles of the students were obtained by using questionnaires. The questionnaires' results showed that nearly 70% of the students worked while studying at the university, and paid the tuition fee with their own salaries without their parents' financial supports. The students also chose the major in information technology because it was regarded as a field with better job opportunities when compared to their current jobs. The average of the students' salary was less than 2 million rupiahs. Additionally, 90% of the students were not aware that mathematical ability was required to learn information technology.

The data of the facilities, supporting instruments, budgets, procedures, and regulations required to achieve the goal was obtained through observations and interviews. On the other hand, the data of the supporting instruments possessed by the students was obtained by questionnaires. The observation results showed that the facility and supporting instrument available in the class during the lessons was LCD projector. Neither computers nor laptops were available. There were also no additional facilities and supporting instruments to increase the students' abilities because of the very low tuition fee. On the other hand, the number of students that owned laptop was less than 30%. The procedures and regulations to fulfill the objective were inputted into the assessment system by the lecturer.

The results showed that: (1) 70% of the inputs were from the lower middle economy class, (2) 90% of the inputs were not aware that mathematical ability was needed to learn information technology, and (3) the facilities and supporting instruments used for learning needed to be improved.

3.3. Process evaluation

The data for process evaluation was obtained through observations, interviews, and document reviews. The results showed that before the lesson started, arithmetic and basic mathematic logic test was conducted in the first meeting. The results of the test showed that the average of arithmetic and basic mathematic logic scores was 55. Because the factual condition did not match the determined pre-requirement, the syllabus was modified since the first meeting by the lecturer. This data was discovered through interview and document review of class schedule report.

The material aimed to be given in the first meeting was only able to be delivered in the third meeting. In the first and second meeting, the lessons were conducted by explaining the use of the learning media. Next, the students were asked to use and re-study the learning media in groups at home. One group consisted of 5 students. The groups were formed by the lecturer, with the students whose score were greater than or equal to 60 chosen as group leaders. Every group was then given a task similar to the problem introduced in the learning media. The evaluation of the assignments submitted showed that every group obtained the score of 100, with the characteristics that every group had the same task.

In the third week, the students were given a quiz which questions were similar to the given assignment, different by only small modifications. The test results showed that only 5% of the students were able to receive the score of 100. On the other hand, the average of the test scores was below 49.

The learning evaluation at the end of the semester showed that 100% of the students were able to answer the questions well when outside stimulus that was similar to test questions existed. The mistakes that were made by the students were: (1) to determine the sequential priority of the arithmetic operations, (2) to estimate the variables used in the transformation processes, (3) to explicitly arrange algebraic equation as a form of input to output transformation, (4) to put the algebraic equation and in equation in the correct flowchart symbols, (5) to determine the truth value of a relation, and (6) the learning objective written in the syllabus was not met.

The findings showed that: (1) the proposed learning plan could not be implemented so that adjustments with students' condition were needed, (2) the learning media used by the lecturer could not increase the students' understanding to a level higher than action.

3.4. Product evaluation

The data for product evaluation was obtained through the document reviews of final scores of the programming algorithm subject and interview with the supporting lecturers.

From the interview results, it was learned that the assessment process was done in two stages. The first stage used the authentic final score. In the second stage, the authentic final score was then readjusted by considering class attendance, activeness during the lessons and honesty during the tests. The document review results showed that the average of an authentic final score was less than 49. Only one student received the authentic final score of 70. The authentic final score was taken from individual midterm test, final test, and quizzes that were given during the lessons. The discussion in this research only used the first stage assessment.

The data of the honesty, hard work, and discipline value were obtained through document reviews. The students that were honest in the test were given the score of 100. On the other hand, dishonest students were given 0. The component of the honesty value was for students to do the midterm test, final test, and quizzes by using their own ability, and did not attempt to ask for other students' help during the tests. In the tests conducted, students were allowed to open and read their notes. The activity during the lessons and assignment scores represented the value of hard work. The students that were actively doing the exercises during the lessons received the score of 100. The students that finished the assignments as instructed also received the score of 100. The assignment score consisted of 2 aspects, which were the correctness of the answer and the discipline value. The assignment scores and punctual attendance represented the discipline value. The students that submitted their assignments on time were given a score of 100. The document review results showed that the honesty score reached 80%, hardworking score reached 30%, and discipline score reached 80%.

These results showed that: (1) the average ability of students to design computational algorithms was 49. This score indicated that the students' abilities were far from what was expected in the syllabus, (2) the honesty score reached 80%, hardworking score reached 30%, and discipline score reached 80%.

4. Conclusion

It can be concluded that there are five steps used in the research design of CIPP implementation on the evaluation of computational algorithm learning program. In the first step, the four components of CIPP, which are the context, input, process, and product evaluations, are determined as the research focus. In the second step, several data are collected. Document reviews and tests data are collected for context evaluation. Questionnaires, observations, and interviews data are collected for input evaluation. Observations, interviews and document reviews data are collected for process evaluation. Interviews and document reviews data are collected for product evaluation. In the third step, the validity of the instruments, which are tests and questionnaires, is tested by using content validation. To test the reliability of the instruments, re-tests are conducted with 3 weeks interval. In the fourth step, the data is analysed by using the average parameter. In the fifth step, the conclusion is drawn based on the data analysis results according to CIPP components. The profile of algorithm learning program according to CIPP is presented as follows. The research focus is based on the CIPP model. The context evaluation result shows that the pre-requirement of the program environment does not correspond with the population characteristics. The input evaluation result shows that 70% of the students come from the lower middle economy class, 90% do not know the importance of mathematical thinking in learning computational algorithm subject, and the supporting facilities and instruments are not sufficient. The process evaluation result shows that the lesson plan is adjusted according to the students' condition and the learning media used has not functioned optimally. The product evaluation result shows that the average of computational algorithm score obtained is 49.80% of the students are honest, 30% of the students are hardworking, and 80% of the students are disciplined. Based on the profile, it is recommended to conduct matriculation program before the actual learning program for students that do not meet the pre-requirements stated in the lesson plan.

The matriculation module consists of the basics of mathematical reasoning, which includes: arithmetic logics, algebraic logics, and mathematical logics.

Acknowledgement
The publication of the paper was made possible because of the full support given by Directorate of Research and Community Service - Directorate General of Development and Research Enhancement - Ministry of Research, Technology, and Higher Education of the Republic of Indonesia. It is also an inseparable part of a research entitled "Development of Mathematical Reasoning Learning Media in the Computational Algorithms Development based on Learning Style and Character Building", which received financial grant in 2017-2018 from Directorate of Research and Community Service - Directorate General of Development and Research Enhancement -Ministry of Research, Technology, and Higher Education of the Republic of Indonesia.

References
[1] Widoyoko P E S 2009 *Evaluasi Program Pembelajaran, Panduan Bagi Pendidik dan Calon Pendidik* (Yogyakarta: Pustaka Belajar)
[2] Asfaroh J A, Rosana D and Supahar 2017 Development of CIPP Model of Evaluation Instrument on the Implementation of Project Assesment in Science Learning. *International Journal of Environmental & Science Education* **12** 1999
[3] Zhang G, Zeller N, Griffith R, Metcalf D, Williams J, Shea C, and Misulis K 2011 Using the Context, Input, Process, and Product Evaluation Model (CIPP) as a Comprehensive Framework to Guide the Planning, Implementation, and Assessment of Service-learning Programs. *Journal of Higher Education Outreach and Engagement 15* 57
[4] Dunn R and Dunn K 1993 *Teaching Secondary Students Through Their Individual Learning Styles, Practical Approaches For Grades 7-12* (Massachusetts: Simon & Schuster) Inc
[5] Prashnig B 2007 *The Power Of Learning Styles. Memacu Anak Melejitkan Prestasi Dengan Mengenali Gaya Belajarnya* (Bandung: Kaifa)
[6] Rahmawati and Hartati S J 2013 The Application of Computer Aided Learning to Learn Basic Concepts of Braching and Looping on Logic Algorithm *The International Journal of Multimedia & Its Applications (IJMA)* **5** 15
[7] Anderson J and Karthwohl 2001 *A Taxonomy for Learning, Teaching, and Assessing* (New York: Addision Wesley Longman, Inc)
[8] Hartati S J 2014 Design of Learning Model of Logic and Algorithms Based on APOS Theory. *International Journal of Evaluation and Research in Education (IJERE)* **3** 109
[9] Asiala M et all 2004 *A Framework for Research and Curriculum Development in Undergraduate Mathematics Education* (Indiana: Purdue University)
[10] DeVries D and Annon I 2004 Solution - What Does It Mean? Helping Linear Algebra Students Develop The Concept While Improving Research Tools *Proceedings of the 28th Conference of the International Group for the Psychology of Mathematics Education* Vol 2 p 55
[11] Dubinsky E 1988 A Theory And Practice Of Learning College Mathematics *Mathematical Thinking and Problem Solving* 221
[12] Tall D and Vinner S 1981 Concept Image and Concept Definition in Mathematics with Particular Reference to Limits and Continuity *Educational Studies in Mathematics* **12** 151
[13] Farrell J 2011 *Programming Logic and Design Introductory, sixth edition* (Canada: Course Technology)
[14] Skemp, Richard R 1982 *The Psychology Of Learning Mathematics* (Great Britain: Hazell Watson &Vney Ltd)
[15] Sulistiowati and Hartati S J 2015 Penerapan TAM pada Pembuatan Aplikasi Multimedia untuk Belajar Logika dan Algoritma Berbasis Gaya Belajar. *Laporan penelitian hibah bersaing DIKTI, tahun ketiga* (unpublish)

Analysis of senior high school students' emotional intelligence in cooperative based mathematics learning

M Ilyas, M Ma'rufi, F Fitriani and S Salwah
Universitas Cokroaminoto Palopo, Jl. Latammacelling No.19, Sulawesi Selatan 91911, Indonesia

Email: muhammadilyas@uncp.ac.id

Abstract. Mathematics learning is a process of interaction between teachers and students in a learning environment deliberately created by teachers involving various elements of mathematics learning to form the mindset, understanding, reasoning, and problem-solving skills of mathematics. A character can be cultivated through habituation including the integration of emotional intelligence in cooperative learning. This article was part of development research of cooperative mathematics learning model by integrating emotional intelligence. The subjects of this research were Year 10 students. This study focused on describing students' emotional intelligence developed during the implementation of cooperative learning in a mathematics classroom. The data was collected through observation of learning and interview. Qualitative analysis including reduction step, categorization, data interpretation, and conclusion was employed. The results showed that students could productively utilize emotions in learning mathematics as indicated by being responsible, focusing on the tasks given, confident, tenacious, skillful in communicating, more attentive in listening to others' opinion, and satisfied with the award given.

1. Introduction

Students' achievement is likely influenced by two factors, internal and external factor. Internal factor is highly dominated by psychological conditions and various students' potential in the form of intelligence, including the intellectual intelligence covering various abilities, such as numerical, spatial, reasoning, abstract thinking, and verbal skills. One of the students' internal factors contribute to the success of students in learning is emotional intelligence, including perseverance, social skills, empathy, patience, sincerity, resilience, and self-belief. Emotional intelligence is the ability to read her/himself and others to be able to put her/himself in the situation of others while at the same time control himself; to manage every action and reaction through self-control, motivation, perseverance, pleasant personality and courtesy leading to success in everyday life or at work [1].

Emotional intelligence is one's ability to create synergy innovation in teamwork to achieve organizational or group goals; one's emotional intelligence skills involve skills in managing themselves and others; social skills as in maintaining working relationships and interactions in other life dynamics. Thus, it is not sufficient for a person to only have feelings instead it is required to learn to recognize, appreciate your feeling and others in pleasantly and accurately accepting opinions. Under normal circumstances, a person's emotional routine activities cannot be observed. However, emotions can arise and be observed through facial expressions, voice stress, facial expressions, and body language. This emotional expression can be identified when a relationship with others occurs and there is a mutual need.

Emotional intelligence plays an essential role for the success of one's career in modern life, in a broader sense, it highly contributes on the success of life. There are several aspects of Emotional Intelligence (EI), namely: emotional self-awareness, managing emotions, productively utilizing emotions, empathy, and building relationships [2]. Emotional self-awareness involves the improvement in recognizing and feeling their emotions, having a better understanding of the causes of feelings arisen, recognizing differences between feelings and actions. Managing emotions concern on higher tolerance for frustration and anger management, reduced verbal mockery, fights and classroom disruptions, being able to better express anger properly, without fighting; reduced temporary entry ban from suspension and aggression or self-destructive behaviour; more positive feelings about themselves, school and family, being able to deal better with mental tension, reduced loneliness and social anxiety. Utilizing emotions productively involves: being more responsible, less impulsive, and more self-controlled as well as focusing more on the task at hand. While, empathy, the emotional reading, involves accepting others' points of view, improving empathy and sensitivity to other people's feelings, as well as being a better listener to others.

Emotional intelligence is defined as the ability to recognize our own and other feelings, the to manage emotions well in relationships with others [3,4]. Building a relationship means improving the ability to analyze and understand relationships, better at resolving and negotiating disputes and issues arising in relations as well as being more assertive and skillful in communicating; being more popular, outgoing, friendly; engaged with peers, needed by peers; being more attentive and considerate toward others; being more concerned with social interests and harmony in groups; preferring to share feelings; being cooperative and helpful, more democratic in socializing. Emotional intelligence as a set of abilities, competencies, and non-cognitive skills affecting a person's ability to succeed in performing the task. It also includes overcoming environmental demands and pressures; adjusting to the people around to achieve goals as well as controlling feelings to support the emotional and intellectual development positively.

Cooperative learning is a model of learning designed in which the teacher plays a significant role as a facilitator serving as a bridge for to a higher level of students' understanding based on their notes. Teachers provide students the opportunity to build knowledge in their minds, to gain hands-on experience in applying ideas, as an opportunity for students to find and apply ideas in learning.

Constructivism theory, in which students must individually discover complex concepts and examine the existing information relating to learning material, underlies the cooperative learning. Cooperative learning encourages students to actively and positively interact in a group, to optimize and generate students' potential and to foster activity and creativity which in turn facilitating the dynamics in the learning process. Cooperative learning model can eliminate the nature of individuality and unhealthy competitiveness among students. In this model, learners routinely work in groups to help each other solve complex problems [5].

Cooperative learning facilitates students to assist each other in learning and ensure that everyone in the group achieves a predetermined goal or task. The success of learning in a group depends on the abilities and activities of group members, both individually and in groups. Cooperative learning provides opportunities for students to interact, work in a team, learn from each and respect other's opinions. Therefore, students are expected to manage their emotions in the learning. This article aimed to describe the students' emotional intelligence in cooperative-based mathematics learning in Year 10 senior high school.

2. Method

This is a Research and Development study. The development model in this study refers to Plomp model [6] consisting of five phases: (a) preliminary investigation, (b) design, (c) construction, (d) test, evaluation, and revision, (e) implementation. This article is part of the process of developing cooperative-based mathematical learning model integrated with emotional intelligence to cultivate students' emotional character and intelligence. The focus of this article is to describe the students' emotional intelligence in mathematics learning. The subjects of this study were 30 Year 10 students.

The research instruments were an emotional intelligence observation sheet and an interview guide. Data are collected through observation during three lessons. Selected students were then interviewed to clarify the unclear information obtained from the observation. The data were then analyzed qualitatively involving the reduction step, categorization, data interpretation, and drawing conclusion. The aspects of emotional intelligence observed in cooperative learning were a productive utilization of emotions and empathy. The indicators of productively utilizing emotion were more responsible, confident, tenacious, and skillful in communicating as well as focusing more on the tasks, accepting the others' perspectives, listening to others, being satisfied with the award given.

3. Result and discussion

The emotional intelligence indicates the ability to read themselves and others so that they can put themselves in other people's situations while at the same time they control themselves. This ability aims to help an individual to manage every action and reaction through the self-control, passion, perseverance, and great personality that may not only bring the success in daily life but also in learning.

The student's emotional intelligence was observed by employing cooperative learning model in three mathematics lessons involving 30 Year 10 students. The teacher divided students into six heterogenous group based on gender and academic achievement. The observer observed the aspect of students' emotional intelligence during the lesson on the topic of drawing the graph of quadratic function. The excerpt of the observation is presented as follows.

The excerpt of lesson 1

Teacher : After determining the coordinates in the table, please proceed to the next step of drawing the graph by plotting the coordinates in the Cartesian plane.
(Student was asked to work as a group while teacher walked around the classroom to carefully supervise and observe the activity in each group. Based on the observation, the teacher was seen to move from one group to another group to give limited guidance to the students who were confused and having difficulty working on the task. The teacher also reminded the student to work together in the group)

Student : Miss. What is the meaning of the negative sign in front of the x variable? Is each number substituting x also negative because of the negative sign?

Teacher : Good question! Well, the negative (-) is a negative sign from the coefficient of negative variable x and the negative sign is not operated with certain x that is substituted. For example: if $x = 2$ then $-x^2 = -(2)^2 = -4$. Do you understand?

Student : Yes, Miss! Thank you!
(Student continued discussing with her group members for the remaining allocated time, the teacher then asked the representation of two groups to present their work in front of the class)

Teacher : Students, discussion time is over! I want two students from different groups to present your group work in front of the class!
(Teacher asked the representation of two groups to present their results in front of the classroom. The two students drew the different graph of the function. Student (a) drew the graph with the curve opening upward and student (b) draw the graph with the curve flipping downward. Teacher asks the other group to pay attention and compare the result of the two students with their works)

Teacher : Ok class. Let see the two results! What are the differences?

Student : The first curve opens upward and the other opens downward, Miss!

Teacher : Correct, we can see here that when the curve open upward, what is the coefficient of a? And if the curve is flipping downward, what is the coefficient of a?

Student : The first graph opening upward means $a > 0$, and the graph flipping downward means $a < 0$!

The excerpt of Lesson 2

- Teacher guided the students to work on the worksheet containing the problems related to drawing graph of the quadratic function by discussing with the members of the group.
- Students discussed in the group while teacher walked around the classroom to supervise and observed the activity in each group in detail. It was seen that the teacher stops at each group to give limited guidance to the students who were confused and having difficulty working on the problems. The teacher kept reminded the students to work together as a team in the group.
- One of the students had a question concerning $-x^2$. The student was confused in understanding the negative sign in front of the x. To solve this problem, the teacher then explained that the negative sign $(-)$ is a sign of the coefficient of variable x with the value of -1. Teacher emphasized that the negative sign is not operated with the certain value of x substituting into the variable x. Once this problem was completed, the teacher asked students to work in their group faster.

In lesson 3, teachers were confident and serious in directing students to build their understanding of learning materials. Besides, teachers also seek to generate students' responses indicated by the emergence of the emotional intelligence aspects of students supporting the achievement of learning objectives. The excerpt of lesson 3 is as follows.

- The teacher provided a problem related to drawing the graph of the quadratic function and asked several students to do it in front of the class.
- Students took turn to complete the table of coordinates (x, y). Each student chose a different value of x and then drew a coordinate. Student drew the function graph using the table to determine the coordinates, drawing the Cartesian coordinate system, determining the coordinates in the Cartesian plane and connecting them to form the graph as a parabola.
- The teacher presented how to draw the function graph by drawing a Cartesian coordinate system and obtained the interception, axis symmetry, and vertex followed by connecting the points to form a parabola. Teacher conveyed that the axis symmetry determined by the formula of $\frac{-b}{2a}$ while for the vertex determined by the formula of $\left(\frac{-b}{2a}, \frac{D}{-4a}\right)$.
- Students solve the problems as a group, investigating the difference between the graph of $y = x^2 + 2x - 3$ and $y = -x + 3x - 2$
- Teacher led students to draw a conclusion that is if the coefficient x^2 positive, the graph opens upward, and conversely, if the coefficient x^2 is negative, the graph flips downward.

Students assigned to come to the class were obedient, while other students also took serious efforts and concentrated to build the understanding of the lesson. The emotional intelligence arised was seriousness, sincerity, tenacious and confident, and skilled in communicating as indicated by student activities during the discussion and completion of tasks in the group. The above results are similar to the research of Myer et al. [6] which also studied the relationship between emotional intelligence and communication skills. A group of Iranian university students, 53 female and 76 male, completed the Emotional Intelligence Scale and the Communication Skills Inventory and the data analysis showed that emotional intelligence was positively associated with communication skills [7].

Teachers rewarded the students when they did the task correctly while providing scaffolding as necessary for students who did not solve the tasks successfully. Teachers also provided students the opportunity to respond to their peers' work or ask questions. In this sense, the empathy of students appeared, that is, the students were responsive and sensitive to the feelings experienced by other students. When one student presented the work, the rest of the students paid close attention. It is in line with the result of [8] stated that the self-perceptions and emotional dispositions are related to the nature of EI, which affect the students' relationship with their school friends. Compared to colleagues who have low EI score, students with high EI score were more likely to be seen as someone having leadership qualities and is cooperative and less likely to be seen as disruptive, aggressive and dependent people.

During the learning activities, teachers should also consider the types of interactions that occur [5] including two-way communication, communication between teachers and students with feedback from teachers. Multiple direction communication can also occur, i.e., communication between teachers and students as well as communication between students. For example, when a student expresses opinions, other students are free to response, while the teacher also contributes to the comments. All learning activities should support the emergence of the emotional intelligence aspects of students, namely: being confident to responding or ask questions, to complete the task, engaged in the learning process, and being optimistic in facing problems, as well as concentrating when necessary.

4. Conclusion

Emotional intelligence is one of the vital intelligence to perform. The implementation of cooperative learning model can facilitate the emergence of several aspects of students' emotional intelligence. Aspects of students' emotional intelligence aspects that arise in cooperative-based mathematics learning are students being more serious and sincere, tenacious and confident, skillful at communicating, being a better listener to others' opinions, satisfied with the rewards given.

Acknowledgment

The researchers would like to thank the Directorate of research and community service of the ministry of research, technology and higher education *(Direktorat Penelitian dan Pengabdian kepada Masyarakat (DRPM) Kementerian Riset, Teknologi, dan Pendidikan Tinggi)* for the great higher education grant.

References

[1] Ilyas M 2014 Pembelajaran matematika berbasis karakter dengan melibatkan kecerdasan emosional mahasiswa *Prosiding* **1** 47
[2] Goleman 2003 *Kecerdasan Emosional untuk Mencapai Puncak Prestos/(Working With Emotional Intelligence)* Alih Bahasa Alex TO Kantjono Wicodo (Jakarta: PT Gramedia Pustaka Utama)
[3] Ginanjar A A 2003 *ESQ Berdasarkan 6 Rukun Islam* (Jakarta: Arga)
[4] Stein S J and Book H E 2003 *Ledakan EC? 15 Prinsip Dasar Kecerdasan Emosional Meraih Sukses* (Bandung: Kaifa)
[5] Ilyas M & Fitriani A 2015 Pembelajaran Matematika melalui Model Kooperatif Tipe Make a Match dan Tipe Scramble Pada Siswa Kelas VII SMP Negeri 4 Palopo *Sainsmat* **3**
[6] Plomp T 1997 *Educational and Training System Design* (Enschede, The Netherlands: University of Twente)
[7] Mayer, J. D., DiPaolo, M., & Salovey, P. (1990). Perceiving affective content in ambiguous visual stimuli: A component of emotional intelligence *Journal of Personality Assessment* **54** (3-4) 772-781
[8] Petrides K V, Sangareau Y, Furnham A and Frederickson N 2006 Trait emotional intelligence and children's peer relations at school *Social Development* **15**(3) 537-547

IOP Conf. Series: Journal of Physics: Conf. Series **1088** (2018) 012083 doi:10.1088/1742-6596/1088/1/012083

Fauzi's cognitive conflict in the development of geometry teaching material: A case study in shifting trapezoidal definition

M T Budiarto
Universitas Negeri Surabaya, Jl. Raya Kampus Unesa, Lakar santri, Surabaya 60213, Indonesia

E-mail: megatbudiarto@unesa.ac.id

Abstract. Since junior high school level, students have been taught about a trapezoidal. It is rectangular which has precisely a pair of parallel sides. This definition remains in a special meaning, for instance, if it finds any rectangle that has a pair of parallel sides, then this rectangular cannot be categorized as a trapezoid. When students are asked to define trapezoid in higher education classes such as at university level, students will experience cognitive conflicts. It was discovered when researchers developed geometry textbooks aimed at improving students' critical thinking skills. In the implementation phase of textbooks, it was found a student who experienced a cognitive conflict. In this paper, we will discuss the existence of cognitive conflict that is caused by shifting the definition of trapezoid. In the preceding time, the student recognizes the trapezoid as a rectangular that has exactly a pair of parallel sides then when the trapezoidal definition is expanded into a rectangular which has a pair of parallel sides; the student cannot readily accept the new definition of trapezoid. This occurs because of the shifting of the trapezoidal definition. Ultimately this shift caused a change to the conceptual map that occurred through assimilation and accommodation process.

1. Introduction

Students have known the definition of trapezoid since junior high school. The given definition is rectangular which has exactly a pair of parallel sides. This definition has been in the student scheme for several years. When students were introduced to a new definition of trapezoid, a rectangular having a pair of parallel sides, it was found cognitive conflicts among students. This fact is obtained when we conduct research on the development of geometry teaching materials that aims to develop students' critical thinking skills. As the materials are developed, specific tasks are designed to stimulate students to think critically about the possible extensions of the trapezoidal definition.

As it is known that someone will have a scheme of what has been learned. Over time, a person will get new information where the new information may be in accordance with a pre-existing scheme or may be incompatible with the existing scheme. When new information matches the existing scheme, then the new information does not undermine the existing scheme. It will be different when the new information obtained does not match the existing schema. The new information will undermine the existing order of the previous scheme. This condition will lead to cognitive conflict. It is a state in which a person realizes the difference between a person's cognitive structure and the environment (external information), or between different components (e.g., conception, belief, substructure, etc.) of a person's cognitive structure" [1]. "Cognitive conflict occurs when a student's mental balance is distracted by experiences (referred to as "anomalous data") that are inconsistent with their current understanding" [2]. The term cognitive conflict in this paper has a meaning similar to the term disequilibrium used by Piaget.

The subject will merge the new information into the scheme it already has. Merging new information takes place through assimilation and accommodation process. If the new information obtained by the subject is inconsistent with the schema the subject has, it cannot be directly attached to the existing scheme through assimilation. Assimilation occurs if new information is obtained according to the scheme. In the Piaget's theory, assimilation can be defined as the process of incorporation of new data into the existing schemes and structures [3]. In addition, "it refers to the process by which a subject incorporates a perceived stimulus into the existing schema" [4]. If the new information obtained does not conform to the existing scheme, then the subject does the accommodation by changing the existing scheme so that it corresponds to the new information obtained. Meanwhile, accommodation is defined as the process of creating new schemes or adjusting the old ones when they can no longer explain new experiences [5]. In addition, it involves changing existing schemes, or ideas, as a result of new information or new experiences [6]. The shift of the trapezoidal definition of a trapezoid is rectangular that has exactly a pair of parallel to rectangular sides that have a pair of parallel sides affecting the change of conceptual map of the rectangular relationship the subject has.

2. Method

This study begins with the specific task may be expressed in a sentence as follows: "What if the definition of a trapezoid is rectangular which has a pair of a parallel side?". Research design undertaken by researchers in developing materials geometry can be described in the form of a diagram as follows.

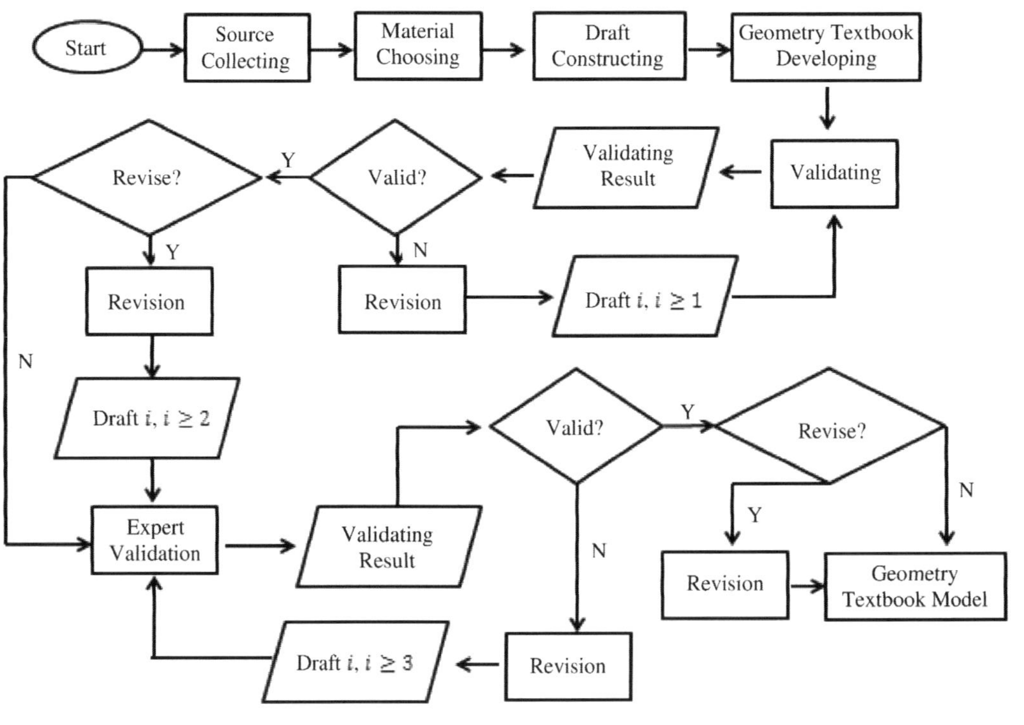

Figure 1. Research design.

3. Results and discussion

A student named Fauzi (the research's subject) has an initial scheme of a parallelogram. It is rectangular which has two pairs of opposite sides parallel. The subject also has another parallelogram definition scheme that is a rectangle which has two pairs of sides equal or parallelogram is a rectangle which has a pair of opposite sides parallel and equal. The three definitions of the parallelogram are equivalent. Another initial scheme is about a rectangle. It is a parallelogram that has one right angle. Whereas, a square is a rectangle that has the same length. A rhombus is a parallelogram whose four sides are equal. The square is a rectangle that has one right angle. The trapezoid is a rectangle that has exactly a pair of parallel sides; a rectangular that has two pairs of adjacent sides of equal length and square is a kite whose four sides are equal. The initial scheme of the rectangular of the subject uses side and angle attributes. If it is based on the initial scheme, then trapezoid is a precise rectangular which has a pair of parallel sides then the set of a parallelogram and the trapezoidal set are disjoint. Whereas if the new scheme (the result of accommodation), an i.e trapezoid is a rectangle which has a pair of parallel sides, then the set parallelogram is a subset of the trapezoidal set. The set of kites and rectangular sets is disjoint. Although the question is repeated and the subject is mediated, the subject remains the stance that the set of kites and the square set are a set of mutually disjoint schemes. The initial scheme of the relationship between the rectangular is shown in Figure 2 below.

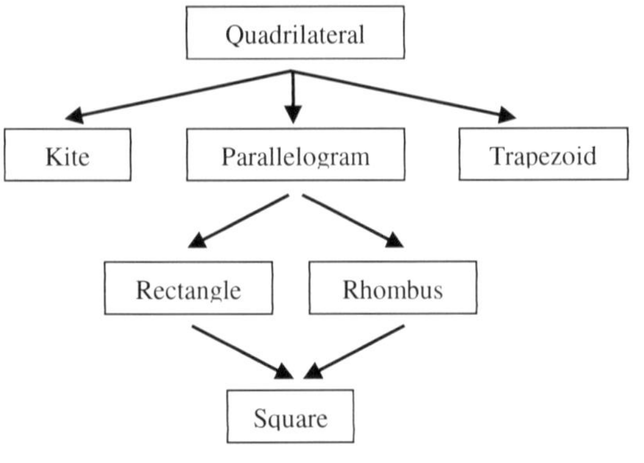

Figure 2. The initial concept map of the rectangular relationship.

Regarding the initial scheme of trapezoidal definitions the subject has, the trapezoid is rectangular which has exactly two pairs of parallel sides, it belongs to the category of analytic definitions with "rectangular" as the genus proximal and "exact two pairs of parallel sides" as differentia specific. "An analytic definition was a definition that mentioned the genus proximum (immediate family) and differentia specific (special distinction)" [6]. The new information the subjects gain is that the trapezoid is rectangular that has two pairs of parallel sides also a trapezoidal definition that belongs to analytic definitions with "rectangular" as the genus proximal and "two pairs of parallel sides." Although both of them are the analytic definition, the differentia specific of both definitions are different. So to incorporate new information into the existing scheme required the process of accommodation.

The shift in the trapezoid definition scheme occurs on the subject through the accommodation process [8]. It is because the new information obtained about trapezoidal definitions is inconsistent with the trapezoid definition scheme that has been specialized in specific differentia. Assimilation is not the only process that occurs when new information is incorporated into existing schemes, but there is also a process of accommodation. Further assimilation and accommodation will complement each other in the process of changing the concept map of the relationship between rectangular based on the new scheme. By using the method of assimilating, subject create a new scheme about the relationship between parallelogram and rectangle. The rectangle is a parallelogram with one angle is equal to 90 degrees. It can be seen in the scheme that the subject has about the concept of rectangular relationships. Based on the scheme, the set of rectangles is a subset of the set of a parallelogram. Thus, through assimilation, in the new concept map scheme of rectangular relationships, a parallelogram is linked to trapezoid.

The relationship between parallelogram and rhombus can be assimilated to the initial scheme. A rhombus is a parallelogram which all four sides are equal. It means the set of a rhombus is a subset of parallelogram set. It can be seen from scheme owned by the subject about concept map of rectangular relationship where parallelogram has a connection with rhombus. Therefore, through assimilation in the new concept map of rectangular relationship, parallelogram still has a connection with rhombus. Subject has a scheme about the relationship between rhombus and square. It can be seen from the initial scheme owned by the subject. There is a relationship between a square and a parallelogram. It means the subject has scheme that the set of a square is a subset of a parallelogram. The new relationship between rhombus and square can be assimilated to the existed scheme. Thus, through assimilation, new scheme about the concept map of rectangular relationship, rhombus still has a connection with squares. Scheme about the relationship between rectangle and square owned by the subject is about the existence of the relationship between square and rectangle. It means the set of squares is a subset of a rectangle set. It can be seen from the scheme owned by the subject about the concept map of rectangular relationship. The new relationship between rectangle and square can be assimilated to the existing scheme. Hence, through assimilation, in the new scheme about the concept map of rectangular relationship, rectangles still has a connection with squares.

Regarding the trapezoidal scheme, it is rectangular which has two pairs parallel side. Hence, the trapezoidal set and the kite set are disjoint. That is, there is no link between the trapezoid and the kite. In preparing a new concept map of the rectangular relationship, the trapezoidal and kite relations are assimilated into the scheme, therefore in the new scheme of concept maps of rectangular relationships, the set of kites and the trapezoidal set remains entirely disjoint. The shifting of the trapezoidal definition scheme allows the subject to also shift in the definition of a parallelogram. By using the new scheme of trapezoidal definition, the subject can accommodate the definition of a parallelogram. The subject formed a scheme about the definition of a parallelogram. It is a trapezoid that has two pairs of the same side of the length. The subject has owned map concept of the rectangular relationship shows there is no link between the kite and the rhombus. It means the set of kite and rhombus set are disjoint. By looking back at the kite and rhombus attribute, the subject to the accommodation process to change the scheme that has been owned in advance. The set of rhombus become a subset of kite set. Therefore, in the new scheme of concept maps of rectangular relationships, the kite has a connection with the rhombus.

Subjects also do accommodation process to the initial scheme of the relationship between the set of kites and the square set. In the initial scheme of the subject, the set of kites and the square set are disjoint. It can be seen from the schemes that the subject has about the conceptual map of the rectangular relationship; the kite has nothing to do with the square. The changes that occur in the scheme of the subject are the subject of changing the initial scheme in which the set of kites and the square set are disjoint then it becomes a subset of the kite set. The subject can accommodate the kites and square attributes that were initially unrelated to have a relation. So in the new scheme of concept maps the relationship of kite has a square relation.

The assimilation and accommodation of the subjects to the new information gained led to a scheme change concerning the conceptual map of rectangular relationships. Assimilation and accommodation also helped the subject in devising a new scheme of conceptualized connection in concept maps such as Figure 3.

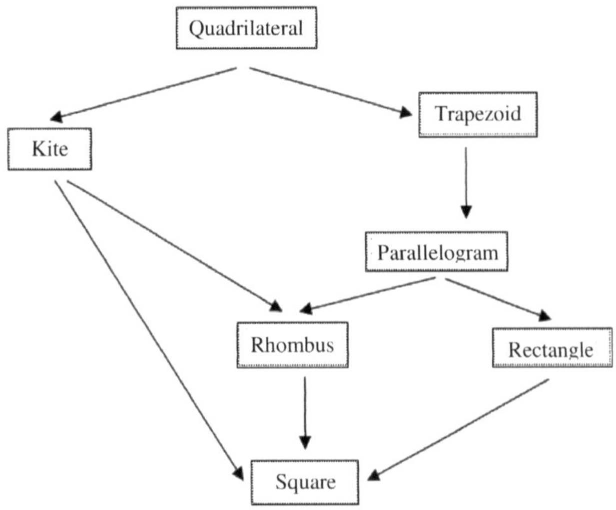

Figure 3. The new concept map of rectangular relationship.

The subject has a preliminary scheme of the general formula of parallelogram, rhombus, rectangle, square and trapezoid. With the newly established scheme, by using assimilation and accommodation of the subject, it can be argued that the general formula of parallelogram, rhombus, rectangle, and square can be derived from the broad formula of the trapezoid image below [8].

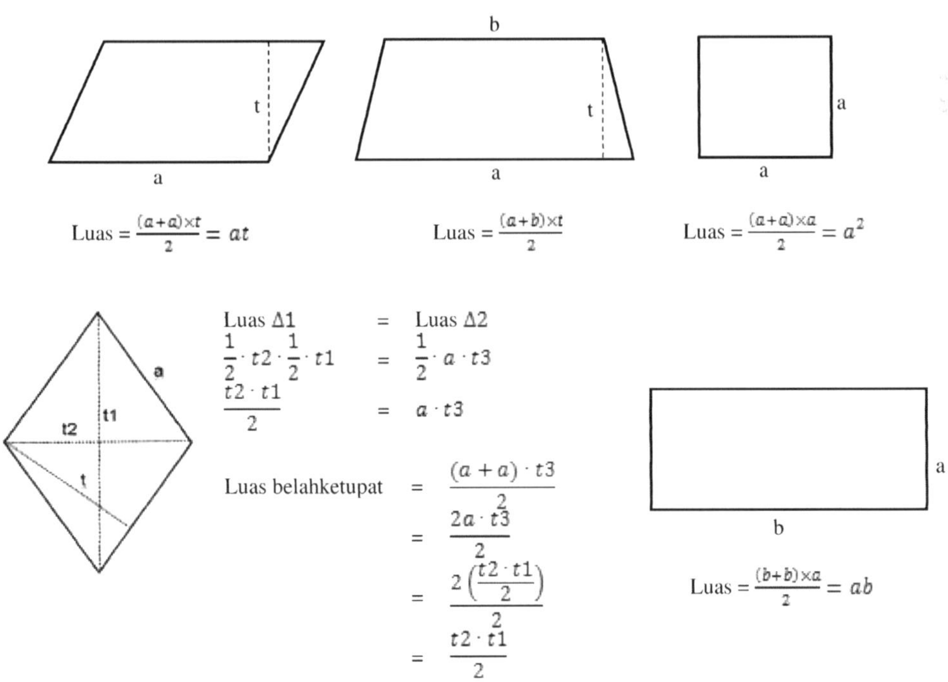

Figure 4. The area pattern of a rectangle.

4. Conclusion

Regarding Fauzi's conflict, he will incorporate new information into the existing scheme. The merger takes place through assimilation and accommodation. His new information conforms to the existing scheme, and then the merger takes place through assimilation. Whereas his new information does not conform to the existing scheme then the merger takes place through accommodation. The shifting of the trapezoid definition in Fauzi's conflict cognitive occurs through accommodation, where new information is obtained coupled to the existing scheme by changing the existing schema. The shifting of the trapezoidal definition causes a change to the conceptual map of the rectangular relationship. The corresponding rectangular relationships with the scheme of the subject will be immediately assimilated, while the unfavorable relationships will be accommodated. Based on the assimilation and accommodation performed, Fauzi composed a new scheme on the conceptual map of the relationship between rectangular. It means the special task of critical thinking in the development of geometry materials is worth developing for other topics. The implications for the results of this study are the learning of geometry in higher education as well as in secondary school needs to be given challenging problems that foster creative thinking.

References

[1] Madu B C and Orji E 2015 Effects of cognitive conflict instructional strategy on students' conceptual change in temperature and heat *SAGE Open* **5** 1
[2] Budiarto M T, Rahaju E B and Hartono S 2017 Students abstraction in recognizing, building with and constructing a rectangular *Educational Research and Reviews* **12** 394
[3] Foster C 2011 A slippery slope: Resolving cognitive conflict in mechanics *Teaching Mathematics and Its Applications* **30** 216
[4] Ribeiro P F H, Stoltz T, de Camargo, D, Blum V C L, Machado J M, de Freitas S P, Dias C M and dos Reis T J 2017 Creation process during learning of gifted students: Contributions from Jean Piaget *Online Submission* **8** 505
[5] Zhang Z 2015 Assimilation, accommodation, and equilibration: A schema-based perspective on translation as process and as product *In International Forum of Teaching and Studies* **11** 84
[6] Budiarto M T, Khabibah S and Setianingsih R 2017 Construction of high school students' abstraction levels in understanding the concept of rectangulars *International Education Studies* **10** 148
[7] Lee G and Kwon J 2001 What do you know about students' cognitive conflict in science education: A theoretical model of cognitive process *Proceedings of 2001 AETS Annual meeting* (Costa Mesa, CA) p 309
[8] Winicki L G 2001 Research of original geometric concepts: some episodes from the classroom *International Journal of Mathematical Education in Science and Technology* **32** 727

Cognitive conflict strategy to minimize students' misconception on the topic of addition of algebraic expression

Irawati[1], C M Zubainur[1] and R M Ali[2]
[1]Syiah Kuala University, Jl. Teuku Nyak Arief Darussalam, Banda Aceh 23111, Indonesia
[2]University Utara Malaysia, Sitok, Kedah Darul Aman Syiah Kuala, Malaysia

E-mail: cutmorinazubainur@unsyiah.ac.id

Abstract. Students' misconception on algebra needs more attention. Implementation of cognitive conflict strategy is needed to minimize student misconception. Learning instruments on algebra operation using a cognitive conflict strategy have been developed using the ADDIE model. The learning instruments consist of a lesson plan, worksheet, and diagnostic test. This paper describes the test result on the addition and subtraction of algebraic expression based on the learning instruments developed. The subjects were 20 students of a junior high school in Lhokseumawe, Aceh, Indonesia. Research instruments used were teacher self-reflection sheet to see how relevant the lesson plan is with the learning implementation. The learning process began with motivating the students through the problems related to their daily activities followed by working on the cognitive conflict test, presenting the result, and discussing the answers. In this phase, the teacher ensured that the students' concept was aligned with the scientific concept. In the next step, the students wrote a self-reflection to find out the differences between the initial concept and the new one. At the end of the learning, they did individual tasks related to the topic. Generally, conceptual changes found are quite good even though few students still faced misconception. However, it can be concluded that the cognitive conflict learning strategy could minimize student misconception on the topic of algebra.

1. Introduction

The misconception is an incorrect understanding of the concept occurring repeatedly and explicitly. The misconception of the mathematic concept raises the concern because it leads the false concept formation and generalization obstructing the learning process [1]. The misconception occurs because before learning the concept, the students have their understanding based on reasoning, intuition, culture, or others. The concept is maintained to explain the symptoms existing surrounding, but the concept is different from the actual concept. The misconception is difficult to be changed to the correct one; so, it can detain the learning process [2].

Many researchers have identified that the students in 6th to 8th-grade face misconception in algebra [3]. The students often merge the algebraic addition (conjoining) incorrectly. For example, the students simplify the algebra of $4 + 3x^2$ become $7x^2$. The students also simplify $4 + 3x$ become $7x$ [4]. Most students also incorrectly solve the fractional operation, e.g., $\frac{x}{2} + \frac{x}{3} = \frac{x^2}{5}$, by multiplying the numerators and denominators. Those examples show that student misconception and misunderstanding are found in using algebraic concept [5].

Misconceptions on algebra are also faced by students in Grade 8 of a junior high school in Lhokseumawe, North Aceh. Based on the initial observation conducted by the researchers, out of 22 students, 17 students faced misconception even though the test was given after the students had just finished learning algebra. The misconception was found in solving the equation of 3x + 2y. The students thought that 3x + 2y is equal to 5xy. Another misconception was found in simplifying $\frac{4}{xy} + \frac{2x}{y}$; the students simplified it into $\frac{6x}{xy}$. They also made an error in solving 3(x-y); their answer was 3x-y.

Generally, the misconceptions faced by the students could be a barrier for their learning. Therefore, it is required the efforts to deal with this problem. One of the efforts that can be undertaken is demonstrating the cognitive conflict [6]. In the learning using cognitive conflict strategy, students are challenged to show the concept they have understood. Thus, students having misconception will present the concept contrary to the actual concept.

Cognitive conflict is the condition in which the students find the contrary between cognitive structure they have and the surrounding condition [2]. The cognitive conflict strategy raises the students' dissatisfaction with the concept they have so that it helps students change their incorrect concept to correct one based on the scientific concept [7]. Cognitive conflict strategy presented by the teacher is expected to make the students more aware of the misconception they made, and finally, the students can construct the concept toward scientific conception. Thus, this article aims to describe algebra learning through cognitive conflict to minimize students' misconception.

2. Method

The learning instruments of algebra using cognitive conflict strategy have been developed using ADDIE (Analysis, Design, Development, Implementation, and Evaluation) development model [8]. The learning instruments used are a lesson plan, student worksheet, and diagnostic test. Those instruments have met the validity criteria based on the recommendation from the experts and practitioners. The data of this study was obtained from the pilot test of algebraic operation learning instruments in the implementation phase based on the ADDIE model. The pilot test aims to measure the practicality and the effectiveness of the instruments. This study involved 20 students of Grade 8 of a junior high school in Lhokseumawe, Aceh, Indonesia. The research instruments used were teacher self-reflection sheet. The reflection sheet was used to see the relevance between the lesson plan and learning implementation. The data were analyzed descriptively.

3. Result and discussion

The learning on algebraic operation using cognitive conflict strategy aims to minimize student misconceptions on algebra addition and subtraction. The learning began with the introduction phase, motivating students through daily problems related to algebraic addition and subtraction as shown in Figure 1. The teacher asked the students a question, "What do you understand about the picture?" Then, the teacher explained the objectives and the steps of the learning process. The learning implementation in this phase was relevant to the lesson plan.

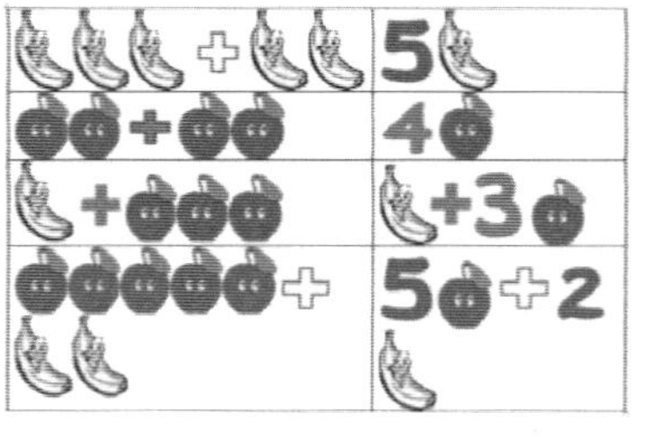

Figure 1. The daily problem used to motivate students.

In the core activity phase, the learning process began with students observe the contextual problems on the worksheet. The students were encouraged to ask questions related to the information

IOP Conf. Series: Journal of Physics: Conf. Series **1088** (2018) 012084 doi:10.1088/1742-6596/1088/1/012084

existing in the problems that they had not understood yet. For example, a student asked, "How to change this problem into the algebra equation? Can we define it with another symbol other than x?"

Furthermore, the students were asked to collect the information related to the algebra addition and subtraction on the textbooks. The information could assist the students in solving the problems. Then, the students were asked to solve questions on the worksheet that could reveal student cognitive conflict on algebraic addition and subtraction (see Figure 2). In this phase, the students completed those questions to find out their initial concepts.

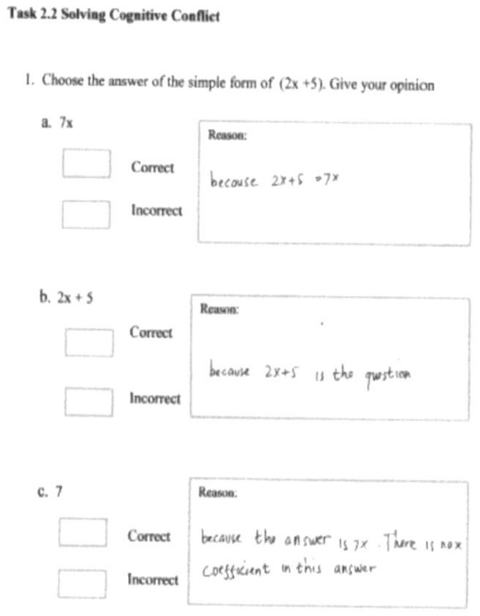

Figure 2. The Cognitive Conflict Questions on algebraic addition and subtraction.

Figure 2 presents an example of student answers related to the cognitive conflict. The students were asked to choose whether the simplified equations are correct or not and give reasons for their response. The options given for the simplified equations of (2x +5) are 7x, 2x +5, and 7. The student chose 7x as the correct answer because the student thought 2x+5 = 7x. Then, the student considered 2x+5 and 7 incorrect. The student reasoned that 2x +5 is the question; it is not the simplified form, and 7 has no variable x; thus, the simplified form is 7x.

Based on the reason given by several students, the students considered that 2x+5 could be changed to 7x. It shows that the students still misunderstood the concept; they added together the unlike terms. The following excerpt shows the conversation between the teacher and the students on the problem.

Teacher: "Please, explain your answer."
Student : "2 is added by 5. So, it becomes 7. Because there is the symbol "x" so it is added by x and finally it becomes 7x."
Teacher: "Are you sure?"
Student : "Yes."
Teacher: "Is it correct that 2x can be added by 7?"
Student : "Yes. So, it becomes 7x."
Teacher: "How about 3x + 4y, what is the simplified form?
Student : "It is 7xy."
Teacher: "If x is apple and y is banana. How many are 3 apples added by 4 bananas?"
Student : (Silent and think)
Teacher: "What is the answer?"
Student : "3 apples and 4 bananas"
Teacher: "Is the answer 7 apples bananas?"
Student : "No"
Teacher: "So, how about 3x + 4y?"

IOP Conf. Series: Journal of Physics: Conf. Series **1088** (2018) 012084 doi:10.1088/1742-6596/1088/1/012084

Student : "3x + 4y"
Teacher: "and how about 2x + 5?"
Student : "it should be 2x + 5"

After the students finished the problems on the worksheet in their groups, each group was asked to present their answers, and other groups were given the opportunity to comment and share their ideas. This activity allowed the students to express and exchange their ideas. Through this phase, it was expected that the students raised the conflict in their minds so that the students encouraged to double check the initial concept they had with the scientific concept. The teacher guided the students to inquiry questions, such as questions started with why and how. These questions help students to correct their misconception.

The next learning phase is that the students summarized the materials they had learned. In this phase, the teacher gave feedbacks to strengthen students' concept. The students were also asked to reflect on their answers to see the difference between the initial concept and the new concept made during the class discussion. Then, the students continued to complete their individual tasks.

The students' answers on the individual tasks show that the students could determine the simplest form of 7a+2b+4ab–4b+2ab+2a. The example of the correct answer is shown in Figure 3.

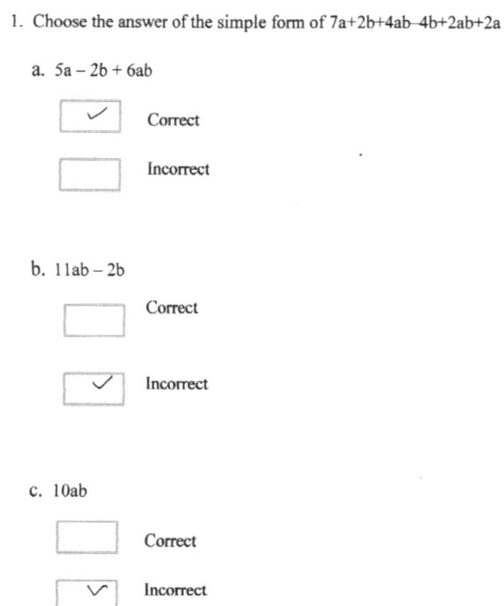
1. Choose the answer of the simple form of 7a+2b+4ab–4b+2ab+2a

a. 5a – 2b + 6ab

Correct

Incorrect

b. 11ab – 2b

Correct

Incorrect

c. 10ab

Correct

Incorrect

Figure 3. The student's answer in the individual task.

Based on the student's answers (see Figure 3), the students no longer had performed misconception in simplifying the algebra equations. The students' answer was confirmed through the interviews as presented in the following excerpt.

Researcher : "Please explain your answer (pointing to the selected answer)."
Student : "I choose the same variable, such as 7a,subtracted by 2a equal to 5a. 2b is subtracted by 4b equals -2b, and 4ab is added by 2ab equals 6ab."
Researcher : "How about 2a + 2a^2?"
NT : "It cannot be calculated."
Researcher : "Why?"
NT : "Because a is not the same as a^2."
Researcher : "Do they use the same variable?"
NT : "Yes. But, the power is different."
Researcher : "So, in your opinion, which formula can be operated?
Student : "the same variable only."

Researcher : "What is the requirement to be the same variables?"
Student : "The same variables have the same power."

Based on the students' answers in the individual tasks and interviews, in general, cognitive conflict strategy can minimize the misconception on the algebraic addition and subtraction concept. Also, giving the additional task, i.e., homework strengthened this activity. Even though most of the students had improved their conceptual understanding, there was a small number of students who still needed additional time and supports from the teacher. The cognitive conflict questions could not help them realize their misconceptions.

The trial test shows that the learning process using the learning instruments could be implemented in the mathematics classroom. The implementation went well as the time allocated in the lesson plan, 90 minutes. Generally, it can be concluded that the learning objective was achieved. The findings show that the learning using cognitive conflict strategy can fix student misconceptions [7]. The cognitive conflict strategy is the conceptual alteration strategy that enables students to change their misconceptions in to the concept aligned with the scientific concept. This change could improve the learning achievement of students [6].

Overall, the students' conceptual alteration was quite good. The students could well develop the understanding of the learning concepts by the conceptual changes. The conceptual understanding includes expanding the concept they already had and fixing the inappropriate concepts.

Also, the test result indicates that some students have not changed their initial concepts as expected. This problem is not because of students' learning weaknesses, but instead the resistant misconception. Therefore, the students need more challenging cognitive conflict tasks and longer time to realize the misconception they faced [9]. The students who made any changes to their concepts through the cognitive conflict tasks will have a strong concept, and the concept will stay longer in the students' mind. [10]. That means that the misconception can be fixed if the right strategy in the longer period is implemented. Thus, it is reasonable when there were several students in this study who still faced misconceptions even though the cognitive conflict strategy had been implemented. This problem happens because it is not easy to change students' misconception due to the resistance.

4. Conclusion

Based on the trial test of the learning instruments using cognitive conflict strategy on the concept of algebra addition and subtraction shows that the learning instruments have been well implemented. The learning process was implemented well as the lesson planned. Generally, students performed conceptual changes through the cognitive conflict task as expected in this study. The task helps students minimize their misconceptions on algebra. Thus, the learning instruments have met the learning objectives.

References

[1] Leinhardt G, Zaslavsky O and Stein M K 1990 Functions, graphs, and graphing: Tasks, learning, and teaching *J. Review of educational research* **60** 1

[2] Lee G, Kwon J, Park S S, Kim J W, Kwon H G and Park H K 2006 Development of an instrument for measuring cognitive conflict in secondary-level science classes. *J. Journal of research in science teaching* **40** 585

[3] Balgamis E 2011 Misconception of Elementary School Students' in Algebra *Conference: ECER 2011 Urban Education*

[4] Allen G D 2007 *What is a misconception* (Texas A & M University: Unpublished manuscript)

[5] Zwelithini D B and Israel K 2014 Grade 9 Learners' Errors and Misconceptions in Addition of Fractions *Journal of Social Sciences* **5** 236

[6] Swedosh P and Clark J 1997 Mathematical misconceptions – can we eliminate them? *Proceedings of the International Conference of Mathematics Education Research Group Australasia - MERGA 20* **2** 492

[7] Baser M 2006 Fostering conceptual change by cognitive conflict based instruction on students' understanding of heat and temperature concepts *Eurasia Journal of Mathematics, Science and Technology Education* **2** 96
[8] Branch R M 2009 *Instructional Design: the ADDIE Approach.* (USA: Springer. Science + Bussiness Media)
[9] Suparno P 2005 *Misconception and Concept Changing in Physical Education* (Yogyakarta: Grasindo)
[10] Özmen H 2011 Effect of animation enhanced conceptual change texts on 6th grade students' understanding of the particulate nature of matter and transformation during phase changes *J Computers & Education* **57** 1114

IOP Conf. Series: Journal of Physics: Conf. Series **1088** (2018) 012085 doi:10.1088/1742-6596/1088/1/012085

Deepening students understanding of triangle topic through 'application' component of ELPSA (Experience, Language, Pictorial, Symbol and Application) framework

B R A Febrilia [1] **and D W Winarti**[2]
[1]IKIP Mataram, Jl. Pemuda No. 59A, Nusa Tenggara Barat 83125, Indonesia.
[2]University of Canberra, 11 Kirinari St, Bruce ACT 2617, Canberra, Australia.

E-mail: rikafebrilia@ikipmataram.ac.id

Abstract. ELPSA (Experience, Language, Pictorial, Symbol and Application) is a learning framework that was introduced as a sequence of a learning process which presents mathematical ideas through lived experience, mathematical conversation, visual stimuli, symbolic notation and the application of applied knowledge. Employing the "application" component of ELPSA framework in the mathematics classroom emphasizes how the understanding of certain material can be applied to a new situation. This study investigates the learning process during the class activity about the triangle. The study was conducted in one of the junior high school in Central Lombok, West Nusa Tenggara. The data were collected from teaching and learning videos and students' work during the learning processes in the classroom. We transcribed and analysed the video as well as looked at the students' works. The result shows that the activity of composing a 2D shape from various types of triangle support students in deepening their understanding about the types of triangles. Also, the activity of composing 2D shape can be a promising foreground for students in their following learning about the area of the composite figure.

1. Introduction

Education agencies and researchers from different countries have been seeing the importance of knowing how to apply the mathematics knowledge that students learn at school not only to assess their understanding of the material delivered in class but also their knowledge on how to apply what they know in the new situation [1,2,3]. For example, Program for International Student Assessment (PISA) 2015 reported that PISA assesses not only the student's ability to extrapolate what they have learned but also whether students can apply their knowledge in unfamiliar setting both in and out of school. This recent demand in education reflects the fact that modern markets values individuals not only for what they know but also for what they can do with what they know [4].

Application of the subject learned is rarely discussed explicitly in mathematics classroom in Indonesia [5]. Based on the interview with mathematics teachers in Central Lombok, West Nusa Tenggara, they stated that the tasks given were mainly the authentic close-end type of problems based on the lesson learned rather than the open-end type of application-based problems. Despite many possible factors causing Indonesia students having the low performance for mathematics in PISA [6], the less task in the open-ended type of application problems could be one of the causes. Therefore, it is critical to include more of the application of the lesson learned within the Indonesian mathematics curriculum.

Numbers of learning framework do not include application in the learning process explicitly [2], meanwhile ELPSA (Experience, Language, Pictorial, Symbol and Application) framework explicitly include "A (Application)" component which help students to develop their ability in utilising the existing knowledge to be applied in various situations [7]. This study focusses on the application component of ELPSA as part of the whole series of ELPSA framework implementation at school for the sub-topic types of triangles. The overall lesson plan consists of three sequences with the duration of 1×40 minutes for each sequence. The first and second sequence focused on "ELPS" components and the third sequence focused on "A" component. This study, however, focused only on the third sequence where the focus is on "A" component of ELPSA, as aforementioned that our concern is on the lack of application implementation in Indonesia mathematics classroom.

This present study, therefore, investigated the learning process during the class activity of composing 2D shape from the various types of the triangle on how it helps students deepening their understanding about the types of triangles. The research aim was then stated in the research question: "How does the activity of composing 2D shape from various types of triangle help student in deepening their understanding about the types of triangles".

2. Methods

2.1 Participants

This qualitative descriptive study involved 24 Year seven students which consist of 12 females and 12 males. During the group-work activity, the students were divided into five groups of 4-5 students. This research conducted in one of the junior high schools in Central Lombok district, West Nusa Tenggara, Indonesia. The school locates in a remote area while the students come from the middle economic family background.

2.2 Data collection and analysis

In order to investigate the research question, the data were collected from teaching and learning videos and students' work during the learning processes in the classroom. The teaching and learning processes were based on the activities that have been designed. We transcribed and analysed the video as well as looked at the students' works. The analysis was focused on the students learning process during the group-work activity in composing 2D figure from various types of triangle.

2.3 Lesson design

The sub-topic types of triangle in mathematics lesson related to this study were designed in three lesson plans. The use of ELPSA framework in designing the lesson was very beneficial because it strongly aligns with how mathematics ideas should be sequenced so that student can understand the lesson easier [8]. The first lesson plan focused on ELP components of ELPSA framework. It contained the main activities to mention the characteristics and similarities of triangles by looking at the sides or angles of the triangles. The second lesson plan focused on the LPS component where the main activities were classifying the triangles and make a concept's map of the types of triangles. On the third lesson plan, we only focused on "A" component through the activity of composing different objects from various type of triangles. In this study, however, we focused solely on the third lesson plan with the structure of the lesson plan described in table 1.

Table 1. The lesson plan description.

Indicator	Activity	Further information
Construct the 2D objects from various types of triangles;	Group discussion: The teacher gave the various type of triangles to the students and asked them to imagine what object can they make from those various triangles given. After that, the students tried to make the composite object, and the teacher asked them to identify the types of the triangles they used and how many triangles (for each type of the triangle) they used for making the object. The teachers asked the students to identify another 2D plane (if available) within the object they formed.	The teacher provides three pieces of triangles for every type of it. There are seven different types of triangle namely: 1. An isosceles triangle with all of the three angles is acute angles. 2. An isosceles triangle with two of its angles is an acute angle, and the other one is an obtuse angle. 3. An isosceles triangle with two of its angle is an acute angle, and the other one is a right angle. 4. A scalene triangle with three of its angle is an acute angle. 5. A scalene triangle with two of its angle is an acute angle, and the other one is an obtuse angle. 6. A scalene triangle with two of its angle is an acute angle, and the other one is a right angle. 7. An equilateral triangle.

3. Results

The third lesson plan designed gave the opportunity for students to deepen their knowledge about the various types of triangles. The activity was making any 2D shape from various types of triangles as shown in figure 1.

Figure 1. Various type of triangle made by the teachers for a class activity

Before the students start making the shape, the teacher asked them to imagine what object will be made from the triangles given. In this part, the students were expected to visualise an object in their mind and the partitions of the object which consist of some triangles.

(a) (b) (c)

Figure 2. The students observed the triangles

IOP Conf. Series: Journal of Physics: Conf. Series **1088** (2018) 012085 doi:10.1088/1742-6596/1088/1/012085

From the video analysis, we captured some activities. The first capture was when the students took two triangles and apparently, she made a comparison between the two triangles (see figure 2a). Once she realized the triangles were not the same, she asked her friend "*Where are the similar triangle*?" while showing the triangle on her hand to her friend. Her friend then looked at the set of the triangles for a moment and took one of it "*Here you are*". She then matched the triangles and got them matched. The second was shown in figure 2b where other student did size matching by measuring the side length of the triangle with a ruler. The third was when the others student also did size matching after he tried to compose some objects from the same type of triangles (see figure 2c). After he did size matching this student then talked to himself "*Two sides of a triangle have the same side length, it is an isosceles triangle with two of its angle is an acute angle, and the other one is a right angle*".

One of the students showed that he could form four different objects from the same types and same numbers of triangles (see figure 3(a), 3(b), 3(c), 3(d)). He kept trying to form the object, and unexpectedly he could find several objects from the triangles composition. The students named the object in figure 3(a), 3(b), 3(c) and 3(d) respectively as sands' clock, parallelogram, trapezoid and fan. In addition, the other student made a frog from various types of the triangle as shown in figure 3(e). The naming of the object showed that the students related their knowledge of the objects to their experience in their daily life inside or outside the school.

Figure 3. Kinds object that students can be made.

Another interesting finding was when the student able to form an object from several triangles as well as able to split the object into other objects. This student first made a trapezoid from three similar triangles. He then called his friend to show if he took one triangle from the trapezoid, the shape became a parallelogram, and if he took one more triangle, the shape became a triangle (see figure 4). From the activity, we can see that this student has the ability of composing and partitioning the shape.

Figure 4. Composing and partitioning triangles.

After collecting the 2D shapes, each group starts putting the result of the object they made on each of the board paper to be displayed during classroom discussion (see figure 5).

IOP Conf. Series: Journal of Physics: Conf. Series **1088** (2018) 012085 doi:10.1088/1742-6596/1088/1/012085

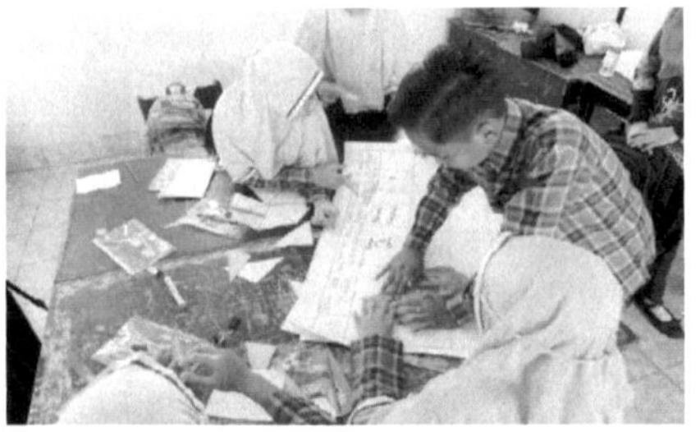

Figure 5. Students paste the objects they have made on paper board

Looking at the collection of students' work (see figure 6) all the objects displayed are typically 2D geometric shape that students learn at schools. The objects they made during the group work activities such as frog, fans or sands' clock were not shown on display.

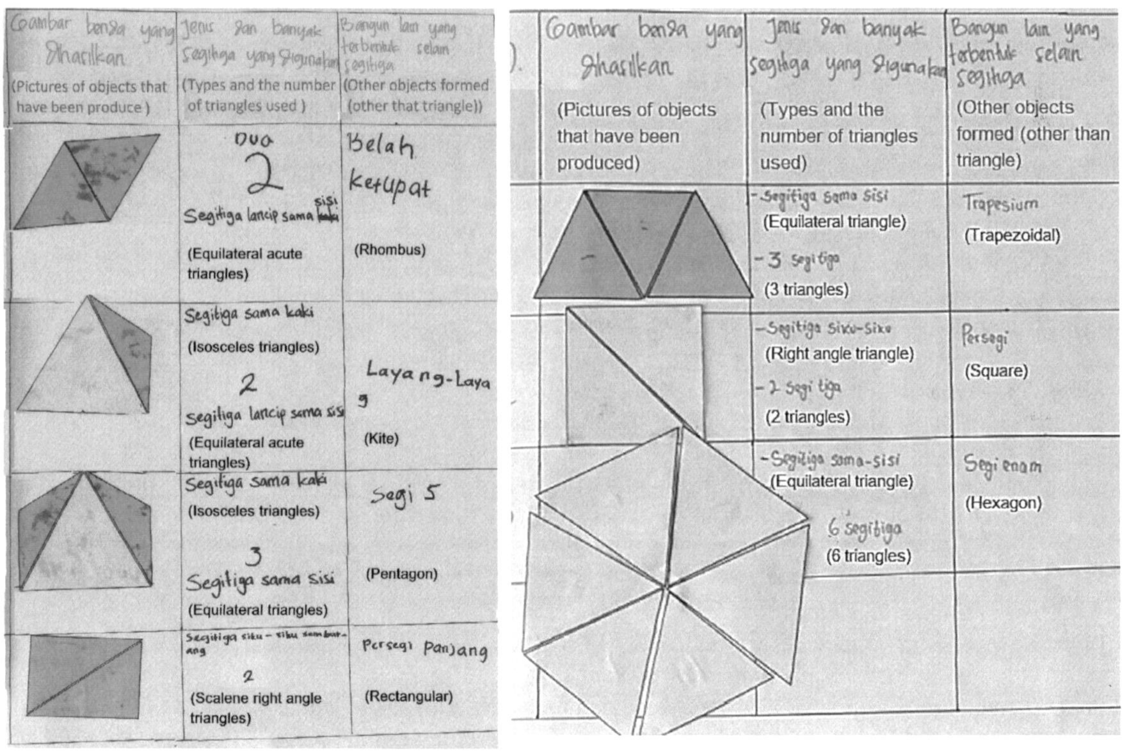

Figure 6. Worksheet of each group.

4. Discussion and conclusion

The activities developed in this lesson gave opportunities for students to deepen what they know about various types of triangles through their ways of constructing the 2D shape as well as through the discussion with their friends. It is described in the result that students have some discussions about the same type of triangles, different types of the triangle and they also consider the side length of the triangle to decide whether certain triangle matched with another triangle. With the "application component of ELPSA, this lesson does not only allow students to apply their knowledge about triangle but also related it to different types of 2D shape that they have already known. It is in line with the description of [8] that the role of application helps students to be able to apply their mathematics knowledge (i.e., triangle) to the new situation (i.e., 2D shape besides a triangle) that help to reinforce concepts and provide opportunities to scaffold understanding.

This activity promotes students' skill of composing an object which also gave a chance for students to learn about the partition of the object. Knowledge of composing the 2D shape will make them think the shape can be cut into multiple shapes (i.e., various triangles). This partitioning skill is one of the foundational concepts that are involved in learning measure area [9]. Therefore this activity can be a good start to help students in developing their understanding about area measurement problems. There are at least five foundational concepts that are involved in learning to measure area [10]. This lesson designed involves three of those five foundational concepts namely partitioning, unit iteration and conservation of the area. The first foundation was shown from student activity in partitioning the trapezoidal into one parallelogram and one triangle or three similar triangles (see figure 4). The 2nd foundation showed from the activity of forming a 2D shape from the same types of triangles into different 2D shape. In this activity, the student arranged the triangles without overlapping and gaps to one another. The arrangement of triangles formed a 2D shape such as parallelogram, trapezoid and rectangle. The 3rd foundation showed from student activity to rearrange two similar triangles into several different objects which show that even though the object looks like different, but it was made from two triangles. This activity urges the students to recognise that, even the object looks different, the area has not changed [11]. These three foundational concepts can be a good start for the student to have the initial concept of the area that they can utilise once they learned about the area measurements.

References

[1] Cheng L P 2013 The design of a mathematics problem using real-life context for young children *Journal of Science and Mathematics Education in Southeast Asia* **36** 23

[2] National Council of Teachers of Mathematics (Ed) 2000 Principles and standards for school mathematics *National Council of Teachers of Mathematics* **1**

[3] Tan J and Gopinathan S 2000 Education reform in Singapore: Towards greater creativity and innovation? *National Institute for Research Advancement (NIRA) Review* **7** 5

[4] Gurria A 2016 Pisa 2015 results in focus *PISA in Focus* **67** 1

[5] World Bank Human Development Department East Asia and Pacific Region 2010 *Inside Indonesia's mathematics classroom: TIMSS Video study of teaching practices and student achievement* (Jakarta: World Bank Office Jakarta)

[6] OECD. (2012). PISA 2012 Results in Focus*: What 15-year-olds know and what they can do with what they know.*

[7] Lowrie T and Patahuddin S M 2015 ELPSA as a lesson design framework *Journal on Mathematics Education* **6** 77

[8] Lowrie T and Patahuddin S M 2015 ELPSA Kerangka kerja untuk merancang pembelajaran matematika *Jurnal Didaktik Matematika* **2** 98

[9] Patahuddin S M, Puteri I, Lowrie T, Logan T and Rika B 2018 Capturing student mathematical engagement through differently enacted classroom practices: applying a modification of Watson's analytical tool *International Journal of Mathematical Education in Science and Technology* **49** 384

[10] Clements D H and Stephan M 2004 Measurement in Pre-K to grade 2 mathematics. *Engaging Young Children in Mathematics: Standards for early childhood mathematics education* 299

[11] Marshall L 1997 Year 7 *Students' Understanding of The Relationship Between Area and Perimeter* Retrieved from http://ro.ecu.edu.au/theses/900

IOP Conf. Series: Journal of Physics: Conf. Series **1088** (2018) 012086 doi:10.1088/1742-6596/1088/1/012086

Mathematical problems of PISA-like with the 200m swimming contexts in Asian Games

D Yansen, R I I Putri and Zulkardi

Universitas Sriwijaya, Jl. Srijaya Negara, Bukit Besar, Palembang, Indonesia

E-mail: ratu.ilma@yahoo.com

Abstract. This study aimed to develop valid and practical mathematical problems of PISA-like with 200m swimming contexts at the Asian Games. The subjects were tenth-grade students of senior high school, the research methodology used was designed research with the type of development study divided into 2 stages, namely preliminary and formative evaluation. The preliminary stage included the student analysis, curriculum analysis, and analysis of PISA problems. Formative evaluation stage included one-to-one that all together with the expert reviews than the small group. Data collection techniques used were documentation, walk through, interview, and observation. This study aimed to produce a valid and practical mathematical problem of PISA-like with 200m swimming contexts at the Asian Games. Valid based on the results of the validator assessment at the expert reviews phase and panel discussion which states that the matter had been well viewed in terms of content, constructs and language and workmanship of students at the one-to-one phase. Practical could be seen from small group phase that students could use the problem device well.

1. Introduction

The curriculum currently applied in Indonesia is the curriculum of 2013. Implementation of the 2013 curriculum [1] requires that competency-based assessments include assessment of attitudes, knowledge, skills that are integrated with the learning process and make the portfolio the main instrument. One that can support the 2013 curriculum is given PISA questions to students.

Programme for International Student Assessment (PISA) is an international study to measure the skills and abilities of students aged 15 years and are held every three years [2]. Indonesia for the first time had become one of the participating countries that participated in PISA since 2003. On PISA 2012 Indonesia ranked 64th out of 65 countries that follow PISA [3, 4], while at PISA 2015 Indonesia was ranked 62nd out of 70 countries [5].

The problem of PISA 2015 was developed based on four content, including change and relationship, Space and shape, Quantity, and Uncertainty and data [2]. One of the PISA content was related to the subject matter of statistics and opportunities of Uncertainty and data. Uncertainty was, therefore, a phenomenon at the heart of the mathematical analysis of many problem situations, and the theory of probability and statistic as well as techniques of data Representation and description had been established to deal with it [2].

The results of the PISA study in 2012 on uncertainty and data content, students were only able to be level 2 to 6 PISA levels [6]. The low PISA result of Indonesian students was caused by the lack of problem-solving abilities of the non-routine or high level, the evaluation system in Indonesia was still using low-level problem, and students were accustomed to acquiring and using formal mathematics knowledge in the classroom [7-9]. Also, the low PISA results were due to students in Indonesia having difficulty in solving PISA-based math problems that were context-based and turn them into math problems [10].

The success of Indonesian students in solving the problems of PISA was determined by the evaluation system and the ability of teachers in developing students' mathematical literacy [11, 12]. One approach that suits the 2013 curriculum is the Indonesia Realistic Mathematics Education Approach, where math should be close to the student and relevant to the daily life situation of the students [13]. The Indonesia Realistic Mathematics Education was a learning innovation that adapt from a similar learning approach as RME (Realistic Mathematics Education) [14, 15]. In the Indonesia Realistic Mathematics Education, learning began by presenting a contextual/realistic problem [16]. Using of context was important because it could motivate students in learning mathematics and help students to understand the topics [17-19]. Context was a situation or phenomenon/natural occurrence associated with the concept of mathematics being studied [20]. The PISA problem was designed to assess, define areas of interest, and use three main characteristics: content, context, and mathematical processes. Indonesia Realistic Mathematics Education and PISA had similarities that both use contextual problems. One context that can be used is the sport at the Asian Games. It is by the proposed by Putri et al. [21] who states that the sport at the 2018 Asian Games can be an activity in mathematics. One of the sports that competed in the Asian Games is swimming. Swimming is very well known even by most students so it can be a real context for them in learning mathematics. This is in accordance with the statement Putri [22] who says that learning mathematics through sports could make students like mathematics, this was because students will quickly adapted because it related to everyday life. Some research that use the context of the Asian Games 2018 sports in designing mathematics learning, such as learning fractions through swimming context for elementary school students [23], *dayung* context in fraction [24], fraction in shot-put: a learning trajectory [25], multiplication of fraction with natural number by using hurdles [26], and sprint context of Asian Games in the division of fractions [27].

From the description above, the formulation of the problem in this study is how the mathematics of PISA-Like uncertainty and data content with the context of a valid and practical 200m swimming. This study aimed to produce a valid and practical mathematical problem of PISA-like with 200m swimming contexts in the Asian Games.

2. Method

The method used in this research is design research method with the type of development studies. This study aimed to generate the mathematics problems of PISA-like as a valid and practical of tenth-grade enrichment program. Validity was shown from the results of the validator assessment at the expert reviews phase and panel discussion which states that the matter had been well viewed in terms of content, constructs and language and also from student comments on the one-to-one phase. The practicality could be seen from small group phase that students could use the problem device well [28, 29].

This research consists of two stages namely preliminary stage which included student analysis, curriculum analysis, and analysis of PISA problems and formative evaluation stage which included self-evaluation, expert reviews, and one-to-one, small group and field test [30, 31]. In this study, researchers only discuss the prototyping stage (formative evaluation) which included self-evaluation, expert reviews and one-to-one and small group.

The subject of this research was the tenth-grade students of Senior High School. Data collection techniques used in the study were documentation, walk through, interview, and observation. The document used was the Curriculum 2013 for high school, PISA framework and the problems of PISA. The walk through was done with experts at the expert reviews phase. Experts provided feedback, comments or suggestions regarding content, constructs, and languages. From the walk through, results were used to revise the prototype 1 to be combined with the one-to-one test to get the prototype 2. Interviews were conducted with students when students worked on the mathematics of PISA-Like at one-to-one and small group phases. Interview results at the one-to-one and small group phases were used as materials to revise the prototype1. Observations were made to students when students worked on the problems of mathematics PISA like at one-to-one and small group phases. In this research, the observation was done to find out students' difficulties in doing the mathematics of PISA like at the one-to-one phase and to know the practicality of the problem when small group phase.

3. Result and discussion

3.1. Preliminary Stage

In the preliminary stage, the researcher conducted student analysis, curriculum analysis, and analysis of PISA questions. Analysis of students was done to know the students who had high ability, moderate, and low and to know student which age 15 years. Curriculum analysis was conducted to determine the *Standard Kompetensi* (SK) and *Kompetensi Dasar* (KD) of statistical materials and opportunities in the curriculum of 2013. Analysis of PISA questions was done to develop the mathematics of PISA like on uncertainty and data content based on the PISA 2015 framework.

3.2. Formative Evaluation Stage

The formative evaluation stage in this research includes self-evaluation, expert reviews, and one-to-one. At the self-evaluation phase, the researchers conducted an evaluation and reviewed their design results about the type of PISA that had been made. The result was called prototype 1. Next, at the Expert reviews phase, the validation of prototype one was checked by an expert regarding content, constructs, and language. The validation process was done by sending the PISA type problem via email to the validator. The validators validating prototype 1 were KS (from Australia), and SA (from Indonesia). In addition to the expert reviews phase, the researchers also conducted a panel discussion at Sriwijaya University and attended by nine peers (Magister Mathematics Education) and three lecturers. Comments and suggestions from expert reviews and discussion panel were used to revise prototype 1.

At one-to-one, prototype one was tested to 3 students who have low, medium, and high ability. The three students were EM, EN, and NA. The result of student's observation and the interview was used to revise prototype 1. The analysis results obtained from expert reviews, one-to-one and panel discussion could be seen on Table 1.

Table 1. Comments from expert reviews, one-to-one and panel discussion

Validation	Comment	Revision
Panel discussion	1. Fix the way of writing. 2. Change the question to make it more meaningful by comparing the fastest time record in Asian games 2014.	1. Fixed the recommended way of writing 2. Change the question to have meaning 3. Added the fastest time record in Asian Games 2014
Expert Reviews	1. Compare with the average time data on Asian games 2014 2. Fix the way of writing to make it easier for students to understand 3. One of the athletes who were attended or all athletes followed the training	
One-to-one	1. The problem was difficult and must repeatedly read it to understand the meaning of the problem 2. Need accuracy and reasoning in answering questions	

IOP Conf. Series: Journal of Physics: Conf. Series **1088** (2018) 012086 doi:10.1088/1742-6596/1088/1/012086

	3. Based on the answer, student 1 could answer the problem according to the instructions about; student 2 had answered the question but was wrong in the settlement; student 3 had answered the question but not finished

Based on the comments/suggestions of expert reviews, panel discussion, and one-to-one conducted in parallel, then the problem on prototype one was improved and revised again. The revised question was called prototype 2 can be seen in Figure 1.

Toward the 2018 Asian Games, Indonesia's 200m individual men's swimming medley athletes plan 20 times of practice. The plan has been implemented 40% and the average time obtained is 116 seconds
Question:
Is it possible that at the end of the exercise the average time the athlete is getting beyond the fastest time in Asian Games 2014 that is 115.34 seconds? Give your reason.

Figure 1. Prototype 2

In the small group phase, prototype two was tested to 6 students who have different ability. Two students low ability, two students medium ability, and two students high ability. The six students were RG, ES, CA, FA, ML, and LA. The small group phase was done to see the practicality of the developed problem. The practical meaning of the problem that was easy to use could be administrated and could be interpreted well by students.

In the small group phase, the results achieved by the students do not very much with the results achieved by students at the phase of one-to-one. This could be seen from the results and the systematic answers of students about looking for average travel time of 200m individual men's swimming. The Figure 2 shows the result of answers from some students.

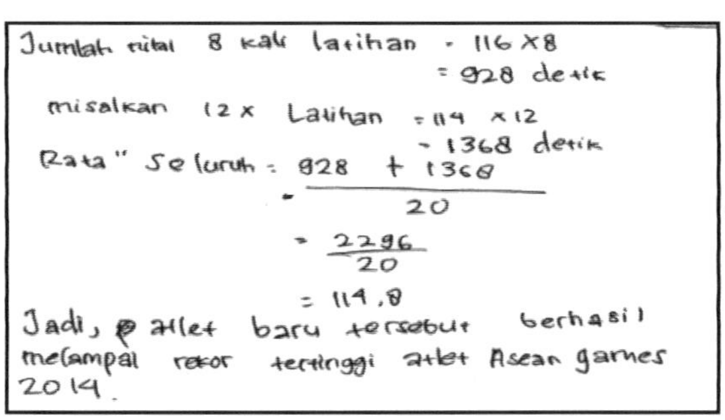

→

The sum of eight times exercise is equal to 116 x 8 = 928 seconds

For example the sum of twelve times exercise is equal to 114 x 12 = 1368 seconds

The average of all exercise is equal to (928+1368)/(8+12) = 2296/20 = 114.8 seconds

So, the new athlete already pass the highest record of ASEAN Games 2014's athlete.

Figure 2. Student's answer

From the student's answer in Figure 2, it appeared that the student had understood the purpose of the problem. In order to surpass the fastest time of athletes in Asian games 2014, students made an equation for the average time of 12 sessions of 114 seconds. So after the calculated as a whole, the average time was obtained is 114.8 seconds. Based on the results and discussion of students at the small group phase, the problems developed could be understood, explanations and clues on the problem were clear and students could answer the problem correctly.

4. Conclusion

This research had produced a PISA-like math problem with a valid and practical 200m swimming context in Asian Games. Validity was shown from the results of the validator assessment at the expert reviews phase and panel discussion which stated that the matter had been well viewed in terms of content, constructs and language and workmanship of students in the one-to-one phase. Practicality could be seen from small group phase that students could use the problem device well.

Acknowledgements

The authors would like to express his gratitude to the Directorate General of Higher Education Indonesia who has funded a post grant research in 2017-2018. Then the authors would like to express his gratitude to `the teacher (Siti Fatimah) and the tenth-grade students of Senior High School number 3 Prabumulih who have helped a lot in this research.

References

[1] Kemendikbud. 2014 Permendikbud No. 59 Tahun 2014 tentang Kurikulum 2013 Sekolah Menengah Atas/Madrasah Aliyah (Jakarta: Kemendikbud)
[2] OECD 2016 *PISA 2015 Assessment and Analytical Framework: Science, Reading, Mathematics, and Financial Literacy* (Paris: OECD)
[3] OECD 2014 *PISA 2012 Results in Focus: What 15-year-old Know and What They Can Do With What They Know* (Paris: OECD)
[4] Putri R I I and Zulkardi Z 2018 Higher-order thinking skill problem on data representation in primary school: A case study *J. of Physics: Conf. Series* **948** 012056
[5] OECD 2016 *PISA 2015 Result (Volume 1): Excellence and Equity in Education (*Paris: OECD)
[6] Zuhra R 2015 *Analisis Strategi Siswa dalam Menyelesaikan Soal PISA (Programme For International Student Assesment) Konten Uncertainty and Data Pada Siswa Kelas VIII Mtsn Model Banda Aceh Unpublished Thesis* (Banda Aceh: Universitas Syiah Kuala)
[7] Stacey K 2010 Mathematical and scientific literacy around the world *J. of Science and Mathematics* **33** 1
[8] Wu M 2012 Using PISA and TIMSS mathematics assessments to identify relative strengths of students in Western and Asian *J. of Research in Education Sciences* **1** 67
[9] Novita R, Zulkardi and Hartono Y 2012 Exploring primary student's problem-solving ability by doing tasks like PISA's question *J. on Mathematic Education* **3** 133
[10] Wijaya A, Heuvel-Panhuizen M v d, Doorman M and Robitzsch A 2014 Difficulties in solving context-based (PISA) *J. The Mathematics Enthusiast* **11** 555
[11] Johar R 2012 Domain soal PISA untuk literasi matematika *J. Peluang* **1** 30
[12] Hendroanto A, *et al* 2018 Mathematics education students solving PISA's mathematics problems: An overview of students' mathematical literacy *International Journal on Emerging Mathematics Education* **2** 129
[13] Putri R I I 2014 *Evaluasi program pelatihan Pendidikan Matematika Realistik Indonesia (PMRI) Bagi Guru Matematika Sumatera Selatan Proc. Seminar Nasional Evaluasi Pendidikan* p 522
[14] Putri R I I 2011 Pembelajaran materi bangun datar melalui cerita menggunakan pendekatan Pendidikan Matematika Realistik Indonesia (PMRI) di sekolah dasar *J. Pendidikan dan Pembelajaran* **18** 234
[15] Risma D A, Putri R I I, and Hartono Y 2013 On developing students' spatial visualisation ability *J. International Education Studies* **6** 1

[16] Revina S and Leung F K S 2018 Educational borrowing and mathematics curriculum: Realistic mathematics education in Dutch and Indonesian primary curriculum *International Journal on Emerging Mathematics Education* **2** 1

[17] Fauziah A, Putri R I I , Zulkardi, and Somakim 2017 Primary school student teachers' perception to Pendidikan Matematika Realistik Indonesia (PMRI) instruction *J. of Physics: Conf. Series* **943** 012044

[18] Arifin S, Zulkardi, Putri R I I, Hartono Y, and Susanti E 2017 Developing Ill-defined problem solving for the context of "South Sumatera" *J. of Physics: Conf. Series* **943** 012038

[19] Rejeki S and Putri R I I 2018 Models to support students' understanding of measuring area of circles *J. of Physics: Conf. Series* **948** 012058

[20] Zulkardi and Putri R I I 2006 *Mendesain Sendiri Soal Kontekstual Matematika Proc. KNM13* (Semarang: UNNES) pp 1

[21] Putri R I I, Zulkardi and Widodo (2015) *Ciptakan Pembelajaran Matematika dengan Asian Games* Available from http://sumselpostonline.com/ciptakan-pembelajaran-matematika dengan-asean-games/ (Accessed 18th September 2017)

[22] Putri R I I 2015 *Penilaian dalam pendidikan matematika di Indonesia*: Lokal, Nasional, dan Internasional. *Paper presented at the special meeting open of senat Sriwijaya University*, Available from http://fkip.unsri.ac.id/index.php/posting/90 (Accessed 20th April 2015)

[23] Putri R I I , Gunawan M S, and Zulkardi 2017 Addition of fraction in swimming context *J. of Physics: Conf. Series* **943** 012035

[24] Nasution M F, Putri R I I and Zulkardi 2017 *Dayung Context In Fraction Proc. 5th South East Asia Development Research (SEA-DR) International Conference 2017* (Banjarmasin: ULM) p 1

[25] Putri, R I I and Zulkardi 2017 Fraction in shot-put: A learning trajectory *AIP Conference Proceedings* **1868** 050005

[26] Rahayu C, Putri R I I and Zulkardi 2017 *Multiplication of Fraction With Natural Number by Using Hurdles Proc. 5th South East Asia Development Research (SEA-DR) International Conference 2017* (Banjarmasin: ULM) p 43

[27] Roni A, Zulkardi and Putri R I I 2017 *Sprint Context of Asian Games in the Division of Fractions Proc. 5th South East Asia Development Research (SEA-DR) International Conference 2017* (Banjarmasin: ULM) p 22

[28] Zulkardi 2002 *Developing A Learning Environment on Realistic Mathematics Education for Indonesian Student Teachers* Published Dissertation (Enschede: University of Twente)

[29] Riyanto B, Zulkardi, and Putri R I I 2017 Mathematical modeling in realistic mathematics education *J. of Physics: Conf. Series* **943** 012049

[30] Tessmer M 1993 *Planning and Conducting Formative Evaluations* (London: Kogan Page)

[31] Zulkardi 2006 *Formative Evaluation : What, why, when, and how* Available from http://www.oocities.org/zulkardi/books.html (Accessed 14th September 2017)

IOP Conf. Series: Journal of Physics: Conf. Series **1088** (2018) 012087 doi:10.1088/1742-6596/1088/1/012087

Designing learning activities on conditional probability

Benidiktus Tanujaya[1], Rully Charitas Indra Prahmana[2] and Jeinne Mumu[1]

[1]University of Papua, Jl. Gunung Salju, Manokwari, West Papua 98314, Indonesia
[2]Universitas Ahmad Dahlan, Jl. Pramuka Kav. 5, Yogyakarta 55161, Indonesia

E-mail: b.tanujaya@unipa.ac.id

Abstract. This study aims to design student-learning activities on the concept of conditional probability using the development of test items. The method used is design research implemented in three stages, namely preliminary design, teaching experiment, and retrospective analysis. The study conducted on 47 mathematics education students of the University of Papua from mathematics statistics lesson. The research produces a learning trajectory, which contains a series of the learning process that students can develop conditional probability test items in three versions, determine sample space, formulate, and use conditional probability formula. The research shows that the development of test items with the approach namely observe, imitate, and modify can help students understand the concept of conditional probability.

1. Introduction

Mathematics is one of the most critical subjects in the education system of Indonesia. The subject studied at all levels of education in Indonesia, from elementary school to university [1]. One of the courses reviewed in the issue is statistics. Students learn several concepts in statistics instruction, including the idea of probability. Probability, a mathematics lesson, is also taught at all education levels including elementary, junior and senior high school.

However, including a subject in the curriculum does not automatically guarantee its correct instruction. The specific characteristics of probability, such as a multidimensional view of possibility or the lack of reversibility of random experiments, are not usually found in other subjects and create specific challenges for teachers and students [2, 3]. Mathematics teachers frequently lack detailed preparations to teach the lesson comprehensively. The teachers do not have particular knowledge and skill to teaching probability [4]. Furthermore, probability in an elementary statistics class is challenging to teach because there is not much time, the concepts and procedures are complicated, and the students do not see the importance of learning it [5]. The situation is even worse for textbooks, and curriculum documents of probability prepared for elementary and high school teachers do not offer adequate support [4].

The mathematics textbooks sometimes provide a too limited understanding of probability (only the conventional approach), and applications at other times limited to games of chance, and the definitions of concepts are inappropriate in several textbooks [6]. In Indonesia, especially in West Papua, mathematics' teachers generally use only one to three mathematics textbooks in mathematics instruction [1]. The books have a standard procedure for teaching probability, i.e., there are only definitions, example, and exercises. This situation leads to insufficient teacher learning resources, especially in teaching about the concept of probability. As a result, mathematics teachers often start the lecture by write down definitions, followed by providing examples and counterexamples [7]. The teachers are not creative in teaching probability. Examples of questions given by those contained in textbooks are of a very limited. As a result, it is urgent to offer the mathematics teachers a better preceding knowledge as well as adequate skills to teach probability creatively.

Probability is one of the subjects in the lesson of Mathematical Statistics in Mathematics Education program at the University of Papua. The mathematics education students' University of Papua, as a mathematics teacher candidate, need to be well prepared for an understanding concept and teaching skill of probability concept. Probability, especially conditional probability concept is an essential concept for mathematics education students. Conditional probability is the basic concept for understanding further concepts in mathematical statistics and other lessons. The idea is a basic so that students can better understand other concepts, such as Bayes theory, random variables, and parameter estimation. While in other lessons, the idea of conditional probability is foundation for the development of queuing theory, decision-making, and games theory. Consequently, it is essential to find effective strategies and methods to teach conditional probability for the mathematics education students, the teacher candidate.

The teachers should be able to predict students' learning difficulties, mistakes, obstacles and problem-solving strategies' in conditional probability lessons to provide some effective strategies of instruction. For example, teachers should know students' difficulties and mistakes in comparing two terms in conditional probability [4]. Moreover, since the teachers want students to build their knowledge actively by solving problems, the teachers should use this same approach in teaching, especially if they want the students' later use a constructivist and social approach in their teaching [1]. The students are required to have more responsibility in their studying and to think creatively and critically. Therefore, the teachers should create suitable atmosphere for students to reflect on their previous beliefs about teaching and discuss these ideas with other classmates [8].

By knowing the students' problems comprehensively, teachers can design various strategies in instruction [1]. The issue of students in learning conditional probability is the inability of students to understand the problem. Examples of test given by teachers are very different from the exam questions. Moreover, Tanujaya [9] stated that there are two main problems of students in solving mathematics problems, that is they seem still face difficulty in understanding the issues, and they have not been able to use the formula that has been studied. The problem is known as the connection in mathematics. Mathematical connections are a bridge where prior knowledge is used to build an understanding of the relationship between mathematical ideas, concepts, or representations [10].

Probability is a subject of mathematics that consists of randomness. The conflict between probability theory and students' view of the world is due at least in part to students' limited contact with randomness. It causes the students to have difficulty in solving the probability problem. Students only memorize the formulas and examples of questions given by teachers without understanding the concept of probability correctly and adequately. Once another matter is given, they cannot solve it. Therefore, the mathematics teachers must prepare the way for studying probability by providing experience with random behavior in the mathematics curriculum [11].

Therefore, developing students' understanding using test item development is proposed to solve those problems. By using test item development, teachers can also find information about students understanding through various instruction activities, comments, explanations, questions, answers, and other students' activities in the classroom. Based on the instruction activities, the teachers can develop students' understanding of the concept of conditional probability [9].

The development of test items can be done with some approach strategy, one of them is through observation, imitation and then modification. These three are the learning strategies that students can do to improve their thinking skills as well as to practice what they think and develop their understanding of a particular concept. It is by the Bandura's social learning theory posits that students learn from one another through observation, imitation, and modeling. The approach is often considered as a bridge between behaviorist and cognitive learning theories. Learning is a change in performance because of practice.

Does it correct that the development of test item can be used to develop students' understand? How to use the development techniques of test item to improve students' understanding of the concept of conditional probability? What procedures and strategies should be undertaken for the development of test item to develop students' thinking skills in learning? The following will describe the methods used in answering these questions.

2. Method

The research conducted in the even semester of the academic year 2017/2018. The method used in this research is design research as a series of approaches that the intent of producing new theories, artefacts, and practices that account for and potentially influence learning and teaching in naturalistic settings [12, 13]. The method has five characteristics namely interventionist nature, process-oriented, the reflective component, cyclic character, and theory-oriented [14]. Primary objective of design research is to develop theories together with instructional materials [15]. Therefore, the purpose of this research is to develop a learning theory of conditional probability using the development of instrument test. This learning theory can be used to improve students' understanding of the concept of conditional probability.

Moreover, three phases of conducting design research are that preparing for the experiment, experimenting in the classroom, and conducting retrospective analyses [15] while the research method also has a cyclic character that consists of three phases: design, teaching experiment and retrospective analysis [16]. Thus, this research used three stages, namely preparation of experiment, experiment design, and retrospective analysis.

2.1. Preparion of experiment

The stage is also called Preparation and Design [17] or Preliminary Design [18]. At this stage, a literature review of conditional probability learning materials was conducted. After that, a suspected strategy and thought of students are in the learning process. Next, continue designing learning trajectory (LT) and the hypothetical learning trajectory (HLT). The LT contains the trajectory of learning material and concept map that students will go through during the instruction. Meanwhile, HLT is a hypothesis of how students' thinking and understanding develop in an instruction [18]. The HLT consists of learning goals, starting points, the description of mathematics activities, conjectures of students' thinking and strategies, and suggestions for teacher regarding what and how to react students' particular response.

2.2. Design experiment

Design experiment comprises pilot and teaching experiment [18]. In the pilot experiment stage, the researcher involves six students. Selected students have different levels of ability, two students with the high knowledge, two students with average ability and two students with low strength. In teaching experiment stage, learning activities designed in the preliminary design stage was tested. At this stage, all the students who take the mathematical statistics courses that amount to 47 people involved in this study.

2.3. Retrospective analysis

Data from the teaching experiment were analyzed in this stage to plan activities and to develop the design of activities in subsequent learning. The study aims to explain how students can generalize from learning activities such as observing and analyzing examples of given questions, developing items to understand the concept of conditional probabilities. The retrospective analysis is to create Local Instructional Theory (LIT). At this stage, HLT compares with actual student learning.

3. Result and discussion

This study designed three activities consisting of activity determining sample space and sample point, and the probability of an event. Here is a description of these activities.

3.1. Preparation of experiment

At this stage, the researcher designed LT and HLT based on the literature review. There is a critical point that the students must pass in studying the conditional probability concept, i.e., the students must be able to define two events appropriately [19]. The two events have members and are not members.

There are two statements that students can use in developing conditional probability instruments test to help students to understand the critical point as follows:

- Determine a sample space (S) of an experiment, with at least two events, e.g., event A and event B, where both events have several corresponding members.
- Develop some items of test instrument about the concept of conditional probability.

Students determine both matters based on the test items obtained, either from textbooks, modules or various other teaching materials. Students then classify the test items into several categories. Based on these categories, students develop new test items, through observations, imitations, and modifications.

3.2. Design experiment

At this stage, the researcher involves six students. The results show that the students understand and carry out the test questions but only imitate examples of problems contained in textbooks. They have not been able to develop more varied test item. It led to the researcher to redesign the learning activity with the following statement:

- Determine a sample space (S) of an experiment, with at least two events, event A, and event B. Both events have several corresponding members.
  For example, it is known that in a container there are five marbles, three are red, two are white, and each marble is numbered. From the container is taken a marble at random. If it is known that the marbles are taken in red, then what is the probability of marbles is numbered 1?
- Determine the two events, where the events are one because of another event, event A, and event B. Thus, there are four events, namely the event A and event, not A, event B and event not B. If an event A occurs, event B is more probable to occur than an event, not B. In other words, if an event, not A occurs, then event not B is more probable than event B.
  For example, based on the records at Rendani Airport of Manokwari, the flight probability of aircraft A departs on time is 0.65; while the flight probability is landing on time is 0.86. For additional information, it also known that the flight opportunity departs and landing on time is 0.57. Calculate how many chances the flight landed on time if known it departed on time.

In teaching experiment stage, learning activities designed in the preliminary design stage tested [20-22]. All of 47 students who take the mathematical statistics courses involved in this experiment. The results of the experiments show that all students are at least able to develop two different test items.

Some students can prepare an instrument test that asks about conditional probabilities, $P(A|B)$, Probability of event A and event B simultaneously, $P(A \cap B)$, and mutually probability of event A, $P(A)$ and event B, $P(B)$. As the result, the students would be able to construct formula of conditional probability.

3.3. Retrospective analysis

Based on results, obtained Local Instructional Theory (LIT) for conditional probability learning, which compared with actual student learning are:

- Students should understand some concepts of probability: sample space, events, independent events, and non-mutually exclusive events.
- Students should be able to determine two independent events and not be independent.
- Students should be able to provide actual examples of sample space that could be countered.
- Students should be able to develop an item of the test instrument, using observation, adaptation, and modification approaches.
- Students should be able to develop conditional probability formulas based on developed sample test questions.

P(A|B) is the probability of A given B to denote a conditional probability [3, 19, 23]. P(A|B) refers to the probability of A when B has already been known. It is assumed that A and B are two events containing a and b outcomes, respectively, and let c denote the number of outcomes in the intersection of A and B. The conditional probability of A given B is the ratio of c to b. However, c/b can be written as the quotient of two other ratios,

$$\frac{c}{b} = \frac{c\,n^{-1}}{bn^{-1}}$$

So, for the particular case,

$$P(A\backslash B) = P(A \cap B)[P(B)]^{-1}$$

4. Conclusion

The research reveals that the development of test items with the approach: observe, imitate and modify can help students understand the concept of conditional probability. Students conduct observation by collecting various test items of conditional probability from different teaching materials, such as textbooks, and modules. The test items collected are grouped into several categories to facilitate observation. Students then develop similar test items by referring to the type of test items that have been obtained. Finally, the student modifies the test item that has been designed to achieve the test item about the conditional probability of his development.

References

[1] Tanujaya B, Prahmana R C I and Mumu J 2017 Mathematics instruction, problems, challenges, and opportunities: A case study in Manokwari regency, Indonesia *World Transactions on Engineering and Technology Education* **15** 287
[2] Batanero C, Chernoff E J, Engel J, Lee H S and Sanches E 2016 *Research on Teaching and Learning Probability* (Hamburg: Springer)
[3] Lee C, Li H C and Shahrill M 2018 Utilising the think-pair-share technique in the learning of probability *International Journal on Emerging Mathematics Education* **2** 51
[4] Ortiz J J, Cañizares M J, Batanero C and Serrano L 2002 An experimental study of probabilistic language in secondary school textbooks *The International Conference on Teaching Statistics 6* Cape Town
[5] Keeler C and Steinhorst K 2001 A new approach to learning probability in the first statistics course *Journal of Statistics Education* **9** 3
[6] Batanero C, Juan D, Godino and Rafael R 2004 Training teachers to teach probability, *Journal of Statistics Education* **12** 1
[7] Mumu J, *et al* 2018 *J. Phys.: Conf. Ser.* **943** 012011
[8] Jaworski B 2001 Developing mathematics teaching: teachers, teacher educators and researchers as co-learners in *Making sense of mathematics teacher education* eds. Lin L and Cooney T J (Dordrecht: Kluwer)
[9] Tanujaya B 2017 Application of assessment as learning in mathematics instruction *Advances in Social Science, Education and Humanities Research* **100** 140

IOP Conf. Series: Journal of Physics: Conf. Series **1088** (2018) 012087 doi:10.1088/1742-6596/1088/1/012087

[10] Hiebert J and Carpenter T P 1992 Learning and teaching with understanding, *Handbook of research on mathematics teaching and learning* ed. Grouws D A 65-97 (New York: Mc Millan)
[11] Moore D S 1990 Uncertainty *On the shoulders of giants: New approaches to numeracy* (Washington: National Academy Press)
[12] Barab S and Squire K 2004 Design-based research: Putting as take in the ground *Journal of the Learning Sciences* **13** 11
[13] Ploom T 2013 Educational design research in Ploom T and Nieveen N (eds.) Educational Design Research- Part A: An Introduction (Enschede: SLO)
[14] Akker J V D, Gravemeijer K, Mckenney S and Nieveen N 2006 *Educational Design Research* (London: Routledge Taylor and Francis Group)
[15] Bakker A 2004 *Design Research in Statistics Education on symbolizing and Computer Tools* (Amersfoort: Wilco Press)
[16] Gravemeijer K and Cobb P 2006 Design Research from A Learning Design Perspective In Akker J V D, Gravemeijer K, Mckenney S, Nieveen N (eds.) *Educational Design Research* (London: Routledge Taylor and Francis Group)
[17] Eerde D V 2013 Design Research: Looking into the Heart of Mathematics Education *The First SEA-DR Proceeding*
[18] Prahmana R C I 2017 *Designing Mathematics Learning Trajectory: An Introduction* (Beau Bassin: Lambert Academic Publishing)
[19] Larsen J L and Marx M L 2011 An Introduction to Mathematical Statistics and Its Applications (New York: Prentice Hall)
[20] Prahmana R C I, *et al* 2017 *J. Phys.: Conf. Ser.* **893** 012001
[21] Setyawan F, *et al* 2018 *J. Phys.: Conf. Ser.* **943** 012004
[22] Hendroanto A, *et al* 2018 *J. Phys.: Conf. Ser.* **943** 012029
[23] Purnami A S, *et al* 2018 *J. Phys.: Conf. Ser.* **948** 012020

IOP Conf. Series: Journal of Physics: Conf. Series **1088** (2018) 012088 doi:10.1088/1742-6596/1088/1/012088

The use of the *wx*Maxima linear algebra module on Gauss elimination lesson for mathematics education students

D Afriza[1], C M Zubainur[1] and A Veloo[2]
[1]Syiah Kuala University, Jl. Teuku Nyak Arief Darussalam, Banda Aceh 23111, Indonesia
[2]North Malaysia University, 06010 Sintok, Kedah Darul Aman, Malaysia

E-mail: cutmorinazubainur@unsyiah.ac.id

Abstract. Students' difficulties in solving problems in Linear Algebra course could be assisted by the use of *wx*Maxima Linear Algebra module. This paper focuses on the learning on Gauss elimination topic using the *wx*Maxima Linear Algebra module. The *wx*Maxima Linear Algebra module has been developed using the ADDIE development model. The participants were 18 mathematics education students of UIN Ar-Raniry Banda Aceh. This study used the lecturer's self-reflection sheet to see the relevance of the lesson plan and the learning implementation. The learning process implemented direct instruction model and the time allocated for the class session was 150 minutes. The learning activities began with the explanation about elementary row operations (ERO) and Gauss elimination (GE) by the lecturer and followed by solving problems manually solving using the paper-pencil method. Furthermore, the solutions obtained manually were checked using the *wx*Maxima software. Learning activities were then continued to the concept of Gauss Jordan elimination method. Although the students spent much time to solve the Gauss elimination problems manually, overall the learning process went well as the time allocated in the lesson plan. The findings suggest that the *wx*Maxima Linear Algebra module could be used for other linear algebra topics.

1. Introduction

Mathematics education department of UIN Ar-Raniry utilizes Indonesian National Qualification Framework-based curriculum. One of the objectives is to educate students as prospective teachers in mastering technology that will be implemented in secondary schools. Students are also required to master mathematics concept during their study in university. One of the courses that must be undertaken by students in mathematics education department is linear algebra.

One of the topics required in a linear algebra class is Gauss elimination (GE). By learning GE, the students are expected to be able to change the enlarged matrix into row echelon form. After that, the students were expected to understand Gauss-Jordan elimination to determine the solution of the linear equation system. However, some students still made mistakes in solving problems related to Gauss-Jordan elimination. Their mistakes generally involved calculation. This is possibly caused by the lack of accuracy in calculating elementary row operations and the limited time to understand the lesson [1]. Some students also had difficulties in solving problems with very large linear equation system since it requires much time to solve. Therefore, it needs a media to facilitate students in solving such problems. Calculation of very large and complex systems require computers to make it easier [2]. This is relevant with the learning objectives expected by the university. That is, the learning should give students the opportunity to learn and use information and communication technology (ICT).

Specifically, the learning should focus students' attention on how to use technology in mathematics appropriately, and provide opportunities for students to apply the technology in mathematics learning [3].

Using software in mathematics learning can improve students' understanding and performance in mathematics [4,5]. One of software that can be used for linear algebra is *wx*Maxima software. *wx*Maxima is a computer algebra system (CAS) that combines graphics, symbols, and numerical capabilities. Professor Schelter was an expert on *wx*Maxima software [6]. The *wx*Maxima is developed by the MACSYMA system in the Massachusetts Institute of Technology. There is other CAS software such as Mathcad, Mathematica, Maple, and Smath Studio are also useful for linear algebra. However, *wx*Maxima is free and easily accessible for students; thus, *wx*Maxima is utilized in this study.

The use of *wx*Maxima to solve linear algebra problems can help students build their own knowledge and develop the desired competencies [7]. Therefore, this article aims to explain the learning process using Linear Algebra module with *wx*Maxima software on Gauss elimination topic.

2. Method

Linear Algebra module with *wx*Maxima software has been developed using ADDIE development model. The ADDIE model consists of five development stages: analysis, design, development, implementation/execution, and evaluation [8]. This study is part of the implementation phase of the first meeting on GE in a large group to describe the learning process on GE. The participants were 18 mathematics education students. The research data is taken from the lecturer's self-reflection sheet to see the relevance between the lesson plan and the implementation. Data were analyzed descriptively about the learning process on GE. The learning activities began with the explanation about elementary row operations (ERO) and GE by the lecturer and followed by solving the problem manually using the paper-pencil method. Then, the solutions obtained manually were checked using the *wx*Maxima software.

3. Results and discussion

The purpose of this learning process was that students could solve problems related to elementary row operations (ERO), solutions of homogeneous linear equation system, and the concept of Gauss Jordan elimination method. GE lesson using *wx*Maxima Linear Algebra module began with reviewing the previous lesson about ERO, then followed by explaining the GE steps to students and completing problems related to GE.

Students were assigned several problems on the *wx*Maxima Linear Algebra module to be solved manually. The assigned problems on *wx*Maxima Linear Algebra module were adopted from Linear Algebra Elementary book on page 16 [2]. One of the problems assigned to students is presented in Figure 1.

Complete the matrix of the following system of equations to be a row echelon matrix using GE.

$$\begin{aligned} x_1 + 3x_2 - 2x_3 + 2x_5 &= 0 \\ 2x_1 + 6x_2 - 5x_3 - 2x_4 + 4x_5 - 2x_6 &= -1 \\ 5x_3 + 10x_4 + 15x_6 &= 5 \\ 2x_1 + 6x_2 + 8x_4 + 4x_5 + 18x_6 &= 6 \end{aligned}$$

Figure 1. One of the problems assigned to students.

The students' answers to the problem were varied. Some students answered correctly with the final answer as shown in Figure 2.

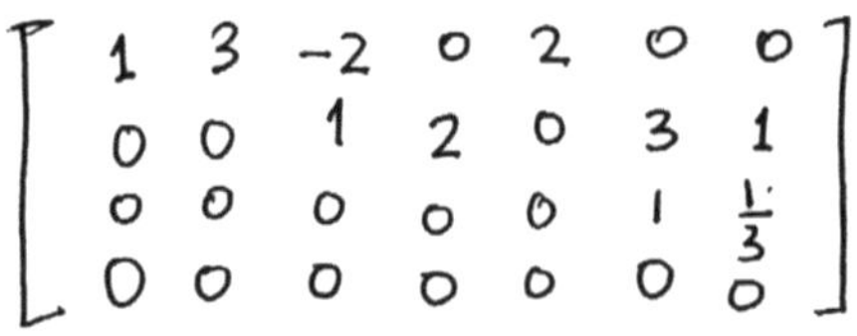

Figure 2. The students' correct answer.

The answer in Figure 2 indicates that the students had understood the concepts and procedures of the GE. The first procedure is that if a row does not consist entirely of zeros, then the first non zero number in the row is a 1, called a leading 1. The second procedure is that if there are any rows that consist entirely of zeros, then the rows are grouped together atthe bottom of the matrix. The third procedure is that in any two successive rows that do not consist entirely of zeros, the leading 1 in the lower row occurs farther to the right than the leading 1 in the higher row [2]. However, some students answered the problem incorrectly as shown in Figure 3.

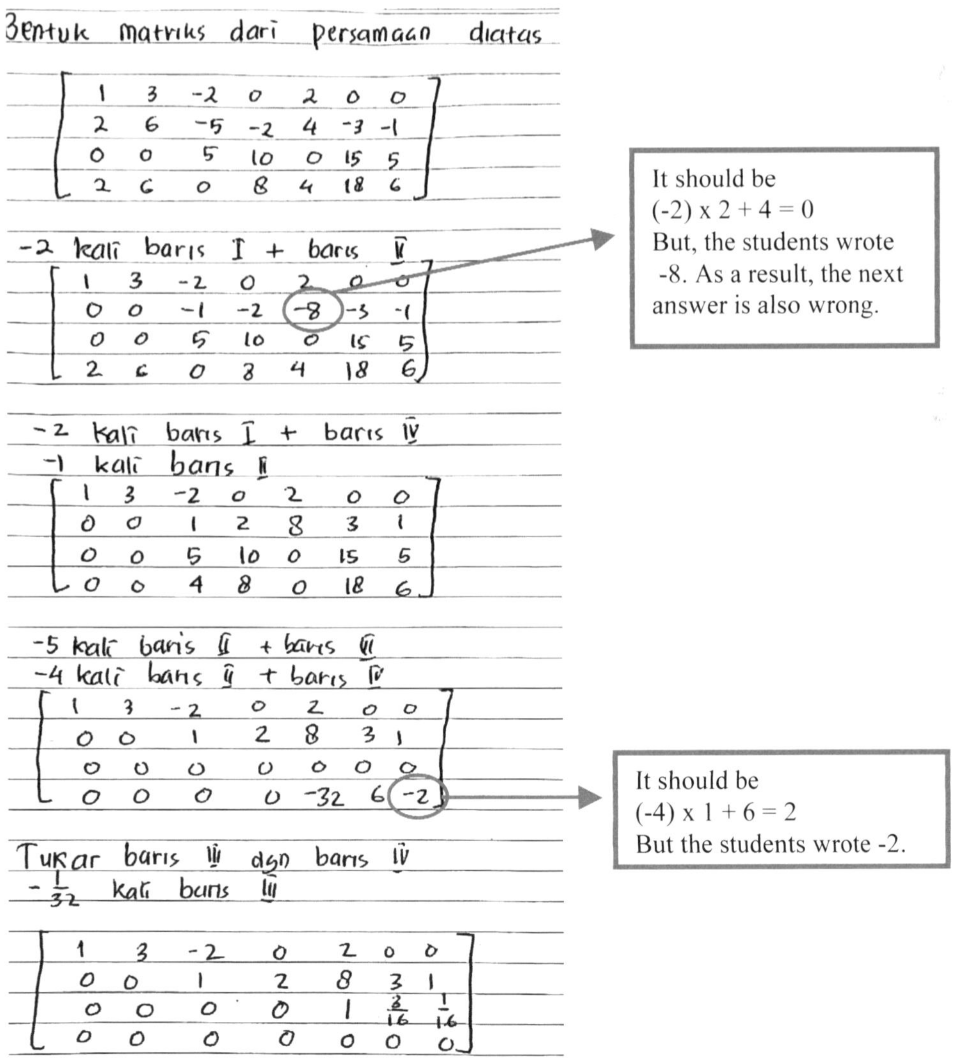

Figure 3. The students' incorrect answer in term of miscalculation.

Figure 3 shows that the students answered incorrectly in determining the multiplication result of the matrix element. The students repeated the mistake in the next step. As a result, the final solution obtained was also wrong. Based on the students' answer in Figure 3, it shows that the students made a mistake out of the GE procedure. The error was in determining the result of GE operation caused by lack of accuracy in calculation. The students were not aware of the mistake they made because the solution is lengthy. The same mistake was also experienced by other students, as shown in Figure 4.

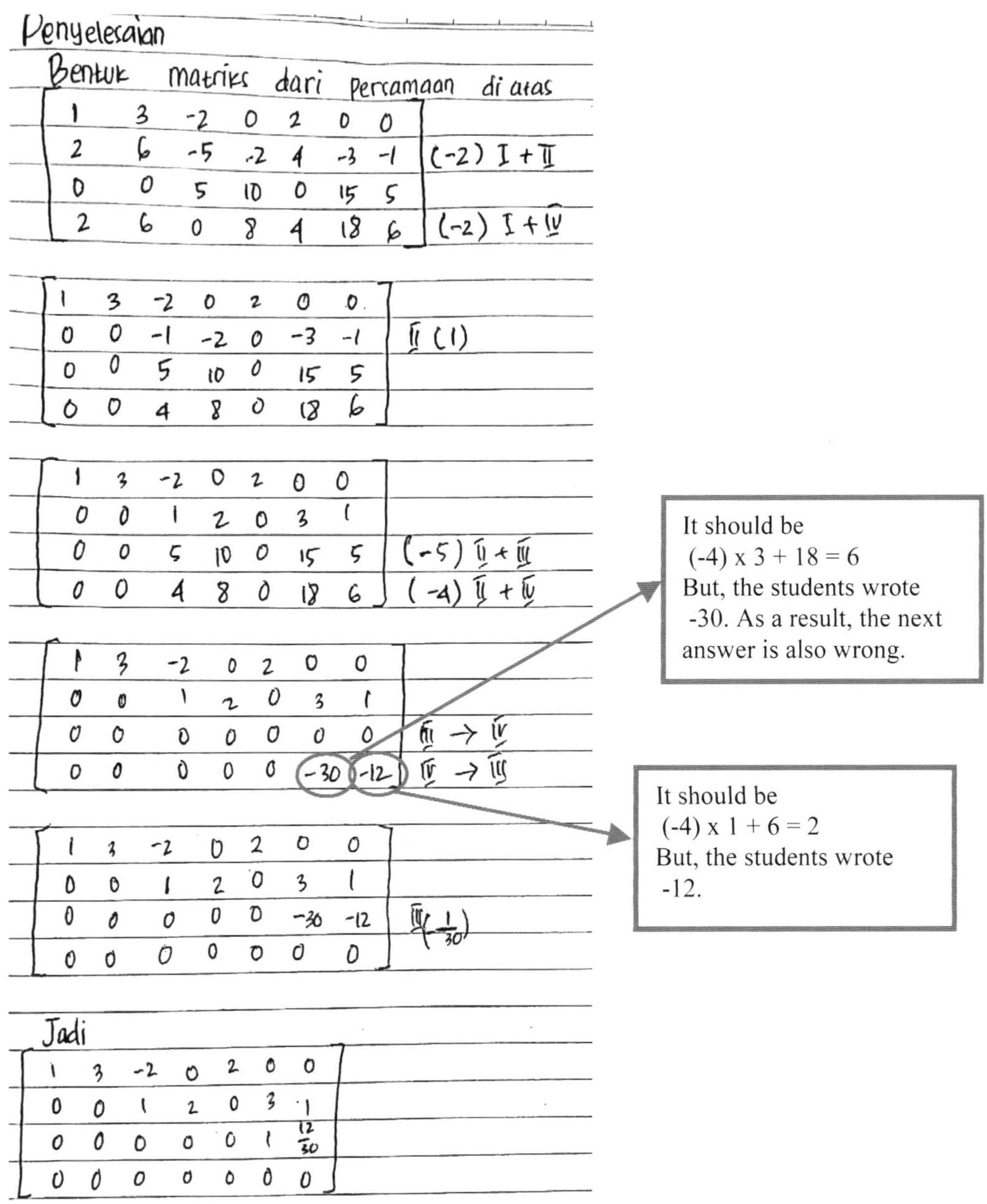

Figure 4. Another students' wrong answer in term of miscalculation.

In addition to the errors in the calculation of the matrix elements, there were also some students who had a mistake in the concept when solving the given problem. The answers given by the students were not based on the GE rules. The students' answer is shown in Figure 5.

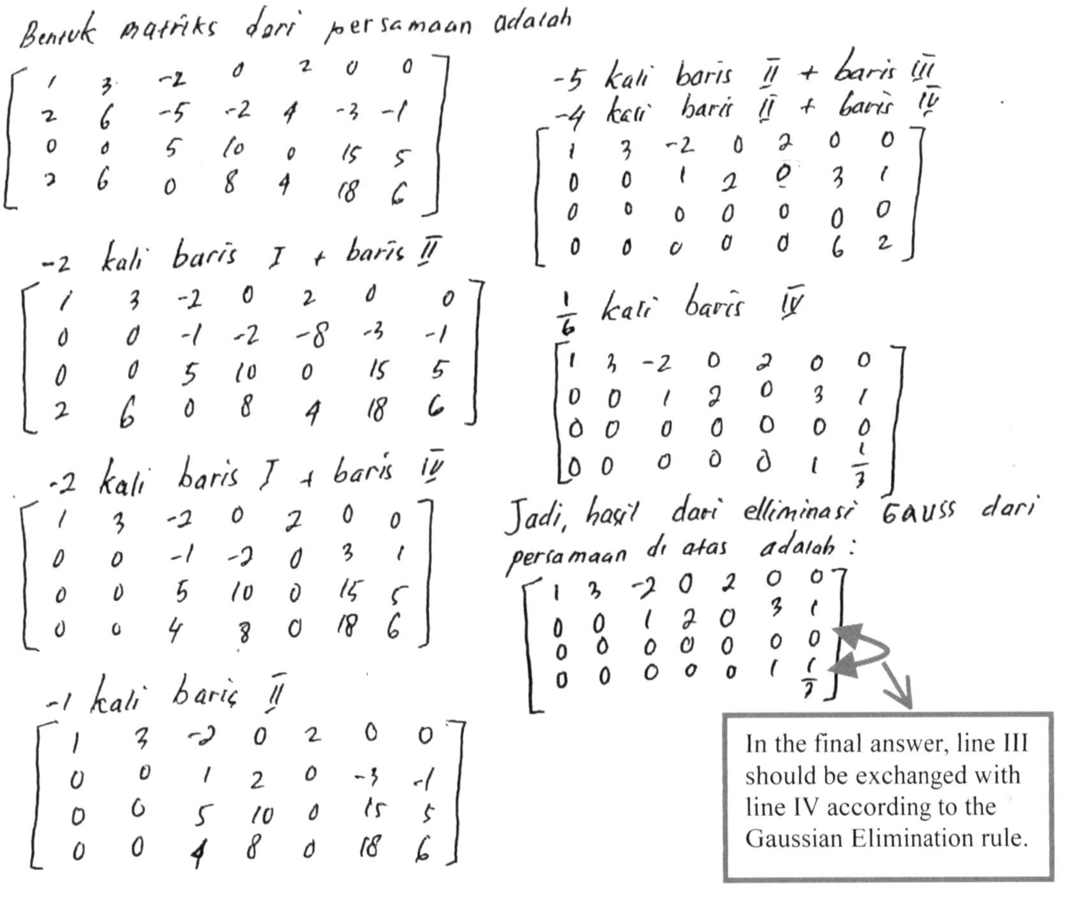

Figure 5. The student's wrong answer in term of misconception.

Figure 5 shows that the student experienced an error concept in the second GE procedure, namely: If there are any rows that consist entirely of zeros, then they are grouped together at the bottom of the matrix. The students should add the next step of exchanging line III with line IV.

When asked to determine the correct answer, all those students could not find it because they assumed that their answers were correct already. Therefore, the students needed to check their answers that had been obtained manually by using the *wx*Maxima software. Checking the calculation steps by using *wx*Maxima software was provided in the module as well. Utilizing the software, the students got a correct answer as shown in Figure 6.

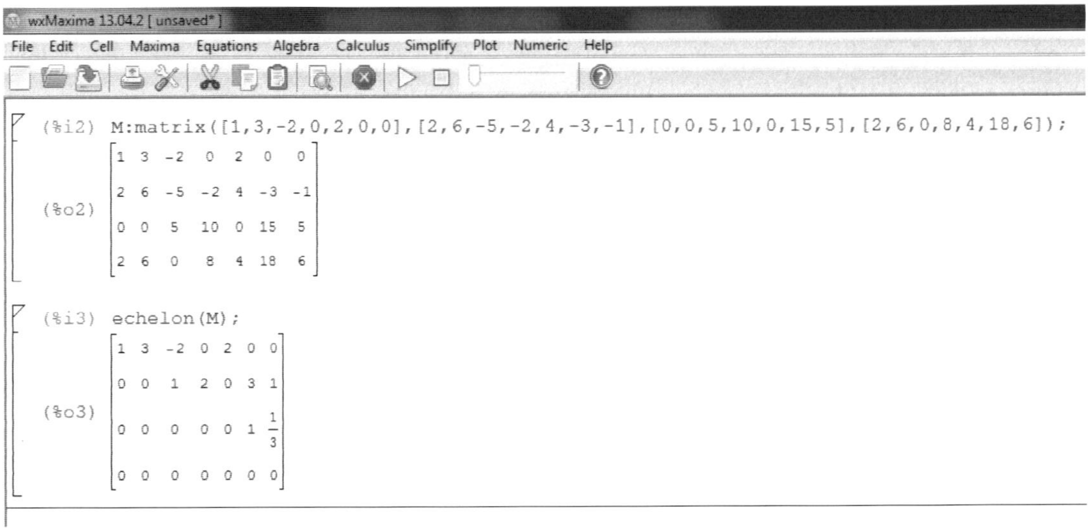

Figure 6. The answer to the problem displayed on the *wx*Maxima software.

Students who answered differently from the answer displayed on the *wx*Maxima software were required to double check their initial answer. At this stage of learning, the students could find out the mistakes they made in their calculation. The same learning procedures were carried out to Gauss Jordan elimination lesson. The Gauss Jordan elimination is a further step of the GE to obtain a reduced row echelon matrix. Therefore, it is essential for students to check the GE steps to improve the answers obtained before they proceed to the next step.

The learning process in the Linear Algebra course especially on GE lesson run well. The students did the exercises and seemed enthusiastic to check their solutions using *the wx*Maxima software. Learning steps in Linear Algebra course especially GE lesson using *wx*Maxima software could be implemented in accordance with the activities on the lesson plan. The time allocated was also well executed in the learning process.

Based on the learning implementation, it shows that the use of *wx*Maxima software could help the students recognize their wrong calculation and misconception. This finding is aligned with the results of a previous study indicating that the use of mathematical software has created a major influence on students' understanding of mathematics and their subsequent performance in learning [9].

During the learning process, the students looked enthusiastic in completing the given task and double-checking their answers using the *wx*Maxima software. This finding is supported by previous research that found that students were active in learning with the use of the software. The use of *wx*Maxima Linear Algebra module could improve student engagement and provide positive effects to students during the learning process [10].

The module also provided assistance for students to follow the learning process. Students could understand the lesson from the module and check their solution based on the solutions presented in the module without waiting for the assistance from the lecturer. Also, the use of the module in learning could attract students' interests, made students more active in learning, and helped students improve their learning achievement [11].

4. Conclusion

Learning linear algebra especially GE topic by using *wx*Maxima software could be implemented in the classroom. The learning activities run well as the activities and the allocated time in the lesson plan. The use of *wx*Maxima software in the learning process also could help students find and notice the wrong calculations and concepts. This is valuable for the students to learn from their mistakes.

IOP Conf. Series: Journal of Physics: Conf. Series **1088** (2018) 012088 doi:10.1088/1742-6596/1088/1/012088

References

[1] Rosita C D, Laelasari L and Noto M S 2014 Analisis Kemampuan Pemahaman Matematis Mahasiswa pada Mata Kuliah Aljabar Linear I *Euclid* **1** 60

[2] Anton H and Rorres C 2002 *Elementary Linear Algebra, 11th editions* (United States of America: Wiley)

[3] Powers R and Blubaugh W 2014 Technology in mathematics education: Preparing teachers for the future *Contemporary Issues in Technology and Teacher Education* **5** 254

[4] Kurz T L, Middleton J A and Yanik H B 2005 A taxonomy of software for mathematics instruction *Contemporary Issues in Technology and Teacher Education* **5** 123

[5] Lavicza Z and Koren B 2015 New Challenges in Developing Dynamic Software for Teaching Mathematics *The Proceedings of the 12th International Congress on Mathematical Education* 621

[6] Kuo C S 2013 Automation of Mathematics Teaching --- Demonstration by Methods of Completing the Square in Solving a Quadratic Equation *International Journal of Computer Science and Mobile Computing* **2** 125

[7] Diaz A, Garcia A and de la Villa A 2011 An Example Of Learning Based On Competences: Use Of Maxima in Linear Algebra For Engineers *International Journal for Technology in Mathematics Education*

[8] Branch R M 2009 *Instructional design: The ADDIE aproach* (USA: Springer Science and Bussiness Media)

[9] Ayub A F M, Tarmizi R A , Bakar K A and Wong S L 2012 *Wx*Maxima Computer Software as an Aid to the Study of Calculus by Students with Different Learning Approaches *Procedia-Social and Behavioral Sciences* 467

[10] Fitria M, Arnawa M and Lufri 2014 Pengembangan Modul Aljabar Linear Elementer Bernuansa Konstruktivisme Berbantuan ICT *EKSAKTA* **1** 34

[11] Nurlainy 2017 Pengembangan Modul Kalkulus II Berbantuan Software *wx*Maxima di Program Studi Pendidikan Matematika Universitas Syiah Kuala (Banda Aceh: Universitas Syiah Kuala)

Students' metacognition on mathematical problem solving through ethnomathematics in Rejang Lebong, Indonesia

D Herawaty, W Widada, T Novita, L Waroka, and A N M T Lubis
Universities Bengkulu, Jl. W.R. Supratman, Bengkulu 38371Indonesia

E-mail: dherawaty@unib.ac.id

Abstract. The structure of knowledge consists of a development scheme of facts. Mathematical activities are based on situations arising from experiences. This study aimed to describe the metacognition of students in solving problems using mathematics problem-based learning in Rejang Lebong ethnomathematics. This research was the preliminary phase of a research and development. This preliminary research was conducted through in-depth interviews, involved six students of a Junior High School in Rejang Lebong, Bengkulu province, Indonesia. The research instrument consisted of an interview guide on problem solving and understanding of mathematical concepts. Data were analyzed qualitatively. The results showed that students could develop the ability to solve problems through self-reflection on planning, monitoring and evaluating the implementation of thinking processes. Students could combine pieces of information about the parts of the traditional house of Rejang Lebong that have similar properties to the mathematical 3-dimensional figures such as pyramids, prisms, rectangular prism, and cubes.

1. Introduction

In mathematics learning process, students often have the misconception. One of the ways to reduce the misconception is by requiring students to explain the way and the results of their thinking, which is widely known as metacognition. Metacognition refers to a strategy of learning, planning, and validation of the learning process [1]. Widada stated that a variety of mental and physical activity undertaken by students could be seen as a collection of structure or model of the students' metacognition [2]. The physical and mental activities of metacognition involve an explicit description of how mathematical concepts/principles of mathematics are developed in students' minds.

In mathematics learning, metacognition has an important role in solving mathematics problems. A development scheme of facts forms the structure of mathematical knowledge, and mathematical activities are based on situations arising from someone's experiences [3]. To obtain optimal results in learning mathematics, solving mathematical problems should involve the well-organized steps. The steps include understanding the problem, making a plan for settlement, implementing the settlement plan, and re-examining solutions that have been resolved. According to Widada, the starting point was the issue of contextual learning mathematics which can enhance students' mathematical abilities to solve mathematical problems successfully [5]. Solving the problem itself was a mental activity which requires an individual's effort to overcome or find the correct solution to a problem. The person needs to manage his thoughts well, take advantage of prior knowledge, control and reflect on their thinking processes and results, to help in resolving a problem. The awareness of the thinking processes was called one of the components of metacognition.

According to Flavell, metacognition was also crucial in other areas, such as communication, reading comprehension, learning, social metacognition, attention, self-checking, memory, self-study, writing and problem-solving [6]. Research carried out by Aini showed that metacognition training could improve mathematical competence and be useful for learners who had difficulties in learning mathematics [7]. According to Herawaty, the students' activities during the implementation of the realistic mathematics teaching model were effective [8]. The ability to understand the concept and troubleshooting results from a class experiment with realistic mathematics learning was better than in control classes that were assigned to conventional mathematics lesson.

The utilization of contextual learning media is appropriate to improve students' abilities to understand and master the concepts and principles of mathematics [9]. Contextual learning could relate the learning activities to local cultures. Meanwhile, the mathematics culture can be applied and used to analyze the innovations. Thus, mathematics can be used as a tool to develop a winning culture. In addition, consciously or not, humans have used mathematics in solving problems.

Mathematical properties tend to be linear and rigid, but when they are integrated to culture, then the properties are flexible, for example, in the architecture area. The building structure involves mathematical principles as well as the aesthetic principles. When thinking about building structures, it not only requires the concept of geometrical shapes (two-dimensional and three-dimensional geometry) but also involves the aesthetic sense. Various cultural products of Rejang Lebong have many heritage and artistic creativity that include mathematical principles. For example, in Rejang Lebong, the design woven cloth forms two-dimensional geometry, ornamental carvings and architectural structures in houses contains geometric patterns. The attribution of mathematics into a culture is better known as ethnomathematics. Therefore, the process of students' metacognition is found in the concept of geometry with problems relating to ethnomathematics becomes important.

Many experts defined the term metacognition in different ways. According to Livingston, metacognition refers to the thinking process about one's thought (thinking about thought) or a person's knowledge about the process of thinking [10]. O'Neil & Brown defines metacognition as a process of how a person thinks about thought to develop a strategy to solve the problem [11]. Wellman stated that "metacognition was a form of cognition, a second or higher order thinking process control, which involved cognitive activity process. It can be simply defined as someone's thinking about thought or as a person's cognition about cognition" [12]. Accordingly, metacognition is the process of thinking two levels or more that involve the control of cognitive activities. Blakey [13] found "metacognition means thinking about thought, as well as knowing what we know and what we do not know", which means that metacognition was awareness of what known, and what unknown. Thus, it can be seen that metacognition was knowledge, awareness, and control of one's thinking process and results. Huitt [14] defined metacognition as one's knowledge of the cognitive system, thinking about though, and essential skills of a person in "learning to learn ". Moreover, Huitt sets out the two components included in metacognition, namely what we know or do not know, and the regulation of how we learn. Therefore, in this research, metacognition refers to the process when someone thinks about the way he thinks about what he has done.

Metacognition has two components, namely the knowledge of metacognition, and the mechanism of self-control, and monitoring of cognitive [11]. The components of metacognition in this research refer to the knowledge of metacognition. Anderson [15] stated that knowledge of metacognition is generally similar to someone's awareness and knowledge of metacognition because metacognition is awareness of what is known is and what is unknown. The reasonable cognitive strategies refer to the ways to increase awareness about the process of thinking and learning that are applied when consciousness is true, and then one can guard the mind by designing, monitoring and assessing what he learned.

According to Pierce [16], to improve the ability of metacognition, students should have three types of knowledge, namely: declarative knowledge, procedural knowledge, and conditional knowledge. Declarative knowledge is factual information that someone understood and expressed orally or in writing. Declarative knowledge was the ability to describe the thinking strategies, procedural knowledge includes knowledge of how to use the strategies, and conditional knowledge was knowledge about the right time to use them [17]. Procedural knowledge is knowledge about how to do things and how to do the steps in a process. Conditional knowledge is the knowledge of when to use a procedure, skill or strategy and when not to use it, why the procedure can be used and under what conditions and why such as a procedure were better than the other.

In this research, metacognition refers to three types of knowledge, namely: declarative knowledge, that refers to knowledge of facts and concepts or factors that affect thinking and attention in solving the problem, procedural knowledge, that is knowledge of how to do things, how to do the steps or strategies in a problem-solving process, conditional knowledge, that refers to a person's awareness of the conditions that affect him in solving the problem is: when a strategy should be implemented, why to implement a strategy and when the strategy used in solving problems. Thus, the problem of this research was how the process of students' metacognition in problem-solving geometry based on the traditional housing ethnomathematics in Rejang Lebong.

2. Method

This research was a development research by Plomp [22], which included only preliminary research conducted by in-depth interviews. The interviews included six junior high school students in Rejang Lebong. The research instrument consists of an interview protocol on problem solving and understanding of mathematical concepts. The researcher was the main instrument. Data were analyzed qualitatively to explain the metacognition of students in finding the concepts and principles of geometry (two-dimensional and three-dimensional geometric figures) based on ethnomathematics in Rejang Lebong.

3. Results and discussion

The traditional house of Rejang Lebong is known as *Cut Umeak Jang*, in which *umeak* means house, *Cut* means artificial, and *Jang* means Rejang. Thus, literally, Umeak Cut Jang means artificial Rejang house. Moreover, the traditional house of Bengkulu is also commonly known as Umeak-An, that means old or older house. Here ethnomathematics traditional house of Rejang Lebong:

Figure 1. Traditional house Rejang Lebong [23]

The process of students' metacognition based on society ethnomathematics Rejang Lebong as the horizontal mathematical process. Mathematics activities that developed students' thinking about the geometrical concepts were related to traditional house culture Rejang Lebong. Students devoted their mind and stated that the traditional houses in Rejang Lebong had been implementing the concept and principles of geometry in the construction of parts of the building traditional houses. The concepts and principles found by other students were geometrical models, including a triangular prism, quadrilateral pyramid, cubes, and blocks.

IOP Conf. Series: Journal of Physics: Conf. Series **1088** (2018) 012089 doi:10.1088/1742-6596/1088/1/012089

3.1. Concepts and principles prism build space triangle

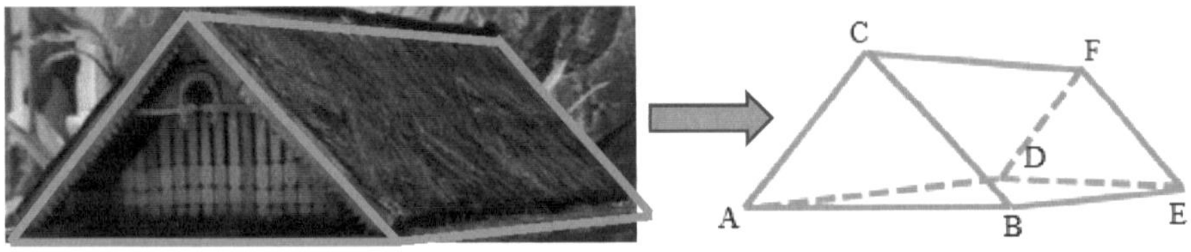

Figure 2. Prism triangle ABC.DEF.

A triangular prism was waking three-dimensional space limited by the base and lid were identical triangular-3 and the sides of the square-shaped upright. Here's the formula volume and surface area of prisms:

$$\text{Volum} = \text{the area of the base of the prism x high from the prism The surface area of the prism}$$
$$= (2 \times \text{the area of the base of the prism}) + (\text{around the triangle} \times \text{high from the prism})$$

3.2. Concepts and principles build pace rectangular space prism

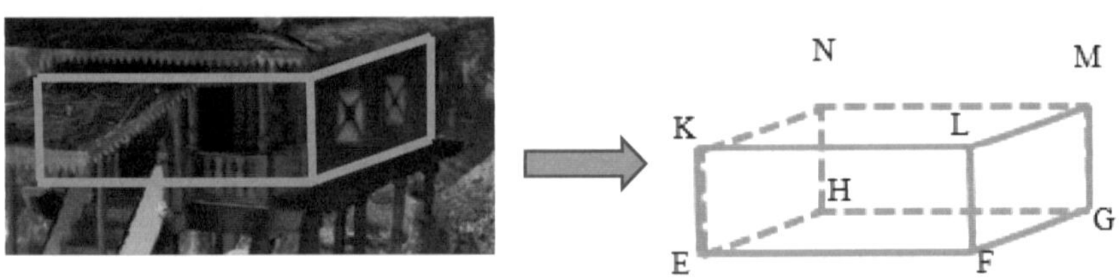

Figure 3. Rectangular prism EFGH.KLMN.

Rectangular prism was waking up three-dimensional space formed by three pairs of square or rectangular, with at least one pair of which were sized differently. Here's the formula volume and surface area of the beam:

$$\text{Volume} = \text{length x width x height}$$
$$\text{The surface area} = (2 \times l \times w) + (2 \times l \times h) + (2 \times w \times h)$$
$$= 2(lw + lh + wh)$$

3.3. Concepts and principles build space rectangular pyramid

Figure 4. Limas quadrilateral T.ABCD.

Limas was a wake bounded by a rectangle as a base and four triangular fields that meet at a peak.

$$\text{Volum} = \frac{1}{3} \times \text{the area of the base x height}$$
$$\text{The surface area} = \text{the area of the base} + \text{L. TBC} + \text{L. TCD} + \text{L. TAD} + \text{L. TAB}$$

3.4. Concepts and principles build space cube.

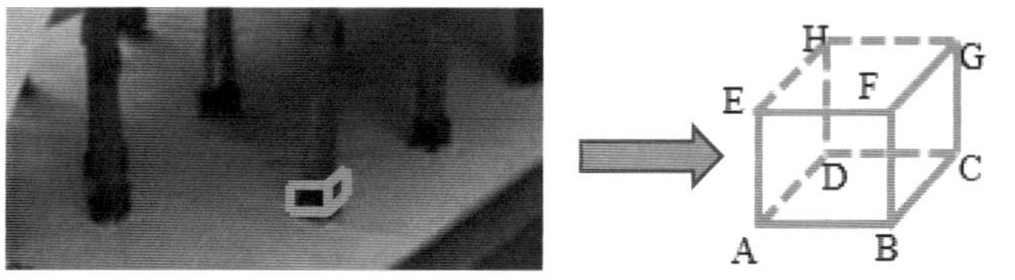

Figure 5. Cube ABCD.EFGH.

Cube was waking three-dimensional space bounded by six areas congruent sides of a square. The cube has six sides, 12 ribs, and 8 vertexes, Here's the formula volume and surface area of the cube:

$$\text{Volum} = s \times s \times s = s^3$$
$$\text{Area} = 6 \times s^2$$

According to Woolfolk [18], cognitive strategies include planning, monitoring, and evaluating processes. NCTM suggested three basic elements of metacognition include to develop an action plan, to monitor the action plan, and to evaluate action plans. In this research, these basic elements were found by students when they related the forms of some elements of traditional house in Rejang Lebong to geometric figures.

Any problems always need to solve. Therefore, the problem solvers should have a contextual understanding. Widada [19] suggested that someone has no understanding of a problem conceptually when the person does not utilize the concepts that have been learned or failed to implement the concept that has been known to solve the given problem. Troubleshooting was an attempt to find the solution to a problem. Polya mentioned that in solving a problem, there were four stages that must be made, including to understand the problem, to plan problem-solving process, to implement the plan of, and to re-check the solution obtained [4]. Santrock [20] stated that solving the problem can be done by finding and developing problems, developing strategies to solve the problem, evaluating the solution, and thinking and redefining the problem and the solution. According to Schoenfeld [21], there are five stages in solving the problem, namely reading, analysis, exploration, planning/implementation, and verification. Artzt & Armor-Thomas has developed step-by-step troubleshooting of Schoenfeld, becoming reading, understanding, analysis, exploration, planning, implementation, and verification [21]. In this research, the students used these steps during their metacognition process to validate the formulas of the 2-dimensional and 3-dimensional geometric figures, including the surface area and volume of pyramids, prisms, rectangular prism, and cubes.

4. Conclusion

Students solved mathematical problems through mathematization process based on ethnomathematics. Students were aware that Rejang Lebong's ethnomathematics was the starting point of horizontal mathematical activity. Just like the traditional house, the culture was a real problem to achieve geometric concepts, such as 2-dimensional and 3-dimensional geometric figures. In particular, students can discover about the surface area and volume of pyramids, prisms, rectangular prism, and cubes. The students' metacognition was used to validate the correctness of the formulas.

Reference

[1] Ozsoy G 2011 An investigation of the relationship between metacognition and mathematics achievement *Asia Pacific Education Review* **12** 227

[2] Widada W 2017 Beberapa dekomposisi genetik siswa dalam memahami matematika *Jurnal Pendidikan Matematika Raflesia* **2** 30

[3] Widada W 2016a Profile of cognitive structure of students in understanding the concept of real analysis *Infinity* **5** 83
[4] Polya G 1981 *How to Solve It* (Princenton University Press)
[5] Widada W 2016b Sintaks model pembelajaran matematika berdasarkan perkembangan kognitif peserta didik *Jurnal Pendidikan Matematika Raflesia* **1** 163
[6] Flavell J H 1979 *Metacognition and Cognitive Monitoring, A New Area of Cognitive-Developmental Inquiry* (American Psychologist)
[7] Aini, Ana N, Wijayanti and Pradnyo 2017 Profil metakognisi siswa dalam memecahkan masalah matematika ditinjau dari kemampuan matematika *Jurnal Ilmiah Pendidikan Matematika* **1**
[8] Herawaty D and Rusdi 2016 Increased capacity of the understanding of the concept and the ability to solve problems through the implementation of the model of teaching mathematics realistic based on cognitive conflict students *Infinity* **5** 109
[9] Widada W 2015 Proses pencapaian konsep matematika dengan memanfaatkan media pembelajaran kontekstual *Jurnal Penelitian Pendidikan Matematika dan Sains* **22** 31
[10] Livingston J A 1997 Metacognition: An Overview http. *www, gse. buffalo, edu/fas/shuell/CEP564/Metaeog. htm*
[11] Syaiful 2011 Metakognisi siswa dalam pembelajaran matematika realistik di sekolah menengah pertama *Edumatica* **1**
[12] Mulbar U 2008 Metakognisi Siswa Dalam Menyelesaikan Masalah Matematika *Tersedia pada: http//www. usmanmulbar. files. wordpress. com.*
[13] Blakey E and Spence S 1990 Developing Metacognition *ERIC Digest*
[14] Huitt W 1997 Metacognition *Educational Psychology Interactive* (Valdosta, GA: Valdosta State University)
[15] Anderson L W, Krathwohl D R, Airasian P W, Cruikshank K A, Mayer R E, Pintrich P R and Wittrock M C 2001 A taxonomy for learning, teaching, and assessing: A revision of Bloom's taxonomy of educational objectives abridged edition (White Plains, NY: Longman)
[16] Peirce W 2004 Metacognition: Study Strategies, Monitoring, and Motivation. A greatly expanded text version of a workshop presented November 17, 2004, at Prince George's Community College
[17] Lestari Y D Metakognisi siswa dalam memecahkan masalah matematika berdasarkan gaya kognitif reflektif dan impulsive [thesis] Surabaya: Unesa University
[18] Woolfolk and Anita 2005 *Education Psychology Ninth Edition* (United States of America: Pearson Education)
[19] Widada W and Herawaty D 2017 Realistic mathematics learning based on bengkulu ethnomatematics to increase cognitive level *Int. Conf. on Science and Education* (Bengkulu, 14-15 Desember 2017)
[20] Cahdriyana R A 2016 Pengaruh Metode Pembelajaran Berbasis Masalah Terhadap Kemampuan Memecahkan Masalah Matematika Siswa SMP Negeri 9 YOGYAKARTA *AdMathEdu: Jurnal Ilmiah Pendidikan Matematika, Ilmu Matematika dan Matematika Terapan* **6**
[21] Lidinillah 2008 Strategi pembelajaran pemecahan masalah di sekolah dasar *Jurnal Pendidikan Dasar* **10** 1
[22] Plomp T J 2013 *Educational Design Research: An Introduction* (Netherlands: Netherlands Institute for Curriculum Development (SLO))
[23] *Source: http: //www.rumahperumahan.com/2016/10/desain-bentuk-Rumah-adat-Rejang-lebong.html*

IOP Conf. Series: Journal of Physics: Conf. Series **1088** (2018) 012090 doi:10.1088/1742-6596/1088/1/012090

Development of learning tools using Treffinger learning model to improve creative thinking

R Handayani[1], Hajidin[1], M Duskri[2], E Maidiyah[1]
[1]Syiah Kuala University, Banda Aceh, Indonesia, 23111
[2] Ar-Raniry Islamic State Islamic University, Banda Aceh, Indonesia, 23111

E-mail: hajidin.gsd@fkip.unsyiah.ac.id

Abstract. The mathematical creative thinking is essential for students. The initial test indicated that students' creative thinking ability is still lacking. Treffinger learning model deals with creativity problems. Based on the needs analysis, Treffinger learning instruments was not available. This study aimed at developing and producing valid and practical learning instruments using Treffinger learning model to improve the junior high school students' mathematical creative thinking. It was a developmental research that refers to the ADDIE development model. The results of this study indicated that the learning instruments satisfy the validity requirements and the results of the validity test by validators showed that instruments were valid and based on the agreement among validators results classified as the good category. Its practicality can be seen from the results of the learning implementation (93%). In addition, the results of teacher responses were classified as 'very good.

1. Introduction

The mathematical creative thinking ability is one of the mathematical skills that must be acquired by students in learning mathematics. In fact, the development of creative thinking ability is one of the focuses of mathematics learning [1]. Thus, the learning applied by teachers should be oriented toward creativity, encouraging students to discover their own solutions from the different point of views [2]. Our societies require creative thinking more and more than in the past [3]. Creativity draws new relationships between experience and knowledge, it also proposes new solutions to problems [4]. Students can develop their creative thinking ability that they can qualify the development of globalization era and compete with others in the future [5].

It is found that students are lacking mathematical creative thinking. This is indicated by the test results of mathematical creative ability administered to 82 junior high school students in Banda Aceh. The test showed that an average of 67.68% of students did not solve the mathematics problems using more than one solution, though the problems provided in the test requires students to solve it in at least two different ways. While, 10.37% of learners have solved the problem in the test providing at least two different ways, yet one of the solutions presented was wrong. Only 20.73% of students answered the questions using at least two ways and they were correctly done. Less than two percent of the students (1.22%) did not answer the problem or all the answers provided were incorrect. Figure 1 presents a problem used to examine the students' mathematical creative thinking ability adapted from Siswono [6].

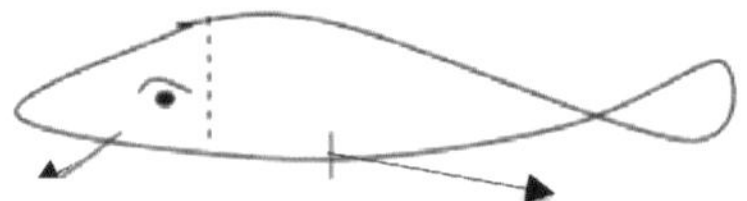

Figure 1. A problem used to examine the students' mathematical creative thinking ability

A fish has three parts of the features, namely head, body, and tail. Each part weight in kilograms.

a. How many kilograms may the weight of the fish be? Write down how to solve it. Are there any other possible answers? If there are, mention at least two possible answers.
b. Check the answers you have obtained. Show two or more different ways to get the answer.

One of the students' answers is presented in Figure 2.

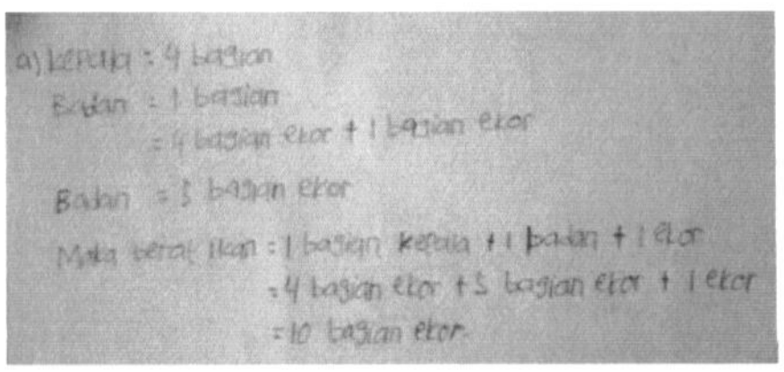

a) Head = 4 parts
Body = 1 part
=4 parts of tail + 1 part of tail
Body = 5 parts of tail
So, the fish weight = 1 part of head +1 body+1tail
= 4 parts of tail +5 parts of tail +1 part of tail
= 10 parts of tail

Figure 2. The result of students' initial test

The student's test result in Figure 2 shows that the students had given the correct answer for the given problem, however, it was only solved using one strategy despite the questions asking the students to solve the problem using at least two different ways. In this case, the researchers also conducted interviews with the students and it is noted that the students were not aware of other ways to solve the problem.

Treffinger learning model is one of the learning models specially designed with the learning stages to trigger students' creative thinking ability involving both cognitive and affective skills. The Treffinger model encouraging creative learning consists of three stages arrangement started from the basic elements and rise to the more complex functions of thinking; students are engaged in skills building activities in the first two stages and later deal with real-life problems on the third phase [7].In addition, the Treffinger model consists of the following steps: level 1 provides the foundation upon which creative learning develops by including a variety of important techniques basic to creative learning. in level 2, the basic cognitive and affective factors from level 1 are extended, higher-level or more complex thinking skills are employed, along with transformation of products and processes, analogies and metaphor, methodological or inquiry skills, dealing with complex feelings and tensions, imagery, and the development of psychological freedom and safety. Level 3, the final level, involves the person in real problems and challenges [8].

One attempt that should be considered to support the learning process triggering the students' mathematical creative thinking ability is the design of learning instruments. All of the teachers in the educational unit is obliged to develop a complete and systematic learning instruments to ensure that the learning is interactive, inspirational, fun, challenging, and motivating the students to be actively engaged, creative, and independence based on their talents, interests, as well as their physical and psychological development [9]. In fact, it is hard to imagine teachers becoming more effective over time without being able to analyze teaching in terms of its effects on student learning [10]. This is confirmed by [11] who stated that teachers need to analyze the concepts of the material and carefully plan the learning before conducting the learning in the classroom.

Based on the needs analysis conducted by the researchers in three junior high schools in Banda Aceh, neither the learning instruments using the Treffinger learning model nor the learning model demanding students in thinking creatively was available. Therefore, there was a need of developing of learning instruments using creative learning model, one of which is Treffinger. The purposes of this study were to describe the development process and produce a valid and practical mathematics learning instruments by using the Treffinger learning model to improve junior high school students' mathematical creative thinking ability.

2. Method

This is a design research. The subjects in this study were 30 Year 7 students of one of the Islamic junior high schools in Meulaboh. The quality criteria of the learning instruments refer to the criteria set by [12], while the design of learning instruments refers to the development model of ADDIE. The data in this study was obtained by observing the implementation of learning activities and examining the teacher's response toward the implementation of learning using the learning instruments of Treffinger learning model. Data analysis of the validity test was carried out by calculating the average score the developed learning instruments assigned by the validator. The validity criteria used is adapted from [13] and are presented in Table 1.

Table 1. Validity criteria of the experts

Average	Validity criteria
$4 \leq TV \leq 5$	Highly valid
$3 \leq TV < 4$	Valid
$2 \leq TV < 3$	Quite valid
$1 \leq TV < 2$	Invalid

The practicality test of the learning instruments was conducted in two ways, namely: observing the implementation of learning instruments and examining the mathematics teacher responses to learning instruments designed. The implementation of learning was carried out by converting the percentage of learning activities (k) into a qualitative value adapted from the qualification of learning implementation used by [14] as shown in Table 2.

Table 2. The percentage of qualification of the implementation of learning instruments

Percentage	Criteria
$80\% \leq k < 100\%$	Excellent
$60\% \leq k < 80\%$	Good
$40\% \leq k < 60\%$	Adequate
$20\% \leq k < 40\%$	Less
$0\% \leq k < 20\%$	Much less

Average score of the practicality test (p) based on the teacher's response to the questionnaire and it was then converted to qualitative criteria [15], as presented in Table 3.

Besides using the average formula, the data analysis of validity and practicality test was also reinforced by inter-rater test. This was conducted to examine the perception or agreement between the validators and the teachers/practitioners toward the Treffinger learning instruments employing the assessment criteria of kappa proposed by [16] as follows KK < 0,4 (categorized as bad), 0,4 ≤ KK ≤ 0,75 (categorized as good), and KK > 0,75 (categorized as excellent).The inter-rater test in this study was obtained by generating kappa coefficients using SPSS 23.0. The learning instruments are said to be valid and practical if kappa is greater than or equal to 0.4, otherwise the learning instruments needs to be revised.

Table 3. The classification guidelines of teacher practicality

Interval Score	Criteria
x>4,2	Excellent
3,4< $x \leq$ 4,2	Good
2,6< $x \leq$ 3,4	Adequate
1,8< $x \leq$ 2,6	Less
$x \leq$ 1,8	Much less

3. Results and Discussion

The process and the results of designing mathematics learning instruments using Treffinger learning model to improve mathematical creative thinking ability is described in the following sections.

3.1 Analyzing stage (analysis)

The activities undertaken at this stage were performing the needs analysis consisting of the analysis of students' mathematical creative thinking ability conducted by administering a creative thinking ability test, the analysis of the availability of learning instruments, and the curriculum analysis.

The result of the needs analysis showed that the mathematical creative thinking ability in three schools was lacking. This was indicated by the test results reported that 71% of the students participated in the test was less fluent in expressing ideas to solve the problem, 24% of the students was rather fluent and only 5% of the students fluently expressed the ideas to solve problems on the creative thinking ability test. For the indicator of flexibility, 76% of the students only provided one way/solution to the given problem even though the instructions required students to answer the problem with more than one way/solution, and only 4% of the students were managed to solve the problem by presenting more than one way/solution. For the indicator of authenticity (originality), it was noted that no student provided different or unique answers compared to other students. Furthermore, there was no lesson using the Treffinger learning model or the learning stages requiring students to think creatively. The activities in the students' worksheet were also found to highly guiding students in solving problems, consequently, students were not given the freedom to explore their ability in various ways/solutions. In addition, the test conducted to evaluate the students' learning outcomes did not allow the students to solve the problem in various ways/solutions or provide the instruction requiring students to present various ways/solutions in solving the problem. Therefore, the development of the learning instruments aiming to improve students' creative thinking ability is required.

3.2 Designing stage (Design)

Once all steps of the needs analysis have been conducted, the initial design of learning instruments was created. The first step carried out at this stage was studying the problem and finding alternative solutions to solve the problems found in the needs analysis. The results obtained at this stage was that the researcher developing the mathematics learning instruments using Treffinger learning model to enhance junior high school students mathematical creative thinking ability. The topic chosen for these learning instruments was rectangular plane geometry, consisting of parallelogram, trapezium, rhombus, and kite taught in the second semester of Year 7. This topic was suitable to be taught by using the Treffinger learning model as well as improving the mathematical creative thinking ability as it can be solved in various ways/solutions.

3.3 Developing stage (Development)

The subsequent stage was to develop the learning instruments that starting with determining the test items needed to measure the ability in accordance with the purpose of learning. This resulted in problems for student worksheet that was tailored to satisfy the indicator of the ability to think creatively using the syntax of Treffinger learning model as activities should be completed by the students. The next step was developing a lesson plan which format was adapted from the ministry regulation number 22 of 2016. The researchers later developed the teaching materials consisting of two parts of the rectangular topic that is the introduction of the rectangular plane geometry as well as the area and perimeter of the rectangular geometry plane. Finally, a test of learning outcomes aiming to evaluate the learning activities was also developed. The products obtained in this step were learning instruments consisting of lesson plans, student worksheets, learning outcome tests and teaching materials.

The learning instruments were validated by the validators. Content and construct validity expressed by the experts and practitioners were used to indicate whether the instruments were valid. The validity aspect of a learning instruments can be identified from (1) content validity, i.e. the instruments was strongly developed based on the theoretical rationale, and (2) construct validity, i.e. there is internal consistency between each component in the instruments [12]. The validation of the learning instruments in this research was carried out twice as the first validation phase of student worksheet and test of learning outcomes were classified as the bad category, and therefore, they should be revised and revalidated. While, the lesson plan and teaching materials were declared as valid in the first validation phase and classified as good category based on the validators' perception. Nevertheless, both teaching materials and lesson plan were also revised in accordance with the suggestions provided by the validators. The revisions made based on the validators' suggestion included the multiple interpretations language or sentence, the stages of learning, and the inappropriate problems and learning activities to improve the mathematical creative thinking ability using Treffingerlearning model. The second validation stage indicated that the learning instruments met the valid criteria based on the validator assessment results and the agreement between the validators. The Kappa values ranged from 0.588 to 1.000 for lesson plan, 0.545 to 1.000 for student worksheet, 0.409 to 0.755 for test of learning outcomes, and 0.462 to 0.611 for teaching materials. This means that learning instruments using Treffinger learning model to improve the mathematical creative thinking ability of junior high school students are both valid in term of content and construct validity.

3.4 Implementing stage (Implementation)

The next stage was to conduct a pilot test (trial). This trial phase consisted of individual trial, small group trial, and field trial. Two criteria of practicality, namely: (1) experts and practitioners declare that the developed instruments can be applied, and (2) the developed instruments can be applied/implemented in the field [12]. Therefore, the learning instruments developed were categorized as practical or easy to implement as indicated by the results of field trials conducted for 30 Year 7 students and teacher responses toward the learning instruments. The observation aimed to examine the implementation of the learning utilizing the developed learning instruments using the Treffinger learning model to improve mathematical creative thinking ability. The analysis results of the learning implementation reported that the implementation of the learning reached to approximately 93%. This means that it is in the excellent category based on the specified criteria.

The data obtained from the teacher questionnaire was teacher general assessment on the developed learning instruments. Two junior high school mathematics teachers participated to complete the questionnaire. The analysis results indicated that the average of the mathematics teachers' assessment of the lesson plan, student worksheet, teaching materials, and test of learning outcomes were 4.57, 4.56, 4.70, and 4.50, respectively, indicating that the assessment of learning instruments was at the excellent category based on predetermined criteria. To measure the practicality more accurately, the relationships between perceptions of teacher I and teacher II were also examined, followed by inter-rater test by calculating the kappa value. The kappa values were obtained using SPSS 23, the value of teacher I and teacher II on the lesson plan, student worksheet, teaching materials, and test of learning outcomes were 0.462, 0.571, 0.545, and 0. 600 respectively, indicating that the perception between the two teachers on the learning instruments was at the good category.

4. Conclusion

It can be concluded that the development process of learning instruments using Treffinger learning model to improve junior high school students' mathematical creative thinking ability consists of five stages, namely: analyzing, designing, developing, implementing, and evaluating. Furthermore, the results show that learning instruments developed are valid as indicated by the validator assessment results and the inter-rater test. Overall, the results indicate that the developed learning instruments satisfy the good criteria and all components in the learning instruments are consistent and support each other. The practicality test results are identified from the field trials results and teacher responses on the developed learning instruments, reporting an average of 93%. The analysis results of the response of two teachers fall into the excellent category. In addition, the inter-rater test also concludes that the perceptions between the two teachers are at the good category. For future research, we suggest to use the learning instruments developed in this research for further investigation such as comparing the conventional learning process with learning by using the Treffinger learning tool.

References

[1] Sudarmat R D *Pengaruh Penggunaan Pendekatan Pembelajaran Differentiated Instruction Untuk Meningkatkan Kemampuan Berpikir Kreatif Matematis Pada Siswa SMP* [Disertasi] Universitas Pendidikan Indonesia 2014

[2] Treffinger D J, Selby E C and Isaksen S G 2008 Understanding individual problem-solving style: A key to learning and applying creative problem solving *learning and individual differences* **18** 390

[3] Kyung-hwa, L 2005 The relationship between creative thinking ability and creative personality of preschoolers *International Education Journal* **6** 194

[4] Türkmen H and Sertkahya M 2015 Creative Thinking Skills Analyzes Of Vocational High School Students *Journal of Educational & Instructional Studies in the World* **5** 74

[5] Adnyani L P W, Kurniawan I and Pinahayu E A R 2018 Development of Creative-Thinking Instrument in Mathematics Problem Solving Based on Logical Mathematics Intelligence *Journal of Research and Advances in Mathematics Education* **3** 1

[6] Siswono T Y E 2007 Desain Tugas untuk Mengidentifikasi Kemampuan Berpikir Kreatif Siswa dalam Matematika *Tersedia: http://tatagyes. files. wordpress. com/2007/10/tatag_jurnal_unej. pdf.[7 januari]*

[7] Shoimin A 2014 *68 Model Pembelajaran Inovatif dalam Kurikulum 2013* (Yogyakarta: Ar-Ruzz Media)

[8] Treffinger D J, Isaksen S G, and Firestien R L 1982 Theorical Perspectives on Creative Learning and Its Facilitation: *An Overview The Journal of creative behavior* **17** 9

[9] Depdiknas 2007 *Peraturan Menteri Pendidikan Nasional RI Nomor 16 Tahun 2007 Tentang Standar Kualifikasi Akademik dan Kompetensi Guru* (Jakarta: Depdiknas)

[10] Hiebert J, Morris A K, Berk D, and Jansen A 2007 Preparing teachers to learn from teaching *Journal of teacher education* **58** 47

[11] Skemp R R 1971 *The Psychology of Learning Mathematics (*England: Penguin)

[12] Nieveen N 1999 Prototyping to Reach Product Quality Dalam Plomp T, Nieveen N, Gustafson K, Branch R M, and van den Akker J (eds) *Design Approaches and Toolsin Education and Training* (London: Kluwer Academic Publisher)

[13] Khabibah *Pengembangan Model pembelajaran Matematika dengan Soal Terbuka untuk Meningkatkan Kreativitas Siswa Sekolah Dasar* [Disertasi] Surabaya Universitas Negeri Surabaya 2006

[14] Fleiss J L 1975 Measuring Agreement between Two Judges on the Presence or Absence of a Trait *Biometrics* **31** 651

[15] Hutami T 2016 Efektivitas Pendekatan Metakognitif dalam Pembelajaran Matematika Terhadap Kemampuan Penalaran Siswa Kelas VIII MTs Negeri Babadan Baru *Jurnal Pendidikan Matematika* **5** 1

[16] Widoyoko S E P 2013) Optimalisasi Peran Guru dalam Evaluasi Program Pembelajaran *Jurnal Pendidikan* **22** 177

Developing worksheets through ISLE-based STEM approach and implementing them on senior high school students

E Rahmayani, I Irwandi and R Rajibussalim
Syiah Kuala University, Jl. T Nyak Arief, Darussalam, Banda Aceh 23111, Indonesia

E-mail: irwandi@unsyiah.ac.id

Abstract. The Indonesian 2013 curriculum demands that a learning process emphasize three domains, namely attitude, knowledge, and skills. One learning environment that supports the curriculum is the *Investigative Science Learning Environment* (ISLE), which is based on Science, Technology, Engineering, and Mathematics (STEM). The ISLE-based STEM approach is a learning approach that encourages students to be active, creative and innovative in conducting observations and experiments on a project. The ISLE-based STEM method can help students solve actual scientific problems through collaborative group activities. The development of worksheets lays an important role in ISLE-based STEM approach. The purpose of this study was to develop worksheets of the ISLE-based STEM by gathering feedback and learning outcomes from grade 10 high-school students in Banda Aceh. This research employed the ADDIE *(Analysis-Design-Develop-Implement-Evaluate)* development model. The result of the analysis of worksheet validation by experts reached 3.91, indicating a valid criterion for a worksheet to be used in a learning process. The percentage of students' responses to learning was 85%, indicating positive students' responses to learning using worksheets developed, while the test of learning outcomes after the implementation of ISLE-based STEM approach reached 86%. These results indicate that ISLE-based STEM worksheets can be used for grade 10 high-school students.

1. Introduction

Physics is a branch of science that is expected to contribute to the development of the quality of human resources because of its importance in the advancement of science and technology. Due to the great role of physics in responding to global challenges and supporting technological advances, it is necessary to improve the process of learning physics so as to improve the quality of education. In fact, the quality of physics learning in schools is currently still low [11]. This fact is evident from the substantial number of learning materials that are not suitable for the teaching time and the learning process in the classroom, thus causing ineffective learning activities. In addition, the process of physics learning in Indonesia is mostly still largely based on "traditional" education system in which educators and students regularly meet face to face at a certain time in the classroom [7].

The Indonesian 2013 curriculum (K-13) requires that a learning process, including physics learning, be focused on three domains, namely attitude, knowledge, and skills [6]. K-13 emphasizes the modern pedagogic dimension in learning and students are expected to be able to dig, process, and present the information through observation, inquiry, and experiment before analysis, reasoning, conclusion, and creation are performed. The development of science and technology in the world has resulted in a thorough change in the world of education. The change is to improve the quality of education. Today there are evolving thoughts about a collaborative discipline known as STEM education, which focuses on the educational process of solving problems in everyday life by integrating into science, technology, engineering and mathematics [9].

Students in Indonesia need a learning process that can help them develop their identity and ability as a scientist. One approach to learning that can support the 2013 curriculum is the ISLE-based STEM. ISLE(*Investigative Science Learning Environment*) is an environmental science investigation approach . According to Etkina, ISLE is an educational framework that can guide instructional design and student learning [4]. It engages students in processes that reflect scientific practice in order to help them learn physics.

Thus, the core of ISLE's approach is to link any learning subject or topic with real life. Drawing the connection can be done in many ways. The material that is being studied can be linked directly to factual conditions or accompanied by illustrations or examples, learning resources, media, etc., which are either directly or indirectly pursued related or there relationships with real-life experiences.

Thus, learning will be more interesting since students can directly feel the direct benefits of what is learned. Most students often complain that the system of teaching and learning in schools is still focused on solving the problems in the books. This concern was expressed by some high school students in Banda Aceh when the researchers made observations in their school. If this condition continues to happen, the reasoning ability of some Indonesian students especially Acehnese is going to decrease. Whereas in order to apply the acquired knowledge, one must have a basic scientific whole where one must be able to apply the knowledge they have in solving everyday problems. In addition, the students do not understand the basic concepts of the topics being studied.

According to Arsyad [1], the use of learning media in teaching and learning process can produce new desires and interests, generate motivation and learning stimuli, and even bring influence to the psychological development of students. Interesting teaching approach can encourage students to learn one of them by using learning media. One of the learning media is student worksheet. Trianto [12] stated that a student worksheet is a student guide used to conduct the investigation or problem-solving activities. It also guides students in developing their cognitive aspect through exercises and guides the development of all aspects of learning. Based on the results of observations and interviews conducted by researchers on high school teachers in Banda Aceh, as preliminary data, it was found that science laboratories are not working properly, physics teachers do the teaching and learning process using conventional approaches whereas physics material requires the use of learning media.

The researcher wanted to do a study with ISLE-*based* STEM approach and selected harmonic motion in a simple pendulum as the topic because the students are only taught the concept without any practicum. Furthermore, most students are not confident and skilled enough in determining the parameters of the quantities in simple harmonic motion. In traditional simple pendulum experiments in high school, students only use a stopwatch to determine the time of a number of swing periods and then calculate the average period, assuming that each period requires the same amount of time. As a result, students get bored and saturated [3].

It is different when ISLE-based STEM approach is used. Teachers collaborate ISLE learning with the factual concept that students encountered. This method of learning focuses on the student's activeness, emphasizes on performance assessment and team cooperation. STEM is integrated with a smartphone, computer/laptop censor accelerometer and tracker application. The combination of ISLE and STEM is expected to be an educational approach which can increase students' quality. Students will be very excited to see the result of the experiment which can be gained easily [5].
The purpose of this study is to find out (1) Validity of student worksheets (2) Students' response to worksheets, (3) and learning outcomes of the grade 10students of high school in Banda Aceh by using worksheets ISLE Based STEM for simple pendulum topics.

2. Method

This research was implemented in grade 10 of 4 senior high schools in the city of Banda Aceh; SMAN 1 Banda Aceh, SMAN 3 Banda Aceh, SMAN 11 Banda Aceh and SMA Lab School. It involved the total of 101 sample students who were selected by *purposive random sampling* technique; that is determining the samples with certain considerations [8]. The schools were selected for the research because (1) they are among favourite schools in Banda Aceh, (2) they have adequate infrastructure (3) unimplemented learning concepts (4) they had granted the permission to conduct the research. As a Research and Development, this research aimed at developing accountable educational products. Product development in this research used ADDIE (*Analysis-Design-Develop-Implement-Evaluate*) models [2].

The data used in this research was quantitative data which were converted into qualitative data through data analysis. The acquired quantitative data consisted of (1) data validation of ISLE-based STEM student worksheet developed based on suggestion, comment and feedback from validators and physics teachers, (2) students' responses (3) data of students' learning outcomes by using ISLE-based STEM worksheets on simple pendulum topics. Stages of this research were:

- Analysis
  Stage analysis is a process of defining what students are learning, i.e. doing *the needs assessment* (needs) and perform analysis tasks *(tasks analyze)* consists of an analysis of the characteristics of students, material analysis, and analysis of the curriculum.
- Design
  Activities undertaken at this stage is designing worksheets ISLE *Based* STEM. This study focuses only on a) preparation of the learning device plan, b) selection of worksheets format, and learning outcomes test. This initial design is based on the results of needs analysis, curriculum analysis, and material analysis.
- Development
  At this stage,worksheets that have been designed will be developed through expert validation and product revision I. The activities are undertaken at this stage are as follows:
  - Expert validation, This stage aims to determine one aspect of product development quality, the aspect of the validity. This is done by testing the validity of product design by experts and subject of physics subject, as well as getting suggestion and criticism from validator to developed product.
  - Product revision stage I, The validation data obtained is then analyzed and revised. The first stage of product revision is development based on expert validation.
    The validation results revealed that the previously invalidated worksheets were not interesting and did not use 4K camera technology, while the already validated and revised worksheets were better and feasible to use because there was an illustration of the story about the development of pendulum. At the next stage, the worksheets were then tested for students to see their responses to the development of worksheets.
    The following (Figure 1) is the revised worksheets section after validation.

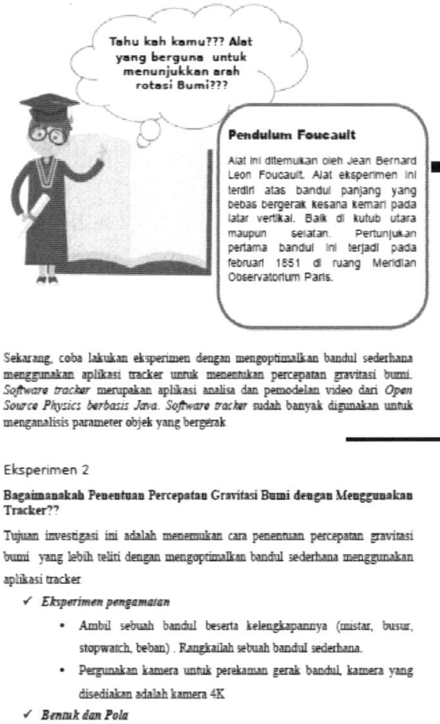
Tahu kah kamu??? Alat yang berguna untuk menunjukkan arah rotasi Bumi???

Pendulum Foucault

Alat ini ditemukan oleh Jean Bernard Leon Foucault. Alat eksperimen ini terdiri atas bandul panjang yang bebas bergerak kesana kemari pada latar vertikal. Baik di kutub utara maupun selatan. Pertunjukan pertama bandul ini terjadi pada februari 1851 di ruang Meridian Observatorium Paris.

Sekarang, coba lakukan eksperimen dengan mengoptimalkan bandul sederhana menggunakan aplikasi tracker untuk menentukan percepatan gravitasi bumi. *Software tracker* merupakan aplikasi analisa dan pemodelan video dari *Open Source Physics berbasis Java. Software tracker* sudah banyak digunakan untuk menganalisis parameter objek yang bergerak.

Eksperimen 2

Bagaimanakah Penentuan Percepatan Gravitasi Bumi dengan Menggunakan Tracker??

Tujuan investigasi ini adalah menemukan cara penentuan percepatan gravitasi bumi yang lebih teliti dengan mengoptimalkan bandul sederhana menggunakan aplikasi tracker.

- ✓ ***Eksperimen pengamatan***
  - Ambil sebuah bandul beserta kelengkapannya (mistar, busur, stopwatch, beban) . Rangkailah sebuah bandul sederhana.
  - Pergunakan kamera untuk perekaman gerak bandul, kamera yang disediakan adalah kamera 4K
- ✓ ***Bentuk dan Pola***

Pendulum Foucault

Jean Bernard Leon Foucault invented this tool. This experimental tool consists of a long pendulum that moves to and fro freely in a vertical background at both north and south poles. The first performance of this pendulum took place in February 1851 in the Meridian Room of the Paris Observatory.

Now, try to experiment by optimizing the simple pendulum using the tracker app to determine the acceleration of Earth gravity. The tracker software is a video analysis and modelling application from Java-based Open Source Physics. Software Tracker is widely used to analyse moving object parameters.

Experiment 2

How to Determine the Acceleration of Earth Gravity by Using Tracker?

The purpose of this investigation is to find a way to determine the acceleration of Earth gravity more carefully by optimizing the simple pendulum using a tracker application.

- ✓ *Observation experiment.*
- Take a pendulum along with its accessories (mister, bow, stopwatch, and load). Prepare a simple pendulum.
- Use a camera to record the movement of the pendulum. The camera provided is a 4K camera

Figure 1. Worksheet after validation and revision.

- Implementation
  At this stage, the revised worksheets were then put on trial. Worksheet trial was conducted on the students of SMAN 1, SMAN 3, SMAN 11 Banda Aceh, and SMA Laboratory Unsyiah.
- Evaluation
  The evaluation stage included worksheet revisions by experts and physics teachers. Physics expert reviewed the data and physics subject matter were analysed by considering the experts' recommendations for revision before stage II started to obtain applicable worksheets.

The instrument used for validity analysis is the validation sheet filled by a validator. Qualitative data of a learning product was obtained through the analysis validation sheets for media experts and content expert. The assessment criteria of the validation sheet for content expert consisted of (1) content, (2) presentation and (3) language, while the media experts' validation sheet included such formatting aspects as (1) clear numbering system, (2) appropriate layout arrangement, (3) readable font size, (4) clear instruction manual, (5) interesting cover design, (6) appropriate animation size, picture, shape and the writing in each page, (7) appropriate color in the media in each page, (8) animation and illustration quality, (9) practical learning media, and (10) manageable learning media. Data validation results by media experts and material experts were tabulated in tabular form, and then the average score from all validators was calculated. Then the average total score was converted into qualitative data based on conversion guidelines in the following table 1:

Table1. Conversion score average guidelines.

Score Interval	Criteria
$\bar{x} > M_i + 1{,}8\ Sb_i$	very good
$M_i + 0{,}6Sb_i < \bar{x} \leq Mi + 1{,}8\ Sbi$	good
$M_i - 0{,}6Sb_i < \bar{x} \leq Mi + 0{,}6\ Sbi$	enough
$M_i - 1{,}8Sb_i < \bar{x} \leq Mi - 0{,}6\ Sbi$	less
$\bar{x} \leq Mi - 1{,}8\ Sbi$	very less

Explanation:
Mi = ideal average = (ideal maximum score + ideal minimum score)
Sbi = standard deviation = (ideal maximum score - ideal minimum score)

The ideal maximum and minimum scores are 5 and 1 respectively. The classification ratings obtained guidance learning device is shown in the following table:

Table 2. Guidelines for classification of learning tool assessment.

Score interval	Criteria
$\bar{x} > 4{,}20$	very good
$3{,}4 < \bar{x} \leq 4{,}20$	good
$2{,}6 < \bar{x} \leq 3{,}4$	enough
$1{,}8 < \bar{x} \leq 2{,}6$	less
$\bar{x} \leq 1{,}8$	very less

Data on students' responses to learning with ISLE-based STEM learning media were analyzed descriptively in percentage. Data on the learning outcomes were analyzed descriptively using the terms of mastery of learning outcomes. Completeness of learning outcomes according to the Department of Education and Culture [15] is achieved if 85% of students achieve a defined score of 75.

Furthermore, the classical completeness is calculated based on the percentage formula, then the percentage of completeness is converted into qualitative data with scoring interval criteria. If the completeness is "very good" if it is > 80; mastery is 'good" if it is ≤ 80 or > 70, mastery is "enough" if it is ≤70 or > 60, the completeness is "inadequate" if it is ≤60 or > 50 and mastery is 'very poor" if it is ≤ 50.

3. Result and discussion

3.1. Results of needs analysis

The development of worksheets was started by the researchers through an observation. This observation began with observing the learning activities conducted in early October 2017 as well as interviewing teachers and students on learning issues they faced before performing needs analysis afterward. This early observation was conducted by the researchers themselves at SMAN 1 Banda

Aceh, SMAN 3 Banda Aceh, and SMAN 11 Banda Aceh. From the analysis stage, the following results were obtained:

- Constraints faced by both the teachers and the students, among others, are inadequate laboratory facilities and unavailable laboratory facilities for learning physics materials. Also, teachers rarely use self-designed worksheets for a specific purpose; as a result, students consider physics a difficult and unenjoyable subject.
- Based on the journals on social media, research has been conducted in Indonesia and especially in Aceh on the development of worksheets, but not using the ISLE-based STEM approach.

3.2. Expert validation result on the student worksheet

The information obtained was then used to design learning media in the form of worksheets (Draft 1) and instruments for feasibility assessment that were prepared at the design stage. Student worksheets that had been prepared at the design stage were consulted with the supervisor. From the consultation, some suggestions for improvements were obtained. The revised worksheets were then validated by the validators to evaluate their quality. The learning product validation was conducted by 4 validators, consisting of 2 expert lecturers, 1 teacher from SMAN 1 Banda Aceh, and 1 teacher from SMAN 3 Banda Aceh.

The purpose of this validation is to obtain input, evaluate the prepared learning products, to determine the validity of the contents and the constructs from the perspectives of physicists and educational practitioners, as well as to determine the feasibility of the learning media.

The results of the validation are tabulated as follows: The result of worksheet validation analysis showed that the worksheet material validity value of 3.91 was within valid criteria, and worksheet validation by media experts reached the value of 3.83, which is within valid criteria. This means that the worksheets are applicable in the learning process. The next stage was that the worksheets were tested on students to see their responses.

3.3. Students' response

The percentage of students' response to learning reached 85%. Based on the Likert scale with 5 criteria of responses (*strongly agree*, *agree*, *neutral*, *disagree*, and *strongly disagree*), this figure showed positive students response to learning using the developed learning media. This means that the learning media could be employed well [14].

3.4 Student learning outcomes

The test of this learning result consisted of five open-ended questions. The test results obtained by the students were briefly presented in the following table:

Table 3. Test of Learning Results.

Description	Test of Learning Outcomes (%)				Average Percentage
	SMAN 1	SMAN 3	SMAN 11	SMAN Lab School	
Completed	85 %	90%	82%	86 %	86 %
Unfinished	15%	10%	18%	14 %	14 %

Table 3 shows the data on the test of students learning outcomes after learning. 86% of students completed the test thoroughly based on the established thoroughness requirements. Thus, the results indicated that students had achieved learning mastery; and therefore the ISLE-based STEM worksheets developed were already good.

Based on the results above, it can be concluded that worksheets which were revised through ISLE-based STEM was applicable and was valid according to the experts and practitioners, so they could already be used in the second stage of evaluation. It can be said that using ISLE-based STEM worksheets, the material on Simple Pendulum can be used to deliver appropriate and creative learning process. According to the theory put forward by Bruner in Tung [16], a learning process will run well and creatively if a teacher gives the students opportunities to discover their own rules through concepts, theories, definitions, etc. ISLE-based STEM worksheets help students discover and learn physics about the simple pendulum. In addition to the valid use by students, ISLE-based STEM worksheets also guide the development of students' ideas through (1) observing phenomena and finding patterns, (2) developing explanations for patterns, (3) using this explanation to make predictions about experimental test results, (4) deciding whether the test results are consistent with the prediction, and (5) revising the explanation if necessary. This encourages students to learn in fun and effective ways since the worksheets enable students to construct their own knowledge of a simple pendulum. This is in line with Bruner's opinion in Suyono and Hariyanto [13] that teachers should guide their students so that they can build their own knowledge base rather than being taught through rote memorization.

Based on the results of field trials conducted by physics teachers in grade X SMAN 1 Banda Aceh, SMAN 3 Banda Aceh, SMAN 11 Banda Aceh and SMA Lab School Banda Aceh, teachers and students could use ISLE-based STEM worksheets well, as shown by 85% positive responses on average from students. This indicates that worksheets helped and facilitated students in understanding the material. Therefore, it can be said that the worksheets have functioned in accordance with the worksheets used in learning [1], 1) as a teaching material that can minimize the role of teachers and enable students, 2) as a teaching material that makes it easier for students to understand materials provided, and 3) as a teaching material that helps the implementation of learning in students.

ISLE-based STEM worksheets gave a big influence on students' learning outcomes. The test results were indicating that as many as 86% of 101 students were able to comprehend thoroughly simple pendulum material available in the worksheets. This is in line with the explanation put forward by Trianto [14], that worksheets contain a set of fundamental activities that must be performed by the students to maximize their understanding to develop basic capabilities in accordance with indicators of the achievement of learning outcomes that must be pursued.

4. Conclusion

Based on the results of this study, it is concluded that the ISLE-based STEM worksheets developed in this study can be used as one of the instruments of learning in learning physics. Based on its implementation, the use of ISLE-based STEM worksheets can also improve students learning outcomes. The assessment of students' skills after implementing the approach in the classroom indicates positive and promising results.

Acknowledgment

This study is funded entirely by NAS and USAID under the USAID Prime Award Number AID-OAA-A-11-00012, and any opinions, findings, conclusion, or recommendations stated in the article are from the author only and do not always reflect the view of USAID or NAS.

Reference

[1] Arsyad 2014 *Learning Media* (Jakarta: PT. Raja Grafindo Persada)
[2] Branch R M 2009 *Instructional Design: The ADDIE Approach* (New York: Springer Science & Business Media)
[3] Chao W 2015 *Pendulum Experiments with Three Modern Electronic Devices and a Modelling Tool* (Beijing: Being Normal University)
[4] Etkina E and Planinšič G 2014 *Light-Emitting Diodes: Exploration of Underlying Physics The Physics Teacher* **52** 212
[5] Etkina E and Heuvelen A V 2007 *Investigative Science Learning Environment: A Science Process Approach AAPT* pp 3-4

[6] Culture K P 2017 *Model Silabus Subject of Middle School Aliyah (SMA/MA) Subject of Physics* (Jakarta: Ministry of Education and Culture)

[7] Farhad S 2001 Distance Education An Introduction *Saba & Associates*

[8] Kothari C R 2004 *Research Methodology: Methods and Techniques* (New Delhi: New Age International (P) Limited Publisher)

[9] National Research Council 2014 *STEM Integration K-12 Education: Status, Prospect and an Agenda for Research* (Washington DC: The National Academies of Science)

[10] Prastowo A 2015 *Creative Guide Creating Innovative Instructional Materials* (Yogyakarta: Diva Press)

[11] Samudra G B, SuastraI W and Suma K 2014 Problems faced by high school students in Singaraja town in studying *e-Journal Physics Graduate Program of Ganesha University of Education*

[12] Scale T M 2013 *LIKERT* (*The Measurement Scale and the Number of Responses in Likert Scale*) **2** 127

[13] Suyono and Hariyanto 2011 *Learning and Learning: Theory and Concepts Basic* (Bandung: Teenagers Rosdakarya)

[14] Trianto 2012 *Designing Innovative-Progressive Learning Models: Concepts, Platforms, and Implementation of Education Unit Level Curriculum* (Jakarta: Kencana Prenada Media Group)

[15] Trianto 2010 *Integrated Learning Model* (Jakarta: Earth Literacy)

[16] Tung K Y 2015 *Learning and Development Learning* (Jakarta: Index of Law of the Republic of Indonesia Number 20 Year 2003 regarding National Education System)

IOP Conf. Series: Journal of Physics: Conf. Series **1088** (2018) 012092 doi:10.1088/1742-6596/1088/1/012092

Assessing conceptual and algorithmic understanding of students in senior high school

Nuzulia, M Hasan and A Ismayani
Syiah Kuala University, Jl. Teuku Nyak Arief Darussalam, Banda Aceh 23111, Indonesia

Email: muhammadhasan.kimia@unsyiah.ac.id

Abstract. Learning chemistry involves algorithmic and conceptual understanding, and students are required to master both of them. This study aims to obtain preliminary data about the algorithmic and conceptual abilities of students in solving chemistry problems. This research is part of development research with Plomp model that focuses on the preliminary investigation stage. The data were collected using test, questionnaire, and an interview involving 50 students of a high school in Banda Aceh. The data were analyzed to find the percentage of students' ability on algorithmic and conceptual understanding. The findings show that 13% of students achieved high algorithmic and high conceptual understanding, 6% of students achieved high algorithmic and low conceptual understanding, and 30% of students achieved low algorithmic understanding. Therefore, it is needed to develop learning materials appropriate to the conditions of students to improve students' algorithmic and conceptual understanding.

1. Introduction

To succeed academically, high school students must be able to solve both algorithmic and conceptual problems. Conceptual problems require students to understand the concepts of a problem, without requiring them to memorize. To solve conceptual questions, students need to have a strong conceptual understanding of the topics involved [1]. Conceptual problem solving requires a good understanding of the concepts of science and the application of conceptual knowledge. Although conceptual understanding is identified as the primary goal of teaching science, instruction in a chemistry course usually emphasizes complex processes for solving algorithmic problems [2]. Algorithmic problems require the application and manipulation of certain mathematical formulas. Science teachers often assume that students' performance in mathematical problem solving indicates the mastery of chemical concepts. Teachers should think further whether the students understand a lesson or not. Also, teachers are expected to know whether students use concepts in problem solving, relate the concepts to daily life, and use critical thinking [3]. On the other hand, many students solve chemical problems using algorithmic strategies but do not understand the chemical concepts behind their algorithmic manipulations. They also did not understand the algorithmic thoroughly; they only follow what their teachers did [4]. Thus, students have problems both the algorithmic and conceptual parts.

The essential indicators in conceptual understanding problems are an explanation of selected answers, predict events, explaining processes, and being able to connect with other topics. Specifically, there are six types of conceptual questions. The first one is the tiered multiple-choice question. It has several interrelated questions between questions; the first question gives a prompt, and the second question gives the reason for the first question. The second type is the question on particle levels that represents chemical situations at the atomic or particle level by using different sizes or colors for circles or spheres representing particles. The third type is the question involving the use of graphs, tables, and other data to forecast or describe phenomena that have occurred in experimental situations. The fourth is the demonstration question, in which students must answer questions after they observe demonstrations, videos or simulations. The fifth is the analogy question. The last one is a series of questions that require students to select an item that will complete the series [1].

On the other hand, the algorithmic indicator utilizes mathematical data and formulas for the solution of problems. The algorithmic troubleshooting strategy requires the application and manipulation of certain mathematical formulas. Many students still lack the conceptual understanding of chemistry. Both teachers and students play an essential role in developing problem-solving skills in chemistry [1].

The US Office of Education and the US National Joint Committee on Learning Disabilities suggests that learning disability is defined as a type of heterogeneous disorder, and contains diverse characteristics among individuals, and student learning outcomes [5]. The effects of learning difficulties include obstacles to listening, speaking, reading, writing, logical thinking, and mathematical ability. Learning difficulties are not always explicit. Therefore, to detect students' learning difficulties, intelligence checking can be performed. Moreover, teachers need to analyze students' skills on a concept before providing a basic concept and correct their misconceptions. Then, the teachers need to give an algorithmic problem and students should be able to develop it. So that, the teachers can understand the needs of their students [3].

The purpose of this study is to obtain preliminary data on students' conceptual and algorithmic understanding to be further used to determine appropriate efforts to assist students' learning. Based on the effort determined, it will be developed learning materials that can assist students in understanding the conceptual and algorithmic understanding. This research is expected to give a starting point for future researchers to conduct more research in this field.

2. Method

This study was part of broader research in the development of learning tools to foster students' algorithmic and conceptual abilities. The broader research followed stages of Plomp model, that is preliminary research, prototyping, and assessment phase [6]. However, this paper focuses only on the preliminary research stage that shows the importance of learning developed instruments.

This research was conducted in a senior high school in Banda Aceh, Indonesia. The instrument of this research is the test, questionnaire, and interview guide sheet. The test consists of problems that measure the understanding of concepts and algorithms. The participants involved 50 students and were categorized into four groups. The first group is the students who have a high algorithmic understanding with a score of 70-100% on algorithmic problems and who have a high conceptual understanding with a score of 50-78% on conceptual problems. The second group is the students who have a high algorithmic understanding with a score of 70-100% on algorithmic problems and who have a low conceptual understanding with a score of 0-33% for conceptual problems. The third group is the students who have a low algorithmic understanding with a score of 0-40% on the algorithmic problems and who have a high conceptual understanding with a score of 50-78% on conceptual problems. Last, the fourth group is the students who have a low algorithmic understanding with a score of 0-40% on the algorithmic problems and who have a low conceptual understanding with a score of 0-33% for conceptual problems [7].

3. Results and discussion

After the students took the test, their understanding of conceptual and algorithmic was analyzed. The findings show that 13% of the students achieved high algorithmic and high conceptual understanding, 6% of the students reached high algorithmic and low conceptual understanding, and 30% of student achieved low algorithmic understanding. The percentage of the finding is shown in Figure 1. Based on the interview result, the first and second group of students preferred algorithmic rather than conceptual. However, the second group of students with higher algorithmic understanding was unable to gain a high conceptual understanding [8].

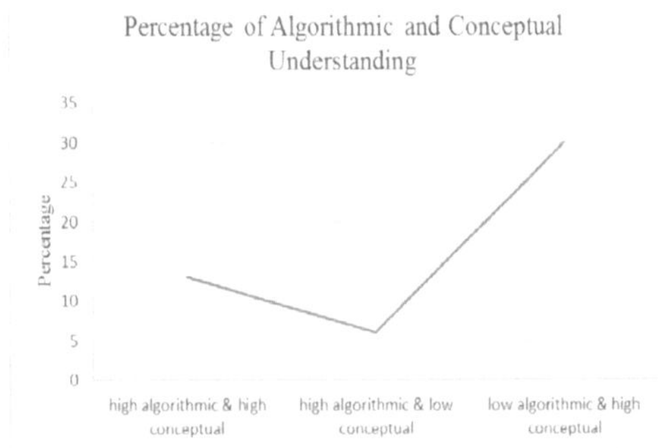

Figure 1. Percentage of conceptual and algorithmic understanding of the students.

Students are considered having high algorithmic or conceptual understanding if they have a score of 40-100 and having low algorithmic or conceptual understanding if they have a score of 0-39 [1]. The students who achieved high algorithmic understanding are not guaranteed success in answering conceptual problems [3]. Moreover, the third group of students preferred conceptual understanding. Nevertheless, the third group of students continued to state that the algorithmic problems were easier than conceptual problems because, sometimes, conceptual problems are more elusive than algorithmic problems. In some cases, there were also some students who had high understanding in concept but were not able to use them in algorithmic problems [9].

One of the algorithmic questions used in this study is the combustion enthalpy of fuel of 5.460 kJ/mol and 5.7 grams of fuel (Mr = 144) is burned then the students need to find the resulting combustion enthalpy. This kind of problem only requires remembering formulas and students should be able to solve them if they know the formula. On the other hand, the conceptual problem requires a deeper understanding of concepts. Students are not able to answer if they only memorize the formula without understanding the concept. Example of the conceptual problems is the problem about the magnitude of the enthalpy price as shown in Figure 2 below.

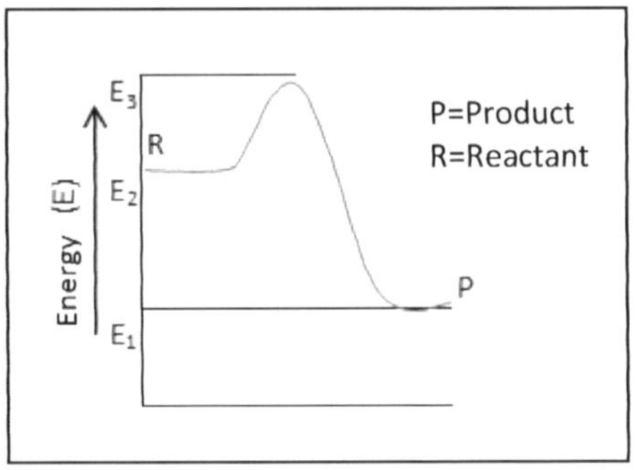

Figure 2. Example of the conceptual problems.

To be able to answer the conceptual problems as in Figure 2, students should understand the concept and be able to read the charts, because the ability to read graphics is part of the understanding of concepts. Many students know that enthalpy change is a product minus reactants, but if they do not understand the graph, then they can not answer the problem. Such a conceptual problem requires a comprehensive understanding. Conceptual problems can be expressed in different forms and types such as pictorial and verbal forms. The problems in the pictorial form are represented using images, graphics while the problems in the form of verbal are expressed without pictures in their descriptions [10].

Students' understanding of thermochemistry might not be improved through the teacher-centered learning. Lack of understanding of students might be caused by the teacher's teaching strategy that was still not appropriate to the students. This is supported by Bowen and Bunce stating that many factors hinder students studying chemistry namely, adequate skills such as algorithmic, hierarchical structure of concepts, textbooks, and learning methods [9]. This will be an indication for educators to seek other alternative teaching and learning approaches to develop students' conceptual understanding of chemistry [1]. This thermochemical topic is quite difficult for students. Based on the data obtained in the last two years, daily test value on the thermochemical topic did not meet the established standards of 75 out of 80.

Students had also been distributed a questionnaire to determine the needs of students on this material. The questionnaire used the Likert scale. The results of the questionnaire indicate a score of 4549. The score in the range 3864 to 5152 is categorized closer to agree that means teachers should balance the delivery between algorithmic and conceptual understanding because both are important. Students should not be focused on the understanding of chemistry only but must be both algorithmic and conceptual to understand the chemistry thoroughly. In thermochemical learning, some teachers still deliver the material concentrate only on understanding chemistry. A balance between conceptual and algorithmic understanding is essential in learning chemistry. One of the topics that require the two understandings is thermochemistry [11].

Another study conducted on students at one of the schools in Ankara, Turkey on the topic of atomic structures and periodical tables, the concept of mole, gas laws and solutions have the similar findings. The study found that students' algorithmic and conceptual abilities were considered high. However, these scores could not be entirely used as a guideline because the participants involved in the study were eleventh grade which had been prepared to enter the university [2]. Another study in Taiwan found that students had higher algorithmic than conceptual abilities [12]. Similarly, 36.8% of high school students in Greece were only able to answer algorithmic problems. A few students were able to answer both with the percentage of 6.4%, and 3.2% of students were able to solve only conceptual problems [6].

Unlike the three studies above, a study in Turkey looked for ways to improve algorithmic skills. The study was conducted on the pre-service teachers in Turkey, and the findings showed that the student-centered learning like Problem-Based Learning (PBL) could improve the algorithmic and conceptual abilities of students in the gas concept topic [13]. The same finding occurred in Malaysia; there were 96% of pre-service teachers who can solve algorithmic questions, and 54% of pre-service teachers who can answer the conceptual question using problem solving approach [1].

The results of this study indicate that students' conceptual understanding was still low. Therefore, it is needed to consider how to develop students' conceptual understanding. There are three important points that should be considered in order to improve the conceptual understanding of students: the emphasis on macroscopic, microscopic, and symbols. The macroscopic understanding is something that can be seen and felt. Microscopic understanding is something at the level of atoms, molecules, and kinetic parts. Symbolic understanding is a representation of chemical symbols, stoichiometry, and equations [14]. Furthermore, it should also be noted that learning styles and teaching strategies contribute to the improvement of students' algorithmic and conceptual abilities. Based on the observations on the teaching strategy, when teachers used teacher-centered learning and provided explanations from the beginning to the end of learning, students' understanding would not have a significant improvement. Also, teachers should connect the topics with the students' experience so that the students' learning in chemistry is more meaningful. Find real examples related to students' everyday life on chemistry topics is needed, especially on the topics that involve the microscopic and symbolic levels.

4. Conclusion

The findings show that 13% of students achieved high algorithmic and high conceptual understanding, 6% of students achieved high algorithmic and low conceptual understanding, and 30% of students achieved low algorithmic understanding. None of the students has a low algorithmic understanding and a low conceptual understanding. Based on the findings, it is necessary to do research on learning strategy and media that can enhance algorithmic and conceptual understanding of students in thermochemistry. The implication of this research is to develop a lesson with the right approach to improve the students' algorithmic and conceptual abilities.

Acknowledgment

The authors would like to thank Syiah Kuala University and the Youth and Sports Office in Banda Aceh for their support in making this study possible.

References

[1] Surif J, Ibrahim N H and Dalim S F 2014 Problem Solving: Algorithms and Conceptual and Open-Ended Problems in Chemistry *Procedia - Social and Behavioral Sciences* **116** 4955
[2] Yilmaz A, Tuncer G and Alp E 2007 An Old Subject with Recent Evidence from Turkey: Students' Performance on Algorithmic and Conceptual Question of Chemistry *J. World Appl. Sci.* **2** 420
[3] Gultepe N, Yalcin C A and Kilic Z 2013 Exploring Effects of High School Students' Mathematical Processing Skills and Conceptual Understanding of Chemical Concepts on Algorithmic Problem Solving *Australian Journal of Teacher Education* **38** 106
[4] Cracolice M S, Deming J C and Ehlert B 2008 Concept Learning Versus Problem Solving: A Cognitive Difference *Journal of Chemical Education* **85** 873
[5] Widayanti C G, Diana R and Siswati 2012 Profil Inteligensi Pada Siswa dengan Kesulitan Belajar di SD Negeri Gisikdrono Semarang *Jurnal Psikologi Undip.* **11** 1
[6] Plomp T 2010 *Educational Design Research: an Introduction* (Netherlands: Institute for Curriculum Development)
[7] Papaphotis G and Tsaparlis G 2008 Conceptual Versus Algorithmic Learning in High School Chemistry: The Case of Basic Quantum Chemical Concept *Chemical Education Research and Practice* **9** 323
[8] Usman H and Purnomo S A 2012 *Pengantar Statistika* (Jakarta: PT Bumi Aksara)
[9] Chiu M 2001 Algorithmic Problem Solving and Conceptual Understanding of Chemistry by Student at A Local High School in Taiwan *Proc. Natl. Sci. Counc.* **11** 20
[10] Bowen C W and Bunce D M 1997 Testing for Conceptual Understanding in General Chemistry *The Chemical Educator* **2** 1
[11] Haláková Z and Proksa M 2007 Two Kinds of Conceptual Problems in Chemistry Teaching *Journal of Chemical Education*, **84** 172
[12] Johnstone A H 2006 Chemical Education Research in Glasgow in Perspective *Review of research*
[13] Chiu M H 2001 Algorithmic Problem Solving and Conceptual Understanding of Chemistry by Students at a Local High School in Taiwan *Proc. Natl. Sci. Counc.* **11** 20
[14] Bilgin I, Senocak E and Sozbilir M 2009 The Effect of Problem-Based Learning Instruction on University Students' Performance of Conceptual and Quantitative Problems in Gas Concepts *Eurasia Journal of Mathematics, Science & Technology Education* **5** 153
[15] Johnstone A H 1991 Why is science difficult to learn? Things are seldom what they seem *Journal of Computer Assisted Instruction* **7** 75

The development of mathematics higher order thinking skills instrument for grade VIII junior high school

Y Yunita, W Wahidin and A Tsurayya
Muhammadiyah Prof. DR. HAMKA University, Jakarta 13830, Indonesia.

E-mail: yunita1996nita@gmail.com

Abstract. 21st-century learning activities require higher order thinking skills including critical and creative thinking. High-level mathematical thinking skills are tools to train and improve higher order thinking skills in mathematics. This study aimed to produce products in the form of a high level mathematical thinking skills instrument of grade VIII based on Bloom's Taxonomy containing C4 (Analyze), C5 (Evaluate), and C6 (Creating) questions. Indicators of higher order thinking include analyzing, organizing, differentiating, and linking. Evaluating is judging, checking and criticizing. While creating is generalizing, designing, and producing. This research was conducted at three junior high schools in Jakarta that consisted of 41, 36 and 36 students respectively. The sample of this research was grade VIII students. The study was conducted during the even semester in 2017. Data collection method used higher order thinking skill test of the eight grade students then analyzed data with the calculation of validity, reliability, difficulty level, and discrimination power. The result showed that the instrument was valid, reliable, the difficulty level was moderate, and the discrimination power was moderate so the problems in this instrument could be used by teachers to measure students' high order thinking skills.

1. Introduction

An international study measuring the student's performance in mathematics and natural science, Trend in Mathematics and Science Study (TIMSS), showed that in 2015, Indonesia ranked 45 out of 50 countries. Based on the data of mathematics content and cognitive domain, the percentage of Indonesian participants with the correct answer in a matter of understanding was higher than that of in the matter of application and reasoning. Moreover, the results of Programme Internationale for Student Assessment (PISA) test showed that in 2015, Indonesian students ranked 62 out of 70 countries in mathematics with an average score of 386. Indonesia's score was 386 in mathematics. This value is below the international average, 493. This shows that Indonesian students' higher order thinking skills in mathematics is still below expectation. The curriculum used in Indonesian education is the Curriculum 2013 Revision 2017 (K13 2017). The K13 2017 intends to promote higher order thinking skill (HOTS), literacy and character education [1]. The aim is to improve students' problem-solving skills so that they are able to overcome difficulties in their learning activities and everyday life. Qualified teachers are the key to achieving this goal [2].

Higher order thinking process is a mental activity that involves multifaceted, creative, and unconstrained thinking that leads to useful solution [3–6]. Learners generally have a variety of abilities. The ability to think creatively and critically are both the ability to create a diverse way of completion and produce the right answer to a difficult problem. These skills result in valid explanation, decision, performance, and product within the context of existing knowledge and experience and they promote continued growth in these and other intellectual skills [7].

The studies on improving critical thinking skills have used a variety of strategies and learning models such as problem-based approach [8], the application of writing strategy [9], collaborative approach work [10], and guided inquiry learning model [11]. Thinking skill is divided into two categories: higher order thinking skill (HOTS) and lower order thinking skill (LOTS). LOTS deals with regular implementation of mathematics formula in solving equations, while HOTS entails strategy and manipulation [12-13]. It is crucial that learners be introduced to problems that require analyzing, interpreting and manipulating of information and producing a correct solution. Bloom states that education should focus on mastery subject and the achievement of higher order thinking as oppose to the use of utilitarian that simply means learning as a means of transferring words [14]. The instrument in this study was formulated based on the Bloom's taxonomy on the ability of higher order thinking skills that requires a high level problem solving skill.

2. Method

This study used research and development method to produce and develop higher order thinking skills instrument for the students. This study was conducted using formative research proposed by Tessmer [15] which was modified as shown in the flowchart.

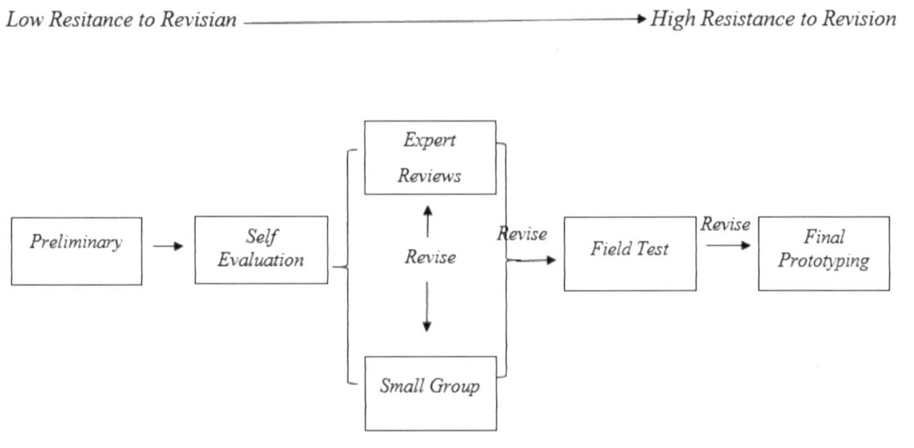

Figure 1. Flowchart of formative research design.

Figure 1 shows that the research method consisted of several stages: (1) preliminary or preparatory phase, (2) self-evaluation including analysis and design stage, (3) expert reviews or properness test of the product, followed by small group or small-scale trials, (4) field test or large-scale trial, and (6) final prototyping instrument. The research subject from small groups in grades VIII was intended to provide advice on higher order thinking skills instrument. The research received the material about mathematics in grade VIII for the field test.

The purpose of this study was to produce and develop instrument, to determine the properness and effectiveness of the product, and also to determine the profile of higher order thinking skills of students on the topics of number patterns, Cartesian fields, relation and function, straight line equations, linear equations with two variables, Pythagoras theorem, circles, wake-up flat space, statistics, and probabilities. Properness test was done by giving product properness questionnaire to the experts and students. Product effectiveness test was done by calculating the validity, reliability, difficulty level, and the discrimination power. The profile of higher order thinking skills was determined by analysing the students' answers to the problems. Data from field tests were used to calculate the feasibility of the instrument.

3. Result and discussion

The results of this study included the design, properness, and effectiveness of instrument, and a profile of the students' higher order thinking skills. The developed higher order thinking skills instrument consisted of 20 problems description. These problems represented the indicators of the higher order thinking skill. The indicators of the higher order thinking skill were as follows: first, student could recognize and distinguish between the causes and the result of a complicated scenario; the student could analyse incoming information and divide it into smaller sections to find patterns and relationships; and the student could identify and formulate questions. Second, the student could describe the idea and the methodology of an evaluation by using match and standard criteria to ensure the effectiveness; the student could hypothesize, criticize, and do the test; and the student could accept or reject a statement based on predetermined criteria. Third, in creation, the student could make a generalization of ideas; the student could determine a way to solve the problem, and the student could organize the parts of something into a new structure that did not exist before [16].

Higher order thinking skills instrument was given to experts and students to measure the properness. The properness questionnaire was prepared with some aspects of the assessment criteria, i.e. the aspect of content, construct, language, time allocation, and instructions. The properness test of higher order thinking skills instrument of students was done by experts in one stage. It was conducted before the higher order thinking skills instrument was tested in a small-scale trial. The result about the properness of higher order thinking skills instrument is shown in table 1.

Table 1. The result of properness higher order thinking skills instrument.

No	Aspects Observed	Evaluator				l_i	V_a
		Validator 1	Validator 2	Validator 3	Validator 4		
1	Content	3	4	4	5	4	3,52
2	Construct	2.78	3.4	3,95	4,3	3,6	
3	Language	3	4	3,3	4,7	3,75	
4	Time Allocation	2	3	3	3	2,75	
5	Instruction	3	3	3	5	3,5	
	Criteria						Quite Valid

The profile of students' higher order thinking skills instrument can be seen from the answers regarding the students' ability to analyze, evaluate, and create. The result for the field test of ability higher order thinking skills instrument is presented in table 2.

Table 2. Profile of higher order thinking skills.

Interval values	Frequency	Percentage (%)	Category
80 < Value ≤ 100	8	7,08	Very good
60 < Value ≤ 80	27	23,89	Good
40 < Value ≤ 60	37	32,74	Moderate
20 < Value ≤ 40	32	28,32	Low
0 < Value ≤ 20	9	7,96	Very Low
Amount of subject	113	100	
The average value	50		Moderate

Table 2 shows that the average value obtained by all of the students was 50 which was in the moderate category, so it could be concluded that in general grade VIII students had moderate ability in solving problems and understanding the problems that were not routine or higher order thinking skills. The score of the students in doing questions of higher order thinking skills is shown in figure 2. Figure 2 shows that the students' scores were normally distributed which meant that students were able to understand the problems that were not routine or higher order thinking skills.

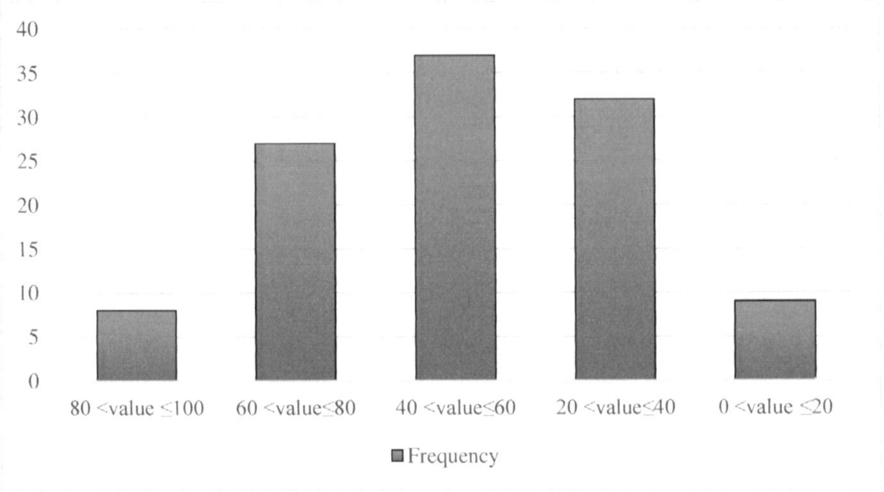

Figure 2. Scores of students in doing questions.

The result of validity test of higher order thinking skills was valid (r_{XY} > r_{table}) as shown in table 3.

Table 3. Validity test of higher order thinking skills.

Item	r_{XY}	r_{table}	Conclusion	Category
1	0, 5193	0, 1555	Valid	Moderate
2	0, 5243			Moderate
3	0, 5626			Moderate
4	0, 4117			Moderate
5	0, 5309			Moderate
6	0, 5841			Moderate
7	0, 6835			Moderate
8	0, 6188			Moderate
9	0, 6043			Moderate
10	0, 6620			Moderate
11	0, 7577			Good
12	0, 7957			Good
13	0, 7314			Good
14	0, 7583			Good
15	0, 7036			Good
16	0, 6476			Moderate
17	0, 6185			Moderate
18	0, 6861			Moderate
19	0, 7161			Moderate
20	0, 5428			Moderate

As shown in table 3, as many as 15 items fell into the moderate category and 5 items fell into good category. As a whole, the instrument could be verified valid.

The result of the reliability test of higher order thinking skills instrument was reliable. The r value was 0.9219 and it fell into very good category.

The result of the difficulty level of the higher order thinking skills instrument showed that one item was hard, seventeen items were moderate, and two items were easy. This is shown in table 4.

Table 4. The difficulty level of the higher order thinking skills instrument.

Item	Difficulty Level	Category
1	0, 3938	Moderate
2	0, 6615	Moderate
3	0, 3368	Moderate
4	0, 5170	Moderate
5	0, 4963	Moderate
6	0, 4904	Moderate
7	0, 3931	Moderate
8	0, 2904	Difficult
9	0, 7887	Easy
10	0, 7777	Easy
11	0, 3374	Moderate
12	0, 5736	Moderate
13	0, 6134	Moderate
14	0, 4729	Moderate
15	0, 4808	Moderate
16	0, 5863	Moderate
17	0, 4882	Moderate
18	0, 4226	Moderate
19	0, 4652	Moderate
20	0, 3285	Moderate

The result of the discrimination power of the higher order thinking skills instrument is shown in table 5.

Table 5. Discrimination power test of higher order thinking skills.

Item	Discrimination Power	Category
1	0, 3226	Moderate
2	0, 4570	Good
3	0, 3629	Moderate
4	0, 2769	Moderate
5	0, 4382	Good
6	0, 5161	Good
7	0, 5672	Good
8	0, 4879	Good
9	0, 3871	Moderate
10	0, 4778	Good
11	0, 6734	Good
12	0, 6754	Good
13	0, 5847	Good
14	0, 5746	Good
15	0, 5833	Good
16	0, 5706	Good
17	0, 4167	Good
18	0, 4879	Good
19	0, 6653	Good
20	0, 3165	Moderate

Table 5 shows that there were five moderate items and fifteen good items. Therefore, it could be concluded that the instrument satisfied the discrimination power criteria.

4. Conclusion

The higher order thinking skills instrument used to measure grade VII junior high school students was moderate and effective. It was a valid, reliable, with moderate level of difficulty and moderate discrimination power. The students were able to understand the problem that was not routine or higher order thinking skills which included analysis, evaluation, and creation. This instrument can be used for junior high school to practice higher order thinking skills and provide references for teacher on the matter of promoting higher order thinking skills. These skills are important for students to compete in the national and international world, especially in mathematics.

References

[1] Latif I A, Saputro D R S and Riyadi 2017 *J. Appl. Phys.* **983** 012129
[2] Collins R 2014 Skill for the 21st century: Teaching higher-order thinking *an Electronic Journal for Leaders in Education* **12**
[3] Resnick L 1987 *Educational and Learning to Think* (Washington D.C.: National Academy Press)
[4] Miri B, David B C and Uri Z 2007 *Research in Science Education* **37** 353
[5] Zohar A 2004 *Higher Order Thinking in Science Classrooms: Students' Learning and Teachers' Professional Development* (Dordrecht: Springer Science+Business Media)
[6] Patton A and Robin J 2012 *Work that Matters: The Teacher's Guide to Project Based Learning* (United Kingdom: Paul Hamlyn Foundation)
[7] Kusuma L S R, Lusia R and Wiryanto 2018 *IOP Conf. Ser. Mater. Sci. Eng.* **336** 012045
[8] Alismail H A and McGuire P 2015 21st Century standards and curriculum: Current research and practice *Journal of Education and Practice* **6** 150
[9] Jackson D 2014 Business graduate performance in oral communication skills and strategies for improvement *The Int. J. of Management Education* **12** 22
[10] Boyles T 2012 21st Century knowledge, skills, and abilities and entrepreneurial competencies: A model for undergraduate entrepreneurship education *Journal of Entrepreneurship Education* **15** 41
[11] Grifin P, Care E and McGraw B 2012 *The Changing Role of Education and Schools: Assessment and Teaching of 21st Century Skills* eds P Grifin, B McGraw and E Care (New York: Springer) chapter 1 pp 1-15
[12] Saido G A M, Siraj S, Nordin A B and Al-Amedy O S 2015 *Malaysian Online Journal of Educational Management* **3** 16
[13] Fisher R 2010 Thinking skill *Learning to Teach in the Primary School (2nd ed)* eds J Arthur and T Cremin (London: Routledge)
[14] Suyono and Hariyanto 2011 *Belajar dan Pembelajaran, Teori dan Konsep Dasar* (Bandung: Rosdakarya)
[15] Tessmer M 1993 *Planning and Conducting: Formative Evaluations* (London: Kogan Page)
[16] Lewy, Zulkardi and Aisyah N 2009 *Jurnal Pendidikan Matematika* **3** 14

IOP Conf. Series: Journal of Physics: Conf. Series **1088** (2018) 012094 doi:10.1088/1742-6596/1088/1/012094

The reflective abstraction profile of junior high school students in solving mathematical problems based on cognitive style of field independent and field dependent

B Panjaitan
Universitas HKBP Nommensen, Jl. Sutomo No. 4A, Medan, Indonesia

E-mail: panjaitan_binur@yahoo.com

Abstract. In solving a problem, students have different characteristics. This research describes students' abstraction profile in solving mathematical problems based on cognitive style. The qualitative method was applied to analyze data taken from one student who has field-independent's cognitive style and one student who has field-dependent's cognitive style. The research finding shows that students' abstraction profile of field-independent's cognitive style is different from students of field-dependent's cognitive style in solving a mathematical problem. The implication is that the teacher should know and understand each student's cognitive style to apply appropriate learning approach.

1. Introduction

Abstraction is the basic process in mathematics. Mathematical thinking is a mental activity that in the process always uses abstraction or/and generalization. Mathematical thinking is defined as "a mental activity that involves abstraction and generalization of mathematical ideas" [1]. This statement means that abstraction is always used in each of mathematical thinking.

Abstraction is defined as "an organization activity of mathematical vertical concept which is constructed previously and becomes a new mathematical structure" [2]. The new mathematical objects are constructed through such formatting relation until finding generalization as a new proof or strategy for solving problems.

Abstraction theory is divided into three parts. Firstly, empirical abstraction that focuses on the way children construct the meaning of object characteristics (focusing on objects and their properties). Secondly, pseudo-empirical abstraction that focuses on the way children construct the meaning of characteristics that involves an object (focusing on action on objects and the properties). Thirdly, reflective abstraction that focuses on the idea of action and operation (focusing on mental objects) [3]. The result of reflective abstraction is the scheme of knowledge for each phase of development and reflective abstraction to extract the scheme from the related form of action.

Reflective abstraction refers to a subject's ability to project and reorganize structure created from activities and the subject's interpretation of the new situation. In line with this, reflective abstraction refers to the subject's ability to project on a new level and organize a structure created from the subject's activities and interpretation" [4].

Levels of reflective abstraction activities were defined as follows: the first level is recognition; the second level is representation; the third level is structural abstraction; and the fourth level or the highest level is structural awareness [4]. The researcher uses the level of reflective abstraction activity that is proposed [4] to describe the abstraction process of junior high school students in solving a mathematical problem.

At the recognition level, the problem solver encounters a new situation and recalls or identifies activity from previous situations as being appropriate. The ability to identify mathematical structure is when one realizes the structure existed and has the ability to solve the problem faced. To solve the problem, they have to recall previous knowledge that has been gained from previous activity and use it for further activity.

The second level is representation. The problem solver utilizes a diagram in resolving a problematic situation to aid reflection. The problem solver is operating at this level if more control over the solution activity is demonstrated or, more precisely, if the solver re-presents this solution activity. This reflective level requires the individual to demonstrate a certain degree of flexibility and control over prior activity in the sense that the activity could mentally be "run through".

At the third level, structural abstraction, students are able to project and reorganize mathematical structure has been created from previous activity and student's interpretation into a new situation. The existing mathematical structure is projected and reorganized so students could gain new knowledge from it.

The fourth level or the highest one is structural awareness. It refers to student's metacognition about activity and organization of its cognitive structure. The problem structure that students comprehend for solving the problem becomes the object in the reflection. Students are able to think such structure as the objects and able to decide without trying or using physical forming or mentality present completion method. When problem takers gain the higher level of reflective abstraction, their thought becomes more flexible.

In the solving problem, the students have different characteristics. Psychology with its branches has indicated many various variables which demonstrated individual differences as well as encourages the learning process such as intelligence, talent, cognitive style, thinking style, adoption style, persistent and commencement ability.

Individual differences in this research are the differences based on the cognitive style. Identifying a student's cognitive style, the educator is helped to understand how somebody organizes and presents information [5]. Psychological dimensions that represent the consistencies of individual's manner in acquiring and processing information [6].

The student who has field-independent's cognitive style is able to abstract the context elements or its background. He/she tends to solve the problem with the analytical approach. Meanwhile, students of field-dependent's cognitive style use problem solving approach globally, he/she thinks all the pictures in the contexts given and has difficulties to separate one item from whole context [7].

Problem solving is defined as the process done by the individual in combining prior knowledge to face a new situation. It means that problem solving is the process that is done by somebody in combining the prior knowledge to solve the problem that has not yet known. The ways of solving problem consisting four steps, they are; (1) understand the problem, (2) devise a plan, (3) carry out the plan, and (4) recheck the result [8].

From these explanations, it can be concluded that solving problems of mathematics is the process of finding mathematics problem which included understanding the problem, planning of solving the problem (devise a plan), solving the problem (carry out the plan), and recheck the result.

Based on the previous background, this research intends to answer the research question "how students' abstraction profile who has field-independent's cognitive style and who has field-dependent's cognitive style in solving mathematical problem?"

2. Method

It is explorative research uses a qualitative approach. The participants were the grade 8 students in one of Indonesian Province. The selective subject is one student who has field-independent's (FI) cognitive style and one student who has field-dependent's (FD) cognitive style. FI and FD subject selection is done in the following; The Group Embedded Figures Test (GEFT) is given to the students, set the students group of FI and FD according to the scores obtained by the students within the time limit available. Students with scores 0-9 are categorized as groups FD and students who score 10-18 are categorized as FI student groups.

Out of 40 students who attended the GEFT test, there were 26 (65%) students who had afield independent cognitive style and 14 people (35%) who had field dependent cognitive style. From the group of field independent cognitive style, one student is selected, who has the highest GEFT test score, and from the group of cognitive style field dependent, one student is selected who has the lowest GEFT test score.

The instruments in this research are GEFT and problem solving task. The problem is "a fond fish in rectangle form has 18 meters in length and 8 meters in width. Around of beyond of it will be built a path of 1-meter scale, which is made of gravels. For each meter quadrate is needed 2 kilograms of gravels. If the price of the gravels is 4.500,00 rupiahs per kilogram, so how much is the cost of buying gravels to cover the scale?"

The recorded data through a deep interview is analyzed by a qualitative method that involved data reduction, displaying data, and taking conclusion. Data reduction phase is the activity of selection, the concentration of attention, simplification, abstraction, and transformation of raw data. The code is then generated so it is known to come from which source. The valid data is analyzed. In data presentation stage, the researcher collected data and organized it; so, it is easier to draw the conclusion from the data. The conclusion step, the researcher draws the conclusions from the data collected and verified the conclusion. Verification is done by reviewing field notes, discussing with colleagues or with teachers from research subjects. This analysis is related to student's appearance of abstraction process in solving a mathematical problem.

3. Results and discussion

This following is found the analysis of the abstraction process of two students in solving mathematical problems.

3.1. Analysis of student' abstraction profile of field independent (subject FI)

To know the abstraction process appear in solving the problem, so the interview is done to Subject 1 (FI). Subject FI has read the item/problem; he gave attention to the problem, so he thought quickly or done abstraction to understand it. After asking of what has been known and asked, he was able to explain rightly in the form of words. It indicates a reflective abstraction level of recognition and representation.

Subject FI state that the information of the item that is the path will be built in around beyond the pond fish is 1 meter. It means that there is a new rectangle whose 2 meters in length and 2 meters in width so will make the planning of new situation (rectangle of 20 meters length and 10 meters width) rather than previous situation (remembering the rectangle of 18 meters length and 10 meters width), recognizing previous activity. The planning has been done before in the form of a rectangle to help reflection. It shows the reflective abstraction level of recognition and representation.

Then the researcher wants to know whether still there is another idea to answer the question of the problem. Subject S1 has gained the insight to measure the width of the path that will be built by gravels. It the whole width or the path width increased by fishpond then decrease by fishpond width.

Subject FI believes that his steps are right, so he completes the idea directly by reflecting previous activity for the new situation; it is to count the path width by decreasing the whole width with the width of pond fish, which is drawn by a rectangle. It shows the abstraction level of representation and structural abstraction.

The ways to recheck the obtained result, subject FI checks the obtained result with the steps done previously by adding the width of each rectangle (there are four rectangles) at around of the pond fish. In this case, the subject FI develops a new strategy or method to count the width of the path. In his work exists crossing which indicates that the subject FI makes the wrongness. Subject FI immediately corrects his work as much as required. Subject FI can give the argument or the true reason for the decisions made, and to summarize his activities while doing the task. It shows the reflective abstraction level of structural abstraction and structural awareness.

3.2. Analysis student's abstraction profile of field dependent (subject FD)

After giving the test sheet to subject FD, he reads the test instruction and writes down his name on the answer sheet provided. To find out the increasing of the abstraction process in mathematical problems, so the researcher (R) did interview to the subject of FD (S2).

Subject FD has read the test/problem. He has a lack of attention to what has been read before, so he thought in a long time to understand it. It can be seen when the subject FD was asked what is known in the test; he is silent. The researcher attempts to understand what has been being thought by the subject by review the same question. Then he answered it incompletely. Meanwhile, in the end, his answer is correct. Besides that, he can describe what is known in the test in the form of words. It shows the reflective abstraction level of recognition and representation.

Subject FD state the information related to the test is the pond fish circumference. He has difficulty to make the plan; he does not understand the previous information because he does not know to differentiate the meaning of "around" and "circumference." The researcher hopes the subject FD can distinguish the around of pond fish and its circumference so the subject can understand the information related to the test.

From the question addressed by the researcher, subject FD is able to count the width of the path by applying one way. Then, the researcher wants to know whether any other ways or not.

Subjects FD state his idea doubtfully. But, through the questions given by the researcher, he can state his idea. Finally, he is able to remember the way to measure the width of the path in another method. He states that the left and right side of each border of the pond fish is 10 meters quadrate. It is said because has a glance the whole circumference of the pond fish and the path is two times 20 meters and add two times 10 meters. It shows that subject FD still cannot differentiate the meaning of words "around" and "circumference." He relates the new situation to the previous situation, so the plan is gotten which is stated in form of words to help reflection. It shows the reflective abstraction level of recognition and representation.

Subject FD writes directly the result of the calculation around of left, right, up, and the bottom side, so he is sure that the steps which are done are the right one. He states that the calculation of the path in the form of words. It shows the reflective abstraction level of representation.

Subject FD checks the result of the path width which will be built and review the wrongness through writing another way of solving the problem then he reviews the necessary one. The review is stated in form of words. It shows the reflective abstraction level of representation.

Subject Field Dependent difficult to interpret the word or information that is in the problem. This is in accordance with the opinion that the subject of the field dependent is difficult to provide a structure of ambiguous information, besides the subject only sees the problem globally [5]. In contrast to the Field Independent subject, before solving the problem, first analyzes the problem and then abstracts it, according to the opinion that individuals who are field independent tend to think analytically [9].

4. Conclusion

Based on the analysis and finding it can be concluded that profile abstraction process of the third class of junior high school in solving the mathematical problems as follows:

4.1. Reflective abstraction profile belongs to a student's cognitive style of field independent

In the step of understanding the problem, the student describes the necessary information based on the problem; he states what has been known and what is asked in the item in form of words. In this case appears reflective abstraction level of recognition and representation. In the step of planning solving problem, the student understands the information at new situation and understand the new information firstly at once he is able to relate the information quickly and uses pictures to help reflection in order to get right planning solving the problem. In this case appears reflective abstraction level of recognition and representation. In the step of solving problem-based on the preparation, the students carry out the plan. He reflects previous activity into a new situation and uses picture representation. In this case appears reflective abstraction level of representation and structural abstraction. In the step of review, the gained result the student uses a new strategy to check the result obtained previously. Giving argument or valid reason for decided decisions and summarizing all activities while solving the problem. In this case appears reflective abstraction level of structural abstraction and structural awareness.

4.2. Reflective abstraction profile belongs to a student's cognitive style of field dependent

In the step of understanding the problem, the student states necessary information based on the problem, stated what is known and asked in the item test in the form of words. In this case appears reflective abstraction level of recognition and representation. In the step of planning solving problem, the student does not understand new information in a new situation and does not remember previous details yet, so he needs longer time to relate the data to the beginning he has faulty in completing the test. In this case appears reflective abstraction level of recognition and representation. In carrying out the planning, the student only does it based on the plan without developing new strategy or method. In this case appears reflective abstraction level of representation. In the step of re-checking the gained result, the student reviews his wrong as much as required or writing another way of completing the test. He is not able to differentiate the result from a different way. In this case appears reflective abstraction level of recognition and representation.

Based on the conclusion the recommendation or implication that the teacher should know and understand each student's cognitive style in order to adapt to the appropriate learning approach. In completing the test, the teacher should be able to create questions which to reach knowledge or prior information to bring he/she up to reflective abstraction level of structural awareness.

References

[1] Wood T, Williams G & Mc Neal B 2006 Children's mathematical thinking in different classroom cultures *Journal for Research in Mathematics Education* **37** 222

[2] Mitchelmore M and White P 2004 *Abstraction in mathematics and mathematics learning* In M J & A B Fuglestad (Eds.) *Proceedings of the 28th Conference of the International Group for the Psychology of Mathematics Education* **3** 329

[3] Gray E and Tall D 2007 Abstraction as a natural process of mental compression *Mathematics Education Research Journal* **19** 2

[4] Goodson-Espy T 2005 *Why Reflective Abstraction Remains Relevant* In Mathematics Education Research In Lloyd G M, Wilson M, Wilkins J L M & Behm S L (Eds) *Proceedings of the 27th annual meeting ofthe North American Chapter of the International Group for the Psychology of Mathematics Education. Applachian State University*

[5] Altun A and Cakan M 2006 Undergraduate students' academic achievement, field dependent/independent cognitive styles and attitude toward computers *Journal of Educational Technology and Society* **9** 289

[6] Ausburn L J & Ausburn F B 1978 Cognitive styles some information and implications for instructional design *Educational Communication and Technology* **26** 337

[7] Riding R and Rayner S 1998 *Cognitive styles and learning strategies: Understanding style differences in learning and behavior* (London: David Fulton Publishers)

[8] Polya G 1973 *How To Solve It Second Edition (New Jersey: Princeton University Press)*

[9] Witkin H, Moore C, Goodenough D & Cox P 1977 Field-dependent and field-independent cognitive style and their educational implications *Review of Educational Research* **47** 1

IOP Conf. Series: Journal of Physics: Conf. Series **1088** (2018) 012095 doi:10.1088/1742-6596/1088/1/012095

Mathematics learning through *Pendidikan Matematika Realistik Indonesia* (PMRI) approach and Adobe Flash CS6

M Zainil, Y Helsa, Y Zainil and W T Yanti

Universitas Negeri Padang, Jl. Prof. Dr. Hamka, Air Tawar, Padang, Indonesia

E-mail: melva_zainil@yahoo.co.id

Abstract. This research based on low cognitive learning outcomes of students and less interesting mathematics learning for students as the result of learning centered on teachers, and many teachers who have not used readable media. This research aims to find out to what extent the influence of PMRI approach assisted media Adobe Flash CS6 to result of learning mathematics at Grade V. It is quasi-experiment research. It was conducted at SDN 2 Marapalam Padang City. It used purposive sampling, Grade V-A as the control group and Grade V-B as the experiment group. The instrument used in this research is the multiple-choice type. Inferential statistical test with a t-test is used to analyze the data. The result shows that the PMRI approach media assisted Adobe Flash CS6 influences the results of mathematics learning Grade V SDN 23 Marapalam, Padang City in the academic year of 2017/2018.

1. Introduction

Mathematics is one of the disciplines studied in educational institutions and offered to students from the primary school level up to the higher levels [1, 2]. Mathematics is the numbers and calculations that are part of human life [3-5]. It indicates that mathematics as a subject has an important role, both scientific mindset in shaping the quality of the students and its usefulness in everyday life. Using the concepts and principles of mathematics can help students to present something logically, creatively, and systematically [6, 7]. Students are invited to use various sources and learning media that can improve students' interaction, logical thinking, creativity and systematic to support in learning mathematics.

The Indonesian realistic mathematics education approach (PMRI) is a mathematical approach developed in mathematics learning which is one of the theories commonly known as the realistic approach [8-10]. The PMRI approach is an approach that places students' reality and experience as the starting point of learning [8]. It is an approach that provides the widest opportunity for students to be active in the learning process in which students are required to build their understanding and understanding of newly learned concepts [9]. PMRI helps for better teaching and learning process where students are more active and creative such as teachers no longer use the chalk and talk method [11-13]. The teacher's role changes from the center of the teaching and learning process to the mentors and resource persons [8].

The learning model and method are not the only supports in the learning process such as instructional media must also be used to achieve better learning [14-16]. The instructional media is constructive for teachers in delivering material to students [15]. It is as an intermediary to read messages from the sender to the recipient to achieve learning objectives to get more satisfactory learning results [17].

One of ICT-based learning media is learning media which use adobe flash CS6 application [16]. Adobe Flash is a program to create 2D vector based animation. A variety of features and its straightforward use of Adobe Flash CS6 application make it a favorite animation program, and it is quite popular. The various plate functions and options, as well as a collection of complete tools, are beneficial in doing interesting animated works. The other benefits of Adobe Flash CS6 are animated graphics software that can create graphical and animated objects so we can directly create object design without having to use graphics software support such as Illustrator or Photoshop. The adobe flash CS6 can make animations that can help us in making learning media interesting [18].

The ICT learning media, especially Adobe Flash CS6 program can be used as a reference for the use of ICT in learning especially the use of Adobe Flash CS6 program on abstract mathematics learning following the material taught in Elementary School. The results of the interviews with the classroom teachers of Grade V-A, V-B, and V-C at SDN 23, Marapalan, Padang city reveal that teachers are less able to use the ICT based learning media with so that children are not interested in learning. Monotonous learning cannot help the teacher to enable students to understand learning materials well. Teachers as one of the main elements in learning are expected to be able to use appropriate learning media and to pay attention to the cognitive developmental stages of students to motivate students to be more active in learning. Innovation in teaching and learning process is needed to improve student achievement as expected.

2. Method

This research is experimental research. The research design used was the non-equivalent control group design. The difference between the students in the experimental group treated with PMRI approach assisted adobe flash CS6 media and control group students treated with PMRI approach can be seen. The population in this research is all students of Grade V SDN 23 Marapalam Padang City. The total number of students in the community is 92 students. The V-B class students were used as the experimental group and the V-A class as the control group. Purposive sampling technique was used in this research that is sampling based on individual consideration [19].

Intensive reading skill test was used to collect the data. The validity test (validity of the grains), reliability test utilizing the Product Moment Correlation formula with rough numbers, the difficulty level of questions and the different questions were used to try out the instrument. Data analysis used a prerequisite test and hypothesis test. The prerequisite analysis test used was the normality test and homogeneity test. Lilifors method was used to test the normality and Fisher method was used to test the homogeneity. T-test (t-test) was used to test the hypothesis.

3. Results and discussion

3.1. Results

Pre-test was given to both control and experimental group before the treatment as shown in Table 1.

Table 1. The result of pre-test

Variable	Pre-test	
	Experiment group	**Control group**
N	31	30
Highest Mark	70	75
Lowest Mark	25	30
Mean	52,58	54,17
SD	12,02	12,80
SD^2	155,7849	163,9368

Table 1 shows that the experimental class with the number of 31 students obtained the highest score of 70 and the lowest score of 25. The experimental group achieved an average rating of 52.58, the standard deviation of 12.02 and the value of variance 155.7849. While the control class with the number of 30 students obtained the highest score of 75 and the lowest score of 30. The average rating for the control group is 54.17, standard deviation 12.80 and the value of variance 163.9368. The description of pre-test results in the above table revealed the cognitive learning outcomes of flash class CS6 treatment using PMRI approach assisted media. Both group then were and then given post-test in where the results can be seen in Table 2.

Table 2. The result of post-test

Variable	Post-test	
	Experiment group	**Control group**
N	31	30
Highest Score	100	95
Lowest Score	55	40
Mean	80,16	71,33
SD	13,15	13,13
SD^2	173,0389	172,2989

Table 2 shows that the experimental group with the number of 31 students got the highest score of 100 and the lowest score of 55. The average rating of the experimental group is 80.16 with the standard deviation of 13.15 and the value of variance 173.0389. While the control class with the number of 30 students obtained the highest score of 95 and the lowest score of 40. The average rating of the control group is 71.33 with the standard deviation, and variance value is 172.2989. Based on the description of post-test results in the table above, it can be concluded that cognitive learning outcomes of the experimental group are higher than the control group. The hypothesis needs to be tested to know the effect of PMRI approach assisted Adobe Flash CS6 media to the result of mathematics learning in Grade V SDN 23 Marapalam, Padang city. Previously, tested prerequisite analysis that is by using normality test and homogeneity test was done. The normality test of the pre-test score can be seen in Table 3.

Table 3. The results of normality test

Group	L_0	L_{table}	n	α	Description
Experiment	0,0735	0,159	31	0.05	Normal
Control	0,09753	0,161	30	0.05	Normal

Table 3 shows that the sample based on the pre-test score has the normal distribution where it can also be seen L0 < L table. Furthermore, the homogeneity test of pre-test can be seen in Table 4.

Table 4. The homogeneity test of pre-test

No	Group	N	α	f count	f table	Description
1	Experiment	31	0,05	1,13	1,85	Homogeny
2	Control	30	0,05			

Based on the homogeneity test in Table 4, F value can be calculated with real level $\alpha = 0.05$. From the F distribution table, it can be seen that the F count ≤ F table is $1.13 \leq 1.85$. It can be concluded that the data of the final test result of both groups have a homogeneous variance at 95%.

Table 5. The results of the normality test calculation based on post test score

Group	L_0	L_{table}	n	α	Description
Experiment	0,10198	0,159	31	0.05	Normal
Control	0,012067	0,161	30	0.05	Normal

Table 5 reveals that the sample based on the post-test score has normal distribution is normal in which it also explains the value of L0 < L table. Furthermore, the result of homogeneity tests of post-test seen in Table 6.

Tabel 6. The results of homogeneity test of pretest

No	Group	N	α	f count	f table	Description
1	Experiment	31	0,05	1,02	1,85	Homogeny
2	Control	30	0,05			

Based on homogeneity test in Table 6, it can be seen the value of F calculation with the real level $\alpha = 0.05$. From F distribution table, it was obtained the value of Fcount ≤ Ftable is 1.02 ≤ 1.85. It can be concluded that the data of the final test result of the sample has a homogeneous variance at 95%.

Table 7. The results of hypothesis testing

No	Group	Numbers	Real level	t_{count}	t_{table}	Description
1	Experiment	31	0,05	2.803	2.000	Significantly different
2	Control	30	0,05			

Based on the hypothesis test in Table 7, it can be seen that at 5% significance level, the result of hypothesis testing of both groups is tcount> ttable (2.803> 2.000) means H0 is rejected. It means that there is an influence of PMRI approach assisted Adobe Flash CS6 media to the results of mathematics learning of Grade V students of SDN 23 Marapalam city of Padang.

3.2. Discussion

The research results show that there is a difference results between the use of PMRI approaches assisted adobe flash cs6 media with traditional learning using PMRI approach only. It is evidenced by the results of hypothesis testing which result in rejecting the H0 at a significance level of 5%.

Based on the results of research done in Grade V SDN 23 Marapalam, learning in the experimental group was treated using PMRI approach with the help of media adobe flash cs6 media reveal that the students were more active and interested in studying [11, 12, 15, 20]. While in the control group students were not motivated to study. So there is an effect of PMRI approach assisted adobe flash CS6 media to student cognitive learning outcomes in identifying the characteristics of shape [16]. They were significant differences in student learning outcomes whose were taught using PMRI approach assisted with adobe flash CS6 media with cognitive learning outcomes of students who were trained using PMRI approach done without adobe flash CS6 media [21]. It can be seen in the final test of the students given to the experimental group and the control group. Using the PMRI approach assisted of adobe flash CS6 media in the experimental class, the level of the students' learning outcomes increase.

Based on the result of the post-test in identifying characteristic shapes the average score for the experimental class was 80.16, while the average rating for the control group was 71.33. Based on data analysis t_count> t_table with 2.803> 2.000 then rejected the H_0 and accepted the H_1. So, it can be concluded that the hypothesis in this study was accepted that there is an influence of PMRI approach assisted with adobe flash CS6 learning media to cognitive learning outcomes in identifying the character shape in Grade V SDN 23 Marapalam, Padang City.

4. Conclusion

The average score of post-test of the experimental group applying the PMRI approach with the help of adobe flash CS6 media is 80.16, and the average score of post-test control group applying PMRI approach is 71.33. Based on the result of hypothesis test using t-test (t-test), it can be concluded that at significant level = 0.05 PMRI approach assisted with adobe flash CS6 media is significantly affected the cognitive learning outcomes in identifying the characteristic of bola in Grade V SDN 23 Marapalam, Padang City in the academic year of 2017/2018.

References

[1] Rofii A, *et al* 2018 Characteristics of students' metacognition process at informal deduction thinking level in geometry problems *International Journal on Emerging Mathematics Education* **2** 89

[2] Ryandi R B, *et al* 2018 Learning combinations through "handshake" *International Journal on Emerging Mathematics Education* **2** 105

[3] Halim A 2009 *Matematika Hakikat dan Logika* (Yogyakarta: Ar-Ruzz Media)

[4] Revina S and Leung F K S 2018 Educational borrowing and mathematics curriculum: Realistic Mathematics Education in the Dutch and Indonesian primary curriculum *International Journal on Emerging Mathematics Education* **2** 1

[5] Tanujaya B, *et al* 2017 Mathematics instruction, problems, challenges, and opportunities: A case study in Manokwari regency, Indonesia *World Transactions on Engineering and Technology Education* **15** 287

[6] Permatasari D and Harta I 2018 The gap between the beginning and the end of algebraic thinking transition period *International Journal on Emerging Mathematics Education* **2** 79

[7] Rofii A, *et al* 2018 Characteristics of students' metacognition process at informal deduction thinking level in geometry problems *International Journal on Emerging Mathematics Education* **2** 89

[8] Hadi S 2017 *Pendidikan Matematika Realistik Indonesia* (Jakarta: Rajawali Pers)

[9] Prahmana R C I, *et al* 2012 Learning Multiplication Using Indonesian Traditional Game in Third Grade *Journal on Mathematics Education* **3** 115

[10] Lestari K E and Yudhanegara M R 2015 *Penelitian Pendidikan Matematika* (Bandung: Refika Aditama)

[11] Wijaya A 2008 Design research in mathematics education: Indonesian traditional games as means to support second graders' learning of linear measurement *Thesis* (Utrecht: Utrecht University)

[12] Revina S, *et al* 2011 Spatial visualization tasks to support students' spatial structuring in learning volume measurement *Journal on Mathematics Education* **2** 127

[13] Zulkardi 2009 The "P" in PMRI: Progress and problems *Proceedings of IICMA 2009 Mathematics Education* (Yogyakarta: IndoMs) p 773

[14] Masniladevi, *et al* 2018 Teachers' ability in using math learning media *J. Phys.: Conf. Ser.* **943** 012059

[15] Muhtadi D, *et al* 2018 The Integration of technology in teaching mathematics *J. Phys.: Conf. Ser.* **943** 012020

[16] Zainil M, *et al* 2018 ICT media design for higher grade of elementary school mathematics learning using CS6 program *J. Phys.: Conf. Ser.* **943** 012046

[17] Mudlofir and Rudydiyah 2015 *Desain Pembelajaran Inovatif* (Jakarta: Rajawali Pers)

[18] Island S 2008 *Panduan Mudah Membuat Animasi* (Jakarta: Media Kita)

[19] Sugiyono 2009 *Metode Penelitian Pendidikan (Pendekatan Kuantitatif, Kualitatif dan R&D)* (Bandung: Alfabeta)

[20] Purnami A S, *et al* 2018 The effect of team accelerated instruction on students' mathematics achievement and learning motivation *J. Phys.: Conf. Ser.* **948** 012020

[21] Shahrill M, *et al* 2018 Processes involved in solving mathematical problems *AIP Conference Proceedings* **1952** 020019

Designing learning trajectory for teaching sequence and series using RME approach to improve students' problem solving abilities

E Gee, A Fauzan and A Atmazaki
Universitas Negeri Padang, Jl. Prof. Dr. Hamka, Sumatera Barat 25173, Indonesia.

E-mail: ahmadfauzan@fmipa.unp.ac.id

Abstract. This study aimed at designing learning trajectory (LT) for teaching Sequence and Series using Realistic Mathematics Education (RME) approach to improve the students' problem solving ability. The LT designed in this research covered real problems that helped students follow the math learning process to build their knowledge in solving mathematical problems. The LT was developed through a design research that consisted of a cyclic process of preparing the experiment, conducting the experiment, and retrospective analysis. The research's subject was 34 grade nine students at a junior high school in Nias Selatan, Indonesia. The data were collected through observation, interviews, checklist, videotaping, and analyzing the students' works. The results showed that the LT for teaching Sequence and Series using RME approach reached the criteria of validity and practicality. The LT also could help students build their own knowledge through problem solving activities to discover sequence and series concepts.

1. Introduction

Problem solving ability was highlighted in many mathematical curricula and has recently become one of the most studied in the field of mathematics education. The National Council of Mathematics Teachers has consistently advocated for problem solving as part of daily math instruction, arguing that solving problems was an important thing to do in learning mathematics [1,2]. Furthermore, Soedjadi stated that mathematical problem solving ability is an ability within the students themselves to use mathematical activities to solve problems in mathematics or other sciences even in everyday life [3]. Beigie also stated that through problem solving, students could learn to improve their understanding of mathematical concepts by working through carefully selected issues which use the application of mathematics in real life problems [4]. The development of mathematical problem solving ability can equip students to think logically, analytically, systematically, critically, and creatively [5]. Therefore, problem solving ability is an important ability that needs to be provided for students to help them learn math as well as solve real life problem.

The development of student problem solving skills is not an easy task. When solving math problems, most students are not able to understand the problem, plan the solution, and apply the solution to the given problem [6]. Other studies also revealed that students poor problem solving ability is caused by their lack of experience in solving nonroutine problems [7]. The same problem was also encountered by the researcher in that students have difficulties in solving mathematical problems due to mathematics teaching that does not connect students' informal knowledge from their lives with their formal knowledge. Mayer defines problem solving as some process that requires the problem solver to find a connection between experience (schemes) that he or she had with the problem he or she is facing and being able to do something to solve the problem [8]. Meanwhile, Gagne defines problem solving as an activity to synthesize between knowledge, rules, concepts, schemes, or experiences they had and the condition they are facing to find a solution [8]. Based on the description, in order to connect students' thinking between real life problems and their formal knowledge in mathematics, a learning trajectory is required to bridge the informal and formal knowledge.

Learning trajectory is a series of activities and tasks that support the development of students' understanding of a particular instructional purpose [9]. Simon included the empirically supported descriptions of how students' thinking evolves over time. Based on research synthesis, clinical interviews, pedagogical experiments and large-scale evaluation data, learning trajectory was empirically defined as descriptions that help determine how students' informal ideas evolve through mathematical understanding [10]. Learning trajectory in learning is guiding students' instructional decisions [11] and improving their ability to use their thinking [12]. In Mathematics learning, students are expected to be the subject of learning. Therefore, teachers should be able to apply the learning trajectory that requires students' involvement and adapt how they think.

The learning trajectory has attracted the attention of some researchers [13-16]. In these studies, the researchers developed the LTs for various mathematics topics. The results of the researches showed that that the LTs could help the students to reinvent mathematical concepts in meaningful ways, so that they can build their understanding of the topics that they learned.

This learning trajectory is based on realistic mathematics education approach because this approach creates student-centered learning as well as directs students to engage in contextual problem solving. The main principle in RME approach is that mathematics is viewed as a human activity and mathematical learning means doing mathematics [17]. This means that in learning mathematics, student involvement is expected as well as directed to solve mathematical problems related to real life. Furthermore, Freudenthal emphasized that in mathematics learning students should be allowed and supported to create their own ideas and use their own strategies. In other words, they must learn mathematics in their own way [18].

The use of the learning trajectory is expected to improve students' problem solving abilities on sequence and series topics. RME learning that emphasizes on skills process of doing mathematics will lead students to do and solve their own problems by taking advantage of informal knowledge from their life. This is in accordance with a learning trajectory that provides predictions and anticipation of students thinking that will help students solve mathematical problems through mathematization process based on RME approach. Through the mathematization process, the informal knowledge of the students will be linked with the formal knowledge to be studied. This will train students to solve problems and have a positive impact on students' mathematical problem solving abilities.

2. Method

This research used design research approach proposed by Gravemeijer and Cobb [19]. We used this approach because design research aims to better understand the interrelatedness between teaching and learning in order to improve teaching [20]. Design research in this study consisted of a cyclic process of preparing the experiment, conducting the experiment, and retrospective analysis. Gravemeijer and Cobb illustrated the cyclic process as can be seen in Figure 1.

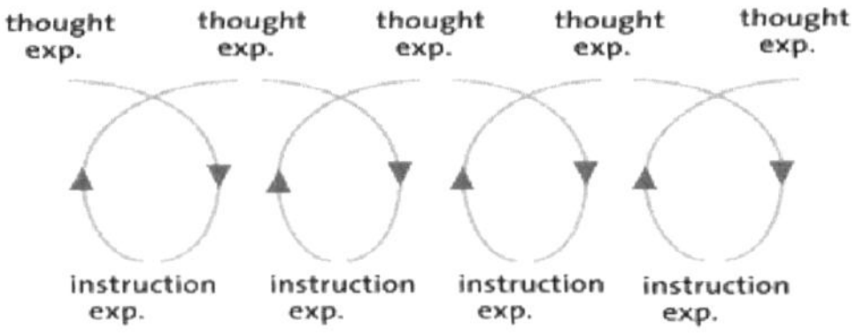

Figure 1. A cyclic thought process and instruction experiment.

In preparing for the experiment, we determined the end point of the instructions. The goals of our sequence and series lessons were for the students to reinvent the concepts of numerical patterns, arithmetic sequences and series, and geometry sequences and series. Considering that in studying sequence and series materials it takes the students' experience to explore the informal way of thinking, we then used the South Nias culture as the starting points of the lesson. After we set the end and the starting points, we designed the HLT that consisted of five main activities and nine sub-activities of solving contextual problems that would facilitate students to do horizontal and vertical mathematizations as well as stimulate students' thinking and reasoning. Besides, we also formulated the predictions of students' thinking and solutions, and their anticipations.

In the experiment phase, we tried out the HLT in two cycles. The first try out was conducted in small group that involved six grade nine students at junior high schools in Nias Selatan, Indonesia. After retrospective analysis and re-design processes, the HLT was tried out to 34 grade nine students at the same school. The retrospective analysis involved the research, a teacher, and an observer. Beside focusing our attention to develop the HLT, we also observed and analyzed the impact of the HLT on the development of students' confidence in using their strategies when solving the contextual problems and development of students' reasoning during try out. The research data were collected through observation, interviews, checklist, videotaping, and analyzing the students' works.

3. Results and discussion

The HLT for teaching the sequence and series topics was validated by three mathematics education experts in Indonesia during preparation for the experiment phase. The result showed that the HLT met the validity criteria [21] with the following characteristics: the activities of solving contextual problems in the HLT had potential to facilitate the students to reinvent the sequence and series concept; the activities were well ordered; the HLT suit the fundamental principles and characteristics of RME; and the components in the HLT were designed and consistent between one and another.

The HLT also satisfied the practicality criteria [22], in which it worked as intended during the tryout. The students understood the contextual problems and the conducted 'doing math' activities without major obstacle. The probing questions that were prepared as the anticipations of students' thinking and solutions also helped the students to achieve the goals of the activities. Besides, the time provided for doing the activities of solving the contextual problem was well planned.

The application of the learning trajectory by using the RME approach to the sequence and series learning creates interactive and student-centered learning. Students who are the subject of learning should be able to build their own knowledge based on experience. This is in line with the principle of RME expressed by Gravemeijer and Freudenthal, that the basic principles of RME are the guided reinvention and progressive mathematizing, didactical phenomenology, and self developed models [23, 24]. So in this learning trajectory, the mathematical problems are made as student activity and stimulate students to find mathematical concepts that are solved in their way based on their cognitive level and experience.

The following example is a contextual problem that can stimulate students' understanding in solving mathematical problems contained in the learning trajectory.

Mr. Fa'a has three children named Nia, Ani, and Adi. The three children will join a singing competition to celebrate Christmas School on January 4, 2018. If that day was November 15, 2017, then there were another 49 days of opportunity to prepare for the clothes that will be worn later. Then Mr. Fa'a gave each of his three children one thousand rupiah coins a day and asked them to keep it. If on the first day Nia obtained 3 coins, Ani 2 coins, and Adi a coin. Then the next day, Nia get 3 coins, while for Ani and Adi each get 2 pieces of coins.

a. Write the pattern of the number of coins received Nia, Ani, and Adi starting on the first day until the 5th day in a sequence of numbers!
b. On the 20th day, what is the number of coins received by Nia, Ani, and Adi?
c. What is the number of coins received by Nia, Ani, and Adi on the 40th day?
d. Can you determine how many coins Nia, Ani, and Adi received for the nth day?

Contextual problems provided greatly helped students' understanding in solving this problem. This was evident from the student's answers, in which they were able to find the concept of numbers pattern. This solution began in an informal way in which the student determined the number of coins received on a given day by the multiplication operation. Then with the help of tiered questions students were assisted and directed to find the formal knowledge of the nth term formula on the numbers pattern. One example of student answers can be seen in Figure 2.

4.

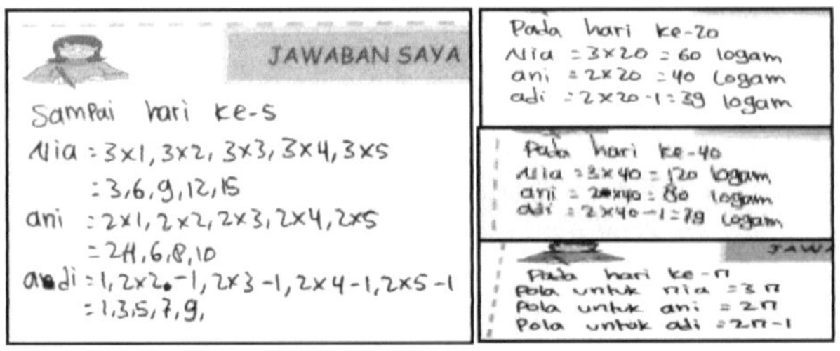

The number of coins the three children received on the fifth day, namely:
1) Nia = 3 × 1, 3 × 2, 3 × 3, 3 × 4, 3 × 5 = 3, 6, 9, 12, 15;
2) Ani = 2 × 1, 2 × 2, 2 × 3, 2 × 4, 2 × 5 = 2, 4, 6, 8, 10;
3) Adi = 1, 2 × 2 - 1, 2 × 3 - 1, 2 × 4 - 1, 2 × 5 - 1 = 1, 3, 5, 7, 9.

On the 20th day, the number of coins received by three children, namely:
1) Nia = 3 × 20 = 60 coins;
2) 2 × 20 = 40 coins;
3) Adi = 2 × 20 - 1 = 39 coins.

Then, on the 40th day, the number of coins received by the three children, namely:
1) Nia = 3 × 40 = 120 coins;
2) 2 × 40 = 80 coins;
3) Adi = 2 × 40 - 1 = 79 coins.

Furthermore, the number of coins received by the three children on the nth day, namely:
1) Nia = 3n;
2) Ani = 2n;
3) Adi = 2n - 1.

Figure 2. Students answers to find the pattern of numbers.

Implementation of the learning trajectory using RME gives a positive response from students because mathematical problems are connected with everyday life problems, making it easier for students to understand the problem. This suggests that the various situations that students experience in everyday life can be utilized in building an understanding of facts, concepts, and principles of mathematics. The imaginary or real-life situation that students gain from experience makes learning mathematics a useful and meaningful activity that emphasizes reasoning rather than mathematical formulas [24]. The results of research by Zulkardi and Ilma states that context is the first step in learning mathematics. In the reinvention approach, problem contexts play a crucial role [25]. Well-chosen contextual problems provide an opportunity for students to solve problems with an informal solution strategy [26].

Students as learning subjects certainly have different experiences and cognitive levels, thus when given mathematical problems, various solution and strategies will emerge. Through the use of the learning trajectory with the RME approach these different ways of thinking in generalizing the knowledge to be built are accommodated. The same was confirmed by Soedjadi that in general the development of cognitive abilities of children starts with concrete things then gradually leads to the abstract. For every child, the journey from real to abstract can be different. Some are fast and some are slow. The fast ones may not require many stages, but for those who are not fast, it is possible that they go through many stages. Thus for every child different learning trajectory may be required [27]. This means that students are able to find their own strategies to be used in solving contextual problems based on their experience and cognitive level, which met one of the RME characteristics called 'students' free production' [28,29]. This is evident from the students answer to several problems in each activity led to some strategies. Finding the right solution is related to the problem solving ability of students in planning and determining the solution of the problem. Through the anticipation questions that teachers have provided in the learning trajectory, differences in the way students think in determining the right solution to the given contextual problems are bridged.

This learning situation highly contributes to mathematical problem solving ability. Learning trajectory that uses the RME approach are oriented to real life problems. This makes it easier for students to understand the given problems. Furthermore, in the learning process students are expected to be able to find the concept of sequence and series through the mathematization process, starting from horizontal to vertical mathematization process. This means that students are not allowed to use mathematical formulas directly, but they are directed to find the formula itself through the given contextual problem. The students' involvement in the mathematization process is a way to practice problem solving skills. The results of the analysis of pretest showed that the average grade of students' initial ability in solving mathematical problems was 48.41. Whereas the result of posttest data analysis which is the average grade of student's mathematical problem solving ability after applying the learning trajectory based on RME reached 74.85.

5. Conclusion

The learning trajectory developed for sequence and series topic meets the criteria of validity, practicality, and effectivity. The result of this design development is beneficial in achieving the learning objectives of sequence and series topics. A learning trajectory that uses a realistic mathematics education approach creates an interactive and enjoyable learning environment. The RME approach is oriented to everyday problems, allowing students to understand the problems and determine the solution of the problem in their own way. Furthermore, this learning activity involved students in the mathematization process from horizontal to vertical mathematization. Through the mathematization process, students' mathematical problem solving abilities are exercised.

References

[1] Ball D, Ferrini-Mundy J, Kilpatrick J, Milgram J, Schmid W and Schaar R 2005 *Am. Math. Soc* ***52*** 1058
[2] Davis R 1992 *J. of Math. Behavior* **11** 241
[3] Soedjadi 1994 *Memantapkan Matematika Sekolah sebagai Wahana Pendidikan dan Penalaran Kebudayaan* (Surabaya: Program Pasca Sarjana Pendidikan IKIP Surabaya)
[4] Beigie D 2008 *Mathematics Teaching in the Middle School* **13** 352-360.
[5] Edy S, Feria A P and Mukhtar 2017 *Contextual Learning Model Journal on Math. Edu.* **8** 85
[6] Seifi M, Haghverdi M and Azismohamadi 2012 *J. of Basic and App. Scie. Res.* **2** 2928
[7] Windari F, Dwina F, and Suherman 2014 *J. Pend. Mat.* **3** 28
[8] Foshay R and Kirkley J 2003 *Principle for Teaching Problem Solving* Technical Paper Plato Learning Inc
[9] Simon, M A 1995 *Journal of Research in Math. Educ.* **26** 137
[10] Confrey J, Maloney A P, Nguyen K H, Mojica G, and Myers M 2009 *Proceedings of the 33rd Confe. of the Psy. of Math. Educ.* **2** 352
[11] Wilson P H 2009 *Teachers' Uses of a Learning Trajectory for Equipartitioning* [Dissertation] North Carolina State University USA
[12] Clements D and Sarama J 2004 *Math Thinking and Learning* **6** 89
[13] Fauzan A, Armiati A, and Ceria C 2018 A Learning Trajectory for Teaching Social Arithmetic using RME Approach *IOP Conference Series: Materials Science and Engineering* **335** 012121
[14] Ramirez RE and Solis AH 2016 Hypothetical learning trajectories that use digital technology to tackle an optimization problem *Int. J. Techno. Math. Educ.* **24** 51
[15] Prahmana R C I and Kusumah Y S 2016 The Hypothetical Learning Trajectory on Research in Mathematics Education Using Research-Based Learning *J. Pedagogika* **123** 42
[16] Webb D C, Kooij H vd and Geist M R 2011 Design Research in the Netherlands: Introducing Logarithms Using Realistic Mathematics Education *J. Math. Edu. Technol. Coll.* **2** 47

[17] Fauzan A, Plomp T and Gravemeijer K 2013 *Educational Design Research Part B: An Introduction ed T Plomp and N Nieveen (Enschede, SLO)* pp *159-178*

[18] Zulkardi 2002 *Development a Learning Environment on RME for Indonesian Student Teachers* [Dissertation] Enschede: University of Twente

[19] Gravemeijer and Cobb 2006 Educational Design Research *Design Research from a Learning Design Perspective* pp 45-85 Routledge

[20] Liljekvist I, Mellroth E, Ollson J and Boesen J 2016 *Proc MADIF10 Development of mathematics teaching: design, scale, effects Linkoping: SMDF* pp 119-127

[21] Nieveen N and Folmer E 2013 *Educational Design Research Part A: An Introduction* ed T Plomp and Nieveen (Enschede: SLO) pp 152-169

[22] Freundenthal H 1991 *Revisiting Math. Educ.* (Dordrecht: Kluwer Academic Publisher)

[23] Gravemeijer K 1994 *Developing RME* (Utrecht: Freudenthal Institute)

[24] Turmudi 2009 *Proc. of IICMA Students' Responses to the Realistic Mathematics Teaching Approach in Junior Secondary School* (Indonesia: Indonesia University of Educ)

[25] Zulkardi and Ilma 2006 Mendesain Sendiri Soal Kontekstual Matematika *Proc. Konferensi Nasional Matematika 13* (Semarang: KNM)

[26] Doorman M, Drijvers P, Dekker T, Heuvel-Panhuizen M, de Lange J and Wijers M 2007 *ZDM Math. Educ.* **39** 418

[27] Soedjadi 2007 *Masalah Kontekstual sebagai Batu Sendi Matematika Sekolah* (Universitas Negeri Surabaya: Pusat Sains dan Matematika Sekolah)

[28] de Lange J 1987 *Mathematics, Insight, and Meaning* (Utrecht: OW & OC)

[29] Streefland L 1991 *Realistic Mathematics Education in Primary Schools* (Utrecht: Freudenthal Institute)

[30] Treffers A 1991 *Three Dimensions: A Model of Goal and Theory Description in Mathematics Education* (Dordrecht: Reidel)

IOP Conf. Series: Journal of Physics: Conf. Series **1088** (2018) 012097 doi:10.1088/1742-6596/1088/1/012097

The quality of mathematics learning material using a modification of Think Pair Share (TPS) model

Maisyura, C M Zubainur and T F Abidin
Syiah Kuala University, Jl. Teuku Nyak Arief Darussalam, Banda Aceh 23111, Indonesia

E-mail: cutmorinazubainur@unsyiah.ac.id

Abstract. Think Pair Share (TPS) modification learning model is expected to strengthen TPS model syntax by changing the number of member in a group, a group of two into a group of three. The researcher has developed learning materials for TPS modification model using Plomp model. This study focuses on the practicability of the learning materials on the assessment phase of Plomp model. The practicality of the learning materials was evaluated by experts and teachers. The data was obtained from the pilot test involving 19 students in grade 9 of a junior high school in Aceh. The research instrument used was an observation sheet to examine the learning implementation. The results show that learning materials using TPS modification model met the practicality criteria. The finding implies that the learning materials using TPS modification model could be applied in the classroom.

1. Introduction

Think Pair Share (TPS) modification model refers to the TPS model by changing one of the learning phases in the TPS model. That is, changing the number of member in a group, group of two into a group of three by considering the third students as a mediator in the group. The group of three students gives many opportunities for each member to share ideas so that the dominance of one person in a group could be avoided [1]. This condition encourages the interaction among the students.

TPS modification phases are oriented on thinking, discussion with the group mates, and sharing activities. The small group discussion in the TPS modification model means that the students discuss in a small group consisting of three students to solve mathematics problems. The group on the TPS modification model consists of students with various capabilities [1].

TPS modification model is expected to give better impacts than TPS model. TPS modification is considered more effective in changing the discussion pattern situation in the classroom; that is, the discussion can provide more time for students to think, respond, and help each other [2]. Moreover, TPS modification also supports students to think about the problem individually then share to other students as in the original TPS model [3]. Similar to the original TPS, TPS modification allows students to think independently, respond to problems given by teachers to find the solution and help other students who have difficulty in solving the problems [4]. Students construct their knowledge to find the solutions to the problems, both individually and in a group.

This paper examines the practicability of mathematics learning materials using TPS modification model. The practicability of the materials is evaluated based on the recommendations of experts and teachers.

2. Method

This study was part of development research on learning materials using TPS modification model. This study used Plomp development model that consists of three phases: the preliminary research, prototyping, and evaluation phases [5]. This paper focuses on the evaluation phase for the practicality aspect. The data to measure the practicability aspect were obtained from a pilot study on the learning materials. The pilot study was conducted for 19 Year 9 students in one of the junior high schools in Aceh, Indonesia. Experts and teachers were asked to assess the practicality of the learning implementation [6].

3. Result and discussion

The pilot study to evaluate the practicability of learning materials with TPS modification model focuses on the topic of solid geometry with the curved surface (non-polyhedrons). The learning implementation consists of four sessions. The first session discussed net and surface area of cylinders. The second session focused on net and surface area of cones. The third session was about the surface area of spheres, and the fourth session was about the volume of solid geometry objects with the curved surface (cylinders, cones, and spheres). Those subtopics were required to learn by junior high school students in grade 9 as mentioned in the basic competencies in Curriculum 2013. That is, generalizing the surface area and volume of solid geometry objects with the curved surface (cylinders, cones, and spheres) and solving a contextual problem related to the surface area and volume of the objects, and the combination of some solid geometry objects with curved surface [7].

The first step of the TPS modification model is giving problems related to solid geometry with the curved surface (non-polyhedrons). The problem given was about net and surface area of cylinders shown in Figure 1.

Aisyah got a cylinder pencil holder from her father. The holder has a hollow surface. Aisyah wants to use it to save her money and put it on her table. However, Aisyah doesn't want the money can be seen from the outside. What should Aisyah do?

Figure 1. The pencil holder problem.

The students were given the opportunity to understand and think about the problem solving strategy individually. The result of a student's individual work is shown in Figure 2. Then, the students discussed the problems in a group of three. In the second phase of the TPS modification model, the students discussed the problem solving with the other members. Every member in a group had the different abilities that consist of high, medium and low. Then, the learning activity was continued to group presentation.

IOP Conf. Series: Journal of Physics: Conf. Series **1088** (2018) 012097 doi:10.1088/1742-6596/1088/1/012097

Penyelesaian:

1. Apa yang dapat dilakukan Aisyah agar uang dalam celengannya tidak terlihat dari luar?
   melapisi alat tulis dengan menggunakan kertas pada semua bagian
2. Berbentuk apa lapisan yang perlu disiapkan Aisyah?
   Berbentuk persegi panjang /jajar genjang , dan 2 lingkaran
3. Gambarkan bentuk kertas yang dibutuhkan Aisyah untuk melapisi wadah alat tulis tersebut!

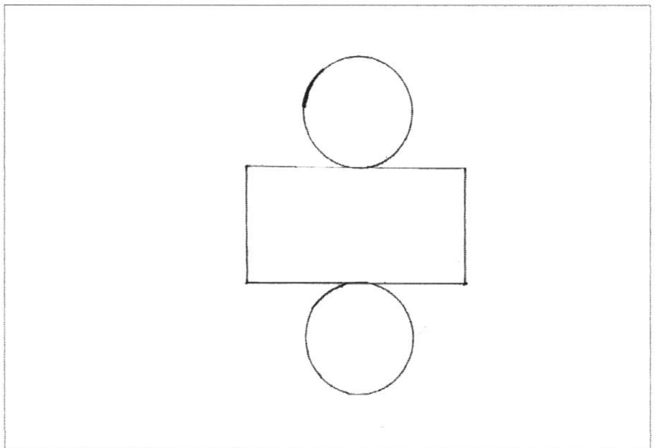

Solution:

1. What can Aisyah do to make the money in the pencil holder not visible from the outside?
   Covering the pencil hoder with paper
2. What shape of the layer does Aisyah need to prepare?
   Rectangular/parallelogram, and 2 circles
3. Draw the layer's shape Aisyah needs to cover the pencil holder!

4. Berapa luas kertas yang dibutuhkan Aisyah untuk melapisi bagian selimut wadah alat tulis?

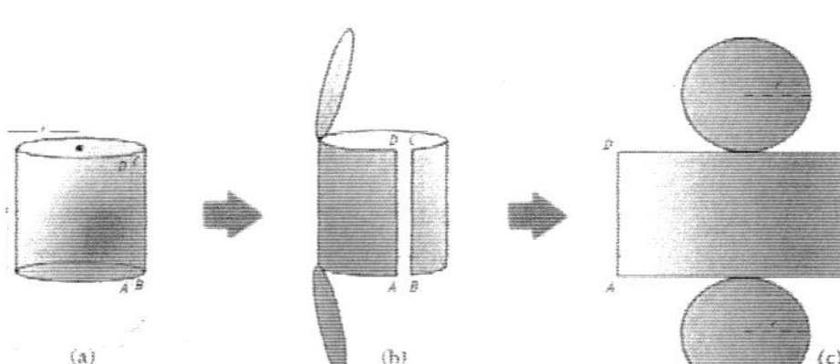

Gambar 1.2

Berdasarkan gambar di atas maka:

a. Panjang selimut tabung (AB = CD) = Keliling lingkaran ($2\pi r$) (1)
b. Lebar selimut tabung (AD = BC) = tinggi tabung (1)
c. Jadi, luas luas kertas yang dibutuhkan Aisyah untuk melapisi bagian selimut wadah alat tulis?
   Luas selimut tabung = $P \times L$
   $= 2\pi r \times t$ (2)
   $= 2\pi r t$

5. Berapa luas kertas yang dibutuhkan Aisyah untuk melapisi bagian alas dan tutup wadah alat tulis?
   $= \pi r^2 + \pi r^2$
   $= 2\pi r^2$ (1)

4. What is the area of paper that Aisyah needs to cover the side/the lateral surface of the pencil holder?

Based on the figure above, then:

a. The length of the lateral surface (AB = CD) = *circumference* ($2\pi r$)
b. The width of the lateral surface (AD = BC) = *the height of cylinder*
c. Thus, the area of paper that Aisyah needs to cover the side/the lateral surface of the pencil holder?
   The lateral surface area = $p \times l = 2\pi r \times t = 2\pi r t$

5. What is the area of paper that Aisha needs to cover the base and the lid of the container?
   $= \pi r^2 + \pi r^2$
   $= 2\pi r^2$

Figure 2. Example of a student's work on the pencil holder problem.

The practicality of the learning materials was evaluated by experts and teachers and based on the observation result of the learning implementation. Based on the evaluation from experts and teachers, the learning materials including a lesson plan, worksheet, and a test could be applied with minor revision. Overall, the quality of the lesson plan was good; only some parts need correction. Similarly, the experts and teachers recommended that the worksheet could be applied, but there was a minor revision. For the test, the errors found in the sentences that did not meet the Indonesian language requirement. Also, most of the typos found in the test were about missing words and ambiguous sentences without the figure explanation. Nevertheless, in general, the test was good enough and can be applied with a small revision.

The learning implementation for every session was evaluated through the observation by two observers, a teacher, and a colleague. Each session of learning has 21 aspects of learning activities. There were six aspects evaluated in the preliminary phase: opening, praying, checking students' attendance list, checking students' prior knowledge, motivating students, delivering the learning objectives, and explaining the learning activities. The observation result shows that all aspects in the preliminary phase were well conducted in the first, second and third sessions.

Moreover, there are 10 aspects evaluated in the core activity. The activities include students observe the problem given related cylinder radius and surface area, students ask questions related to the problems given, the teacher distributes a worksheet about cylinder radius and surface area, the students collect the information related to the problems in worksheet, students solve the worksheet, students are assigned a group, students discuss their solutions in a group, the teacher guides each group, group presentation, and whole class discussion. The finding shows that generally the aspects of the core activity from the first session to the fourth session could be implemented well. Out of the 10 aspects, only one aspect did not occur. That is the activity in the whole class discussion. The other groups did not respond to the presentation delivered by the presenter group because the answers of all groups were almost the same. Moreover, the students were still shy to deliver ideas publicly in the class. On the contrary, beyond the learning process, the students were more courageous to ask questions of the other students who were more capable and to teachers about the material that they did not understand.

Furthermore, in the final phase, there were five aspects evaluated that consists of guiding students to conclude the topic that had been learned, giving reinforcement about net and surface area of cylinder, having students reflect their learning process, evaluating students through a test, providing information about topics that will learn in the next session. All activities in the final phase could be applied well. In detail, the result of observation of the learning implementation is shown in Table 1.

Table 1. The observation results of the implementation of learning.

The measured aspect	Implementation	
	Observer 1	Observer 2
Greeting, praying, and checking the attendance list of the students, open the learning activities.	√	√
The teacher checks the prior knowledge of the students by asking some prerequisite materials.	√	√
The students answer the question and give response related to prerequisite materials	√	√
The teacher motivates the students about the importance to understand cylinder radius and surface area	√	√
The teacher delivers the learning objectives	√	√
The teacher explains the learning activity that the students will conduct	√	√
The students observe the problem given in power point about cylinder radius and surface area	√	√

IOP Conf. Series: Journal of Physics: Conf. Series **1088** (2018) 012097 doi:10.1088/1742-6596/1088/1/012097

The students ask about the problems given	√	√
The teacher distributes a worksheet of cylinder radius and surface area	√	√
The students collect the information to solve worksheet about cylinder radius and surface area	√	√
The students complete the worksheet about cylinder radius and surface area	√	√
The students are divided into several heterogeneous groups (3 members)	√	√
The students are asked to discuss together with the group member to discuss the worksheet given in the previous phase.	√	√
The teacher guides every group facing the problems in completing the worksheet.	√	√
The group representative presents the discussion result in *Pair Plus Phase* to other groups	√	√
Other groups give a response to the presenter group	-	-
The teacher guides the students to conclude the material learned	√	√
The teacher strengthen the students in solving the problem of net and surface area cylinder	√	√
The students reflect the learning process conducted	√	√
The teacher asks the students to do the evaluation test/quiz	√	√
The teacher gives information about the next topic.	√	√

The findings show that the learning materials using TPS modification model met the requirements of the practicability criteria. Experts and teachers declared that the learning materials could be applied in the classroom, and the level of implementation of the learning materials is categorized very high [6]. TPS modification model encouraged students to be more active in the discussion and more confident in delivering their ideas. A small group discussion provided more opportunities to discuss the topic with their partners. Low-achieving students were more confident in discussing the topic with moderate-achieving students. Similarly, moderate-achieving students were motivated to talk with the high achieving students. The learning outputs from the implementation of TPS modification model have the same impact as the original TPS model. This is consistent with the results of some studies that found the TPS model encouraged the students in the classroom to discuss actively, and improve safety environment for students to construct their thoughts and ideas [8,9]. Also, cooperative models, e.g., TPS can provide the feedback to respond and help each other in the learning process [10]. The students who have difficulty in constructing their knowledge can be assisted by other group members to find a solution to their difficulty [10]. The interaction processes occurring in the group can help the students to develop thinking and the idea so that the learning process is more active.

4. Conclusion
Overall, the learning materials using TPS modification model met the criteria of practicality. Based on the pilot study on the mathematics learning materials using TPS modification model, it can be concluded that the experts and teachers generally recommended lesson plan, worksheet, and test developed can be applied in the classroom with minor revision. Also, based on the observation, all aspects were done well. Thus, the practicality of learning materials using TPS modification model was considered good.

References
[1] Lie A 2004 *Practicing Cooperative Learning in the Classroom* (Jakarta: Gramedia Widiasarana Indonesia)

[2] Arends I R 2008 *Learning To Teach* (Yogyakarta: Pustaka Belajar)

[3] Tint S S and Nyunt E E 2015 Collaborative learning with think pair share technique *Computer Applications: An International Journal (CAIJ)* **2** 1

[4] Sumarni S 2016 *Proceeding The 2nd International Conference On Teacher Training and Education* vol 2 (Surakarta: Universitas Sebelas Maret) p 783

[5] Plomp T and Nieveen N 2013 *Educational Design Research: An Introduction* (Enschede: The Netherlands SLO)

[6] Nieveen N 1999 *Prototyping to Reach Product Quality* In Plomp T, Nieveen N, Gustafon K, Branch R M and Van Den Akker J (Eds.) *Design Approaches and Tools in Education and Training* (Dordrecht: ICO Cluwer Academic Publisher)

[7] Kemendikbud 2013 *Salinan Lampiran Permendikbud No. 103 Tahun 2014 Tentang Pembelajaran Pada Pendidikan Menengah* (Jakarta Badan Pengembangan Sumber Daya Manusia Pendidikan dan Kebudayaan dan Penjamin Mutu Pendidikan)

[8] Azlina N N 2010 CETLs: Supporting collaborative activities among students and teachers through the use of think pair share techniques *International Journal of Computer Science Issues (IJCSI)* **7** 18

[9] Sampsel A 2013 Finding the Effects of Think-Pair-Share on Student Confidence and Participation *Honors Projects* 28

[10] Cooper J L and Robinson P 2000 Getting started: Informal small-group strategies in large Classes *New Directions for teaching and learning* **81** 17

IOP Conf. Series: Journal of Physics: Conf. Series **1088** (2018) 012098 doi:10.1088/1742-6596/1088/1/012098

Student's understanding of numbers through the number sense strategy

N F Helmy[1], R Johar[1] and Z Abidin[1,2]
[1]Syiah Kuala University, Jl. Teuku Nyak Arief Darussalam, Banda Aceh 23111, Indonesia
[2]National Islamic University Ar-Raniry, Jl. Ibnu Sina No. 2, Banda Aceh 23111, Indonesia

Email: rahmah.johar@unsyiah.ac.id

Abstract. Many students have difficulties to do arithmetic operations with fraction. One way to overcome student difficulties is to implement number sense strategy. Therefore, it is necessary to develop a learning instrument that consists of the lesson plan, worksheet, and test. This study is qualitative research that only discusses student understanding after implementation of number sense strategy. The participants are 19 students of year 7 in one of the secondary schools in Meurah Mulia, North Aceh, Indonesia. The test used in this study are number sense problems and unstructured interview guidelines. The tests were conducted before and after the lesson. The result of the research shows that the students' understanding of the numbers has increased.

1. Introduction

Understanding numbers is not just about knowing arithmetic, but it is beyond that, namely having a good intuition about numbers, understanding well the properties of numbers, and knowing the relations between numbers. Students have started studying the numbers since the beginning of school, and fraction began to be taught in Year 3. Unfortunately, many students still have difficulties to do the arithmetic operation. This is according to Sutrisno that states that students still do not understand the standard procedures of the fractional operation [1].

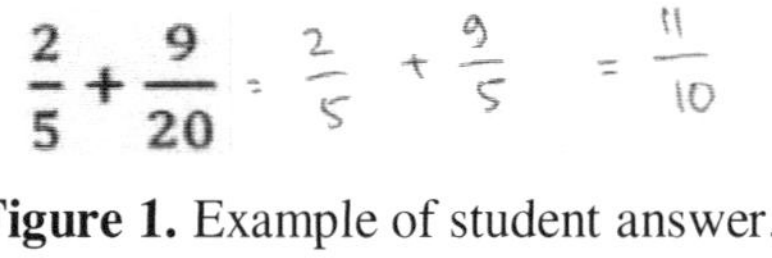

Figure 1. Example of student answer.

This student made a mistake in adding fractions, he changed $\frac{9}{20}$ to $\frac{9}{5}$ and he sums up both the numerator with the numerator and the denominator with the denominator, so the result is wrong. Some previous research also mentions that students still make mistakes in calculating operations, such errors including procedural errors, concept errors, misinterpretation, error due to carelessness, application error, and learning errors [2, 3]. Witri, Putra, and Nurhanida [4] mentioned that students had more difficulty to solve tasks on fractions than integers. Usher [5] says that if someone has a problem to understand the fractions and we are not doing anything, then he will get lost. Therefore, efforts are needed to help students improve their understanding.

One effort that can be conducted to help improve students' understanding is to introduce students to number sense strategies. Yang [6] states that the number sense refers to one's understanding of numbers and the ability to handle daily problems associated with numbers. Number sense is not a method of calculation; number sense is just a kind of sensitivity or awareness to numbers. A person with a good number sense will have the ability to process numbers. This ability is not only used in mathematical calculations but also in daily life. The importance of studying the number sense was expressed in several studies [6-11].

Number sense can be well taught if supported by appropriate learning materials. Therefore, the development of learning materials and tests based on the number sense strategies is needed to improve students' understanding on fraction. This paper only examines the development of students 'understanding of the numbers using learning instruments, so that the problem in this research is how does students' understanding about fraction develop after learning based on the number sense strategy?

2. Method

This study is a qualitative research that aims to describe the students' understanding of numbers through the number sense strategies after learning processes of remedial I. The participants were 19 students from year 7 of junior high school in Meurah Mulia, North Aceh, Indonesia. This study used learning materials and tests that have been developed in the previous study. The test was designed to meet the number sense indicators: 1) to understand the density of numbers, 2) to use the benchmark, 3) to estimate and assess the reasonability of the calculation, 4) to know an operation effect, 5) to recompose/reconstruct the number [9]. The test given to the students consisted of nine questions: five multiple choice and four essay questions (Examples of question can be seen in student interviews on results and discussion). Students given the test were also interviewed to see which strategies they used.

The materials had been assessed and validated by four validators: a learning expert, a mathematics teacher, a teacher who has researched number sense and a friend who have attended research and learning mathematics workshops, so the developed materials have “worth using with revision” criteria. Validation results are revised based on validator suggestions and comments provided so that the device is eligible for use. In addition, the test used in the study also met valid and reliable criteria [12].

The data of this research were student answer sheets and interview results. Data collection was conducted in two stages. The first stage was conducted before the learning process took place to determine the initial ability of students and the second stage was implemented after learning processes to know the development of students' abilities. The learning conducted in this research was remedial learning on fraction. The learning processes were conducted for two cycles, namely remedial I and remedial II. Each remedial was designed for two meetings. The first meeting, students were taught about the density of numbers, comparing numbers, addition and subtraction of fractions including decimal fractions, and on second meeting students were taught about the multiplication of fraction. After following the learning processes, students were given a test to see their understanding and progress. Students who did not pass in Remedial I must follow Remedial II which was also designed for two meetings with different worksheet and test that are given at remedial I.

3. Result and discussion

Pre-test and interview results show that no students used number sense strategy to answer the question and the highest score obtained by students is 42.5%. Many students made a lot of mistakes when trying to do an appropriate calculation, for instance, on the task about (adding fractions). Figure 2 is an example of students’ mistake.

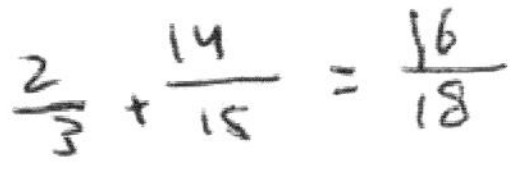

Figure 2. Student's written answer on adding fraction.

This student sums up the fractions without equating the denominator, but he instantly sums up both the numerator with the numerator and the denominator with the denominator, which is wrong. The students were focused only on solving the given problem by using a standard procedure, so students did not understand the meaning underlying the calculation or operations [4, 13].

Students are given a test after remedial I to see their number sense development. Here are some examples of interviews with students:

3.1 Multiple choice test result

Each multiple choice item contains an indicator of estimating the result of a calculation operation. However, each question could be conducted using another indicator. Problem 1 and 5 were given to test students' ability to estimate multiplication results. The following is an example of interview with a student.

S06 : By using the estimation, the location of the decimal point in the following multiplication is ... 4.346 x 3.125 = 1358125
P : The answer?
S06 : B. 13.58125. Because the initial numbers lie in the multiplication is 4 and 3, if they are multiplied, they will produce a two-digit number that is 12, but if all numbers are multiplied, they will be equal to the result above.

The student rounded each number (4.346 and 3.125) into one digit decimals (4 and 3), and then he tried to estimate the correct answer by looking the number closed to 12. So, he used one of number senses strategies called recomposing numbers. The similar idea was also presented by eight students.

Problems 2 and 3 are designed to train students' ability to estimate the sum. Many students answered this question using a number sense strategy correctly. They used benchmark $\frac{1}{2}$ to answer this question. An excerpt of an interview with one of the students is showed as follows.

S01 : Without doing a calculation, the following fractions which result in more than 1 are ...

a. $\frac{5}{11}+\frac{3}{7}$
b. $\frac{7}{15}+\frac{5}{12}$
c. $\frac{1}{2}+\frac{4}{9}$
d. $\frac{5}{9}+\frac{8}{15}$

P : Why did you choose d?
S01 : It is because $\frac{5}{9}$ is greater than $\frac{1}{2}$, $\frac{8}{15}$ also greater than $\frac{1}{2}$. So, greater than $\frac{1}{2}$ + greater than $\frac{1}{2}$ equal greater than 1
P : What about option a? Why don't you choose a. $\frac{5}{11}+\frac{3}{7}$?
S01 : It is less than 1
P : less than1? Why is it less than 1?
S01 : Because $\frac{5}{11}$ is less than $\frac{1}{2}$, $\frac{3}{7}$ is less than $\frac{1}{2}$.
P : So?
S01 : It is less than 1

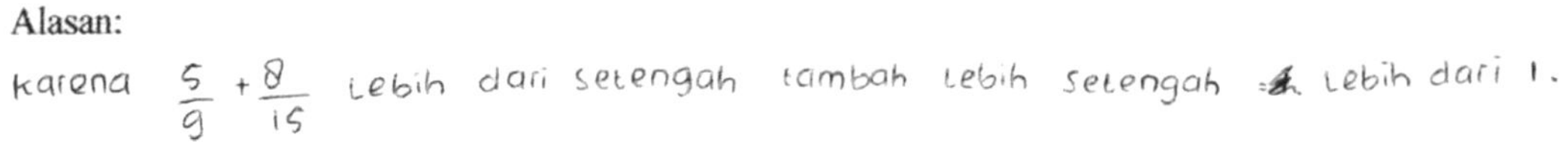

Figure 3. S01's written answer.

This student knows that $\frac{5}{9}$ and $\frac{8}{15}$ is a fraction of which each value is greater than $\frac{1}{2}$, so if they are summed up $(\frac{5}{9} + \frac{8}{15})$ = greater than $\frac{1}{2}$ + greater than $\frac{1}{2}$ = greater than 1. She used one of number senses strategies called benchmark. The benchmark used in solving this problem is $\frac{1}{2}$. This similar idea also presented by 11 students.

Question number 4 was given to know the student's ability to estimate the result of fraction subtraction. Six students answered this question correctly, but only one student answered with number sense.

3.2 Essay test result

The essay test consisted of four questions. The first question was given to test the students' understanding of the density between the two decimal numbers. The following dialogue is an excerpt of interviews with students who understood the density of numbers.

P	:	How many decimal numbers are between 1.21 and 1.22? Please, Explain clearly!
S02	:	Too much
P	:	Why is so?
S02	:	It is because between two numbers there are infinitely decimal numbers.
P	:	Please, Explain your reason!
S02	:	It is because there are 1,211 ; 1,2111; etc, also 1,12; 1,1211; etc. Regardless we mention the numbers, they remain less than 1.22

Students who answered the questions and gave reasons above understood the density of numbers and knew that between two numbers there are infinite numbers. The correct answer was also given by 11 students.

Essay question number 2 and number 3 were given to know the ability of students in comparing two numbers. Eighteen students answered this question correctly. The following is the excerpt from the interview.

P	:	$A = \frac{5}{8}$. $B = \frac{8}{9}$, which one is greater?
S06	:	$\frac{8}{9}$
P	:	Why?
S06	:	Because it is closer to 1, so $\frac{8}{9}$ is greater than $\frac{5}{8}$. Because b also has fewer remaining namely $\frac{1}{9}$, so b is the answer.

This student used 1 as a mental reference and decided that $\frac{8}{9}$ is closer with 1 so $\frac{8}{9}$ is greater of both. In addition, these students also used residual theorems, in which a fraction that has less residue is larger than another one.

Essay question number 4 was given to know the ability of students to understand the effect of the multiplication operation. Eighteen students answered this question correctly (nine of them using the number sense strategy), there were also students who did the formal calculation or short multiplication method to answer the question. An interview excerpt is showed as follows.

S01	:	Consider the following multiplication! 3 x 0,2 3 x 0,5 Without doing the calculations, which one of the multiplications produces a greater result?
P	:	What do you think? Which one is greater?
S01	:	3x 0,5
P	:	Why is 3x 0,5 greater?
S01	:	Because they are multiplied by a larger number. Here, there is 3 in both questions, so 0.5 is greater than 0.2.
P	:	So?
S01	:	3 x 0,5 is greater than the two of them.

This student knew that a multiplication operation would produce a larger number when it was multiplied by larger numbers. The student was aware that the two given operations contain three, so the student must determine a larger multiplier, 0.5 or 0.2. Then he determined whether a larger or smaller multiplier had a larger result.

The problem with the right answer at least for the right reasons was essays question number 1, 2, and 3 about the density of numbers and comparing fractions. However, the students' understanding of numbers has shown progress in determining the right calculation strategies and also students have begun to use number sense strategies to solve the given problems. Number sense strategies used include re-composing numbers, using benchmarks, understanding the value of the place (number magnitude), assessing the reasonableness of the results, and knowing the effects of the operation. In addition, the highest score that students got after learning is 90%. This finding was in accordance with the study [14] which stated that number sense learning could provide positive results on student learning outcomes. These results also corresponded to those found in previous studies [6, 15] which suggested that number sense could be a starting point for students who wanted to understand the concept deeply.

Based on the result above, it can be seen that students have experienced the development of understanding. However, it was also found that most students had difficulty in subtracting and comparing fractions. During the learning process, these difficulties could be overcome by guiding students when working on material or answering questions. In addition, the need for improvement is to add a special task for students who have difficulties with certain materials so that they can continue learning. Student's worksheet and other assessment instruments have reached the desired indicator so that it can be used without revision.

4. Conclusion

The test results and student interviews showed progress toward students' understanding of numbers. The results of the first test showed that most students could not determine the appropriate calculation strategy to answer the question and no one student used number sense strategy. However, after the learning was completed, students began to show progress where students have been able to determine the appropriate algorithm to answer the problem and also already used the number sense. The number sense indicators used were re-composition of numbers, using benchmarks, understanding the value of the place (number magnitude), assessing the reasonableness of the results, and knowing the effects of the operation.

Learning materials developed were good in terms of preparing the learning steps and division of the group, but need a little revision on the part of the special assignment for students who experienced obstacles which were expected to help students could continue the learning process. In addition, student worksheets and other assessment instruments can be used without revision.

References

[1] Sutrisno 2015 *Aksioma* **6** 1
[2] Ramlah, Bennu and Paloloang 2016 *Jurnal Ilmiah pendidikan Matematika* **1** 182
[3] Nuraini, Suhartono and Yuniawatika 2016 *Sekolah Dasar Kajian Teori Dan Praktik Pendidikan* **25** 168
[4] Witri G, Putra Z H and Nurhanida 2015 *Proceeding: 7th International Seminar on Regional Education* **2** 755
[5] Usher E L 2009 *American Educational Research Journal* **46** 275
[6] Yang and Hsu C J *International Electronic Journal of Mathematics Education* **4** 92
[7] Kroesbergen E H, Noordende J E and Kolkman M E 2012 *Reading, Writing, Mathematics and the Developing Brain: Listening to Many Voices* **6** 295
[8] Westwood P 2008 *What Teacher Need to Know About Numeracy* (Australia: Acer Press)
[9] Alsawaie ON 2011 *International Journal of Science and Mathematicsematic Education* **10** 1071
[10] van Nes F 2009 *Young Children's Spatial Structuring Ability and Emerging Number Sense* (Fisme: Universiteit Utrecht) chapter 2 p 17
[11] Aghileri J 2006 *Teaching Number sense* (London: Continuum International Publishing Group)
[12] Helmy F H [Tesis] Banda Aceh, Aceh: Magister Universitas Syiah Kuala 2018
[13] Purnomo Y W and Alyani F 2014 *International Education Studies* **7** 73
[14] Saputra WS, Rukayah and Syarifuddin M 2013 *Number sense Jurnal Dikdaktika Dwija Indria* **2** 1
[15] Soot S and Mackey M 2015 *International Journal of Humanities Social Sciences and Education* **2** 14

Patterns playing for early childhood education: Mathematics learning for early childhood education

Y Yuhasriati and A Yuriansa
Universitas Syiah Kuala, Jln. Teuku Nyak Arief Darussalam, Banda Aceh 23111, Indonesia

E-mail: yuhasriati@unsyiah.ac.id

Abstract. Mathematics learning for children should be conducted through playing because children's interests are highly on games and playing is one of the main ways how children learn. Thus, games should be creatively designed to develop children's potential. This research focused on developing mathematics learning models to improve children's competencies in creating patterns. This study aimed to develop patterns playing model for five to six years old learners. The research was conducted for a year at one of the kindergartens in Aceh Besar, Indonesia. The study carried out in this phase was to develop the models conceptually. The findings of this study showed that patterns playing model for learners of the ages of five to six were the patterns of shapes, colors, sizes and numbers. The proposed activities were arranging objects, coloring, sticking, and making beaded bracelets. The types of the patterns playing that were developed included repetitive patterns, growing patterns and symmetrical patterns. The developed model of patterns playing can be used in learning to develop children's cognitive aspect, namely the ability to create patterns.

1. Introduction

Early childhood is considered as golden period (golden age), specifically children whose age of 0 to 6 years old. According to Hurlock [1] the development of children's intelligence at this age reaches 80% and therefore it should be carefully dealt with and well-planned to achieve the optimal development. As Santrock [2] explained that children grow in a sensitive period in their early years, in addition, children can easily receive certain stimulations at this age.

Education for early childhood in Indonesia is known as *Pendidikan Anak Usia Dini (PAUD)* which can be implemented both formally and informally. The regulation of the Ministry of Education and Culture of the Republic of Indonesia Number 137 Year 2014 mentioned that the implementation of PAUD is classified into 3 categories based on the children's ages: 0 to 2 years old is categorized as childcare, named as *Taman Penitipan Anak (TPA)*; 2-4 years old is categorized as playgroup, named as *Kelompok Bermain (KB)*; and 4-6 years old is categorized as kindergarten named as *Taman Kanak-Kanak (TK)* [3]. Early childhood education is one of the implementations of education focusing on improving six aspects including religion and moral, social emotional, language, physic and motoric, cognitive, and art.

Children's ability to create patterns is one of the cognitive developments. The development may occur when it is stimulated by fun activities in the classroom. One of cognitive abilities for early childhood learners of age 5 to 6 is problem solving skill, as listed in the regulation of the Ministry of Education and Culture of the Republic of Indonesia Number 137 Year 2014 about children's development achievement standard known as *Standar Tingkat Pencapaian Perkembangan Anak (STPPA),* which includes the ability to create patterns.

Developing children's ability to create patterns can be done through play. Ahmad [4] stated that children learn while playing. Children's ability to create patterns can assist them to understand the relations between objects, shapes, and numbers [5]. Early childhood learners should be trained to form patterns as this can improve their abilities of sorting, classifying, identifying shapes, and creating graphs. The children's ability to create patterns can be one of their basic skills in mathematics.

Preliminary studies at TK IT Azkia Cadek Aceh Besar showed that the problem that occurred at the institution was lack of mathematical games in developing early childhood cognitive abilities, especially in creating patterns. Based on this reason, the researchers were interested in investigating mathematics learning for early childhood education to improve the patterns playing model for 5 to 6 years old learners.

The research questions were as follows.

1. What were the activities of patterns playing in the learning process of early childhood learners?
2. What were the possible patterns playing models to develop children's ability to create patterns?

1.1. Mathematics learning for early childhood learners
Early childhood learners are children of age 0 to 6 years old. Suyanto [6] mentioned the characteristics of the learners are: 1) highly curious, 2) having unique personality, 3) enjoying fantasies and imaginations, 4) having potential period of learning, 5) having egocentric behavior, 6) having short concentration range, and 7) part of society. Based on these characteristics, Suyadi [7] asserted that early childhood learning should be done through meaningful games designed by teachers considering the concepts and the learning process.

Mathematics is one of cognitive aspects that should be introduced to children in their early years to achieve the standard cognitive development known as *Standar Tingkat Pencapaian Perkembangan Anak (STPPA)* as listed in the regulation of the Ministry of Education and Culture of the Republic of Indonesia Number 137 Year 2014. As Hurlock [1] explained that the early five years of a child's life seem to be the most essential period to their future development. This statement is in line with Bloom's suggestion that children's intelligence reaches 50% at the age of 4 and 80% at the age of 8 [8]. Therefore, children's potentials, including mathematics, need to be developed in their early years.

Educative games can be employed to introduce mathematics to children as they learn by playing and trying. The main principle of early childhood learners is learning by playing or playing by learning, as Risaldy [9] affirmed that teachers need to design playing activities by incorporating educative elements in order to ensure learners learn and to create a fun learning. Mathematics learning should be conducted using games allowing the learners to work and learn both by themselves and in group. Mathematics learning could be carried out in an integrated way using certain themes related to the learners' life and experiences. Mathematics concepts for 5 to 6 years old learners are about shapes, colors, sizes, and patterns. The achievement levels are: 1) classifying objects based on their shapes, colors, or sizes; 2) Introducing patterns AB- AB, ABC-ABC and ABCD-ABCD; 3) Arranging objects in five different sizes and colors, 4) creating patterns.

1.2. Patterns playing
Educative games play an important role in the implementation of learning activities in the early childhood education as children learn from playing [9]. The playing activities designed to improve their abilities are related to the concepts of shapes, colors, sizes and patterns conducted by the patterns playing activities. The patterns that are created include repetitive patterns, growing patterns and symmetrical patterns [10].

Repetitive patterns are obtained from repeating objects based on their colors, shapes, or sizes. The ABC-ABC patterns are from the objects repetitions based on colors, such as red, yellow, green, red, yellow, green, red, yellow, green, etc. The shapes of ABC-ABC patterns are from repeating the objects based on shapes, for example triangle, circle, square, triangle, circle, square, triangle, etc. The shapes of ABC-ABC are created from repeating sizes, for example arranging pencils based on sizes, short, medium, long, short, medium, etc.

Growing patterns are obtained from the objects growth based on their colors, shapes, and sizes. For example, gradation or degradation colors, such as arranging objects from large to small sizes; short to long; little amount to many and vice versa.

Symmetrical patterns are created from reflections of objects based on colors, shapes, and sizes. The symmetrical patterns based on colors, for example red, yellow-yellow-red (ABBA). The patterns based on shapes, for instance leaf and wing shapes. Based on sizes, for example large, small, small, and large. The activities of patterns playing can be applied by involving learners individually or in group to continue the sequencing based on colors, shapes, and sizes. The activities can be conducted by arranging objects, coloring, sticking, and making beaded bracelets which in turn lead to repeating, growing patterns and symmetrical patterns.

2. Method

Research and Development method was applied in this study. This method was chosen in accordance with the aims of the study which was to develop patterns playing model to be implemented in early childhood's learning.

Patterns playing model developed in this research was conceptual and procedural. However, the focus was only on the conceptual model. The phase of development model used in this research was Constructivist Instructional Design (CID) [10] and the steps for developing the model referred to Borg and Gall [11]. The development of this playing model was based on the constructivism theory designed by Piaget and Vygotsky (social constructivism). The development process is known as 4D (Divine, Design, Development, and Dissemination).

The research was conducted in one of early childhood schools in Aceh Besar, Indonesia. The school was chosen mainly because of its lack of mathematics games designed for classroom activities. This study was carried out for one year and the subjects for this study were 5 to 6 years old learners in the school. Data collected for this research were the result of discussion with expert team, the result of the design of patterns playing, and the test result of the effectiveness of patterns playing model. For data analysis, descriptive qualitative was implemented.

3. Result and discussion

The research findings were in accordance with the aims of this study: developing patterns playing model for 5 to 6 years old learners. Based on the results, some findings were identified: (1) the assessment instrument could be used in the activity of creating patterns for 5 to 6 years old learners, and (2) the designed model of patterns playing was for 5 to 6 years old learners.

3.1. Model adaptation

The model development of patterns playing was conducted by adapting learning activities which were normally implemented in the research location (PAUD). They were: coloring, arranging objects, sticking and making beaded bracelets. The four activities were developed into patterns playing model to enhance learners' cognitive aspects. This model is one of alternatives for integrated activities to develop other aspects of learners' development.

Furthermore, educative patterns playing were organized in this study. The playing activities needed to fulfill certain criteria namely: (1) having specific aims, (2) incorporating learning activities, (3) using didactic methods, (4) using media/educative equipments, (5) having playing procedures and (6) having assessment instrument. The sixth criteria were used as reference to analyze the activities of patterns playing.

The first criterion was related to the aims of study as listed at the regulation of the Ministry of Education and Culture of the Republic of Indonesia Number 137 Year 2014 about learner's development achievement standard known as *Standar Tingkat Pencapaian Perkembangan Anak (STPPA).* The second criterion described the contents of learning (referring 2013 curriculum) that should be inserted in the playing patterns model activities. The learning contents consisted of religious moral values, literacy, science, mathematics, social emotional, and arts. The third criterion was related to implementation of didactic method and it was linked to some learning methods including fun, contextual, and meaningful learning. The fourth criterion was related to the media. The patterns playing model needed several media and equipment to support the activities. The fifth criterion emphasized that the activities of patterns playing should have clear and specific procedures. The procedures were the technical instructions for the subjects implementing the activities. The last criterion described that patterns playing model should assess both the cognitive development and the activities of the learners.

3.2. The model development of patterns playing

The model obtained satisfied the pre-defined criteria: (1) having specific aims, (2) incorporating learning activities, (3) using didactic method, (4) using media/educative equipment, (5) having clear procedures and (6) having assessment instruments. The activities were referred to the patterns which were normally implemented in the research location (PAUD): coloring, arranging objects, sticking and making beaded bracelets.

The patterns playing activities of coloring were supported by several tools such as papers, crayons or coloring pens, pictures of pearl necklace, pearl bracelet and house fences. The patterns resulted from the activities were coloring and repeating patterns.

The patterns playing activities of organizing objects were conducted by using geometry, parts of twigs, and small stones. The patterns resulted from the activities were patterns based on sizes: large to small and vice versa.

The patterns playing activities of sticking was carried out by using colorful papers in five sizes. Some equipments were required for sticking activities, such as ice cream sticks, glue and cardboards. The results of these activities were size and color patterns. The types of pattern obtained were repeating and growing patterns: large to small size and vice versa.

The patterns playing activities of making beaded bracelets were conducted by using three kinds of flowers, three colors of beads, and threads. The patterns of shape, size, color and numbers were obtained from these activities. The patterns formed were repeating and growing patterns: large to small size and vice versa.

The materials used for patterns playing in this study were both artificially and naturally created. The artificial ones were solid geometry, beads, ice cream sticks, and papers. Meanwhile, the researcher also took several natural materials such as stones, twigs, and leaves. Patterns playing model with various activities was developed to improve learners' cognitive skills and fine motor abilities and to train learners' patience.

The activities of patterns playing developed in this research were the activity of coloring, arranging objects, sticking, and making beaded bracelets. The patterns playing models developed were a solitaire game model, play in groups, and competition.

The model of pattern playing was designed to develop children's cognitive abilities. The developed model can be used to play alone or play in groups because children's cognitive development is influenced by social activities. This is consistent with Vygotsky's statement as cited by Kuntjojo [13] that children's cognitive development cannot be separated from social and culture activities

Pattern playing is done through coloring, composing, sticking, and dancing. These activities are useful to stimulate the development of early childhood multiple intelligence, cognitive abilities, fine motor skills, patience and art. Therefore, learning process in PAUD institutions must be carried out in an integrated manner. This is as stated in the Minister of Education Regulation Number 146 of 2014 concerning 2013 Early Childhood Education Curriculum that learning for early childhood is carried out in a planned and integrated manner [14].

The developed model was designed specifically for children aged 5-6. The activities included the shape patterns, color patterns, size patterns and number patterns. This is in accordance with Piaget's theory that the stage of cognitive development of children aged 2-7 years is still in the Pre-Operational period which means that the child still tends to focus on the most interesting characteristics of a stimulus, the child has not been able to reflect on and integrate various characteristics of the stimulus and the child is not yet able to reason rationally [15].

4. Conclusion

Based on the results, it could be concluded that: (1) the development of patterns playing model was conducted by adapting the routine activities of 5-6 years old learners such as arranging objects, coloring, sticking, and making beaded bracelets. (2) The materials used by 5-6 years old learners for patterns playing activities were solid geometry, ice cream sticks, papers, crayons, coloring pencils, twigs, flowers, and leaves. (3) The types of patterns developed in the patterns playing for 5-6 years old learners were repeating, growing and symmetrical patterns. (4) The patterns model for 5-6 years old learners were the patterns of shapes, colors, sizes and numbers. (5) Patterns playing models with various activities were developed to improve learners' cognitive skills and fine motor abilities, and to train learners' patience.

References

[1] Hurlock E B 1998 *Child Development* 6th edition (Jakarta: Erlangga) pp 76–87
[2] Santrock J W 2005 *Child Development* (New Jersey: McGraw Companies) p 46
[3] Peraturan Menteri Pendidikan dan Kebudayaan Republik Indonesia No 137 tahun 2014 *Standar Nasional Pendidikan Anak Usia Dini*
[4] Ahmad, Anizar, Yuhasriati and Nurchasanah S 2017 Introducing numbers to early childhood children by using number cards in PAUD Negeri 2 Banda Aceh *Indonesian Journal of Early Childhood Education Studies* **6** 94–9
[5] Samuelsson I P and Fleer M 2008 *Play and Learning in Early Childhood Settings: International Perspectives* (Melbourne: Springer) p 84
[6] Suyanto 2005 *Konsep Dasar Anak Usia Dini* (Jakarta: Departemen Pendidikan Nasional) p 9
[7] Suyadi 2010 *Psikologi Belajar PAUD* (Yogyakarta: Pedagogia) p 16
[8] Suryadi 2015 *Teori Pembelajaran Anak Usia Dini dalam Kajian Neurosain* (Bandung: Rosdakarya) p 33
[9] Risaldy S 2014 *Bermain, Bercerita, dan Menyanyi* (Jakarta: Luxima) p 7
[10] Lestari 2011 *Konsep Matematika untuk Anak Usia Dini* (Jakarta: Kementerian Pendidikan Nasional) p 83
[11] Willis J 2000 The maturing of constructivist instructional design: Some basic principles that can guide practice *Educational Technology* **40** 5–16
[12] Borg and Gall 1983 *Educational Research: An Introduction* (New York: Longman Inc.) p 772
[13] Kuntjojo 2013 *Strategi Pembelajarn untuk Anak Usia Dini* Available from: https://ebekunt.wordpress.com/2010/07/27/strategi-pembelajaran-untuk-anak-usia-dini/
[14] Peraturan Menteri Pendidikan Nomor 146 tahun 2014 *Kurikulum 2013 Pendidikan Anak Usia Dini* (Jakarta: Kementerian Pendidikan dan Kebudayaan)
[15] Hijriati, 2017 Pengembangan model pembelajaran anak usia dini *Bunayya: Jurnal Pendidikan Anak* **3** 74-92

IOP Conf. Series: Journal of Physics: Conf. Series **1088** (2018) 012100 doi:10.1088/1742-6596/1088/1/012100

Teachers' activities during designing Higher-Order Thinking Skills (HOTS) mathematical questions through teacher assisting program

Wahyuni[1,2], R Johar[2] and M Duskri[3]
[1]Educational Quality Insurance Institution, JL. Banda Aceh Medan Km 12,5, Aceh Besar 23361, Indonesia
[2]Syiah Kuala University, Jl. Teuku Nyak Arief Darussalam, Banda Aceh 23111, Indonesia
[3]Ar-Raniry State Islamic University, Jl. Syeikh Abdul Rauf , Banda Aceh, 23111, Indonesia

E-mail:rahmah.johar@unsyiah.ac.id

Abstract. Support for primary school teachers when designing mathematical questions would be crucial since many teachers were not able to develop mathematical questions that could train students to reach higher order thinking skills. This study aims to provide teachers with guidancein designing HOTS questions as a part of the development process of teaching instruments by using Plomp model. This paper only explains teachers' activities during the assisting programs in the prototyping phase. The assisting programs were conducted in four meetings and involved ten primary school teachers in Banda Aceh, Indonesia. The findings showed that teachers' activities during the first assisting program in designing HOTS questions were not as effective as had been intended, due to the lack of teachers'attention to the tutor's instructions during the program. Meanwhile, the second assisting program showed an effective result on all components of teachers' activities. This program could potentially be applied to other mathematics topics involving many teachers, which is the main program in the next phase of this development research.

1. Introduction

HOTS involves a complicated thinking process, which does not offer a specific algorithm to solve a particular mathematical problem, it is unpredictable and uses different approaches from the previously existing problems [1]. Moreover, higher-order thinking is nonalgorithmic, that is, the path of action is not fully specified in advance, tends to be complex, often yields multiple solutions, each with costs and benefits, rather than unique solutions, involves nuanced judgment and interpretation, and efforts. There is considerable mental work involved in the kinds of elaborations and judgments require [2]. In other words, HOTS could train someone to think critically and creatively. The characteristics of HOTS are to solve mathematical questions without algorithm and to produce multiple solutions [1].

Many teachers were not able to develop higher order thinking skills. A study showed that when teachers were required to write a question representing the Higher-Order Thinking (HOT) category, 60% of the teachers were only able to write a question from the Lower Order Thinking (LOT) category [3]. Hence, an assisting program to help teachers in designing HOTS mathematical questions would be necessary. The assisting program aims to provide enforcement and supports for teachers to improve their teaching professionalism in designing HOTS mathematical questions. The assisting program means to create a strategy of "making the best of the client's resources" [4]. While several studies were conducted on designing HOTS questions [1,5,6,7], it seems that no study has been conducted to assist teachers in designing HOTS mathematical questions. This paper only explains the activities of teachers during the assisting programs. It is important to observe teachers' activities during the assisting programs to yield an active learning environment. Also, observing teachers' activities could monitor the effectiveness and efficiency of particular activities, as well as could optimize the educational goals that are intended for those activities [8]. In this study, the teachers' activities include the activities of listening to the explanation, designing mathematical questions, paying attention to the colleagues' explanations, asking questions, responding to questions, making conclusions and assessing irrelevant attitudes during the program [9].

2. Method

This study used a Research and Development (R&D) approach, involving a model of Plomp that consists of three main phases: the preliminary research, prototyping phase and assessment phase [10]. The preliminary research phase involved an analysis of the teaching requirements for the mathematical topics that would be taught. The result of this phase suggested that most of the teachers were not able to create the indicators for mathematical questions of the topics. This paper specifically only explains the prototyping phase. In this phase, teaching instruments were developed, including tests, lesson plans, powerpoint slides, and observation sheets on teachers' abilities. This phase involved ten primary school teachers who were teaching at Year 5 in Banda Aceh, Indonesia. The teachers were selected, only those who were able to design mathematical questions on the level of C1 (knowledge) and C2 (comprehension) of Bloom's taxonomy. The participating teachers had the initials of AW, NA, RW, AI, IW, NM, YA, RN, SI, and ID. The teachers were between 25-45 years old and had teaching experiences at least five years.

Table 1. The criteria of the ideal times for teachers' activities during the assisting program.

No	The observed activities	Relevance percentage (P)	
		Ideal time	Tolerance level 5%
1.	Listening/paying attention to the tutor's explanation	25%	$20\% \leq P \leq 30\%$
2.	Designing HOTS questions as instructed and guided by the tutor	25%	$20\% \leq P \leq 30\%$
3.	Listening/paying attention to colleagues' explanation	10%	$5\% \leq P \leq 15\%$
4.	Asking question(s)/ expressing idea(s) to the tutor or colleagues	15%	$10\% \leq P \leq 20\%$
5.	Answering/responding to questions	15%	$10\% \leq P \leq 20\%$
6.	Making a conclusion about particular concepts or procedures	10%	$5\% \leq P \leq 15\%$
7.	Irrelevant attitudes during the assisting program	0%	$0\% \leq P \leq 5\%$

The assisting programs consisted of two different programs and were conducted in four meetings, wherein each program for two meetings. At the end of each program, the teachers who were in "Fair" and "Poor" categories were interviewed, as a reflection on the teachers' activities when designing HOTS questions. The interview result was used to improve the next meeting's program. Teachers' activities during the assisting program were observed by using observation sheets, which consisted of seven categories as shown in Table 1. There was one observer who examined the dominant activities performed by the teachers during the program. The observer noted the relevant codes and numbers of the teachers' activities every 10 minutes. The observation data were analyzed using percentages. An activity of teachers was effective once teachers could complete the activity at the time that was allocated during the initial training, with a tolerance level of 5%. Table 1 shows the relevance of teachers' activities to the proper allocated time for each activity during designing HOTS mathematical questions.

One of the activities during the first meeting in the assisting program was provided the participating teachers with a pre-test to evaluate their prior knowledge. The tutor then presented the teachers with the explanation of the assessment process of the knowledge domain using Bloom' taxonomy from C1 to C3. The process included the assessment technique, assessment instruments, classification of questions types, procedures of writing questions, and grading guidelines. After the presentation, a 30-minute-question-and-answer-session was held for clarifications on the topic of formulating questions on the knowledge domain. The teachers were then provided with a worksheet to write questions for a written test, which also included the basic competency, topics, indicators of competency accomplishment, indicators of the questions, question types, and the number of questions. Each teacher chose one of the basic competencies from Year 5 textbook to design the questions. The teachers then formulated the indicators of the questions based on the chosen basic competencies. The tutor guided the teachers who found difficulties when formulating the indicators of the questions. At the end of the first meeting, the tutor interviewed the teachers with difficulties in designing the questions. The interview result suggested that some of the teachers were not able to differentiate between the indicators of the questions and the indicators of competency accomplishment.

At the second meeting, the tutor conducted a presentation regarding the previous studies of HOTS and designing HOTS questions by teachers. The presentation also showed the results of the national examination of teachers' competencies and the differences between LOTS and HOTS questions. After the presentation, the tutor grouped the teachers, two teachers for each group, to discuss how to design HOTS questions. At the end of the meeting, a post-test was posed to the teachers and the reflection of the first program was also discussed together.

3. Results and discussion

Teachers' activities during the first assisting program on designing HOTS mathematical questions were not fully effective. This situation was because the teachers' activities in designing HOTS questions were not performed as intended and as explained by the tutor. Besides, the teachers did not possess enough capability of designing HOTS questions. Meanwhile, teachers' capability would be the main factor in the success of designing HOTS questions. The results of teachers' activities during the assisting program are shown in Table 2a and 2b.

Table 2a. Teachers' activities during the first assisting program.

No	Observation category	Percentage of teachers' activities during the assisting program (%)		Percentage of ideal time (%)	Average (%)	Effectiveness based on ideal time
		First meeting	Second meeting			
1	Listening/paying attention to the tutor's explanation	28,9	24,4	25%	26,6	Effective
2	Designing HOTS questions as instructed and guided by the tutor	17,8	20	25%	18,9	Not effective

Table 2b.

3	Listening/paying attention to colleagues' explanation	13,3	14,4	10%	13,8	Effective
4	Asking question(s)/ expressing idea(s) to the tutor or colleagues	18,9	18,9	15%	18,9	Effective
5	Answering/respondi ng to questions	11,1	14,4	15%	12,7	Effective
6	Making a conclusion about particular concepts or procedures	7,8	7,8	10%	7,8	Effective
7	Irrelevant attitudes during the assisting program	2,2	0	0%	1,1	Effective

Based on Table 2 the data suggested that the teachers' activities in designing HOTS questions were not fully effective. Further, the unsatisfying findings of the first assisting program were revised and improved for the second assisting program. The revision of the first assisting program is shown in Table 3.

Table 3. Findings and revision of the first assisting program.

No	Reflection	Findings	Revision
1.	Program I	Some teachers did not fully pay attention during the assisting program in designing HOTS questions. Thus, the teachers' activities did not run as efficiently as intended and instructed by the tutor.	The tutor encouraged and motivated all teachers to be able to design HOTS questions as had been instructed and guided before. The tutor also re-grouped the teachers and reset the sitting plan of the teachers, in order to encourage teachers to be more engaging during the assisting process

Teachers' activities during the first assisting program on designing HOTS questions were effective at level 85.7%. In other words, most of the activities have fulfilled the ideal time requirement. Only 14.3% of the teachers' activities were not effective, in which the activity to design HOTS questions did not run as instructed and guided by the tutor. This ineffectiveness was caused by some teachers who did not fully pay attention to the assisting program in designing HOTS questions.

The second assisting program was conducted twice at the third and fourth meetings. At the third meeting, the tutor explained the classification of the complex thinking (Higher-Order Thinking). The discussion session was also held afterward, to discuss how to differentiate and use the operational verbs of Bloom's taxonomy according to the cognitive level, as well as to relate the verbs to the indicators of HOTS questions. The last session at this meeting was that all the teachers tried to design HOTS questions. Furthermore, the second post-test was conducted at the fourth meeting, in which the teachers were asked to design HOTS questions consisted of level C5 of the Bloom's taxonomy. After the post-test, reflection, and evaluation were held among the teachers.

The findings of the second assisting program showed that teachers' activities in this program were effective, as the details could be explained in Table 4.

Table 4. Teachers' activities during the second assisting program.

No	Observation category	Percentages of teachers' activities during the assisting program (%)		Percentage of Ideal time (%)	Average (%)	Effectiveness based on ideal times
		First meeting	Second meeting			
1	Listening/paying attention to the tutor's explanation	31,1	17,8	25%	24,5	Effective
2	Designing HOTS problems as instructed and guided by the tutor	20	32,2	25%	26,1	Effective
3	Listening/paying attention to colleagues' explanation	13,3	16,7	10%	15	Effective
4	Asking question(s)/ expressing idea(s) to the tutor or colleagues	18,9	15,6	15%	17,2	Effective
5	Answering/responding to questions	10	11,1	15%	10,6	Effective
6	Making a conclusion about particular concepts or procedures	6,7	6,7	10%	6,7	Effective
7	Irrelevant attitudes during the assisting program	0	0	0%	0	Effective

According to Table 4 above, the observation findings showed that the teachers' activities during the second assisting program on designing HOTS questions were effective. This effectiveness was indicated by the fulfillment of the ideal times requirement when the teachers designed HOTS questions. The involvement of the teachers during the assisting activities was believed as one of the factors that supported the improvement of teachers' competency when designing HOTS questions at the second assisting program. During the assisting program, the teachers not only followed the tutor's instructions but also designed the HOTS questions based on the academical theories they acquired from the tutor. As suggested by a study [11], to achieve success in a training program, the learning activities in training are supposed to be supported by involvement of the participants in all activities, including in planning, conducting, and assessing the learning activities of the training. Also, the use of the participative model in teacher training is believed to be more effective, because the main objects of the training are teachers who generally already have the prior knowledge.

This study suggested that the assisting programs that were conducted in two cycles seem likely to improve the teachers' abilities to design HOTS questions. This idea is also relevant to the findings of a study [12], which showed an improvement in teachers' abilities to conduct a classroom action research, after following an assisting program on a competency-based reflective classroom action research. Also, another study [12], showed that the abilities to design a classroom action research of a training's participants could be optimized through tutorials, guidance, and assistance.

In this study, the observation of teachers' activities during the assisting program was conducted by an observer. The recapitulation of the effectiveness of teachers' activities during both of the assisting programs is shown in Table 5.

Tabel 5. The recapitulation of teachers' activities during the first and second assisting programs.

No	Observation category	Effectiveness based on the ideal times	
		Program 1	Program 2
1	Listening/paying attention to the tutor's explanation	Effective	Effective
2	Designing HOTS problems as instructed and guided by the tutor	Not effective	Effective
3	Listening/paying attention to colleagues' explanation	Effective	Effective
4	Asking question(s)/ expressing idea(s) to the tutor or colleagues	Effective	Effective
5	Answering/responding to questions	Effective	Effective
6	Making a conclusion about particular concepts or procedures	Effective	Effective
7	Irrelevant attitudes during the assisting program	Effective	Effective

Based on Table 4, only 85.7% of teachers' activities during the first assisting program were effective, which suggests that not all activities on designing HOTS questions were successful as intended by the tutor. Meanwhile, at the second assisting program, the tutor emphasized on teachers' participation in designing HOTS questions, by using the drilling method on the provided worksheets. This emphasis aimed to optimize the activity of designing HOTS questions as instructed and guided by the tutor, which was not effective in the first program.

The results of teachers' activities during the second assisting program were relevant to the intended indicators, in which all of the teachers' activities were effective. This relevance resulted from the active involvement of the teachers in designing HOTS questions. Also, the teachers were focussed on the operational verbs of the Bloom's taxonomy for the cognitive level C4 to C6, when designing the indicators of the HOTS questions. These findings were relevant to a study [11], that suggested that learning or assisting process should be designed to yield potential circumstances for the participants to create effective learning activities. Moreover, group activities during the programs could allow the participating teachers to discuss, interact, and share their ideas with each other. The success of the second program also resulted from the interest, motivation, and determination of all teachers to improve their abilities to design HOTS questions. Some of the factors that could influence someone's success include intelligence, interest, attention, and readiness [12]. Besides, the factor of teachers' prior knowledge, that teachers usually acquire from participating in seminars or training, might also be the supporting factor of the teachers' success in designing HOTS questions.

4. Conclusion

In conclusion, based on the assisting programs that were conducted for four meetings, teachers' activities during the first assisting program were ineffective at category designing HOTS questions as instructed and guided by the tutor. This ineffectiveness was caused by the lack of teachers' attention to what was instructed by the tutors when designing HOTS questions in the first assisting program. Meanwhile, in the second assisting program, all activities of the teachers were effective consequently teacher's ability in designing HOTS questions had also improved.

Reference

[1] Thompson T 2008 Mathematics teachers' interpretation of high-order thinking in blooms taxonomy *International Electronic journal of Mathematics Education* **3** 96

IOP Conf. Series: Journal of Physics: Conf. Series **1088** (2018) 012100 doi:10.1088/1742-6596/1088/1/012100

[2] Arends R I 2008 *Learning to Teach* (New York: Mc GrawHill Company)
[3] Harpster D L *A study of possible factors that influence the construction of teacher-made problems that assess higher-order thinking skills* [Dissertation] Montana State University-Bozeman 1999
[4] Payne M 1986 *Social Care in The Community* (London: MacMillan)
[5] Wahyuni D E and Alimufi A 2015 Implementasi pembelajaran *scientific approach* dengan soal *higher order thinking skill* pada materi alat-alat optik kelas X di SMA Nahdatul Ulama 1 Gresik *Jurnal Inovasi Pendidikan Fisika (JIPF)* **4** 32
[6] Rosnawati R 2009 *Enam Tahapan Aktivitas dalam Pembelajaran Matematika untuk Mendayagunakan Berpikir Tingkat Tinggi Siswa* (FMIPA UNY)
[7] Kusuma M D, Rosidin U, Abdurrahman and Suyatna A 2017 The development of higher order thinking skill (HOTS) instrument assessment in physics study *Journal of Research and Method in Education (IOSR-JRME* **7** 26
[8] Trimo 2013 Peningkatan aktivitas guru dan kemampuan mengolah nilai melalui pelatihan berbantuan microsoft excel pada guru-guru SDN 1 Magelung *Jurnal Progress PKPI2 FAI Universitas Wahid Hasyim* **1**
[9] Mukhlis *Pendekatan Matematika Realistik untuk Materi Pokok Perbandingan di Kelas VII SMP Negeri Pallangga* [Tesis] tidak diterbitkan, Universitas Negeri Surabaya 2005
[10] Plomp T 2013 *Educational Design Research: an Introduction,* Ed T Plomp and N Nieveen (Enschede: SLO) p 10
[11] Slameto 2010 *Belajar dan Faktor yang Mempengaruhinya* (Jakarta: PT Asdi Mahasatya)
[12] Sudjana D 1993 *Metoda dan teknik pembelajaran partisipatif* (Bandung: Nusantara Press)

IOP Conf. Series: Journal of Physics: Conf. Series **1088** (2018) 012101 doi:10.1088/1742-6596/1088/1/012101

The development of algebra teaching materials to foster students' creative thinking skills in higher education

M Mursalin[1], N L S Nuraini[2], H Purnomo[2], N W Damayanti[3], D Kristanti[4], A Rohim[5], R Widyastuti[6], Y O Wulandari[3], H Saleh[7], S N Mayangsari[3], M Fonna[1], R Rohantizani[1], I Muhammad[1], H Nufus[1], R Sulastri[8], R Amalia[9], N Nuraina[1] and M Muliana[1]

[1]Universitas Malikussaleh, Jl. Cot Tengku Nie Reuleut, Aceh Utara, Indonesia
[2]State University of Malang, Jl. Semarang No. 5, Sumbersari, 65145 Malang, Indonesia
[3]Wisnuwardhana University, Jl. Danau Sentani Raya No. 99, Madyopuro, 65139 Malang, Indonesia
[4]STKIP Bina Bangsa Meulaboh, Jl. Lintas Barat Sumatera, Meureubo, 23681 Meulaboh, Indonesia
[5]Universitas Islam Darul Ulum Lamongan, Jl. Airlangga 03, Lamongan, 62253 Jawa Timur, Indonesia
[6]Politeknik Kediri, Jl. Mayor Bismo No.27, Mojoroto, 64121 Kediri, Indonesia
[7]Universitas Madura, Jl. Raya Panglegur No. Km. 3,5, Barat, Panglegur 69317 Pamekasan, Indonesia
[8]Universitas Serambi Mekkah, Jl. Unmuha, Batoh, Banda Aceh 23245 Aceh, Indonesia
[9]Universitas Samudra, Jl. Meurandeh, Langsa 24354 Aceh, Indonesia

E-mail: mursalin@unimal.ac.id

Abstract. Students' creative potential must be developed. The purpose of this study is to develop algebra teaching materials to foster students' creative thinking skills at universities that are valid, practical and effective. The model of Plomp was used in this study that consisted of several stages including validation by experts, validation by practitioners and field trials. Based on the validation results, the algebra materials were valid. While from the observation results, the implementation test was considered practical and effective. Thus, algebra teaching materials can be used by lecturers in algebra courses in the mathematics education department in higher education.

1. Introduction

Every individual has natural ability to think creatively. This ability needs to be developed otherwise it will only be an inactive potential. Creative potential of an individual will be latent if not developed and formed [1,2]. The ability of creative thinking can also be referred to as divergent thinking [3]. Fong [4] said that creative thinking is a capability based on available data or information to find many possible answers to a problem where the emphasis is on the quantity, usability and diversity of the answers. Creative thinking can also be interpreted as a combination of logical thinking and divergent thinking based on conscious intuition [5].

Creative thinking is both authentic and reflective and produces a complex product [6,7]. The ability to think involves the activities of synthesizing ideas, building new ideas and determining their effectiveness. It also involves the ability to make decisions and produce new products. Creative thinking can be indicated as (1) a product of mental activity that has novelty and subjective value subjectively and objectively; (2) process of new thinking, i.e. transforming an initial idea and deciding whether to receive or to reject it; (3) the thinking process is characterized by a strong and stable motivation, and can be observed over time with high intensity [7,8,9]. The ability to think creatively like this is very important in learning mathematics, especially in problem solving.

The development of students' creative thinking is very important in the era of global competition considering the level of complexity of problems in all aspects of modern life is higher [10]. Creative thinking needs to be developed through active and creative learning, in order to guide students to practice solving problems from different perspectives so that they are able to deal with complex situations in their surrounding community [11,12]. To promote creative thinking, Kandemir suggested that the learning applied by teachers should be oriented to creativity that invites students to find their own solutions from various angles of view, the goal is to train the ability to think [2].

Basic thinking skills in learning mathematics are usually formed through activities that are convergent, that is the process of seeking the single most appropriate answer. These activities generally tend to be algorithmic, mechanistic, and routine. While the ability to think is divergent creative process to various directions and generates many alternative solutions [13]. The ability to think creatively can be useful for dealing with possibilities, and this ability has the characteristics most likely to be developed through mathematics learning [14,15,16]. Therefore, teachers who teach mathematics are expected to be able to help students in developing reasoning, logical thinking, logical systematic, creative, intelligent, sense of beauty, open attitude and curiosity. The success of learning, especially in developing students' creative thinking ability, is also very dependent on the teaching materials used.

Teaching materials are a set of learning resources that enable students and teachers to do learning activities [17,18,19]. The materials are systematically arranged to create an environment or atmosphere that allows students to learn independently, critically, creatively, and analytically [18]. One of the teaching materials that is intended in this study is algebra teaching materials used in linear algebra and elementary algebra courses for students at the mathematics education department in higher education.

Algebra is a branch of mathematics that deals with properties of operations and the structures which these operations are defined on [20,21]. The Elementary Algebra Course follows the study of arithmetic. This course is mostly preoccupied with operations on sets of whole and rational numbers and solving first and second order equations [22]. Algebra is a suitable subject to train students' creative thinking skill. Therefore, this study aimed at developing teaching materials that support students' creative thinking, because so far there has been no similar research conducted for a special college level on the development of algebra teaching materials in college.

The development of the teaching material was based on the development of the digital era today, where students were required to be able to be creative in dealing with real problems. Creativity is a manifestation of creative thinking ability. Therefore, the researchers were interested in conducting research by developing algebra teaching materials for elementary algebra courses that could foster students' creative thinking skills in higher education.

2. Method

The type of this research was a development research with reference to the development model developed by Plomp. The development steps consisted of (1) Preliminary research, (2) Prototyping phase and (3) Assessment phase. The research was conducted at three departments of mathematics education in Universitas Malikussaleh, State University of Malang and Wisnuwardhana University. The participants were 12 students from each university who were enrolled in the odd semester of the 2017/2018 academic year. The students came from a class with heterogeneous ability. The research also involved Algebra lecturers at the university who acted as observers for the implementation of the algebra teaching materials.

The quality criteria of the algebra teaching materials included valid, practical, and effective aspects. After the algebra materials were validated, they were then tested on the students at the Mathematics Education Department of Malikussaleh University. The types of data collected in this study were qualitative and quantitative. The qualitative data were obtained from the feedback and suggestions from validators and students. The quantitative data were obtained from the scores of test of students achievement.

Data analysis used descriptive qualitative and descriptive quantitative methods. Data analysis was performed to process the data from validation sheet, observation sheet, student response sheet and student ability test. The data were in the form of scores on the student's student validation sheet, observation sheet, student response questionnaire and the students' test results. The results of this analysis were used to determine whether the algebra teaching materials met the valid, practical, and effective criteria.

3. Result and discussion

3.1. Result of preliminary research phase

The algebra teaching material was developed using Plomp development research design which consisted of 4 stages: preliminary research phase, prototyping phase, assessment phase and dissemination phase. In this study, the dissemination phase was not conducted due to time and financial constraints. At the preliminary stage, the research group conducted direct observations at several departments of mathematics education in Indonesia including Universitas Malikussaleh, State University of Malang, Wisnuwardhana University, STKIP Bina Bangsa Meulaboh, Islamic University of Darul Ulum Lamongan, Kediri Polytechnic, Madura University, Serambi Mekah University and Samudra University.

During this phase, it was deduced that students' creative thinking skill was low with the average score below 60. Additionally, it was also found that students' motivation in learning algebra was also low and the availability of algebra books was limited. Based on these data, the researchers conducted this study in order to produce algebra teaching materials that could foster students' creative thinking skills in higher education. At this phase, the researchers designed plans and collected the materials to be organized into algebra teaching materials for the Elementary Algebra Course. The materials included content material, worked examples and exercises.

3.2. Result of prototyping phase

At the prototyping phase, the algebra teaching materials were validated by two professors of mathematics. Figure 1 shows a comment from one of the validators.

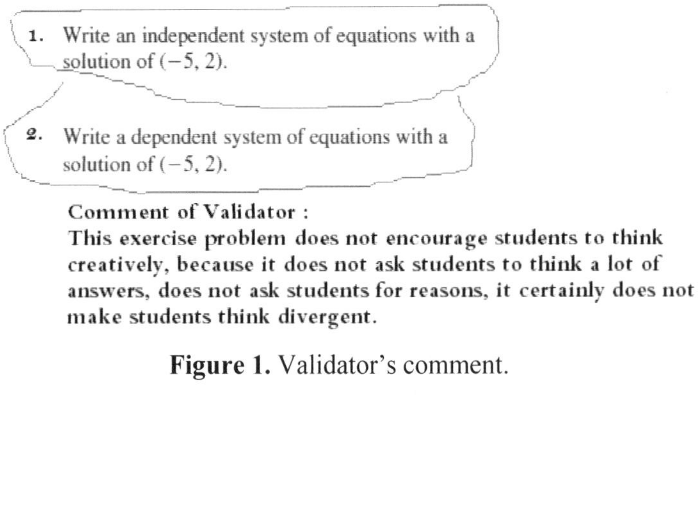
1. Write an independent system of equations with a solution of $(-5, 2)$.
2. Write a dependent system of equations with a solution of $(-5, 2)$.

Comment of Validator :
This exercise problem does not encourage students to think creatively, because it does not ask students to think a lot of answers, does not ask students for reasons, it certainly does not make students think divergent.

Figure 1. Validator's comment.

As shown in figure 1, the validator suggested revising the algebra problem because it did not promote students' creative thinking. Based on this comment, the researchers revised the problem. Figure 2 shows the result of the revision.

1. How many solutions does the system of equations graphed on the right have? Give three of the solutions. Is the system consistent or inconsistent?

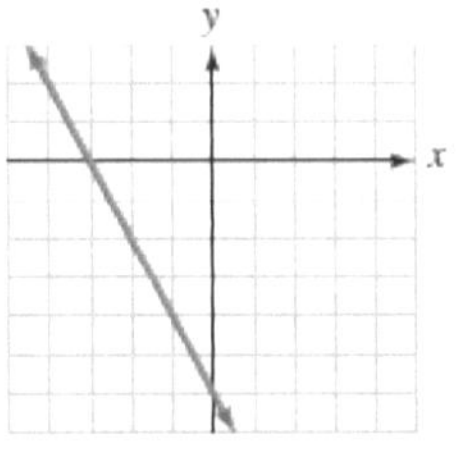

2. Can a system of two linear equations have exactly two solutions? Why or why not?

Figure 2. Result of revision.

The algebra teaching materials was revised three times in accordance with the comments of the validators. Finally, the validation result for the final revision was 82.15%, so it could be established that the algebra teaching materials were valid because the result was more than 70%.

3.3. Result of assessment phase

Trials were conducted to assess the practicality and effectiveness of the algebra materials. The trials were conducted at three departments of mathematics education in Universitas Malikussaleh, State University of Malang and Wisnuwardhana University. Each trials involved 12 students. The trials covered eight topics which were conducted in eight class meetings and one meeting for final test.

The class meetings ran from October 3, 2017 to December 23, 2017. During the class meetings, two observers were present in the classroom. The purpose of the observation was to observe the implementation of the algebra materials and activities in the class in using the algebra materials. Based on the results of field observations, the level of the implementation was in high category with the average score of 86.13%. Therefore, it could be concluded that the algebra teaching materials met the criteria of practicality.

Based on the results of the test of creative thinking skills and students' mastery of the learning content, the average score was 87.25%. Therefore, it could be concluded that the algebra teaching materials met the criteria of effectiveness.

3.4. Discussion of final product

The algebra teaching materials could be considered feasible if they met the criteria of validity, practicality, and effectiveness. The validity test measured the instruction, content, construction, conceptual, and language aspects. Based on the validation result of the algebra teaching materials, the average value was 82,15%. Accordingly, the algebra materials were included in valid criteria.

The algebra teaching materials met the criteria of practicality if the implementation was more than 70%. Data obtained from the observation showed that the average value of the implementation of the materials was 86.13%. Thus, the materials could be said to be practical.

The algebra teaching materials could be categorized effective if 75% of the students passed the final test. The data obtained from the test results of creative thinking skills showed that 87.25% students passed the test. Thus, it could be concluded that the algebra teaching materials were effective.

There were three aspects of learning mastery indicators on the algebra teaching materials: (1) creative thinking ability test as well as algebra content mastery test, (2) competency test and (3) group work result. All these indicators were achieved. Therefore, it could be concluded that algebra teaching materials were valid, practical, and effective.

Based on the result, it could be stated that the algebra teaching materials could be used by lecturers of mathematics in Indonesian universities because they have fulfilled the criteria of validity, practical and effectiveness. Nevertheless, not all learning methods are applicable with the algebra teaching materials that we developed. This study was limited to product design, it did not cover dissemination. This gap allows other researchers to conduct study on the materials' time effectiveness, suitable learning methods and more.

4. Conclusion

Based on the research, it could be concluded that the algebra teaching materials to foster students' creative thinking ability in higher education were feasible to use because they were proven valid, practical, and effective. The average value of the validity was 82.15%. Thus, the algebra teaching materials can be used without further revision. The average value of the practicality was 86.13%. The average score of the effectiveness test 87.25%. Therefore, the conclusion of this research was that the final product of algebra teaching materials had fulfilled the valid, practical, and effective criteria.

Acknowledgement

The authors would like to thank the Dean of the Faculty of Teacher Training and Education, Universitas Malikussaleh, Wisnuwardhana University, and Samudra University. The authors would also thank Prof. Dr. Bansu I Ansari, M.Pd. as validators and Head of Department of Mathematics Education, Universitas Malikussaleh who have facilitated this research, and all those who have assisted this research.

References

[1] Ahmar A S, Rahman A, Minggi I, Mulbar U, Upu H and Tiro M A 2017 The profile of creativity and proposing statistical problem quality level reviewed from cognitive style *J. Phys. Conf. Ser.* **954**

[2] Kandemir M A and Gür H 2009 The use of creative problem solving scenarios in mathematics education : views of some prospective teachers *Procedia - Soc. Behav. Sci.* **1** 1628–35

[3] Tabach M and Friedlander A 2016 Algebraic procedures and creative thinking *ZDM* (July)

[4] Fong N S 2004 Developing algebraic thinking in early grades : Case study of the Singapore primary mathematics curriculum *Math. Educ.* **8** 39–59

[5] Henriksen D, Richardson C and Mehta R 2017 Design thinking: A creative approach to educational problems of practice *Think. Ski. Creat.* **26** 140–53

[6] Yuli T and Siswono E 2011 Level of student's creative thinking in classroom mathematics *Educ. Res. Rev.*

[7] Fonna M and Mursalin M 2018 Role of self-efficacy toward students' achievement in mathematical multiple representation ability (MMRA) *Jurnal Ilmiah Peuradeun* **6** DOI 10.26811/peuradeun.v6i1.174

[8] Arnon D S 1988 Geometric reasoning with logic and algebra *Artificial Intelligence* **37** 37–60

[9] Mursalin M 2016 Pembelajaran geometri bidang datar di sekolah dasar berorientasi teori belajar Piaget *DIKMA Jurnal Pendidikan Matematika* **4** 250–258

[10] Runco M A 2012 Creative and imaginative thinking *Encyclopedia of Human Behavior* pp. 602–5.

[11] Valéria Š, Rumanová L and Pavlovi G 2014 Support of pupil's creative thinking in mathematical education *Procedia - Soc. Behav. Sci.* **116** 1715–9

[12] Bolden D S, Harries T V and Newton D P 2010 Pre-service primary teachers' conceptions of creativity in mathematics *Educational Studies in Mathematics* **73** 143–57

[13] Jeffrey B and Craft A 2004 Teaching creatively and teaching for creativity: Distinctions and relationships *Educ. Stud.* **30** 77–87

[14] Unal H and Demir İ 2009 Divergent thinking and mathematics achievement in Turkey: Findings from the programme for international student achievement (PISA-2003) *Procedia - Soc.*

Behav. Sci. **1** 1767–70

[15] Barak M and Levenberg A 2016 A model of flexible thinking in contemporary education *Think. Ski. Creat.* **22** 74–85

[16] Innabi H and El Sheikh O 2006 The change in mathematics teachers' perceptions of critical thinking after 15 years of educational reform in Jordan *Educ. Stud. Math.* **64** 45–68

[17] Norton S J and Irvin J 2007 A concrete approach to teaching symbolic algebra *Proc. 30th Annual Conf. Math. Educ. Res. Gr. Australas (Hobart)* vol 2 (Sydney: MERGA) pp 551–60

[18] Ansari S A and Rahman A 2017 Development of teaching material using an android *Glob. J. Eng. Educ.* **19** 72–6

[19] Susac A, Bubic A, Vrbanc A and Planinic M 2014 Development of abstract mathematical reasoning: The case of algebra *Front. Hum. Neurosci.* **8** 679

[20] Hall W 2005 *The Teaching of Algebra* pp 1–8

[21] Kaput J J 2008 What is algebra? What is algebraic reasoning? *Algebra in the Early Grades* 5–17.

[22] Ellis A B 2011 Algebra in the middle school: Developing functional relationships through quantitative reasoning *Early Algebraization: A Global Dialogue from Multiple Perspectives* pp 215–38

Effectiveness of the development of the inquiry-based learning model to improve students' psychomotor achievement

R A Siregar
Institut Pendidikan Tapanuli Selatan, Jl. Sutan Moh. Arif, Batang Ayumi Jae, Sumatera Utara 22733 Indonesia

E-mail: rabiyatuladawiyah8620@yahoo.co.id

Abstract. Students often consider chemistry as a difficult subject. Therefore, an engaging learning model is important in chemistry learning. This study aimed to produce an effective learning model to improve students' psychomotor achievement. This study used research and development using ADDIE model. The research was conducted in two senior high schools in Padang Sidempuan, North Sumatera. The participants were divided into an experimental class and a control class. The effectiveness of the learning model was investigated by assessing students' learning achievement, especially on the psychomotor aspect. The result showed that the inquiry-based learning model was effective. This result implies that the model can be used to improve students' psychomotor achievement.

1. Introduction

Chemistry studies specific phenomena that occur in substances and everything related to substances, namely the composition, structure and properties, transformation, dynamics and energetics of substances. Chemical materials are macroscopic, microscopic and symbolic. Judging from its content, learning chemistry involves skill and reasoning. There are two things related to the process of learning chemistry that cannot be separated, that is chemistry as a product and chemistry as a process (scientific work). Therefore, the learning process of chemistry should pay attention to the characteristics of chemistry as a process and product.

The goal of chemistry learning in high school is for students to gain experience in applying scientific method through experiments where students perform hypothesis testing by doing experiments, data retrieval, processing and interpreting data, and communicating experimental results orally and in writing. Students gain experience of scientific method by using inquiry-assisted learning model that includes presenting questions or problems, making hypotheses, designing experiments, experimenting to obtain information, collecting and analyzing data, and concluding. These syntaxes are only suitable for studying macroscopic chemistry with laboratory activities, but not suitable for microscopic and symbolic chemistry [1].

The learning process in the 2013 curriculum for all levels is implemented using a scientific approach which includes observing, questioning, trying/gathering information, reasoning and communicating. Scientific approach is a learning approach that adopts the steps of scientists in constructing knowledge through scientific method. This is in line with one of the goals of chemistry learning in high school which is to gain experience in applying scientific method through experiments where learners perform hypothesis testing by doing experiments, retrieving data, processing and interpreting data, and communicating experimental results orally and in writing.

Not every high school in Padang Sidempuan has adequate laboratory. Initial observation found that 50% of state high schools in Padang Sidempuan do not have their own chemical laboratory. This situation causes difficulties in performing chemically-related activities. The schools that do have lab equipment and chemicals have not made good use of them. This information was obtained from a survey conducted on 6 to 20 April 2015. The survey reveals that 50.00% of high school chemistry teachers in Padang Sidempuan carry out chemical practicum activities only 1 to 2 times in a year. This finding is supported by a research conducted by Festiyed [2] who found that about 19% of high school chemistry teachers made use of the laboratory more than 6 times in a semester, while 47% have not used laboratory. Learning chemistry outside of laboratories can affect students' scientific attitude, problem solving skills, scientific investigations and scientific acquisition skills [3].

The importance of conceptual understanding in the teaching and learning process greatly influences attitudes, decisions and ways of solving problems. Previous studies describe some problems in learning chemistry: students' skills in laboratory activities are still low, students' thinking ability is still low which causes low learning achievement, students have difficulty in explaining chemical terms, difficulty in understanding chemical concepts, and difficulty in solving problems related to calculation [4,5]. Other studies suggest that the causes of these difficulties are: poor understanding of prerequisite concept, teacher-oriented learning, the learning process focus more on memorization of information due to time constraints, and the teaching materials do not fit the characteristics of students [6,7].

Another problem is the teaching approach. Most teachers start teaching by explaining a topic, followed by giving questions and answers, and then end it with exercise. This monotonous routine does not encourage scientific attitude. In addition to that, in terms of assessment of learning achievement, teachers usually focus only on the cognitive domain. They often forget about psychomotor domain in conducting practicum activities. In order to improve students' chemical learning activities, it is necessary to develop a learning model that can collaborate learning approaches and models. The learning model developed is the inquiry-based learning model.

The use of inquiry-based learning model is expected to improve students' learning achievement in the psychomotor and affective aspects. Students are invited to actively conduct experiments or investigations and construct their understanding of a topic. This model can train students to observe macroscopic, microscopic and symbolic chemicals through practicum activities.

Inquiry-based learning is a learning model that provides adequate instructions for students through complete procedures and directive questions during the process of inquiry [8]. Teacher asks many questions during learning process to guide students to come to a conclusion on their own. The type of inquiry that is suitable for high school level is guided inquiry, since guided inquiry provides more directions for students who are not ready to solve problems with inquiry without assistance due to lack of experience and knowledge or have not reached the level of abstract cognitive development [9]. Through guided inquiry, a teacher can provide guidance and direction to students to conduct investigations. The formulation of the hypothesis is based on the information that the students obtain through listening to information from books and the internet.

Inquiry-based learning model is very helpful for teachers in implementing the 2013 curriculum because it uses scientific method. Therefore, this study aimed to develop an inquiry-based learning model to improve students' psychomotor achievement in chemistry.

2. Method

This study applied Research and Development (R&D). The product was an inquiry-based learning model that was designed to improve students' psychomotor achievement in chemistry. The study was conducted at senior high schools in Padang Sidempuan, North Sumatera, Indonesia. The ADDIE model was used in the development of the product. ADDIE stands for analysis, design, development, implement and evaluation. This paper focuses on the results of the evaluation phase.

The population in this study included all public high schools in Padang Sidempuan, North Sumatra consisting of 8 schools. The test subjects were chemistry teachers who taught in grade XI and the students. Selection of the test subjects was based on the criteria of schools' accreditation. Three high schools were chosen to participate in the study, they were SMA Negeri 2 Padang Sidempuan as high category, SMA Negeri 5 Padang Sidempuan as medium category and SMA Negeri 7 Padang Sidempuan as low category.

The effectiveness of the product was assessed by investigating students' psychomotor achievement. Friedman test was used to assess students' achievement. Data analysis used a normalized gain test. The hypothesis proposed was $P \neq 0.05$, meaning there was an improvement of psychomotor learning achievements with practicum activities in every meeting.

3. Result and discussion

This section discusses the results of the development of the inquiry-based learning model. The model used computer laboratory and simulation to improve students' learning achievement. The development model was the ADDIE model which included five steps: Analysis, Design, Development, Implementation and Evaluation. The analysis phases covered analysis of curriculum and students' characteristics. The result of the analysis phase showed that almost all chemistry lessons for grade XI could be done through lab work activities, except for Hydrocarbon and Petroleum. The analysis of students' characteristics showed that there was a need to improve the students' interests in learning chemistry. 0

A teacher needs to facilitate students to develop their thinking skills through their interactions with the teacher, with themselves and with the environment. After the implementation of the inquiry-based learning model, there was an improvement of students' psychomotor achievement with an average value of 0.35 as shown in figure 1.

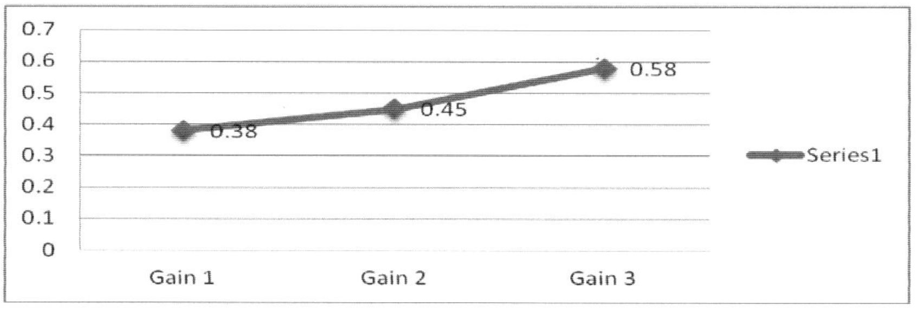

Figure 1. Students' psychomotor achievement.

Figure 1 shows that there was improvement in students' psychomotor achievement for every meeting. The assessment of the students' psychomotor achievement was done by using performance appraisal. The assessment covered the ability to prepare lab work, activities during the lab work, and the ability to deliver experiment results.

The significance of the students' psychomotor achievement for each meeting through practicum activities was analyzed using Friedman test. The value was 0.000. If this result was compared, then it was obtained that $P \neq 0.05$. Therefore, it could be concluded that there was a difference between the control group and the experiment group. Therefore, it could be stated that there was an improvement in students' psychomotor achievement using the inquiry-based model.

This result is supported by the results of research conducted by Naaman and Barne [9] who found that laboratory activities could facilitate students' psychomotor achievement. Furthermore, Hofstein, Shore and Kipnis [10] stated that the activities in the laboratory could provide an opportunity for students to learn in an authentic environment and build their knowledge of chemical-related phenomena. In addition, by conducting experiments in the laboratory students were able to practice inquiry skills such as posing questions, formulating hypotheses, and conducting further investigations of questions that have been asked.

The use of appropriate learning model in each learning was expected to produce a fun learning process and could improve students' competence or achievement. A teacher must be able to choose the appropriate learning model for their students. The application of the inquiry-based learning model had good impact on students' psychomotor domain. It could be seen that the value of the process of conducting practicum activities in collecting data was the highest value. The reason for this was because the students were highly motivated in conducting the experiment or practicum. Usually the students only listened to their teacher's explanation and then asked to answer the questions that the teacher gave. Now they had practicum activities. The chance to fulfill their self-actualization needs made the students to be in high spirit to learn and they were happy doing the tasks assigned by the teacher. One way to meet students' self-actualization needs was by creating meaningful learning associated with the students' real life and involving the students in the activity of knowledge discovery.

As for the students' ability in presenting experiment result, the score was lower. The reason was because the students paid less attention to the questions that hade been answered when analyzing the data. Furthermore, the students did not work collaboratively in making conclusions; they relied on their clever friends.

The components of the inquiry-based learning model which was developed in this research consisted of syntax, social system, principles, support system, and the instructional impact. The syntax consisted of: 1) presenting a phenomenon/problem; 2) formulating the problem; 3) proposing hypothesis; 4) collecting data; 5) testing the hypothesis; 6) elaborating knowledge relating to aspects (macroscopic, microscopic and symbolic); and 7) concluding. This syntax integrated scientific approach (observing, asking, collecting data/information, reasoning and communicating). The social system in the implementation of the inquiry-based learning model was that the teacher acted as a mentor and facilitator who helped the students to construct their understanding of chemistry to promote the students' sense of responsibility both individually and in groups. The principle reaction that emerged in the inquiry-based learning model was the teacher acted as a guide in the learning group in the discussion and practical activities that helped each other. The teacher acted as a facilitator who supported every learning activities and motivated the students to be able to construct information through learning activities using scientific measures. The support system in inquiry-based learning model developed in the form of inquiry-based learning model book, teacher book of inquiry-based learning model and student book of inquiry-based learning model. The instructional impact of the application of inquiry-based learning model was the success of student learning related to the improvement of cognitive, psychomotor and affective areas. The accomplishment of the application of inquiry-based learning model was the students' increasing interest in researching, self-confidence, openness, reading ability, learning motivation and cooperative spirit.

Overall achievement of student learning in the psychomotor domain was increased. This result was reinforced by Friedman's test analysis of 0.000. The value of significance was ≤ 0.05 then it could be said that there was a difference in students' psychomotor achievement between the students who were taught using inquiry-based learning model and the control group.

In learning with inquiry-based learning model, the students were required to conduct scientific investigation through formulating problem, making hypotheses, collecting data through experiments, analyzing data and concluding. All these activities required the students to play an active role in their learning process. This is in accordance with the theory constructivism which suggests that a good learning is a learning that actively involves the students in knowledge construction.

The impact of this activity was that the students were specifically required to be able to construct their own knowledge, attitudes and ability, and to use all their senses to seek or find answers to a question. The students was responsible directly to solve problems that the teacher provided. There was also an improvement in the students' performance and learning achievement especially in the psychomotor domain.

4. Conclusion

It was concluded that the inquiry-based learning model that had been developed was effective in improving students' psychomotor achievement.

References

[1] Eggen P and Kauchak D 2012 *Strategies and Models for Teachers: Teaching Content and Thinking Skills* (Boston: Pearson)

[2] Festiyed 2015 *Studi Pendahuluan Pengimplemetasian Kurikulum 2013 dalam Mengintegrasikan Pendekatan Saintifik melalui Model Inkuiri dan Authentic Assesment dalam Pembelajaran IPA di Kota Padang* Paper presented at: SEMIRATA 6-9 May 2015 Pontianak: Universitas Tanjung Pura

[3] Akani O 2015 Laboratory teaching: Implication on students' achievement in chemistry in secondary schools in Ebonyi State of Nigeria *Journal of Education and Practice* **6** 206

[4] Yakina, Kurniati T and Fadhilah R 2017 Analisis kesulitan belajar siswa pada mata pelajaran kimia kelas X di SMA Negeri 1 Sungai Ambawang *Ar Razi Jurnal Ilmiah* **5** 287

[5] Bayram Z, Oskay O, Erdem E, Ozgur S D and Sen S 2013 Effect of inquiry based learning method on students' motivation *Procedia-Social and Behavioral Sciences* **106** 988

[6] Suyono 2009 *Model Pembelajaran Kimia Berbasis Multipel Representasi dalam Meningkatkan Penguasaan Konsep Kinetika Kimia dan Keterampilan Berpikir Kritis Siswa* (Surabaya: UNS)

[7] Pratiwi D M, Saputro S and Saputro A N C 2015 Pengembangan LKS berbasis inkuiri terbimbing pada pokok bahasan larutan penyangga kelas XI IPA *Jurnal Pendidikan Kimia* **4** 32

[8] Gormally C, Brickman and Peggy 2011 Lessons learned about implementing an inquiry-based curriculum in a college biology laboratory classroom *Journal of College Science Teaching* **40** 45

[9] Rachel M N and Barnea N 2012 Laboratory activities in Israel *Eurasia Journal of Mathematics, Science and Technology Education* **8** 49

[10] Hofstein A, Shore R and Kipnis M 2004 Providing high school chemistry students with opportunities to develop learning skills in an inquiry-type laboratory: A case study *International Journal of Science Education* **26** 47

IOP Conf. Series: Journal of Physics: Conf. Series **1088** (2018) 012103 doi:10.1088/1742-6596/1088/1/012103

Developing mathematics training modules to enhance teachers' content knowledge

R Sulastri[1], R Johar[2], M Duskri[3], M Ikhsan[2] and H Meutia[1]

[1]Serambi Mekkah University, Jl. Unmuha, Batoh, Banda Aceh 23245, Indonesia
[2]Syiah Kuala University, Jl. Teuku Nyak Arief Darussalam, Banda Aceh 23111, Indonesia
[3]State Islamic University of Ar-Raniry, Jl. Syeikh Abdul Rauf, Darussalam, Banda Aceh, 23111, Indonesia

E-mail: rini.sulastri@serambimekkah.ac.id

Abstract. Teacher competence test is one way of measuring teachers' ability in pedagogic and content knowledge. The result of the competency test shows the low category because of teachers' difficulty in understanding and teaching several topics of mathematics. The purpose of this study was to develop six mathematics-training modules concerning the content knowledge of mathematics teachers satisfying the valid and practical criteria. This research employed the research and development approach of Borg and Gall model in ten phases. However, the development in this research was conducted only up to the eighth phase. The results showed that six mathematics-training modules satisfied the valid and practice criteria. The validity was indicated by the conformity of the materials and concepts for mathematics with teachers' problems, and the curriculum. The practical criteria were based on the conformity of the implementation of the training modules designed; the efficiency, usefulness, and attractiveness as the modules were designed into several different parts included the instructions and estimation time; and the legibility and the easy to use aspects.

1. Introduction

The Ministry of Education and Culture of Indonesia measures the academic professionalism of teachers the administering the teacher competency test (TCT) that is conducted every year. The TCT started in 2012 only for teachers who will participate in the certification and starting from 2015 the TCT is conducted to measure the professionalism of all teachers. The test aims to determine the level of individual teacher competence and map the teachers' mastery of pedagogical and professional competences. Thus, the TCT focuses on identifying teachers' weaknesses in the mastery of the pedagogical and professional competencies [1]. The TCT enable a teacher to obtain the information concerning his/her ability, and the results can be a trigger to improve his/her quality as a professional teacher.

Professional competence or content knowledge defined as knowledge of the subject matter or subject that is usually learned or taught [2]. It is further explained that the content includes knowledge of concepts, theories, ideas, the framework of thinking, methods of proof and its evidence, as well as their components [3]. Thus, professional competence is tightly intertwined with the ability of teachers to acquire their subject matter. In other words, pedagogical knowledge in learning is the teachers' method to teach and construct a mathematical concept following the purpose of learning [4].

Overall, the results of the TCT in 2015 were poor, but there was an increase in 2016. Aceh is one out of 34 provinces in Indonesia showing some changes. Aceh is ranked 23rd nationally in 2016. This showed improvement compared to the previous year being in the last third position [5]. Although, in general, the ranking of Aceh has increased, one of the districts, Pidie district, is still experiencing problems with 86.34% of teachers do not master the subject matter and cannot teach well [6]. The result of this TCT is used as a consideration of policy in the provision of coaching and development programs as well as awards and appreciation for teachers.

The results of the discussion with five mathematics teachers from different junior high school schools in Banda Aceh and the district of Pidie, Aceh, identified that most teachers had performed poorly in the TCT since 2015. One of the reasons is the materials in the pedagogical and professional competence test being too broad so that teachers find it challenging in understanding and solving the mathematics problems. This is in line with some obstacles in the TCT, included the type of problems, the language of the problem being difficult to understand, the advanced content of the question, and the lack of socialization about the implementation of TCT [7,8].

Based on the results of the TCT, the government has conducted various training to improve the quality of teachers related to their ability and learning. Some activities stop at the training phase without following-up on the results of the training. It is, therefore, necessary to continue to follow up the training activities for the short and long-term impact, such as the results of the TCT test in the following year and the final students learning outcomes.

Based on the issues previously mentioned, an effort is required to improve the pedagogical and professional competences of teachers by developing a mentoring model to be applied in the learning process. In developing the model, a module to be used in the training phase is required. Thus, this study is limited to obtaining a valid and practical training module to be used by Junior High School Mathematics teachers to improve their professional competence.

2. Method

This research was development research conducted to obtain a mathematics training module satisfying the valid and practical criteria. The development model used referred to the Borg & Gall model. Development research is oriented to develop and validate the products used in education [9]. The development consists of ten stages: research and information collection, planning, preliminary form of product development, preliminary field testing, main product revision, main field testing, operational product revision, operational field testing, a final product, revision, and dissemination and implementation. However, this research only conducted the first eight phases as this stage has obtained a product meeting the valid and practical criteria following the objectives of the study. The valid product was obtained from the validation carried out by three experts in the field of mathematics content and evaluation, while the practical product was derived from the usefulness and appearance of products used by users.

This research was conducted in Pidie district, Aceh, Indonesia involving ten junior high schools. The school was selected by purposive sampling method, based on the willingness of the school to participate in the study as well as the school level quality (medium or low). Two or three mathematics teachers represented each school.

3. Result and discussion

3.1. Research and information collection

This activity is related to the initial activity including the analysis of mathematics learning materials perceived to be difficult for students and teachers, literature study or review of previous research or reports, and field studies directly related to the products developed. Therefore, the steps taken are determining the subject of the study, the experts as the validator, the place and time of the study, as well as the subject for the pilot study. For the needs analysis, the District Education Department of Pidie, Aceh contributed in releasing the consent, determining the number of schools, teachers and

IOP Conf. Series: Journal of Physics: Conf. Series **1088** (2018) 012103 doi:10.1088/1742-6596/1088/1/012103

school supervisors as subjects of the study as well as providing the literature related to the score of Teacher Competency Test (TCT).

Regarding the schools involved, all schools recommended by the Education Department were directly followed up to confirm their involvement in the research and some schools were not willing to participate. Thus, it took about three months to obtain a desired number of schools to be involved. At this stage, we also received the results of TCT from the Education Department as one of consideration in conducting the study. The Education Department was supportive and helpful starting from selecting the school as the objects of the study to inviting the participants to each activity undertaken.

3.2. Planning

The planning conducted consisted of preparation of test items for pedagogical and professional competence in mathematics. This test aimed to determine teachers' initial ability and examine the materials that were challenging for teachers and difficult to teach to the students. The mathematics test was adapted from the problems of National Exam from the previous years. The problem used was a problem considered difficult and requiring reasoning to examine teachers' level of mathematics materials and concepts understanding.

At this stage, test and the Focus Group Discussion (FGD) were conducted by involving 45 participants consisting of supervisors, principals, and mathematics teachers, as well as the research team and the expert team of lecturers. The test was administered for three hours for two categories, followed by four hours of FGD located in one of the schools in Pidie district, Aceh. Based on the results of the test and the discussion, six mathematics topics were selected for additional training because they were difficult to be taught to students and difficult to understand by the teachers. Thus, a review of the literature related to the topics and problem identification was conducted, and the problems obtained were also summarized for the planning of training module design.

3.3. Develop preliminary form of product

The activities undertaken were designing the initial product developed namely the training module related to the six mathematics topics selected and considered to be the most challenging. The six topics were Arithmetic sequence and series, Geometry sequence and series, Comparison, Triangle, Plane geometry, and Statistics. Besides, a module validation and an observation guideline were also designed.

There were several stages of designing the training modules included: the introduction discussing the background of training and problems occurred in learning from both teachers and students point of view; aim; general guidance; sources, materials and tools; time; general outline of the activities; and the detail steps of the activity included time spent. The instructions and some notes provided for the facilitator related to the materials and the implementation of the activities were expected to facilitate the facilitator or user in using the module. In addition, there is also a worksheet in the training module that is tailored to the materials (Figure 1).

Revisions were made regarding the content and conformity of the instructions based on the validation results of the modules. Content-related revisions included the misplaced topics such as the Geometry sequence and series located in the Arithmetic sequence and series. Also, there is a problem in the worksheet considered to be less challenging, for example, the Arithmetic sequence problem concerning a sequence of numbers to determine the next term as this types of problem can directly be solved using the formula without a reasoning process.

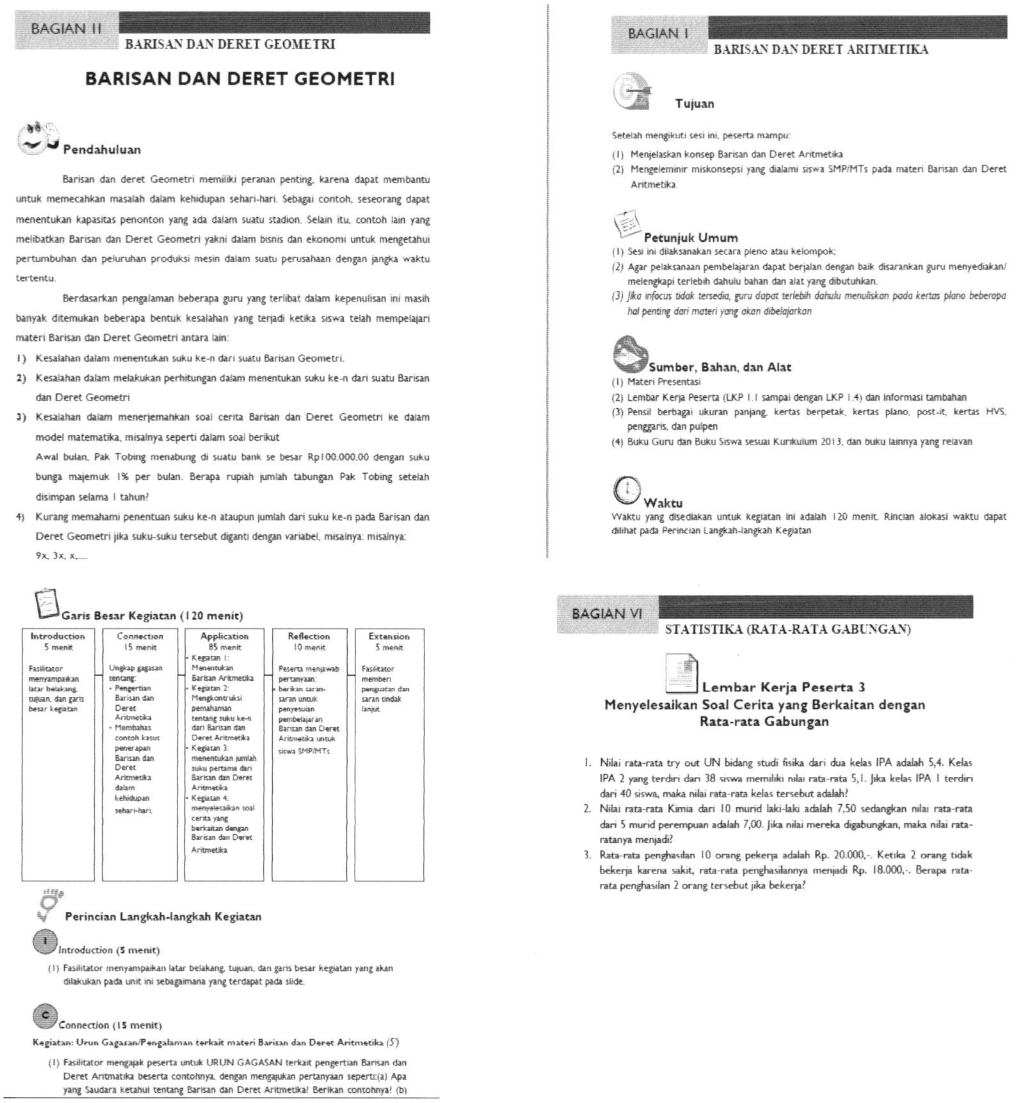
BAGIAN II
BARISAN DAN DERET GEOMETRI

BARISAN DAN DERET GEOMETRI

Pendahuluan

Barisan dan deret Geometri memiliki peranan penting, karena dapat membantu untuk memecahkan masalah dalam kehidupan sehari-hari. Sebagai contoh, seseorang dapat menentukan kapasitas penonton yang ada dalam suatu stadion. Selain itu, contoh lain yang melibatkan Barisan dan Deret Geometri yakni dalam bisnis dan ekonomi untuk mengetahui pertumbuhan dan peluruhan produksi mesin dalam suatu perusahaan dengan jangka waktu tertentu.

Berdasarkan pengalaman beberapa guru yang terlibat dalam kepenulisan ini masih banyak ditemukan beberapa bentuk kesalahan yang terjadi ketika siswa telah mempelajari materi Barisan dan Deret Geometri antara lain:

1) Kesalahan dalam menentukan suku ke-n dari suatu Barisan Geometri.
2) Kesalahan dalam melakukan perhitungan dalam menentukan suku ke-n dari suatu Barisan dan Deret Geometri
3) Kesalahan dalam menerjemahkan soal cerita Barisan dan Deret Geometri ke dalam model matematika, misalnya seperti dalam soal berikut
Awal bulan, Pak Tobing menabung di suatu bank se besar Rp100.000,00 dengan suku bunga majemuk 1% per bulan. Berapa rupiah jumlah tabungan Pak Tobing setelah disimpan selama 1 tahun?
4) Kurang memahami penentuan suku ke-n ataupun jumlah dari suku ke-n pada Barisan dan Deret Geometri jika suku-suku tersebut diganti dengan variabel, misalnya: misalnya: 9x, 3x, x,....

Garis Besar Kegiatan (120 menit)

Introduction 5 menit	Connection 15 menit	Application 85 menit	Reflection 10 menit	Extension 5 menit
Fasilitator menyampaikan latar belakang, tujuan, dan garis besar kegiatan	Ungkap gagasan tentang: - Pengertian Barisan dan Deret Aritmetika - Membahas contoh kasus penerapan Barisan dan Deret Aritmetika dalam kehidupan sehari-hari.	- Kegiatan 1: Menentukan Barisan Aritmetika - Kegiatan 2: Mengkonstruksi pemahaman tentang suku ke-n dari Barisan dan Deret Aritmetika - Kegiatan 3: menentukan jumlah suku pertama dari Barisan dan Deret Aritmetika - Kegiatan 4: menyelesaikan soal cerita yang berkaitan dengan Barisan dan Deret Aritmetika	Peserta menjawab pertanyaan: - berikan saran-saran untuk penyesuaian pembelajaran Barisan dan Deret Aritmetika untuk siswa SMP/MTs	Fasilitator memberi penguatan dan saran tindak lanjut

Perincian Langkah-langkah Kegiatan

Introduction (5 menit)
(1) Fasilitator menyampaikan latar belakang, tujuan, dan garis besar kegiatan yang akan dilakukan pada unit ini sebagaimana yang terdapat pada slide.

Connection (15 menit)
Kegiatan: Urun Gagasan/Pengalaman terkait materi Barisan dan Deret Aritmetika (5')
(1) Fasilitator mengajak peserta untuk URUN GAGASAN terkait pengertian Barisan dan Deret Aritmatika beserta contohnya, dengan mengajukan pertanyaan seperti:(a) Apa yang Saudara ketahui tentang Barisan dan Deret Aritmetika? Berikan contohnya? (b)

BAGIAN I
BARISAN DAN DERET ARITMETIKA

Tujuan

Setelah mengikuti sesi ini, peserta mampu:
(1) Menjelaskan konsep Barisan dan Deret Aritmetika
(2) Mengeleminir miskonsepsi yang dialami siswa SMP/MTs pada materi Barisan dan Deret Aritmetika.

Petunjuk Umum
(1) Sesi ini dilaksanakan secara pleno atau kelompok;
(2) Agar pelaksanaan pembelajaran dapat berjalan dengan baik disarankan guru menyediakan/melengkapi terlebih dahulu bahan dan alat yang dibutuhkan.
(3) *Jika infocus tidak tersedia, guru dapat terlebih dahulu menuliskan pada kertas plano beberapa hal penting dari materi yang akan dibelajarkan*

Sumber, Bahan, dan Alat
(1) Materi Presentasi
(2) Lembar Kerja Peserta (LKP 1.1 sampai dengan LKP 1.4) dan informasi tambahan
(3) Pensil berbagai ukuran panjang, kertas berpetak, kertas plano, post-it, kertas HVS, penggaris, dan pulpen
(4) Buku Guru dan Buku Siswa sesuai Kurikulum 2013, dan buku lainnya yang relavan

Waktu
Waktu yang disediakan untuk kegiatan ini adalah 120 menit. Rincian alokasi waktu dapat dilihat pada Perincian Langkah-langkah Kegiatan

BAGIAN VI
STATISTIKA (RATA-RATA GABUNGAN)

Lembar Kerja Peserta 3
Menyelesaikan Soal Cerita yang Berkaitan dengan Rata-rata Gabungan

1. Nilai rata-rata try out UN bidang studi fisika dari dua kelas IPA adalah 5,4. Kelas IPA 2 yang terdiri dari 38 siswa memiliki nilai rata-rata 5,1. Jika kelas IPA 1 terdiri dari 40 siswa, maka nilai rata-rata kelas tersebut adalah?
2. Nilai rata-rata Kimia dari 10 murid laki-laki adalah 7,50 sedangkan nilai rata-rata dari 5 murid perempuan adalah 7,00. Jika nilai mereka digabungkan, maka nilai rata-ratanya menjadi?
3. Rata-rata penghasilan 10 orang pekerja adalah Rp. 20.000,-. Ketika 2 orang tidak bekerja karena sakit, rata-rata penghasilannya menjadi Rp. 18.000,-. Berapa rata-rata penghasilan 2 orang tersebut jika bekerja?

Figure1. The layout of some revised training module.

3.4. Preliminary field testing

Before conducting the pilot study, the second FGD was conducted involving some experts in mathematics, evaluation, and education learning. The experts were limitedly involved in providing feedback to the designed modules. The second FGD results enabled some improvements to the module such as adding contextual questions to assist teachers in solving a problem. Later, a preliminary test was conducted on a limited scale involving three mathematics teachers at one of the junior high schools in Pidie district, Aceh. The test results show that most of the contents of the module were correct and appropriate with the mathematics topics, however, more PISA-like problems or problems requiring reasoning were required to be added.

A preliminary test was conducted on a limited scale involving three mathematics teachers in one of the junior high schools in Pidie, Aceh, Indonesia. The test results showed that most of the contents of the module are correct and in line with the mathematical content, however, the reasoning and PISA-like problems (Program of International of Student Assessment) required to be added to the modules.

3.5. Main product revision

The mathematics training module for six topics was revised based on feedback and suggestions from the preliminary study results. One of the revisions was to add PISA-like questions in the last section of the worksheet for each material. Furthermore, FGD was conducted to refine the developed product involving all research teams including the expert teams and validators to examine the development of the training modules designed. The discussion results suggested that a test should be conducted at the end of the training to determine teachers' ability to understand and solve PISA-like problems.

3.6. Main field testing

A limited test involving ten mathematics teachers from three secondary schools in Pidie, Aceh, Indonesia for the design of mathematics training module developed was conducted. It aimed to obtain information regarding the application of the product and the level of difficulty experienced by teachers in the trial period. The test results showed that the module was impressive with the challenging and varied problems ranging from the low level to PISA level. This leads the teachers to recall their understanding of the initial concepts related to the topics and solve the high-level questions requiring reasoning.

3.7. Operational product revision

Based on the results of the limited trial, the revision was made to the module by adding the final test of reasoning and PISA level problems. The FGD was conducted by involving the research team to obtain improvements and feedback to the module before the product was expanded. Based on the discussion results, it was agreed to add the tests related to the PISA-like problems at the end of the lesson, as well as correct some typos to obtain a valid and targeted training module. This was a second revision of the modules.

3.8. Operational field test

The revised product was further tested to 25 teachers from ten junior high schools in Pidie, Aceh, Indonesia. The training was conducted for three days involving four lecturers who were expert in the subject matter, mathematics learning, and education evaluation. The activity took place the whole day with different presenter and modules. In addition to product implementation during the training, there were also discussions and interviews with mathematics teachers and school supervisors to obtain the feedback of the training and examine the ability level and understanding of the materials. Most participants provided positive responses of the training because they rarely had the opportunity to participate in the training and they can also learn directly to the experts such as learning geometry material from the expert lecturer of geometry. Also, participants gained new knowledge related Higher Order Thinking Skills (HOTs) and the utilization of the problems. This stage was also conducted to determine the practicality of the module.

The development of the mathematics training modules was only conducted to the stage of obtaining valid and practical modules. The process did involve not only validation but also discussion through FGD involving the expert team, user, and research team. It was aimed to obtain the module designed following the problems faced by the mathematics teacher. These modules dealt with professional competences. According to the Government Regulation Number 74 the Year 2008, professional competencies include the competence for mastering the subject matter in breadth and depth in line with the standard contents of the educational unit program, subject, and/or group of subjects handled. It also covers the relevant scientific concepts and method and technological or relevant artistic that are conceptually coherent with the educational unit program, subjects, and/or groups of subjects handled. Thus, it is crucial to improving the professional competence.

Improving teacher quality is the main priority for the government through its policies. The activities conducted include providing learning resources and manipulative, improving schools' facilities, as well as teacher training [10]. Teachers can initiate to enhance their professional competence by reading educational books, writing scientific papers, following the actual news from the media, and participating in the training [11]. In this case, the training should not only focus on the technical training of teaching but also regarding the classroom governance at school, curriculum, and classroom management. Besides, teachers may involve in other activities such as attending a class working group, conducting classroom action research, and actively participating in professional organizations.

This module was developed to assist the government improves the competence of teachers, especially the professional competence. The development of this module satisfied the valid and practical criteria. The validity of the module is used to examine whether the module is adequate and in line with the expected objectives [12]. The analysis of module validation was done by experts including the lecturers and mathematics teachers with several revisions to obtain a valid module. The practicality was indicated by the application of the module during the training. Thus, the module met the needs of the teachers as a user.

4. Conclusion

The development of training modules related to mathematics content knowledge refers to the Borg & Gall development model including ten phases. However, this study was conducted only until the eighth stage as the valid and practical training modules have been obtained. The module considered to be valid based on the mathematics topics and concepts matching both the problems experienced by the mathematics teachers and the applied curriculum. The practical criteria are based on the conformity of the implementation of the training and the designed modules. It is also assessed regarding efficiency, usefulness, and attractiveness, as there are several different sections in the design as well as the instruction and estimated implementation time. Besides, also it is viewed based on the aspect of the legibility of the module and easy to use of the modules. Thus, this module is ready to be used on a large scale to help improve the content knowledge of mathematics teachers.

References

[1] Dinas Pendidikan Kabupaten Pidie 2014 *Data Hasil Uji Kompetensi pada 9 Kabupaten Mitra USAID Prioritas Provinsi Aceh untuk Kompetensi Pedagogik dan Profesional*

[2] Koehler M J and Mishra P 2009 What is technological pedagogical content knowledge (TPACK)? *Contemporary Issues in Technology and Teacher Education* **9** 60

[3] Shulman L S 1986 Those who understand: knowledge growth in teaching *Educational Researcher* **15** 4

[4] Subramaniam K 2013 Prospective secondary mathematics teachers ' pedagogical knowledge for teaching the estimation of length measurements (Crites 1993). http://doi.org/10.1007/s10857-013-9255-2

[5] Yusuf N *Peringkat Uji Kompetensi Guru di Aceh Naik ke Ranking 15 Nasional* Serambinews. Kutaraja [Internet] 2018 Agustus 4 [cited 2018 Agustus 4] Available from: http://aceh.tribunnews.com/2018/03/09/peringkat-uji-kompetensi-guru-di-aceh-naik-ke-ranking-15-nasional

[6] Murthalamuddin *Pelatihan Manajemen Sekolah Modul 3 USAID PRIORITAS jenjang SMP/MTs Kabupaten Pidie* Dinas Pendidikan Aceh. Banda Aceh 2016 Jan 7 [cited 2016 Jan 7] Available from: http://Disdik.Acehprov.Go.Id/Node/2047

[7] Yuswono L C, Martubi S and Agus B 2014 Profil kompetensi guru SMK teknik kendaraan ringan di daerah istimewa Yogyakarta *Prosiding Konvensi Nasional Asosiasi Pendidikan Teknologi dan Kejuruan (APTEKINDO) ke 7 FPTK Universitas Pendidikan Indonesia*

[8] Warganegara N S, Berchah P and Hermi Y 2013 *Persepsi Guru terhadap Pelaksanaan Uji Kompetensi Guru di SMA Negeri 3 Bandar Lampung Tahun Pelajaran 2012/2013*

[10] Jalal F, Samani M, Chang M C, Stevenson R Ragatz A B and Negara S. 2009 *Teacher Certification in Indonesia: A Strategy for Teacher Quality Improvement* Departemen Pendidikan Nasional Republik Indonesia

[11] Asmarani N 2014 Peningkatan Kompetensi Profesional Guru di Sekolah Dasar *Jurnal Administrasi Pendidikan* **2** 505

[12] Telaumbanua Y N, Sinaga B, Mukhtar M and Surya E 2017 Development of mathematics module based on metacognitive strategy in improving students mathematical problem solving ability at high school *Journal of Education and Practice* **8** 73

Student's representation of fraction through ELPSA framework

E Juliangkary[1] and R Johar[2]
[1]IKIP Mataram, Jl. Pemuda No. 59A Mataram, NTB 83125, Indonesia
[2] Syiah Kuala University, Jl. Teuku Nyak Arief, Darussalam, Banda Aceh 23111, Indonesia

E-mail: rahmah.johar@unsyiah.ac.id

Abstract. Representation has an important role in developing students' mathematical ability. Therefore, learning materials need to be developed. One of them is by implementing experience, language, pictorial, symbol, and application (ELPSA) framework. The aim of this research is to analyze students' mathematical representation on fraction material after implementing ELPSA framework. There were 20 students of Year 7 from one of the junior high schools in West Lombok, Indonesia who joined the lessons. However, only six students were involved in the interview section to investigate their answers about the representation of fraction. This research shows that the students still have difficulties in the representation of fraction. Errors in representation on fractional material also occur due to students' mistakes in understanding fraction unit. Students' representation of fraction should be conducted through exercise continually.

1. Introduction

Representation can be interpreted as a process of modeling concrete matter in the real problem into abstract concept or symbol [1]. Mathematical representation is used to plan, monitor, and control the mathematical problem-solving process [2]. While Lowrie and Patahuddin state that concrete representations supported students' learning as they move towards more abstract concepts and ideas [3]. Interpreting the given image is called encoding. The decoding process in which learners are given problems in the form of words and learners need to interpret the problem and present it in the form of sketches, diagrams or drawings.

In relation to the fractional concept, there are seven types of representations according to [4] as follows: (1) part group, congruent part; (2) part-whole, congruent parts; (3) part group, non-congruent parts; (4) part group, comparison (comparing the number of members or objects of two sets); (5) Number line, a line of numbers containing a line segment with a length of one unit; the segment of the line is divided into b equal parts of length; (6) part whole, comparison; and (7) part whole, non-congruent parts.

Representation of fraction using number line is more difficult for students to understand than others [5]. However, according to Wong and Evans, most elementary school students can use symbolic representation but find difficulties in representing images in the topic of fractional [6]. Therefore, students' ability in representing fraction is necessary to be developed. Several studies have been conducted by [4,7,8]. However, there is no research conducted in developing learning material through the ELPSA Framework to improve students' representation ability on fraction.

ELPSA framework is a learning approach consisting of elements experience, language, pictorial, symbol, and application. This framework is based on constructivism learning theories and its social nature. The ELPSA framework views learning as an active process in which students construct their own way of understanding something through individual thought processes and social interactions with others [3]. Pictorial components in ELPSA are components that are closely related to mathematical representation. This paper discusses the students' representation ability after the implementation of the learning materials of ELPSA framework.

2. Method

This research is design research to obtain learning materials namely lesson plan, students' worksheet and mathematical representation test through ELPSA framework. Process of lesson materials development is started by conducting (1) preliminary investigation by organizing an assessment of curriculum and fraction topic, (2) prorotyping phase by designing and validation process of lesson plans, worksheets, and mathematical representation test, (3) assessment phase by implementing the learning materials and test the representation skill of the students [9]. This paper only discusses the phase of the learning materials especially related to assessing the ability of students' mathematical representation after implementation of ELPSA framework. Before implementation, authors have designed the teaching materials. Two experts were involved in this study in assessing the validity of teaching materials. According to them, the teaching materials were valid. There are seven lessons of fraction that have been implemented. Each lesson takes 80 minutes. This study only discusses the students' mathematical representation after implement the first lesson about basic concepts of fraction. One of the authors has discussed and performed (as simulation) some activities of the first lesson to the teacher before implementation of the teaching materials. Authors and teacher have joined several Professional Development workshop about ELPSA framework at Mataram, Indonesia. These workshops for mathematics teachers were conducted by Government Partnership for Development (GPFD) Project between the University of Canberra, Australia and IKIP Mataram, Indonesia (see www.elpsa.org).

The participants of this research were 20 students (one female and nine males) of Year 7 of one of junior high school in West Lombok, Indonesia. The student's mathematical representation is measured by three indicators: (1) solve problems by involving mathematical expressions; (2) create an image to clarify the problem and facilitate its completion; and (3) create a problem situation based on the data or representation given and write it down [10]. The test for 40 minutes was performed after the students learned the fraction through ELPSA Framework. After doing the test and reviewing the answers of the students, the authors conducted interviews with six students to trace their responses to three questions of mathematical representation. Table 1 shows the indicator and the item test.

Table 1. Mathematical representation indicator and item test.

Mathematical Representation Indicators	Item Test
Solve problems by involving mathematical expressions	1
Create an image to clarify the problem and facilitate its completion	2
Create a problem situation based on the data or representation given and write it down	3

3. Result and discussion

Student's answer to the test of mathematical representation is described as follows.

Question 1: Please divide the following square into four equal parts in different ways.

IOP Conf. Series: Journal of Physics: Conf. Series **1088** (2018) 012104 doi:10.1088/1742-6596/1088/1/012104

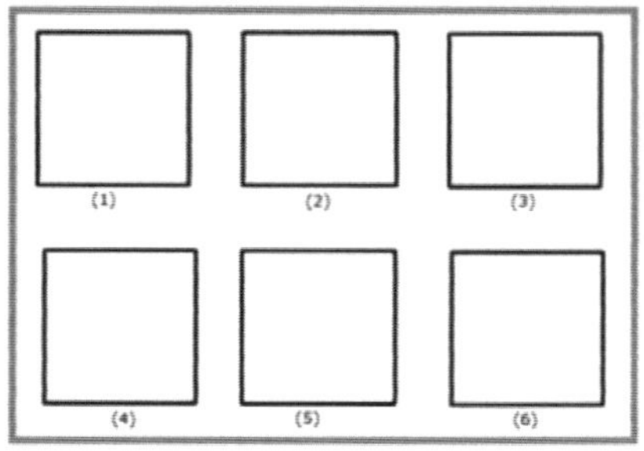

Figure 1. Question 1

The students were expected to be able to answer six pictures on the provided answer sheet. Generally, the students drew such fraction as follows.

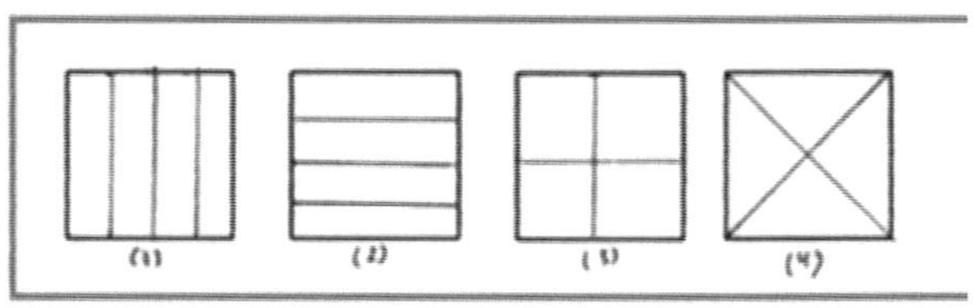

Figure 2. General question answered by the students for Question 1.

From 20 students only one student who answered five correct answers, 15 students answered four correct answers, and three students answered three correct answers. Students' wrong answer was described as follows.

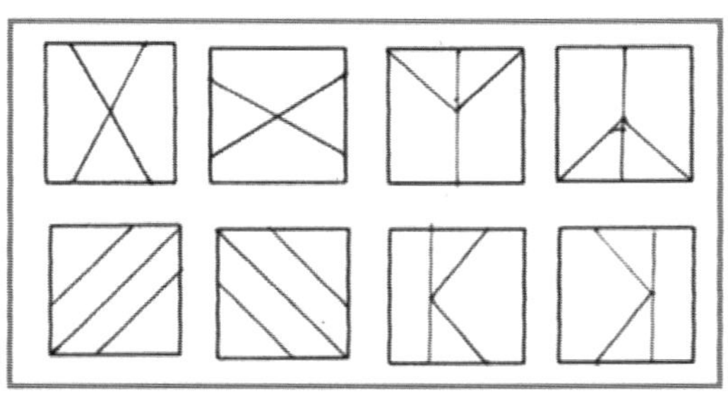

Figure 3. Students' wrong answers to Question 1.

Therefore, the researchers conducted the interview with students to confirm their response. Excerpts of interviews between the researchers and Student 1 (S1) are as follows.

Researchers	:	Please explain how did you obtain this answer? (Researchers pointed drawing created by S1
Student 1 (S1)	:	For this one, (appointed drawing 1) I just crated line from top to bottom, Mam!, for this one (appointed drawing 2) I created line sidelong, for this one (appointed drawing 3) I created line from the top to the bottom and sidelong, for this one (appointed drawing 4), I created line from the side corner to this side edge, for this one (appointed drawing 5), I created cross line, and for this one (appointed drawing 6) I just did the same as Question 5 however I rotated it, Mam !
Researchers	:	Please read the question carefully.

Student 1 (S1) : (Reading the question)
Researchers : What about this one? (Appointed drawing 1, 2, 3, and 4).
Do they have the same size?
Student 1 (S1) : Yes, Mam! Yes, they do.
Researchers : How about this one? (Appointed drawing 5 and 6).
What do you think?
Student 1 (S1) : All of them are not the same, however, two parts which face to face has the same size.
Researchers : Are you sure that is right?
Student 1 (S1) : I am pretty sure Mam.

It should be six different representations to state fraction $\frac{1}{4}$ as follows.

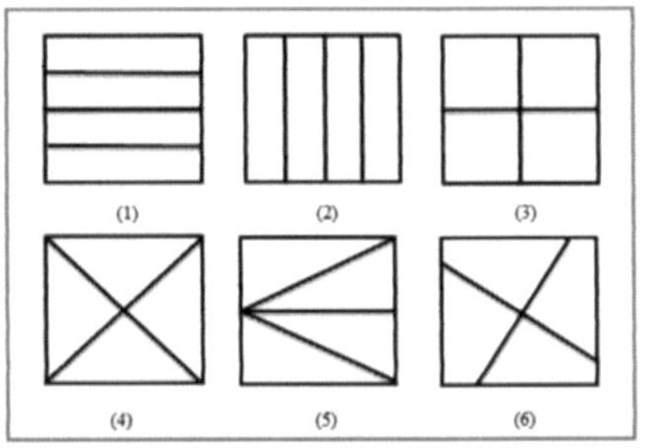

Figure 4. Answer key Question 1.

Question 2: Please represent fraction $\frac{1}{3}$ in various ways!
There was 90% of the students answered it by representation as follows.

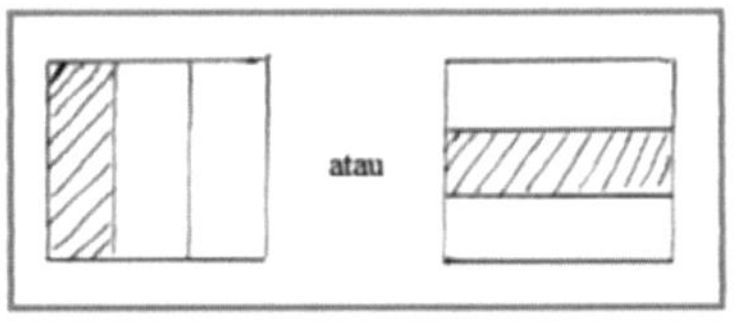

Figure 5. Students' representation in answering Question 2.

The researchers conducted interview with the students to confirm their answer. Here are excerpts of the interview between the researchers and Student 2.

Researchers : Please take a look at your answer for question no.2
Could you explain how did you obtain this answer?
Student 2 (S2) : Rechecking his answer. Yes, Mam, this is a form of fraction $\frac{1}{3}$
Researchers : Is there any other representation?
Student 2 (S2) : Yes, there is. However, I can't remember it. This is all I can remember.

This matter provides information that 90% of students have limited ability in declaring fraction $\frac{1}{3}$ in various representations. There are only 10% of students who answered by using three representations to declare fraction $\frac{1}{3}$ as described in Figure 4 as follows.

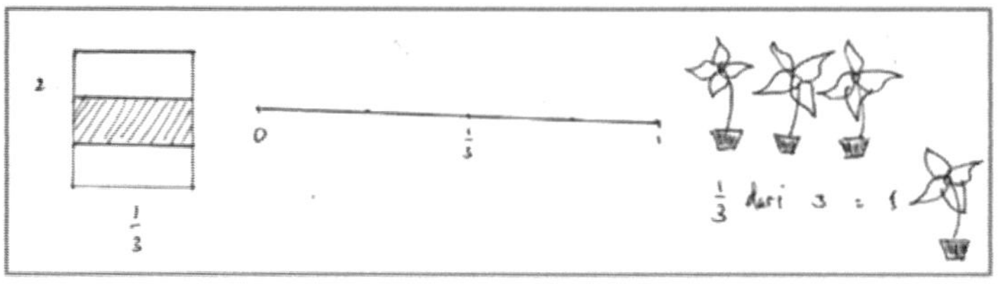

Figure 6. Students' representation in answering Question 2.

Here are excerpts of the interview between the researcher and student 3 (S3):

Researchers	:	Could you explain how did you obtain this answer? (Pointed S3's answer sheet).
Student 3 (S3)	:	Firstly, I created a square divided into 3 areas, but I shaded one of the areas which equals to $\frac{1}{3}$, secondly, I used number line. Third, I drew 3 flowers, if $\frac{1}{3}$, of the 3 flowers is one flower Mam.
Researchers	:	Please take a look at the number line! Pay attention to the position of the $\frac{1}{3}$.
Student 3 (S3)	:	I was wrong Mam, it should be here (writing the position which was deemed correct by S3.

However, during the teaching and learning process, the teacher has explained in various ways to represent fraction $\frac{1}{2}$. In this research, the students still have difficulties in mathematical representation. This is supported by opinion [6] who stated that most of the Elementary students able to use representation symbolically and have difficulties to present it in the form of drawing or fraction materials.

Question 3: Please answer this question!

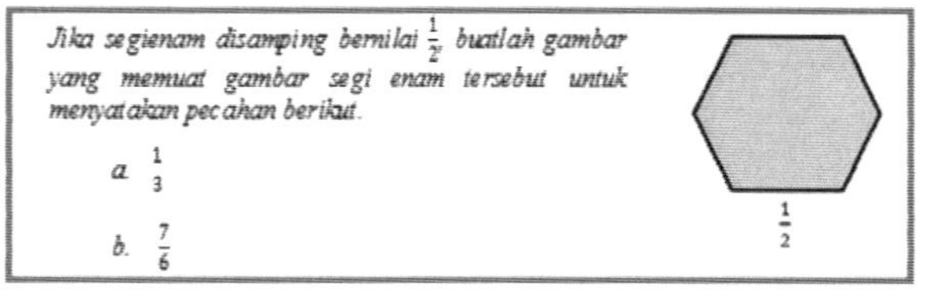

If this hexagon equals to $\frac{1}{2}$, could you please make drawing represent the fraction as below?

Figure 7. Question 3.

For question no. 3 point a, there were 80% of the students answered with figure 8(a). This is the correct answer to find out fraction equals to $\frac{1}{2}$ and $\frac{1}{3}$.

Fraction equals to $\frac{1}{2}$ is: $\frac{1}{2} = \frac{2}{4} = \frac{3}{6} = \frac{4}{8} = \frac{5}{10} = \frac{6}{12}, \ldots$

Fraction equals to $\frac{1}{3}$ is: $\frac{1}{3} = \frac{2}{6} = \frac{3}{9} = \frac{4}{12}, \ldots$

$\frac{1}{3} = \frac{1}{12} + \frac{1}{12} + \frac{1}{12} + \frac{1}{12} = \frac{4}{12} = \frac{1}{3}$

There were 10% of the students provided the wrong answer as stated in Figure 8(b) and 5% of the students provided the wrong answer as stated in Figure 8(c), also 5% of them provided the wrong answer as stated in Figure 8(d).

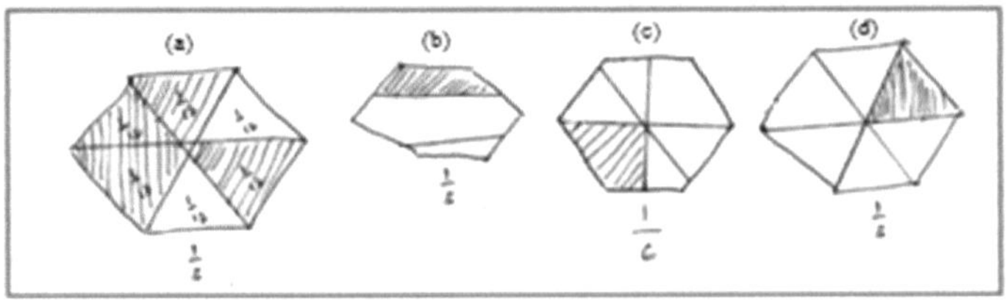

Figure 8. Students' representation in answering Question 3.

For question no.3 point b, there were 50% students provided correct answer as shown in figure 9(a) and 5% was appropriate to the expected answer because this based on answer 3a. Due to, 1 hexagon equals $\frac{1}{2}$ therefore for 2 hexagon equals to 1 and $\frac{1}{6} = \frac{2}{12}$, therefore it was obtained $\frac{7}{6}$. There were 5% of the students provided the wrong answer as shown in Figure 9(b) and 45% of them provided the wrong answer as shown in Figure 9(c).

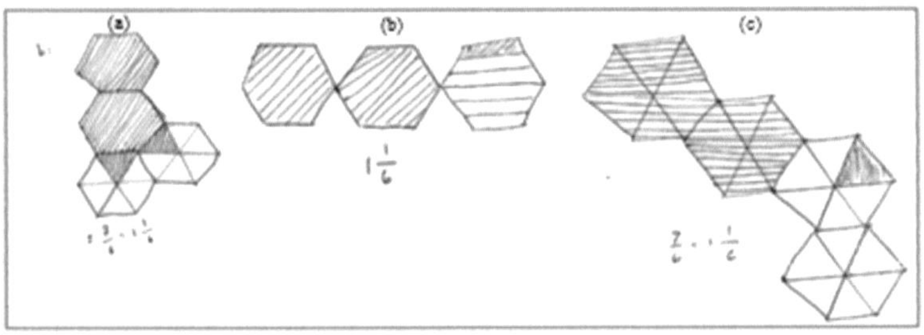

Figure 9. Students' representation in answering Question 3b.

The researchers conducted the interview to find out the students' reason and answer as shown in Figure 9 (b) and (c). Here is the excerpt from interviews between researchers and S2 and S3.

Researchers	:	Please pay attention to your answer for question no.3b. Could you please explain how did you obtain this?
Student 2 (S2)	:	Question 3b. The question is $\frac{7}{6} = 1\frac{1}{6}$. It has been known that 1 hexagon equals to $\frac{1}{2}$ if its value is 1, therefore, there are 2 hexagons which are shaded. The rest is $\frac{1}{6}$. Then I divided 1 hexagon become 6 then I shaded into 1 area, Mam.
Researchers	:	Could you explain how did you get the answer for Question 3b?
Student 3 (S3)	:	It was from 3a Mam, I divided hexagon 3a becomes 6, so every part equals to $\frac{1}{12}$ therefore $\frac{12}{12} = 1$. I drew another which equals to 1, so there were 4 hexagons. The rest should be $\frac{1}{6}$, I shaded another 1 part, Mam, therefore, it equals to $\frac{7}{6} = 1\frac{1}{6}$

The problem occurred on question no. 3 because the students forgot the information on that question stated that the hexagon equals to $\frac{1}{2}$. This occurred as well on [7] concerning the students' mistakes in understanding fraction unit on the number line. He stated that mistakes based on the understanding $\frac{3}{4}$ part of a fraction unit.

The ELPSA framework is a cyclical learning design approach, through daily life experiences, mathematical conversations, visual stimuli, symbol notation, and applications [3]. The teacher has implemented the learning materials of ELPSA framework to support the students' representation skill of fraction, especially through the pictorial component of ELPSA framework. Nevertheless, the students still have difficulties in solving mathematical representation about fraction.

During the lesson, students have learned about dividing the squares into two equal parts in different ways. Based on observation, most of the students in the classroom found the correct different ways. On the other hand, when students were asked to divide the square into four equal parts in different ways, only one of 20 students who answered five correct answers, 15 students answered four correct answers, and 3 students answered three correct answers. It means, dividing the square into four equal parts is more difficult for them than into two equal parts in different ways for students.

Most of the students' representations about fraction $\frac{1}{3}$ are only used part of a whole representation. This representation is familiar to Indonesian students [11]. The most difficult problem of students is the question no. 3b. There were only 50% of students provided the correct answer to represents fraction $\frac{7}{6}$ when the representation of fraction $\frac{1}{2}$ is given. All three questions never presented in Indonesian mathematics textbook so that these questions are non-routine for students. Students have some errors and difficulties in solving non-routine problems [12] and they need some informal strategies to solve them [13].

One meeting of the lesson is not enough to develop the students' representation. It is necessary for the students to obtain exercise any longer. In conducting mathematical representation, the students are required to have such a high level of creativity. Creative thinking is a process used when we produce a new invention [14]. Teachers have the important role to support students' mathematical representation.

4. Conclusions

Three indicators measure the student's mathematical representation. The most difficult indicator is creating a problem situation based on the data or representation given. Student mathematical representation on fraction should be support continuously and required more exercises. One meeting is not enough to develop representation. In conducting mathematical representation, students are also required to have high-level creativity. Errors in the representation of fraction also occur due to students' mistakes in understanding fraction units.

References

[1] Hwang W Y, Chen N S, Dung J J and Yang Y L 2007 Multiple representation skills and creativity effects on mathematical problem solving using a multimedia whiteboard system *Journal of Educational Technology & Society* **10**

[2] Goldin G A and Kaput J J 1996 A joint perspective on the idea of representation in learning and doing mathematics *Theories of mathematical learning* 397

[3] Lowrie T and Patahuddin S M 2015 ELPSA as a lesson design framework *Journal on Mathematics Education* **6** 77

[4] Novillis C F 1976 An analysis of the fraction concept into a hierarchy of selected subconcepts and the testing of the hierarchical dependencies *Journal for Research in Mathematics Education* 131

[5] Larson C N 1980 Locating proper fractions on number lines: effect of length and equivalence *School Science and Mathematics* **80** 423

[6] Wong M and Evans D 2007 Students' conceptual understanding of equivalent fractions *J. Watson & K. Beswick Mathematics Essential Practice* **2** 824

[7] Hannula M S 2003 Locating fraction on a number line *International Group for the Psychology of Mathematics Education* **3** 17

[8] Surya E 2010 Peningkatan representasi visual thinking matematika siswa SMP N 11 Medan dengan melatih ketrampilan menggambar dan pendekatan kontekstual *Jurnal Title*

[9] Plomp T 2013 *Educational Design Research: an Introduction*, Ed T Plomp and N Nieveen (Enschede: SLO) p 10

[10] Sulastri S, Marwan M and Duskri M 2017 Kemampuan representasi matematis siswa smp melalui pendekatan pendidikan matematika realistik *Beta Jurnal Tadris Matematika* **10** 51
[11] Johar R, Patahuddin S M and Widjaja W 2017 Linking pre-service teachers' questioning and students' strategies in solving contextual problems: a case study in Indonesia and the Netherlands *The Mathematics Enthusiast* **14** 101
[12] Ulu M and Akar C 2016 The effect of visuals on non-routine problem solving success and kinds of errors made when using visuals *Educ. Res. Rev* **11** 1871
[13] Arslan C and Altun M 2007 Learning to solve non-routine mathematical *Problems Elementary Education Online* **6** 50
[14] Pehkonen E 1997 The state-of-art in mathematical creativity *ZDM* **29** 63

Improving students' higher order thinking skills in thermochemistry concept using worksheets based on 2013 curriculum

R Verdina, A Gani and Sulastri
Syiah Kuala University, Jl. Tgk. Chik Pante Kulu No 5, Banda Aceh 23111, Indonesia

E-mail: sulastri@fkip.unsyiah.ac.id

Abstract. Higher order thinking skills could facilitate students' thinking to find a new knowledge from the observation of phenomena or natural phenomena that exists around. The purpose of this research was to improve students' higher order thinking skills for teaching and learning thermochemistry concept using student worksheets (SWS) based on curriculum 2013. The research and development method developed by Borg and Gall's model was used in this study. The samples were 25 students of grade 11 from a secondary high school and 15 chemistry teachers in Banda Aceh, Indonesia. The higher order thinking skills were measured using the essay test. Based on the results of data analysis, the N-gain value was 79% which is in the high category. It can be concluded that the developed SWS could improve students' higher order thinking skills. Therefore the developed SWS could be used as the material for teaching and learning of Thermochemistry.

1. Introduction

Associated with the issue of educational development at the international level, Curriculum 2013 in Indonesia is designed with various improvements. One of them is the refinement of assessment standards. Assessment of learning outcomes is done by facilitating learners to improve higher-order thinking because high-level thinking can encourage learners to think broadly and deeply about the subject matter.

The ability to think is a cognitive domain that according to Bloom's taxonomy consists of six levels. In the 2013 curriculum, this domain refers to a form that has been revised by Anderson & Krathwohl. The domain level consists of the ability to remember, understanding, applying, analyzing, evaluating, and creating [1]. Human thinking skills can be classified into two major groups; low-order thinking skills (LOTS), and higher-order thinking skills (HOTS). LOTS consists of first three aspects of taxonomy bloom including remembering, understanding, and applying. HOTS consists of the last three aspects of taxonomy bloom namely analyzing, evaluating, and creating.

The important goal of science education is to improve students' higher order thinking skills (HOTS) such as critical, reasoning, reflective and science process skills. The students' HOTS enables them to face the challenges of daily life [2]. In the other words to define the problem around them, observe, analyze, hypothesize, experiment summarize, and apply their information. The easiest learning habit is through activities guided by student worksheets (SWS) to train students' HOTS.

HOTS is thinking level that higher than simply memorizing. HOTS cannot be unmeshed from the levels of learning; they involve interdependent, multiple components and levels. In real life, students will learn content in both community and school experiences. The concepts that they learn in the prior year will help them to improve HOTS and new content in the next year. Furthermore, HOTS involves a variety of thinking processes applied to complex situations and having multiple variables [3].

Based on the results from the 2017 National Exams (UN) for Senior High School subjects in Aceh, the ability of the students in chemistry, especially in thermochemistry was still low. It is showed by the average score of 54%, which was lower than the acceptable target score [4]. That result indicated to us that the ability of the students to effectively understand the concepts was still low. The participants had difficulties to understand the concepts of chemistry such as thermochemistry because they only calculated problems and used the formulations without any motivation to understand thermochemistry rather than to link those concepts with real life [5].

Teaching and learning activities can be done effectively and efficiently if the learning materials designed by the teachers can support a conducive learning atmosphere. Therefore, teachers are expected to be creative in deciding the model, the method, the media and the learning materials in the teaching and learning processes. The most frequent teaching aid used is SWS. The SWS is one or more sheets of paper that are prepared or provided by the teacher as learning aids to be used by the students. SWS is used to support active learning and to increase students' enthusiasm to learn, as well as to be used for teacher evaluation. A well prepared SWS will have a positive effect on students' learning outcome [6].

Student worksheets developed by teacher rarely improved HOTS. As a consequence, students' thinking skills will be decreased. Based on the description of facts and expectations discussed above, it is necessary to improve students' HOTS through SWS. The SWS should be developed both theoretically and empirically valid.

2. Methods

The research method used was the Research and Development (R&D) method by adapting Borg and Gall's model. Type of this research is both quantitative and qualitative. The SWS development was done qualitatively and the result of using the SWS (through pre-test and post-test calculation) was done quantitatively.

Random sampling was used to select the sample for this study. The sample selected was 25 students on grade 11 from one secondary high school and 15 chemistry teachers in Banda Aceh, Indonesia. The higher order thinking skills were measured using the essay test and analyzed by calculating the pre-test and post-test (N-Gain score).

3. Results and discussion

The development of the SWS for learning about thermochemistry to improve HOTS was done based-on the model of Borg and Gall [7] as follow:

3.1. Research and information collection

A study of relevant literature was done to find information which could be used as the basis for developing the SWS, which was developed in accordance with the revised 2013 curriculum. In addition, information was gathered about previous researches from various journals, textbooks and teaching materials as material to support the development of the SWS for teaching and learning. The choice of materials was done based on the previous low score results from the 2016/2017 National Examinations (UN). Based on the observations of previous materials used for teaching and learning chemistry, the modules or the SWS did not yet have any material for improving HOTS. As a result, the students did not have the ability to solve the problems which required them to do analysis and evaluation and to use creativity as a part of cognitive domains. These domains are concerned with HOTS. In [8], they have said that the skills improvement in HOTS can be achieved through teaching and learning theory or practice using aspects of analysis, evaluation and creative design.

3.2. Planning
The SWS was prepared in accordance with the syllabus. The thermochemistry material consisted of two basic competencies for the cognitive skills and two basic competencies for the psychomotor skills. Based on those competencies, indicators were prepared with the integration of HOTS. The time allocated in the syllabus for studying thermochemistry was 6 sessions.

3.3. Development of a preliminary form of the product (SWS)
The SWS has developed HOTS for several meetings. Each meeting aimed to achieve one or more indicators in accordance with the syllabus. The various components of the SWS included the cover, the introduction, the content, the summary, the tests and the list of references. These components are shown in Table 1.

Table 1. Components of the SWS

No	SWS components	Materials included in each component
1.	Cover	SWS title and drawing connected with Thermochemistry.
2.	Introduction	Introduction, contents list, basic competency, indicators, purpose, concept mind-map, and the activity.
3.	Main contents	Information about HOTS components (analysis, evaluation and creative thinking).
4.	Summary	A Summary of the Thermochemistry material studied.
5.	Test of understanding	Tests of the Thermochemistry materials based on HOTS.
6.	References	Sources of information used in preparing the Thermochemistry SWS based on HOTS.

The SWS which was developed must be able to increase the motivation of the students, one way of increasing motivation is to display drawings which are contextual and interesting so that the students will more easily understand its relevance in daily life [9].

3.4. Preliminary field testing
Components which were evaluated in testing the first draft included the drawings prepared for the SWS, the materials provided, the language used such as correct spelling, the ease of reading and the usefulness of the material for the users of the SWS. The initial testing to validate the SWS was done by two expert lecturers from the Chemistry of the Education and Teacher Training Faculty at Syiah Kuala University and one expert from the Institute for Teaching Quality as independent evaluators. The validity score of the SWS was 89%. This result showed that the SWS was valid to be used after the revisions proposed by the validator.

3.5. Main product revision
The results of the SWS validation by the experts would be valid to use after modifications had been made. Their recommendations for modifications to the SWS draft included how the drawings must illustrate by analogy and be contextual to help student to understand the concepts of thermochemistry in their daily life. The design of the SWS must assist students to expand their thinking ability using HOTS while being in accordance with their needs.

3.6. Main field testing
Field tests were done by giving the SWS to chemistry teachers. The teachers selected for this test were five chemistry teachers from two favorite senior high schools in Banda Aceh. Those two schools were chosen based on similar vision, mission, and level of results in the 2016-2017 National Exams.

Therefore, the feedback from those teachers was used as the material to evaluate the SWS for further improvement.

3.7. Operational product (SWS) revision

The response of the two chemistry teachers from two favorite senior high schools has added extra examples of questions in the column "Let's Practice". In addition, those teachers needed to make the positive (+) and the negative (-) signs clearer in order to avoid misunderstanding of the concepts in the sub-section about changes in enthalpy in exothermic and endothermic reactions. Thus, the revision of SWS product was shown by extra questions in the column "Let's Practice". The signs (+) and (-) were made to show whether the reaction was exothermic or endothermic. While the sign got bigger, it showed larger number of calories generated exothermically for the sign (+) or absorbed endothermically for the sign (-).

3.8. Operational field testing (the SWS)

Trials to further improvement of the SWS were done by distributing the revised SWS to various Chemistry Teachers in the Chemistry Teachers Forum in Banda Aceh. The recommendations from these teachers became the final input of the final SWS distribution to schools.

3.9. Final product revision

The revisions of SWS were to provide examples that are more contextual to open, closed and isolated systems, such as chicken eggs get rotten when enclosed in their shell. In addition, at the summary part, it is recommended to add information related the benefits of studying thermochemistry in our lives. This revision was the final step to complete the SWS product to be used in thermochemistry teaching and learning processes. The SWS product is shown in Figure 1, 2 and 3.

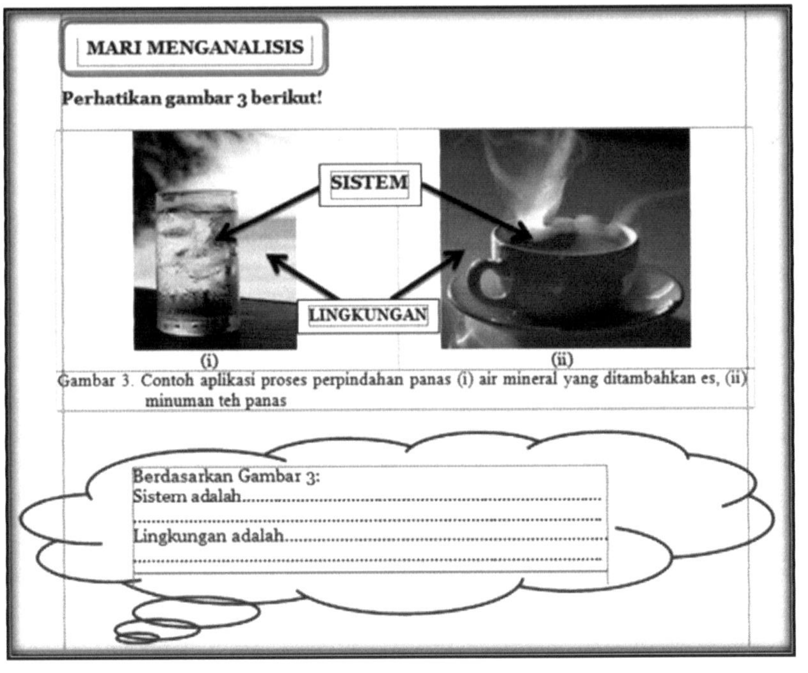

Figure 1. Worksheet based on HOTS (Analyzing)

Figure 2. Worksheet based on HOTS (Evaluating)

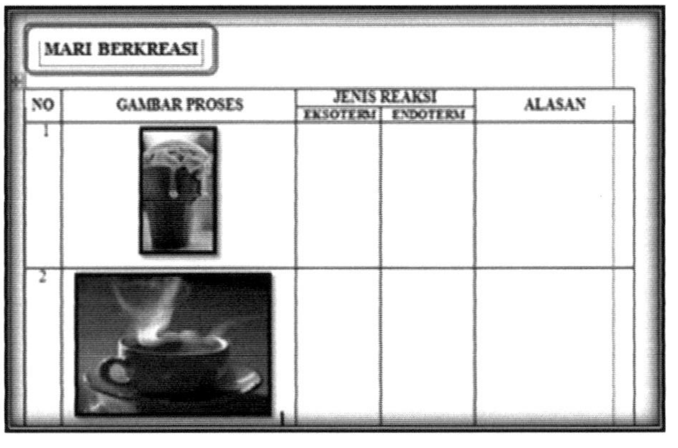

Figure 3. Worksheet based on HOTS (Creating)

Figure 1, 2 and 3 show that worksheets were consists of three components of hots including analyzing, evaluating and creating. In the analyzing section, there were images that stimulate students to analyze the system and environment. In the evaluation section, there were images that stimulate students to evaluate how the characteristics of energy and mass transfer occur in closed, open and isolated systems. While in the creating section, there were images that stimulate students to be creative in providing reasons for images that include exothermic or endothermic reactions.

3.10. Dissemination and implementation

The final revision of the SWS was then implemented to one secondary high school in Banda Aceh, Indonesia. The teaching and learning program was run in accordance with lesson plan that had previously been prepared to socialize the SWS product in supporting teaching and learning of thermochemistry.

The quality of this SWS product was evaluated using a questionnaire that accessed three important aspects including (a) compatibility of the SWS with the students, (b) compatibility of the SWS with the aims of the lesson and (c) the technical quality, and those three aspects showed values of 82%, 86% and 85%. These values showed that the SWS product prepared was satisfactory. Furthermore, the

SWS distributed could improve HOTS through the questions being asked and give positive effect on the development of cognitive, affective and psychomotor. The similar finding was supported by the behaviorist theory which said that the process of teaching-learning depends on interaction between the stimulus and the response [10].

The HOTS was measured by using questions in the form of analysis which was included in the SWS. However, before measuring the HOTS, trials were done to check the validity, reliability, and the power differentiation towards the questions difficulty level. The results showed that 10 of the 12 questions were suitable to be used.

Based-on the analysis, N-gain score was at the high-level category with the value of 0.79 or 79%. It means that the test score was increased from pre-test to post-test with the average of 36 to 87 (which was a very good result). Similar finding was found by [11] that students, whom were always given questions which required higher order thinking skills, tended to have high conceptual understanding in accordance with Bloom's taxonomy. Another study was done by [12], it showed that teaching and learning that could improve HOTS was able to increase the ability in problems solving and critical thinking.

4. Conclusion

The SWS is theoretically and empirically feasibility worthy. This conclusion is based on the high value of N Gain of students' HOTS. So, it can be concluded that the SWS developed could improve students' higher order thinking skills. Therefore, this developed SWS could be used as the material for Thermochemistry teaching and learning.

References

[1] Moore B and Stanl T 2010 *Critical thinking and formative assessments*: Larchmount (New York: Eye on Education Inc.)

[2] Zachariades T, Christou C and Pitta-Pantazi D 2013 Reflective, systemic and analytic in real numbers *Educational Studies in Mathematics* **82** 5

[3] King F J, Goodson L and Rohani F 2000 *Higher Order Thinking Skills* (Center for Advancement of Learning and Assessment Florida State University)

[4] Puspendik 2017 *Laporan Hasil Ujian Nasional Tahun Ajaran 2016/2017* (Jakarta: Departemen Pendidikan Nasional)

[5] Cigdemoglu C and Geban O 2015 Improving students' chemical literacy levels on thermodynamics concepts through a context-based approach *Chemistry Education Research and Practice* **16** 302

[6] Oren S F 2012 An application about pre-service teachers' development and use of worksheets and an evaluation of their opinions about the application *Educational Sciences: Theory and Practice* **12** 263

[7] Borg W R and Gall M D 1989 *Educational research:* An Introduction (New York: Longman)

[8] Hugerat M and Kortam N 2014 Improving higher order thinking skills among freshmen by teaching science through inquiry *Eurasia Journal of Mathematics, Science & Technology Education* **10** 447

[9] Lee C D 2014 Worksheets usage, reading achievement, classes' lack of readiness, and science achievement: A cross-country comparison *International Journal of Education in Mathematics, Science, and Technology* **2** 96

[10] King L A 2007 *The Science of Phsychology* (New York: Mc-Graw Hill)

[11] Jensen J L, McDaniel M A, Woodard S M and Kummer T A 2014 Teaching to the test... or testing to teach: Exams requiring higher order thinking skills encourage greater conceptual understanding *Educational Psychology Review* **26** 307

[12] Soltis R, Verlinden N, Kruger N, Carrol A and Trumbo T 2015 Process-oriented guided inquiry learning strategy enhances students' higher level thinking skills in a pharmaceutical sciences course *American Journal of Pharmaceutical Education* **79** 11

IOP Conf. Series: Journal of Physics: Conf. Series **1088** (2018) 012106 doi:10.1088/1742-6596/1088/1/012106

The effectiveness of guided inquiry-based student worksheets on students' generic science skills

M Faradilla, M Hasan and Sulastri
Syiah Kuala University, Jl. Teuku Nyak Arief Darussalam, Banda Aceh 23111, Indonesia

E-mail: sulastri@fkip.unsyiah.ac.id

Abstract. The generic skill of chemistry is the ability to think and act according to the chemistry obtained. The purpose of this study was to determine the effectiveness of guided inquiry-based student worksheets on students' generic science skills. The effectiveness of learning outcomes is seen in the acquisition of the N-Gain value of learning outcomes. The approach used experiment was using one group pretest-posttest design. The sample consisted of 62 students year eleven from two secondary schools in Banda Aceh, Indonesia. The data were collected through generic science skill observation sheets and learning results test. The results showed that generic science skills of the first school were developed on indicators of direct and indirect observation, logical consistency and logical inference. As for the second school were developed on direct observation indicators, logical consistency, and logic inference. Both schools had generic science skills indicators with good criteria. The value of N-gain learning outcomes of the first school was 0.54 and the second school was 0.47 with the medium category and t-test results were significantly different. Thus, the guided inquiry-based student worksheets effectively could be used as an alternative teaching material on acid-base material practice.

1. Introduction

Generic science skill (GSS) is a terminology given for the certain skills in the learning process. Every country might have different basic skills which according to the country's conditions [1]. In Indonesia, basic skill is also called the generic science skill with ten skills indicators: direct and indirect observation, scale awareness, symbolic language, logical framework, logical consistency, causality, mathematical modeling, concept constructing and abstraction [2].

Curriculum changes occur around the world, the development of generic skills has become an important issue for educational success [3,4]. There are several important levels of generic skills that are considered important, such as communication skills followed by integrity and professional ethics, and teamwork [5]. In addition to these skills, there are other skills such as the ability to summarize and analyze, teamwork, problem-solving, proficiency in information and technology, entrepreneurship, creative and critical, leadership, and self- qualities [6]. In other words, students will have a significant advantage in developing generic skills for their studies and professional lives in helping to develop their common skills [7].

People realize that GSS is important, but its implementation is not as expected. Particularly in Aceh, based on observations at two secondary schools in Banda Aceh, Indonesia, students' generic science skills have been not developed well because students are rarely trained to practice GSS, including on the topics of acid-bases. Those chemistry topics are still problematic because many concepts of the topics are failed to master, particularly in the interpretation of pH indicator tables [8]. The lack of generic skills of students in learning is also influenced by less varied learning models [9]. One of the efforts to improve the students' GSS is by applying the guided inquiry learning model.

GSS can be trained using guided inquiry learning model. Furthermore, the results of usage of inquiry models can improve students' learning outcomes of cognitive aspects [10]. The advantages of this learning model for learners are to improve their ability to write, analyze and find the truth about a problem systematically and directed according to the purpose; then the learners can also develop their critical thinking to have good reasoning ability [11]. Based on research that has been conducted shows that the guided inquiry student worksheets are very feasible to be applied [12]. Further, the results of another study also showed that the development of guided inquiry student worksheets could train the students' skills in conducting experiment [13]. Based on these descriptions, the researchers consider it necessary to do a study with the title of the effectiveness of guided inquiry-based student worksheets on students' generic science skills.

2. Methods

This research was a research application based on research and development that has been done before. Development model used is the research model developed by Borg and Gall's [14]. The resulting product of development was a guided inquiry-based student worksheet. The percentage of eligibility value to the guided inquiry-based student worksheet was averaged 86.67%; it was considered eligible to implement in a class. The validated guided inquiry-based student worksheets were implemented in chemistry classes of regular and private high schools. The effectiveness of the guided inquiry-based student worksheet was applied to the experimental method using one-group pretest-posttest. This design research diagram can be seen in table1 [15].

Table 1. The study designs one-group pretest-posttest.

O_1	X	O_2
Pretest	Treatment	Posttest

The sample of the study was chosen by purposive sampling technique which was determined based on the students' ability. School research was chosen based on the value of national examination of students according to students' absorbance of acid-base material. The sample of this study consisted of 30 students of year eleven from regular secondary school (the first school) and 32 students year eleven from private secondary school (the second school). The research instrument was used generic science skill observation sheets and learning results test. The data collected in the study were analyzed by calculating the normalized N-gain values of the pretest and posttest results. Generic science skill observation data were analyzed by calculating the overall percentage of the generic science skill indicators of each student.

3. Results and discussion

Activities undertaken by learners during a practicum were reading and executing instructions in student worksheets, investigating information on natural substances as acid-base indicators, designing work procedures, and discussing accomplishing the tasks in student worksheets. The activities of students during learning can be seen in Figure 1. Furthermore, learning by using student worksheets based guided inquiry is expected to train generic science skills of learners to understand the various acid-base indicator of natural materials through acid-base material practice.

(a) (b)

Figure 1. The activity of students to use student worksheets that are based on guided-inquiry practice: (a) students learn the student worksheet instructions, and (b) the students conducted an Acid-Base material experiment practice.

Figure 1 shows the students activities as follow: (a) the students read and study the worksheet practice guideline to determine any solution that can be used as Acid-Base indicator according to the colour of the solution; (b) the students conduct experiment to determine the acid-base indicator and write the observation results obtained into a table that have been provided in the student worksheet.

The GSS score of students is presented in table 2.

Table 2. GSS scores of students at two secondary schools in Banda Aceh, Indonesia.

No	The indicator of generic science skills	Score (%)	
		The first school	The second school
1.	Direct observation	98.75	91.80
2.	Indirect observation	82.08	79.30
3.	Cause and effect	64.17	58.59
4.	Logical consistency	92.50	88.02
5.	Logical inference	85.42	82.04
	Total	422.92	394.27
	Average	84.58	78.85

Table 2 shows that an assessment of GSS of students at two secondary schools in Banda Aceh, Indonesia, the scores were 84.58% and 78.85% on average and was categorized as a good category. This suggests that the practical implementation can build GSS of learners. Then, with the application of student worksheets, it can also build participation of learners so that learners' generic science skills increased. The previous research mentioned that their generic skills in participatory learning strategies could put the students at the center of the learning process and also the community [16].

The GSS scores of students at the first school were different from the second school; significant assay results indicate that significant 0.000<0.05 indicates a significant difference. This is because the learners both school prepared to have generic skills are conceptualized as skills that help them understand the advantages and disadvantages, helping to independent, analyze the situation from a different perspective and can take advantage of the resources at their disposal to address the problems with the challenges that exist in future [17]. So that, the implementation of this generic learning skills

IOP Conf. Series: Journal of Physics: Conf. Series **1088** (2018) 012106 doi:10.1088/1742-6596/1088/1/012106

will support students in the higher education curriculum. The reason that supports shows that students that have the generic science skills will be better, the capabilities needed in a variety of job requirements, and the discipline will assure itself as the best graduate [18].

The results acquired were supported by a study showed that the average gain in the experimental class <g> on direct observation indicator of 0.38, which was consistent and logical framework indicators 0.34 [19]. Then, the results of data analysis questionnaire showed that nearly all learners had good skills in the application of guided inquiry learning. The logical inference indicators show that learners were able to deliver the results that have been studied. Research that has been conducted shows that the average yield of 88.50 percentage inference logic was a very good category. So, it can be interpreted that the use of student worksheets based guided inquiry can improve the ability of logical inference learners. Learners were able to explain in accordance with the reference, and also can draw the conclusion properly [20]. Skills convey this information is not restricted to mass communication. The key to effective oral communication is to know how to do, listen to, and requires intelligence to the overall process of communication activity [21].

The effectiveness of the implementation of student worksheets based guided inquiry tested the *N-Gain* and t-test after pretest and posttest the practical implementation of acid-base materials at two secondary schools in Banda Aceh, Indonesia. As for some forms of student worksheets that have been filled can be seen in Figure 2.

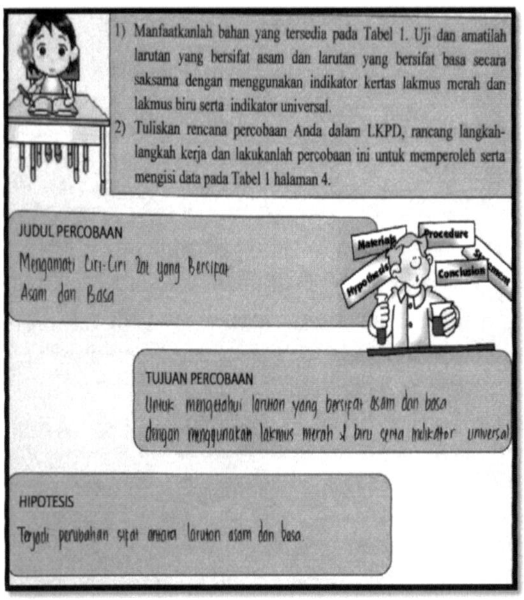

(a)

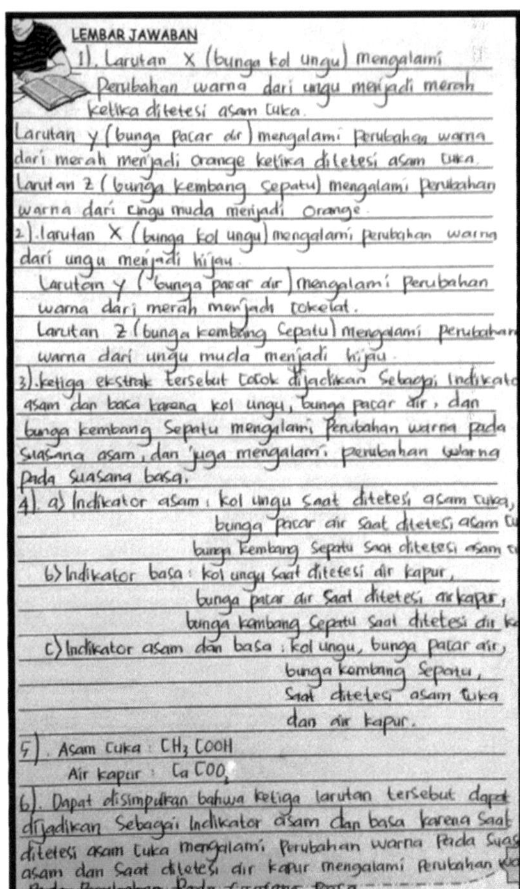

(b)

Figure 2. Representative image of student worksheets based on guided inquiry, (a) Student activity guide sheets; and (b) answer sheets and student discussions report.

Figure 2 part (a) shows practice guideline of a student worksheet that containing some important aspects including problem topics, the title of the experiment, experiment objectives, experimental hypotheses, and the availability of tools and materials according to learning indicators. The problems given in the student worksheet are the characteristics of guided inquiry learning. Based on the guided inquiry learning, the students will analyse the problems provided by teachers using self-made experimental design and procedures [22]. While the Figure 2 part (b) shows student answer sheet consisting of all students' answers based on the questions given in the student worksheet. This sheet contains the discussion of student's activities related to the solution colour's description after the experiment, the characteristics of each solution colour obtained, acidic and basic solutions grouping, the chemical formulas for each ingredient used, and the summary of the experiments results.

Test *N-Gain* had an increase in the study of students. A t-test was conducted to determine the significance of the study of students. Figure 3 *N-Gain* in both researched schools is shown here.

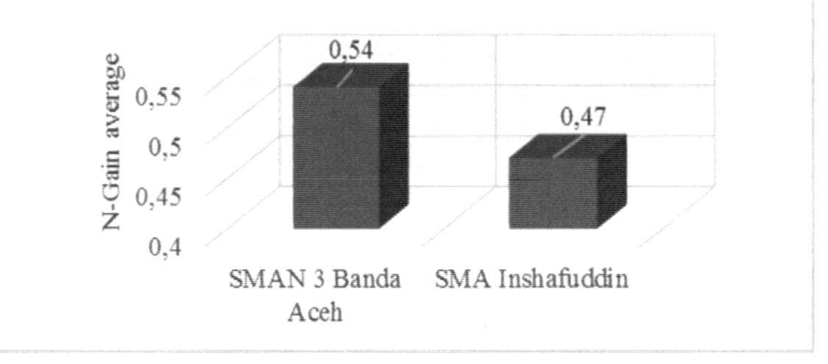

Figure 3. Average graph of N-Gain.

Based on Figure 3, it shows that the average N-Gain in both schools were 0.54 and 0.47 with medium criteria. This indicates that learners at two secondary schools in Banda Aceh had different capabilities. a The results of the analysis of the t-test for equality of two average values pretests had significant results 0.096 > 0.05 meaning that the average value of pretest ability learners in the first school was not significantly different from the second school. However, after the implementation of student worksheets based guided inquiry, the results of t-test analysis for differences in the average value of two posttests showed significant 0.036 <0.05 meaning the ability of learners in both schools on acid-base materials in terms of practical implementation using student worksheets based guided inquiry had significantly different capabilities, but both of the schools had different school passing standard grades in chemistry subject chemical learning (KKM) that was 75 in the first school and 70 the first school. So, based on hypothesis testing concluded that the implementation of learning chemistry lab on acid-base materials using student worksheets based guided inquiry was effective in improving student learning outcomes. This confirmed the results of this study support previous research showing that colloidal module based guided inquiry can effectively improve cognitive learning outcomes of students shown by the analysis of N-*Gain* from acquisition pretest and posttest, score capabilities cognitive learning outcomes of learners at 0.688 categorized as a gain medium [23]. Furthermore, other studies show that learners had a better understanding of the basic concepts of acids and bases as well as a more positive attitude to the application of guidelines on teaching materials based guided inquiry [24]. In addition, by using a model of guided inquiry is able to maximize the learning process that shows an increase in cognitive abilities of students in the cycle I and II both classical and average class [25]. As well as the relevant results showed that the analysis of the effectiveness of assisted instructional materials show that can improve generic skills such as oral and written communication, teamwork [26].

4. Conclusions

Learning outcomes of students at two secondary schools were in medium level with both N-Gain is 0.54 and 0.47, science generic skill values were 84.58% and 78.85% with a good category. Thus, student worksheets effectively can be used as an alternative teaching material on acid-base material practice.

Acknowledgment

We would like to thank the people who have contributed to this research including research instrument validate, teachers at two secondary schools who had participated in this research.

Reference

[1] NCVER 2003 *Defining generic skill-At a glance*. Australian National Training Authority. This publication was produced by the National Centre for Vocational Education Research Ltd. ABN 87 007 967 311
[2] Khabibah E N, Masykuri M and Maridi 2017 *ASSEHR.* **158** 251
[3] Green W, Hammer S and Star C 2009 *Higher Edu. Res. Develop.* **28** 17
[4] Horsburgh M 1999 *Quality in Higher Educ.* **5** 9
[5] Singh P, Thambusamy X R and Ramly A M 2014 *Proc. Soc. Behavioral Sci.* **123** 315
[6] Rahman S, Mokhtar B S, Yasin M R and Hamzah M I M 2011 *Proc. Soc. Behavioral Sci.* **15** 3713
[7] Freudenberg B, Rimble M and Cameron C 2011 *Asia-Pac. J. Cooperative Edu.* **12** 79
[8] Adlim M, Wilyta I and Hasan M 2017 *J. Pencerahan.* **11** 15
[9] Harahap R P S, Sani A R and Simanjuntak P M 2017 *IOSR J. Res. Method Edu. (IOSR-JRME)* **7** 60
[10] Vlassi M and Karaliota A 2013 *Proc. Soc. Behavioral Sci.* **93** 494
[11] Hamdani 2011 *Strategi Belajar Mengajar* (Bandung: Pustaka Media)
[12] Rahmi R, Hartini S and Wati M 2014 *Berkala Ilmiah Pendidikan Fisika* **2** 240
[13] Rokhmah A and Madlazim 2015 *J. Inovasi Pend. Fisika (JIPF)* **4** 88
[14] Borg W R and Gall M D 1989 *Educational Research* (London: Longman Group)
[15] Fraenkel J R, Wallen N E and Hyun H H 2011 *How to Design and Evaluate Research in Education (eighth edition)* (New York: Mc.Graw-Hill Book)
[16] George R 2011 *Int. J. Fundamental Psycho. Soc. Sci.* **1** 14
[17] Chan W S C 2010 *Proc. Soc. Behavioral Sci.* **2** 4815
[18] Crebert G, Bates M, Bell B, Patrick C J and Cragnolini V 2004 *Higher Edu. Res. Develop.* **23** 147
[19] Yuniarita F 2014 *J. Pengajaran MIPA* **19** 111
[20] Diba F P, Wardani S and Sudarmin 2017 *J. Innovative Sci. Edu.* **6** 1
[21] Urios I M, Rangel R E, Tomàs B R, Salvador T J, García C F and Piquer F C 2015 *J. Technol. Sci. Edu.* **5** 107
[22] Meador G 2010 *Inquiry physics: A modified learning cycle curriculum* (Oklahoma: Bartlesville High School)
[23] Novilia L, Iskandar M S and Fajaroh F 2016 *Int. J. Edu.* **9** 17
[24] Bilgin I 2009 *Scientific Res. Essay.* **4** 1038
[25] Karsilah, Febriastuti D Y and Siswanto 2017 *Indo. J. Sci. Edu.* **1** 49
[26] Abdullah S R S, Takriff M S, Ismail M, Kalil S M, Rahman A R, Daud W R, Mohammad W A, Mohammad B A, Anuar N, Abdullah N, Harun S, Kamarudin S K, Tasirin M S, Yaakob Z, Jahim J, Kofli T H N, Markom M, Rahman A N, Rahim A A R R R and Rahaman A S M. 2012 *Proc. Soc. Behavioral Sci. (UKM Teaching and Learning Congress 2011)* **60** 512

The association between conceptual understanding and reasoning ability in mathematics: An analysis of DNR-based instruction models

MT Bakar[1,2], DSuryadi[1], Darhim[1], W STonra[2] and M S Noto[3]

[1]Universitas Pendidikan Indonesia, Jl. Dr. Setiabudi No. 229, Bandung 40154, Indonesia
[2]Universitas Khairun Indonesia, Jl. Bandara Babullah Akehuda Ternate 97751, Indonesia
[3]Universitas Swadaya Gunung Djati, Jl. Perjuangan No.1 Cirebon 45132, Indonesia

E-mail: marwiatamrin@student.upi,edu

Abstract. Mathematical reasoning is a tool used to understand mathematical concepts. Mathematical understanding is the basic ability in learning mathematics which is a mental construction process that connects, makes generalizations and synthesizes internal and external representations of mathematical objects. It is suspected that these two abilities have a close relationship. Based on these assumption, this study aimed to describe and reveal if there was an association between mathematical understanding and mathematical reasoning of college students who were taught using the Duality Necessity Repeated reasoning (DNR)-Based Instruction in the Basic Mathematics Course. 107 prospective primary education students participated in the study. Data was collected using a test consisting of four problems related to two-variable linear equations. Association analysis was assessed using Spearman test. The result showed that there was a strong and positive association between the ability of mathematical concept understanding and mathematical reasoning. This implies that college students' mathematical reasoning abilities are strongly influenced by their understanding of mathematical concepts.

1. Introduction

Mathematical understanding is an ability to understand the basic nature of mathematical knowledge [1]. One of the basic natures of mathematics is havingabstract objects. Understanding abstract objects of mathematics cannot be done just by memorizing. As Idris stated [2] that mathematical understanding is not only remembering mathematical concepts or procedures, but it requires deep understanding. The definition of further mathematical understanding is put forward by Wiggins & Mc Tighe [3] as a mental construction and abstraction; being able to communicate, solving problems and demonstrating reasoning [4]; being able to prove relationships and the representation of concepts in a situation or problem solving [5]. This corresponds to Harel's remark that mathematics is *ways of thinking* (WoT) and *ways of understanding* (WoU); duality that has a reciprocal relationship where both of them are mental actions such as interpreting, summing up, proving, explaining, generalizing and so on [6]. Therefore, a person is said to have the ability of mathematical understanding when he or she is capable of performing the mental acts as mentioned earlier. This is in line with Hoosain's opinion [7] that mathematical understanding includes mathematical facts and mathematical proofs.

Mathematical reasoning is the ability to think logically. It has something to do with making connection between concepts and situations and the ability to make conclusions. The study of mathematical reasoning has been extensively researched. For example, English's [8] study found that mathematical reasoning is imaginative, and Russel's [9] study stated that reasoning is a tool of understanding abstraction. Research findings on mathematical concept understanding and reasoning havealso been widely practiced. For example, a research conducted by Hoosain [7] recommended educators to set students' understanding as their goal in teaching mathematics. A theoretical study conducted by Kieran [10] suggested that mathematical concept understanding is sustainable and continuous. Hence, it can be concluded that there are some stages involved in achieving mathematical concept understanding.

Previous research on pre-service teachers' mathematical ability show that there are at least five weaknesses possessed by students: (1) lack of prerequisite knowledge, (2) lack of understanding of the basic concepts of mathematics, (3) lack of focus in learning, (4) inability to identify problems, and (5) poor logical reasoning ability [11]. This is similar with a study conducted by Fuadi, Johar and Munzir [12] involving a number of eight grade students which was aimed to determine the improvement of mathematical concept understanding and reasoning after applying contextual learning model. The result shows that students who are taught by contextual learning model perform better than those who are taught with conventional learning. From some of these studies, there has been no research that aims to see the association between the ability of mathematical concept understanding and mathematical reasoning using the DNR-Based Instruction. Therefore this is what seems to be the novelty of this research. For that reason, it is necessary to develop an instrument to measure students' mathematical concepts understanding and mathematical reasoning.

The DNR-Based Instruction model was developed by Harel [13] who suggested that the goal of mathematics learning is to help students constructing ways of thinking and ways of understanding [6]. DNR-Based Instruction consists of Duality, Necessity and repeated Reasoning, which are also the principles in Harel learning [14]. The duality principle is the fundamental concept of DNR. The principle of necessity is that what teacher wants to convey is what they need. This is called intellectual need [15]. This highly supportive model characteristic is what this research conducted.

2. Method

This study aimed to describe students' mathematical concept understanding and mathematical reasoning ability and to determine whether there is an association between the conceptual understanding and mathematical reasoning of students who were taught with DNR - Based Instruction learning. Then, an instrument test related to mathematical conceptual understanding indicators and mathematical reasoning indicators was designed.

The indicators of mathematical conceptual understanding included being able to relate a concept to another concept, represent a mathematical situation in various ways, determine better representations. The indicators of mathematical reasoning included being able to perform calculations based on mathematical formula and draw general conclusions based on the mathematical process/concepts involved. This instrument test was developed based on validation process by four experts. The experts were two lecturers of mathematics education, an evaluation lecturer, and a linguistic lecturer. This instrument had been tested on 35 students at a university in Ternate, Indonesia.

The participants of this study were 107 students at the University of Khairun Ternate, Indonesia, who were enrolled in the Basic Mathematics Course. The scores of the students' conceptual understanding and mathematical reasoning were obtained from the test given at the end of the lesson. The test consisted of four problems as presented in table 1.

Tabel 1. Problem description.

No	Problems
1	Mr. Amir bought five recreational park tickets for two adults and three children for Rp105,000.00. Meanwhile Pak Iksan bought three tickets for adults and five tickets for children for Rp165,000.00. State this situation in the most appropriate form youcan think of to determine the price of each recreational park ticket.
2	A convenience store sells two types of rice. Rice type I and rice type II. The price of rice type I is Rp10,000.00 and the price of rice type II is Rp13,000.00. The sales on that day were Rp2,935,000.00 and the amount of rice sold was250 liters. a. Which type of rice sells better? Explain youranswer. b. Suppose that the amount of rice soldwas 370 liters and the sales were Rp4,450,000. How many litters was each type of the rice sold? c. What is your conclusions based on these two problems?
3	The circumference of an isosceles triangle is 20cm. If the length of both feet added by 3 cm and the length of the base is two times the original length, then the circumference is 34 cm. Represent your answers using more than two ways to calculate the lengths of all three sides of the isosceles triangle.
4	There is a rectangular object with a circumference of 22 cm. If the length is changed into three times the original length and the width twice the original width, then the circumference of the object becomes 58 cm. Find the area of the rectangle and explain how you get your answer.

Descriptive analysis was used to obtain the description of the students' ability of mathematical concept understanding and mathematical reasoning, and to predict the association between the two variables using Spearman's rhotest. The interpretation of the Spearman association coefficient is presented in table 2.

Table 2. Interpretation of Spearman's association coefficients.

Associationcoefficient	Interpretation
0.00 to 0.20	There is almost no association
0.21 to 0.40	Weak association
0.41 to 0.60	Moderate association
0.61 to 0.80	High association
0.81 to 1.00	Perfect association

3. Result and discussion

In general, the average scores of the students' mathematical conceptunderstanding and mathematical reasoning were 78.055 and 72.98, respectively. This data showed that the students' mathematical concept understanding was higher than their mathematical reasoning. This result is in accordance with Kieran's study [10] which stated that understanding of mathematical concepts is the basis of mathematical mastery and the foundation for learning more advanced mathematical concepts. Students' existing knowledge plays an important role in connecting new ideas and information [16].

Before conducting statistical tests to determine whether there was an association between the two capabilities, data normality was tested using Kolmogorov Smirnov Z test. Normality test result showed thatone of the data was not normally distributed. Therefore the association test used Spearman non-parametric test instead. The hypotheses were:

H_0: There is no association between the ability of mathematical concept understanding and mathematical reasoning ability.
H_1: There is an association between the ability of mathematical concept understanding and mathematical reasoning ability.

The testing criterion based on Spearman association test of the hypothesis was that if the significance value was less than the degree of $\alpha = 0.05$ then H_1wouldaccepted. Otherwise H_0 would be accepted.

The results of Spearman's association test between the ability of mathematical concept understanding and mathematical reasoning are shown in table 3.

Table 3. Association between students' mathematical understanding and mathematical reasoning.

			Mathematical concept understanding	Mathematical reasoning
Spearman's rho	mathematical concept understanding	Correlation coefficient	1.000	.801**
		Sig. (2-tailed)	.	.000
		N	107	107
	mathematical reasoning	Correlation coefficient	.801**	1.000
		Sig. (2-tailed)	000	.
		N	107	107

Based on table 3, it could be seen that the correlation coefficient based on Spearman test between the ability of mathematical concept understanding and mathematical reasoning was 0.801, and the value of significance (p value) was 0.000. Therefore, based on the criteria of the association, it couldbe interpreted that the ability to understand mathematical concepts and mathematical reasoning were highly and positively correlated. This meant that the ability to understand mathematical concepts and mathematical reasoning had a positive relationship with high closeness. In other words, it couldbe interpreted that the higher ability mathematical concept understanding was, the higher mathematical reasoning. Students with high understanding of mathematical concepts tended to have high reasoning abilities. Thus it could be said that learning with DNR-Based Instructions developed the ability to understand mathematical concepts and mathematical reasoning. This finding is in line with Harel's research [13] that the DNR-Based Instructions learning model helped students to devise ways of thinking and improvedtheir understanding of mathematics, thus ensuring a positive impact on their mathematical concept understanding and reasoning.

4. Conclusion

Based on the findings, it could be concluded that the students' mathematical concepts understanding ability was higher than their mathematical reasoning ability. There was a strong and positive association between the ability to understand mathematical concepts and mathematical reasoning.

Acknowledgments
Thank you to Yulyanti Harisman and Dr. Karman La Nani, M.Si who have helped to review this article.

References

[1] Godino J D 1996 Mathematical concepts, their meanings and understanding *Proc. PME Conf.*vol 2 (Universidad de Valencia) p 2-417

[2] Idris N 2009 Enhancing students' understanding in calculus through writing *Int. Electronic Journal of Mathematics Education* **4** 1

[3] Wiggins G and McTighe J 2012 *Understanding by Design* (Alexandria, VA: ASCD.org)

[4] Baig S and Halai A 2006 Learning mathematical rules with reasoning *Eurasia Journal of Mathematics, Science and Technology Education* **2** 15

[5] Machin M C, Diaz J P and Trigo M S 2012 An exploration of students' conceptual knowledge built in a first ordinary differential equation course (part 1) *The Teaching of Mathematics XV* pp 1–20

[6] Harel G 2008 What is mathematics? A pedagogical answer to a philosophical question *Proof and Other Dilemmas: Mathematics and Philosophy* ed R B Gold and R Simons (Washington: Mathematical American Association)

[7] Hoosain E 2001 What does it mean to understand Mathematics? *Humanistic Mathematics Network Journal* **25** 9

[8] English L D 1997 Analogies, metaphors and images: Vehicles for mathematical reasoning *Mathematical Reasoning Analogies* (Lawrence Erlbaum Associates Inc)

[9] Russel S J 1999 Mathematical reasoning in the elementary grades *Developing Mathematical Reasoning in Grades K-12, 1-12* eds Lee V Stiff and F R Curcio (Virginia: NCTM)

[10] Kieran C 1994 Doing and seeing things differently: A 25-year retrospective of mathematics education research on learning *Journal for Research in Mathematics Education* **25** 1

[11] Wahyudin *Ability of math teachers, math teacher candidate and students in mathematics subjects* [Dissertation] UPI Bandung (Indonesia) Not published 1999

[12] Fuadi R, Johar R and Munzir S 2016 Improving the ability of mathematical understanding and reasoning through contextual approach *Journal Didaktika Matematika* **3** 1

[13] Harel G 2002 The development of mathematical induction as a proof scheme: A model for DNR-Based instruction *Learning and Teaching Number Theory: Research in Cognition and Instruction* p 185

[14] Harel G 2008 DNR perspective on mathematics curriculum and instruction Part II: With reference to teacher's knowledge base *Zentralblattf'ur Didaktik der Mathematik* **40** 1

[15] Harel G 2013 Intellectual need *Vital Direction for Mathematics Education Research* ed K R Leatham (New York: Springer)

[16] Stylianides A J and Stylianides G J 2007 Learning mathematics with understanding: A critical consideration of the learning principle in the principle and standards for school mathematics *The Mathematics Enthusiast* **4** 103

Students' understanding of charts: the study of PISA's problem-solving in the content of data

I K Sari, A Nasriadi and M Salmina
STKIP Bina Bangsa Getsempena, Tanggul Krueng Aceh Baru Street 34, Banda Aceh, 23112, Indonesia

E-mail: intankemalasari00@gmail.com

Abstract. This study aimed to analyze students' ability to interpret data that were presented in charts. The analysis was based on the students' problem-solving of PISA problems regarding charts. The participants were 50 grade-seven students from 3 different schools in Banda Aceh. From the four types of charts namely scatter chart, line chart, bar chart and pie chart, it was found that the students' tended to understand problems that were presented on only one observational dimension. The students' explanation about the charts focused solely on either the x-axis or y-axis. The students seemed to find it difficult to understand data which were presented in bar charts and pie charts. Basically, 13-14-year-oldstudents, according to Piaget's cognitive development theory, are moving towards the stage of formal development so they mainly think of a concept from a single point of view. Therefore, it is necessary for teachers to bridge students' thinking from informal stage to formal stage. Through the results of this study, we expect to enrich teachers' references regarding the development of learning materials.

1. Introduction

Program for International Student Assessment (PISA) is an international program to measure the educational success of a country. The survey measures students' literacy and problem-solving skills. Particularly in mathematics, the PISA test investigates students' mathematical literacy that can prepare the students to socially involved in a modern society [1]. The purpose of PISA in mathematics is to assess students' knowledge in solving problems in daily life. With the increasing challenges of today's life, students need to have good understanding and reasoning skills that can be used in their life.

Many of the weaknesses of Indonesian students' mathematical skills are revealed in PISA results. The weaknesses include inability to reason, lack of reading comprehension, and tend to only "receive" information and then forget about it. This condition indicates that the teaching and learning of mathematics in Indonesian classroom needs improvement in order to promote students thinking [2]. The mathematics topics covered in PISA test were also taught in Indonesian classroom such as uncertainty and data.

The regulation from the Indonesian Education Minister No. 22 Year 2006 asserts that the topic of statistics is one aspect in mathematics subjects that must be taught to students in junior high school level. Even before that, statistics has been introduced to students since the 6th grade of elementary school. This means that statistics is not a new concept for junior high school students. But, the real condition is that many junior high school students consider statistics as a difficult subject, especially data presentation such as pie chart [4]. In addition to that, students still find difficulty in interpreting variables placed horizontally and vertically on histograms [5].

According to Piaget's constructivist theory, cognitive development is a process where learners actively build their understandings using experiences and interactions [6]. This theory explains how children adapt to and interpret objects or events around them. This development follows certain stages from simple to complex. Junior high school students grade7 are students with the age of 13-14 years old. At this age, students have entered the formal operational stage, which is the top level of cognitive structure development. Adolescent students are able to think logically for all kinds of hypothetical problems and verbal problems, and they can use scientific reasoning and can accept the views of others. They can think flexibly because they can see all the elements and possibilities that exist and can think effectively because they can see which thoughts are suitable for a certain problem. Viewed from biological factors, this stage occurs at puberty (when other major changes occur), marking the entry into the adult world physiologically, cognitively, moral reasoning, psychosexual development, and social development.

Some people do not fully progress until this stage, so they do not have thinking skills like adults and keep using reasoning from the concrete operational stage. Moreover, junior high school students are students who are just starting the stage of cognitive development. In learning mathematics, according to Freudenthal, mathematical process is a key element for two reasons. The first reason is that mathematical process is not only an activity for mathematicians, but also for students through the application of mathematics in everyday-life situations. The second reason is that the final stage of mathematics learning is the formation of axioms. Thus, Freudenthal suggests that mathematics be organized in the process of rediscovery(reinvention) where students construct mathematical concepts through a process that is similar to what experienced by the experts [7].Similarly, in learning mathematics, students need to be introduced first to the concept of the real world before building his/her personal concept into an abstract formal concept. This is in line with Sari's [8] remark that learning activity is one of the major supporting elements in a learning process. Without that activity, the teacher will only be transferring knowledge and the students will only be passive recipients without having true understanding. Activities can build a conducive learning atmosphere. Building the learning activities that are relevant to learning materials can be done by creating and applying learning trajectory.

Based on the background above, it is clear that data presentation is not new for junior high school students of age 13-14 years old. However, students still face difficulties in understanding bar charts and pie charts. For this reason, the researchers conducted a study to investigate students' problem-solving strategies in solving problems related to data presentation. This study aimed to describe students' abilities in understanding scatter charts and line charts in solving PISA problems.

2. Method

The study was conducted using a qualitative research method based on the phenomenological method. The sample of this study was 50 seventh-grade students (age of 13-14) from three different junior high schools in Banda Aceh. The schools were different in terms of academic ability level according to Banda Aceh Educational Government ranking.

The steps undertook in the study consisted of three stages. The first stage was selecting PISA questions or developing PISA related questions about charts. In the second stage, the selected PISA problems were used to examine students' strategies for solving problems. In the third stage, interview was conducted to clarify students' answers. Data analysis was conducted in two ways: analysis of the data related to the students' answer based on selected PISA problems and analysis of the data related to the interview.

3. Result and discussion

This research was carried out by giving PISA problems to 50 students of grade7 in three different junior high schools in Banda Aceh. The given questions consisted of scatter plot materials, line charts, bar charts and pie charts. Figure 1 is one of the questions in the test.

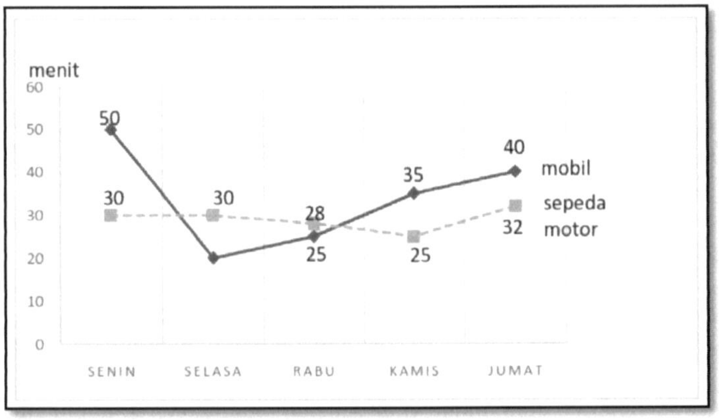

Figure 1. A line chart about two kinds of vehicle.

As shown in figure 1, the x-axis showed the days from Monday to Friday. The y-axis showed the time from 0 to 60 minutes. The black line referred to the car and the broken line referred to the motorcycle. The questions related to figure 1were: (1) On what day is the shortest time for the car and for the motorcycle? and (2) On what day are the two vehicles traveling relatively at the same time? The problem shows data presented in two dimensions of time and day involving two objects, a motorcycle and a car. Thus, this chart has 4 dimensions of observation. Nevertheless, this problem is still relatively easy and requires simple reasoning.

Out of50 students who answered these questions, 18 students answered both questions correctly: Tuesday for the car and Thursday for the motorcycle. 10 of the students answered the first question correctly and the second question incorrectly. Their answers were: 20 minutes for both vehicles. Six students answered the first question incorrectly and the second question correctly, and 16 students answered the second question incorrectly. That means 22 students or 44% had incorrect answers for the first question and 26students or 52% of students had incorrect answers for the second question.

During interview with the students who had incorrect answers, it was discovered that the students misinterpreted the line chart. For the first question, they only looked at the bottom of the horizontal line without paying attention to which vehicle had the shortest time. As for second question, the students raised the number 25 for every aspect of motorcycle and car observations, which showed a 25-minute travel time. This implies that in this case the students paid attention only to the information displayed from the *y*-axis or horizontal line so they did not realize the integration with the vertical line which indicated the time. This means that majority of the students only looked at the presentation of the data from one dimension only.

The second problem was related to line charts, where students were asked to make statements and reasons based on figure 2. The students were asked whether they agreed if firm A was the company that offered the cheapest transportation service than firm B and firm C. The chart gave two dimensions of observation: the cost of car rental and the number of passengers.

From 50 students who answered this question, 37 students answered correctly and 13 students answered incorrectly. Two of the students who answered correctly used answers that matched the manipulation of numbers and mathematical formulas to find the right number of rents for each passenger. Interestingly, 39 students reasoned this problem based on their observations of the horizontal line. Since firm A had the longest line with the highest point, they assumed that firm A had the most expensive cost of car rental.

In figure 2, the x-axis showed the number of passengers and the y-axis showed the rate of the rent starting from 0 to 16 thousand rupiahs.

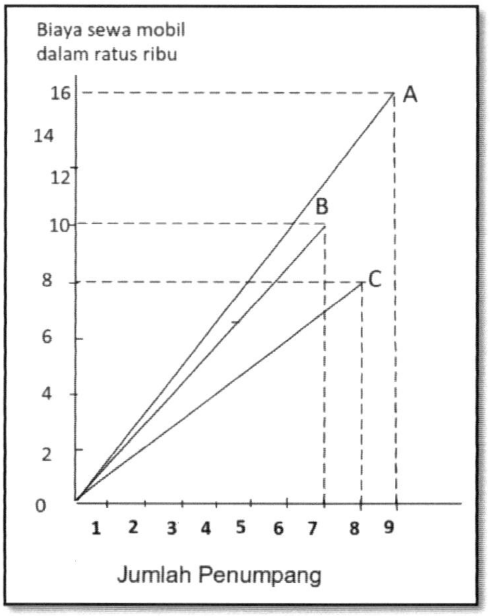

Figure 2. A line chart about car rentals.

Apart from that, 8 students disagreed with the statement in the question. They reasoned that although the price was in the highest position, the number of passengers was also many, this was based on observations from the vertical line so they assumed that the more passengers, the rent would be cheaper. The other 3 students who answered wrongly did not give reasons for their answers.

In the third and fourth questions, the students were asked to create bar charts and pie charts based on previous questions in tabular form. For this session, the students were asked to work in groups of four so that they could collaborate by sharing ideas and get feedback from their friends.

Out of 15 groups of the students who participated in the study, only one group could make a bar chart with information from the vertical and horizontal sides. For the pie chart problem, no group answered correctly. This showed that the students were still using observations from only one dimension. Table 1 shows students' answers to the PISA questions.

Table 1. The number of students' responses based on the answer types.

Number of Problem	Answer Type 1[a]	Answer Type 2[b]	Answer Type 3[c]	Answer Type 4[d]
1.a	7	21	0	22
1.b	7	17	0	26
2	7	30	0	13

[a]True answer and Two-dimensional viewpoint.
[b]True answer and One-dimension viewpoint.
[c]Wrong answer and Two-dimensional viewpoint.
[d]Wrong answer and One-dimensional viewpoint.

According to the Piaget's theory of cognitive development, the students have entered the formal operational stage. This means that the students should have been able to build their own mathematical concepts through the principles of hypothesis and deductive reasoning. In the further development, students aged 13-14 years old begin to develop abstraction skills using symbols or notation [12].

However, this study indicated that some students aged 13-14 years old in Banda Aceh had not reached the appropriate cognitive development.

Whereas, the ability of students who only observed an event through a single point of view according to Piaget's Theory was a cognitive development of level 2 which was the pre-operational stage. This level was actually a stage of cognitive development of children aged 2-7 years old. According to Piaget, children's perceptions in this stage of development were generally restricted to one aspect or dimension of an object. Piaget tested the concept of conservation by pouring the same amount of liquid into two similar containers. When the liquid from one container was poured into a third, wider container, the level was lower and children thought there was less liquid in the third container. Thus, the children were using one dimension, height, as the basis for their judgment of volume.

This was the same case as the findings of this study. When the students were given a problem involving two dimensions, they only paid attention to one dimension. This indicated that in considering two interrelated things, the students tended to look at one side only. This finding was confirmed through the interview. 37 out of 50 students, or equal to 74%,exhibited that they consciously and repeatedly interpreted the line chart on one dimension only. The students' errors in interpreting the scatter chart and line chart made them faced difficulties in interpreting or presenting data in the form of bar charts because the three types of the diagrams were actually the same, just different in shapes. Pie chart was even more challenging for the students because of its shape. It required the students to be able to convert data into degrees. It meant that mathematical problem-solving in junior high schools in Banda Aceh still needed attention from teachers or educators [13].

4. Conclusions

Students aged 13-14 at some junior high schools in Banda Aceh had not reached the suitable cognitive development stage yet. Students in grade 7 should have entered the formal operational stage, but this study found that the students were still at the pre-operational stage according to Piaget's theory. A pre-operational stage is a condition for children aged 2-7 years old. It means that some junior high schools in Banda Aceh fall behind two stages from the normal stage. This can be seen from the students' perception in interpreting problems of presenting data in the form of a chart where they only paid attention to one dimension, either the x-axis or y-axis. This condition must be improved otherwise the curriculum content will always be difficult for the students. This also explains why Indonesian students' achievement in PISA survey has not improved.

Acknowledgments

We would like to thank the Indonesian Ministry of Research and Technology of Higher Education (Kemenristekdikti) who has given us great support so that we were motivated to conduct research that were beneficial for Indonesian education.

References

[1] OECD 2009*Learning Mathematics for Life a View Perspective* (OECD Publishing)
[2] Wardhani S 2011 *Program Bermutu Better Education through Reformed Management and Universal Teacher Upgrading* (Yogjakarta: Kementerian Pendidikan Nasional Badan Pengembangan Sumber Daya Manusia Pendidikan Penjaminan Mutu Pendidikan)
[3] OECD 2013 *PISA 2012 Assessment and Analytical Framework* (OECD Publishing)
[4] Wadhiyantini T 2008 *Permasalahan Pembelajaran Statistika - Peluang SMP dan Alternatif Pemecahannya* (Yogyakarta: Pusat Pengembangan dan Pemberdayaan Pendidikan Tenaga Kependidikan Matematika)
[5] Saiman 2016 Analisis kesulitan siswa dalam belajar statistic khususnya pada histogram *Aksioma Jurnal Pendidikan Matematika* **5** 231
[6] Trianto 2011 *Model-Model Pembelajaran Inovatif Berorientasi Konstruktivis* (Jakarta: Prestasi Pustaka)

[7] Treffers A 1991 Didactical background of a mathematics program for primary education *Realistic Mathematics Education in Primary School on the Occasion of the Opening of the Freudenthal Institute* ed L Streefland (Utrecht: CD – ß Utrecht University)
[8] Sari I K 2016The Students' Learning Trajectory of Transformation Geometry *Proc. of ICMSE* **3** 120
[9] OECD 2009 *Take the Test Sample Questions from OECD's PISA Assessments* (OECD Publishing)
[10] Hayat B and Yusuf S 2010 *Mutu Pendidikan* (Jakarta: BumiAksara)
[11] Stacey K 2010 Mathematical and scientific literacy around the world *Journal of Science and Mathematics Education in Southeast Asia* **33** 1
[12] Tessmer M 1993 *Planning and Conducting – Formative Evaluations* (London: Philadelphia Kogan Page)
[13] Ojose B 2008 Applying Piaget's theory of cognitive development to mathematics instruction *The Mathematics Educator* **18** 26
[14] Sari I K 2015 Profil Pemecahan Masalah Matematis Siswa Usia 14-15 Tahun di Banda Aceh *Numeracy* **3** 73-86

IOP Conf. Series: Journal of Physics: Conf. Series **1088** (2018) 012109 doi:10.1088/1742-6596/1088/1/012109

Questioning skill of science teacher from the students perscpective in senior high school

A Halim[1], Yusrizal[1], H Mazlina[1], Melvina[1,2] and Zainaton[3]
[1]Syiah Kuala University, Jl. Teuku Nyak Arief Darussalam, Banda Aceh 23111, Indonesia
[2]Syracuse University, Syracuse, United States
[3]National University of Malaysia, Kuala Lumpur, Malaysia

E-mail: abdul.halim@unsyiah.ac.id

Abstract. One of the most important emphasis in curriculum of 2013 is to produce graduates who have higher order thinking skill. Implicitly, it proved that graduates students of senior high school still have lack of development of higher order thingking skill. Several theoritical studies showed higher order thingking like creative thinking can be stimulated by teachers who have good questioning skill. This study aims to describe the activities of the questioning skills of physics teacher based on developed instruments. This research used qualitative and quantitative approach with survey technique. The sample was six physics teachers from six different senior high school in Banda Aceh chosen based on working period. Data were collected using questionnaire to indicate students responses. Data analysis with descriptive statistics explained that questioning skill of physics teachers being on moderate category. According to this research, in order to improve questioning skill in physics learning, the teachers need to distibute questions evenly, and the questions asked should be traceable to investigate to what extent for the students to be able to understand all concepts what they have been learned. The influence of teaching experience will be reached on the components of advanced questioning skill based on questions asked.

1. Introduction

Communication is important for learning to improve students participation in teaching process [1]. Teaching science plays important role in improving to make students interesting in asking question [2]. In case, learning science especially in physics is tend to create students to be more pasive thus their thingking skill will be decreased [3,4]. One of communication methods to influence students contribution in the class in order to to create effective teaching and learning through questioning [5]. Questioning is one of learning strategies which is believed will be able to facilitate in developing thingking skill to solve the problems, activate the participation in learning, create the interest and motivat them to find or grab new knowledge [6-9].

Questioning activity in education has very important role to improve learning effectivity. The effective question is depend on the use of teachers skills when provided the questions. The teachers who asked the questions must appreciate students responses well and it should be presented with full attention, spirit, polite and does not create confusion and restrict the creative thingking of students [10,11]. Hence, the teachers must mastery the strategies for discussion effectively thus they will be able to submit higher order questions [12]. The higher order question is believed to be able to help analysis and diagnosis for the students and able to evaluate learning progress of learning [13,14].

Therefore, self assessment to the discussion strategies for teachers is very crutial to improve effectivity and teachers success in physics learning [15,16].

One of discussion strategies for teachers that is able to facilitate critical thingking skills of students is provided by questions with analysis information based on basic fact completely [17]. Creating infromation in context to fulfill students necessary is can be done by submitting questions in form of 5 "W" + 1 "H" which are what, who, where, when, why and how [18]. But, the effectivity of those questions supported to learning achievement is still less disclosed. It is in line with Bay who stated that the questions which are provided by teachers and freequently asked on low level for memorizing, understanding, and applied questions [19]. Other than that, many teachers do not provide questions emphasized on critical thingking [20]. Therefore, science teachers are required to present questions in learning process in order to motivate students during learning and to evaluate their progress in learning [21].

According to explanation and introduction above, there are a few studies in reviewing the questioning skill of teachers in physics learning. Trhough this research, it is expected that teachers are able to apply questioning skill in learning so that the students will motivate to be more active. Besides that, this research data will be very important to find teachers profile who have good ability and skills. This study aims to understand how is teachers ability in applying questioning skill in physics learning in senior high schools in Banda Aceh for academic year of 2017/2018.

2. Method

This research used quantitative approach with survey method. It was conducted for 6 physics teachers in state senior high school number 2,3,7,14,15 and 16 in Banda Aceh which were chosed randomly withouth proportion for working time (disproportionate statified random sampling) based on school acreditation with category A and B.

Instrument which was used for this research is questionnaire with Likert sclae which has four multiple choise. It was developed according to validation process by 3 experts, which were 1 physics lecture as physicist, 1 physchology lecture who expert in questioning skill, and 1 physics teacher. It was arranged using 2 indicators of teachers questioning skill and it was the development result which is have been used by Uno [22]. According to the assessment from 3 validators, it can be concluded that the instrument is ready for field test. All sections of trial test have been conducted for 64 students in one of Senior High School in Banda Aceh to test validity and reliability. Technique for validity and reliability instrument refers to method which is had been used by previous researchers [23]. The trial test showed that the instrument has good score for reliability and validity (0.90 and 0.25). The research was done by distributing quetionnaire of questioning skill for students when the teacher conducted the learning process for one meeting. Then, the data were analyzed using descriptive statistics with percentage.

3. Result and discussion

3.1 Analysis for teacher questioning skill

According to analysis for the whole result on the component of basic questioning skill and anvanced questioning skill, the percentage of average score for teacher achievement of questioning skill in learning physics can be seen in Figure 1.

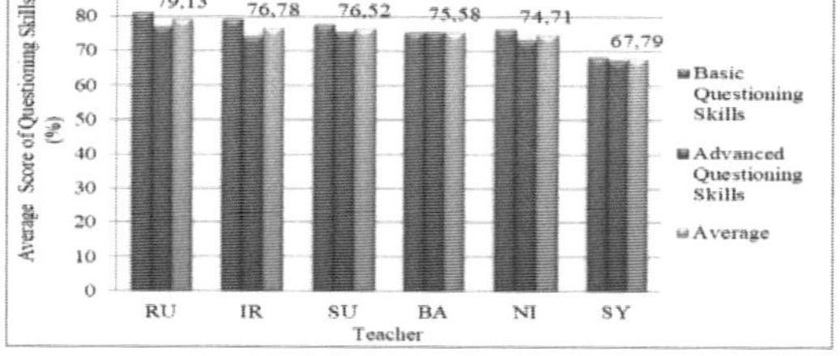

Figure 1. Percentage of teachers questioning skill in learning physics in state senior high schools in Banda Aceh

Based on Figure 1, the teachers questioning skill in learning physics was categorized by good achivement for RU (79.13%), IR (76.78%), and SU (76.52%). Then, the skill categorized with pretty good is obtained by NI (74.71%), BA (75.58%), and SY (67.79%). The analysis for six teachers especially for SU and BA, they did not have good ability in distributing questions, but it was provided for the whole class, hence it caused the students can not be participated well. The result is in line with Ermasari [24] who explained that the uneffectivity of teachers questioning skill are shown by the question distribution which is not evenly, providing response which is not good and the teachers habbits who directly answered the question asked by teachers.

According to Figure 1, it also result that the teachers still not optimal for advanced questioning component because of quation type which is submitted by teachers on low level which is memorizing question, understanding and applied question. As for, the type of question used by teachers in submitting questions in physics learning can be seen in Table 1.

Table 1. Question forms submitted by teachers in physics learning

Teacher	Question forms					
	What	Who	Where	When	Why	How
SY	√ (what is the conclussion form conducted experiment?)	-	-	√ (when compass needle shifted to the left?)	√ (why the compass needle can be shifted to the right or to the left?	-
NI	√ (what is differences between acceleration and deceleration?)	-	-	-	-	-
RU	√ (what is your impression to the existance of electricity?)	√ (who is famous scientist declared about static electricity?)	-	-	√ (why the different charge can be attracted?)	-
SU	√ (what thing that can be attracted by magnet?)	-	-	√ (when magnetic field can be exist?)	√ (why magnetic field can create magnetic forces?)	-
BA	√ (what is differences between conduction and convection?)	-	-	-	-	√ (how is the heat transfer can be happen by convenction ?)
IR	√ (what is static fluids?)	-	-	-	√ (why the woods can be floated on the water surface?)	-

Based on Table 1, it obtained that generally teachers used question type with word "what" by category for memorizing (C1). Only one teacher (E) who provide question "why". The result showed that the teachers still not able to submit higher order question in order to facilitate the development of students higher order thinking skill in learning process [19]. Besides that, category of different question type will encourage reflection and build students knowledge [26].

3.2 The influence of teaching experience to the questioning skill

3.2.1 Basic questioning skill

The influence of teaching experience to the basic questioning skill can be seen in Figure 2.

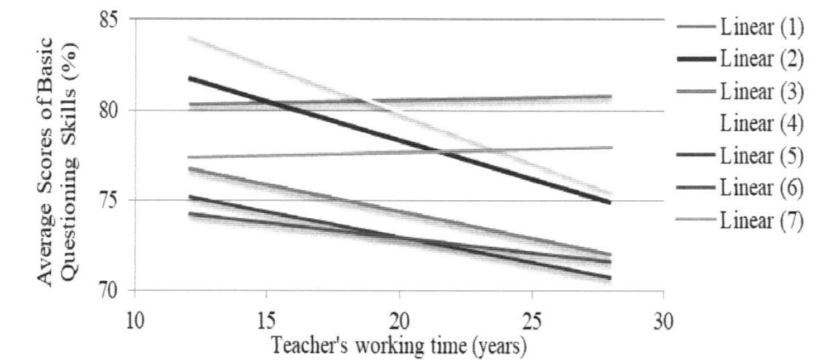

Figure 2. The Influence of Teaching Experience to the Basic Questioning Skill

Information of Indicators:
(1) Disclosure the questions clearly and concisely
(2) Giving references
(3) Centralization towards the requested answer
(4) Replacement turn replied
(5) Spreading questions
(6) Giving time for thingking
(7) Giving guidelines

According to Figure 2 the influence of teaching experience only exist on indicators of spreading questions, giving references and giving guidelines. From the video observation, it obtained that SU and BA were more success on indicators of giving references with helping the students who feel difficult to answer the questions. This finding proved that basically attitude change which is showed by students must be influenced by educational background and teachers experience [22].

3.2.2 Advanced questioning skill

The influence of teaching experience to the advanced questioning skill will be shown in Figure 3.

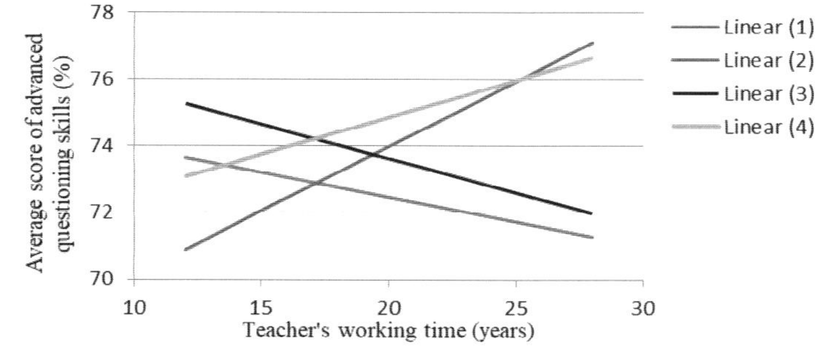

Figure 3. The Influence of Teaching Experience to the Advanced Questioning Skill

Information of indicators:
(1) Changing the demands of cognitive level of the questions
(2) Question chronology
(3) Giving questions for tracking
(4) Encourage the interaction among students

According to Figure 3, there is the influence of teaching experience on indicators (2) questions chronology and (4) encourage the interaction among students. It means the more experience obtained by teacher in teaching, so it will create questioning skill regularly and be able to motivate students for doing the interaction among them. Based on the graph in Figure 3, it can be understood that the addition of teaching experience will decrease the ability in figuring out the students capabilities. The result of video observation obtained by SY, NI, RU, SU, and BA used question chronology by starting comprehension questions and then will be closed by applied question through question problem solving. This finding showed that the teachers who have many experiences will be able to provide higher order questions with cognitive chronology appropriated to the questions [13].

4. Conclussion

The result of teachers questioning skill in physics learning in several state senior high school in Banda Aceh is presented by percentage of 75.04% with pretty good category with basic questioning skill by percentage of 76.37% (good) and advanced questioning skill by percentage of 73.83% (pretty good). Other than that, this research indicated that the influence of teaching experience will be reached on the components of advanced questioning skill based on questions asked.

References

[1] Ma X 2008 The Skills of Teacher's Questioning in English Classes. *International Education Studies J* **14** 92

[2] Ozer D Z and Gungor S N 2017 Analysis of Middle School Students' Views and Impressions about a Science Center *Turkish Science Education J* **14** 108

[3] Mahardika I K, Maryani dan Murti S C 2012 Penggunaan Model Pembelajaran *Creative Problem Solving* Disertai LKS Kartun Fisika Pada Pembelajaran Fisika Di SMP *J Pembelajaran Fisika* **12** 231

[4] Halim A, Suriana dan Mursal 2017 Dampak Problem Based Learning terhadap Pemahaman Konsep Ditinjau dari Gaya Berpikir Siswa pada Mata Pelajaran Fisika *J Penelitian dan Pengembangan Pendidikan Fisika* **31** 1

[5] Mauigoa L 2006 Enhancing Teachers' Questioning Skills to Improve Children's Learning and Thinking in Pacific Island Early Childhood Centres. *New Zealand J of Teachers' Work* **31** 12

[6] Philips N and Duke M 2001 The Questioning Skills Of Clinical Teachers and Preceptors: A Comparative Study *J of Advanced Nursing* **33** 523

[7] Shi-ying X 2011 The Present Situation of English Teacher's Questioning in Senior Middle School and Positive Strategies *Asia-Pacific Science and Culture J* **13** 1

[8] Yang H A 2017 Research on The Effective Questioning Strategies in Class *Science J Education* **54** 158

[9] Wangru C 2016 The Research on Strategies of College English Teachers Classroom Questioning *International Education Studies J* **98** 144

[10] Kira E, Komba S, Kafanabo E and Tilya F 2013 Teachers' Questioning Techniques in Advanced Level Chemistry Lessons: A Tanzanian Perspective *Australian J of Teacher Education* **12** 66

[11] Tofade T, Elsner M S J and Haines S T 2013 Best Practice Strategies for Effective Use of Questions as a Teaching Tool *American J of Pharmaceutical Education* **77** 1

[12] Saeed T, Khan S, Ahmed A, Gul R and Cassum S 2012 Development of Students' Critical Thinking: The Educators' Ability to Use Questioning Skills in The Baccalaureate Programmes in Nursing in Pakistan *J of the Pakistan Medical Association* **62** 200

[13] Wang A, Chai C S and Hairon S 2017 Exploring The Impact of Teacher Experience on Questioning Techniques in A Knowledge Building Classroom *J Computer Education* **4** 27

[14] Yusrizal and Halim A 2017 The Effect of the one-tier, two-tier, and three-tier diagnostic test toward the students' confidence and understanding *Unnes Science Education J* **6** 1583

[15] Davoudi M and Sadegh N A 2015 A Systematic Review of Research on Questioning as a High-level Cognitive Strategy. *English Language Teaching J* **8** 76

[16] Carthy P M, Sithole J P C and Gyan E 2016 Teacher Questioning Strategies in Mathematical Classroom Discourse: A Case Study of Two Grade Eight Teachers in Tennessee, USA *J of Education and Practice* **7** 80-

[17] Yuhui L and Sen D 2015 The Influence of Teachers' Questioning and Feedback On Students' Affect in The Interaction of Efl Classroom *International J of Arts & Sciences* **8** 467

[18] Hart G J 1996 The Five W's: an Old Tool for The New Task of Audience Analysis *J Technical Communication* **43** 139

[19] Bay D N 2016 The Question Asking Skills of Preschool Teacher Candidates: Turkey and America Example *J of Education and Training Studies* **4** 161

[20] Nevers S T and Melrose S 2016 Commentary: Posing Questions to Support and Challenge – A Guide for Mentoring Staff *Internet J of Allied Health Sciences and Practice* **14** 1

[21] Yip D Y 2004 Questioning Skills for Conceptual Change in Science Intsruction *J of Biological Education* **38** 76

[22] Uno H B 2010 *Orientasi Baru dalam Psikologi Pembelajaran* (PT Bumi Aksara: Jakarta)

[23] Halim A dan Yusrizal 2009 Developement and validation of an Instrument to Access the Lecturers Performance in the Education and Training Duties *Jurnal Pendidikan Malaysia* **34** 33

[24] Ermasari G, Subagia I W dan Sudria I B N 2014 Kemampuan Bertanya Guru IPA Dalam Pengelolaan Pembelajaran. *e-J Program Pascasarjana Universitas Pendidikan Ganesha Program Studi IPA* **4** 1

[25] Engin M 2013 Questioning to Scaffold: an Exploration of Questions in Pre-service Teacher Training Feedback Sessions *European J of Teacher Education* **36** 39

IOP Conf. Series: Journal of Physics: Conf. Series **1088** (2018) 012110 doi:10.1088/1742-6596/1088/1/012110

Designing computer-based fraction worksheets for junior high school

R Amalia[1], S Saiman[1], S Sofiyan[1] and M Mursalin[2]
[1]Universitas Samudra, Jl. Meurandeh, Langsa Lama, Kota Langsa 24354 Indonesia
[2]Universitas Malikussaleh, Jl. Cot Tengku Nie Reuleut, Muara Batu, Aceh Utara, Indonesia

E-mail:rizkiamalia@unsam.ac.id

Abstract. In mathematics learning, the use of student activity worksheets is still limited.In addition, the use of computers as one of the media for learning mathematics is still very minimal. The concept of fractions, as one of mathematics subjects, is quite complex but fundamental to be mastered by students. The purpose of this study was to develop valid, practical, and effective computer-based worksheets on the topic of fractions for grade 7 junior high school. This study used modified 4-D research development model. The result showed that the students responded positively towards theimplementation of the computer-based student activity worksheets.Therefore, the computer-based worksheets can be useful in mathematics learning, especially on the topic of fractions.

1. Introduction

Mathematics plays an important role in various disciplines and human life in general. Learning mathematics aims to connectknowledge learned at schools to everyday life, provide skills acquisition, prepare students for their future, and cultivate mathematical thinking [1]. Mathematics is often considered to be the most difficult subject, boring and unrelated to everyday life. The concept of fractions is one of mathematics subjectsthat many students find to be difficult. Fractions have been taught since elementary to junior high school level, but many junior high school students and college students cannot solve fractional problem seven though fractions are very closely related to everyday life. Empson and Levi argued that "the study of fractions of algebra on the fundamental mathematical relationships that constitute the core of algebra ... [that] govern how addition, subtraction, multiplication, and division work in algebra as well as arithmetic" [2]. More broadly, fractions applications is useful in various professions, such as in the field of medicine for example in determining the level of drug delivery; pilots need fractions in reading the direction of a flight radar; traders use fractions in calculating the percentage of profit or loss; and farmers use fractions in determining the appropriate fertilizer for maximum plant growth.

The Delors' Report marked the beginning of UNESCO's 21st-century learning competencies which include learning to live together, forming the four pillars of learning [3]. Teachers need to be aware of the four pillars of education in the current era of globalization. Teachers as learning agents need to master and apply information and communication technology (ICT) in the learning of mathematics at school. The National Council of Teachers of Mathematics argues that technology is an important tool for teaching and learning mathematics. Viewed as an integrated part of learning tools, technology can broaden the scope of course material that students can learn and can expand their knowledge [4].

The implementation of technology in teaching fractionscan help teachers to assessonline materials, interactive media, software, games and so on. Previous studies suggest that exploratory games that implicitly support mathematics knowledge can improve students' understanding of mathematics and their classroom engagement [5]. Researchers have developed some effective learning strategies to improve students' understanding of fractions. Recent research on interventions to support students at-risk and/or those with learning disabilities, report that they have been successful in promoting students' understanding of fractions using: a) mnemonic devices for teaching addition and subtraction of fractions [6-7]; b) manipulatives with pictures to solve problems involving fractions [8]; and c) direct instruction model [9-10]. Other researchers have reported success with similar strategies with a broader range of student abilities. These techniques include direct instruction using guided practice and focused feedback [9], and cue cards [11].

Teacher-designed learning in each mathematics topic will have an effect on students' mastery of mathematical concepts. Therefore, teachers must prepare appropriate learning tools. One means to increase student activity during learning is by using student activity worksheets (SAW). Teacher-made SAWsshould be integrated with ICT. Teachers must be able to design mathematical tasks using an attractive, varied display of images, stimulating curiosity and understanding of the material. Teachers can also use other media such as calculators, games, interactive whiteboards and other audiovisual aids, together with various software packages. The Internet can also be used to design effective learning tasks such as problem-solving simulations. Therefore, computer-based student activity worksheets should be developed. Based on the above description, this study aimed to produce computer-based student activity worksheets on fractions that is valid, practical, and effective for students of junior high school grade 7.

2. Method

The development model used in this research was the Four-D Model. The product was a computer-based student activity worksheet. Thiagarajan's 4-D model consisted of 4 development stages: Define, Design, Develop and Disseminate [12]. In the Define stage, there were 5 activities to be done: curriculum analysis, student analysis, task analysis, concept analysis and objective learning analysis. While in the Design stage thesteps to be done were: format selection (i.e. selection of existing resource formats and setting the format of teaching materials), and making the initial design according to the selected format. The Develop stage included evaluation by experts. Experts' suggestions were then used to revise the material and the design of the learning instruments. Developmental testing was also conducted to test the product on real targets. At the time of this trial, perceptions, reactions or comments from the target userswere collected. Test results were used to revise the product. After the product was revised, it was re-tested to obtain effective results. The last phase, Disseminate, was done by way of disseminating teaching materials through distribution for a limited number of teachers and learners [13]. But in this development research, disseminate stage was not done because of time constraints.

The procedure of development in this research consisted of (1) preparation of research, (2) implementation of research, and (3) data analysis. The steps taken by researchers in the preparation of research were as follows:

- Developing research instruments. Instruments in this research were validation sheet, computer-based student activity worksheets, competency test, and student response questionnaire.
- Submitting prototype computer-based student activity worksheets to the validator.
- Revising the prototype of computer-based student activity worksheets that have been validated to produce student worksheets that were ready to be tested (draft-1).
- Asking for permission of the school to be used as the place of study, as well as consulting with mathematics teachers to determine the timetable of the study.

At the time of conducting the research, the researchers conducted the test using mathematics student worksheets on fractions for students in grade 7 in one of the junior high schools in Kota

Langsa, Indonesia. The study took place in the academic year 2017/2018. After the implementation of the research, researchers analyzed the data using appropriate analytical techniques. After that, researchers compiled a report of the study. The subjects of this trial were all students in one of the classesata junior high school in Kota Langsa, which consisted of 20 male and 10 female students.

3. Result and discussion

The results of each stage of the development that has been done were as follows.

3.1. Defining stage

This stage consisted of 5 phases. The first phase was curriculum analysis. At this phase, the thing to note was that the curriculum was applicable in the school where the research took place. The school usedCurriculum 2013 for grades7 and 8. The approach of the research in developing computer-based student activity worksheets,which contained a scientific approach, grouping, and utilization of ICT, was very compatible with Curriculum 2013. In the development of computer-based student activity worksheets we focused only on one basic competency, namely knowing about fractional numbers and performing fractional operations in problem-solving.

The second phase was student analysis. When viewed from the background of students' knowledge, in general, the students of grade 7 have mastered some prerequisite materials that have been taught before studying fractions, such as simple fractions and integers. These existing knowledgewere important in studyingfractions. As for the learning conditions of the students, after interviewing with teachers of mathematics at that school, it was revealed that the teachers rarely used ICT or computer-based student activity worksheets, and the learning activity was usually concentrated on the teacher.

The third phase was task analysis. Computer-based student activity worksheets on fractions was designed as a teaching material that would be used for learning fractions by utilizing ICT so that learning would be more interesting and fun. Assessment of students' achievement was done by examining students' activities while they dothe tasks. The tasks were assigned both individually and in groups according to sub-materials. Individual tasks consisted of guiding questions in embedding fractional concepts through guided discovery. While the group taskswere aimed to examine students' performance in solving more difficult problems or questions that required exchanges of opinion through group discussions.

The fourth phase was concept analysis. The concept analysis was done by analyzing the concepts of fractions, focusing on determining the meaning of fraction, determining the value of a fraction, determining the way to simplify fractions, determining and comparing two Fractions,and determining fractions between two fractions. The computer-based student activity worksheets encompasseddefinitions, worked examples, problems and practice questions, online learning resources, softwares, and varied educational games.

The fifth phase was analysis of learning objectives.The analysis of learning objectives was done by elaborating basic competencies into the indicators of achievement in accordance with sub-material of fractions. In elaborating the basic competencies into indicators of achievement, we also took into account the results of concept and task analysis.

3.2. Design stage

The design stage consisted of 2 phases. Phase 1 was the selection of worksheets' format.The selection of student worksheet formats was intended to design the content of learning, strategies, approaches, learning methods, and learning resources, and ICT. The material on the student worksheet was prepared by the researchers by adapting the materialsavailable in books, including from the mathematics book of Curriculum 2013.

Phase 2 was initial design of student worksheets. After choosing the format of student worksheet, the next step was to develop the initial design of computer-based student activity worksheets which was called the prototype of student worksheet. This prototype was then validated by experts. The validated product was referred to as draft 1.

3. 3. Development Stage

The development stage aimed to produce student worksheets that had been validated and revised by experts. The experts were Mr. Muhammad Zaki, S.Pd.I., M.Pd. and Ms. Fadillah, SP, S.Pd., M.Pd., lecturers of Mathematics Education at Samudra University, and Mrs. Masrah Nasution, S.Pd., a mathematics teacher at the school where the research took place.

There were three aspects to be validated using a five-point rating scale: 1 = very poor, 2 = poor, 3 = fair, 4 = good, and 5 = very good. The validation result of the computer-based student activity worksheets can be seen in table 1.

Table 1. Results validation of computer-based student activity worksheets.

Number	Aspects rated	Validator			Average	Average of Each Aspect	Average Amount
		1	2	3			
I	Format:						4,74
	1.Instructions.	4	5	5	4,67	4,67	
II	Language:						
	1. Goodness of language.	4	4	4	4	4,56	
	2. Ease students in understanding the language.	4	5	5	4,67		
	3. Clarity of sentence structure.	4	5	4	4,33		
	4. Sentences do not contain double meanings.	4	5	5	4,67		
III	Content:						
	1. Material correctness.	5	5	5	5	5	

There were some suggestions of improvement from the validators for the worksheets. The validators suggested that not all the material should be strictly guided. Rather, they suggested to provide somesub-materials that the students could find the concepts by themselves. Following this advice, for the sub-content of determining the fractions between two fractions, the students were asked to learn independently by themselves. Various online learning resources were made accessible for the students.

The next step was the limited trial. After the limited trial, we obtained data about the validity, reliability, sensitivity of the item andstudents' perceptions on their learning completeness. The calculation of the validity of the problems in the worksheet was done by correlating the existing score of each item with the total score by using the formula of product moment correlation. Then we obtained the validitycoefficient of less than 0.80 and more than 0.60, meaning the worksheets had high validity.The calculation of the reliability of the description was done using the Alpha formula. The result was 0.62.

As for students' achievement, the students were assumed to have achieved mastery of a concept if they earned the minimum score of standard accomplishment which was established by the school. The minimum score was 75. Table 2 shows the students'learning outcomes.

Table 2. Students' learning outcomes.

Categoryof Achievement	Number of students	Percentage (%)	Average
Pass (Score ≥ 75)	24	80	
Fail (Score < 75)	6	20	86,03
Amount	30	100	

Based on table 2, it could be concluded that majority of the students had achieved mastery learning with an average score of 86.03.

Students' questionnaire was aimed to investigate students' perceptions on the implementation of the computer-based student activity worksheets. The results of the student response questionnaire are as follows: the percentage on the worksheet display statement is interesting that is 87.5%, has clear instructions 83.3%, worksheet instructions are easy to understand 85%, the language used on the worksheet is easy to understand 85%, Worksheet allows students to understand the topic of fractions 80 % and worksheet can make students more motivated 80%. Thus, the average percentage of students' response questionnaires from 30 respondents to computer-based student activity worksheets was 83.472%. The respondents' questionnaire value is very strong.

Based on the assessment of the expert validators, the test of learning outcomes and mastery of learning, and the result of a questionnaire of student response, computer-based student activity worksheets on fractions was valid, practical, and effective for students of junior high school grade 7 semester 1.

Fractions are essential foundational skills for future mathematics success [14]. Students' mastery of fractions can be achieved through the management of learning by the teacher. One of them is by utilizing ICT in learning. The implementation of ICT in mathematics enables students tolearn integers, geometry, problem-solving and working with data using technology [15]. This is in line with research conducted by Tomljenovic and Zovko (2016) who found that using ICT in teaching leads to better learning and knowledge acquisition in primary schools [16].

Developing meaningful learning of fractions in the classroom depends on how teachers connect these fraction concepts to their students' own learning [17].The teacher can make a guided discovery of the material through a worksheet designed by the teacher. This is because the teacher know their students better, the teacher knows how to explore the knowledge and curiosity of their students.A worksheet can be designed to help students identify and analyze important mathematical concepts. A worksheet should include definition, the web of attributes, examples and non-examples [18].

The computer-based student activity worksheet was made by adjusting the competency standards, basic competencies and learning objectives on fractions. The learning included blended learning, which combined face-to-face learning and the deepening of online material directed in the worksheet. The worksheet consisted of an introduction, sample questions, questions, competency tests and online source materials.

The source materials also came with online games that could be downloaded and played by students. Fraction gamesprovided opportunities for teachers and students to think about mathematical contexts and to engage in ongoing reflection about the teaching and learning of mathematics [19]. Games provided a situation that naturally combined issues of practice and theory, and reflection on those relationships and provided opportunities for discussion, reflection, and collaboration within a meaningful context [19].

So, the computer-based student activity worksheetoffered an interesting and fun learning alternative through worksheets that combined offline and online learning. The guided discovery of offline material on the worksheet consisted of an introduction, sample questions, questions, competency tests and some topics. The fractional materials could be accessed through various online learning resources, virtual manipulatives, online games, software, and more.

4. Conclusion

The validity test for the computer-based student activity worksheets was based on three aspects, namely content feasibility, language, and presentation. The validation score was 4.74. This score fellinto very valid category. The practicality of the worksheets was viewed from general assessment of all validators. Based on the general assessment of all validators, the computer-based student activity worksheets could be used with little revision. Thus, it could be concluded that the student worksheets were practical.The effectiveness of the student worksheets was judged by the students' learning outcomes.80% of the students passed the test with an average score of 86.03. So it could be concluded that computer-based student activity worksheet was effective.

The questionnaire used to investigate the students' perceptions on the implementation of the computer-based student activity worksheetsshowed that the students responded positively towards the worksheets. The average score of the questionnaire was 83.472% which fell into very strong category.

Acknowledgment
We gratefully thank thelecturersofMathematics Education of Samudra University, mathematics teachers and all students of a junior high school in Kota Langsa who participated in our study.

References

[1] Sinay E and Nahornick A 2016 *Teaching and Learning Mathematics Research Series l: Effective Instructional Strategies* (Research Report No. 16/17-08) (Canada: Toronto District School Board).
[2] Bruce C, Chang D and Flynn T 2013 *Literature Review Foundations to Learning and Teaching Fractions: Addition and Subtraction* (Trent University Shelley Yearley).
[3] Griffin Pand Care E 2014 *Developing learners' collaborative problem solving skills* (Assessment Research Centre, Melbourne Graduate School of Education, European Schoolnet Academy).
[4] Ball L and Stacey K 2005 Teaching strategies for developing judicious technology use *Technology-supported mathematics learning environments* eds W J Masalski and P C Elliot (Reston, VA: National Council of Teacher of Mathematics).
[5] Gresalfi M S, Johnson B R, Loehr A and Nichols I 2017 Design matters: explorations of content and design in fraction games *Educational Technology Research and Development* **66** 579
[6] Joseph L M and Hunter A D 2001 Differential application of a cue card strategy for solving fraction problems: Exploring instructional utility of the cognitive assessment system *Child Study Journal* **31** 123
[7] Test D W and Ellis M F2005 The effects of LAP fractions on addition and subtraction of fractions with students with mild disabilities *Education & Treatment of Children* **28** 11
[8] Butler F M, Miller S P, Crehan K, Babbitt B and Pierce T 2003 Fraction instruction for students with mathematics disabilities: Comparing two teaching sequences *Learning Disabilities: Research and Practice* **18** 99
[9] Flores M M and Kaylor M 2007 The effects of a direct instruction program on the fraction performance of middle school students at-risk for failure in mathematics *Journal of Instructional Psychology* **34**84-95
[10] Scarlato M C and Burr W A 2002 Teaching fractions to middle school students *Journal of Direct Instruction* **2** 23
[11] Bruce C D and Ross J 2009 Conditions for effective use of interactive on-line learning objects: The case of a fractions computer-based learning sequence *The Electronic Journal of Mathematics and Technology* **3**
[12] Sumaji 2015 *Pengembangan Perangkat Pembelajaran Matematika dengan Model Pembelajaran Pemecahan Masalah untuk Meningkatkan Kemampuan Pembelajaran Matematis* (Kudus: Universitas MuriaKudus Gondang Manis Bae)
[13] Trianto 2012 *Model Pembelajaran Terpadu* (Jakarta: PT Bumi Aksara)
[14] The National Mathematics Advisory Panel 2008 *Reports of the Task Groups and Subcommittees* (Washington, DC: U.S. Department of Education)
[15] Sivakova D, Kochoska J, Ristevska M and Gramatkovski B 2017ICT - the educational programs in teaching mathematics *TEM Journal* **6** 469

[16] Tomljenovic K and Zovko V 2016 The use of ICT in teaching mathematics: A comparative analysis of the success of 7th grade primary school students *Croatian Journal of Education* **18** 215
[17] Fisher J 2009 Fractions: Partitioning and the part-whole concept *SET* **2** 12
[18] Toumasis and Charalampos 1995 Concept worksheet: An important tool for learning *The Mathematics Teacher* **88** 98
[19] Kafai Y B, Franke M L, Ching C C and Shih J C 1998 Game design as an interactive learning environment for fostering students' and teachers' mathematical inquiry *International Journal of Computers for Mathematical Learning* **3** 149

The potential of jangka beaches as a natural laboratory for learning the concept of biodiversity

A Abdullah, L Fitriana, M Ali Sarong and C Nurmaliah
Syiah Kuala University, Jl. Teuku Nyak Arief Darussalam, Banda Aceh 23111, Indonesia

E-mail: abdullah@unsyiah.ac.id

Abstract. Utilizing the surrounding environment as a laboratory will be more impressive for passive learners to be active in learning. This study aimed to analyze the ecological criteria that include bird species diversity, richness of flora and fauna found in the area and the response of teachers to the suitability Jangka coastal areas as a natural laboratory. Data collection for the diversity of bird species was conducted using observation *line transect* techniques. The data have been analyzed using the formula Shannon-Weaver diversity index. Bird diversity index is $\bar{H} = 3.6913$. Data collection for the wealth of flora and fauna was conducted by observation. The percentage obtained for the rich flora and fauna, which is 62%. The conclusion of this study is that there is bird diversity in the area of Jangka Beach Alue Ue is high with $\bar{H} = 3.6913$ and for the wealth of flora and fauna are also included in the high category with the percentage obtained by up to 62%. The survey results revealed that the majority (83%) of teachers answer (YES) for a potential Jangka coastal areas as a natural laboratory for materials biodiversity, while the teacher's response stating (NO) is obtained (17%).

1. Introduction

The learning process is the interaction between teachers and students through reciprocal communication that takes place in a situation effectively to achieve the learning objectives. Teaching and learning process, teachers and learners are involved in an interaction with learning materials as a medium. The interaction in the learning process demanded more active learners rather than teachers. The interaction is said to be maximal, in case between teachers and learners, between learners and teachers, the learners to the learners, learners with learning materials and media, but remain within the framework of achieving the goals set together.

According to student learning activities is a thing done to achieve the learning outcomes obtained from observation, the experience of working as well as discussions with colleagues. Teaching in schools should involve students in learning, both mentally, physically, and socially. Learning emphasis should be focused on training skills, attitudes, and understanding of the concept. The phenomenon of teaching less directly involve students in learning activities causing psychomotor and cognitive abilities of students lacking. Students rarely discuss and cooperate with other students that lead students to become passive so that students do not develop [3].

The results of observation the of the existing schools in the districts of Bireuen show student learning outcomes are still low. Not maximal learning process is carried out as the lack of laboratories and experiments to reinforce the concepts learned. Other information indicates that teachers are only emphasis on students' mastery of concepts and give less to the learning exercise directly or practice.

Matter of biodiversity is one of the materials that are difficult to understand by students at a school in Bireuen district. Due to students only get an understanding of material from books, was a teacher has not been able to put the students as a whole to understand what actually happens in the real environment. This resulted in learners decreased interest because of lack of understanding of the material, thereby adversely affects the student learning outcomes.

The results of interviews with 30 biology teachers of class XI SMA in Bireuen, obtained a description and explanation that the mastery of learning outcomes on the concept of biodiversity has not reached the maximum completeness criteria (KKM) the process of learning biology in schools that matches the nature of science in ideal conditions cannot be fully implemented. Teachers are still a source of learning, students are not involved in the process of finding the concept. Teachers rarely apply direct media in the learning process. Several observations have been made, it was found that among the students with the other students in the process of activeness learning activities they have not been seen clearly, the interaction between student one with other students has not happened dynamically. Students have less discussion with other students which results in students being passive so that learning outcomes are low.

The way to overcome this gap will require a learning process in the form of direct media and interactive learning models and can assist students in understanding the learning materials greater emphasis on the search process of knowledge rather than knowledge transfer. One such study is by exploiting the natural surroundings as a medium of learning as a natural laboratory, especially for materials *biodiversity*. Utilizing the surrounding environment as a laboratory would be more memorable for learners who are passive to active learning, independent also able to come up with ideas and creative ideas and be able to interact in asking questions to teachers and peers which in turn can improve student learning outcomes [5].

[6] States that media-based learning is a learning environment that uses learning objects like a real experience, observe directly, obtaining data accurately and be able to learn independently or in groups. In the school environment is a source of useful learning, especially in the biodiversity study material.

Based on preliminary observations in the area of Jangka Beach in Bireuen district, it appears that the Coastal Region suitable for use as a media person or a natural laboratory in the learning process materials biodiversity. Because of the diversity of flora and fauna found in the coastal region was found with a number of very varied and can be used as a learning resource optimally. Jangka Alue Ue Territory has an area that is still very beautiful and beautiful, away from the noise of vehicles and other disturbances. Thus, it is suitable as a learning media in the form of a natural laboratory for the students because the atmosphere is calm and support the learning process takes place. In addition, its strategic location and easy to reach into its own added value for the region as a natural laboratory location for students to use as a medium in the learning process.

Biodiversity in the area Jangka Alue Ue, if used as a natural laboratory, will have a positive impact. Students will be easier to understand the subject matter by linking the learning materials are being studied to real life that they get in a natural laboratory. In addition, to generate awareness and a sense of community awareness of conservation efforts for the species of birds, flora, and fauna found in the region. The public will be aware that nature's biodiversity besides giving beauty also supports education. So it is necessary to examine more deeply about the potential of Jangka Alue Ue Beach neighborhood Jangka District of Bireuen District as Laboratarium term nature.

2. Method

The method used in this research is descriptive qualitative with ecological criterion view and observe using observation sheet. The parameters of this study are ecological criteria (diversity of bird species, a wealth of flora and fauna). Data collection to determine ecological criteria (diversity of bird species) from coastal areas of Jangka Alue Ue of Bireuen District using observation method with transect line technique. The transect line is used to observe the birds at the time of movement from one point to the next counting point [8].

The next step was the researcher who conducted the data collection methods of documentation about the conditions actually existing in the study, as proof of valid supporting the research that has been done [1].

Data from observation in this study were analyzed by using the formula diversity index Shannon-Wiener:

$$\bar{H} = -\Sigma \text{ pi ln pi}$$

Where:
$\bar{H}$: Diversity Index Species
pi : The Importance of Species All i = $\frac{ni}{N}$
ni : Number of Individuals Species All i
N : Total Individuals entire Species

With Conditions:
$\bar{H}$ >3 high diversity index
$\bar{H}$ 2-3diversity index was
$\bar{H}$ <2 low diversity index

Analysis of the data used in processing this data is descriptive analysis using observation sheets filled by observers (researchers). The data have been taken subsequently collected, and to describe the contents of each of the elements of ecological criteria coastal area Jangka Alue Ue. Analysis of data from observation using percentage analysis. The scores obtained by each ecological criteria were summed and the result is called the total score. Furthermore, the percentage of the average value by dividing the total score by the maximum score is taken from the total number of statements and then multiplied by 100%.

Table 1. Categories observed values ecological criteria.

Assessment Score	Category
81-100%	Very High
61-80%	High
41-60%	Medium
21-40%	Low
0-20%	Very Low

Data analysis to determine the response of 30 teachers to the conformity of the conservation area as a laboratory Jangka nature by using a questionnaire in the form of statements on the level of biodiversity possessed by the region. Determination of scores for positive and negative statements as in the following table:

Table 2. Determination of score questionnaire.

Alternative answers	Score	
	Expressions of the positive	Expressions negative
Yes	1	0
No	0	1

Data were analyzed by percentage.

IOP Conf. Series: Journal of Physics: Conf. Series **1088** (2018) 012111 doi:10.1088/1742-6596/1088/1/012111

Category responses of teachers based on Table 3.4 below:

Table 3. Guidelines score questionnaire criteria.

No.	Score	Category
1	$77 \leq q \leq 100$	High
2	$51 \leq q \leq 76$	Medium
3	$25 \leq q \leq 50$	Low

3. Results and discussion

Observations were made on 19 January to 18 February 2018, obtained by bird species found in the Jangka Alue Ue Jangka Subdistrict Bireuen District 45 species of birds from 21 families in the neighborhood of Jangka Alue Ue Beach Bireuen District. Most species are the families Nectariniidae and Ploceidae, while the least species are from the families Apodidae, Campephagidae, Dicaedae, Hirundinidae, Paridae, Pittidae, rallidae, Silviidae, Sturnidae, Turdidae, and Turnicidae.

Data analysis has been done, it is known bird species diversity index in Region Jangka Alue Ue Jangka Subdistrict of Bireuen District is high with $\bar{H}$ = 3.6913. Index of bird species diversity in Region District of Jangka Alue Ue can be seen more clearly in Table 4.1:

Table 4. Bird diversity index in Jangka Alue Ue of Bireuen district.

No.	**Station**	$\bar{H}$	**Level Biodiversity**
1	I	2.9095	Medium
2	II	2.5498	Average
3	III	3.0844	Height
4	Overall	3.6913	High

Data analysis results obtained in the diversity index of birds Jangka Alue Ue overall contained in the study site is $\bar{H}$= 3.6913. This suggests that the presence of birds with high levels of diversity. Bird species that are easy to find means the population is quite high in the area, this is due to the ability of birds to interact with vegetation and food. The Columbidae family has a large rate that is widespread throughout the world, especially fruit and grain eaters with a slightly plump body, solid, high forehead, and a distinctive short beak. It sounds rattling repeatedly. The Nectariniidae family is a very many-membered bird, widespread in the world. These birds are very diverse in color patterns but are generally of medium size, round heads, rather short legs, pointed beaks and wide wings. These birds feed on insects, often in places filled with fertile flowers and fruits. Many of these family members have a sweet voice. In line with the study [4] that the extent of the existence of different bird species in a place depends on how well the area fits as a nest and as a place where they depend for life.

Tourist area Jangka Alue Ue District of Jangka Bireuen district comprised mostly of farms population, residential and farm/plantation residents are allowed to be used as a suitable habitat for various species of birds. The high vegetation in a place is also a matter that affects the presence of bird species in an area. The existence of such diverse vegetation Jangka Alue Ue Regional District of Jangka very conducive to bird life, both to get food, rest and breed so that the birds do not migrate again to find a new habitat.

In line with the research [7] which states that the process of adaptation of a species habitat is intended to preserve the life of an individual in a given environment. Similarly, the birds, the presence of a residence with the availability of sufficient feed will maintain a bird species to inhabit a location in the long term. The success of a species of bird defends itself strongly influenced by its success in choosing a suitable habitat and special for it. In the process of obtaining a suitable habitat depends on the ability of the bird species interactions with habitat and vegetation that inhabit the region. The observation that obtained from observation directly, encountered several species of flora and fauna contained in the Jangka Beach area Alue Ue. Species that are along the beach area Jangka Alue Ue namely Bamboo yellow *(Bambusavulgaris)*,of palm *(Arengapinnata)*,Pinus *(Cupressus lusitanica),* coconut *(Cocus nucifera),* banana *(Musa paradisiaca)* and tree fig *(Ficuscarica)*is a species of plant like many species of birds.

The fauna species seen along the Jangka Alue Ue area are several types of animals including family crustaceae such as *Ocypoda* sp, some insects such as *Delias fruhstorferi*, *Aesha* sp, *Cantharis,* and *Leptocorisaacuta*. One of the bird species refers to rare animals namely *Haliastur indus* and *Heliacetus leticogaster* as well as some small birds, and reptiles such as the green snake *Timeresurus albolabris* and *Lacerta agilis* and amphibians such as *Fejervarya cancrivora*. This is in line with the study [9] which states that the tourist attraction has the potential to have high value if it has a high diversity of flora and fauna in the region.

In the development potential of coastal areas as a natural laboratory response, Jangka of teachers is absolutely necessary. To optimize it, keep in mind also the public opinion against the development potential of Jangka Beach as a natural laboratory that has been implemented by filling out the questionnaire. The results can be seen in Table 4.2 below:

Table 5. Analysis of response teachers against region.

No	Object	Answer	Persentase
1	Perceptions of neighborhood	1.Yes	18 %
		2. No	2 %
2	bird species diversity	1.Yes	22 %
		2. No	0 %
3	Diversity of flora and fauna	1.Yes	17 %
		2. No	2 %
4	The public perception	1.Yes	13 %
		2. No	8 %
5	Learning	1.Yes	13 %
		2. No	5 %

The survey results show that the majority (83%) of teachers answered (YES) to the Jangka Alue Ue area as a suitable coastal area as a natural laboratory for the concept of biodiversity. Much of the flora and fauna diversity found in the region and the air environment is very natural from cold air without pollution, thus bringing a comfortable feeling for students to study in the region. The teacher's response stating (NO) is obtained (17%), from a questionnaire distributed to teachers can know the things that make them do not agree with the term coastal areas as a natural laboratory that is due to lack of time in school student study hours as well as the absence of a special school transportation to bring children into a natural laboratory for the study. Although a small percentage of responses (NO) is, but it should be a serious concern and can be addressed. This is in line with research [2] which states that to assess the potential of a region to assess three important indicators of the type of object area, the amount of biodiversity in it, the traditions and attitudes of the local community.

4. Conclusions

Jangka Alue Ue beach area District of Bireuen based on ecological criteria that include (bird species diversity, richness of flora and fauna) have high levels of bird diversity, richness of flora and fauna are also obtained in the high category. The Alue ue area has the potential to serve as a natural laboratory as a learning medium for students.

References

[1] Arikunto S 2006 *Research Procedures A Practice Approach* (Jakarta: Rineka Copyright)

[2] Dhayita T 2014 Potential Development of community based area in Semawa Rawa Pening area *Journal of Engineering* **3** 146
[3] Fathurrohman P 2007 *Teaching and Learning Strategy* (Bandung: PT Rafika aditama)
[4] Iskandar S 2004 Study of Water Type Utilization in North Beach Indramayu West Java Journal of Biology **10** 354
[5] Istikomah A 2010 The use of group investigation learning model to foster students' scientific attitude *Journal of Physics Education Indonesia* **2** 242
[6] Juairiah 2014 Environmental Based Learning To Improve Student Learning Outcomes In Biodiversity Concepts *Journal of Education Biology Edition* **13** 178
[7] Priyono Yesser 2012 Development of Bukit Tinggi Ecotourism Area *Journal of Architectural Perspectives* Yogyakarta Faculty of Forestry Universitas Gadjah Mada **21** 567
[8] Tang Y 2017 Protecting the biodiversity in the fast urbanization age anecotourism zone in Hainan China *Journal Procedia Engineering* **198** 419
[9] Yosevita L 2007 Study on Regional Potential and Ecotourism Development in Tual Kabupaten Maluku Tenggara *Journal of Agroforestry* **2** 444

IOP Conf. Series: Journal of Physics: Conf. Series **1088** (2018) 012112 doi:10.1088/1742-6596/1088/1/012112

STEM learning in regular and vocational high schools on the topic of scientific menu card fabrication

R P Sari, M Adlim and A Gani
Syiah Kuala University, Jl. Teuku Nyak Arief Darussalam, Banda Aceh 23111, Indonesia

Email: adlim@unsyiah.ac.id

Abstract. Only a small percentage of high school graduates in Indonesia successfully enter university and the rest returned to their community without life skill. Therefore, it is crucial to stimulate entrepreneurial practices among students by integrating with science and technology. This study was initiated by developing STEM students' worksheet and subsequently tested the effectiveness for regular and vocational high school students. The worksheet was previously developed with R and D method following STEM syntax on the topic of scientific-menu-card fabrication. The STEM project related to student's innovation on fabricating of scientific menu card that containing nutritional data and the efficacy of fruit juice beverage. Students respond to STEM learning process were collected based on product assessment sheets, and questionnaire. The vocational students have a higher score in design card which general high school students were better in menu card presentation. There was not significantly different score among students except on indicator of motivation to accomplish the project, entrepreneurship motivation, and vision broadening. Vocational students have better responds on these indicators. STEM worksheet was recommended by students & teachers to use as an alternative teaching material in entrepreneurial learning.

1. Introduction

Indonesia has more regular high schools which reached 70% of 4,135,975 students and 30% of which was vocational school high students. From the total high school graduates, it was only 29% successfully enter university and the rest returned to their community without life skill [1]. Therefore, it is crucial to stimulate entrepreneurial practices among students, by integrating with science and technology. Learning model that accommodate science, technology, and innovation for entrepreneurship is known as STEM learning. Science, technology, engineering, and mathematics (STEM) is a learning approach that integrating content and skill of science, technology, engineering and mathematics in real life situations simultaneously [2,3].

Learning with STEM approach from several researches demonstrated a positive influence [4]. STEM-based learning can improve motivation since the students actively explore their ideas and learning experience. It also provides experience in the engineering process. It motivates more students in learning and participation in the class [5,6]. However, very few STEM module commercially available and barely any STEM learning practice especially in high schools. This because STEM is considered a new model, it will take time to introduce for educators.

This study aimed to explore teacher response and student capability to use STEM worksheets to integrate science and technology as well as the effect on student learning motivation and entrepreneurship awareness among students. The implementation was carried for chemistry classes in regular and vocational schools on the topic of fabrication scientific menu cards.

2. Methods

This study was initiated by developing STEM module using R and D approach with the model of ADDIE (analysis, design, develop, implementation and evaluation). Based validity assessment the STEM module with the nearly maximum score (91.15%), it was considered eligible to implement in a class. The validated STEM module was implemented in chemistry classes at regular and vocational high schools. The project was conducted by 26 students; 6 males and 20 females from a regular high school in Aceh Besar. It is suburb school and most of the students' parents were farmers. The other class was 32 students; 23 males and 9 females from a vocational high school in Banda Aceh). The school located in Banda Aceh, a capital of Aceh and the student parents were not dominantly farmers.

Students' worksheet was developed to train students to fabricate scientific-menu card of juice type beverage. Students were guided to explore literature and to sort list fruit or leaf extract that have nutritional data and the efficacy for health from scientific sources as well as familiar foods. There were several criteria for fruit choice; it should be a familiar fruit juice for Indonesians, it has been reported the nutritional contents in scientific journal. Subsequently, the students prepared the beverage, served for the taste test, took photos and fabricated the scientific-menu-card. The teachers assessed the project process and product with several rubrics. Before and after class activities, students were asked to fill in a questioner to express their impression on STEM learning and entrepreneurship motivation.

3. Result and discussion

Before STEM project started, the students fill in the questioner (pre-test) to answer what their expectation the STEM classes. Will the STEM class motivate them to learn? Will it improve their creativity? Will it provide a source of knowledge? Will it explore their self-potential? Will it provide information they need? The Similar questioner was given after STEM project accomplished (post-test). The students' responses have been compared between pre-test and post-test at the end of this section. The Students read the STEM module and the worksheet containing several information on fruit juice including the nutritional data and health benefit quoted from scientific journals.

Students have learnt some of STEM component skills after applying STEM worksheets in their learning experience. Science part was represented by knowledge on fruit that has nutritional and health benefit. Students have learnt to create their own recipe of fruit juice beverage with nutritional and health benefit consideration. They learnt the chemical composition of several fruit and in its role in nutritional and health. They learnt how to prepare and serve the beverage that looks attractive and nice taste. Technology aspects were developed by students when they were composing a scientific menu card containing scientific information and attractive beverage pictures, which are not common uses in the restaurant. Engineering skills were practiced when students used food photography technique & processing, as well as choosing the right material for menu cards. They used a small cardboard equipped with lighting as a mini studio for their food photography. Mini studio was prepared by teacher but the photography technique was based on student creation after they learnt it from google searching. The mathematical calculation was involved in cost processing, nutritional doses for each menu and potential revenue gain and the size of the market. Activities of students during STEM project learning are displayed in Figure 1. These practiced skills have confirmed the STEM learning aspects that are found in the literature [7].

Menu card and fruit juice drink that served and presented by students are further assessed by the teachers. The results of the assessment are presented in Table 1. Table 1 showed that student vocational high school has the higher score in creativity aspect in menu card design and better taste of their fruit juice drink. Therefore, STEM worksheet was effective to develop student creativity. This finding seems similar to previously reported [8]. Students from the vocational school have a better score than their counterpart might have a correlation with their experience in their own school that is trained to be skilful workers with entrepreneurship vision and such skill is not experienced by regular school students.

Figure 1. Activities of students during STEM project learning.

Table 1. The results assessment of the menu card and fruit juice presentation.

Student-assignment performance indicators		Scores	
		Regular high school[a]	Industrial vocational high school[b]
Menu card	Physical appearance	78.75	78.75
	Content	87.50	85
	Creativity	62.50	81.25
Juice	Taste	85.00	90
Average		80.00	84

[a]6 males and 20 females
[b]23 males and 9 females

Gender as shown in Table 1 also might be influenced, vocational students dominating by male students would have more enthusiastic to use the computer to creatively design the menu card compared to students in the regular high school where more female [9,10]. This is consistent with studies that find that men are superior in technology than women so they prefer a career in technology while women are superior in the field of science [11,12]. However, from the physical aspect of the menu card, the score was not significantly different between the group. However, in the aspect of the image, content and layout on the menu card, the regular high school presented more attractive presentation. There were 8 menu cards that were fabricated by students, each class had 4 groups of students and each group presented one menu cards. The representatives of menu cards are displayed in Figure 2.

(a-1)

Translation the figure legend :

Tamarillo Juice
Rich of antioxidant content

Nutrition content:
Carbohydrate : 8.25 g
Protein (1.03 g)
Fat (0.36 g)
Vit. C(29.5 mg)
Vit.B1(0.043 mg)
Vit. B3 (0.271 mg)
Vit B6 (0.198 mg)
Vit B9 (4 µg)
Vit. A (189.17 mg)
Vit. E (2.08 mg)
Ca (10 mg)
Mg (20.6 mg)
K (321 mg)
Na (3 mg)
Zn (0.8 mg)
P (38.9 mg)

(a-2)

Translation figure the legend :

Health benefit :
Increase endurance
(Asih et al, 2005
Reducing blood glucose & cholesterol
(Asvita & Berawi 2016)
Preventing obesity
(Kadir et al, 2015)
Reducing cancer risk
(Mutalib et al, 2017)
Reducing high blood pressure (Fabon, 2016)

(a-3)

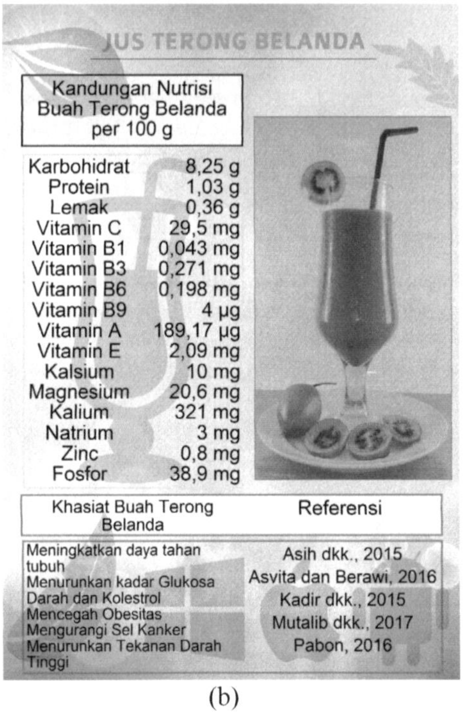

(b)

(c)

Figure 2. Representations of menu cards fabricated by (a-1) vocational school student, (a-2 and (a-3) are representative zoom legend in the menu (English Translation). Figure 2 (b-c) by menu cards fabricated by regular high school students

The teachers from both schools were asked to respond on STEM worksheets by filling in a questionnaire consisting of 10 questions. The summaries of the questions including would STEM worksheets be used as an alternative material for student creativity and entrepreneurship workshops? Did it match with learning objectives, did it contain STEM skills and students' abilities?, was the content easy to understand and easily to teach to students, was it systematics writing?, did the guideline have completed instructions, was it attractive to the view, and was it increasing to learn?. All teachers gave 100% positive response. Student responses were also recorded by giving a questionnaire and the findings are presented in Table 2.

Table 2. Students' response to STEM worksheets.

No.	Indicators	Percent of answers; "Yes"	
		Regular schools	Vocational school
1.	Compliance with learning objectives	96.15	90.6
2.	Easy to comprehend the elaboration	92.31	96.9
3.	Broadening vision	96.15	100.00
4.	Guiding students to search for additional literature	73.08	81.30
5.	Appropriate writing composition	88.46	90.60
6.	Containing a systematic instruction	92.31	93.80
7.	Interesting worksheet for students	92.31	93.80
8.	Enhancing motivation in entrepreneurship	88.46	100.00
9.	Presenting attractive image to motivated students	88.46	96.90
10.	Stimulating students to accomplish the STEM project	92.31	100.00
Average		90.00	94.38

Table 2 showed that the average positive responses from regular and vocational high school were 90.00% and 94.38% respectively. This finding is a line with the previous study that learning using STEM made students innovative, gain knowledge better and stimulated entrepreneurship [13]. In addition, STEM learning can enhance students' ability in science and innovate in technology products to compete globally [14-17]. Through STEM learning, students gain impressed experience following the learning stages and it generates motivation and interest in learning [18-19].

Students' responses on entrepreneurial learning before and after learning using STEM worksheets were also different and data are presented in Table 3.

Table 3. The students' response to the entrepreneurial learning.

No.	Indicator	Percent of answers "Yes"			
		Regular school		Vocational school	
		Pretest	Posttest	Pretest	Posttest
1.	Students are motivated	65.38	88.46	87.50	96.88
2.	Supports creativity	69.23	84.62	84.38	90.63
3.	Use of teaching materials	38.46	84.62	59.37	93.75
4.	Developing self-potential	84.62	88.46	81.25	100.00
5.	In accordance with the wishes	73.08	88.46	78.12	93.75
Average		66.15	86.92	78.13	95.00

Table 3 showed that students' responses from both schools on entrepreneurial aspect improved after learning STEM project. This was in line with their response on STEM worksheets. Before the implementation of STEM learning, the average positive response of students from regular school was lower than from vocational school although STEM objective had been explained at the beginning of the class. Probably beginning regular high school students have low expectation STEM model until they experienced directly they thought it would be just like a common class. Vocational school students have higher score even at beginning compare to the regular students because of the STEM project matched with their learning goal that was trained to be an entrepreneur or skillful worker. The improvement on the positive response was higher in regular schools.

4. Conclusion

STEM module and students' worksheets effectively drilled STEM knowledge and skills. It generated students' creativity and innovation as well as entrepreneurial motivation. It was accepted by teachers and students as interesting learning material and practical works. Vocational students who were most were male show more skill in using computer compared to regular high school who were dominantly female. There was not significantly different score among students except on indicator of motivation to accomplish the project, entrepreneurship motivation, and vision broadening. Vocational students have better responds on these indicators. Vocational students have better responds on these indicators because they realized that entrepreneur will be one of their careers.

Acknowledgment

We would like to thank regular high school of SMA Negeri 1 Darul Imarah and the vocational school of SMK SMTI Banda Aceh who has participated in this research.

Reference

[1] Adlim M, Samingan and Hasibuan R A 2014 *J. Turk. Sci. Edu.* **11** 111
[2] Siew N M, Amir N and Chong C L 2015 *Springer Plus* **4** 1
[3] Cinar S, Pirasa N and Sadoglu G P 2016 *Universal J. Edu. Res.* **4** 1479
[4] Adlim, Hanum L and Toa A A M 2015 *Proc. 5th Annual Int. Conf.*, Syiah Kuala University, Banda Aceh, 9 - 11 September 2015
[5] Karahan E, Bilici S C and Unal A 2015 *Euras J. Edu. Res.* **60** 221
[6] Chittum J R, Jones B D, Akalin S and Schram A B 2017 *Inter. J. STEM Edu.* **4** 1
[7] Ismayani A 2016 *Indo. Digital J. Math. Edu.* **3** 264
[8] English L D and King D T 2015 *Inter. J. STEM Edu.* **2** 1
[9] Iswanto P, Prihantoro C R and Avianti R A 2015 *Int. J. Vocational Edu. Training Res.* **1** 1
[10] Marniati 2016 *Proc. Int. Mechanical Engineering and Engineering Education Conf.*, Malang, 7-8 October 2016
[11] Hango D 2013 *Statistics Canada Insights on Canadian Society* **75-006-X** 1
[12] Diekman A B, Brown E R, Johnston A M and Clark E K 2010 *Psycho. Sci.* **21** 1051
[13] Zokowski P, Geramita K, Ashdown J, Brooks B and Thompkins A 2016 *Proc. IEEE Integrated STEM Education Conf.*, Princeton University, New Jersey, 5 March 2016
[14] Quang L X, Hoang L H, Chuan V D, Nam N H, Anh N T T and Nhung V T H 2015 *British J. Edu. Soc. Behav Sci.* **11** 1
[15] Vahedi S and Yari M 2014 *European Online J. Natural Soc. Sci.* **3** 572
[16] Khaeroningtyas N, Permanasari A and Hamidah I 2016 *J. Pend. IPA Indo.* **5** 94
[17] Kelley T R and Knowless J G 2016 *Int. J. STEM Edu.* **3** 1
[18] Spanias A, Banavar M K, Braun H, Spanias P and Zhang Y 2016 *Proc. IEEE Integrated STEM Education Conf.*, Princeton University, New Jersey, 5 March 2016
[19] Afriana J, Permanasari A and Fitriani A 2016 *J. Inovasi Pend. IPA* **2** 202

IOP Conf. Series: Journal of Physics: Conf. Series **1088** (2018) 012113 doi:10.1088/1742-6596/1088/1/012113

Validation of prototype instruments for implementing higher order thinking learning using the IMPROVE method

BI Ansari[1], Saminan[2] and R Sulastri[3]
[1,3]Serambi Mekkah University, Jl. Unmuha, Batoh, Banda Aceh 23245 Indonesia
[2]Syiah Kuala University, Jl. Teuku Nyak Arief, Darussalam, Banda Aceh 23111 Indonesia

E-mail: bansuiriantoansari@serambimekkah.ac.id

Abstract. Indonesian education, as stated in the National Curriculum of 2013, aims to develop higher order thinking skills (HOTS). HOTS are thinking processes that involve higher cognitive level of Bloom's Taxonomy, namely analysis, evaluation and creation.This study aimed to produce a valid, practical and effective prototype of HOTS learning using IMPROVE method. This study employed the Research and Development approach using ADDIE model (Analysis, Design, Development, Implementation and Evaluation). Initial stage of the development of the prototype was carried outby analysing, designing, developing and revising the instrument. Next activity was the validation of the prototype instrument by experts. Based on the validator's opinion, it was indicated that the prototype of higher order thinking learning using IMPROVE method was valid and practical. It was concluded that the prototype was ready to be piloted at schools to examine itspracticality and effectiveness.

1. Introduction

Indonesian education, in the National Curriculum of 2013, aims to develop critical and creative thinking skills, both skills are higher order thinking skills (HOTS). HOTS are thinking process involving the higher cognitive level of Bloom's Taxonomy, namely analysis, evaluation and creation. The domain of cognitive processes belonging to HOTS is the domain of analysis, evaluation and creation [1]. Analysis is the ability to identify problems systematically, evaluation is the ability to provide reasons for the answers provided, and creation is the ability to design and formulate word problems into a mathematical model to solve the problem.

HOTS is more than simple memorizing or understanding, instead it involves various cognitive processes such as creating, generating ideas, exploring consequences, reviewing, monitoring progress and so on [2]. HOTS tends to involve complex problems having many solutions [3].

Higher order thinking is a thinking process that requires students to use brilliant ideas to apply previously learned knowledge. Thus, this ability supports students to think critically, creatively and reflectively. Students are often confused when they face a complex problem because their teachers do not usually give higher order thinking problems. Most schools tend to emphasize lower order thinking skills in their teaching and learning. The consequence of this habit is that students forget easily and havepoor learning outcomes [4].This is in line with the 2011 Trends in International Mathematics and Science Study (TIMSS) and the 2012 Program for International Student Assessment (PISA)survey that Indonesian students have poor achievement in the international mathematics contests presenting non-routine problems [5]. Furthermore, it is revealed that Indonesian students are poorly ranked in the

TIMSS in terms of understanding complex information,theory, analysis and problem solving,using procedures in problem solving, and investigating [6].

The factors that cause Indonesian learners to incorrectly solve high order thinking problems are being less meticulous in solving the problems, poor initial mathematics ability, ineffective learning, and lack of understanding of the problems [7]. This condition indicates that our educational condition is still not encouraging HOTS. Students' thinking ability in solving mathematical problems is mainly at the lower order thinking. This is actually the root of the problems that needs to be addressed, like a chronic disease that needs to be immediately treated. This means students' higher order thinking skills, especially in solving mathematics problems, need to be developed. Students need to be equipped with higher order thinking skills, namely the ability to remember more relevant concepts to a problem and to present alternative solutions before choosing the most appropriate one. In addition, the ability to systematically compile facts in the problems by forming patterns and relationships leads to students being able to simplify complex information usinga correct model to solve the problem [8].

One of the promising methods for developing HOTS is the IMPROVE method. IMPROVE is an acronym of Introducting new concept, Metacognitive questioning, Practicing, Reviewing and Reducing difficulties, Obtaining mastery, Verification, and Enrichment. This method develops HOTS by metacognition questions and peersinteraction. The metacognition questions are built in four stages of the problem solving process, namely: orientation and problem identification, organization, treatment, and evaluation [9].

The metacognition questions include comprehension questions, strategic questions, connection questions, and reflection questions [10]. It is reported that the average score of the cooperative learning group undergone the metacognitive exercise is significantly higher than the cooperative group without the exercise and groups of individuals undergone the metacognitive exercises on the students' mathematical representation skills [11].

It is expected that the metacognitive questions, one of the important components in IMPROVE method, can develop students' HOTS. Therefore, it is necessary to develop a learning of HOTS using the IMPROVE method to produce valid, practical and effective products.

2. Method

Based on the previously mentioned background and conceptual model, the research questions were presented as follows: 1) What were the processes to develop a valid HOTS learning for senior high school students using IMPROVE method? 2) What were the results of the development of a valid HOTS learningfor senior high school students using IMPROVE method?

This study employed an R & D Approach based on the ADDIE development model[12], which involved analysis, design, development, implementation and evaluation. The first year of this study focused on assessing the validity of the developed products based on the initial stages of development (analysis, design and development). The analysis phase began with analyzing of the school curriculum, user needs analysis (teacher), and students' characteristics to obtain information on whether HOTS learning model with IMPROVE method for targeted senior high schools was needed. The second stage was design: determining the basic idea or the design of HOTS learning based on the cognitive orientation, organization and elaboration pattern using the IMPROVE method. The third stage was development which involved the validation ofexperts and revision.

This research was conducted at three high schools in Banda Aceh. The subject of this pilot study were nine mathematics teachers from the three schools. Initial data collection was done in three stages: (1) Analysis stage, which was also called pre-planning. In this stage, the instruments needed were the curriculum analysis sheet, teacher's need analysis sheet, and student characteristic analysis sheet. (2) Design stage, the activity undertaken was designing the prototype based on the needs analysis conducted in the first stage. This stage was a systematic process of designing a prototype by settingsubject matter and learning objectives, designing lesson plan (LP), discussion worksheet (DW), assignment worksheet (AW), assessment tools and guidelines for using the IMPROVE methods in learning. The instrument required in this stage was the validation sheet. Next, the preparation of lesson plan started with determining the components of lesson plan such as basic competence, indicators of competence achievement, linkages between subject matter, indicators of competence achievement and time, as well as linkages between scientific approach, IMPROVE method, settings, communicative language. (3) Development was the stage of products realization. In this stage, learning/prototypes consisting of LP, DW, AW, assessment tools as well asthe guidelines for using the IMPROVE methods in learning were developed. The products were then consulted with the partner teachers and subsequently validated by two expert lecturers, and two teachers (excluding partner teachers). The validation aimed to determine the quality of the products/prototypes. The validation process was carried out until the developed products were declared feasible to be used in the learning process. The quality of the products were based on the aspect of validity which was obtained at this stage.

There were six instruments involved in this initial stage study, namely: curriculum analysis sheet, requirement analysis sheet, student characteristic analysis sheets, LP validation sheets, DW & AW validation sheets, evaluation tools and guidelines for using the IMPROVE method, and feedback sheet of experts and teachers. Data obtained from the analysis and design stages was then analysed descriptively, while data analysis from the development stage was aimed to examine the validity of the products. To investigate the feasibility of these products, formative evaluation was required at each phase of development, including the phase of expert validation and revision.

Expert validation was aimed to examine one aspect of the products development quality. It was done by expert lecturers and teachers. The validators' recommendations were necessary to examine the feasibility of the products based on the indicators proposed by Nieveen: 1) at least three out of the four experts (validators) stated that the HOTSlearning prototype using the IMPROVE method was based on a strong theoretical basis, and 2) at least three out of the four experts stated that the components of the productswere consistently related [13]. Furthermore, product revision phase I (prototype), for field trial in the second year of this research was carried out after the products had been declared to be feasible by the experts.

3. Results

This section is reported in three parts. Part 1 is the analysis stage, part 2 is the design stage, and part 3 is the development stage.

3.1. Analysis stage

This research was started by administering a questionnaire concerning the curriculum, the needs of teachers, and the characteristics of the students to the nine partner teachers. The results were then analyzed and it wasrevealed that the students were required to have HOTS even though the teachers did not design topics related to HOT due to lack of literature and lack of understanding of the skills. The teachers expected thatthe learning model to solve non-routine problems could be used easily with clear steps to encourage students to be more active in order to achieve expected learning goals. Materials needed to be developed were for the topics of algebra, probability, geometry, and trigonometry.

Based on the teachers' needs analysis on the learning model needed to improve students' thinking skills, the teachers mentioned that one of the problems in their school today was limited number of books and teaching materials, especially books containing HOTS. In addition, the teachers have never attended training on designing HOTS problems and how to teach them. Thus, they needed additional teaching materials related to HOTS.

Further analysis was carried out to analyze the students'characteristics including mathematics skills and attitudes towards mathematics learning, especially towards HOTS problems. The teachers mentioned that the students were less enthusiastic when they were provided with non-routine problems. The students tended to only listen to what the teacher said. They did not build their understanding of a concept independently. Usually only few students were active in solving the problems given by the teacher. Most of the students were silent or talked among themselves and in the end they could not solve HOTS problems. Based on the three analysis, it could be concluded that the teachers needed a learning model to promote students' HOTS using the IMPROVE method.

3.2. Design stage

After the analysis phase had been completed, the next stage was to create the design based on the results of previous analysis. The activities undertaken at this stage were designing LP, guidelines for using the IMPROVE method, DW, AW and its evaluation, and the instruments.

At this stage we designed nine LP together with the DW, AW, training and evaluation tools for nine meetings. LP I, II and III focused on algebra and probability; LP IV, V and VI discussed about geometry; LP VII, VIII and IX discussed about trigonometry.

3.3. Development stage

This stage involved the development of HOTS learning (product) and the validation of the instrument. The developed products included LP, DW, AW, exercises and evaluation tool. The learning was developed with regards to the principles and structure of the preparation of LP, DW, AW and its toolsby paying attention to eleven criteria/indicators (see Table 1, 2 & 3). The development result was called the prototype 1 design.

The initial product (prototype 1) was then consulted with the partner teachers and some feedbacks were obtained. Validation was done by the validators after the revision to examine the quality of the products. The validation process, which was content validity, was conducted by two expert lecturers (material experts), and two teachers (not from partner teachers). The validation results from these experts were used as a reference to revise and refine the prototype. In tables 1, 2, and 3 below, we can see the prototype validation result.

Table 1. The validation results of lesson plan (LP).

No	Aspect	Criteria/Indicators	Validators				Mean
			1	2	3	4	
1	Contents	Basic competences (KD) & Competence Achievement Indicators (CAI)	4	4	5	4	4,25
		Conformity of subject matter, CAI and time	4	4	3	5	4,0
		Conformity of Scientific Approach and the IMPROVE Method	4	5	4	5	4,5
2	Format	Setting	4	5	5	5	4,8
3	Language	Communicative	5	3	5	5	4,5
		Average validator	4,2	4,2	4,6	4,6	4,4

Table 2. Validation results of DW, AW and exercises.

No	Criteria/Indicators	Validators				Mean
		1	2	3	4	
1	Content Eligibility	4	3	4	4	3.75
2	Presentation Eligibility	3	4	4	4	3.75
3	Language Eligibility	4	4	4	4	4
4	Graphic Eligibility	4	5	4	5	4.5
	Average validator	3.75	4	4	4.25	4

Table 3. Validation results of the evaluation tool.

Aspect of problems	Criteria/Indicators	Validators				Mean
		1	2	3	4	
Construction	Problem in accordance with solution demands	4	5	4	4	4.25
	Rubrics	4	3	4	4	3.75
Language	Communicative problems	5	5	5	4	4.75
Average validator		4.25	4.25	4.5	4	4.25

The results of the instrument validity presented in tables 1, 2 and 3 show that the validity of the prototype fell in the valid criteria with the average score of 4.22.

4. Discussion

The procedures undertook in developing the prototype to produce a qualified product started from the analysis stage, analyzing the curriculum, teachers needs and characteristics of students, followed by designing the prototype and the validation as well as revision.

4.1. Analysis stage

In the analysis stage, it was found that the HOTS learning using IMPROVE method was important to be implemented. In this stage, it was also revealed that HOTS learning using the IMPROVE method was never done before or it was limited to C4 level with group discussion method. The Directorate of High School Development said that there were three criteria of teaching materials: (1) the availability of materials according to curriculum demands, meaning that the development of learning materials must be in accordance with the curriculum, (2) characteristics of the target, meaning that the developed learning should be tailored to the characteristics of students as target, covering the social, cultural, geographical and developmental stages of students, (3) the development of learning should be able to solve problems or difficulties in learning [14].Thus, the prototype that has been developed in this study could satisfy the three criteria of teaching materials.

4.2. Design stage

The activity undertook at this stage was the selection of initial formats and HOTS learning. Some of the difficulties encountered in this stage included the lack of teaching materials consisting HOTS problems; it wasdifficult to distinguish the problems in the cognitive domain of C4, C5 and C6; and it was difficult to find references concerning the IMPROVE method. Nevertheless, the design of this HOTS learning could be designed well with solid teamwork.

4.3. Development stage

The development stage was the products/prototypes realization. A product based on the framework from the design stage was developed. The products/prototypes were LP, DW, AW, exercise, and evaluation tools as well as learning guidance ofthe IMPROVE method. These products met the valid criteria based on the result of lecturers and teachers' assessment.Although they had achieved good classification, as shown in tables1, 2 dan 3, the material distribution aspect had a low score compared to the other aspects. In order to develop the lesson plans better, the feedbacks given by the assessor was adjusted in the material distribution to the allocated learning time.Meanwhile, based on the validation sheet, the average score was 4.22, this indicated that the prototypes met both criteria: based on a strong theoretical basis and had consistently interrelated components. This is in accordance with the indicators of validity analysis developed by Nieveen.Such studies have been conducted byLeris et al. about "validation of indicators for implementing an adaptive platform for MOOCs". The score of the internal consistency, which was statistically evaluated using Cronbach's alpha, was greater than 0.8, indicating good internal consistency [15].

5. Conclusion

The HOTS learning was developed based on ADDIE theory consisting of five stages, namely Analysis, Design, Development, Implementation and Evaluation. For this year, the process of developing the prototype was conducted in the first three stagesof ADDIE model, they were: (a) the stage of analysis, determining the information and needs of the HOTS learning. In this stage, data on the curriculum description used in the schools, teacher needs analysis and characteristic analysis of students. (b) The design stage, obtainingoverview of the prototype to be designed based on the needs analysis conducted in the first stage. This stage begins by designing LP, DW, AW, evaluation tools, guidelinesof the IMPROVE method, and validation sheets for feedbacks of the designs; (c) The development stage, developing prototypes in the design phase, based on the validation results from the experts to be tested at the partner schools. Based on the development process of three stages of the ADDIE model, it was concluded that the prototype of HOTS learning using the IMPROVE method was valid and practical.

References

[1] Krathwohl D R 2002 A revision of Bloom's taxonomy: An overview *Theory Into Practice* (The Ohio State University: College of Education) p 132

[2] Wang S and Wang H 2011 Teaching higher orderthinking in the introductory MIS course: A model directed approach *Journal of Education for Bussines* **86** 208

[3] Widana I W 2017 *Modul Penyusunan Soal Higher Order Thinking* (Jakarta: Direktorat Jenderal Pendidikan Dasar dan Menengah, Departemen Pendidikan dan Kebudayaan) p 74

[4] Bassham G, Irwin W, Nardone H and Wallace J M 2007 *Critical Thinking* (USA: A Student)

[5] PISA 2012 *Result: What Student Know and Can Do: Student Performance in Reading, Mathematics, and Science* **1**(OECD)

[6] Kemendikbud 2011*Kemampuan Matematika Siswa SMP Indonesia* (Jakarta: Badan Penelitian dan Pengembangan) p 17

[7] Gais Z and Afriansyah E A 2017 Analisis kemampuan siswa dalam menyelesaikan soal higher order thinking ditinjau dari kemampuan awal matematis siswa *Mosharafa* **6** 255

[8] Ansari B I 2016 *Komunikasi Matematik, Strategi Berpikir dan Manajemen Belajar* 978-602-1620-43-4 (Banda Aceh: Yayasan PeNA) p 90

[9] Mavarech Z R and Kramarski B 1997 IMPROVE: A multidimensional method for teaching mathematics in heterogeneous classroom *American Educational Research Journal* **34** 365

[10] Kramarski and Mizrachi N 2000 Metacognitive discourse in mathematics classroom *Europen Research in Mathematics Education* **III** 11

[11] Kramarski and Mizrachi N 2001 Enhancing mathematical literacy with the use of metacognitive guidance in forum discussion *Proc. of the 28th Conf. of the International Group for Psychology of Mathematics Education* **3** 171

[12] Branch R M 2009 *Instructional Design, the ADDIE Approach* (USA: Springer Scoence+Bussiness Media)

[13] Nieveen N 1999 Prototyping to reach product quality *Design Approaches and Tool in Education and Training* eds Van den Akker, Branch R M, Gustafson, N Nieveen and Plomp (Nederlands: Kluwer Academic Publishers) pp 125-35

[14] Direktorat Pembinaan Sekolah Menengah Atas 2008 *Panduan Pengembangan Bahan Ajar* (Jakarta: Departemen Pendidikan dan Kebudayaan) p 11

[15] Leris D, Sein-Echaluce M L, Hernandez M and Bueno C 2017 Validation of indicators for implementing an adaptive platform for MOOCs *Computers in Human Behavior* **72** 783

Developing guided-inquiry-student worksheets to improve the science process skills of high school students on the heat concept

M Mahyuna, M Adlim and I Saminan
Syiah Kuala University, Jl. Teuku Nyak Arief Darussalam, Banda Aceh 23111, Indonesia

E-mail: adlim@unsyiah.ac.id

Abstract. Conventional student worksheets applied in schools had not been effectively improve the student creativity because the students were not trained science process skills in most lab practical activities. Practical works were only confirmation of lab procedure. The purpose of this research is to compose guide inquiry-based student worksheets using R & D methods with ADDIE model. The validated student worksheet was implemented at public senior high school in Banda Aceh by randomly sample the class from five parallel classes. After implementation, a test for basic science process skills was performed by students and students' respond were collected using questioners. The finding showed that the student process skills were significantly improved after the guided-inquiry learning was implemented using the effective student worksheet. Most students recommended the guide inquiry students worksheet to apply in other classes because easy to understand, contextual and motivating students to do science investigation.

1. Introduction

Previously study had been explored remoteness of school location caused lower in quality of learning [1]. Later we found also nearly similar problems at public high school in the capital of Aceh. The school still uses conventional student-worksheets in science subjects. The conventional worksheet has not accommodated constructivism learning. The students just follow the fix laboratory procedure without activity of drilling students to do science skill process even in basic level. The basic science skill process that are hypothesizing, identifying tools and materials, analyzing data, and summarizing the results of the investigation.

One of the learning methods that accommodate the science process skills were guided inquiry and it is recommended as a learning method in 2013-national curricula [2]. Guided inquiry is a learning method that helps students to learn and acquire knowledge by themselves discovering what is being studied [3]. In guided inquiry, students are given a problem but the teacher will not provide the experimental procedure for the students, they have to develop by themselves under guidance of teachers. Unlike confirmation inquiry, the conclusion in guided inquiry cannot easily answered theoretically, it must come from the experimental work. The model also includes discovery of meaning, organization, and ideas, gradually they learn how to organize and conduct research to achieve the learning objectives. Guided inquiry student worksheets have been reported as an effectively method to improve learning outcomes, knowledge, attitudes and students skills [4-7]. Similar studies on other subject has also reported that guided-inquiry learning method significantly improve the students' science process skills, but teachers usually reluctant to perform due to time consuming and tide teaching schedule [8-11]. Guided inquiry module and the students' worksheet have not been widely published commercially especially in high school physics of heat concept. Therefore, the purpose of this study is to develop guide-inquiry student worksheets and the impact on science process skills of student improvement.

2. Method

The guided inquiry student worksheet was developed by using R & D with ADDIE (Analysis, Design, Develop, Implementation and Evaluation) model. The implementation was carried out at public senior high school in Banda Aceh and 27 students on grade 11 were the participants. The students group (class) were randomly selected from five parallel classes in this target school based on similarity of previous student test scores. All stages of ADDIE procedure started from need assessment, analyses (A), designing (D), developing (D), implementation (I) and up to evaluation (E) were carried out. The student guided-inquiry worksheets were validated by 5 experts previously before implemented. Revision was conducted several times based on expert suggestion and respond from students and teachers following ADDIE steps. Instrument for basic science skills were modified from previously reported [12] and validated and test the reliability. Based on Spearman-Brown reliability tests, all test items were reliable but 11 out of 20 test items were removed due to low validity. Prior to worksheet implementation, pretest was administered to the students to examine their initial science process skills and posttest was carried out after implementation.

3. Result and discussion

3.1. Development of student guided-inquiry worksheet with ADDIE model

The need assessment found that the school has not have guided-inquiry module and the student worksheet. It was confirmed by interviewed, documented and direct observation. In design step, the revision has been done according to reviewers' suggestion including adjustment on the font, color, illustration, inquiry learning syntax, adaptable language almost in entitle draft. Almost similar types of correction with some additional citation and references were performed on the step of develop. At implementation stage, the validator gave appropriateness score of 84.00, 83.33 and 80.00 for content, presentation and language of worksheet respectively. Four students' worksheets were composed that were heat experiment, Joseph Black principle, conduction, convection and radiation. The list of contents was the title, the objective of project, the problem statement, material and equipment, hypothesis, experimental procedures, data analysis, conclusion, questions. After material and equipment section, there was instruction for students to read the problem statement, the objective, the material and equipment and a blank of observation table & experimental diagram given in experimental procedure before they wrote their experimental procedure.

3.2. Student respond on student guided-inquiry worksheets

Students' response collected from on student worksheets were dominantly on positive response at the stage of strongly agree (19.4%), agree (62.9%), disagree (14.5%), strongly disagree (3.08%) toward several statements in questioners as describe the characteristic of guided inquiry worksheet. The worksheets were on experiment of heat and heat transfer for 5 groups of students with different projects: (a) heat experiment (*kalor*), Joseph Black principle (*asas* Black), conduction (*konduksi*), convection (*konveksi*) and radiation (*radiasi*). A representative quote of student worksheet and activities are described in Figure 1. The students were instructed to read the project problem statement, to observe the experimental diagram (pictures) in worksheet and to note the design-observation-data-sheet containing experimental variables. An experimental simulation was demonstrated before students start their project. Students, then, were asked to write their hypothesis, to write experimental procedure, to run project (collecting data), to analyze data and to write conclusion and to present the project accomplishment. Figure 1 was a representative of experimental work conducted by students. The students compared the different of temperature elevation of fluid in the different color of containers (a black and a white tin). They recorded in observation sheet, analyzed and wrote the conclusion.

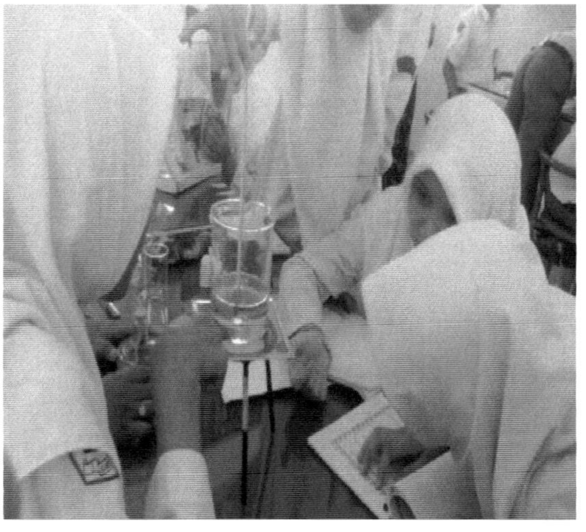

(a)

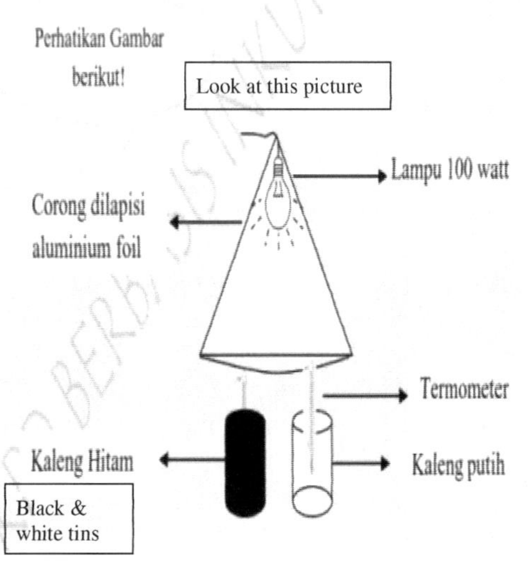

(b)

Figure 1. (a) Student conductedthe experiment at high school laboratory, (b) fragmentof student guide-inquiry student-worksheet containing anexperimental diagram to guide student tobe able to compose their own experimental procedure.

3.3. Science process skills data analysis

The student science process skills data analysis is provided in Figure 2. It shows that the post-test scores increased significantly after applying the guided inquiry project where students were guided by the inquiry student worksheets. Guided inquiry learning model emphasizes the process of discovery of a concept that appears scientific attitude on students and can be designed for use by teachers according to the level of intellectual development of students [13]. The pre-test scores were only 55.5% and after treatment the post-test score was 90.2% and the gain was up to 34.7%. The students' activities during the class were composing hypothesis, designing experimental, using tools, interpreting data, applying concepts and asking questions, classifying and stating the conclusion. Students were actively involved in class activities, so class was not boring and they enthusiastic to run experiment and confirming the previously [14]. This finding aligns to the previous study that inquiry learning in the laboratory gave significant effect on student science process skills [15,16].

IOP Conf. Series: Journal of Physics: Conf. Series **1088** (2018) 012114 doi:10.1088/1742-6596/1088/1/012114

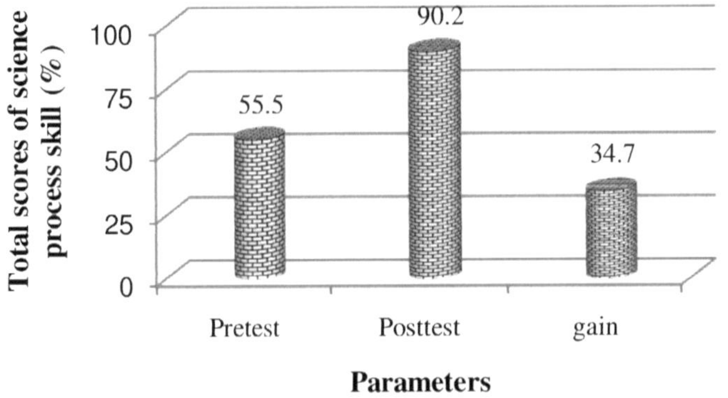

Figure 2. Comparison of pre- and post-tests before and after project implemented.

Further analysis of student science process skills for each indicator component was presented in Figure 3. Generally, for all indicator components, post-test score was higher than the pre-test. Student skill on asking question represents as a large gap between pre- and post-test. Each indicator shows high score and this consistent with previous study in different subject that implementation of inquiry learning let to improve the basic science process skill of students [17,18].

Giving a problem to the student will arise his curiosity, how to use the tool to solve the problem, the concepts needed for the solution and what method is appropriate for its completion. It will encourage them to use the knowledge and tools they already have and look for what needs to be known to solve the problem and teacher played important role on stimulate students to run the project [19].

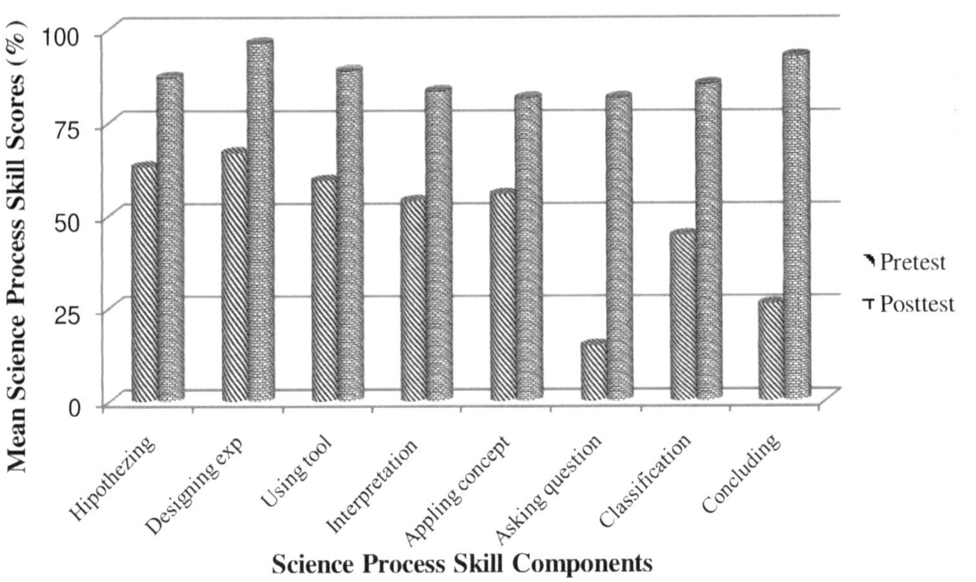

Figure 3. Science process skills analysis categorization based on the indicator clusters.

4. Conclusion

Guided inquiry students' worksheets have been developed and considered valid by experts. The validation score was 84 and highly appropriate in implementation for high school students as shown by high positive respond from students with distribution score of 82.3%. The students show their high capacity to use the worksheet in learning of heat transfer. Applying the worksheet in guided inquiry learning gave significant improvement on science process skills of students especially on this topic. The post-test score was 90.2% and the N gain was up to 34.7% after implementation of student worksheets.

Acknowledgment

We would like to thank public high school of SMAN 4 Banda Aceh who have participated in this research. High appreciation was addressed to instrument validators who gave significant contribution in this study.

Reference

[1] Adlim M, Soewarno S, Khairil, Usman, Hasbi, Hasmunir, Armia, Ishak andYasin B 2014 *Inter. J. Sci. Math. Educ.* **12** 817
[2] Kuhlthau C 2010 *School Libraries Worldwide* **16** 17
[3] Wahyudi L E and Supardi Z A I 2013 *J. Pend. Fisika* **2** 62
[4] Annafi N, Ashadi and Mulyani S 2015 *J. Inkuiri* **4** 21
[5] Abdi A 2014 *Universal J. Edu. Res.* **2** 37
[6] Alameddinea M M and Ahwalb H W 2016 *Proc. Social Behavioral Sci* **232** 332
[7] Kitota A K A, Ahmada A R and Seman A A 2010 *Proc. Social Behavioral Sci.* **7** 264
[8] Gormally C, Brickman P, Hallar B and Armstrong N 2009 *Inter. J. Scholarship Teach. Learn.* **3**
[9] Astuti Y and Setiawan B 2013 *J. Pend. IPA Indo.* **2** 88
[10] Yunus S R, Sanjaya I G M and Jatmiko B 2013 *J. Pend. IPA Indo.* **2** 48
[11] Aristini N K D, Sudarman I K and Riastini P N 2017 *e-J. PGSD Universitas Pendidikan Ganesha* **5** 1
[12] Herman 2015 *J. Sains Pend. Fisika* **11** 120
[13] Dewi N L 2013 *J. Pend Dasar.* **3** 1
[14] Chebii R, Samwuel W and Joel K 2012 *Sci. Res.* **3** 1291
[15] Khan M and Iqbal M 2010 *J. Strength for Today and Bright Hope for Tomorrow* **11** 169
[16] Rahmawati R, Haryani S and Kasmui 2014 *J. Inovasi Pend. Kimia* **8** 1390
[17] Kitot A K A, Ahmad A R and Seman A A 2010 *Proc. Social Behavioral Sci.* **7** 264
[18] Yasmin N, Ramdani A and Afriana A 2015 *J. Pijar MIPA X* **5** 69
[19] Rauf R A A, Rasul M S, Mansor A N, Othman Z and Lyndon N 2013 *Asian Social Sci.* **9** 47

IOP Conf. Series: Journal of Physics: Conf. Series **1088** (2018) 012115 doi:10.1088/1742-6596/1088/1/012115

Characteristics of pre-service teachers' performance in problem posing

M Masriyah, I Kurniasari and E L W Palupi
Universitas Negeri Surabaya, Jl. Raya Kampus Unesa, Lidah Wetan, Surabaya 60213 Indonesia

E-mail: masriyah@unesa.ac.id

Abstract. Problem posing is as much important as problem solving. In order to enhance problem solving skill, students need to be trained to pose a question or problem. Hence, teachers need this ability. This study was a descriptive research which aimed to describe the characteristics of pre-service teachers' performance in problem posing. The research participants were 34 pre-service mathematics teachers who were enrolled in the Assessment Course in 2017/2018 academic year. The research instrument was a validated problem posing worksheet. Data were collected through a test. The results showed that the students preferred to make free essay problems than the structured ones, and only a few made open problems. All of the students' problems were resolvable. The difficulty level of the problems was varied from high (dominated by analytical problem), medium, to low (dominated by understanding) level. Related to the type of the problems, the ability of pre-service teachers in posing questions with the type of post-solution was the lowest, and the type of within-solution was the highest. This results implicate that pre-service teachers need to be accustomed to creating open and good problems for their future students.

1. Introduction

The school mathematics curriculum has not paid much attention to students' problem posing instead puts more emphasis on solving mathematical problems. Basically, the topic of problem posing is a long-standing issue. Singer, Elerton, and Cai suggested that the new issue of problem posing is the awareness of various parties regarding the need for content of problem posing in the mathematics curriculum at all levels, both as a learning tool and as an object of learning [1]. Kilpatrick argued that problem posing is an essential content in mathematics and the essence of mathematical thinking, and an important part of mathematical problem solving [2]. A person cannot solve the problem if it is not formulated or filed properly by the problem compiler. Shriki pointed out the importance of problem posing that "Problem formulation is more essential than the solution" [3]. Raising new problems, new possibilities, looking at old problems from a new point of view requires creative imagination and marks real progress in science.

The importance of problem posing in solving mathematical problems, among others, was described in Polya's suggestion to assist students in overcoming their difficulties when solving problems. Polya suggested that we can help students in solving a problem by: a) providing questions that lead students to solve problems, b) helping students to explore their knowledge and compile questions by themselves, c) providing meaningful cues to solve problems but not solution of the problem, d) helping students solve their own difficulties [4]. The importance of the study of mathematical problem posing was also embodied in the statement of Ellerton and Clarkson that the development of mathematical ability requires the ability of creative mathematical imagination that, among other things, is developed when raising new questions, creating new opportunities, and looking at old questions from a new point of view [5]. The rationale of the need and importance of developing the ability to apply mathematical problems to students, pre-service teachers, and mathematics teachers is implied in the advice of the National Council of Teachings of Mathematics (NCTM) which suggested that students should be trained to formulate problems or questions based on various situations presented either in or beyond mathematics [6]. Students should also be given the opportunity to develop their own problems or questions, find conjectures, and generalize and expand the problems by compiling the next question of the existing question and then solving the problems.

In any mathematics learning approach, mathematics teachers should present problems or mathematics questions that encourage students to learn, stimulate students' critical and creative thinking. For example, in contextual and problem-based learning approaches, mathematics learning begins with the presentation of contextual problems that match with the mathematical concepts to be taught and in accordance with students' pre-existing mathematical knowledge. Then during the lesson, the teachers must also select mathematical problem or task that is relevant to the mathematical ability to be developed. If the mathematical problem does not encourage students to think further then the students will only have procedural or mechanical knowledge and not used to apply high order mathematical thinking.

Silver argued that the posing of mathematical problems helps develop mathematical thinking, problem-solving skills, attitudes and self-esteem in learning mathematics and broadening the understanding of mathematical concepts [7]. Cai suggested that submission of mathematical problems is one of the key components of mathematical exploration [8]. Therefore, allowing the students to pose mathematical problems is a way to investigate the students' way of thinking in terms of perspectives.

Mathematics education students are teacher candidates who are required to have good understanding of mathematics content and teaching strategies. They need to know how to teach according to curriculum requirement. In addition, pre-service teachers should also have good skills in conducting an authentic assessment in order to assess the success of a learning process. This is in accordance with the competence of teachers that must be owned by mathematics education students as stated in the curriculum of Mathematics Education. "The four teacher competencies that must be possessed by teachers are pedagogical, personality, professional, and social competencies" [9]. Pedagogical competence is the ability to understand students, the design and implementation of learning, evaluation of learning outcomes, and the development of students to actualize the various potentials they have.

In conducting assessment activities, there are several capabilities that must be possessed by pre-service teachers. One of the abilities is the ability of problem posing and problem solving, especially in mathematics. Because this ability is needed in order to be able to explore and evaluate the extent to which learners understand the material being taught. In addition, the teacher's ability to pose problems can assist students in reducing their dependence on textbook and help students to become more involved in learning activities [3].

The curriculum applied in mathematics education at Universitas Negeri Surabaya has been revised to meet the current curriculum requirements of primary and secondary schools. The lecturers understand important aspects that must be done and applied in the learning process at the university. One important aspect that must be considered by the lecturers of Assessment is the aspect of students' ability in raising mathematical problems, because this is a prerequisite ability that must be possessed by pre-service teachers. A good teacher must be able to design a learning well, carry out learning with a method or model of learning that fits well, and also able to assess the efficiency of the learning. The objective of this research was to describe the characteristics of pre-service teachers' performance in problem posing.

2. Problem Posing

According to Brown and Walter, the term problem posing was officially recognized by the National Council of Teachers of Mathematics (NCTM) as part of the National Program for Re-Direction of Mathematics Education [10]. The formal enactment of the term mathematics problem is related to education reform by NCTM in 1991 [10].

Meanwhile, according to Law Number 14 Year 2005 regarding teachers and lecturers, pedagogic competency is a skill that is related to students' understanding and educational and dialogical learning. Substantially, this competency is a teacher's ability to evaluate students' learning outcomes. This is in accordance with the opinion of Masriyah which stated that one of the competencies teachers must have is the ability to conduct assessments, both in the process and product of learning [11]. Thus, all pre-service teachers must have the ability of problem posing for each learning activity that they do.

In any mathematics learning approach, mathematics teachers should be able to construct questions that encourage students to learn, stimulate students' critical and creative thinking. During the lesson, they must be able to choose appropriate mathematics problems or tasks that are relevant to the mathematical ability to be developed.

According to Siswono, problem posing has several advantages, they are: (1) provides opportunities for students to achieve a broader understanding and analyse more deeply about a topic, (2) motivates students to learn more, (3) provides an opportunity for students to develop creative, responsible, and independent attitudes, (d) knowledge which is gained from experimental learning which has close relation with students' daily life will be well remembered by students [12].

Silver proposed that the term problem posing applies to three different forms of mathematical cognitive activity [7]:

- Pre-solution posing, i.e. a student makes a problem of the situation held.
- Within-solution posing, a student re-formulates the problem as it has been completed. That is, students can create a question that can lead to the answer to the problem created previously.
- Post-solution posing, i.e. a student modifies the goal or condition of the completed problem to create a new problem. That is, students create new problems similar to those that have been resolved.

Judging from how to present a question, the form of essay questions can be grouped into two kinds, namely: structured essay and free essay. The structured essay form is presented in detail of several sections that are mutually supportive. Problems are usually arranged from the simplest to the most complex. The question of a free essay is presented globally, not in detail. In answering it, students asked questions are allowed to do part of the answer freely. Single problem have generally included as free essay form [13].

The difficulty level of a problem can be categorized into high, moderate, and low. The difficulty level is high or high order thinking if it measures the achievement of indicators for analysis, evaluation or create category. The level is moderate if the question measures the achievement of indicators for the applying category, and it is low category when measuring the achievement of indicators for remembering or understanding category. In addition, a problem is said to be open, if the problem has several ways of settling or many solutions [14].

3. Method

This study was a descriptive research. In this article we described the characteristic of pre-service teachers' performance in problem posing. The participants of this research were 34 pre-service teachers who were enrolled in the Assessment Course. The steps that researchers conducted in this research were (1) preparing research proposal, (2) developing research instruments, (3) conducting validation of research instruments, (4) revision of research instruments to be used in research (if needed), (5) after the research instruments have been declared valid by the validators, the next step is giving a brief information about the problem posing, (6) giving the task of filing the problem to all research subjects, (7) conducting data analysis, and 8) preparing research report.

This research was conducted using a validated problem posing worksheet. Data analysis was done by 1) separating the results of the students' tasks in the filing of the problem for pre-solution, within-solution and post-solution; 2) supervising the results of the task of filing the problem while taking into account: (a) the kind and type of questions, (b) the structure of the language/ problem sentence, (c) whether the problem could be solved, (d) whether the problem posed contained open-ended questions, (e) correctness of the problem, and (f) the level of difficulty of the questions made by the pre-service teachers.

4. Result and discussion

Based on the study, the pre-service teachers' performance in problem posing could be expressed that 85% of them made free essays and 15% of them made structured essays. Only 5.88% of them made open-ended questions with many ways of solution, and only 11.76% of them made open-ended questions with many solutions. The structure of the sentences/language used was in good category (73.52%) and fair category (26,48%). All of the problems that the students made could be solved. The difficulty level of the designed problems could be categorized into high (29.41% dominated by the problems that belonged to the analysis category), medium (79.41%), and low (17.64% dominated by understanding category). Almost all of the students (94.12%) could solve the problems they made. Only 14.71% of the students posed problems of post-solution type, 23.52% of pre-solution type, and 88.24% of within-solution type.

This study showed that pre-service students preferred making the within-solution problem type. This was because they just needed to compose some questions related to the answer of the previous problem that have been solved. In addition, they preferred making free essays rather than structured essays. This was understandable, because a free essay could be presented globally and did not need to be structured in detail.

Only a few of the participants made open-ended questions with many ways of solution. This was understandable because the students were not used to dealing with solving open-ended problems when they were in elementary, junior or senior high school. While the structure of the sentences/language used can be categorized as good or good enough. This was understandable because they were pre-service teachers.

All of the problems that were made could be solved because the problems they made were at junior or senior high school level. Almost all of the pre-service teachers could solve the problems they made. This was understandable because the problems were for junior or senior high school students

Furthermore, the difficulty level of the problems was varied from high (dominated by analytical problem), medium, to low (dominated by understanding) level. Related to the type of the problems, the ability of pre-service teachers in posing questions with the type of post-solution was the lowest, and the type of within-solution was the highest. In general, problems are made with moderate difficulty. This is in accordance with the criteria of questions that should be made by the teacher in making the problem should be a comparison of problems made between difficult, medium and easy questions have a ratio of 3:5:2 or 2:5:2 [11].

When viewed from the type of problem created, the ability of pre-service teachers in posing problems with the type of post-solution was the lowest, and the type of within-solution was the highest. This is similar with the finding of a research was conducted by Kurniasari et al that the ability of posing problems of post-solution was the lowest with the average of 56.62%, the type of within-solution was 62.75%, and the the type of post solution is the highest with the average of 82.84% [15].

5. Conclusion

Based on the results of the study, it could be concluded that pre-service teachers could pose good quality problems. The researchers suggest to the lecturers of the Assessment Course to train pre-service teachers to make open-ended questions with many solutions or many ways so that they can apply this knowledge when they become teachers at a school. In addition, students need to be trained to make problems with good quality, they should always be accustomed to check the truth of a problem by making the key answer to the problem or alternative solutions, and check whether the problems they make can be solved.

References

[1] Singer F M, Ellerton N and Cai J 2013 *Educ. Stud. Math* **83** 1

[2] Kilpatrick J 1987 *Cognitive Science and Mathematics Education* ed A H Schoenfeld (Hillsdale: Lawrence Erlbaum Associates) pp 123-147

[3] Shriki A 2013 *Creative Education* **4** 430

[4] Polya, G 2014 How to Solve It: A New Aspect of Mathematical Method (USA: Princeton)

[5] Ellerton N F and Clarkson P C 1996 *International Handbook of Mathematics Education* ed A J Bishop, K Clements, C Keitel C, J Kilpatrick and C Laborde (Netherlands: Kluwer) pp 987-1033

[6] National Council of Teachers of Mathematics 2010 *Why is Teaching with Problem Solving Important to Student Learning* Available from: https://www.nctm.org/uploadedFiles/Research_and_Advocacy/research_brief_and_clips/Research_brief_14_-_Problem_Solving.pdf

[7] Silver E A 1994 *For the Learning of Mathematics* **14** 19

[8] Cai J 2002 *Int. Journal of Mathematical Education in Science and Technology* **34** 719

[9] Tim 2015 *Kurikulum Program Studi S1 Pendidikan Matematika (KKNI) Unesa* (Surabaya: UNESA Press)

[10] Brown S I and Walter M I 1983 *The Art of Problem Posing Philadelphia* (PA: Franklin Institute Press)

[11] Masriyah 2018 *Asesmen Proses dan Hasil Belajar* (Surabaya: Unesa University Press)

[12] Siswono T Y E 1999 *Metode Pemberian Tugas Pengajuan Soal (Problem Posing) dalam Pembelajaran Matematika Pokok Bahasan Perbandingan di MTs Negeri Rungkut Surabaya* (Surabaya: Thesis PPs UNESA)

[13] Silver E A and Cai J 1996 *Journal for Research in Mathematics Education* **27** 521

[14] Branca N A 1980 *Problem Solving in School Mathematics* ed Krulik and R E Reys (Reston, VA: National Council of Teachers of Mathematics) pp 3–8

[15] Kurniasari I, Masriyah and Evangelista L W P 2017 Identifikasi kemampuan problem posing mahasiswa pada matakuliah asesmen *Prosiding Seminar Nasional Matematika dan Pembelajarannya* (Malang: Jurusan Matematika FMIPA UM) pp 1043-53

IOP Conf. Series: Journal of Physics: Conf. Series **1088** (2018) 012116 doi:10.1088/1742-6596/1088/1/012116

Students' thinking process in solving mathematical problems based on the levels of mathematical ability

A Sanjaya[1], R Johar[1], M Ikhsan[1] and L Khairi[1]
[1] Syiah Kuala University, Jl. Teuku Nyak Arief Darussalam, Banda Aceh 23111, Indonesia

E-mail: rahmahjohar@fkip.unsyiah.ac.id

Abstract. Problem-solving is one of the main skills in mathematics learning. In Indonesia, problem-solving skill has been embedded in the national education goals. Nevertheless, many students still have difficulties in solving mathematical problems that involve higher order thinking process. This study aimed to analyze students' thinking process when solving mathematical problems according to the levels of mathematical ability. This study is descriptive research and uses a qualitative approach. The participants of the study were undergraduate students in one of the universities in Aceh province, Indonesia. The data were collected by a triangulation method, involving mathematical problem-solving test and semi-structured interviews. The data were analyzed by using by Miles and Huberman's method, including the data reduction, data presentation, and conclusion drawing stages. The findings showed that the three groups of students had different thinking processes. While students with high mathematical ability thought conceptually when solving mathematical problems, students with low ability had computational thinking. Meanwhile, students with moderate ability used semi-conceptual thinking during the problem-solving process. The findings of this study suggested that the students' mathematical ability might lead to different ways of students' thinking processes when solving mathematical problems and different levels of communication skills. Thus, mathematics educators are supposed to consider this aspect when designing teaching instructions.

1. Introduction

Thinking ability is a crucial element during a learning process to understand and acquire the new knowledge. Thinking is a natural process in an individual's mind that can help to solve various problems and improve the quality of life. Thinking involves a process of mental representation that is formed through the transformation of information with a complex interaction of mental attributes, such as assessment, abstraction, logic, imagination, and problem solving[1]. Furthermore, a student's thinking process during a learning process is supposed to be optimized in order to help the student to improve problem-solving skill. As suggested by Gagne, that "the central point of education is to teach people to think, to use their rational powers, to become better problem solvers" [2].

In mathematics education context, problem-solving has become an inseparable part of mathematics learning. Problem-solving skill is considered strongly related to the success of mathematics education so that the integration of problem-solving during the learning process should be necessary [3]. Also, problem-solving is generally regarded as the most critical cognitive activity in the everyday learning process [4]. In addition, it is believed that problem-solving skill is the only skill that significantly contributes to the development of students' thinking ability, while other skills are only the parts of problem-solving [5]. Nevertheless, students often find mathematical problems challenging, and they cannot immediately find the steps or procedures to solve the problems [6]. A student might find mathematical questions problematic when the student does not know the requirements to get the answer of the questions, or the student is not challenged to answer the questions [7]. Besides, studies mentioned that many Indonesian students showed the poor understanding of the fraction topics [8] [9]. Whereas, fraction in mathematics is one of the topics that involve mathematical thinking skill and related to many other mathematical issues, such as decimal, the comparison of fractional functions, scale, inheritance, and measurement. These issues indicate that many students might still have the poor quality of mathematical thinking, which might contribute to the low problem-solving skill.

Most of higher education includes the mathematics subject as a compulsory course in their programs, even though the programs are not the mathematics-related programs. The aspect of problem-solving skill has been embedded into the main goals of the Indonesian mathematics curriculum, in most of higher education. The government expects that all university graduands in Indonesia could possess the problem-solving skill through mathematical thinking to meet the demands of the 21st-century skills in the globalization era. However, the studies on the quality of mathematical thinking and problem-solving skill among the university students, who are not from mathematics-related programs, have rarely been conducted. Therefore, there is a need to investigate the university students' thinking process when solving mathematical problems, which could picture the quality of students problem-solving skill. This study focused on students with different levels of mathematical abilities since the intellectual ability of students could be instrumental in the mastery of facts and concepts [10]. The analysis on the different groups aimed to examine the how the relations between the mathematical conceptual understanding with students' thinking process when solving mathematical problems. This study aimed to provide a better understanding about to what extent the levels of mathematical abilities could influence students thinking process when they solve mathematical problems. The findings of this study were expected to provide educators or researchers with the insight about the aspects that could contribute to the quality of students' problem-solving skills, which they need to focus on when design teaching instructions or conducting further studies.

2. Method

This study is descriptive research that refers to a study directed to describe the symptoms, facts or events systematically and accurately on the nature of research subjects [11]. Further, the qualitative approach was used in this research to understand the thinking process of university students whose different levels of mathematical abilities when solving mathematical problems. This study involved 49 students of Forestry Faculty of Agriculture Studies Program in one of the universities in Aceh. The participants were categorized into three groups, based on their final scores on mathematics course. The first group was students with high mathematical ability, the second group was students with medium mathematical ability, and the third group was students with low mathematical ability. This classification aimed to analyze the differences of students' thinking process according to different levels of mathematics ability. Further, the problem-solving skill of students was analyzed through the quality of stages when they solve the questions, including when understanding the problem; preparing a problem-solving plan; implementing a problem-solving plan; and re-examining the problem solving [12]. This study focussed on fraction topics, which are related to other materials such as decimal, the comparison of fractional functions, scale, inheritance, and measurement.

The data were collected by the triangulation method, in this case, by using two different research instruments for the same participants to strengthen the validity of the data. The first data collection method was a mathematical test on fractions topics, in which the test questions were developed by the researchers and were validated by mathematics education experts as Figure 1. The semi-structured interviews were conducted to get the further descriptions about students' thinking process when solving mathematics problems. The interviews were conducted with three students who represented each level group. The three students were then given initial names based on the represented groups. MKT for the student with high level of mathematical ability, MKS for the student with the medium level of mathematical ability, and MKR for the student with low mathematical ability.

Question 1

Mr. Said has a garden. Half of the garden area is planted with spinach in raised beds, while the water spinach is planted in half as much as the spinach area. The remaining area of the garden is planted with coffee trees.

a. What is the part of the garden area that is planted with water spinach?
b. What is the total part of the garden area that is planted with vegetables?
c. What is the part of the garden that is planted with coffee trees?
d. How many raised beds can be made to plant spinach if each bed has the area 1/20 of the garden area?

Question 2

Mr. Amru has a rectangular garden that has the area 90 m^2 and the length 15 m. To provide the water for the plants in the garden, he made a circular lake with radius 2 m in the middle of the garden as shown in the following picture.

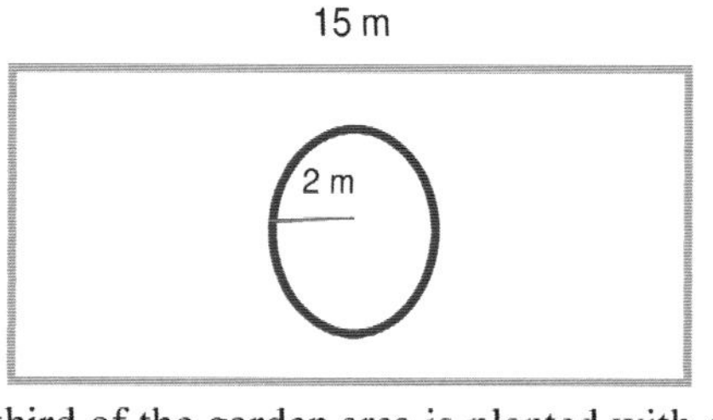

A third of the garden area is planted with coffee trees, while in the opposite side, chocolate trees are planted as much as half of the area planted with coffee. The remaining garden area is planted with potatoes. Find each garden area that is planted with chocolate trees and the potatoes!

Question 3

If the following picture has the value of $\frac{1}{2}$, draw a $\frac{7}{3}$ picture.

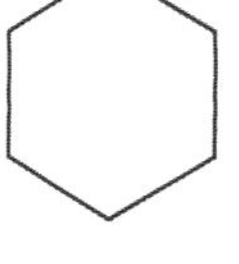

Figure 1. The questions of the problem-solving test

The data were then analyzed by Miles and Huberman's method that included the data reduction, data representation, and conclusion drawing [13]. The data reduction stage refers to summaries of the data, by selecting the main points and focussing on the important information, to be analyzed the common themes of the data. Regarding the interview data, the interview results were transcribed first before being reducing. The data representation stage involves the displays of the data, which includes grouping the data based on the themes and patterns obtained from the previous stage. The last stage is to draw the conclusion based on the groups of the data, in this case, about students' thinking process when solving mathematical problems.

3. Results and Discussion

The data in this study showed the students with different levels of mathematical ability had different thinking processes. The data in this study showed that the students with high mathematical ability were able to analyze the test questions in details, including what was known and unknown in the questions. Also, they could solve the questions correctly. During the interviews, they were also able to re-explain the procedures they performed when solving the questions. They admitted that they knew the procedures from the concept of fractions they had learned before. When solving mathematics questions, students who are able to understand the problem with their own words correctly and able to see the relationship between the known and unknown information from the question are students who use their conceptual thinking [14]. These students tend to apply the concepts they have learned and are able to reveal the steps they have taken when solving the problem. In other words, the students with high ability in this study were students whose conceptual ability when solving mathematics education.

Meanwhile, students with medium mathematical ability in problem-solving were able to identify what was known from the questions. However, some information was missed in their descriptions. Similarly, when they were trying to figure out what was being asked from the questions, even they could determine the asked points, there was also some missing information from the participants of this moderate ability group. Also, they could not use the detailed procedures of the prior concepts they have learned before when solving the questions. Furthermore, during the interviews, some of them they explained every step of their strategies incompletely and incorrectly. This group of students has semi-conceptual thinking when solving mathematical questions, in which they are less able to express with their own words about what is known and unknown of the posed problem [14]. In other words, these students could not fully understand when and how to apply the mathematical concepts they have learned when solving mathematical problems.

With regards to students with low mathematical skill, the students were unable to re-explain the problem with their own language, as well as were unable to identify what was known and unknown in the questions. They expressed the information of the questions incompletely and incorrectly. During the interview, the student with low mathematical skill was unable to justify her answers when answering the questions. Besides, her response to the interviewer sounded like she had guessed the procedures she had used to solve the questions. These students could be categorized in the group of students with computational thinking. A student with computational thinking is usually not able to express mathematical ideas in their own sentence [14]. They tend to forget the concept that has been learned, and not able to explain the steps taken in answering the question.

These findings imply that students' thinking process could be influenced by the levels of the students' mathematical ability. A strong understanding of the mathematical concepts may lead to conceptual thinking, which has a strong correlation with the ability to solve mathematical problems, as well as could effectively communicate the procedures how to get the solutions of the mathematical problems. In contrast, students with low mathematical skill were unable to analyze the questions and only apply the computational thinking skill when dealing with mathematics questions. Students with low mathematical skill might find difficult to identify the information of the questions correctly. The students could have misconception due to the poor quality of their conceptual understanding. The lack of conceptual understanding might also have led to a poor quality of students' skill to communicate mathematical ideas. In other words, besides contributing to the thinking process, mathematical ability, which is indicated by the quality of conceptual understanding, might also influence students' mathematical communication skill. Considering this importance, there appears a need for teachers to pay attention to the level of students' conceptual understanding when designing mathematics teaching instructions. Teachers are supposed to build the learning instruments that could foster students' mathematical thinking and communication, at the same time could respond to the needs of students at all levels. Improving the quality of mathematics tasks could be the starting point for teachers since the mathematical task is considered as important vehicles to build student capacity for mathematical thinking [15]. The factors of the task features and cognitive demands, such as the problem variations, number and kind of representations, the connection to the mathematical concept, and communication requirements of the task could be the basic considerations for teachers when developing the mathematical tasks [15]. Furthermore, the mathematical tasks could be effective once the tasks are understandable and challenging for students to answer, and cannot easily be solved by routine procedures [16]. Thus, it would need the efforts of teachers to actively explore the insight about the mathematical tasks that are suitable to be posed to their students, which could match the different levels of mathematical abilities and various student needs.

4. Conclusion

This study showed that students with different levels of mathematical ability tend to have different thinking processes. The students with high mathematical ability tended to think conceptually when solving the mathematical problem. They were able to analyze the information and requirements of the questions, including to determine what is known and unknown, to solve the questions correctly, as well as to re-explain the steps to solve the questions verbally. Moreover, the students with moderate mathematical ability tended to have semi-conceptual thinking. They had a middle level of ability to generate the mathematical information from the questions. They sometimes had the misconception when re-explaining their procedures how to solve the questions. Meanwhile, the students with low mathematical ability tended to have computational thinking when solving mathematical knowledge. They were not able to express with their own sentence what was known and asked in the questions. Also, the students with low mathematical ability tended to forget the basic concepts that are prerequisite of the questions. The findings indicated that it is crucial to consider the levels of students' mathematical ability when teachers develop teaching instructions. Enhancing the quality of the mathematical tasks could be the possible aspect that teachers should focus on.

References

[1] Solso R L 1995 *Cognitive Psychology. Boston.* Allyn and Bacon.
[2] Gagne R M 1980 *The condition of learning* (New York: Holt, Rinehart, & Winston) p 85
[3] Shadiq F 2004 *Pemecahan Masalah, Penalaran dan Komunikasi Masalah disajikan Instruktur/ Pengembangan Matematika SMA Jenjang Dasar. PPPG Matematika* Yogyakarta
[4] Jonannsen D H2000 Toward A Design Theory of Problem Solving *ETR&D* ***48*****(4)** 63-84
[5] Carso J 2007 A ProbemWith Problem Solving: Teaching Thinking Without Teaching Knowledge *The Mathematics Educator* **17(2)** 7-14.
[6] Hudojo H 1979 *PengembanganKurikulum Matematika dan Pelaksanaanya di Depan Kelas*. Surabaya: Usaha Nasional.
[7] Rizal M 2011*Proses Berpikir Siswa Sekolah Dasar Melakukan Estimasi dalam Pemecahan Masalah Berhitung Ditinjau dari Kemampuan Matematika dan Jenis Kelamin*. Disertasi. Surabaya: PPs-UniversitasNegeri Surabaya.
[8] Netriwati 2016 Analisis Kesulitan Mahasiswa tentang Pembelajaran Pecahan pada Kitab Faroid. *Al-Jabar: Jurnal Pendidikan Matematika* **7(2)** 21-28.
[9] Sugiatno 2010 Model Sajian Verbal-Model-Abtrak dan Model-Verbal-AbstrakUntuk Materi Penjumlahan Dan Pengurangan Pecahan di Sekolah Menengah Pertama. Pontianak. Jurnal Cakrawala Kependidikan: FKIP-PMIPA, Universitas Tanjungpura.
[10] Suryanto S 2005 *Dasar-dasar Pendidikan Anak Usia Dini*. Yogyakarta: Hikayat Publishing
[11] Riyanto Y 2008 *Paradigma Pembelajaran.* Surabaya: Unesa University Press.
[12] Polya G 1973 *How to solve It*. New Jersey: Princeton University Press.
[13] Miles M B, Huberman A M 1992 *Analisis Data Kualitatif*. Terjemahan oleh Tjetjep R. Rohidi. Jakarta: Universitas Indonesia (UI-Press).
[14] Maulana 2008 *Dasar-dasar Keilmuan Matematika*. Subang: Royyan Press

[15] Stein M K, Grover B W and Henningsen M 1996 Building student capacity for mathematical thinking and reasoning: an analysis of mathematical tasks used in reform classrooms *American Educational Research Journal.* **33** 455-488
[16] Hudojo H 2001 *Pengembangan Kurikulum dan Pengembangan Matematika* Malang: Universitas Negeri Malang

IOP Conf. Series: Journal of Physics: Conf. Series **1088** (2018) 012117 doi:10.1088/1742-6596/1088/1/012117

Designing PISA-like mathematics problems using the context of Karawang

I N Aini[1], Zulkardi[2], R I I Putri[2] and Turmudi[3]
[1]Universitas Singa Perbangsa Karawang, Jl. HS. Ronggo Waluyo, Jawa Barat 41361, Indonesia
[2]Universitas Sriwijaya, Jl. Srijaya Negara, Palembang 30128, Indonesia
[3]Universitas Pendidikan Indonesia, Jl. Setiabudhi No 229, Bandung 40154, Indonesia

E-mail: indrie.nooraini@staff.unsika.ac.id

Abstract. The improvement of students' mathematics ability has been the major attention of educators in many countries, including Indonesia. The purpose of this research was to generate PISA-like mathematics problems using the context of Karawang. This research was a development research, and the evaluation flow was conducted by using a formative study including self-evaluation, expert review and field test. The discussion of this article was on the expert review phase and one-to-one. The participants were three students of a junior high school in Karawang. The data collection techniques were walk through and interview. The walk through technique was based on expert reviews to obtain valid PISA-like problems in content, construction, and language aspects. The interview was one-to-one interview. Once collected, the data were analyzed by using descriptive analysis method. Based on expert reviews as well as students' responses, valid PISA-like mathematics problems using the context of Karawang were obtained. The results of this study provide implications for the development of PISA-like mathematics problems using the context of Karawang.

1. Introduction

Program for International Student Assessment (PISA) is an international survey that assesses the skills and abilities of 15-year-old students [1]. PISA focuses on assessing students' skills and abilities in mathematics, reading, and science [1,2]. As a large-scale assessment forum, PISA examines students' mathematical literacy. According to PISA, mathematical literacy is "an individual's capacity to identify, and understand the role that mathematics play in the world, to make well-founded judgments and to use and engage with mathematics in ways that meet the needs of that individual's life as a constructive, concerned, and reflective citizen" [3]. Mathematical literacy is the ability of a student to identify and understand mathematical rules and their application in everyday life.

The question of how to improve students' mathematics ability has been a significant concern of educators worldwide, including Indonesia. PISA results since 2000 to 2015 show that Indonesian students were at the 10th lowest rank of all participating countries in PISA [4].

Students' low achievement in PISA cannot be separated from the teaching and learning process in schools. Students are not accustomed to solving contextual problems, and only do problems exemplified by teachers without truly understand the meaning of mathematics in everyday life. In fact, Indonesian students are particularly not accustomed to solving high-level contextual problems because mathematics learning in Indonesian schools do not use high-level problems [5]. Therefore, students need a better learning activity that promotes their higher order thinking skill.

One of the efforts to familiarize students with PISA-based mathematical problems is by introducing the problems as early as possible or at the beginning of students' middle school experience. Hence many studies have attempted to develop PISA model in junior high schools with various focuses.

Formal mathematics problems cause students to consider that mathematics is not easy and fun. Some students even think that mathematics does not have implementation in their everyday life [6]. Therefore, putting mathematics problems in context is very important [7]. It is also necessary to use local contexts in order to help students to understand mathematical phenomena from their daily-life experiences. One context that can be applied is the context of Karawang.

Karawang is one of the districts in West Java province, Indonesia. There are many local contexts of Karawang district including the aspect of heritage buildings such as *Kebulatan Tekad* monument, Soekarno's Exile House, Walahar Dam and many more. Previous research has developed many PISA-like problems [8]. However, the context of Karawang has never been used. This study aimed to produce valid PISA-like mathematics problems using the context of Karawang. The main reason for selecting the context of Karawang for the PISA-like problems wasto enable the students who participated in the study to understand the problems more easily since the context was familiar to them.

The PISA-like problems developed in this research were at levels three and four considering that students in Karawang were not yet familiar or never heard of PISA problems. The problems that their teacher usually presented in the classroom or available in the text books and student worksheet are of level one or level two [9].

According to the Organisation for Economic Co-operation and Development (OECD), level three mathematical competence on PISA problems requires students to be able to perform procedures well, including procedures that require consecutive decisions [2]. Students can choose and apply simple problem-solving strategies. Students at this level can interpret and use representations based on different sources of information and rationalize the reasons. The competence of mathematics at level four is that students can work effectively with concrete models in complex situations. In this level, students can choose and integrate different representations and relate them to real situations. Furthermore, they can use their skills well and present reasonsas well as views that are flexible according to the context. Students can explain their interpretations and actions.

The question addressed in this study was formulated as follows: How is the designing of valid PISA-like mathematics problems using the context of Karawang. The purpose of this research was to develop valid PISA-like mathematics problems utilizing the context of Karawang.

2. Method

This research applied research development or development studies research design [10]. This study was conducted in three steps: preliminary or preparatory step, the prototyping step, and assessment step [11]. In the prototyping step, the evaluation used formative evaluation which included self-evaluation, expert review and one-to-one, small group, and field tests. The assessment step was conducted to analyze the preliminary and the prototyping steps [12,13].

In the self-evaluation step, the researchers evaluated the PISA-like mathematic problems using the context of Karawang. The design process was based on three characteristics: content, construct, and language as shown in table 1. These three characteristics were validated by the researchers. The results were analyzed to form the first prototype.

Table 1. Examination of the characteristics of the prototype.

No.	Characteristics	Aspects Examined
1.	Content	The problems are following the characteristics of PISA
2.	Construct	The problems are suitable for junior high school level The problems can be developed In accordance with the PISA framework
3.	Language	The problems are in line with the general guidelines of Indonesian spelling Unambiguous Questions The language of the questions and the questions are clear

The researchers developed three PISA-like problems using the context of Karawang. The success criteria of this research was when the problems were declared valid. The validity was obtained from mathematics-expert validators. The validation process of the first prototype required the expertsto review the steps that were executed in three ways including face-to-face review, mail/review (mails review), and panel review [12]. In conjunction with that, the researchers tested students individually (one-to-one). The results of the expert reviews and one-to-one were used to revise the product.

3. Result and discussion

Expert reviews and one-to-one steps were implemented to produce a product which was then tested on three students. The one-to-one step was in line with the formative evaluation guidelines to produce substantial changes to PISA-like mathematic problems using the context of Karawang [11,13]. The three students were referred to as MS, NA, and AC respectively. The purpose of this trial was to find out the responses and difficulties that the students encountered when reading or answering the questions. The observation focused on the legibility and clarity of the matter. The problems covered the topics of quadrilateral, circle, and Pythagoras. The students were asked to read and answer the problems. Interview was conducted to see the students' responses and understanding [14,15]. Table 2 shows the questions for the interview.

Table 2. Interview questions.

Number	Question
1	Do you know the meaning of the question?
2	Using your own words, can you tell me what the problem is in the question?
3	Do you think the information in the question can be used to solve the problem?
4	Have you ever encountered a problem like this before?

The result of one-to-one showed students' opinions that the PISA-likeKarawang's context were good and couldencourage the students' reasoning skills. All the three students were able to answer the questions correctly andexplaintheir answers. However, there were several words in the problems that were poorly understood by the students. Besides, the students felt unfamiliar with the illustrations in the problems. They confessed that they had to read over and over again to understand them. Thus, the explanation must be more specific. Based on the student's answers, it was concludedthat the students could interpret and use representations based on existing sources of information and explain the reasons.

Then, the researcher asked the students to give comments or suggestions on the PISA-like mathematic problems in the context of Karawang. Table 3 shows the comments and suggestions provided by the students.

Table 3. Students'comments.

No	Name	Comments
1.	MS	"I do not understand, and I am not accustomed to working with problems like this. It takes me a long time to finish it. And there is a word that is not clear in the problem.
2.	NA	Too many words that I have to read repeatedly in order to understand the question, to see the relation between the problem and the picture provided in the problem.
3.	AC	I know more about the history of the relics in Karawang.

It was revealed from the students' answers and comments that they understood and were able to answer the given problem even though not perfectly. In the one-to-one stage, it was known that the students' ability to read problems and interpret them into mathematical expressionswas good enough, but it took a long time. This was an input for the researchers in revising the problems to make the problems clearer so that students could understand the problems more easily. The students attempted to solve the problems by guessing and trying to find the pattern. This indicated that the students generally understood the PISA-like mathematics problems. This meant that PISA-like mathematics problems could serve as a alternative form of assessment to be used by teachers in the classroom.

During the expert review, the researchers requested for the opinions of mathematics education experts including Prof. PY, a lecturer at postgraduate program of Pasundan University, Dr. DL, a lecturer at the University of Singaperbangsa Karawang and MF, a junior high school teacher. The experts' feedbacks and suggestions are shown in table 4.

Table 4. Experts' comments.

No	Name	Comments
1.	Prof. YP	The issues presented should be tailored to the learning indicators in junior high school. The language should follow the general Indonesian spelling guidelines.
2.	Dr. DL	It is still better to present an interesting question that fits the context you want to research.
3.	MF	Text writing should follow the general guidelines for Indonesian spelling. The content and construct of the problems are in line with the framework of PISA problems. It all leads to the content of shape and space, and the general context of Karawang. Similarly, the questions are appropriate for junior high school level. Furthermore, the basic competencies presented in this instrument are in line with the materials studied at junior high school level. So, the essence and the content are appropriate.

The PISA-like mathematics problems were developed by researchers and validated through one-to-one and expert reviews. From both steps mentioned, the researcher revised the problems in accordance with the general guidelines of Indonesian spelling, eliminated irrelevant information and pictures and fixed the layout and display of thequestions.

The results from expert reviews and one-to-one showed that the designed problems could be categorized as valid PISA-like mathematics problems.

4. Conclusion

The results of this research were valid PISA-like mathematics problems using the context of Karawang. The first prototype of the development was declared valid based on expert reviews and one-to-one test. The prototype was valid after the content, construction, and language revisions. There were three PISA-like mathematics problems produced.

References

[1] OECD 2009 *Take the Test: Sample Questions from OECD's PISA Assessments* (Paris: OECD Publishing)
[2] OECD 2003 *Literacy Skill for the World of Tomorrow: Further Results from PISA 2000* available from: http://www.oecd.org/edu/preschoolandschool/2960581.pdf.
[3] OECD 2010 *PISA 2012 Mathematics Framework* available from: http://www.oecd.org/dataoecd/8/38/46961598.pdf.
[4] OECD 2013 *PISA 2012 Assessment and Analytical Framework: Mathematics, Reading, Science, Problem Solving and Financial Literacy* (Paris: OECD Publishing)
[5] Novita, Zulkardi and Hartono 2012 *Indo MS. J.M.E* 133-150
[6] Arifin S, Zulkardi, Putri R I I, Hartono Y and Susanti E 2017 *IOP Conf. Ser: Journal of Physics* **943** 012038
[7] The Cornerstone Tech Prep 1999 *Teaching Mathematics Contextually* (Texas: CORD)
[8] Charmila, Zulkardi and Darmawijoyo 2016 *Jurnal Penelitian dan Evaluasi Pendidikan* **20** 198
[9] Zulkardi Z and Kohar A W 2018 *IOP Conf. Ser.: Journal of Physics* **94 7**012015
[10] Akker J 2006 *Educational Design Research* (London and New York: Routledge)
[11] Nieveen N and Plomp T 2007 Formative evaluation in educational design research *An Introduction to Educational Design Research* (Enschede: SLO)
[12] Tessmer M 1993 *Planning and Conducting Formative Evaluation* (Philadelphia: Kogan)
[13] Zulkardi 2006 *Formative Evaluation: What, Why, When, and How* available from: http://reocities.com/zulkardi/books.html
[14] Putri R I I and Zulkardi 2017 *AIP Conference Proceedings* **1868** 050005
[15] Putri R I I and Zulkardi 2018 *IOP Conf. Ser. Journal of Physics* **948** 012056

Optimalization of student's learning outcomes and learning activities in physics using open ended model

A Hamid and Nofiza
Syiah Kuala University, Jl. Teuku Nyak Arief Darussalam, Banda Aceh 23111, Indonesia

E-mail: Abdulhamid_fkip@unsyiah.ac.id

Abstract. This study aims to optimize the learning outcomes and physics learning activities of senior high boarding school in Banda Aceh, Indonesia on the subject of Newton's law about gravitation using an open-ended learning model. The study was conducted by involving 29 students of grade 11st. Data collected include data of learning result, data of observation result of student activity, and student response questionnaire data. The data collected is then analyzed using descriptive and quantitative statistics. Based on the results of processing and data analysis shows that the results of learning and physics learning activities of senior high boarding school in Banda Aceh, Indonesia has been optimal.

1. Introduction

In the learning activities at school, teachers are not only required to master the teaching materials, but also expected to know the characteristics of the students because each student has a difference in learning activities and problem solving. With regard to problem solving activities, learning needs to be designed so as to accommodate the diversity of student characteristics so that they can develop optimally according to their potential [1]. The failure of the teacher to convey the subject matter is not only because he lacks the mastery, but it can be because he does not know how to teach certain subject matter properly and correctly.

According to Yulianti [2] in the context of physics learning at school, students learn not only how to understand the physics concept but are also expected to develop the skills to ask questions, make explanations, and test explanations and rewrite them using their own language.

According to Popov [3], there are several solutions to address current physics education issues, including (1) teachers need to identify each student's individual abilities and challenge him in the proper way, (2) teachers need to have the ability to explain physics in different ways when meeting with students of different characteristics, and (3) teachers who convey the desire to learn have often lessons relate to reality, and they engage students in challenging discussions and show how theoretical knowledge can be used in practice.

Furthermore, it is said that successful learning depends on two-way communication and discussion between teacher and students. Therefore, if students are aware of their ability to learn then they can also plan lessons with teachers. However, the increased desire to learn does not automatically lead to students achieving better academic performance, but the development and progress in the field of interest will give potential to increase students' desire to learn. Increased desire to learn can create opportunities to achieve better learning outcomes in physics [3].

The outdoor physics approach broadly uses open-ended authentic problems, and hence, it is necessary to classify problems and find suitable problem-solving strategies for open-ended problems. All of problems can be classified with three variables; *data*, *method* and *goal*. The fewer variables that are known to the solver the more open the problem will be [3,4] and therefore, open-ended problems will have a variety of solutions and options of different solving methods are possible. Problem solving can be seen an essential part of physics learning and usually students solve problems, where most of the problems are constructed with the aim to find a suitable formula and then apply given numerical values into that formula [5].

Based on the results of interviews with several teachers and students in high school shows that during physics learning activities in school students tend to be passive. At the time the learning activities are ongoing students only pay attention to the explanation of teachers and have not asked or responded many questions and have not been involved in the learning process, consequently the students are less active and creative in solving problems given teachers.

One of the alternative activities that can be done to improve the process of physics learning is to use a learning model that provides an opportunity for students to develop their mindset in solving a problem. The processes undertaken by the student in selecting, organizing and integrating new knowledge, behaviors and ideas will influence his motivation and attitudes [6].

To optimize the process and the results of physics learning in students, it is necessary to use open-ended learning model. Open ended learning model can be used in physics learning because it can help students think creatively and support the understanding of physics concepts, and this model of learning gives individuals the freedom to develop different ways and problem-solving strategies according to their respective abilities [7].

Furthermore, the open ended learning model is a learning that presents a problem that has method or the correct solution more than one [8]. According to Takahashi [9], the open ended problem is a matter of having multiple solutions or settlement strategies. In addition, the use of open questions can stimulate creativity, original thinking skills, and innovation in learning [10], while the advantage of using open-ended problems is to encourage students to develop creativity and think analytically [11]. Meanwhile, according to Nohda [12], one of the objectives of open matter in learning physics is to encourage students' creative activities in solving problems.

According to Takahashi [9], there are some benefits of using open-ended questions in learning, among others (1) students become active in expressing their ideas, (2) students have more opportunities to comprehensively use their knowledge and skills, (3) students have rich experience in the process of finding and receiving approval from other students of their ideas. In studying physics, open questions can be given to students early in the learning process. Furthermore, the learning process is developed based on the variations of answers that appear. In this way, students have experience in finding something during the problem-solving process.

2. Method

This action research was conducted on students of grade 11st senior high boarding school in Banda Aceh, Indonesia involving 29 students as subjects and the object is open ended learning on Newton's law about gravitation.

Techniques used to collect data in this study are using test questions, student activity observation sheets, and student response questionnaires. The data obtained during this research is processed using descriptive and quantitative statistics.

The stages for this study began by preparing the test set for Newton's law about gravitation, student activity observation sheets, and student response questionnaires.

3. Result and discussion

3.1. Implementation of lesson plan
Based on the result of the observer's evaluation during the learning process, there was an increase of the average percentage of lesson plan implementation, from cycle I of 72,92%, second cycle was 84,75% and cycle III was 95,50% (see table 1). Increasing the implementation of the lesson plan is possible because the teacher is very concerned about each step listed in the lesson plan that he has designed and seeks to improve it for each cycle. Increasing the implementation of this lesson plan shows that physics learning activities for Newton's law subject about gravitation are highly optimized.

Table 1. Percentage of lesson plan implementation per cycle.

Activities learning	Cycle I		Cycle II		Cycle III	
	%	Category	%	Category	%	Category
Introduction	75.00	Good	91.75	Very good	95.75	Very good
Main part	68.75	Good	75.00	Good	97.00	Very good
Closing	75.00	Good	87.50	Very good	93.75	Very good
Average per cycle	72.92	Good	84.75	Very good	95.50	Very good

3.2. Student activity
Student activity data during physics study using open ended learning model can be seen in table 2. During the process of physics learning using open ended learning model, students are given the opportunity to learn through real life activities by presenting the natural phenomenon openly. The form of presentation of this open phenomenon can be done through learning that is oriented to the problem or open matter.

Table 2. Percentage of student activities per cycle.

No.	Student Activity	Observation result (%)		
		Cycle I	Cycle II	Cycle III
1	Hear the teacher's explanation	83.3	83.3	100
2	Answer the teacher's question for initial knowledge	85.7	85.7	85.7
3	Reading (gathering information and so on)	80.0	85.7	92.3
4	Perform critical analysis of problem solving	88.2	88.2	93.3
5	Reviewing and reviewing the issues raised	85.7	92.8	93.3
6	Summing up learning	83.3	83.3	100
	Average	84.4	86.5	94.1

There are a number of benefits of using open-ended questions on open ended models, including (1) students participating more actively in learning and expressing their ideas more intensively, (2) students have more opportunities to use their knowledge and skills comprehensively, (3) students have more opportunities to develop their reasoning [13]. Increased student learning activity during the use of open-ended learning model because this model provides an opportunity for students to express ideas and investigating strategies to solve problems, either individually or in groups. Students are encouraged to learn to build, construct and maintain argumentative and reasonable solutions.

3.3. Student learning outcomes
The results of physics learning that students achieve during the course of action using open-ended learning model can be seen in table 3.

Table 3. Learning outcomes achieved by students each cycle.

Research cycle	The number of students		Classical completeness
	Completed	No completed	(%)
Cycle I	10	19	34.48
Cycle II	19	10	65.51
Cycle III	26	3	89.65

The students learning outcomes can be observed through the number of students who achieve learning classical completeness. In the first cycle, the number of students who achieve mastery in the classical reach 34.48% with the category less good, in cycle II, the number of students who reaches the value of classical completeness reached 65.51% with good category and in cycle III, the number of students who achieve the test results above or equal to 75 reached 89.65% with very good category so that in this cycle can be said that learning has been completed in a classical. This shows that the provision of action with the improvement has been able to optimize student learning outcomes from the category of less good to the category very well and expressed in a classical manner.

Students' learning completeness in class shows that open ended learning model can improve their learning result. This is in accordance with the important characteristic of the open ended model is the flexibility of students to use a number of methods and all possibilities that are considered appropriate to solve problems and improve learning outcomes.

3.4. Students' response

Students' responses to the learning process can, among other things, be assessed from the aspect of likes, interests, desires, and so on can be seen in table 4, which shows good results. Students who are interested in a particular subject tend to give greater attention to the subject. The form of attention can be seen from the participation of students in learning activities.

Table 4. Percentage of student interest responses.

No	Aspect	Student response	
		Percentage	Category
1	Preferences	100	Very good
2	Interest	100	Very good
3	Desire	82.76	Very good
4	A passion for the subject of physics	65.52	Good
5	Help understand the material	68.96	Good
6	Clarity of concept learned	65.52	Good
7	Worksheet language	75.86	Good
	Average	84.4	Good

4. Conclusion

Based on data and analysis of research results, it can be concluded that the results of learning and physical learning activities in students for Newton's law of gravitation can already be optimized using open-ended learning model. The implementation of lesson plan is excellent, learning activities are categorized very good, student learning outcomes are very good, and student response is good. Recommendations from the results of this study is open ended learning model can be used and developed for the material or other subject contained in the subject of physics.

References

[1] Arend R I 2008 *Learning to Teach (Edisi ke-7)* ed Terjemahan Helly Prajitno Soetjipto (Yogyakarta: Pustaka Pelajar)

[2] Yulianti L 2010 *Model-model Pembelajaran Fisika* (Malang: LP3 UM)

[3] Popov O 2006 Developing outdoor activities and a website as resources to stimulate learning physics in teacher education *Journal of Physics Teacher Education Online* **3** 18

[4] Reid N and Yang M J 2002 Open-ended problem solving in school chemistry: A preliminary investigation *International Journal of Science Education* **24** 1313

[5] Benckert S and Pettersson S 2008 Learning physics in small-group discussions *Eurasia Journal of Mathematics, Science and Technology Education* **4** 121

[6] Anthony G 1996 *Classroom Intructional Factors Affecting Mathematics Student Strategics Learning Behavior,* ed Philips C. Clarkson (Australia: Mathematics Education Research Group of Australia)

[7] Suherman 2003 *Strategi Pembelajaran Matematika Kontemporer* (Bandung: JICA)

[8] Shimada S and Becker J P 1997 *The Open-ended Approach: A New Proposal for Teaching Mathematics* (Virginia: NCTM)

[9] Takahashi A 2006 Characteristics of Japanese mathematics lesson *Tsukuba Journal of Education Study in Mathematics* 37

[10] Livne N L, Livne O E and Wight C A 2008 Enhanching Mathematical Creativity Through Multiple Solution to Open-Ended Problems Online 1

[11] Surif J, Ibrahim N H and Dalim S F 2013 Prosedia-social and behavioral sciences **116** 4955

[12] Nohda N 2000 Learning and Teaching Through Open-ended Approach Method *Proceeding of the 24th of the International Group for the Psychology of Mathematics Education* (Hiroshima: Hiroshima University)

[13] Heinemann 2008 *Why Use Open-Ended Question*?

IOP Conf. Series: Journal of Physics: Conf. Series **1088** (2018) 012119 doi:10.1088/1742-6596/1088/1/012119

The development of a module with Microsoft Excel-based interactive media on the topic of buffer solution

Anizar[1,2], A Gani[2], I Khaldun[2] and M Bahi[2]
[1]Islamic Senior High School Ruhul Islam Anak Bangsa, Aceh Besar 23352, Indonesia
[2]Syiah Kuala University, Jl. Teuku Nyak Arief Darussalam, Banda Aceh 23111, Indonesia

E-mail: anizaransari@yahoo.com

Abstract. Topic of Buffer Solution is considered as complex and abstract subject in chemistry because it covers the macroscopic, the microscopic and the symbolic levels. Several studies have shown that interactive simulations using computer technology could help learners to understand abstract concepts. In this research, a module with *Microsoft Excel-based* interactive media was developed as an alternative to real lab practice on the topic of Buffer Solution. The method used in this research was Research and Development method with ADDIE model. The effectiveness of this media was tested to the second graders of natural science program of a senior high school as the samples with a direct experiment and a response questionnaire regarding the developed media. The samples were chosen randomly. Based on the results, it is indicated that the module with Microsoft Excel-based interactive media received positive responses from the teachers (88%) and the students (86%). The results suggest that the improvement of the students' conceptual understanding was in high category (N-gain = 0,98) and the improvement of their analytical thinking skills was in intermediate category (N-gain = 0,5). Further research with larger class experiment is advisable to be conducted with respective module.

1. Introduction

The rapid development of science and technology resulted in a change of educational paradigm followed by improvements in curriculum, media, and technology. Nowadays, multimedia technology has become a necessity in the efforts made to build and develop learning media. According to Hendra and Siagian [1], one of the examples is an interactive learning media that is based on the development of Information and Communication Technology (ICT).

A multimedia-based learning approach has an ability to move a concept from a static textbook to a more interesting learning style [2]. In addition to increasing interest in learning, it is also able to foster creativity and innovation of learners in producing new technologies for the purposes of learning and education. Improved learning outcomes might be obtained by adding multimedia contents into teaching materials that will become multimedia teaching materials [3]. A computer-based teaching material used in teaching and learning process is also an attempt to improve thinking skills [4].

Several case studies indicated that some students faced difficulties in understanding chemistry concepts. The complexity of chemistry science requires a teaching and learning process including the macroscopic, the microscopic and the symbolic levels [5]. In addition, Laksono, et al.,[6] suggests that chemistry teaching and learning process should involve activities in which the learners interact with the materials, communicate, and use their mind. One of them is the material of Buffer Solution. The learning process which is still teacher-centered and focuses only on memorization with limited learning resources and the absence of science laboratory facilities in schools make buffer solution become a boring topic.

The topic of Buffer Solution is one of the subjects discussed in chemistry textbooks including theories and practices. The prerequisite for understanding the concept of Buffer Solution is that the learners should have mastered the concept of acid-base, pH, acid-base stoichiometry and chemical equilibrium. The inability to understand the basic concepts leads to many more problems, such as poor learning outcomes and assumption that the topic of Buffer Solution is difficult to understand [7]. Furthermore, Harun et al. [8] stated that most teachers in Banda Aceh and Aceh Besar regions said that the topic of Buffer Solution is rarely delivered with practicum. This happened because of several reasons, such as the target that should be achieved by the teacher according to the syllabus and the unavailability or the incompleteness of laboratory facilities in the schools [9 , 10]. Furthermore, Tuysuz [11] added that the obstacles faced by the teachers were (1) the expensive tools and materials, so that they are not affordable for underprivileged schools; (2) a quite long time needed for the preparation and execution of the practicum; and (3) too many students that are hardly monitored by the teacher.

Various kinds of innovative learning media, such as animations, modules, concept maps, comics, virtual labs and others can be used by teachers in order to attract students' attention in the classroom [12]. Good learning media interpret theoretical concepts into easy-to-understand concepts and simulate materials that are difficult to be delivered in classroom[10,13].

The method attempted to overcome the problems faced by the students and the chemistry teachers on the topic of Buffer Solution is using a module as an interactive media of virtual lab. According to Chen et al. [14], using a teaching material developed in accordance with the needs of the learners might improve the learners' achievement. It is because the module presents adequate exercises for the cognitive development of the learners [15]. This additional teaching materials can support the process of achieving the learners' competence so that learning materials can be delivered effectively and efficiently. If the module is developed in accordance with the needs, the learning outcomes may be more satisfying.

According to Turkoguz [16] and Alexander et al. [17], by following technological developments, modules can be developed and combined with other teaching materials in the form of audio or video as well as other alternative teaching materials facilitating students in understanding the contents. A module as an interactive media of virtual lab is selected to replace experiments due to some constraints. The use of interactive chemistry multimedia modules, as reported in Azziz et al. [18], showed a positive response and improved the conceptual understanding on the topic of Stereochemistry. Another research conducted by Oktiarmi et al. [19] also obtained the same result, a positive response on the use of interactive multimedia teaching materials. Moreover, it was accepted by learners with different levels of capabilities.

Another advantage of interactive media is by using only a computer program, we are able to simulate a series of experiments without performing the direct activities [20]. Interactive media also helps the learners to interact directly with the experiments, not only by seeing and listening but also by repeating them when needed [21]. With this media, the students can experiment with buffer not only in school laboratories but also at home. Experiments can repeatedly be done without the use of any chemicals. The media makes the experiments cheaper, safer and more suitable to be used and explored by learners who have visual learning styles based on their needs and speeds [22].This study aims to develop viable module with Microsoft Excel based interactive media which can increase understanding concept and analytical thinking skills of the students.

2. Method

Research and Development Method with Analysis, Design, Development, Implementation, and Evaluation (ADDIE) model [23] was employed in this research as follow:

IOP Conf. Series: Journal of Physics: Conf. Series **1088** (2018) 012119 doi:10.1088/1742-6596/1088/1/012119

2.1. Analysis
The analysis is a crucial stage in ADDIE model [23]. The process consists of identifying the needs and problems found and setting the goals to be achieved by the development as well as the knowledge and skills needed in the process of obtaining the final product.Feasibility study was done in order to find out the needs and problemsat an Islamic Senior High School of Aceh Besar Regency.

2.2. Design
The design stage is in which things related to the desired final product are identified. The module with Microsoft Excel-based interactive media was designed based on the needs for the concept of calculation on the topic of Buffer Solution which is quite complicated. The specific documents of learning objectives, assessment instruments, exercises and content were designed.

2.3. The development
The actual creation of the module is completed in the development phase. A module with Microsoft Excel-based interactive media was developed to make it easier for the students to understand the concept of buffer solution, buffer solution components, buffer solution in acid-base titration graphs and the calculation of the pH of buffer solutions, and its application in daily life.Before the learning tools were used, the validity and reliability of the tools were tested and reviewed by the content experts and instructional media specialists. Then, the tools were then revised. The validity and reliability of the module was tested and reviewed by two content experts and two instructional media specialists prior to its implementation in classroom.

2.4. Implementation of the module with microsoft excel-based interactive media
When the tools was reliable, it was implemented in a small-scale experiment in order to see the response and the effectiveness of the product by distributing questionnaires. Twenty-five students in science program at an Islamic Senior High School of Aceh Besar Regency were selected randomly. The response questionnaires were distributed to five teachers and twenty-five students. The responses were considered in revising the media being developed. The experiment was also designed to identify the students' conceptual understanding and analytical thinking skills on the topic of buffer solution. The data were collected with pretest and posttest.To measure the improvement of conceptual understanding and analytical thinking skills before and after the use of learning media, the g-factor formula (N-gain) with Hake formula [24] was used.

2.5. *Evaluation*
After delivery, the effectiveness of the module is evaluated. Formative evaluation runs parallel to the learning process and to evaluate the quality of the module and their reception by the students. Summative evaluation is to prove, once the experiment is finished, that the module had a positive effect.

3. Result and discussion
Feasibility study was done in order to find out the needs and problemsat an Islamic Senior High School of Aceh Besar Regency. A module with Microsoft Excel-based interactive learning media was designed to find the best solution for the problems encountered according to the feasibility study.Based on the observation by questionnaires done in the school, an interactive virtual lab media is needed as an alternative to the unavailable real laboratory as well as a complement of limited learning media, and lack of understanding on the topic of Buffer Solution. This is in line with Sutrisno [25], which states that virtual labs might efficiently and significantly improve the learning process and create new experiences providing stimulation to the cognitive dimension of the concepts being learned.

The average score of experts' judgment on the aspect of the module assessment was 3.42 of 4 maximum score. Based on the score, the module is considered to be implemented. The validity of tested questions was found 0.991 and the reliability was 0.998 with high category. On the basis of statistical analysis, it is assumed that the implemented instrument is valid and reliable.

After several revisions the final display of the interactive media can be seen in Figure 1 and in Figure 2 is shown the cover page of the module.

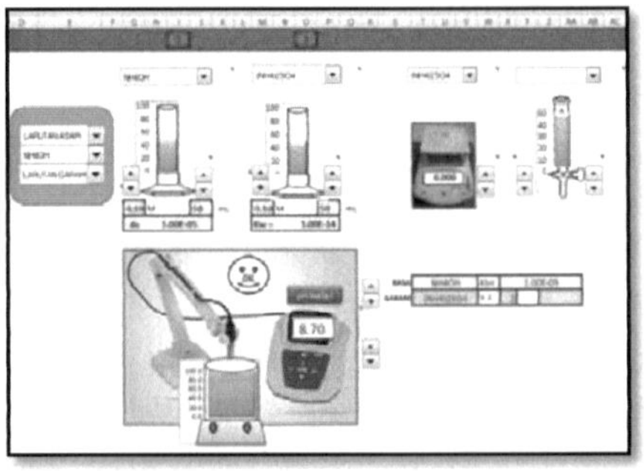

Figure 1. Interactive media

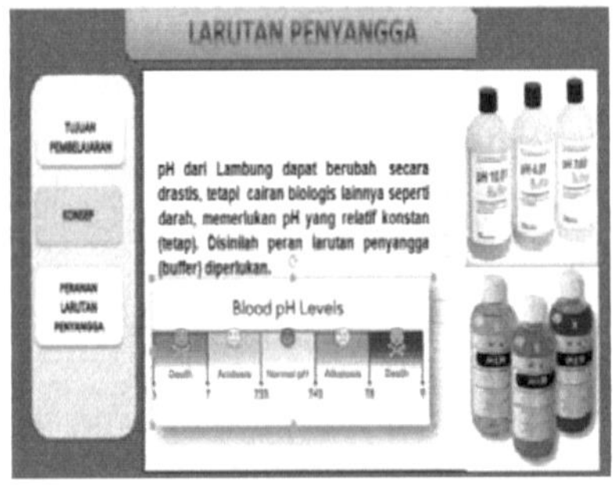

Figure 2. Module

According to our study, it was also found that most of the teachers gave positive responses (88%) indicating that the product was useful in helping the children to understand the topic of buffer solution and increase the students' learning motivation. In addition, positive responses were also given by the students (86%). Some of students also found that the topic of buffer solution became more simple, interesting, and detailed with easily-solved problems.Hence, the product has been expected to enhance the students' analytical thinking skills. The improvement of the students' conceptual understanding and analytical thinking skills before and after applying the trial module is shown in Figure 3.

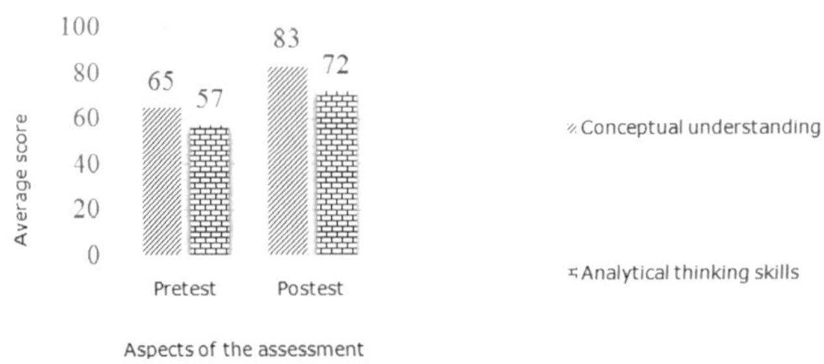

Figure 3. The comparison of the average score of pretest and posttest on the students' conceptual understanding and their analytical thinking skills

In addition, Figure 3 above has also revealed the difference between pretest and post-test average scores of the students' conceptual understanding and their analytical thinking skills. The students' average posttest score is higher than their average pretest score. It was because the module with interactive media was equipped with figures and brief materials which could be easily understood. This result is in line with Arsyad [26] who states that giving module and media is a treatment that can help the learners to learn independently. More detailed results of the experiment on the students' conceptual understanding are listed in Table 1.

Table 1. The improvement of conceptual understanding on the topic of buffer solution.

No.	Score	Average	*N-gain*	Category
1	Pre-test	65	0,99	High
2	Post-test	73		

Based on the analysis of pretest and post-test scores, there was a difference between the students' average score in the test regarding their conceptual understanding before and after the use of the module with interactive media. The average posttest score is 0.99 point (category: high) higher than the average pretest score, which was taken before any treatment was given. Tapilouw [27]and Irianto [28] suggests the same finding in which the conceptual understanding of learners with the use of the module with interactive media is higher than those without any use of the module. The module with interactive media is a tool used to attract the learners' sympathy and interest in an active and independent process of understanding a particular concept.

Moreover, Sukiman [29] has reported that the module with interactive media can be developed based on the development of science and technology by presenting the material in a sequence that facilitates the learners in understanding the concept. In addition, Chomsin [30] has also reported that by developing a module with interactive mediaa teacher can overcome the limitations of time, space and sensory power, of both the learners and the teachers. In addition, a research conducted by Trian et al.[31] also concluded that modules and media designed with the principles of increasing motivation would foster the learners' character. The differences between the pretest and the post-test scores of the students' analytical thinking skills regarding the trial of the module are shown in the following Table 2.

Table 2. The improvement of analytical thinking skill indicators.

No.	Indicators of analytical thinking skill	Average score		Category		(*N*-gain)
		Pretest	Posttest	Pretest	Posttest	
1	Arguing	55	57	Medium	Medium	0,2 (Low)
2	Drawing inferences and conclusions	75	85	High	High	0,2 (Low)
3	Identifying problems	57	72	Medium	High	0,5 (Medium)

Table 2 shows the results of an experimental analysis of the module with interactive media based on several indicators of critical thinking skills. There was an increase in the average posttest score with the N-gain percentage of 0,5, (category: medium) while the average pretest score got N-gain percentage of 0,2 (category: low). Even though it was not significant, there was an increase which indicates that the module helped the learners to be actively involved in the classroom activities and use their analytical thinking skills. Another study conducted by Nisai [32] obtained the same result. It was found that the direct teaching method with the module trained the learners to learn independently either in the school or at home, so they could optimize their analytical thinking skills in solving various problems provided in the module.

The increase of the students' critical thinking skills in the three indicators is generally categorized as low-medium. In line with this result, Ismienar [33] also states that analytical thinking skill is a complementary component which is hard to stimulate. It takes a quite long time to stimulate the students' analytical thinking skills. Table 2 above shows the improvement of several indicators of analytical thinking skill.

The increasing scores of the three indicators of analytical thinking skills in the trial did not show noticeable changes. This probably happened because the learners were rarely trained to develop their thinking skills and to conduct experimental activities [34]. A similar result was also reported by Rahmawati et al. [35], one of the reasons for low level of analytical thinking skills in a school is lack of practice of using thinking skills. Learners were only trained to answer questions by rote so that their high-level thinking was not honed. Teachers play crucial roles in teaching thinking skills during science learning process in the classroom. Analytical thinking skill is a mental process in solving problems. It can be trained through experimental activities in the laboratory, and a module with interactive media might be an alternative to a real experiment [35].

4. Conclusion

Based on the results of our research, it can be concluded that the module with Microsoft Excel–based interactive media is feasible and practical as a teaching material on the topic of Buffer Solution.

Positive responses have been shown with high score of statistical analysis by both teachers and the students toward the module in this research. The trial of the module showed that the module was able to improve the students' conceptual understanding and analytical thinking skills. In addition, this interactive media-based module can be used as an alternative teaching-learning approach to deal with the topic of buffer solution, and as one of virtual laboratory applications in chemistry education.

References

[1] Hendra and Siagian 2014 *J. Teknologi Informasi dan Komunikasi dalam Pendidikan* **1** 87
[2] Jamaluddin H and Zaidatun T 2003 *Multimedia dalam pendidikan* (Selangor :PTS Publication)
[3] Setiawan M A, Dasna I W and Marfuah S 2016 *J. Pendidikan* **1** 746
[4] Shahsavar Z 2014 *J. International Scientific Publications* **3** 99
[5] Fitria, Priatmoko S and Kasmui 2016 *J. Inovasi Pendidikan Kimia* **1** 41
[6] Laksono W E, Rohaeti E, Suyanta and Irwanto 2017 *J. Kependidikan* **1** 100
[7] Mustofa, Pikoli M and Suleman N 2013 *J. Entropi* **8** 551
[8] Harun M, Khaldun I, Abdullah, Taher A, Yusrizal, Yusuf R, Bambang R M, Raihani and Heriansyah H 2011 *Laporan Penelitian* (Universitas Syiah Kuala)
[9] Harun M, Khaldun I, Abdullah, Taher A, Yusrizal, Yusuf R, Bambang R M, Raihani and Heriansyah H 2012 *Laporan Pengabdian Masyarakat* (Universitas Syiah Kuala)
[10] Yuniarti F, Dewi P and Susanti R 2012 *Unnes J. Biologi Education* **1** 86
[11] Tuysuz C 2010 *International Online J. Educational Science* **2** 37
[12] Kusnadi, Masykuri M and Mulyani S 2013 *J. Inkuiri* **2** 163
[13] Yusuf I, Widyaningsih SW and Purbawati D 2015 *Pancaran* **4** 189
[14] Chen Y T, Chen T J and Tsai L Y 2011 *International J. Physical Sciences* **6** 31
[15] Dhamija N and Kancha 2014 *J. Education Confab* **3** 26
[16] Turkoguz S 2012 *Chemistry Education Research Practice* **13** 401
[17] Alexander M H, Hall GE and Dagdigan PJ 2011 *J.Chemical Education* **88** 38
[18] Azziz S S S A, Suhairun A A, Siais S, Talib O, Zain Z N, Shariman T P N T, Tajudin N M, Bakar N A and Jusoff K 2013 *Asia Pasific J. Educators and Education* **28** 53
[19] Oktiarmi P, Rusdi M and Asrial 2014 *Edu-Sains* **3** 6
[20] Wahyuni E S, Sudarisman S and Karyanto P 2013 *J. Inkuiri* **2** 269
[21] Kusumaningrum A C and Sukarmin 2014 *J. Chemical Education* **3** 75
[22] Dobrzanski L A and Honysz R 2010 *J. Achievements in Materials and Manufacturing Engineering* **42**1 96
[23] Danks S 2011 *ASQ Primary and Secondary Education Brief* **4** 5
[24] Hake R 1998 *Amer. J. Physic,* **66** 165
[25] Sutrisno 2012 *Kreatif Mengembangkan Aktifitas Pembelajaran Berbasis TIK* (Jakarta:Referensi Anggota IKAPI)
[26] Arsyad A 2011 *Media Pembelajaran* (Jakarta: PT Raja Grafindo Persada)
[27] Tapilouw S and Setiawan W 2008 *J. Pendidikan dan Komunikasi* **1** 19
[28] Irianto E S 2008 *Widyatama* **6** 31
[29] Sukiman 2012 *Pengembangan Media Pembelajaran (*Yogyakarta: Pedagogia)
[30] Chomsin W and Jasmadi 2008 *Panduan Menyusun Bahan Ajar Berbasis Kompetensi* (Jakarta: PT. Alex Media Komputindo)
[31] Trian E A, Haryani S and Sedyawati SMR 2013 *Unnes Science Education Journal* **2** 269
[32] Nisai K 2016 Pengaruh model pengajaran langsung dengan bantuan modul terhadap kemampuan berpikir analitis dan hasil belajar siswa kelas XI IPA SMA N 14 Bulukumba pada materi larutan penyangga *eprints*.unm.ac.id/3635/
[33] Ismienar 2009*Thinking (*Malang: Universitas Negeri Malang)
[34] Irwanto, Rohaeti E, Widjajanti E and Suyanta 2017 Student's science process skill and analytical thinking abilityin chemistry learning *AIP Conference Proceeding* **1868** 030001
[35] Rahmawati A, Amiruddin and Buranda J P 2013 *J. Pendidikan Geografi* **2** 1

IOP Conf. Series: Journal of Physics: Conf. Series **1088** (2018) 012120 doi:10.1088/1742-6596/1088/1/012120

The development of student worksheets based on metacognitive approach to improve students' mathematical representation ability

A Murni, R D Anggraini and Sakur
Riau University, Jl. Kampus Bina Widya, Simpang Baru, Tampan, Kota Pekanbaru, Riau 28293, Indonesia

E-mail: murni_atma@yahoo.co.id

Abstract. Mathematical representation ability is an important mathematical skill. It can be trained by using a student worksheet as a learning resource. This research aimed to develop and produce valid, practical, and effective student worksheets which were based on metacognitive approach to improve students' mathematical representation ability. This research used the 4-D model (Define, Design, Development, and Disseminate). This research was conducted at a secondary school in Pekanbaru, Indonesia. The research participants were seventh-grade students at the school. Data collection instruments were questionnaires and mathematical representation ability test. Data were analyzed using descriptive statistical analysis technique. The result showed that the worksheets were very valid, very practical and very effective. The validity rate was of 81.71%, the practicality rate was 82.30% and the effectiveness rate was 83.33%. Hence the worksheets can be used in mathematics learning to improve student's mathematical representation ability.

1. Introduction

The abilities to be achieved in the process of learning mathematics according to National Council of Teachers of Mathematics (NCTM) are reasoning, communicating, connecting, representing and problem solving [1]. The ability of mathematical representation is very important because it can be used to represent a situation or problem in order to facilitate the search for a solution. The success of problem solving depends on: (1) the ability to represent problems such as constructing and using mathematical representation in words, graphs, tables and equations; (2) problem solving skill; and (3) the ability to manipulate symbols [2-3]. Some representations are more concrete than others. A concrete representation can be used as a reference for a more abstract concept and as a tool in problem-solving [4-5].

Mathematical representation is everything that a student make to internalize and show their work [4]. Student-made representations are expression of their understanding of a mathematical concept or idea that they use in finding a solution to a mathematical problem [1]. A student can communicate his or her ideas for a mathematical strategy or solution in various ways of representation: (1) symbolically (numerical and/or algebraic symbols); (2) verbally; and (3) visually in the forms of diagrams, graphs or tables [2].

The quality of learning needs to be improved by applying various approaches. One of the recommended approaches for mathematics learning is the metacognitive approach. Learning with a metacognitive approach is a learning that instills awareness to students about a process of how to design, monitor, and evaluate the activities undertaken to determine the solution of a problem. Metacognitive approach can be done by: (1) asking questions and facilitating the students so that the students can control their thinking process and activity, choose appropriate problem-solving strategies, assess their work, not give up easily, reflect and re-check their problem-solving activities; (2) asking the students to analyze their peers' answers; (3) giving rewards to students for their performances; (4) asking the students to keep a diary or journal to record their learning experience; and (5) modeling metacognitive behavior in learning [6-8].

According to Wara [9], metacognitive approach in mathematics learning can be applied by adopting the Meyer model through the following three stages.

1.1. First stage: Initial discussion
At this stage, the teacher explains the general purpose of the topic. The students are encourage to think by answering metacognitive questions listed in the student worksheet. Students are guided to instill awareness by asking themselves when answering the questions. At the end of understanding the concept, the students are expected to understand all the material descriptions and be aware of what they have done, how they did it, which parts they did not understood, what questions are left unanswered and how to find solutions to them.

1.2. Second stage: Students work independently to solve problems
The students are given a set of problems asked to do it individually. The teacher walks around the classroom and provides interpersonal feedback. The teacher helps the students keep an eye on their thinking, not just giving the right answers when the students make mistakes.

1.3. Third stage: conclusion
The inference made by the students is a recapitulation of what is done in class.

Learning process is an important thing that should lead to students' mastery of learning objectives. To facilitate learning, the teacher can facilitate with teaching materials such as a student worksheet. Unfortunately, student worksheets are not yet widely used by teachers in most secondary schools in Pekanbaru. Therefore, it is necessary to develop a student worksheet to be used in Pekanbaru in order to create a meaningful learning process and in accordance with the provisions of the curriculum. This will give students the opportunity to construct knowledge by engaging in active thinking activities.

Based on the explanation above, the researchers conducted a research related to the development of student worksheets which were based on metacognitive approach to improve the mathematical representation of secondary school students. The question addressed in the research was: How were the validity, practicality, and effectiveness rates of the student worksheets which were developed based on metacognitive approach to improve students' mathematical representation ability in the topics of one-variable linear equations and linear inequalities for grade 7 students in Pekanbaru?

2. Method
The type of this research was research development using the 4-D development model. The 4-D model was developed by Thiagarajan and it consisted of Define, Design, Development and Dissemination [10]. In this research, the dissemination stage was not done because of time and cost constraints. The participants of the study were grade 7 students at a secondary school in Pekanbaru. The products were student worksheets focusing on the topics of one-variable linear equations and linear inequalities. The worksheets were developed to improve students' mathematical representation ability. This research was conducted during the odd semester of the academic year 2017/2018.

At the validation stage, the student worksheets were validated by mathematics experts, learning design experts and teachers. After being declared valid, the worksheets were piloted into a small group consisted of 12 students which were divided into three groups. These three groups were heterogeneous in terms of ability and gender. They learned using the student worksheets. After the small group trial, revisions were made. Then, the worksheets were tested in large groups. Large group trials were conducted in another class consisted of 36 students. This trial was done to determine the practicality level of the worksheets. More revisions were made after the large group trial. The effectiveness test was done with yet another class. This class was divided into two groups, one group was taught using the designed student worksheets and the other group without the worksheets.

The instruments in this research were validation sheet, questionnaires for practicality test and a set of tests. The validation sheet was designed to validate the products. For the practicality test, we used questionnaires which were administered to investigate students and the teachers' perceptions on the implementation of the products. The effectiveness of the products was assessed using a set of tests in order to gain data about the students' mathematical representation ability after using the worksheets. Data were analyzed by using descriptive statistical analysis method.

3. Result and discussion

The result of the development research was student worksheets for grade seven on the topics of one-variable linear equations and linear inequalities. The development model that was used in the development of the products was the 4-D model. The stages were as follows.

3.1. Define

This stage aimed to define the terms of the implementation of learning using student worksheet that would be developed. The steps consisted at this stage included curriculum analysis, student characteristics analysis, material analysis and objectives.

At the stage of curriculum analysis, the researchers established basic competencies that were required by the curriculum, they were: students able to explain the definition of a one-variable linear equation and a linear inequality; and students able to solve problems related to one-variable linear equations and linear inequalities.

In the analysis phase of the students' characteristic, researchers reviewed the level of students' cognitive development and looked at the skills of the students both individually and in groups. The findings of this stage were: (1) the students' mathematics ability could be classified as medium, (2) the teaching and learning in the classroom tended to be passive where the students, most of the time, listened to teacher's explanation; (3) when the students were divided into groups, they chose their group members which led to monogenous groups; (4) students tended to memorize the formula without understanding the concept so they usually found difficulties in solving problems. These data indicated that the students were qualified to participate for the limited and field trials.

In the material analysis stage, the researchers analyzed the scope of the material based on the required basic competencies as stated in the curriculum. The subtopics of linear equations and linear inequalities were: closed and open sentences; one-variable linear equation; the equivalent form of a linear equation of one variable; and one-variable-linear inequalities. These subtopics were then formulated into several indicators of learning achievement as follows: (1) able to explain the meaning of closed sentences and open sentences; (2) able to distinguish between closed sentences and open sentences; (3) able to determine open or closed sentences of everyday problems; (4) able to find the concept of one-variable linear equation; (5) able to solve problems related to one-variable linear equations; (6) able to determine the equivalent form of a one-variable linear equation; (7) able to solve problems related to one-variable linear equation and linear inequality; (8) able to find the concept of one-variable linear inequality; and (9) solve problem of one-variable linear inequality.

3.2. Design

At the design stage, the researchers compiled four student worksheets to be delivered in four meetings. The student worksheets were structured to include five components: title, discourse, study guide, work steps, and task. Student worksheet was designed in such a way in accordance with the metacognitive approach. The first worksheet was referred to as Student worksheet (SW-1) as shown in figure 1. Figure 1 shows 1 out of 5 pages of SW-1.

IOP Conf. Series: Journal of Physics: Conf. Series **1088** (2018) 012120 doi:10.1088/1742-6596/1088/1/012120

Student Worksheet (SW-1)

Close and Open Sentences

Name : ....
Class : ....
Group : ....

Guidelines:
1. Read the instructions carefully.
2. Discuss the questions in your group.
3. If you need help, you can ask the teacher after trying your best.
4. Make a conclussion for each activity in the provided columns.
5. You will be asked to present your finished work in front of the class.

Initial Discussion

Read the following dialog carefully.

Toman : Who is the first president of Indonesia?
Rizky : The first president of Indonesia is Soekarno.
Rizky : Who is the composer of the song Indonesia Raya?
Toman : The composer is Kusbini.
Rizky : What is two plus five?
Toman : Two plus five is seven.
Rizky : What is six minus one?
Toman : Six minus one is ten.

The sentences in the dialog can be classified into three groups:

Group 1 consists sentences that cannot be said to be true of false.
Group 2 consists true statements.
Group 3 consists false statements.

The sentences belong to group 1 are:

....

The sentences belong to group 2 are:

....

The sentences belong to group 3 are:

....

Now, to better understand about close and open sentences, let's do the following activity.

Figure 1. Student worksheet-1.

Figure 1 is page 1 of SW-1. It contains the initial discussion stage. In this stage the students were asked to pay close attention to a dialog and determine whether the sentences were true, false or neither. This discussion was intended to lead the students to construct their understanding about close and open sentences. This stage served as a basis for the next independently and conclusion stages.

3.3. Develop
At the develop stage, the researchers conducted the validity, practicality, and effectiveness test.

3.3.1. Validity test. The student worksheet that had been designed were then validated by the validators using validation sheet and through discussion. Validity test by material experts were conducted to examine the material aspects, objectives, and metacognitive approach. Validity test by design and learning experts was aimed to assess the writing construction aspect, language used, worksheet display and image placements. The validators' suggestions were used to revise or improve the worksheets. Table 1 shows the validation result.

Based on the validation results as shown in table 1, the validity of the worksheets fell into very valid criteria with the average percentage of 81.71%. This indicated that the worksheets were feasible to be used in learning and could be used in the practicality test.

Table 1. The validation result of the student worksheets.

No	Aspect	Validity score	Criteria
1	Presentation of goals	79.85%	Valid
2	Metacognitive approach	80 %	Valid
3	Material truth	81.5%	Very valid
4	Mathematical representation ability	84%	Very valid
5	Construction	83.57%	Very valid
6	Language	81.67%	Very valid
7	Display	81.43%	Very valid
	Average percentage	81.71%	Very valid

3.3.2. Practicality test. Practical test was done after the validation process had been completed and the worksheets had been declared valid. The practicality test was conducted to find out whether the worksheets were practical or easy to use by students as users. The practicality test involved two classes of students at a secondary school in Pekanbaru.

3.3.2.1. Practicality test on small group (limited). A small group test was conducted to see if there were errors and deficiencies in the worksheets found by the students. The small group test involved 12 students from class 1. Table 2 shows the results of the practicality test during small group testing.

Table 2. Practicality test result on small group.

No	Variable of practicality	Practicality score	Criteria
1	Students' interest	84.26%	Very practical
2	Student worksheet display	85.15%	Very practical
3	Process of use	81.16%	Very practical
4	Metacognitive approach	80%	Practical
5	Students' mathematical representation ability	79.81%	Practical
	Average	82.08%	Very practical

Based on table 2, it could be seen that the overall percentage of the students' ratings suggested that the worksheets were very practical. The overall practical score was 82.08%

3.3.2.2. Practicality test on large group. The student worksheets based on metacognitive approach which had been declared practical during the small group test, were then tested in a large group test (field trial). This involved 36 students of class 2. After being taught using the worksheets, the students were asked to give their perceptions on the worksheets using a questionnaire. Table 3 shows the result of the practicality test on large group trial.

It could be seen in table 3 that the overall percentage of the students' ratings suggested that the worksheets were very practical. The overall practical score was 82.52%.

Table 3. Practicality test result on large group.

No	Variable of practicality	Practicality Score	Criteria
1	Students' interest	85.36%	Very practical
2	Student worksheet display	84.15%	Very practical
3	Process of use	82.14%	Very practical
4	Metacognitive approach	81.12 %	Very practical
5	Students' mathematical representation ability	79.81%	Practical
	Average	82.52%	Very practical

3.3.3. Effectiveness test. Analysis of effectiveness was based on students' test results after being taught using the worksheets. The worksheets were designed to train the students' individual noticing ability to solve problems and to be able to answer metacognitive questions. These questions were meant to become a trigger for the students' metacognitive activity in problem-solving. Teacher gave feedback interpersonally to students so that they could correct their mistakes in solving problems. The test results showed that 83.33% students passed the test.

4. Conclusion

The result showed that the worksheets which were based on metacognitive approach were very valid with the validity rate of 81.71%, very practical with the practicality rate of 82.30%, and very effective with the effectiveness rate of 83.33%. Therefore, this research had produced student worksheets which were valid, practical, and effective.

References

[1] National Council of Teachers of Mathematics (NCTM) 2000 *Principles and Standards for School Mathematics* (Reston: NCTM)

[2] Neria D and Amit M 2004 Students preference of non-algebraic representation in mathematical communication *Proc. of the 28th Conf. of the International Group for the Psychology of Mathematics Education* vol 3 p 409

[3] Putri H E 2015 The influence of concrete pictorial abstract (CPA) approach to the mathematical representation ability achievement of the pre-service teachers at elementary school *International Journal of Education and Research* **3** 113

[4] Minarni A, Napitupulu E E and Husein R 2016 Mathematical understanding and representation ability of public junior high school in north Sumatra *Journal on Mathematics Education* **7** 45

[5] Supandi S, Waluya S B, Rochmad R, Suyitno H and Dewi K 2018 Think-talk-write model for improving students' abilities in mathematical representation *International Journal of Instruction* **11** 77

[6] Ozsoy G and Atamn A 2017 The effect of metacognitive strategy training on mathematical problem solving achievement *International Electronic Journal of Elementary Education* **1** 67-82

[7] Costa A L 1991 *Developing Minds: A Resource Book for Teaching Thinking* vol 1 3rd edition (Virginia USA: Association for Supervision and Curriculum Development)

[8] Radmerhr F and Drake M 2018 Students'mathematical performance, metacognitive experiences and metacognitive skills in relation to integral-area relationships *Teaching Mathematics and Its Applications: An International Journal of the IMA*

[9] Wara P H 2009 *Peningkatan Kemampuan Penalaran dan Pemecahan Masalah Matematis Siswa SMA Melalui Pembelajaran dengan Pendekatan Metakognitif* [Thesis] UPI Bandung: not published

[10] Mulyatiningsih E 2011 *Metode Penelitian Terapan Bidang Pendidikan* (Bandung: Alfabeta) p 195

AUTHOR INDEX

Abdullah, A 696
Abidin, T F 315, 365, 610
Abidin, Z....114, 197, 616
Adlim, M....702, 714
Adnan, A....28
Afriadi, J....260
Afriza, D....555
Agussalim....254
Ahmad, A....354, 492, 508
Aini, I N 730
Ainy, C 347
Ali, R M 531
Amalia, R....634, 689
Amini, R....371
Andriani....480
Andriani, A 486
Anggoro, A Y....334, 428
Anggraini, R D 328, 454, 746
Anizar 740
An'nur, S....294
Ansari, B I....365, 434
Ansari, BI....708
Anwar 133
Arianti, H....465
Arsad, N M 1
Artika, W....139
Asiah, M D....300
Atmazaki, A....604
Awang, M I 81
Azrul 471
Badaruddin, D N B P....42
Bahi, M 740
Bahri, F 321
Bahrun 248
Bakar, MT 672
Budiarto, M T....401, 525
Daimah, S....421
Damayanti, N W....634
Daqiqil, I 28
Darhim....672
Demitra 126
Desyandri....447
Dewi, I 486
Dolk, M 9
Duskri, M....354, 568, 627, 645
Effendi, K N S....215
Ekawati, R 166
Elizar, E 229
Espiritu, M J B 56
Fandanu, N R 248
Faradilla, M....666
Fauzan, A....260, 604
Febrianto, R....447
Febrilia, B R A 537
Fitri, D....508
Fitria, Y 267
Fitriana, L....696
Fitriani, F 520
Fitriati, F....101, 190
Fonna, M....634
Gani, A 660, 702, 740
Gee, E 604
Gradini, E 321
Hajidin 280, 568
Halim, A....389, 683
Halim, L 1
Hamid, A....735
Handayani, R....568
Harisman, Y....68
Hartati, S J....514
Hartini, S....294
Hartono, S....166
Hasan, M....222, 583, 666
Hasbi, M....382
Helmy, N F 616
Helsa, Y 267, 371, 447, 471, 599
Helsa, Yullys 504
Herawaty, D....182, 562
Hidayat, M 365
Hizir....465
Hobri 174, 413
Husna, M....280
Husna, U....434
Ibrahim 81
Idris, K 210
Ikhsan, M 76, 114, 237, 248, 254, 480, 645, 724
Ilyas, M....341, 520
Iqbal, M....197
Irawati....531
Irwandi....88, 389
Irwandi, I 575
Ismayani, A 583
Isnanda, M F....294
Iwannitona 273
Jannah, R D 119
Johar, R....160, 229, 254, 280, 440, 492, 508, 616, 627, 645, 652, 724
Juliangkary, E 652
Julie, H 287, 334, 428
Jumila....34
Kamid....360
Kartasasmita, B G....203
Khairi, L....724
Khairina....237
Khairunnisak, C 229, 480
Khaldun, I....222, 740
Kohar, A W....166
Kristanti, D 634
Kristanti, F....347
Kurniasari, I 719

AUTHOR INDEX

Kusnandi, K. 68
Kusumah, Y S. 68
Low, J. 95
Lubis, A N M T 182, 562
Mahtari, S 294
Mahyuna, M 714
Maidiyah, E 568
Mailizar 160, 280
Maisyura 610
Manoy, J T. 107
Mardhiah, MZ. 382
Ma'rufi 341
Ma'rufi, M 520
Marwan 81, 160
Masniladevi 504
Masriyah, M 719
Mauliana 76
Maulina, S. 498
Mawaddah 354
Mayangsari, S N. 634
Mazlina, H 683
Meilantifa, M. 401
Melvina 389, 683
Mentari, M. 88
Meutia, H 645
Miaz, Y 447
Misbah, M. 294
Mislinawati, M 376
Muhajir 514
Muhammad, I. 634
Mulia, I S 88
Muliana, M 634
Mumu, Jeinne 549
Munzir, S 480
Murni, A 328, 454, 746
Mursalin, M. 634, 689
Mursawal, A 300
Musdi, E. 260
Nasriadi, A. 677
Nazareth, E. 174, 413
Nilawati 222
Nirwana, H 267
Nizar, H. 406
Nofiza 735
Noto, M S 672
Novita, R 101, 190
Novita, T 562
Nufus, H 634
Nuraina, N 634
Nuraini, N L S 634
Nurfadhillah 492
Nurliana 395
Nurmaliah, C. 696
Nurmasyitah, N 376
Nuzulia 583
Oktavia, R 88, 389
Palupi, E L W 719
Pangaribuan, F 459
Panjaitan, B 594
Perera, J S H Q. 56, 95
Permatasari, R. 62
Prahmana, R C I 42, 56, 95, 153, 549
Pratiwi, I 308
Prihandoko, A C. 413
Purnomo, H. 634
Putra, M 101
Putra, Z H 15
Putri, R I I. 62, 119, 145, 215, 308, 406, 543, 730
Rahaju, E B. 475
Rahayu, L. 34
Rahmadhani, E. 395
Rahmayani, E 575
Rahmi, U 471
Rajibussalim 88, 389
Rajibussalim, R 575
Ridha, M 160
Ritawati 504
Rofiki, I 133
Rohantizani, R 634
Rohati 360
Rohim, A 634
Romlah, S. 174, 413
Rosayanti, E. 101
Rosjanuardi, R. 203
Rosyidi, A H. 166
Roza, Y 28
Rudhito, M A 334, 428
Safitri, J. 174, 413
Sahat, N 153
Saiman, S. 689
Sakur 328, 454, 746
Salam, S. 28
Salasi, R. 382
Saleh, H 634
Salmina, M. 677
Salwah 341
Salwah, S 520
Saminan. 708
Saminan, I 714
Sanjaya, A 724
Sanjaya, F 334, 428
Saputra, M 365
Saputri, M. 139, 300
Saputro, B A 203
Sarbain 49
Sari, D M. 114
Sari, I K. 677
Sari, R P 702
Sarong, M Ali. 696
Sayidah, N 514

AUTHOR INDEX

Shahrill, M 56, 95
Shoffa, S 347
Siregar, N 486
Siregar, R A 640
Siregar, S N 28
Siswono, T Y E 166
Sofiyan, S 689
Sofyan, H 34
Soh, T M T 1
Sthephani, A 243
Subianto, M 76
Suhartati 237, 273
Sulastri 660, 666
Sulastri, R 634, 645, 708
Sulisworo, D 126, 421
Suparno 440
Supriyatno 300
Suryadi, A 421
Suryadi, D 203, 672
Susanto 174
Tanujaya, Benidiktus 549
Tengah, K A 42, 153
Toifur, M 421
Tonra, W S 672
Tsurayya, A 588
Turmudi 730
Umam, K 498
Usmeldi 371
Utami, T P 229
Veloo, A 555
Verdina, R 660
Vitoria, L 376
Wahidin, W 588
Wahyuni 627
Wandari, A 360
Waroka, L 562
Wati, M 294
Widada, W 182, 562
Widiati, I 243
Widyastuti, R 634
Wijayanti, P 166, 475
Winarti, A 49
Winarti, D W 21, 537
Winsløw, C 15
Wulandari, Y O 634
Yamin, M 49
Yaniawati, P 215
Yansen, D 543
Yanti, W T 599
Yuhasriati, Y 622
Yunita, Y 588
Yuriansa, A 622
Yusoff, N 81
Yusrizal 197, 683
Yustinaningrum, B 395
Zainaton 683
Zainil, M 599
Zainil, Y 599
Zaura, B 254
Zubaidah, T 440
Zubainur, C M 273, 315, 434, 465, 531, 555, 610
Zuhra, S F 315
Zulfikar 300
Zulkardi 62, 119, 145, 215, 308, 406, 543, 730
Zulkarnaini, A P 267

Institute of Physics
Dirac House, Temple Back
Bristol BS1 6BE UK

ISSN: 1742-6588
ISBN 978-1-5108-7372-8

International Conference PhysicA.SPb/2017

Journal of Physics: Conference Series Volume 1038

St. Petersburg, Russia
24 - 26 October 2017

IOP
Institute of Physics